Bio and Nanoremediation of Hazardous Environmental Pollutants

Editors

Fabián Fernández-Luqueño
Sustainability of Natural Resources and Energy Program
Cinvestav-Saltillo, Coahuila de Zaragoza, Mexico

Fernando López-Valdez
Agricultural Biotechnology and Agronanobiotechnology Group
Research Centre in Applied Biotechnology (CIBA), Instituto Politécnico Nacional
Tepetitla de Lardizábal, Tlaxcala, Mexico

Gabriela Medina-Pérez
Instituto de Ciencias Agropecuarias
Universidad Autónoma del Estado de Hidalgo
Tulancingo, Hidalgo, Mexico

CRC Press is an imprint of the
Taylor & Francis Group, an **informa** business
A SCIENCE PUBLISHERS BOOK

Cover credit:
The picture and micrographs of the cover correspond to studies by Ana Lucia Flores-Hernández (photographed by Nayelli Azucena Sigala-Aguilar), Selvia García-Mayagoitia, or Gabriela Medina-Pérez, all of them led by Fabián Fernández-Luqueño, in the greenhouse and laboratory of plant-environment interactions, at Cinvestav Saltillo, Mexico.

First edition published 2023
by CRC Press
6000 Broken Sound Parkway NW, Suite 300, Boca Raton, FL 33487-2742

and by CRC Press
4 Park Square, Milton Park, Abingdon, Oxon, OX14 4RN

Library of Congress Cataloging-in-Publication Data (applied for)

ISBN: 978-0-367-51237-8 (hbk)
ISBN: 978-0-367-51239-2 (pbk)
ISBN: 978-1-003-05298-2 (ebk)

DOI: 10.1201/9781003052982

Typeset in Times New Roman
by Radiant Productions

Preface

Soil, water, air, and the environment are indispensable for life and human well-being. However, these have been neglected and polluted during the last decades despite the efforts of some governments, scientists, technicians, and international decision-makers to preserve, protect, manage and harness these natural resources and ensure sustainable development. This book compiles 14 chapters, divided into two sections covering exciting and updated topics, concerns, and troubleshooting regarding the Bio and Nanoremediation of Hazardous Environmental Pollutants. The first section is about the importance of the soil, the relevant concerns, changes in paradigm of soil remediation, microbial interactions, and nanotechnologies relevant for agriculture. In the second one, some topics are presented about soil, water, wastewater, and air remediation and environmental concerns, such as phytoremediation with genetically modified plants, studies of science and technology frontiers such as nanomaterials: for example, studies of nanoparticles for bacterial disinfection and virus inactivation in the environment, synthesis and green synthesis of nanomaterials for bio- and nano-remediation of soil, water or air, including hazardous environmental pollutants, chemical and biochemical processes via bio- and nano-remediation for hazardous contaminants degradation, use of fungi and novel nanomaterials for a high-degradation capacity of hazardous contaminants, and nanocomposites technologies for the environmental pollutants detection.

Editors hope readers can learn or strengthen their abilities and technological knowledge via this book to create better, cheaper, greener solutions for sustainable development by conserving and recovering degraded environments.

The editors
Fabián Fernández-Luqueño
Fernando López-Valdez
Gabriela Medina-Pérez

Contents

SECTION 1

CHAPTER 1

The Jeopardized Soil

Mariana Miranda-Arámbula,[1] *Ada María Ríos-Cortés,*[1]
Gabriela Medina-Pérez,[2] *Fabián Fernández-Luqueño,*[3]
Pedro Antonio López[4] and *Fernando López-Valdez*[1,*]

Soil holds life, the sun generates it, and water the powerfulness, catalyzses it, and all of them are in perfect equilibrium, in exquisite harmony, and in constant dynamics.

F. López-Valdez
February, 2021

1. Introduction

Soil is a natural resource that covers almost all continental regions. This vital structure holds life over the globe and produces food and raw materials, among dozen of soil services and benefits for the organism, including human beings. Soil covers almost the entire surface of the Earth, but it must be taken into account that there are some places where soil does not exist due to the genesis of soil or anthropogenic activities (Nortcliff, 2009). Soil comprises biotic material (5%) and minerals (45%), plus air (25%) or water (25%). The soil's last two components change constantly depending on the soil's humidity. Biotic materials are the biomass of living organisms and the residues or wastes of dead organisms.

Regarding water and the percent of air , it could change according to the humidity or water saturation of the soil. For example, if the soil is completely humid (water-saturated soil, flooded), the percent of air could be zero or close to that value, while if the soil is thoroughly dried, the water content will be zero or close to that value.

Soil is also considered as a natural system composed of a complex mixture of minerals and organic matter, from organic wastes of several forms of life, under weather conditions and biological processes, allowing nutrients to be recycled for other forms of life organized in niches, keeping a thermodynamic equilibrium state (Nortcliff, 2009).

[1] Agricultural Biotechnology and Agronanobiotechnology Group. Research Centre in Applied Biotechnology (CIBA), Instituto Politécnico Nacional. Tepetitla de Lardizábal, Tlaxcala. 90700. Mexico.
[2] Instituto de Ciencias Agropecuarias, Universidad Autónoma del Estado de Hidalgo. Tulancingo, Hidalgo. 43600. Mexico.
[3] Sustainability of Natural Resources and Energy Program, Cinvestav-Saltillo, Coahuila de Zaragoza. 25900. Mexico.
[4] Colegio de Postgraduados, Campus Puebla. San Pedro Cholula, Puebla, México.
[*] Corresponding author: flopez2072@yahoo.com, flopezva@ipn.mx

A question that one should keep in mind is: Why is soil essential for us? The answer is simple: it provides support, water and several ecosystem services for humans and live organisms. Frequently, the soil is just a thin layer that can stand for plants, animals and microorganisms. As Ding et al. (2018) mentioned, soils can provide various essential ecosystem services to produce healthy and affordable food, energy, water, climate change abatement and biodiversity, particularly in densely populated countries. According to Adhikari and Hartemink (2016) and Novák et al. (2020), soils can also play an essential role in ecosystem functioning (providing ecosystem services) as water retention, regulating services such as carbon sequestration and provisioning services such as nutrient cycling or habitat provision.

As mentioned earlier, soils are fundamental for living organisms; however, organic matter plays another critical role, as stated by Jackson et al. (2017). They claimed that soils and organic matter are the foundation of terrestrial life and that dynamic accumulation and distribution of organic matter arise from stabilization and destabilization processes that, at the same time, are influenced by biotic, abiotic and anthropogenic factors.

The soil is a very complex system; for example, it is a ternary system involving three phases: liquid (by water), gaseous (air and volatile compounds) and solid (soil, organic matter and many solids from several origins). And if that is not enough, the soil presents a series of gradients such as temperatures, humidity, composition, characteristics or properties that make it a nonlinear system. Now, that one has a better idea about the soil's importance and the factors which make it so challenging to study. Nevertheless, as humans, our tendency is to complicate the circumstances of soil. How does one do that? By making it difficult and assuming nothing is occurring. One should realize that all activities involve some residues or wastes and contamination. Thus, many severe contaminants have an evident final, finishing into the water or soil matrix. While it is true that soil has an incredible capacity to show essential physical-chemicals capacities such as absorption, adsorption, resilience or buffer, sometimes the soil seems to have a higher capacity to safeguard the effects of many contaminants and factors of degradation. However, the truth is that this is not always the case, at least not without one's help. The soil should be seen as an entity with physical, chemical, and biological characteristics working together in harmony in diverse niches and complex relationships of organisms, plants-organisms or soil-plant-organisms. Soil is not a box to extract services, materials, organisms or money.

The main aim of this chapter is to raise awareness about soil, its transformations, its loss, its importance and its pollution to make better decisions for the near future. In the end, Earth is the only planet one now has.

2. Background of the Status of Soil

Increased and intensified agricultural production and land-use changes by human beings have pushed the soil to its limits in many areas worldwide, resulting in degradation, erosion and loss of natural soils, and agricultural soils. Soil degradation may result from loss of soil structure, chemical imbalances (e.g., salinity), nutrient loss or soil loss by erosion, a biophysical process exacerbated by socio-economic and political factors resulting from inadequate management (Lal, 2010, 2001). Degradation in soil quality and loss of nutrients present within the soil may lead to poorly achieved crops by malnutrition, reducing the quality and quantity of food available for a rapidly growing population. A simple way to note the degree of soil degradation is by the estimated and observed quality by changes in agricultural sustainability, declined productivity, non-maintenance of soil or without possibilities for soil restoration by itself or assisted, and the biotic functions, according to the case (Oldeman and Lynden, 1996).

Regarding the condition of soil in the world, different factors that impact it must be considered, such as agricultural activities, salinization, acidification, extraction of soil due to mining activities, contamination by chemical substances, urban area expansion or land-use change, among others.

The above processes causes oil degradation (decrease in quality and capacity to produce food and agricultural and environmental goods and services) and desertification (a process in which the soil totally or partially loses its production potential. Lal (2010), Oldeman and Lynden (1996) described four levels to determinate the soil degradation processes: slight, moderate, severe, and extreme grades, where a slight grade of soil degradation is considered when the biotic function is conserved; in a moderate grade, it is partially destroyed; in the severe grade, the biotic function is largely destroyed; and for the extreme grade, the biotic function is entirely destroyed. The last implies that agricultural activities are jeopardized, putting environmental sustainability at risk. Lal (2010) and Oldeman and Lynden (1996) reported that agricultural soil with good biological activity lost 2,000 million hectares (Mha) of soil, classifying it as extreme (irreversible degradation). In 1993, 1,950 Mha of agricultural, forest and pasture land were tagged with some grade of degradation: a slight degradation, 750 Mha; moderate, 900 Mha; and severely degraded, 300 Mha. This degradation was an effect of agricultural activities (63%), deforestation (37%) and industrial pollution (2%). In 1994, soil degradation by wind erosion was 474 Mha; water erosion, 915 Mha; physical degradation, 50 Mha; and chemical degradation, 213 Mha (Oldeman and Lynden, 1996). It was estimated that soil degradation was 4.32 billion hectares (Bha) due to population density, highlighting that 1.1 Bha corresponds to populations with high density (more than 41 people per km^2) (Lal, 2010). Bai et al. (2008) estimated that 23.5% of the world's surface presents some level of soil degradation, of which 1 to 2.3 Mha of these arable soils were classified within the extreme level.

3. Natural and Anthropogenic Threats to the Soil Quality

The biological activity of the planet is a natural cause of changes in the structure and quality of soil; the rest of the causes are created by activities of human beings, so these activities and their contaminants as natural or anthropogenic bases can be grouped. The soil quality depends on those natural or anthropogenic activities, some examples of anthropogenic threats can be land-use changes, agriculture, rangeland, forest use, mining, chemical and petrochemical spills, urbanization and communication routes and also, the production of ubiquitous and recalcitrant pollutants. On the other hand, inundations, earthquakes and tsunamis, acid rain, forest fires, volcanic eruptions, species introduction, weathering and erosion as natural threats. These important topics to understand why soil is a jeopardized resource will be briefly addressed next.

3.1 Soil quality

There are many reasons for considering soil quality to assess the soil's capability to maintain agricultural productivity, ecological and environmental quality. At the same time, many programs for soil quality at several governmental levels are undertaken on an individual basis. In other cases, there is great planning of programs, but they have not been implemented for several reasons yet (Drewry et al., 2021). Another important aspect is the trace elements in monitoring soil quality; these chemical elements are not always present in monitoring soils (Drewry et al., 2021). These elements can accumulate over time through pesticides, veterinary drugs, fertilizers and manure. The most common variables measured as key for adequate soil monitoring have been considered: depth of humus/organic layer, bulk density, soil texture, porosity, particle density, soil organic matter and organic carbon, pH, electrolytic conductivity, nutrients depending on several factors and the decision of experts for a given soil (Mahmood et al., 2021). According to these arguments, soil quality can define the conditions for maintaining and covering the essential services that sustain life. Which means that all human activities can be included, but with corresponding responsibility, control for preserving the equilibrium of the entire ecosystem.

3.2 Land-use changes

Land-use change is a significant way of anthropogenic changes that have reformed or transformed the Earth's surface, affecting all the ecological functions of the Earth (Hasan et al., 2020). They also stated that understanding these impacts of land-use change is essential for mitigating the consequences of human-environment interactions (Hasan et al., 2020). As stated by Solomon et al. (2019), forest ecosystems, as ecosystems that provide the best healthy soils, have been changing into other land-use systems over the past decades across the world, and their importance as the primary source of services that support the livelihoods of millions of people.

Some remarkable land-use changes in the world are caused by urbanization, such as urban land use and land cover changes (Zheng et al., 2021), a second factor could be agriculture, rangeland, forest use and mining. Some of them will be briefly explored next.

3.2.1 Agriculture

First, to understand the term soil health, Yang et al. (2020) stated that soil health is the capacity of soil to function within ecosystem boundaries, sustain crop and animal productivity, maintain or enhance environmental sustainability and improve human health worldwide. Besides, they highlighted that soil microorganisms are critical factors in soil health. As it is known, agricultural practice is an anthropogenic activity that can alter soil health by inadequate management of chemical fertilizers, preferred cropping practices (monocropping, for example) and intensive land-use management, among others; which can make a substantial impact on soil function (Yang et al., 2020). Therefore, Miner et al. (2020) declared that it is imperative to prevent land degradation that occurs via soil erosion, nutrient losses and losses of ecological integrity. Soil is constantly exposed to erosion and severe commercial exploitation, mainly due to a lack of integral and adequate management, leading to soil matrix loss. Consequently, soil is a living entity that must have time to regenerate itself and with our support.

3.2.2 Rangeland use

Soil is essential for many human activities, such as grazing, where cattle are one of the primary livestock for food industries or businesses. However overgrazing may lead to soil degradation. For example, the change in land use, from the use of forests to grassland to grazing, has cleared areas by excessive grazing activity. Over time, erosion, loss of organic matter, and loss of fertility, lead to the loss of the soil itself. Swette and Lambin (2021) mentioned that livestock grazing on natural rangeland vegetation is one of the most extensive land uses on Earth, with important implications for livelihoods, food security and the environment. In addition, Swette and Lambin (2021) commented that rangelands are of crucial importance in global land use dynamics, even though they generally receive less attention than forest, agricultural and urban land uses. To avoid these problems on rangeland use and into soil, some strategies could be delaying the entrance of livestock until the desired level of growth of the vegetation, closing areas to grazing to avoid soil erosion, reducing grazing pressure on riparian vegetation or reducing grazing during a period of drought, as proposed by Swette and Lambin (2021). Regarding the laws for forest and rangelands protection, in a strict sense to move soils of forest and rangelands for diverse applications, there are not laws or some kind of protection or regulation by the main governmental entities in most countries. Roudgarmi and Mahdiraji (2020) declared no criminalization in forest and rangeland protection laws. They stated there are not acts and regulations for the destruction, pollution, sale and transfer of soil in forest and rangeland areas.

It should be noted that grazing by wild species in some areas is vital for the regeneration of the pasture itself and the benefit of biodiversity, as is the case with migratory species such as deer, zebras or goats, among others. Only a certain number of particular species do not put the diversity of an ecosystem at risk.

3.2.2.1 Bad farming practices

Other important topics are the harmful farming practices due to lack of training or technologies. Livestock activity is also a polluting factor for the soil. The primary role of livestock is to provide a food source for the population and to be continuously growing to contribute to food security; many countries have led to an intensification of livestock farming and a more globalized food market, which has caused enormous changes in land use. The soil is one of the main impacts since livestock waste is not treated correctly. Urine, feces that can contain parasites and veterinary substances that are applied to livestock can be accumulated in the soil. One of these substances is veterinary antibiotics used worldwide in livestock and poultry to control diseases and promote fast growth. Many of these antibiotics are poorly absorbed in the intestines of animals, resulting in 30 to 90% being excreted via urine and through feces or excreta, contaminating the soil as recalcitrant or persistent chemical compounds that they are. Excreta are then used in many countries as fertilizer; some studies have indicated that these can be taken by plants which is a potential risk to human health (Li et al., 2013; Pan et al., 2011).

Similarly, livestock have contributed to the following points. (a) Pollution by excreta: where the primary pollutants are nutrients (N and P), organic matter, bacteria, other entero-pathogens, residual drugs and even heavy metals. These pollutants can reach the water by specific routes and other unknown ones. (b) Residues from the processing of livestock products: slaughterhouses are an essential source of local pollution, and tanneries emit a wide range of organic and chemical pollutants. (c) Pollution from animal feed production: the primary sources are nutrients from mineral fertilizers, pesticides and sediments caused by erosion. (d) Impact on the water cycle: intensive grazing and land-use conversion alter the water cycle (Pérez-Espejo, 2008), and finally soil is the final receiver of all waste due to lack of wastewater treatment, laws (legislation and expeditious law application) and poor farming practices.

3.2.3 Forest use

As mentioned above, laws do not always protect or regulate anthropogenic activities such as the exploitation of natural resources, in particular, forest resources like wood, via illegal and excessive logging of tropical rainforests and temperate forests around the world. They expose the soil to several types of damage or jeopardy, such as erosion, loss of organic matter and the soil by itself, changes of the micro- and macro-flora and the micro- and macro-fauna, land-use change as urbanization, rangeland or agriculture. Ahirwal et al. (2021) stated that anthropogenic land-use change affects soil quality and the global carbon pool. Land-use change is a potential threat to forest ecosystems because it can alter the soil biome and increase the emission of greenhouse gasses.

By eliminating the cover of forests and sparse savannas both by human activities and by natural phenomena, the soil and the multiple networks of interactions that take place, there are directly impacted. The main adverse effects are decreased richness and abundance of biological diversity and loss of soil due to wind and water erosion, representing a considerable ecological cost (Bragg et al., 2020). On the other hand, excessive timber or logging residues (due to fallen trees in contact with the ground due to uncontrolled logging) represent an additional problem since they favor conditions for forest fires. And in turn, it can allow the establishment of different invasive species putting healthy species at risk (Bragg et al., 2020).

As soil recovery strategies, different plant species of the genera *Miscanthus*, *Salix*, *Populus*, *Eucalyptus*, *Casuarina*, *Robinia*, *Cynara*, and *Leucaena* are considered as multi purposes since they can be used for different goods and services such as soil restorers, for construction, pulp for the paper industry, etc. In addition, some trees, such as *Leucaena* incorporate nitrogen into the soil through symbiotic association with bacteria (*Rhizobium*). It is a short rotary crop to recover soil and improve soil quality. Soil fertility and carbon sequestration were improved using short rotation cropping of *Leucaena* species, and harvested annually can produce up to 39 $Mg \cdot ha^{-1}$ $year^{-1}$ of total dry biomass, which can generate up to 201 $MW h \cdot ha^{-1}$ $year^{-1}$ energy (Fernández et al., 2020). The recovery of forests favors improving the conditions of deforested soils. In some cases, schemes that include invasive species for specific periods are needed to recover the soil and the diversity that allows the establishment of tree species again (Bragg et al., 2020).

3.2.4 Mining

Mining is an activity that generates economic development in countries that practice it; however, it has one of the most significant environmental impact. On a large scale, highly mechanized processes that are generally inclined for efficient use and high production are associated with large volumes of waste and environmental impacts (Feng et al., 2019). On a small scale (artisanal way), heavy metals create havoc on miners' health due to their bioaccumulationin the body (Bravo et al., 2016). There are four types of mines, open-pit or surface mines, underground mines, drilling shafts,underwater or dredging mines, from which various raw materials classified as fuels, metallic and non-metallic are extracted and that undoubtedly pollute according to the type of deposit, extraction process and beneficiation.

In mining, two tasks are distinguished: extraction, which corresponds to excavation operations and processing tasks (metallurgy), in which they separate the sought after metal from the rest of the material (Folchi, 2005). Therefore, at an environmental level, soil is one of the most disturbed since, during the extraction, physical and biological changes that are caused on the surface, eliminating the vegetation, causing geographical and landscape changes and altering the richness and bacterial diversity of the soil. Besides the mining interrupting the hydrological cycles and during them, chemical contamination is generated since toxic elements are created in the purification process, affecting ecosystems, flora, fauna, water sources and agricultural soils (Kumar et al., 2015; Shi et al., 2017).

3.2.4.1 Coal mining

As illustrative examples are gold and coal mining, in last one is essential since this mineral, in addition to organic matter, contains water and several minerals that in the purification process of them could be pollutants or residues so, polluted the environment. This mineral carbon is used for the generation of electrical energy, steel production, the manufacture of cement and liquid fuel generation (Finkelman et al., 2019). Due to its great demand during its extraction and beneficiation, substantial environmental impacts have been caused in addition to the change in land use. These can be considered as the radical transformation of the landscape, the loss of soil structure, increased density, low water infiltration rate in the soil, compaction and erosion, directly affecting the microbial community and also causing acidification that not only affects the soil. In addition, according to Quadros et al. (2016), since coal, like any other rocky material, undergoes physical and chemical changes when it is in conditions different from those that prevailed during its formation (Montero and Martínez, 2014). Regularly after mining, the soil surface becomes contaminated with pyrite, which oxidizes to sulfuric acid when exposed to oxygen and water, resulting in high levels of soil acidification (Worrall and Pearson, 2001).

3.2.4.2 Gold mining

Gold mining, like carbon mining, causes significant damage due to its extraction, contributing to the disturbance of the natural landscape, water pollution and ecosystems destruction. Mining operations and their waste disposal methods are considered one of the leading causes of degradation of environmental health, substances such as mercury, cyanide and other toxic substances are regularly released into the environment through the extraction of gold (Orimoloye and Ololade, 2020). Cyanide is one of the most toxic and dangerous substances known in the world, and is prohibited in various countries. Cyanide is not the only toxic waste associated with gold mining because when rocks are mined and exposed to rain and air, they can contain sulfides that react with oxygen and become sulfuric acid. This acid condition also releases heavy metals such as cadmium, lead and mercury. The latter one is easily transformed into methyl-mercury by the action of microorganisms. This mercury compound is dangerous since its accumulation occurs largely in aquatic biotics, resulting in the contamination of fish, therefore affecting other levels in the chain food. Once released into the atmosphere, mercury undergoes a series of chemical reactions that cause the deposit of Hg (II) compounds in the soil in the short or middle term, which is harmful to people and fish, even at low concentrations (Gavilán-García et al., 2008; Hidayati et al., 2009). Currently, the number of artisanal miners in many developing countries is increasing exponentially due to the price of gold, which with the COVID-19 pandemic eliminated millions of jobs, causing that more and more people to seek and extract gold at handcraft way. The situation has inspired the new small-scale gold rush; consequently, the amount of mercury emitted and released into the environment is increasing proportionally (Martinez et al., 2021). Recently (in March 2021), the news of "gold fever" became relevant as gold was discovered in the Republic of Congo, Africa, changing and impacting the soil in this place at a minimal scale (artisanal miners), where probably after that, the scale of industrial mining could complete the job with high probabilities of soil and environmental impacts.

Historically, mining and metallurgy have been linked to occupational and environmental health problems due to exposure of chemicals and waste typical of this activity. When good planning and correct management implementation are carried out, impacts are prevented or significantly reduced. However, in developing countries, there are health problems linked to modern mining (Plumlee and Morman, 2011). Several environmental problems are reported that are related to issues of physical instability (breakdown of dikes and landfills) and leaching of heavy metals that lead to the contamination of streams and lakes. The mining industry is an essential sector in the economic development of countries.

3.2.5 Chemicals and petrochemicals, spills, and wastes

Soils can be contaminated through agricultural and industrial activities through the atmospheric deposition of chemicals, oils and fuels, which are hardly degradable residues that, in small proportions, are capable of contaminating large areas of the soil, disturbing the natural balance of the environment and living species (Singh et al., 2021).

Although Polycyclic Aromatic Hydrocarbons (PAHs) are primarily air, water and soil pollutants, soil and water are the final depositories of these chemicals. PAHs are a class of compounds with at least two aromatic rings; familiar sources of PAHs are natural or anthropogenic. The first ones are due to forests and grassland fires, oil seeps, volcanic eruptions and exudates from trees. Anthropogenic sources include fossil fuel use and burning, bituminous coal or black coal, wood, garbage, lubricating oil spill, municipal solid waste incineration and oil spills (Haritash and Kaushik, 2009). These compounds are widely distributed and have harmful biological effects, toxicity, mutagenicity and carcinogenicity, and they are present everywhere, recalcitrant, bioaccumulate and have carcinogenic and teratogenic activities.

Consequently, they represent a significant threat to human and soil health and, in particular, those with higher molecular weight increase their persistence, which is why they have generated significant concerns due to their presence in all components of the environment (da Silva Júnior et al., 2013; Haritash and Kaushik, 2009). The case of creosote is a persistent chemical mixture composed of approximately 85% PAHs, 10% phenolic compounds and 5% N, S and O-heterocyclics and has been widely used throughout the world as a wood preservative. This has led to widespread soil and groundwater contamination near wood-treatment plants (Muckian et al., 2007). Crops can absorb pollutants in sufficient quantities to cause some health problems for consumers. For example, the absorption of PAHs by plants varies considerably and is affected by several factors such as concentration, plant species and the microbial population in the soil (Khan et al., 2008).

One of the significant effects on soils is oil hydrocarbon spills, which can lead to the deterioration of the environmental quality. Especially concerning the negative influence on soil and water properties due to their chemical stability, intrinsic resistance to different types of degradation and high toxicity for microorganisms, animals and plants, also act as a source of groundwater contamination (Serrano et al., 2009). The type of soil, its texture and the amount of organic matter determine the final disposal of these hydrocarbons and the damage they will cause. Therefore, Serrano et al. (2009) conducted an experiment searching the phytotoxicity of garden cress (*Lepidium sativum* L.) in agricultural soil polluted with simulated diesel fuel spill at a dose of 1 L $\cdot$ m^{-2}. They reported that a stress period of 18 d following the spill led to a decrease in soil biological activity, reflected by the soil microbial biomass and soil enzymatic activities, which increased again. In addition, the soil's germination activity was recovered 200 d after the spill (Serrano et al., 2009).

Contamination of the food chain is one of the most frequent ways for these toxins to enter the human body, i.e., in countries that have a shortage of fresh water and/or where there is not legislation that regulates the use of wastewater or even surveillance and immediate execution thereof, so that, these domestic or industrial effluents are used as irrigation for agricultural production (Chen et al., 2005). Even in the 21st century, lack of awareness and ignorance persist. For instance, Sayo et al. (2020) conducted a study on the levels of copper, zinc, cadmium and lead in wastewater effluent, in vegetables and in soil irrigated with this effluent. The results showed a mean concentration of 0.48–1.83 mg L^{-1}, 1.43–4.61 mg L^{-1}, 0.015–0.35 mg L^{-1}, 0.011–2.123 mg L^{-1} for copper, zinc, cadmium and lead, respectively, were obtained in the wastewater that were above the levels allowed by the WHO in wastewater for irrigation. Due to the continued use of wastewater for irrigation, a gradual accumulation of heavy metals in the soil may occur, which could eventually lead to increased absorption of heavy metals by growing vegetables, which occurs in many countries.

Contamination of soil by these compounds exerts adverse effects on plants; in the case of seed germination, it has been observed in soils contaminated with diesel, where germination is mainly dependent on the plant species. Some plants are remarkably tolerant, and others intolerant; the inhibitory effect on germination can be attributed to the physical limitations induced by diesel fuel remaining in the soil on the seed since it forms an oil film around the seed that acts as a physical barrier reducing oxygenation (Adam and Duncan, 2002). Grifoni et al. (2020) studied the effect that hydrocarbons have on *Zea mays* L. plants growing in soils affected by fuel spills, both in a laboratory and in the field, showing a decrease in plant yield, the results of which may be due more to a reduction in soil fertility that can be attributed to the accumulation of hydrocarbons in plant tissues, this is due to a physical barrier effect that avoids the intake of nutrients, negatively affecting plant growth. Adieze et al. (2012) studied four tropical plants in a greenhouse; they observed that the height of shoots and the biomass of the plants are reduced when the concentration of hydrocarbons increases. However, some plants, such as *Panicum maximum* and *Centrosema* sp. tolerated pollution stress, plants with potential for bioremediation.

Another pollutant source is plastics. Microplastics are a diverse group of polymer particles of a large variety and range available in the soil environment due to their overexploitation and unplanned practices (Kumar et al., 2020). They can be divided into *macro* (> 25 mm), *meso* (< 25 to 5 mm), *micro* (5 mm to 0.1 μm) and *nano* (< 0.1 μm) plastics that are also a polluting factor for the soil, which is increasing and persisting at a global level (Golwala et al., 2021). Several factors affect the deposition, retention and transport of microparticles in the soil, such as anthropogenic activities, physical characteristics of the plastic particles, climatic conditions and topography (Kumar et al., 2020). In agricultural soils, the main causes for microplastics' contamination have come from the application of biosolids, compost, irrigation with wastewater, mulching films, polymer-based fertilizers and pesticides and atmospheric deposition (Bläsing and Amelung, 2018). Microplastics can affect the biophysical properties of soil, but they can also generate abiotic changes in the plant-soil interaction (de Souza Machado et al., 2019). Soil is affected by microplastics since, during their manufacture or processing, a range of chemicals and additives are used to improve the properties and ensure the applications of the final products. After prolonged environmental exposure, photochemical degradation occurs with adverse effects on soil microbial diversity. Nonetheless, these microplastics may present some advantages as a result of their physical properties on the properties of soil, helping in soil aeration and root penetration by reducing the bulk density of soil and promoting plant growth (for example, onion bulbs and their roots) with an increase in total crop biomass (Kumar et al., 2020). The later results could lead to positive effects. Nevertheless, there are gaps in the knowledge on microplastics, as mentioned by Kumar et al. (2020), where there are no standard protocols for isolating, quantifying and characterizing microplastics from the soil environment and these microplastics are recognized as emerging persistent pollutants that reach different trophic levels along the food chain; so, it should be remembered that polymeric compounds are strange or exogenous compounds for soil and water microbiology and they will not disappear by a magical act.

3.2.6 Urbanization and communication routes

Urbanization has contributed enormously to environmental pollution, and soil has been one of the primary resources affected. The extraction of firewood and deforestation as a mechanism for agricultural and urban expansion and the use of various materials for the construction of houses, buildings and roads, cause impacts on the drainage pattern, waterproofing and canceling the functions of the soil (Cotler et al., 2007).

On roads for transport, in addition to emissions from combustion, soil near roads acts as sinks for pollutants, which are transported by air or through infiltration by road runoff and spray water, this type of pollution is associated with high environmental and human health risks (Zehetner et al., 2009). Emissions from highways are of different types, such as metals, polycyclic aromatic hydrocarbons and de-icing salts (according to higher latitudes and altitudes). For instance, Zehetner et al. (2009) studied soil samples along a transect perpendicular to the highway. They analyzed the soils for road salt residues (Na), total and mobile heavy metals (Pb, Cd, Cu, Zn, Ni, Cr) and PAHs, and their results were that roadside soil pollution was highly heterogeneous. All contaminants showed an exponential-like decrease with a distance from the road, reaching background levels at 5 to 10 m from the road curb (Zehetner et al., 2009). These results are consistent with Werkenthin et al. (2014), who stated that the influence of traffic on soil contamination decreased with increasing soil depth and distance to the road. The metal concentration patterns in the soil solution were independent of concentrations in the soil matrix. At a 10-m distance, elevated soil metal concentrations, low pH and low percolation rates led to high solute concentrations. Nikolaeva et al. (2021) analyzed soils from 1, 2, 7, 25 and 50 m from the edge of the road, immediately on finding a ditch with trees and later agricultural land. Besides, they reported soil contamination by total and phyto-available heavy metals, total petroleum hydrocarbons and de-icing salts in the zone from 0 to 25 m and contamination with polycyclic aromatic hydrocarbons up to 50 m. These results are consistent with world averages for roadside soils for total petroleum hydrocarbons, de-icing salts and heavy metals. As the highest levels were found in tree-lined ditches, the authors proposed that the tree-lined ditch can intercept road runoff and splashes, providing a defense barrier for agricultural soils behind it. It should be noted that despite soil contamination, low toxicity was found in higher plants.

In a similar way, daily activities in urbanized areas bring contamination to soil due to the high consumption of plastics, household chemicals for personal care, the use of DDT to control insects and lead-based paints that contaminate the soil.

3.3 Persistent organic pollutants

Thousands of chemicals are persistent or recalcitrant to the environment. Some are known as Persistent Organic Pollutants (POPs) and have a long half-life in soil, sediments, air or biota. They could have a half-life of years or decades in the soil and several days in the atmosphere. POPs are hydrophobic or lipophilic; this combination of resistance to metabolism and lipophilicity leads to POPs accumulating in food chains (Jones and de Voogt, 1999). Among the most important classes of POPs are many families of chlorinated and chromium aromatics, including polychlorinated biphenyls, dibenzoparadioxins and polychlorinated furans (PCDD/Fs), polybrominated diphenyl ethers (PBDEs) and different organochlorine pesticides (for example, DDT and its metabolites, toxaphene, chlordane, etc.) (Jones and de Voogt, 1999). Some are by-products of combustion or the industrial synthesis of other chemicals (such as PCDD/F) that are not deliberately produced. Many have been synthesized for industrial uses, such as polybrominated diphenyl esters and different organochlorine pesticides (Jones and de Voogt, 1999). Some types of these compounds also have high mobility. In their gaseous phase, they can be highly soluble in water during hot weather or volatilize from soils up to the atmosphere. This can lead to subsequent deposition miles away from their point of release, as temperatures drop, as million tons of polychlorinated biphenyls (PCBs) produced from 1929 through the mid-1970s (Schmidt, 2010). Most of the deposition of POPs by the environment reside in soils and sediments, where they have a high affinity for organic matter. Soils are the main environmental reservoir for these persistent pollutants. These pollutants form a stable bond with the organic matter of soil, where they remain in a non-extractable form. However, changes in the environmental conditions of soil can modify the rates of the division of POPs in the soil, causing them to become easily extractable. Cold temperatures favor the deposition of POPs, and forest soils can accumulate them for long periods due to their high organic carbon content (Komprda et al., 2013).

3.4 Natural threats

3.4.1 Inundation, Earthquakes and Tsunamis

Floods, earthquakes and tsunamis are natural meteorological phenomena that, when exceeding the frequencies, become natural disasters, bringing with them multiple environmental disorders, soil being one of the main affected. Soil damage is one of the most severe threats to global resources.

3.4.1.1 Inundation (flooding)

Inundations are partial or total water occupations that occur on a surface where there is usually no water, where its frequency and duration have increased due to land use and the differences in climate change. It represents about 40% of natural disasters on Earth (Euripidou and Murray, 2004). The effect of a flood on the soil varies according to its history, composition, how long the water lasts on it and the environment of the flooded soil. Floods can be generated by rainwater or the rise in the water table, affecting the soil differently. In the case of soil flooded with rainwater, a decrease in edaphic salinity is caused, where water-avoiding respiration processes replace the air in the matrix of the soil, so biological activity is affected. In the case of floods due to the rise of groundwater, the effect will depend on the concentration of mineral salts it contains, which will determine the potential for salinization in soils. This is generally accompanied by alkalization or sodicity processes. Inundations can remove a considerable amount of soil due to dragging by strong currents when the flood finds a way to flow or drain precipitously or change the microenvironmental conditions of the soil. For instance, from aerobic to anaerobic conditions or changes in chemical, physical or biological soil properties; in other cases, soil may remove from the roots of plants or trees as a washed effect by these natural threats.

It has been observed that, in alluvial plains, the process of flooding and drying causes biogeochemical changes, such as the reduction of the redox potential that influences the pH of the soil, which is the main factor that influences the structure of bacterial communities in a variety of soils and ecosystems (Shen et al., 2021).

Wang et al. (2015) proposed the effect of seawater floods generated by climate change on wetlands, observing the changes that it could generate on invasive and native plants, where it was observed that the more significant floods favored invasive species. In urban areas, floods in addition to causing material damage, generate pollution by the mobilization of chemical products or their remobilization to the environment. This danger can increase when industrial or agricultural lands are adjacent to residential lands. Floods that cause chemical contamination clearly affect the morbidity and mortality pattern after floods (Euripidou and Murray, 2004). Several studies have shown that when a flood occurs, water that mixes with sewer water induces the spread of bacterial pathogens, such as fecal coliforms or enteropathogenic bacteria, resulting in contamination of the soil with exogenous bacteria, causing an increase in genetic diversity, for example, antibiotic resistance genes (Pérez-Valdespino et al., 2021).

3.4.1.2 Earthquakes

Earthquakes, quakes or tremors are synonymous with the natural phenomenon generated by the land's vibrations caused by the propagation inside or on its surface. The damages caused by this natural threat are from movement, floor rupture, landslides, tsunamis and soil mass liquefaction. The primary danger in an earthquake is the rupture of the surface; this may be caused by vertical (trepidatory) and horizontal (oscillatory) movements. Land displacement can affect large areas, producing severe damage to structures, roads, vegetation, etc. They can also cause soil displacement that lead to rocks and debris to fall, causing damage to the environment and significant impacts on the soil, as well as human losses and all kinds of lives (Earle, 2019).

Liquefaction is a significant cause of earthquake-related destruction (even more so than the direct action of waves on buildings). Soil liquefaction is a phenomenon in which soils, due to water saturation and particularly recent sediments such as sand or gravel, lose their firmness and flow as a result of the stresses caused on them due to earthquakes (Uyanık, 2020). In other words, liquefaction is capable of displacing, sinking or even overturning an infrastructure such as houses, buildings or other heavy structures. Therefore, all work before being built must have previous and detailed studies that characterize the soil type and the statement for corresponding use in urbanization or construction.

With the 1985 earthquake in Mexico city, several investigations were opened due to the consequences that occurred on the urban area. It concluded that the seismic risk came from the subduction of plates and the severe damage was attributed mainly to the effects of the local site due to amplification and elongation of seismic waves trapped duration in the sediment of shallow lakes (former lake of Texcoco). The high amplification potential of seismic waves was due to clay soil of Mexico city (Mayoral et al., 2019).

3.4.1.3 Tidal waves/tsunamis

A tsunami is a natural phenomenon that consists of a series of waves (huge ocean waves are known as tsunamis) caused by long-period seismic activity in the ocean by rapid large-scale disturbances such as tidal waves, volcanic eruptions and rarely meteors. These waves move at high speed with very long wavelengths. In the open sea, a tsunami is barely perceptible (González González et al., 2012).

The coastal areas hit by tsunamis present human and material losses and suffer various environmental damages, such as damage to mangroves, wetlands and floods in croplands. In this case, well-drained lands are flooded for only a few hours, in contrast to areas that remain covered with seawater for several weeks, generating salinity as well as outbreaks of animal diseases due to the damage caused to the water and sanitation systems. The effect of tsunamis on agricultural soils causes salinization, sodicity, an increase in soil organic matter, contamination by heavy metals, an increase of Na^+, K^+, Ca^{+2}, Mg^{+2}, Cl^- and SO_4^{-2} ions, increasing the pH of the soil. Changes in the topographic and hydrological surface are associated with erosion and sedimentation, physical degradation or soil removal (McLeod et al., 2010).

3.4.2 Acid rain

Acid rain is a phenomenon that consists of the release of sulfur and nitrogen oxides by various mechanisms from several sources, such as anthropogenic activities, which by reacting with the different components of the atmosphere and, on many occasions, caused and catalyzed by light energy and acidify rainwater with a pH of fewer than 5.6 units. Acid precipitation has the most significant impact on soils that are scarce in cations and has a slight buffering effect, such as podsolic soils located in coniferous forests. Acid precipitation might have adverse effects over time since it increases the leaching of calcium, magnesium and potassium from soil and replaces these cations with protons, further increasing the soil's acidity (Granados-Sánchez et al., 2010). In the same way, acid rain can change the soil's microbiology by inhibiting the activity of fungi and bacteria. It can also reduce the microbial population and negatively affect the metabolic function of an ecosystem (Xu et al., 2015).

3.4.3 Forests fires

Fire is a natural factor in the development of the ecosystem (Dymov et al., 2018). According to Certini (2005), the physical, chemical, biological, even mineralogical soil properties are affected by forest fires. For instance, severe fires (wildfires) generally show adverse effects on soil due to significant removal of organic matter, loss of vegetation cover and covering of ash in soil, a considerable loss of nutrients through volatilization by higher temperatures, deterioration of structure and porosity, leaching and erosion, marked alteration of both quantity and specific composition of microbial and soil-dwelling invertebrate communities (Certini, 2005). The effects on forest ecosystems are diverse and increase if the event occurs with high frequency; since there is a significant loss of organic matter from the soil by combustion of all organic elements of soil, hydrophobic surfaces are formed due to the formation of water-repellent organic substances.

The heating of soil produces variations in some physical and chemical properties. The pH and electrolytic conductivity typically increase due to the contribution of carbonates, basic cations and oxides from ashes. The recovery time for initial (original) pH is varied and is considered to be more or less rapid depending on the time that the ashes remain in the soil matrix. This contribution of ashes also enriches the soil with increased nutrients (Ca, Mg, K, Na and P, mainly). However, some nutrients are lost with wild fire smoke and are volatilized, such as nitrogen. Likewise, there is a risk of loss of fundamental nutrients by the action of the wind, erosion or leaching, especially when the vegetation is non-existent (Zavala et al., 2014).

3.4.4 Volcanic eruptions

As stated by Hernández et al. (2020), volcanism is a primary process of land formation. According to Chen et al. (2020), volcanic eruptions are natural disturbances that provide a model to explore the effects of their disturbances on soil microorganisms. When large amounts of pyroclastic materials are expelled, those vast cover areas, the enormous coarse sands, are deposited near the emission site. In contrast, the finest ones (from fine sands to silts) travel to a greater distance. In the same way, the fallen ash layer can be very variable according to several factors as electrostatic interactions between particles under climatic conditions. These ashes can show a moderately acidic to neutral pH, the electrolytic conductivity of the suspension of these in water may be low, and the concentration of nutrients that they contribute to the soil in an available form is low. Therefore, they are slowly incorporated into soil by biotic or abiotic weathering.

Regarding microorganisms, bacterial communities in soils are affected by volcanic eruptions. In contrast to fungi, they are more stable after an eruption (Chen et al., 2020). Volcanic eruptions have favored the robust adaptability of organisms to environmental disturbances. For instance, a finding highlighted by Chen et al. (2020) was that volcanic eruptions selectively retained some particular microorganisms (i.e., *Conexibacter*, Agaricales and Gaiellales), showing strong adaptability to the environmental disturbances, enhanced metabolic activity for Na^{+} and Ca^{+2} reabsorption, and increased relative abundances of the lichenized saprotrophs.

After a volcanic eruption, the microbiota (pioneer bacteria and archaea) plays an essential role in the initial transformation of damaged soil and simultaneously allows new niches for other microorganisms. According to Hernández et al. (2020), the archaeal communities, such as Thaumarchaeota, were detected in similar amounts in all soils analyzed, but Euryarchaeota was rare in older soils; the youngest soils showed high abundances of Chloroflexi (37%), Planctomycetes (18%) and Verrucomicrobia (10%). The microbial profiles were uncommon for the youngest soils, with a high abundance of bacteria (for example, order Ktedonobacterales (Chloroflexi)) that decreased from 37, 18, and 7%, over time.

These studies revealed that these phenomena could change the soil, vegetation, and microbial communities that dwell there in a radical way, and the passage of time will not necessarily be the same again.

3.4.5 Species introduction and species migration

Migration or introduction of species can change the conditions of soils; for example, the introduction of eucalyptus is a remarkable example. When human beings introduce anexogenous species, the area that hosts native species of plants and trees can be displaced by eucalyptus due to a series of secondary metabolites produced by this species that execute an allelopathic action. At the same time, the soil condition change by the microbial and eucalyptus action, displacing native species and changing the quality soil, exposing the soil to erosion, loss of organic matter, change of pH, among others. Some species are considered invasive (or exotic) species due to their high capacity to expand in a new area, appropriating it, such as *Panicum miliaceum*, *Sorghum halepense*, *Cynodondactylon* and *Digitaria* spp., establishing a grassland kind of species. However, they have become exotic weeds in many countries (Christoffoleti et al., 2007). To clarify, not at all exogenous species would have adverse effects on soil, plants or animal species, such as *Ficusmacrocarpa*.

Some species, such as bovine and caprine (by species migration), can devastate natural grasslands, displacing native grasses and other plant species, exposing the soil to weathering and erosion, changing or displacement of flora and microflora, pH soil change and local fauna and microfauna are also affected. Some species, like deer, due to migration, do not seem to show a high negative impact on grassland; thus, the soil can recover and preserve its properties. On the other hand, without migration, these species may carry out soil to devastate in moderate to extreme grades, losing the soil and its properties.

3.4.6 Weathering and erosion

Weathering is a natural process of depletion and transformation of rocks and their minerals to residues that remain on the Earth's surface or move to a variable depth. At the same time, these rock fragments and minerals are always under erosive forces giving place to new forms of minerals by sunlight, rain or wind actions, which in turn allow chemical processes or reactions (reacting all materials, minerals and organic compounds, to each other producing new compounds or materials). As natural processes of soil, constant change is part of the soil's natural dynamics, i.e., the mineral matter is by nature a vibrating matter by their atoms that holds a natural molecular vibration, that additionally, the rain and wind accelerate to change in the matter.

3.4.6.1 Mechanical weathering

Temperatures effects. The main effect of temperature is the disintegration of stone due to frequent variations between heating and cooling. Where minerals are aggregated (forming a stone) and each one has different coefficients of expansion, inducing several levels of micro-fragmentation in the inner layers of stone composition by temperature variation.

Water and wind effects. Another type of erosion by weathering is caused by water from rain? hitting the soil or by the course of water currents. Another important erosive agent from water is ice with a high capacity as mechanical weathering. At the same time, wind is another crucial weathering by transport of particles due to the high capacity to move, drag or carry mineral material involving abrasive forces, this type of weathering is called Eolic erosion.

3.4.6.2 Biological weathering.

Vegetation effects. Primary forms of plants (complex systems of life) as lichens and mosses, can grow on rocks forming a thin biofilm, creating a micro-habitat, with help from other types of erosions and powders (mineral residues from the wind), that at the same time serve as nutriments of these forms by vegetation and weathering for rocks. The rest of the plants can exert erosive effects on the rock.

Microbial effects. Similar to lichens and mosses, microorganisms as bacteria, mainly, can produce several acids that have the goal of dissolving the mineral material. Rhizobacteria (PGPRs) and mycorrhiza can produce organic acids to solubilize mineral phosphorus (toward available phosphorus) for plants, changing the soil's structure. Additionally, many microorganisms can be associated with lichen and mosses for mutual association, using all their resources for a sustainable community and transforming the soil environment.

3.4.6.3 Chemical weathering

Chemical weathering alters the rock material composition in the direction of surface minerals, for example, clays. Usually, one can find reactions such as carbonation, hydration, hydrolysis or oxidation. The frequent chemical reactions are oxidation (redox) and the other may be by hydrolysis.

Hydrolysis. It can be promoted by acid rain, where acid water can lead the order of the molecules in the stone surface structure interchanging or releasing cations, i.e., acid rain reacts with rock forming minerals such as feldspar to produce clay and salts that are removed in solution. From a chemical point of view, water as a natural component is effective at introducing and moving chemically active agents by way of fractures and causing rocks to crumble piecemeal, called *decomposition reaction* and creating new molecules or materials.

Oxidation. Chemical oxidation refers to the use of oxidation reagents which include the reactions where oxygen is involved with a metal element on a rock surface. A typical reaction could be an iron element in the presence of oxygen forming its corresponding oxide with its characteristic red color (as hematite or magnetite).

4. Modern Strategies to Recover Polluted or Degraded Soils

The leading conventional technologies to remediate soils contaminated with heavy metals have been soil acid washing, dilution, mixing, removal and replacement. However, these strategies do not always cover the entire contaminated area. They can alter the physical-chemical and biological properties, lose soil fertility and be costly strategies for prolonged periods (Xiao et al., 2021; Yeh, 2021). Therefore, ecologically viable remediation processes with a sustainable approach include various plant species, considering that these strategies,such as phytoremediation, can adsorb, absorb and accumulate heavy metals from contaminated soils (Yeh, 2021). These biotechnologies for remediation can be divided into: *in situ* and *ex situ*, and although this strategy is promising, it has not been widely used to date. Additionally, several processes can be categorized by their main function: phytoextraction, phytodegradation, phytostabilization, phytovolatilization, phytostimulation and rhizofiltration (Yeh, 2021).

4.1 Remotion of metals

Soil contamination with cadmium (Cd) by anthropogenic activities such as mining, disposal of hazardous waste and sewage in confined soils for agricultural purposes represents risks to humans, animals, plants and the heath of soil through the trophic chain (Wang et al., 2021). Considering this condition, one of the strategies to remove Cd is through immobilization in its insoluble form, which allows it to accumulate in crops (Wang et al., 2021) and be removed later. The work of Wang et al. (2021) in a soil cultivated with amaranth, showed that the immobilization and the remobilization of Cd in contaminated soils and supplemented with different concentrations of fertilizers (ammonia, NH_4^+-N; nitrate, NO_3^--N), can be a potential alternative strategy since, in the rhizosphere, the abundance of some bacterial phyla (Bacteroidestes, Cyanobacteria and Proteobacteria) was increased, which favor the mobilization of Cd in the contaminated soil and surprisingly identified that the treatments with NO_3^--N, were the more efficient since they showed greater extraction of Cd (7–8 $\mu g \cdot kg^{-1}$) in the soil of the rhizosphere and lower concentration (6 mg DW $\cdot$ kg^{-1}) of it in the culture tissues. Another promising strategy is the use of microorganisms such as *Rhodobacter sphaeroides*; according to Peng et al. (2018), these microorganisms in soils contaminated with zinc (Zn) and cadmium (Cd) favored the pollutants to be in more stable forms. Therefore, bioavailable for bioremediation in wheat crops with a reduction in the soil of 100% for Zn and 30.7% for Cd. In addition, the levels of Cd in roots registered a reduction of 47.2% and a more significant effect in leaves with a reduction of 62.3%. However, they also observed a correlation between a higher concentration of Zn and Cd in soil with *R. sphaeroides* and the lower removal of these metals (Peng et al., 2018).

4.2 Other strategies

Another promising procedure could be the combination of strategies, as the study proposed by Xiao et al. (2021) showed promising results with the combination of animal remediation (*Eisenia fetida*) and microbial remediation (*Bacillus megaterium*). This study evaluated the removal of Cd (2.5 mg $\cdot$ kg^{-1}) in artificially contaminated soil. They evaluated four treatments: soil with Cd as a control treatment, T1; soil with Cd + earthworms [160 individuals per experimental unit], T2; soil with Cd + earthworms + 2% biochar, T3; and soil with Cd + earthworms + *Bacillus megaterium* [2×10^9 CFU $\cdot$ mL^{-1}], T4, for 35 d. Their results recorded that treatment 4 was the most efficient for the removal of Cd in the soil with 35.7%, followed by T3 (34.4%) and finally, T2 (30.5%), with respect to the control treatment. These results are similar to those presented by (Peng et al., 2018) (mentioned above) for removing Cd in soil via phytoremediation. However, the separation of fauna (earthworms) from the soil must be considered, and this may involve additional costs, some strategies such as mechanical separation or the external application of heat or light can facilitate the extraction of the earthworms (Xiao et al., 2021).

On the other hand, an innovative plan was presented by Kumar and Prasad (2021). They proposed the biosynthesis of selenium nanoparticles (Se-NPs) with the bacterium *Vibrio natriegens* (which shows high production rates of Se-NPs) as one of the new strategies to treat soils contaminated with heavy metals (Se, Cd, Zn, Cu and Hg). It was considered that Se-NPs adsorb these heavy metals and act as effective remediators of the soil. Besides, they also have other outstanding properties, such as antidiabetic, anticancer, antioxidant and antimicrobial activity, which is an advantage for the treatment of contaminated soils (Kumar and Prasad, 2021).

To conclude, it is essential to consider that when selecting the appropriate technologies for the remediation of soils contaminated with heavy metals, it is vital to contemplate the different factors that influence optimal bioremediation. The most relevant factors to consider are the source of contamination, the pollutants' physical and chemical properties, the site's hydrogeological capacity, the space and capacity to treat large volumes, sustainable processes and the availability of financial resources. On the other hand, heavy metals or pollutant molecules are primarily complex due to their synthetic nature and hence, exogenous and recalcitrant in many cases. Under these conditions, it is crucial to examine more than one strategy for bioremediation, considering that each strategy ensures or seeks to improve the site and, at the same time, can be more secure for microbial native soil communities (Yeh, 2021).

5. Some Final Considerations

Several anthropogenic activities, such as agriculture, mining, industry, urbanization and deforestation, among others, have negatively impacted our finite resource: soil. According to Orgiazzi and Panagos (2018), it is estimated that about 1.4 tonnes · (ha year)$^{-1}$ of soil is formed (in Europe), and about 2.8 tonnes · (ha year)$^{-1}$ is lost through erosion. Losing soil affects the richness (number of taxa per m^2) and abundance (number of individuals per m^2) of local biodiversity (Orgiazzi and Panagos, 2018). This is the onset of the risk of lack of sustainability for life in general. In soil, there is a network of interactions where the physical-chemical properties can influence the biological diversity and the soil itself. In this binomial, the negative impact of our anthropogenic activities have caused, and the risk that this represents for the soil must be considered. This is why the chapter addresses some impacts and strategies for soil recovery. It is important to emphasize that this will not be enough, and it is necessary to constantly develop new proposals for the recovery of soils and, at the same time, develop procedures to avoid the negative impact on the structure, physicochemical properties and the biological diversity of the soil. Finally, it is considered that the strategies to recover, conserve, protect and manage soil should be based on a complete study that identifies the richness and abundance of the organisms. Besides, considering the economic and environmental costs to implement the remediation procedures and final disposal of pollutants (depending on the case), besides assuming political and social responsibilities with critical factors to solve integrally.

6. Conclusions

Soil is always exposed to many anthropogenic activities and natural phenomena, i.e., this is a complex topic due to the multiple factors involved. Humanity has more challenges than answers, equipment, treatment approaches and strategies. Nevertheless, the effective primary strategy is and has been the *will* and the *responsibility* (of several sectors, such as politics, academic, economic and social) with the environment and ourselves.

During the last decades, several practical and low-cost technologies have been launched for social services to dissipate pollutants and alleviate the stressed environment. However, the easier and cheaper procedure is the management of pollutants, residues and wastes, not the remediation technologies. Threats to the soil could be tackled with functions that jeopardize soil-based ecosystem services required for healthy organisms, including humans.

Besides, the health of the soil is endangered by inadequate tools and approaches to soil quality assessment. For example, suitability for crop growth, productivity, environmental or human health is not enough in today's conditions and necessities. Today, multi-functionality, ecosystem services, resistance and resilience are the main objectives of soil quality assessment through curring-edge digital, analytical and visual tools.

There are jeopardized soils worldwide, so global debates, decisions and actions are required to shape sustainable development. There is a challenge here for the soil science community, and its active involvement is necessary. A holistic or global vision should consider the interactions between soil, water, atmosphere, plant and anthropogenic activities. Decades have been spent developing sectional, partial or short-term studies, but the dynamic soil behavior should currently be studied in an interdisciplinary and trans disciplinary context through real conditions and long-term studies, and also Big Data and the use of geographical information system technology.

Acknowledgment

This research was founded by 'Ciencia Básica SEP-CONACyT' project 151881, the Sustainability of Natural Resources and Energy Program (Cinvestav-Saltillo), and Cinvestav Zacatenco. 'FONCYT-COECYT Fund-Call 2021-C15, Bio- and nano-remediation of soils and water, contaminated with heavy metals, at San Juan de Sabinas, Coahuila de Zaragoza, Mexico (COAH-2021-C15-C095)', 'FONCYT-COECYT Fund-Call 2022-C19, Identification and Characterization of Microplastics in the Municipality of Saltillo, Coahuila de Zaragoza, Mexico: Bases for risk management, (COAH-2022-C19-C088)', and to Instituto Politécnico Nacional for financial support (Projects: SIP-20161934, 20172065, and 20181453). G. M.-P., F. F.-L., M. M-A, and F. L.-V. received grant-aided support from 'Sistema Nacional de Investigadores (SNI)', Mexico.

References

Adam, G. and Duncan, H. 2002. Influence of diesel fuel on seed germination. Environ. Pollut. 120: 363–370. https://doi.org/10.1016/S0269-7491(02)00119-7.

Adhikari, K. and Hartemink, A.E. 2016. Linking soils to ecosystem services—A global review. Geoderma 262: 101–111. https://doi.org/10.1016/j.geoderma.2015.08.009.

Adieze, I.E., Orji, J.C., Nwabueze, R.N. and Onyeze, G.O.C. 2012. Hydrocarbon stress response of four tropical plants in weathered crude oil contaminated soil in microcosms. Int. J. Environ. Stud. 69: 490–500. https://doi.org/10.1080/00207233.2012.665785.

Ahirwal, J., Kumari, S., Singh, A.K., Kumar, A. and Maiti, S.K. 2021. Changes in soil properties and carbon fluxes following afforestation and agriculture in tropical forest. Ecol. Indic. 123: 107354. https://doi.org/10.1016/j.ecolind.2021.107354.

Bai, Z.G., Dent, D.L., Olsson, L. and Schaepman, M.E. 2008. Proxy global assessment of land degradation. Soil Use Manag. 24: 223–234. https://doi.org/10.1111/j.1475-2743.2008.00169.x.

Bläsing, M. and Amelung, W. 2018. Plastics in soil: Analytical methods and possible sources. Sci. Total Environ. 612: 422–435. https://doi.org/10.1016/j.scitotenv.2017.08.086.

Bragg, D.C., Hanberry, B.B., Hutchinson, T.F., Jack, S.B. and Kabrick, J.M. 2020. Silvicultural options for open forest management in eastern North America. For. Ecol. Manag. 474: 118383. https://doi.org/10.1016/j.foreco.2020.118383.

Bravo, M.L., Luna, J.S., Abad, C.Q., Osorio, M.S. and Rodriguez, J.P. 2016. Actividad minera y su impacto en la salud humana/Themining and itsimpacton human health. Cienc. UNEMI 9: 92–100.

Certini, G. 2005. Effects of fire on properties of forest soils: A review. Oecologia 143: 1–10. https://doi.org/10.1007/s00442-004-1788-8.

Chen, J., Guo, Y., Li, F., Zheng, Y., Xu, D., Liu, H., Liu, X., Wang, X. and Bao, Y. 2020. Exploring the effects of volcanic eruption disturbances on the soil microbial communities in the montane meadow steppe. Environ. Pollut. 267: 115600. https://doi.org/10.1016/j.envpol.2020.115600.

Chen, Y., Wang, C. and Wang, Z. 2005. Residues and source identification of persistent organic pollutants in farmland soils irrigated by effluents from biological treatment plants. Environ. Int. Soil Contamination and Environmental Health 31: 778–783. https://doi.org/10.1016/j.envint.2005.05.024.

Christoffoleti, P.J., Carvalho, S.J.P., Nicolai, M., Doohan, D. and VanGessel, M. 2007. Prevention strategies in weed management. *In*: Non-Chemical Weed Management: Principles. Concepts and Technology. CABI, Oxfordshire, pp. 1–16.

Cotler, H., Sotelo, E., Dominguez, J., Zorrilla, M., Cortina, S. and Quiñones, L. 2007. La conservación de suelos: un asunto de interés público. Gac. Ecológica 5–71.

da Silva Júnior, F.M.R., Silva, P.F., Garcia, E.M., Klein, R.D., Peraza-Cardoso, G., Baisch, P.R., Vargas, V.M.F. and Muccillo-Baisch, A.L. 2013. Toxic effects of the ingestion of water-soluble elements found in soil under the atmospheric influence of an industrial complex. Environ. Geochem. Health 35: 317–331. https://doi.org/10.1007/s10653-012-9496-5.

de Souza Machado, A.A., Lau, C.W., Kloas, W., Bergmann, J., Bachelier, J.B., Faltin, E., Becker, R., Görlich, A.S. and Rillig, M.C. 2019. Microplastics can change soil properties and affect plant performance. Environ. Sci. Technol. 53: 6044–6052. https://doi.org/10.1021/acs.est.9b01339.

Ding, K., Wu, Q., Wei, H., Yang, W., Séré, G., Wang, S., Echevarria, G., Tang, Y., Tao, J., Morel, J.L. and Qiu, R. 2018. Ecosystem services provided by heavy metal-contaminated soils in China. J. Soils Sediments 18: 380–390. https://doi.org/10.1007/s11368-016-1547-6.

Drewry, J.J., Cavanagh, J.-A.E., McNeill, S.J., Stevenson, B.A., Gordon, D.A. and Taylor, M.D. 2021. Long-term monitoring of soil quality and trace elements to evaluate land use effects and temporal change in the Wellington region, New Zealand. Geoderma Reg. e00383. https://doi.org/10.1016/j.geodrs.2021.e00383.

Dymov, A., Abakumov, E., Bezkorovaynaya, I., Prokushkin, A., Kuzyakov, Y. and EYu, M. 2018. Impact of forest fire on soil properties (Review). Theor. Ecol. 2018. https://doi.org/10.25750/1995-4301-2018-4-013-023.

Earle, S. 2019. Physical Geology - 2nd Edition. BCcampus.

Euripidou, E. and Murray, V. 2004. Public health impacts of floods and chemical contamination. J. Public Health 26: 376–383. https://doi.org/10.1093/pubmed/fdh163.

Feng, Y., Wang, J., Bai, Z. and Reading, L. 2019. Effects of surface coal mining and land reclamation on soil properties: A review. Earth-Sci. Rev. 191: 12–25. https://doi.org/10.1016/j.earscirev.2019.02.015.

Fernández, M., Alaejos, J., Andivia, E., Madejón, P., Díaz, M.J. and Tapias, R. 2020. Short rotation coppice of leguminous tree Leucaena spp. improves soil fertility while producing high biomass yields in Mediterranean environment. Ind. Crops Prod. 157: 112911. https://doi.org/10.1016/j.indcrop.2020.112911.

Finkelman, R.B., Dai, S. and French, D. 2019. The importance of minerals in coal as the hosts of chemical elements: A review. Int. J. Coal Geol. 212: 103251. https://doi.org/10.1016/j.coal.2019.103251.

Folchi, M. 2005. Los efectos ambientales del beneficio de minerales metálicos: un marco de análisis para la historia ambiental. Varia Hist. 21: 32–57. https://doi.org/10.1590/S0104-87752005000100003.

Gavilán-García, I., Santos-Santos, E., Tovar-Gálvez, L.R., Gavilán-García, A., Suárez, S. and Olmos, J. 2008. Mercury speciation in contaminated soils from old mining activities in Mexico using a chemical selective extraction. J. Mex. Chem. Soc. 52: 263–271.

Golwala, H., Zhang, X., Iskander, S.M. and Smith, A.L. 2021. Solid waste: An overlooked source of microplastics to the environment. Sci. Total Environ. 769: 144581. https://doi.org/10.1016/j.scitotenv.2020.144581.

González González, R., Ortiz Figueroa, M. and Montoya Rodríguez, J.M. 2012. Tsunami: Un problema matemáticamente interesante. Rev. Matemática Teoría Apl. 19: 107–119.

Granados-Sánchez, D., López-Ríos, G.F. and Sánchez-Hernández, M.Á. 2010. Acid rain and forest ecosystems. Rev. Chapingo Ser. Hortic. XVI. https://doi.org/10.5154/r.rchscfa.2010.04.022.

Grifoni, M., Rosellini, I., Angelini, P., Petruzzelli, G. and Pezzarossa, B. 2020. The effect of residual hydrocarbons in soil following oil spillages on the growth of Zea mays plants. Environ. Pollut. 265: 114950. https://doi.org/10.1016/j.envpol.2020.114950.

Haritash, A.K. and Kaushik, C.P. 2009. Biodegradation aspects of Polycyclic Aromatic Hydrocarbons (PAHs): A review. J. Hazard. Mater. 169: 1–15. https://doi.org/10.1016/j.jhazmat.2009.03.137.

Hasan, S.S., Zhen, L., Miah, Md.G., Ahamed, T. and Samie, A. 2020. Impact of land use change on ecosystem services: A review. Environ. Dev., Resources Use, Ecosystem Restoration and Green Development 34: 100527. https://doi.org/10.1016/j.envdev.2020.100527.

Hernández, M., Calabi, M., Conrad, R. and Dumont, M.G. 2020. Analysis of the microbial communities in soils of different ages following volcanic eruptions. Pedosphere 30: 126–134. https://doi.org/10.1016/S1002-0160(19)60823-4.

Hidayati, N., Juhaeti, T. and Syarif, F. 2009. Mercury and cyanide contaminations in gold mine environment and possible solution of cleaning up by using phytoextraction. HAYATI J. Biosci. 16: 88–94. https://doi.org/10.4308/hjb.16.3.88.

Jackson, R.B., Lajtha, K., Crow, S.E., Hugelius, G., Kramer, M.G. and Piñeiro, G. 2017. The ecology of soil carbon: pools, vulnerabilities, and biotic and abiotic controls. Annu. Rev. Ecol. Evol. Syst. 48: 419–445. https://doi.org/10.1146/annurev-ecolsys-112414-054234.

Jones, K.C. and de Voogt, P. 1999. Persistent organic pollutants (POPs): State of the science. Environ. Pollut. 100: 209–221. https://doi.org/10.1016/S0269-7491(99)00098-6.

Khan, S., Aijun, L., Zhang, S., Hu, Q. and Zhu, Y.-G. 2008. Accumulation of polycyclic aromatic hydrocarbons and heavy metals in lettuce grown in the soils contaminated with long-term wastewater irrigation. J. Hazard. Mater. 152: 506–515. https://doi.org/10.1016/j.jhazmat.2007.07.014.

Koch, A., McBratney, A., Adams, M., Field, D., Hill, R., Crawford, J., Minasny, B., Lal, R., Abbott, L., O'Donnell, A., Angers, D., Baldock, J., Barbier, E., Binkley, D., Parton, W., Wall, D.H., Bird, M., Bouma, J., Chenu, C., Flora, C.B., Goulding, K., Grunwald, S., Hempel, J., Jastrow, J., Lehmann, J., Lorenz, K., Morgan, C.L., Rice, C.W., Whitehead, D., Young, I. and Zimmermann, M. 2013. Soil security: Solving the global soil crisis. Glob. Policy 4: 434–441. https://doi.org/10.1111/1758-5899.12096.

Komprda, J., Komprdová, K., Sáňka, M., Možný, M. and Nizzetto, L. 2013. Influence of climate and land use change on spatially resolved volatilization of persistent organic pollutants (POPs) from background soils. Environ. Sci. Technol. 47: 7052–7059. https://doi.org/10.1021/es3048784.

Kumar, A. and Prasad, K.S. 2021. Role of nano-selenium in health and environment. J. Biotechnol. 325: 152–163. https://doi.org/10.1016/j.jbiotec.2020.11.004.

Kumar, M., Xiong, X., He, M., Tsang, D.C.W., Gupta, J., Khan, E., Harrad, S., Hou, D., Ok, Y.S. and Bolan, N.S. 2020. Microplastics as pollutants in agricultural soils. Environ. Pollut. 265: 114980. https://doi.org/10.1016/j.envpol.2020.114980.

Kumar, S., Maiti, S.K. and Chaudhuri, S. 2015. Soil development in 2–21 years old coalmine reclaimed spoil with trees: A case study from Sonepur-Bazari open cast project, Raniganj Coalfield. India. Ecol. Eng. 84: 311–324. https://doi.org/10.1016/j.ecoleng.2015.09.043.

Lal, R. 2001. Soil degradation by erosion. Land Degrad. Dev. 12: 519–539. https://doi.org/10.1002/ldr.472.

Lal, R. 2010. Managing soils and ecosystems for mitigating anthropogenic carbon emissions and advancing global food security. BioScience 60: 708–721. https://doi.org/10.1525/bio.2010.60.9.8.

Li, Y., Zhang, X., Li, W., Lu, X., Liu, B. and Wang, J. 2013. The residues and environmental risks of multiple veterinary antibiotics in animal faeces. Environ. Monit. Assess. 185: 2211–2220. https://doi.org/10.1007/s10661-012-2702-1.

Mahmood, H., Saha, C., Paul, N., Deb, S., Abdullah, S.M.R., Tanvir, Md.S.S.I., Bashar, A., Roy, S., Rabby, F., Ahmed, S.N. and Ali, Md.H. 2021. The soil quality of the world's largest refugee campsites located in the Hill forest of Bangladesh and the way forward to improve the soil quality. Environ. Chall. 3: 100048. https://doi.org/10.1016/j.envc.2021.100048.

Martinez, G., Restrepo-Baena, O.J. and Veiga, M.M. 2021. The myth of gravity concentration to eliminate mercury use in artisanal gold mining. Extr. Ind. Soc. 8: 477–485. https://doi.org/10.1016/j.exis.2021.01.002.

Mayoral, J.M., Asimaki, D., Tepalcapa, S., Wood, C., Roman-de la Sancha, A., Hutchinson, T., Franke, K. and Montalva, G. 2019. Site effects in Mexico City basin: Past and present. Soil Dyn. Earthq. Eng. 121: 369–382. https://doi.org/10.1016/j.soildyn.2019.02.028.

McLeod, M.K., Slavich, P.G., Irhas, Y., Moore, N., Rachman, A., Ali, N., Iskandar, T., Hunt, C. and Caniago, C. 2010. Soil salinity in Aceh after the December 2004 Indian Ocean tsunami. Agric. Water Manag. 97: 605–613. https://doi.org/10.1016/j.agwat.2009.10.014.

Miner, G.L., Delgado, J.A., Ippolito, J.A. and Stewart, C.E. 2020. Soil health management practices and crop productivity. Agric. Environ. Lett. 5: e20023. https://doi.org/10.1002/ael2.20023.

Montero, R.L. and Martínez, M. 2014. Composición físico-química de aguas ácidas procedentes de dos minas de carbón: lobatera, Estado Táchira, Venezuela. Rev. Fac. Ing. Univ. Cent. Venezuela 29: 55–66.

Muckian, L., Grant, R., Doyle, E. and Clipson, N. 2007. Bacterial community structure in soils contaminated by polycyclic aromatic hydrocarbons. Chemosphere 68: 1535–1541. https://doi.org/10.1016/j.chemosphere.2007.03.029.

Nikolaeva, O., Karpukhin, M., Streletskii, R., Rozanova, M., Chistova, O. and Panina, N. 2021. Linking pollution of roadside soils and ecotoxicological responses of five higher plants. Ecotoxicol. Environ. Saf. 208: 111586. https://doi.org/10.1016/j.ecoenv.2020.111586.

Nortcliff, S. 2009. The Soil: Nature, Sustainable Use, Management, and Protection—An Overview. GAIA - Ecol. Perspect. Sci. Soc. 18: 58–68. https://doi.org/10.14512/gaia.18.1.14.

Novák, T.J., Balla, D. and Kamp, J. 2020. Changes in anthropogenic influence on soils across Europe 1990–2018. Appl. Geogr. 124: 102294. https://doi.org/10.1016/j.apgeog.2020.102294.

Oldeman, L.R., Lynden and van, G.W.J. 1996. Revisiting the GLASOD Methodology (No. 96/03). ISRIC, Wageningen.

Orgiazzi, A. and Panagos, P. 2018. Soil biodiversity and soil erosion: It is time to get married. Glob. Ecol. Biogeogr. 27: 1155–1167. https://doi.org/10.1111/geb.12782.

Orimoloye, I.R. and Ololade, O.O. 2020. Potential implications of gold-mining activities on some environmental components: A global assessment (1990 to 2018). J. King Saud Univ.-Sci. 32: 2432–2438. https://doi.org/10.1016/j.jksus.2020.03.033.

Pan, X., Qiang, Z., Ben, W. and Chen, M. 2011. Residual veterinary antibiotics in swine manure from concentrated animal feeding operations in Shandong Province, China. Chemosphere 84: 695–700. https://doi.org/10.1016/j.chemosphere.2011.03.022.

Peng, W., Li, X., Song, J., Jiang, W., Liu, Y. and Fan, W. 2018. Bioremediation of cadmium- and zinc-contaminated soil using Rhodobactersphaeroides. Chemosphere 197: 33–41. https://doi.org/10.1016/j.chemosphere.2018.01.017.

Pérez-Espejo, R. 2008. El lado oscuro de la ganadería. Probl. Desarro. Rev. Latinoam. Econ. 39. https://doi.org/10.22201/iiec.20078951e.2008.154.7734.

Pérez-Valdespino, A., Pircher, R., Pérez-Domínguez, C.Y. and Mendoza-Sanchez, I. 2021. Impact of flooding on urban soils: Changes in antibiotic resistance and bacterial community after Hurricane Harvey. Sci. Total Environ. 766: 142643. https://doi.org/10.1016/j.scitotenv.2020.142643.

Plumlee, G.S. and Morman, S.A. 2011. Mine wastes and human health. Elements 7: 399–404. https://doi.org/10.2113/gselements.7.6.399.

Quadros, P.D. de Zhalnina, K., Davis-Richardson, A.G., Drew, J.C., Menezes, F.B., Camargo, F.A. de O. and Triplett, E.W. 2016. Coal mining practices reduce the microbial biomass, richness and diversity of soil. Appl. Soil Ecol. 98: 195–203. https://doi.org/10.1016/j.apsoil.2015.10.016.

Roudgarmi, P. and Mahdiraji, M.T.A. 2020. Current challenges of laws for preservation of forest and rangeland, Iran. Land Use Policy 99: 105002. https://doi.org/10.1016/j.landusepol.2020.105002.

Sayo, S., Kiratu, J.M. and Nyamato, G.S. 2020. Heavy metal concentrations in soil and vegetables irrigated with sewage effluent: A case study of Embu sewage treatment plant, Kenya. Sci. Afr. 8: e00337. https://doi.org/10.1016/j.sciaf.2020.e00337.

Schmidt, C. 2010. How PCBs are like grasshoppers. Environ. Sci. Technol. 44: 2752–2752. https://doi.org/10.1021/es100696y.

Serrano, A., Tejada, M., Gallego, M. and Gonzalez, J.L. 2009. Evaluation of soil biological activity after a diesel fuel spill. Sci. Total Environ., Thematic Papers: Selected papers from the 2007 Wetland Pollutant Dynamics and Control Symposium 407: 4056–4061. https://doi.org/10.1016/j.scitotenv.2009.03.017.
Shen, R., Lan, Z., Rinklebe, J., Nie, M., Hu, Q., Yan, Z., Fang, C., Jin, B. and Chen, J. 2021. Flooding variations affect soil bacterial communities at the spatial and inter-annual scales. Sci. Total Environ. 759: 143471. https://doi.org/10.1016/j.scitotenv.2020.143471.
Shi, P., Zhang, Y., Hu, Z., Ma, K., Wang, H. and Chai, T. 2017. The response of soil bacterial communities to mining subsidence in the west China aeolian sand area. Appl. Soil Ecol. 121: 1–10. https://doi.org/10.1016/j.apsoil.2017.09.020.
Singh, S., Kumar, V., Datta, S., Dhanjal, D.S., Parihar, P. and Singh, J. 2021. Chapter 8 - Role of plant–microbe systems in remediation of petrochemical-contaminated water and soil environment. pp. 79–88. *In*: Kumar, A., Singh, V.K., Singh, P. and Mishra, V.K. (eds.). Microbe Mediated Remediation of Environmental Contaminants, Woodhead Publishing Series in Food Science, Technology and Nutrition. Woodhead Publishing. https://doi.org/10.1016/B978-0-12-821199-1.00008-0.
Solomon, N., Segnon, A.C. and Birhane, E. 2019. Ecosystem service values changes in response to land-use/land-cover dynamics in dry afromontane forest in northern ethiopia. Int. J. Environ. Res. Public. Health 16. https://doi.org/10.3390/ijerph16234653.
Swette, B. and Lambin, E.F. 2021. Institutional changes drive land use transitions on rangelands: The case of grazing on public lands in the American West. Glob. Environ. Change 66: 102220. https://doi.org/10.1016/j.gloenvcha.2020.102220.
Uyanık, O. 2020. Soil liquefaction analysis based on soil and earthquake parameters. J. Appl. Geophys. 176: 104004. https://doi.org/10.1016/j.jappgeo.2020.104004.
Wang, J.-F., Li, W.-L., Li, Q.-S., Wang, L.-L., He, T., Wang, F.-P. and Xu, Z.-M. 2021. Nitrogen fertilizer management affects remobilization of the immobilized cadmium in soil and its accumulation in crop tissues. Environ. Sci. Pollut. Res. https://doi.org/10.1007/s11356-021-12868-z.
Wang, W., Wang, C., Sardans, J., Tong, C., Jia, R., Zeng, C. and Peñuelas, J. 2015. Flood regime affects soil stoichiometry and the distribution of the invasive plants in subtropical estuarine wetlands in China. CATENA 128: 144–154. https://doi.org/10.1016/j.catena.2015.01.017.
Werkenthin, M., Kluge, B. and Wessolek, G. 2014. Metals in European roadside soils and soil solution—A review. Environ. Pollut. 189: 98–110. https://doi.org/10.1016/j.envpol.2014.02.025.
Worrall, F. and Pearson, D.G. 2001. The development of acidic groundwaters in coal-bearing strata: Part I. Rare earth element fingerprinting. Appl. Geochem. 16: 1465–1480. https://doi.org/10.1016/S0883-2927(01)00018-X.
Xiao, R., Liu, X., Ali, A., Chen, A., Zhang, M., Li, R., Chang, H. and Zhang, Z. 2021. Bioremediation of Cd-spiked soil using earthworms (Eisenia fetida): Enhancement with biochar and Bacillus megatherium application. Chemosphere 264: 128517. https://doi.org/10.1016/j.chemosphere.2020.128517.
Xu, H., Zhang, J., Ouyang, Y., Lin, L., Quan, G., Zhao, B. and Yu, J. 2015. Effects of simulated acid rain on microbial characteristics in a lateritic red soil. Environ. Sci. Pollut. Res. Int. 22: 18260–18266. https://doi.org/10.1007/s11356-015-5066-6.
Yang, T., Siddique, K.H.M. and Liu, K. 2020. Cropping systems in agriculture and their impact on soil health—A review. Glob. Ecol. Conserv. 23: e01118. https://doi.org/10.1016/j.gecco.2020.e01118.
Yeh, T.Y. 2021. Current status of soil and groundwater remediation technologies in Taiwan. Int. J. Phytoremediation 23: 212–218. https://doi.org/10.1080/15226514.2020.1803202.
Zavala, L.M., Celis, R. de and Jordán, A. 2014. How wildfires affect soil properties. A brief review. Cuad. Investig. Geográfica 40: 311–332. https://doi.org/10.18172/cig.2522.
Zehetner, F., Rosenfellner, U., Mentler, A. and Gerzabek, M.H. 2009. Distribution of road salt residues, heavy metals and polycyclic aromatic hydrocarbons across a highway-forest interface. Water. Air. Soil Pollut. 198: 125–132. https://doi.org/10.1007/s11270-008-9831-8.
Zheng, Q., Weng, Q. and Wang, K. 2021. Characterizing urban land changes of 30 global megacities using nighttime light time series stacks. ISPRS J. Photogramm. Remote Sens. 173: 10–23. https://doi.org/10.1016/j.isprsjprs.2021.01.002.

CHAPTER 2

Isoproturon Herbicide Interaction with Crops and Biodegradation

Himani Singh, Gurminder Kaur, Shubhra Khare* and *Vijaya Yadav*

1. Introduction

Biological systems are progressively influenced by degradation from herbicides, which are extensively used for weed control in different rural programs. Herbicides can improve the demand of food by expanding crops to fulfil the current needs of the population from the developing world. The fast increase and continuous use of dangerous herbicides can lead to severe effects on plants and their products and cause major ecological harm. Isoproturon [3-(4-isopropylphenyl)-1, 1-dimethylurea] is a phenyl urea-derived systemic herbicide widely used, for managing pre- and post-growth of broad-leaved weeds and annual grasses in crop fields (Yin et al., 2008). Its use in the agricultural field has increased greatly, thereby creating an urgent the need for the introduction of isoproturon-resistant plants in agricultural fields (Dhawan et al., 2005). Isoproturon (IPU) in general, is applied to control pre- or post-emergence annual grasses and broad leaved weeds like bluegrass, foxtail, bent grass, ryegrass, chickweed, parthenium and lady's mantle (Rigaud and Lebreton, 2004). According to an official survey by the Government of India in 2009–10 (http://www.pesticideinfo.org), it was reported that IPU is amongst the most consumed pesticide in the country. In the UK, around 3.0 million hectares of rural land were placed on roughly 3300 tons of isoproturon in 1997. It is a selective, systemic herbicide which transfers into the target plant through the roots as well as leaves and directly inhibits the process of photosynthesis (Ducruet, 1991; Fedtke and Duke, 2005; Williams et al., 2009). IPU is highly water soluble, moderately hydrophobic and weakly absorbed by soils (Ding et al., 2011). In daily agricultural practices, IPU is applied to the soil, but most of it is not absorbed by targets which results in the accumulation of its remnant residue in soils and crops. Moreover, due to its high-water soluble property, ground water gets contaminated and serves as the main source of IPU being transferred to adjoining agricultural fields, which are especially irrigated through pumped water on the surface of soil. The expanding use of heavy metals, pesticides and organic chemicals is becoming a major concern and posing a threat for the environment. It either increases or decreases the number of bacterial and fungal propagules in the rhizosphere and has detrimental effects on the

Institute of Bioscience and Technology, Shri Ramswaroop Memorial University Lucknow.
Emails: gurminder.ibst@srmu.ac.in; khare.shubhra27@gmail.com; vijayasbyadav@gmail.com
*Corresponding author: himanisingh.ibst@srmu.ac.in

microorganisms present in the soil and ultimately leads to a large reduction in soil fertility (Mudd et al., 1983). IPU has mutagenic, carcinogenic, teratogenic and genotoxic properties which impairs reproduction and disrupts endocrine inorganism (Behera and Bhunya, 1990; Badawi et al., 2009). IPU can also directly affect plant growth and germination of non-weed crop seeds, viz., wheat and beans (Yin et al., 2008; Liang et al., 2012). The impacts of IPU on the health of different plants are summarized in Table 2.1.

Table 2.1. Isoproturon uptake and their effects on plant.

S. No.	Dosage	Model Plant	Mode of Application	Effects	References
1	10 µg L^{-1}	*Elodea densa*	Water column/In nutrient medium	growth inhibition	Feurtet-Mazel et al., 1996
2	1,000 g/ha	*Triticum astivum*	Soil	grain yield decreased	Chhokar and Malik, 2002
3	124 mgg^{-1}	Green gram	Pre-emergence treated soil Soil	minimum grain protein	Khan et al., 2006
	200 µg kg^{-1}			stimulated the chlorophyll content	
4	2.5 Kg ha^{-1}	*Zea mays*	Foliar sprays	reduced fresh and dry weights of shoots and roots as well as chlorophylls and carotenoids contents inhibitions in activities of the glutathione S-transferase (GST) isoforms	Nematalla et al., 2008
5	30 µg L^{-1}	*Scenedesmus obliquus*	Batch culture/In nutrient medium	fluorescence yield was at J transient increased	Dewez et al., 2008
	> 30 µg L^{-1}			fluorescence yield at P transient was decreased	
6	2–20 mg/kg	*Triticum astivum*		decreased the growth rates of both root and shoot SOD, APX activity decreased roots and maximum accumulation of TBARS in leaves	Yin et al., 2008
7	0.014 mg/L	*Myriophyllum spicatum*	Sprayer on the water surface	reduction in photosynthesis	Knauert et al., 2010
8	2.00 kg/ha	*Triticum astivum*	Foliar spray	grain yield enhanced	Ding et al., 2011
9	2–4 mg kg^{-1}	*Triticum astivum*	Soil	reduced the chlorophyll accumulation growth stunt, oxidative damage, higher levels of transcripts of Cu/Zn-SOD and heme oxygenase-1 increase expression of some glucosyltransferases genes	Liang et al., 2012
10	35–50 µg L^{-1}	*Chlamydomonas reinhardtii*	Blue Green-11 medium/In nutrient medium	cell number decreases growth, reduces maximum TBARS accumulation CAT activity was increased HO1 transcripts maximal expression in cells	Bi et al., 2012
11	1.50 kg ha^{-1}	Weed	Soil	reduced weed density, and dry weight of weeds	Amare, 2014
12	1000 g ha^{-1}	Weed	Foliar spray	reduction in weed dry weight increase in grain yield	Elahi et al., 2011
		Triticum astivum			
13	2.5 l ha^{-1}	*Triticum astivum*	Foliar spray	decreased shoot height, fresh and dry weights, Carotenoids, chlorophylls, anthocyanin, activities of d-aminolevulinate dehydratase phenylalanine ammonia lyase and tyrosine	Alla and Hassan, 2014

Table 2.1 contd. ...

...Table 2.1 contd.

S. No.	Dosage	Model Plant	Mode of Application	Effects	References
				ammonia lyase were significantly inhibited, increased malondialdehyde decreased the contents of glutathione and ascorbic acid reduced the activities of superoxide dismutase catalase and ascorbate peroxidase	
14	4 mg kg^{-1}	*Triticum astivum*	Soil	electrolyte leakage in leaves and roots increased activities of GTsgene CD876318 promoted in shoots	Lu et al., 2015
15	10 mM	*Pisum sativum*	Soil	inhibited growth variables reduced carotenoids, chlorophylls, protein activity of nitrate reductase and antioxidants were inhibited	Singh et al., 2016

Metabolites of IPU may be more toxic than IPU itself for fresh water algae, aquatic invertebrates, animals, plants, microbial activities and humans (Remde and Traunspurger, 1994; Mansour et al., 1999; Dosnon-Olette et al., 2010). Recently a number of major IPU metabolites: 3-(4-isopropylphenyl)-1-methylurea (MDIPU), 3-(4-isopropylphenyl)-urea (DDIPU), 4-isopropyl-aniline (4IA), 3,4-dichloroaniline (3,4-DCA), 3,4-dichlorophenylurea (DCU) and N-(3,4-dichlorophenyl)-N-methylurea (DCMU) have been found in the environment (Perrin-Ganier et al., 2001; Sorensen and Aamand 2003). They were seen to be persistent in nature and contributed in the contamination of surface and ground water (Schuelein et al., 1996; Johnson et al., 1998). MDIPU is considered to be the main metabolite which accumulates periodically during the biodegradation of IPU in different soils (Mudd et al., 1983; Cox et al., 1996; Juhler et al., 2001). 4-IA is found to be 600 times more lethal than IPU (Tixier et al., 2002). However IPU herbicides are intentionally manufactured to affect the viability of plants , therefore it is not surprising that high concentrations of IPU and its derivatives have a detrimental effects on algae (Pérés et al., 1996; Mostafa and Helling, 2001), aquatic organisms like amphibians (Greulich et al., 2002), invertebrates and ciliated protozoan (Perrin-Ganier et al., 2001). Its application also reduces the activity of microbes to almost half (50%) of the methanogenesis activity (Attaway et al., 1982; Remde and Traunspurger, 1994) and have severe genotoxic effects (Behera and Bhunya, 1990; Hoshiya et al., 1993). High carbon content increases microbial growth on a contaminated site which in turn, accelerates the degradation rate of organic pollutants (Scow and Hicks, 2005). However, insufficient nutrients reduce the catabolic pathway and the absence of oxidoreducatse chemicals slow down the process of degradation of pesticides (Pandey and Bajpai 2019). Several techniques like precipitation, adsorption, electrodialysis, ion exchange, phytoremediation, microbiological approaches, membrane filtration and nanotechnology have been applied to improve pollutant transformation by natural processes. Herbicides such as atrazine or isoproturon have been often used to treat soil weeds that contribute to the contamination of the soil. The application of polarized electrodes has also been seen to increase microbial mineralization by 20-fold as compared to electrode-free controls of 14C-labelled atrazine and isoproturon in soil (Wang et al., 2020).

A number of scientific experiments have been conducted in recent years on IPU, related to the ability of the growth of plants, transportation and toxicity of IPU and it has been observed that the phytotoxicity and adaptive mechanism for IPU biodegradation has not yet been discovered and needs further investigation in this field. Thus, this chapter takes the first step to target the aforesaid gaps in earlier studies and unravels the IPU accumulation, biotransformation and biomagnification of IPU and their metabolites at higher tropic levels. This work also emphasizes the physiological effect to uncover the real phytotoxicity of isoproturon and their residual effect.

2. Effect of IPU on Crop

2.1 Photosynthesis

Herbicides after being absorbed by plant roots or leaves are translocated to other parts of the plants like shoot tissues, leaf meristem and vascular tissues and then closely interact with the plant content. Pigments are the most efficient and important component of all the compounds. IPU blocks the hill reaction by interrupting the forward flow of the electron through PS II, and thus blocks the transfer of excitation energy from chlorophyll molecules to the PS I, hence disorganizing grana and intergrana, and reducing the pigments content and RuBP carboxylase activity (De Felipe et al., 1989; Kleczkowski 1994; Pascal-Lorber et al., 2010). IPU is commonly used as a photosynthesis inhibiting the herbicide and induces toxicity by electron flow and by generating Reactive Oxygen Species (ROS) (Singh et al., 2016). IPU reduces pigment content of chlorophyll and blocks the flow of electron from PSII to PSI (Kleczkowski, 1994). It has also been found that IPU binds to the D1 protein of the photosynthetic apparatus and affects their stability, thus inhibiting the overall ATP and NADPH production (Nadasy et al., 2000). In a study the alternation of photosynthetic fluorescence parameters was observed in isoproturon-treated *Scenedesmus obliquus* cells. It was also seen that the exposure of *S. obliquus* to 30 g L^{-1} of isoproturon, increased the fluorescence yield at the J transient, showing an inhibition of electron transport at the acceptor side of PSII; however, when *S. obliquus* was exposed to isoproturon at higher than 30 g L^{-1}, the fluorescence yield at P transient decreased, suggesting that the herbicide induced inhibition at the water-splitting level. In some similar experiments, IPU exposure was also seen to cause reduction in the photosynthetic efficiency of algae (Dewez et al., 2008), seaweeds, phytoplankton (Knauert et al., 2008; Dosnon-Olette et al., 2011) and aquatic periphyton (Laviale et al., 2010; Laviale et al., 2011), submerged macrophytes (Knauert et al., 2010). Grouselle et al. (1995) also reported that changes in d-aminolevulinate dehydratase (ALA-D) activity by IPU could result in substantial changes in chlorophyll biosynthesis. Nemat et al. (2008) reported an IPU induced photo-destruction of chlorophylls in maize seedlings due to inhibition of carotenoids (Nemat Alla and Hassan, 2014). A small amount of IPU showed higher bio-concentration and a saturation point in aquatic macrophytes due to this specific binding (Feurtet-Mazel et al., 1996; Crum et al., 1999). While high concentration of IPU had no specific binding to D1 hence showing minimum bioaccumulation in *Elodea densa* (Grouselle et al., 1995). Wheat exposed to 2 mg kg^{-1} of IPU decreased chlorophyll accumulation and induced oxidative stress (Liang et al., 2012). Application of 10 μg L^{-1} IPU in *Scenedesmus obliquus* and *Scenedesmus quadricauda* decreased its photosynthetic efficiency and inhibited its growth rate (Dosnon-Olette et al., 2010). IPU also reduces the chlorophyll content with 25 g L^{-1} IPU in 2 d after IPU treatment (Bi et al., 2012).

2.2 Oxidative stress

Various unsuitable conditions, such as drought, heavy metal toxicity, soil salinity and herbicide contamination cause the production of Reactive Oxygen Species (ROS) which leads to oxidative stress in plants. To maintain normal physiology and homeostasis of the plant cell's antioxidants play an important physiological role in the development of plant growth and can also regularize the state of plants under different stresses. IPU disrupts the electron transport which leads to recombination of charges and formation of triplet chlorophyll to combine with O_2 to form singlet oxygen (Rutherford and Krieger-Liszkay, 2001). The transfer of electrons to O_2 forms reactive oxygen species (ROS) such as superoxide radical (O_2^-), hydrogen peroxide (H_2O_2) or hydroxyl radical (HO^-). ROS have the ability of uncontrolled oxidation of different cellular components which in turn leads to oxidative damage of the cell (Hassan and Nemat Alla, 2005). IPU-induced ROS affect various macromolecules in the cell and cause peroxidation of lipid and protein,nucleic acids and enzymes (Yin et al., 2008). The impact of IPU on membrane leakage and enzyme inactivation causes severe necrosis and chlorosis in plants (Pascal-Lorber et al., 2010). Isoproturon-induced oxidative stress enhances the accumulation of MDA content which is a biomarker for lipid peroxidation and H_2O_2 generation in maize and wheat (Nemat Alla and Hassan, 2007; Nemat Alla and Hassan, 2014). Bi et al. (2012) in their study reported that algal cells accumulated 3.43-folds higher TBARS than the control at 50 g L^{-1} concentration of IPU. Plants have a ROS-scavenging system with enzymatic and non-enzymatic antioxidants (Mittle, 2002). Glutathione (GSH), a non-enzymatic antioxidant has been seen to repress IPU action in wheat (Nemat Alla and Hassan, 2014). The activity of soil enzyme dehydrogenase (DHA) has also been reported to decrease due to the negative impact of IPU on soil microbial communities (Lu et al., 2015). IPU exposure also increases the activity of antioxidant enzymes such as, SOD, APX and POD due to oxidative stress (Liang et al., 2012). They also suggested a parallel alteration in gene expression of Cu/Zn-SOD. IPU at 20 mg/kgconcentration in soil decreased SOD activities, caused as a result of destruction of enzymes by accumulation of H_2O_2 in wheat plants (Yin et al., 2008). In this study the authors tested the activity of Glutathione S-Transferase (GST) and observed the activity stimulated by isoproturon at 2–10 mg/kg. The defence system includes secondary metabolites, called phytochemicals having a large role in ROS detoxification and accumulates during defence responses such as phenolic compounds, and the generation of enzymes for the synthesis of lignin and phytoalexin. These important compounds are required for the viability of plant cells under stress conditions (Sunohara and Matsumoto, 2004). The alteration in anthocyanin component and two secondary metabolites activities Phenylalanine Ammonia Lyase (PAL) and Tyrosine Analine Lyase (TAL) enzymes were reported to change due to ROS accumulation. This switches to the secondary metabolism for the production of terpenoids, isoflavonoids, lignins, tannins, anthocyanin, etc., to cope with stress generated due to excessive uses of IPU (recommended doses) (Nemat Alla and Hassan, 2014).These metabolites also increase under various abiotic stresses to buttress the defence system of plants (Gottstein et al., 1991). IPU induces non-enzymatic activities in plant systems (Nemat Alla and Younis, 1995). IPU exposure reduced anthocynanin content in plants, and anthocyanin is related with PAL and TAL activities (Scarponi et al., 1998). Uncharacterized glycosyl transferase (IRGT1) confers plant resistance to isoproturon–acetochlor, which instigates cellular injury and reduces rice toxicity under isoproturon–a cetochlor stress (Su et al., 2019).

3. IPU Resistance in Crop

IPU resistance in weeds follows two primary pathways referred to as 'non-target site' resistance, reduced absorption or translocation and 'target site' resistance (mutation). Control of resistant weeds requires more than 10 to 12 times higher rate of IPU than susceptible population (Singh et al., 1996; Singh et al., 1997). *Phalaris minor* is controlled only by IPU, but regular use of this chemical causes resistance against IPU. The acquired resistance of IPU is due to a point mutation in the nucleotide sequence of psbA gene of D1 protein (Tripathi, 2003; Tripathi et al., 2005). D1 protein is a QB binding protein encoded by chloroplast psbA gene which is highly conserved (Mordern and Golden, 1989). Resistance to IPU due to alteration in QB binding site decreases proximity with the herbicide to shift plastoquinone and electron flow from QA to QB (Tripathi et al., 2005). In *Phalaris* and purple bacteria psbA, single site mutation in its gene, causes resistance against the herbicide (Trebst, 1991; Singh et al., 2004). It has been reported that resistance takes place due to four-point mutations in which two mutations are found in thylakoid lumen (Ser and Ala residues) 3rd in stromal transmembrane (Leu residue) and the 4th mutation is silent (Singh et al., 2012). However, resistance may also be due to modification of secondary structure of D1 protein which alters the binding position, loss of hydrogen bonds and hydrophobic interaction and displaced binding site. Resistance of weeds against IPU may be endorsed to increased herbicide detoxification by P-450 mono oxygenase activity (Singh et al., 1998). In a recent study it was reported that widely used chlorinated pesticides including linuron, chlorotoluron, atrazine and isoproturon could be completely oxidized by the cytochrome P450s (Kawahigashi et al., 2007). Khanom et al., 2019 reported the involvement of ginseng-derived CYP736A12 in chlortoluron and isoproturon tolerance when over expressed in *A. thaliana*. Höfer et al., 2014 illustrated that the CYP76C1 gene conferred tolerance to chlorotoluron and isoproturon, when overexpressed in native *A. thaliana* plants.

According to the obtained results, transgenic CYP1A2 *A. thaliana* plants showed high efficiency toward isoproturon herbicide degradation and can be recommended for phytoremediation of correspondingly contaminated sites (Azab et al., 2020). In tolerant soybean plants, isoproturon (IPU) was detoxified to monodesmethyl-IPU, 2-hydroxy-IPU, and 2-hydroxy-monodesmethyl-IPU. While in the case of wheat, the major metabolic pathway degraded the isoproturon (IPU) to 2-hydroxy-IPU as a primary metabolite, then to 2-hydroxy-monodesmethyl-IPU, 2-Hydroxy-IPU and an olefinic metabolite (isopropenyl-IPU) (Glassgen et al., 1999). *CYP76B1* was isolated from Jerusalem artichoke (*Helianthus tuberosus*) and was found to metabolize herbicides belonging to the phenylurea class (Robineau et al., 1998; Didierjean et al., 2002). Homology based molecular docking of PgCYP76B9 showed more possible interaction with chlortoluron rather than isoproturon (Jang et al., 2020).

4. Uptake and Translocation of IPU in Crops

The interactions between pesticides and plants have gained more attention in recent years. However the systemic uptake, translocation and metabolism of IPU by plants still remains unclear and needs to be clarified. This work is progressively more recognized by the scientific community and studies on plant–IPU interaction are playing a major role in improving rural livelihoods. It has been observed that after cuticle penetration, foliar-applied herbicides may translocate apoplastically in the transpiration stream or symplastically in the phloem or remain immobilized in the treated leaf. Achhireddy and Kirkwood (1986) reported that the amount of IPU translocated through xylem was 10 times higher as compared to translocation via phloem. This study suggested that isoproturon is predominately an apoplastically translocated herbicide. Blair's experiments (1978) showed that IPU action occurs mainly by penetration by roots. IPU with 17 to 105 d of half-life is moderately mobile in soil and pollutes the environment and water resources through diffusion via spraying (Johnson et al., 2001; Sorensen and Aamand, 2003). Sometimes this compound also tends to bind with organic matter residues or humus particles in the soil. Therefore,in order to establish a significant relation, testing is required to determine whether they are either "Dead dogs" or rather "Time bombs". Schroll and Kühn (2004) reported that IPU is metabolized to plants available and reactive compounds in rhizosphere soil. In the agricultural field, its distribution is affected by plant uptake, leaching, adsorption and abiotic and biotic biodegradation processes and run-off from the field to different areas of the surrounding environment (Gooddy et al., 2007). IPU is carried in the soil atmosphere depending on the partition coefficient (*K*d). *K*d is assessed as pesticides between the solid and the water phase of the soil (Walker et al., 1999). IPU has *K*d between 1.5 to 10 L kg^{-1} which means that it is poorly retained in soil (Sorensen and Aamand, 2003). Adsorption of IPU is affected by the moisture content, cations like Ca^{2+} and K^{+} and temperature (Coquet, 2003; El-Arfaoui et al., 2010). The elevated temperature limits the rate of adsorption of IPU (Amita et al., 2005). Organic acids (low molecular weight)and dissolved organic matter in soils influence leaching out of IPU (Ding et al., 2011). IPU has 50% reduction between 17 to 235 d after application, therefore, has low affinity and average persistence to adsorb on the soil (Chhokar and Malik, 2002; Tian et al., 2010). It has been found that some soil organisms (earthworms) also take part in pesticide leaching. The association between worms and soil enables earthworms to absorb pesticides (Van Straalen and Gestel, 1993). Leaching occurs due to water flowing through burrow linings of worms and mechanical cracks (to ploughing or shrinking) of soil (Dorfler et al., 2006; Dousset et al., 2007).

5. Biodegradation of IPU

In order to assess the impact of a polluting agent, studies of biotransformation need to be reviewed. The pollutant could undergo transformation yielding compounds which are more toxic than the parent molecule. In the environment, organic pollutants can be degraded by different biological, chemical or photochemical processes. Wastewaters containing heavy metals and pollutants are broadly discharged from industries and enter into nourishment chains thus resulting in mutagenesis, carcinogenicity and disabilities in living organisms. A number of technologies are executed to remediate both inorganic and organic contaminants from waste waters. Chemical mechanisms like hydroxylation, oxidation and hydrolysis play an important role in transforming IPU into metabolites (Acero et al., 2007; Gangwar and Rafiquee, 2007b; Reddy et al., 2010). IPU exposure to sunlight causes photochemical decomposition of the herbicide in demethylated metabolites found in water and the soil surface (Parra et al., 2000; Rubio et al., 2006). IPU degradation is caused by solarization and photo-fenton photocatalytic mechanism in aquatic and agriculture fields (Rubio et al., 2006; Navarro et al., 2009; Lopez-Munoz et al., 2013). TiO_2 as a photocatalyst induces photocatalytic degradation of IPU and its major metabolites under exposure of solar radiation (Haque and Muneer, 2003; Reddy et al., 2011). TiO_2 when used with other additives such as mesoporous Al-MCM-41, H-mordenite and HY zeolite shows better photocatalytic activity (Sharma et al., 2008a; Sharma et al., 2008b). IPU forms different transformation products during alkaline and acid hydrolysis (Gangwar and Rafiquee, 2007a; Gangwar and Rafiquee, 2007b). Soil contamination with organic pollutants and their derivatives is a major concern in industrialized countries. Exploiting the metabolic capability of microorganisms to transform pollutants is a promising approach for bioremediation. Microbes playa key role in the biodegradation of herbicides (Bending et al., 2003; El-Sebai et al., 2007; Hussain et al., 2011). In addition to microbes, other organisms like plants, algae, animals and earthworms also biotransform IPU to metabolites (Greulich et al., 2002; Paris-Palacios et al., 2010). It has been reported that *Lemna minor* and *Scenedesmus* phytoremediate from 25 and 58% of IPU (Bottcher and Schroll, 2007; Dosnon-Olette et al., 2011). Microorganisms have a significant role in biodegradation and dissipation of IPU in soil (Johnson et al., 1998; Sørensen and Aamand, 2001). Microbes mineralize more than 40% of the previously added IPU into carbon dioxide in agricultural fields (Bending et al., 2001; Sørensen et al., 2001). A bacterial mixed culture found in soil, when treated with IPU has the ability to mineralize IPU and its metabolites in contaminated soil (Sørensen and Aamand, 2001). Different strains of *Sphingomonas* sp. when co-cultured enhanced the mineralizing activity of IPU by providing amino acid to each other (Sørensen et al., 2002; Hussain et al., 2009).

Sun et al. (2009) described that genes encoding for catechol metabolites are present in *Sphingobium* strains YBL1 and YBL3. The three strains of *Sphingomonas* sp., i.e., YBL1, YBL2, YBL3 and another strain Y57, which was examined from soil collected from Changzhou. These strains were isolated from various soils containing different herbicide-manufacturing plants, however all three soils had been exposed to IPU for a long time (Sun et al., 2006). The isolation of these strains confirmed that members of the genus *Sphingobium* characteristically play a significant role in the organic waste degradation (Dai and Copley, 2004; Prakash and Lal, 2006; Sharma et al., 2006). These studies suggested that soil with these strains when exposed to herbicides increases the activity and abundance of those bacteria responsible for degradation of herbicides (Ostrofsky et al., 1997; Pussemier et al., 1997). 16S rRNA gene sequencing, physiological and phenotypic analysis proved that of these three isolated strains of *Sphingobium* sp., have an IPU biodegradation capability. The first type isoproturon-degrading strain was isolated by Sorensen and Aamand (2001) and represented by the strain *Sphingomonas* sp. SRS2, which can also act on some closely related dimethylurea herbicides, such as diuron and chlorotoluron, starting by *N*-demethylation preceding by phenyl structure degradation. Other strains reported from this group such as YBL2 and YBL3 can also degrade chlorotoluron, diuron, fluometuron and 4IA (Sun et al., 2009). The second strain type was *Methylopila* sp. TES (El-Sebai et al., 2004), which degraded only IPU and caused 4IA mineralization. However, it was unable to act on other phenylurea herbicides. The third type of strain degrading phenylurea herbicide are, strains *Arthrobacter globiformis* D47 and *Variovorax* sp. WDL1 in contrast to SRS2 strain of *Sphingomonas* sp. (Turnbull et al., 2001; Dejonghe et al., 2003). This group is able to transform various phenylurea herbicides to their corresponding aniline derivatives, but then unable to degrade the corresponding aniline further.

Strain YBL1 are different from other strains because it could biodegrade of most of the phenylurea herbicides derivatives including dimethylphenylurea and methoxymethylphenylurea (linuron), though this strain was also able to degrade 4IA and aniline (Sun et al., 2009). Some other bacterial strains taking part in degradation of phenylurea herbicides,especially IPU are *Mycobacterium brisbanense* JK1, *Pseudomonas aeruginosa* JS-11 and *Sphingomonas* sp. SH (Khurana et al., 2009; Dwivedi et al., 2011; Hussain et al., 2011).

Pseudomonas aeruginosa strain JS-11 acts as a plant growth stimulator and controls disease; it is isolated from wheat rhizosphere, when grown on phenylurea herbicide, it partially degrades IPU into aniline derivatives (Dwivedi et al., 2011). During IPU biodegradation several phenotypes of bacterial strains are found which play an important role in degrading metabolites like MDIPU and DDIPU (Roberts et al., 1998). Bacterial isolates, viz., *Sphingobium* and *Sphingomonas* bacteria are widespread, bear a catabolic pathway for IPU mineralization and utilize carbon and nitrogen for their growth (Fredrickson et al., 1999; Sorensen and Aamand, 2001; Bending et al., 2003; Sun et al., 2009; Hussain et al., 2011). Different fungal strains of *Zygomycetes*, *Basidiomycetes* and *Agonomycetes* transform IPU into hydroxylated or demethylated products (Khadrani et al., 1999; Ronhede et al., 2005; Hangler et al., 2007). A white rot fungus *Phanerochaete chrysosporium* has manganese peroxidase enzyme activity which is also related with IPU biodegradation (Castillo et al., 2001). *Rhizoctonia solani* and *Bjerkandera adusta* when degraded have shown to add around 84 and 80% of IPU respectively (Vroumsia et al., 1996; Khadrani et al., 1999). The microbial strain *Sphingomonas* sp. within this microbial community can only conduct IPU mineralization because these strains use IPU as a carbon source (Kiesel, 2014).

Biotransformation of IPU and its derivatives are enhanced by natural and additives of low cost, such as Salicylic Acid (SA) in crops (Liang et al., 2012; Lu et al., 2015). These transformed products further degrade into vacuole or peroxisome (Cole and Edwards, 2000; Brazier et al., 2002). However, detailed studies are not reported on biodegradation and detoxification of toxic organic compounds by SA regulated mechanism. In recent studies SA is capable of reducing accumulation of IPU in wheat plants (Liang et al., 2012; Mutlu et al., 2009). This is due to degradation of IPU by SA and not because of SA-mediated inhibition in intake of IPU in wheat plants which is responsible for decreased accumulation of IPU. Cellular toxicity induced by IPU is significantly reduced when plants are treated with SA, this reduced toxicity indicate that SA is capable of degradation of IPU through a detoxification pathway (Lu et al., 2015). Enzymes that regulate function of SA are glycosyltransferases (GTs). GTs belong to large group of enzymes that plays a crucial role in SA regulations and these are *O*-glucosyl-glucosyltransferase (*O*-GTs), *N*-glucosyl-glucosyltransferase (*N*-GTs). Some genes are also involved in this process such as (CD876318), this gene is not only induced by SA but also by IPU in stems (Desmond et al., 2008). Recent studies suggested that SA-mediated regulation causes glycosylation of IPU-derivatives through activation of enzymes, GTs (Lu et al., 2015). This study is potentially useful for enhancement of IPU degradation in wheat crops grown in mild IPU-contaminated soils showing potential inhibition in accumulation of IPU when treated with SA and lowers the adverse effect of IPU on health.

Methyl-jasmonic acid has been reported to degrade IPU in wheat crops when grown in an isoproturon-polluted soil (Li et al., 2018). Transformed products conjugate with glucosyl moieties and enhance their solubility in cells (Coleman et al., 1997; Sun et al., 2009; Pascal-Lorber et al., 2010). Application of exogenous SA in wheat roots ameliorates IPU intracellular catabolism and induces organic acids (LWMOAs) efflux which affects rhizosphere microbial activity due to biodegradation of IPU residue (Lu et al., 2015). Lu et al., 2015 also reported that the population of Gram−bacteria is higher in IPU+SA compared to IPU as in the presence of SA more carbon source is accessible. It is evinced that natural growth regulator/allelochemical has the potential to overcome the adverse effects of IPU (Singh et al., 2016). SA exposure causes reduction in heme oxygenase-1 expression and mitigates the toxicity of IPU (Liang et al., 2012) (Fig. 2.1). OsPAL (phenylalanine ammonia lyase) as an important SA synthetic component exhibits an essential role in the biodegradation of IPU in paddy crops.

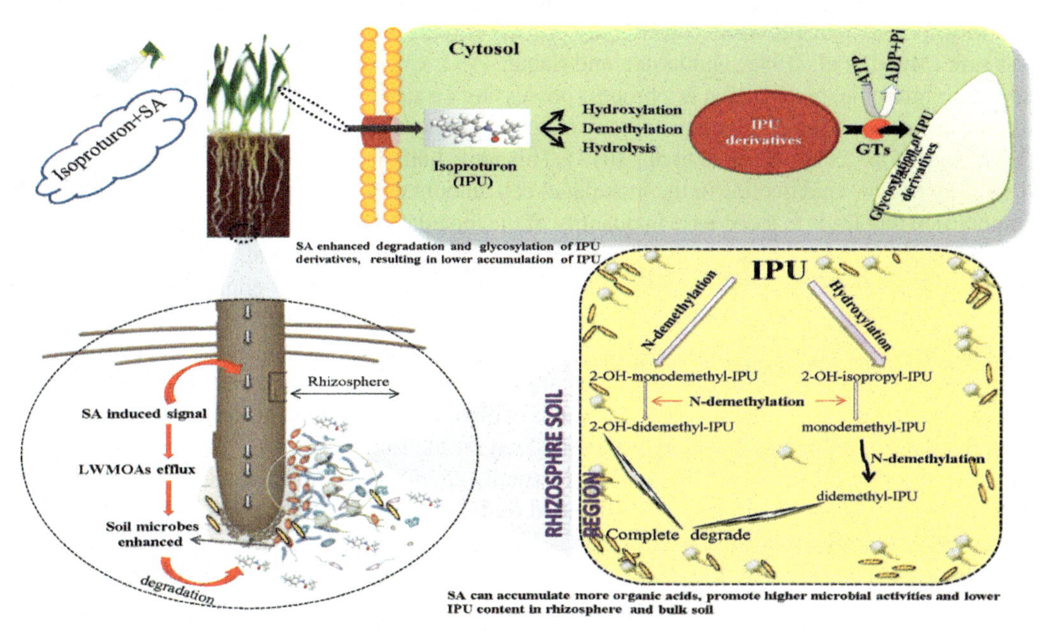

Fig. 2.1. Pathway of leaf and root giving general idea and principle of uotaken, translocation and bio-transformation of IPU by plant. Foliar IPU in a natural environment are uptake through stomata or direct penetration through aerial surface. Soil IPU is absorbed and uptake by root hairs. This pathways also represent the role of rhizophere in IPU degradation and their metabolites into cell organelles.

Some genes such as *pdm*AB, *Pdm*A and *Pdm*B in *Sphingobium* sp. code the enzymes responsible for IPU biodegradation. The above reported genes are highly conserved along with their transposable elements. IPU is found to induce aniline dioxygenase (ADO) and catechol 2,3-dioxygenase (CD-2,3) activity in some strains of *Sphingobium* sps. The phenyl ring of IPU may be degraded through ortho-cleavage of catechol (Hussain et al., 2011). The gene *pdm*AB has been characterized in *Sphingobium* sp. YBL2 is also accountable for the initial N-demethylation of the phenylurea herbicide (Gu et al., 2013). *Pseudoxanthomonas japonensis* a novel bacterium, which has an IPU metabolizing capacity and can be a good option for the remediation of IPU-contaminated sites (Giri et al., 2016). Some strategies are used by crop plants to degrade a wide range of xenobiotics such as pesticides, while it is reported that mechanisms of xenobiotics degradation and detoxification are necessary for plant growth. In IPU degradation, there is a transformation of IPU from one form to another by the process of hydroxylation and *N*-demethylation. Conjugation of IPU and its derivatives to glucose and glutathione have resulted in degradation of IPU. Based on recent studies, for complete degradation these modified compounds are exported into vacuole or peroxisome (Brazier et al., 2002; Cole and Edwards, 2000). As shown by Time-of-Flight Mass Spectrometry (TOF-MS/MS), isopropenyl-IPU can be formed only in aerial parts as an immediate product from 2-OH-isopropyl-IPU. In plants conjugation of xenobiotics to glucosyl, malonyl-glucosyl and acetyl-glucosyl moieties are found (Sun et al., 2009; Pascal-Lorber et al., 2010). This mechanism increases the solubility of xenobiotics in cells that results in detoxification of artificial substances (Coleman et al., 1997).

Phytoremediation innovation can help in cleaning up contaminated zones. Phytoremediation is considered a viable green innovation and low-cost arrangement for water and soil defilement. Remediation of phenylurea isoproturon herbicide using transgenic *Arabidopsis thaliana* plants showing a human cytochrome P450-1A2 has also been explored in some scientific studies. P450-1A2 has been seen to metabolize the herbicide isoproturon (Azab et al., 2020). Phytodegradation is also known as phytotransformation, that is able to arrange omnipresent natural contaminations with care and possibly poisonous natural compounds, through using plant species to take-up and debase toxins (Newman and Reynolds, 2004).

In situ bioremediation is an eco-friendly process which is a cost-effective, less labour-intensive and safe (Muud et al., 1983; Gaillardon and Sabar, 1994; Cox et al., 1996). Bioaugmentation also known as microbial remediation is a process possessing versatile catabolic capabilities and has been extensively used for the elimination of many organic pollutants by inoculation of degrading microbes (Elkhattabi et al., 2007; Hussian et al., 2013). This remediation has some disadvantages such as its instability due to rapid decline in the inoculated cell amount during its indigenous microorganism's competition that leads to a decrease in quantity of the inoculated cell which causes its instability and it does not show much impact on pollutants residing in deep sites (Sun et al., 2009; Bending et al., 2003). Phytoremediation is a renewable and autotrophic method used for elimination of pollutants by rapidly growing plants which are environmentally well-adapted (Abhilash et al., 2009). Some mechanisms are adapted by plants for uptake of pollutants from deep sites by developing their root system extensively and these pollutants are then translocated to the plants tissues for further metabolized processes. Some supplementary aids of phytoremediation are the production of biofuels, stabilization of soil and carbon sequestration (Doty et al., 2000). Unlike, microbes the versatile catabolic capacity for intractable pollutants are usually absent in plants. Therefore, significant genes responsible for pollutant elimination is intended to enhance the catabolic ability of plants that are transferred from microbes to plants. Enzymes like bacterial pentaerythritol tetranitrate reductase, nitroreductase, cytochrome P450, extradiol dioxygenase (DbfB), haloalkane dehalogenase (DhaA) and naphthalene dioxygenase have been successfully isolated from *Arabidopsis*, tobacco and rice plants. These enzymes have been observed to play a prominent role in detoxification, remediation and degradation nitroglycerin (French et al., 1999). Many compounds such as 2,4,6-trinitrotoluene (TNT) (Hannink et al., 2001; van Dillewijn et al., 2008), cyclotrimethylenetrinitramine (RDX) (Rylott et al., 2006), 3-dihydroxybiphenyl (2,3-DHB), 1-chlorobutane (1-CB) (Uchida et al., 2005) and aromatic hydrocarbons (Peng et al., 2014a; Peng et al., 2014b) are involved in the detoxification of compounds. Studies on these bioactive compounds provide new insight on the utility of phytoremediation methods in contaminated sites by transgenic plants which express similar catabolic enzymes reported in bacterial populations. However, for target pollutant mineralizing, it is generally hard to allocate the whole cluster of a catabolic gene to the plants, irrespective if a single catabolic gene is transferred to plants and could release some intermediates which can induce phytotoxicity and cause other environmental issues. To overcome these problems more studies on new strategies are required to improve the efficiency of phytoremediation. In recent experiments the prominent role of rhizospheric microbes is observed in phytoremediation that is successfully used for degradation of pollutants (Kuiper et al., 2004; Weyens et al., 2009; Barac et al., 2004). There is a synergetic association between microbes and plants, this relationship is mutual and could be beneficial for both of them (Abhilash et al., 2012). Thus, to solve the difficulties that appear by the application of both distinct phytoremediation and bioaugmentation practices, the use of plants in combination with microbes has several advantages and could serve as a captivating process (Yan et al., 2018).

Expression of N-demethylase PdmAB of bacteria in a transgenic *Arabidopsis* plant is formed to induce the initial degradative process by reducing N,N-dimethyl-substituted PHs. Principal-coordinate analysis (PCoA), a PdmAB ferredoxin substitute of bacteria is belongs to bacterial communities of weighted Unifrac distances when exposed to different treatments given in soils (Scheunert and Reuter, 2000). Recent studies suggested that not only bacteria but many fungi also play a crucial role in the degradative process of IPU (Hussain et al., 2017). Strains of Agonomycetes, Rhizoctonia and Zygomycetes were found to have an IPU degradation ability. However, from all these strains, Rhizoctonia strains were reported to be more effective to degrade about 84% of IPU which was initially added within 2 d than the others (Vroumsia et al., 1996). Furthermore, no herbicide metabolites accumulation was reported. Steiman et al. (1994) also found the rapid degradative ability of *Rhizoctonia* sp. for IPU herbicides. IPU were also transformed into demethylated and hydroxylated products by some other strains belonging to the genus Basidiomycetes, Ascomycetes and Zygomycetes out of which hydroxylation was reported to be a major pathway of transformation by these fungi (Ronhede et al., 2005).

Hydroxylation reaction increases IPU mineralization in soil (Ronhede et al., 2007). Similarly, it was examined that JS/2 of *Cunninghamella elegans* was capable of producing many metabolites from IPU (Hangler et al., 2007). The addition of IPU in soil initiated degradative responses of fungal strains and many other enzymes also participated in the degradative process of IPU. A soil fungal strain such as Gr4 strain of *Mortierella* sp. used two consecutive processes to degrade IPU, these two procedures involved, urea side chain N-demethylation and isopropyl ring constituent's hydroxylation (Badawi et al., 2009). These reports suggested that pathways utilized to degrade IPU by bacteria were different from the ways followed by different fungal genus. Some freshwater macrophytes such as *Elodea densa* and *Ludwigia natans* can accumulate IPU up to 60-fold (Fuertet-Mazel et al., 2000). However, due to massive inhibition in the growth of these plants this process of bioremediation cannot be used for longer periods. Phenylureas compounds have a high organic carbon amount that tends to persist in soils. Even if they are biodegradable, an ample amount may remain in the soil in the form of chloroanilins which is also persistent in soil and causing toxicity (Coleman et al., 2002). Specific features of most cereals crops show them as tolerant and good targets for phytoremediation. However, facilitation of the isoproturon uptake into the root of the plant has yet to be done.

Bioavailability of IPU in contaminated lands depends upon its use in the plant parts as well as different varieties of these factors are chiefly responsible for removal of IPU through the phytoremediation process and by the uptake into the plants. The metabolites formed after degradation of IPU by bacteria were poorly utilized by plants. However, studies have shown the harmful effects of these metabolized products such as chloroanimins have very damaging effects to the biosphere. In water, there might be a net uptake of the isoproturon observed depending on the conditions set.

6. Metabolic Pathways of Isoproturon

The outcome of pesticides in the subsurface is of great interest to the public, industry and regulatory authorities. Earlier studies have reported that the concentration of MDIPU is higher in agricultural fields and in bacterial and fungal culture as compared to the parent molecule (Berger, 1999; Perrin-Ganier et al., 2001; Hussain et al., 2011; Zhang et al., 2012). MDIPU is formed by demethylation of the dimethylurea side chain of IPU (Sorensen and Aamand, 2001, 2003). Bacterial strain, viz., *Sphingomonas* and *Sphingobium* and fungal strain *Mortierella* Gr4 transform IPU into MDIPU, DDIPU and 4-IA metabolites. This work also suggested that4-IA is formed directly by MDIPU in bacterial culture (Sørensen et al., 2001; Zhang et al., 2012; Badawi et al., 2009).

There are some other alternative pathways which are also found in the *Mortierella* sp. Gr4 and other diverse bacterial cultures which show the hydroxylation of the first and second position of IPU side chain, lead to the formation of 1-OH-IPU and 2-OH-IPU (Lehr et al., 1996; Sorensen and Aamand, 2003; Badawi et al., 2009). Lehr et al. (1996) recognized that 1-OH IPU is a very deadly toxic metabolite synthesized during the culture. C, N and H isotope fractionation of the IPU reflects different modes of microbial transformation (Penning et al., 2010). Hydroxylation of the isopropyl group by fungi was found to be associated with C and H isotope fractionation. In contrast, hydrolysis by *Arthrobacter globiformis* D47 caused strong C and N isotope fractionation. Metabolization of aniline compounds takes place by the development of important intermediate catechol by oxidative deamination steps (Sørensen et al., 2003). 4-IA induces activity of aniline dioxygenase in *Sphingobium* sp. and transformsit into 4-isopropylcatechol by a catechol ring arrangement and cleavage at *ortho* and meta position (Sun et al., 2009; Zhang et al., 2012).Gene *catA* encodes catechol 1,2-dioxygenase which is found in different strains of *Sphingobium* (Sun et al., 2009) and *Sphingomonas* sp. (Hussain et al., 2011). Other residues 4-isopropyl-cis and cis muconic acid developed due to IPU biodegradation by catechol (*ortho* cleavage) (Zhang et al., 2012; Hussain et al., 2013). IPU exposure enhanced Catechol-2,3-dioxygenase enzyme activity and *meta*-cleavage of aromatic ring resulting in the transformation of catechol to 4-isopropyl-hydroxymuconic semialdehyde residue (Zhang et al., 2012). Several other aniline compounds have also been reported to endure a range of chemical reactions in soils to form a variety of compounds (Pieuchot et al., 1996; Perrin-Ganier et al., 2001) summarized in Fig. 2.2.

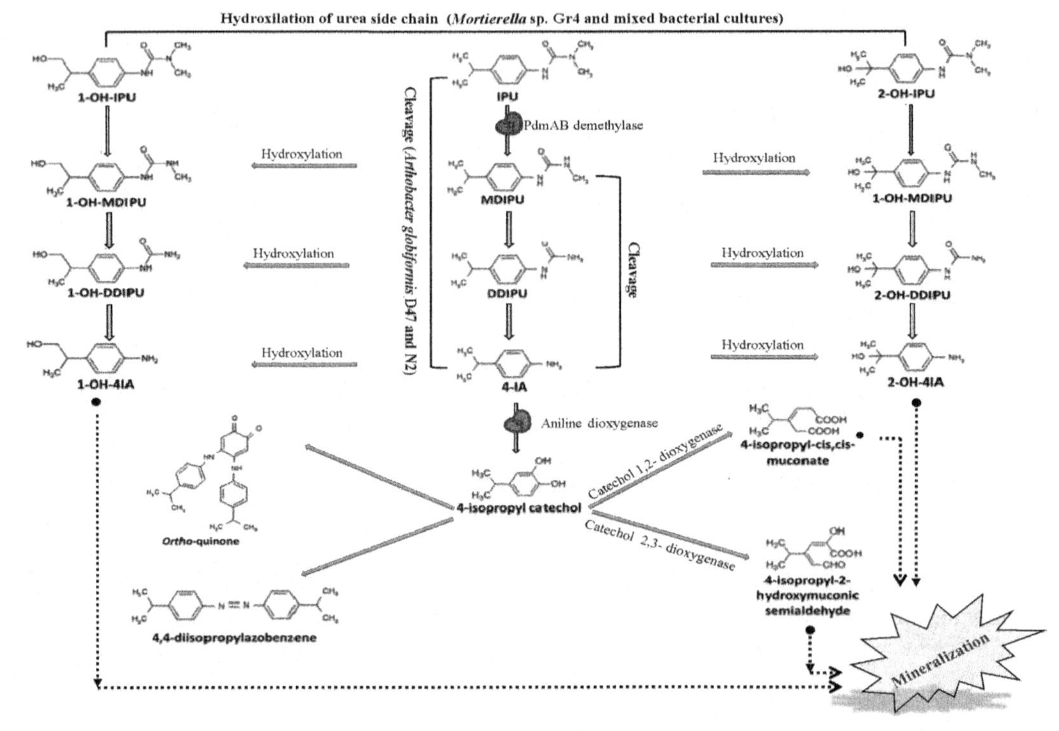

Fig. 2.2. Pathway of IPU biodegradation. Various biochemical processes governing the environmental fate of IPU in agricultural field. The recent advances in the microbial aspects of their degradation in the field as well as in pure culture (Hussin et al., 2015).

7. Factors Affecting Biotransformation of Isoproturon

The adaptation of crops and soil towards degradation of IPU has been experienced in agronomic fields worldwide (Walker et al., 2001; Bending et al., 2006; Hussain et al., 2013). The climatic conditions of soil and the chemical properties of pesticides influence the biodegradation of pesticides (El-Sebai et al., 2007). These factors affect the diversity, abundance, availability of pesticides and biodegradability and pesticide-degrading contagious populations (Hussain et al., 2011). The biodegradation of IPU is affected by pedoclimatic factors as well, viz., pH, moisture content, profundity, microbe population, oxygen accessibility, cation exchange capacity and organic matter of soil (Bending and Rodriguez-Cruz, 2007; El-Sebai et al., 2011; Fenlon et al., 2011; Grundmann et al., 2011; Hussain et al., 2013). IPU transformation is correlated with the soil microbe's abundance which depends on soil pH (range 7 to 7.5) (Bending et al., 2003). Soil pH is a restrictive factor for biodegradation of IPU in the field. The optimal temperature is 30°C for IPU mineralization by *Sphingobium* sps. under laboratory conditions (Sun et al., 2009).The rate of IPU degradation is affected by the depth of the soil as well. The horizon region of 0–20 cm exhibits a higher rate of degradation in comparison to sub-surface horizon 20–50 cm (Issa and Wood, 2005). The biodegradation activity of the subsurface range is low due to the small population of microorganisms (Bending and Rodriguez-Cruz, 2007). Adsorption and soil bound residues of IPU may affect its availability to microorganisms which results in decreased rate of IPU deprivation (Johannesen et al., 2003; Alletto et al., 2006). The biodegradation rate of IPU and formation of its derivatives depend on soil compression which reduces availability of oxygen to microorganisms (Mamy et al., 2011). Various other factors like agricultural practices, organic matters, municipal waste, sewage and charcoal, etc., affect the biotransformation of IPU (Sun et al., 2009; Si et al., 2011). Composts added to the soil also increases IPU biodegradation but heavy metal toxicity reduces mineralization rate of IPU (Suhadolc et al., 2004; Coppola et al., 2011).

8. Conclusion

In summary, the relationship between ecosystem flora (plants) and IPU is complex. Herbicides have organic and financial benefits; however, they can accumulate within the biosphere and cause long-term unfavourable impacts on the encompassing environment. Rapid progression with respect to mechanisms like phytotoxicity, translocation and metabolism of IPU it is still an emerging field and numerous questions related to it need to be dealt with. More light on understanding and exposure of IPU by plants is required. Earlier studies related to IUP, have shown both positive and negative impacts of IPU on plant growth, development, grain quantity and productivity. However, there is need for critical data to justify the mode of action of IPU and their relationships with other cell metabolites and in gene expression. The varied plant responses to IPU are yet to be explored. Many questions like the effect of the vascular bundle of crops in the translocation and biotransformation of IPU, still need to be explored and answered. IPU and their residue at higher tropic levels and the biomagnifications of residue should also be described. Exact translocation mechanism of IPU via modes like apoplast and other pathways such as symplast inside a plant system requires authentic evaluation by the scientific community. Finally, where and how IPU are biotransformed into metabolites inside plant tissues is as yet not explored much. Several bacterial and fungal strains involved in its mineralization are yet to be discovered. Inoculation with IPU-degrading microbial community not only enhances the mineralization, but also improves the biodegradation quality and thus reduces the possible environmental risks associated with IPU and its degradation products. The molecular mechanism of IPU catabolism has not yet been elucidated much. Therefore, further studies will enhance our understanding of the microbial mineralization of IPU and evolutionary molecular mechanism of the IPU-catabolic pathway and thus reduce the possible environmental risks associated with IPU and its biodegradation products.

Acknowledgement

The authors have not received any funding.

References

Abhilash, P.C., Jamil, S. and Singh, N. 2009. Transgenic plants for enhanced biodegradation and phytoremediation of organic xenobiotics. Biotechnol. Adv. 27: 474–488. https://doi.org/10.1016/j.biotechadv.2009.04.002.

Abhilash, P.C., Powell, J.R., Singh, H.B. and Singh, B.K. 2012. Plant-microbe interactions: Novel applications for exploitation in multipurpose remediation technologies. Trends Biotechnol. 30: 416–420. https://doi.org/10.1016/j.tibtech.2012.04.004.

Acero, J.L., Real, F.J., Benitez, F.J. and Gonzalez, M. 2007. Kinetics of reactions between chlorine or bromine and the herbicides diuron and isoproturon. J. Chem. Technol. Biotechnol. 82: 214–222. https://doi.org/10.1002/jctb.1660.

Achhireddy, N.R. and Kirkwood, R.C. 1986. The foliar uptake of isoproturon by wheat and Alopecurus myosuroides huds. Pest. Sci. 17(1): 53–57. https://doi.org/10.1002/ps.2780170107.

Alla, M.M.N. and Hassan, N.M. 2014. Alleviation of isoproturon toxicity to wheat by exogenous application of glutathione. Pesticide Biochemistry and Physiology 112: 56–62.

Alletto, L., Coquet, Y., Benoit, P. and Bergheaud, V. 2006. Effects of temperature and water content on degradation of isoproturon in three soil profiles. Chemosphere 64: 1053–1061. https://doi.org/10.1016/j.chemosphere.2005.12.004.

Amare, T. 2014. Effect of weed management methods on weeds and wheat (*Triticum aestivum* L.) yield. Afr. J. Agri. Res. 9(24): 1914–1920. http://www.academicjournals.org/AJAR.

Amita, B., Anjali, S., Srivastava, A., Bali, R., Srivastava, P.C. and Govindra, S. 2005. Effect of temperature on adsorption-desorption of isoproturon on a clay soil. Indian J. Weed Sci. 37: 247–250.

Attaway, H.H., Paynter, M.J.B. and Camper, N.D. 1982. Degradation of selected phenylurea herbicides by anaerobic pond sediment. J. Environ. Sci. Heal. B 17: 683–699. https://doi.org/10.1080/03601238209372350.

Azab, E., Hegazy, A.K., Gobouri, A.A. and Elkelish, A. 2020. Impact of transgenic arabidopsis thaliana plants on herbicide isoproturon phytoremediation through expressing human cytochrome P450-1A2. Biology 9(11): 362. https://doi.org/10.3390/biology9110362.

Badawi, N., Ronhede, S., Olsson, S., Kragelund, B.B., Johnsen, A.H., Jacobsen, O.S. et al. 2009. Metabolites of the phenylurea herbicides chlorotoluron, diuron, isoproturon and linuron produced by the soil fungus Mortierella sp. Environ. Pollut. 157(10): 2806–2812. https://doi.org/10.1016/j.envpol.2009.04.019.

Barac, T., Taghavi, S., Borremans, B., Provoost, A., Oeyen, L., Colpaert, J.V., Vangronsveld, J. and van der Lelie, D. 2004. Engineered endophytic bacteria improve phytoremediation of water-soluble, volatile, organic pollutants. Nat. Biotechnol. 22: 583–588. https://doi.org/10.1038/nbt960.

Behera, B.C. and Bhunya, S.P. 1990. Genotoxic effect of isoproturon (herbicide) as revealed by three mammalian *in vivo* mutagenic bioassays. Indust. J. Experi. Biol. 28: 862–867.

Bending, G.D. and Rodriguez-Cruz, M.S. 2007. Microbial aspects of the interaction between soil depth and biodegradation of the herbicide isoproturon. Chemosphere 66: 664–671. https://doi.org/10.1016/j.chemosphere.2006.07.099.

Bending, G.D., Lincoln, S.D. and Edmondson, R.N. 2006. Spatial variation in the degradation rate of the pesticides isoproturon, azoxystrobin and diflufenican in soil and its relationship with chemical and microbial properties. Environ. Pollut. 139: 279–287. https://doi.org/10.1016/j.envpol.2005.05.011.

Bending, G.D., Lincoln, S.D., Sørensen, S.R., Morgan, J.A.W., Aamand, J. and Walker, A. 2003. In-field spatial variability in the degradation of the phenyl-urea herbicide isoproturon is the result of interactions between degradative Sphingomonas spp. and soil pH. Appl. Environ. Microbiol. 69: 827–834. https://doi.org/10.1128/AEM.69.2.827-834.2003.

Berger, B.M. 1999. Factors influencing transformation rates and formation of products of phenylurea herbicides in soil. J. Agric. Food Chem. 47: 3389–3396. https://doi.org/10.1021/jf981285q.

Bi, Y.F., Miao, S.S., Lu, Y.C., Qiu, C.B., Zhou, Y. and Yang, H. 2012. Phytotoxicity, bioaccumulation and degradation of isoproturon in green algae. J. Hazard. Mater. 243: 242–249. https://doi.org/10.1016/j.jhazmat.2012.10.021.

Blair, A.M. 1978. Some studies on the sites of uptake of chlortoluron, isoproturon and metoxuron by wheat, Avena fatua and Alopecurus myosuroides. Weed Res. 18: 381–387. https://doi.org/10.1111/j.1365-3180.1978.tb01178.x.

Bottcher, T. and Schroll, R. 2007. The fate of isoproturon in a freshwater microcosm with Lemna minor as a model organism. Chemosphere 66: 684–689. https://doi.org/10.1016/j.chemosphere.2006.07.087.

Brazier, M., D.J. Cole and Edwards, R. 2002. O-glucosyltransferase activities toward phenolic natural products and xenobiotics in wheat and herbicide-resistant and herbicide-susceptible black-grass (Alopecurus myosuroides) Phytochemistry 59(2): 149–156. https://doi.org/10.1016/S0031-9422(01)00458-7.

Brazier-Hicks, M., Offen, W.A., Gershater, M.C., Revett, T.J., Lim, E.K., Nowles, D.J., Davies, G.., and Edwards, R. 2007. Characterization and engineering of the bifunctional N-and O-gulcosyltransferase involved in xenobiotics metabolism in plants. Proc. Natl. Acad. Sci. USA 104: 20238–20243. https://doi.org/10.1073/pnas.0706421104.

Castillo, M.D.P., Wiren-Lehr, S.V., Scheunert, I. and Torstensson, L. 2001. Degradation of isoproturon by the white rot fungus Phanerochaete chrysosporium. Biol. Fertility Soils 33: 521–528. https://doi.org/10.1007/s003740100372.

Chen Lu Yi, Zhang, S., Miao, S.S., Jiang, C., Huang, M.T., Liu, Y. and Yang, H. 2015. Enhanced degradation of herbicide isoproturon in wheat rhizosphere by salicylic acid. J. Agric. Food Chem. 63: 92–103. https://doi.org/10.1021/jf505117j.

Chhokar, R.S. and Malik, R.K. 2002. Isoproturon-resistant littleseed canary grass (Phalaris minor) and its response to alternate herbicides1. Weed Technol. 16: 116–123. https://doi.org/10.1614/0890-037X(2002)016[0116:IRLCPM]2.0.CO;2.

Cole, D.J. and Edwards, R. 2000. Secondary metabolism of agrochemicals in plants. pp. 107–154. *In*: Roberts, T.R. (ed.). Agrochemicals and Plant Protection. John Wiley and Sons Press, London.

Coleman, J.O., Frova, C., Schröder, P. and Tissut, M. 2002. Exploiting plant metabolism for the phytoremediation of persistent herbicides. Environ. Sci. Pollut. Res. 9(1): 18–28. https://doi.org/10.1007/BF02987314.

Coleman, J.O.D., Randall, R. and Blake-Kalff, M.M.A. 1997. Detoxification of xenobiotics in plant cells by glutathione conjugation and vacuolar compartmentalization: A fluorescent assay using monochlorobimane. Plant, Cell & Environment 20(4): 449–460.

Coppola, L., Pilar Castillo, M.D. and Vischetti, C. 2011. Degradation of isoproturon and bentazone in peat- and compost-based biomixtures. Pest Manage. Sci. 67: 107–113. https://doi.org/10.1002/ps.2040.

Coquet, Y. 2003. Sorption of pesticides atrazine, isoproturon, and metamitron in the vadose zone. Vadose Zone J. 2: 40–51. https://doi.org/10.2113/2.1.40.

Coquet, Y. and Barriuso, E. 2002. Spatial variability of pesticide adsorption within the topsoil of a small agricultural catchment. Agronomie 22: 389–398. https://doi.org/10.1051/agro:2002017.

Cox, L., Walker, A. and Welch, S.J. 1996. Evidence for the enhanced biodegradation of isoproturon in soils. Pesticide Science 48: 253–260. https://doi.org/10.1002/(SICI)1096-9063(199611)48:3%3C253::AID-PS466%3E3.0.CO;2-V.

Crum, S.J.H., Van Kammen-Polman, A.M.M. and Leistra, M. 1999. Sorption of nine pesticides to three aquatic macrophytes. Arch. Environ. Con. Tox. 37: 310–316. https://doi.org/10.1007/s002449900519.

Dai, M. and Copley, S.D. 2004. Genome shuffling improves degradation of the anthropogenic pesticide pentachlorophenol by Sphingobium chlorophenolicum ATCC 39723. Appl. Environ. Microbiol. 70: 2391–2397. http://doi:10.1128/AEM.70.4.2391 2397.2004.

De Felipe, M.R., Golvanao, M.P., Lucas, M.M., Lang, P. and Pozuele, J.M. 1989. Differential effects of isoproturon on the photosynthetic apparatus and yield of two varieties of wheat and L. Rigidum. Weed Res. 28: 85–89.

Dejonghe, W., Berteloot, E. and Goris, J. 2003 Synergistic degradation of linuron by a bacterial consortium and isolation of a single linuron-degrading variovorax strain. Appl. Environ. Microbiol. 69: 1532–1541. doi:10.1128/AEM.69.3.1532-1541.2003.

Desmond, O.J., Manners, J.M., Schenk, P.M., Maclean, D.J. and Kazan, K. 2008. Gene expression analysis of the wheat response to infection by Fusarium pseudograminearum. Physiol. Mol. Plant Pathol. 73: 40–47. https://doi.org/10.1016/j.pmpp.2008.12.001.

Dewez, D., Didur, O., Vincent-Héroux, J. and Popovic, R. 2008. Validation of photosynthetic-fluorescence parameters as biomarkers for isoproturon toxic effect on alga Scenedesmus obliquus. Environ. Pollut. 151: 93–100. https://doi.org/10.1016/j.envpol.2007.03.002.

Dhawan, R.S., Dhawan, A.K., Kajla, S. and Moudgil, R. 2005. Assessment of variation in isoproturon susceptible and resistant biotypes of Phalaris minor by RAPD analysis. Ind. J. Biotechnol. 4: 534–537. http://hdl.handle.net/123456789/5764.

Didierjean, L., Gondet, L., Perkins, R., Lau, S.M., Schaller, H., O'Keefe, D.P.R. and Werck-Reichhart, D. 2002. Engineering herbicide metabolism in tobacco and Arabidopsis with CYP76B1, a cytochrome P450 enzyme from Jerusalem artichoke. Plant Physiol. 130: 179–189. https://doi.org/10.1104/pp.005801.

Ding, Q., Wu, H.L., Xu, Y., Guo, L.J., Liu, K., Gao, H.M. and Yang, H. 2011. Impact of low molecular weight organic acids and dissolved organic matter on sorption and mobility of isoproturon in two soils. J. Hazard. Mater. 190: 823–832. https://doi.org/10.1016/j.jhazmat.2011.04.003.

Dorfler, U., Cao, G., Grundmann, S. and Schroll, R. 2006. Influence of a heavy rainfall event on the leaching of [14C] isoproturon and its degradation products in outdoor lysimeters. Environ. Pollut. 144: 695–702. https://doi.org/10.1016/j.envpol.2005.12.034.

Dosnon-Olette, R., Couderchet, M., Oturan, M.A., Oturan, N. and Eullaffroy, P. 2011. Potential use of Lemna Minor for the phytoremediation of isoproturon and glyphosate. Int. J. Phytorem. 13: 601–612. https://doi.org/10.1080/15226514.2010.525549

Dosnon-Olette, R., Trotel-Aziz, P., Couderchet, M. and Eullaffroy, P. 2010. Fungicides and herbicide removal in Scenedesmus cell suspensions. Chemosphere 79: 117–123. https://doi.org/10.1016/j.chemosphere.2010.02.005.

Doty, S.L., Shang, T.Q., Wilson, A.M., Tangen, J., Westergreen, A.D., Newman, L.A., Strand, S.E. and Gordon, M.P. 2000. Enhanced metabolism of halogenated hydrocarbons in transgenic plants containing mammalian cytochrome P450 2E1. Proc. Natl. Acad. Sci. USA 97: 6287–6291. https://doi.org/10 .1073/pnas.97.12.6287.

Dousset, S., Thevenot, M., Pot, V., Simunek, J. and Andreux, F. 2007. Evaluating equilibrium and non-equilibrium transport of bromide and isoproturon in disturbed and undisturbed soil columns. J. Contam. Hydrol. 94: 261–276. https://doi.org/10.1016/j.jconhyd.2007.07.002.

Ducruet, J.M. 1991. Les inhibiteurs du photosysteeme II. pp. 79–114. *In*: Scalla, R. (ed.). Les herbicides: Mode d'Action et Principes d'Utilisation. INRA, Paris, France.

Dwivedi, S., Singh, B.R., Al-Khedhairy, A.A. and Musarrat, J. 2011. Biodegradation of isoproturon using a novel Pseudomonas aeruginosa strain JS-11 as a multifunctional bioinoculant of environmental significance. J. Hazard. Mater. 185: 938–944. https://doi.org/10.1016/j.jhazmat.2010.09.110.

Elahi, M., Cheema, Z.A., Basra, S.M.A., Akram, M. and Ali, Q. 2011. Use of allelopathic water extract of field crops for weed control in wheat. Int. Res. J. Plant Sci. 2(9): 262–270. http://www.interesjournals.org/IRJPS.

El-Arfaoui, A., Boudesocque, S., Sayen, S. and Guillon, E. 2010. Terbumeton and isoproturon adsorption by soils: Influence of Ca2+ and K+ cations. J. Pestic. Sci. 2: 131–133. https://doi.org/10.1584/jpestics.G09-48.

Elkhattabi, K., Bouhaouss, A., Scrano, L., Lelario, F. and Bufo, S.A. 2007. Influence of humic fractions on retention of isoproturon residues in two Moroccan soils. J. Environ. Sci. Health B 42: 851–856. https://doi.org/10.1080/03601230701555104.

El-Sebai, T., Devers-Lamrani, M., Lagacherie, B., Rouard, N., Soulas, G. and Martin-Laurent, F. 2011. Isoproturon mineralization in an agricultural soil: impact of temperature and moisture content. Biol. Fertility Soils 47: 427–435. https://doi.org/10.1007/s00374-011-0549-1.

El-Sebai, T., Lagacherie, B., Soulas, G. and Martin-Laurent, F. 2007. Spatial variability of isoproturon mineralizing activity within an agricultural field: Geostatistical analysis of simple physicochemical and microbiological soil parameters. Environ. Pollut. 145: 680–690.

El-Sebai, T.E., Lagacherie, B. and Soulas, G. 2004. Isolation and characterization of an isoproturon-mineralising Methylopela sp. TES from French agricultural soil. FEMS Microbiol. Lett. 239: 103–110. doi:10.1016/j.femsle.2004.08.017.

Ertli, T., Marton, A. and Foldenyi, R. 2004. Effect of pH and the role of organic matter in the adsorption of isoproturon on soils. Chemosphere 57: 771–779. https://doi.org/10.1016/j.chemosphere.2004.07.

Fedtke, C. and Duke, S.O. 2005. Herbicides. pp. 247–330. *In*: Hock, B. and Elstner, E.F. (eds.). Plant Toxicol. New York, NY: Marcel Dekker.

Fenlon, K.A., Jones, K.C. and Semple, K.T. 2011. The effect of soil:water ratios on the induction of isoproturon, cypermethrin and diazinon mineralisation. Chemosphere 82: 163–168. https://doi.org/10.1016/j.chemosphere.2010.10.027.

Feurtet-Mazel, A., Grollier, T., Grouselle, M., Ribeyre, F. and Boudou, A. 1996. Experimental study of bioaccumulation and effects of the herbicide isoproturon on freshwater rooted macrophytes (Elodea densa and Ludwigia natans). Chemosphere 32: 1499–1512. https://doi.org/10.1016/0045-6535(96)00058-6.

Fredrickson, J.K., Balkwill, D.L., Romine, M.F. and Shi, T. 1999. Ecology, physiology, and phylogeny of deep subsurface Sphingomonas sp. J. Ind. Microbiol. Biotechnol. 23: 273–283. https://doi.org/10.1038/sj.jim.2900741.

French, C.E., Rosser, S.J., Davies, G.J., Nicklin, S. and Bruce, N.C. 1999. Biodegradation of explosives by transgenic plants expressing pentaerythritol tetranitrate reductase. Nat. Biotechnol. 17: 491–494. https://doi.org/10.1038/ 8673.

Fuertet-Mazel, A., Grollier, T., Grouselle, M., Ribeyre, F. and Boudou, A. 2000. Experimental study of bioaccumulation and effects of the herbicide isoproturon on freshwater rooted macrophytes (Elodea densa and Ludwigia natans). Chemosphere 32: 1499–1512.

Gaillardon, P. and Sabar, M. 1994. Changes in the concentrations of isoproturon and its degradation products in soil and soil solution during incubation at two temperatures. Weed Res. 34: 243–250. https://doi.org/10 .1111/j.1365-3180.1994.tb01992.x. 3.

Gangwar, S. and Rafiquee, M. 2007a. Kinetics of the acid hydrolysis of isoproturon in the absence and presence of sodium lauryl sulfate micelles. Colloid Polym. Sci. 285: 587–592. https://doi.org/10.1007/s00396-006-1596-2.

Gangwar, S.K. and Rafiquee, M.Z.A. 2007b. Kinetics of the alkaline hydrolysis of isoproturon in CTAB and NaLS micelles. Int. J. Chem. Kinetics 39: 39–45. https://doi.org/10.1002/kin.20213.

Geissbuhler, H., Martin, H. and Voss, G. 1975. The substituted urea. pp. 209–291. *In*: Kearney, P.C. and Kaufman, D.D. (eds.). Herbicides, Marcel Dekkev, New York.

Giri, K., Pandey, S., Kumar, R. et al. 2016. Biodegradation of isoproturon by Pseudoxanthomonas sp. isolated from herbicide-treated wheat fields of Tarai agro-ecosystem, Pantnagar. 3 Biotech 6: 190. https://doi.org/10.1007/s13205-016-0505-8.

Glassgen, W.E., Komassa, D., Bohnenkamper, O., Haas, M., Hertkorn, N., May, R.G., Szymczak, W. and Sandermann, H.J. 1999. Metabolism of the herbicide isoproturon in wheat and soybean cell suspension cultures. Pestic. Biochem. Physiol. 63: 97–113. https://doi.org/10.1006/pest.1999.2394.

Gooddy, D.C., Mathias, S.A., Harrison, I., Lapworth, D.J. and Kim, A.W. 2007. The significance of colloids in the transport of pesticides through Chalk. Sci. Total Environ. 385: 262–271. https://doi.org/10.1016/j.scitotenv.2007.06.043.

Gottstein, D., Gross, M., Klepel, M. and Leham, H. 1991. Induction of phytoalexins and induced resistance to Phytophthora infestans in potato tubers by phenylureas. Pestic. Sci. 31: 119–121.

Greulich, K., Hoque, E. and Pflugmacher, S. 2002. Uptake, metabolism, and effects on detoxication enzymes of isoproturon in spawn and tadpoles of amphibians. Ecotoxicol. Environ. Safety 52: 256–266. https://doi.org/10.1006/eesa.2002.2182.

Gross, D., Laanio, G., Dupuis, G. and Esser, H.O. 1979. The metabolic behaviour of chlorotoluron in wheat and soil. Pestic. Biochem. Physiol. 10: 49. https://doi.org/10.1016/0048-3575(79)90007-5.

Grouselle, M., Grollier, T., Feurtet-Mazel, A., Ribeyre, F. and Boudou, A. 1995. Herbicide isoproturon-specific binding in the freshwater macrophytem Elodea densa a single-cell fluorescence study. Ecotoxicol. Environ. Saf. 32: 254–9. https://doi.org/10.1006/eesa.1995.1111.

Grundmann, S., Doerfler, U., Munch, J.C., Ruth, B. and Schroll, R. 2011. Impact of soil water regime on degradation and plant uptake behaviour of the herbicide isoproturon in different soil types. Chemosphere 82: 1461–1467. https://doi.org/10.1016/j.chemosphere.2010.11.037.

Gu, T., Zhou, C., Sørensen, S.R., Zhang, J., He, J., Yu, P., Yan, X. and Li, S. 2013. The novel bacterial N-demethylase PdmAB is responsible for the initial step of N, N-dimethyl-substituted phenylurea herbicide degradation. Appl. Environ. Microbiol. 79: 7846–7856. 10.1128/AEM.02478-13.

Guzzella, L., Capri, E., Di Corcia, A., Barra Caracciolo, A. and Giuliano, G. 2006. Fate of diuron and linuron in a field lysimeter experiment. J. Environ. Qual. 35: 312–323. https://doi.org/10.2134/jeq2004.0025.

Hangler, M., Jensen, B., Ronhede, S. and Sorensen, S.R. 2007. Inducible hydroxylation and demethylation of the herbicide isoproturon by Cunninghamella elegans. FEMS Microbiol. Lett. 268(2): 254–260.12. https://doi.org/10.1111/j.1574-6968.2006.00599.x.

Hannink, N., Rosser, S.J., French, C.E., Basran, A., Murray, J.A.H., Nicklin, S. and Bruce, N.C. 2001. Phytodetoxification of TNT by transgenic plants expressing a bacterial nitroreductase. Nat. Biotechnol. 19: 1168–1172. https://doi.org/10.1038/nbt1201-1168.

Haque, M.M. and Muneer, M. 2003. Heterogeneous photocatalysed degradation of a herbicide derivative, isoproturon in aqueous suspension of titanium dioxide. J. Environ. Manage. 69: 169–176. https://doi.org/10.1016/S0301-4797(03)00143-9.

Hassan, N. and Nemat-Alla, M. 2005. Oxidative stress in herbicide-treated broad bean and maize plants. Acta Physiologiae Plant. 27: 429–438.

Heppell, C.M., Burt, T.P., Williams, R.J. and Haria, A.H. 1999. The influence of hydrological pathways on the transport of the herbicide, isoproturon, through an underdrained clay soil. Water Sci. Technol. 39: 77–84. https://doi.org/10.1016/S0273-1223(99)00321-2.

Höfer, R., Boachon, B., Renault, H., Gavira, C., Miesch, L., Iglesias, J., Ginglinger, J.-F., Allouche, L., Miesch, M., Grec, S. et al. 2014. Dual function of the cytochrome P450 CYP76 family from Arabidopsis thaliana in the metabolism of monoterpenols and phenylurea herbicides. Plant Physiol. 166: 1149–1161. [CrossRef] https://doi.org/10.1104/pp.114.244814.

Hoshiya, T., Hasegawa, R., Hakoi, K., Cui, L., Ogiso, T., Cabral, R. and Ito, N. 1993. Enhancement by nonmutagenic pesticides of GST-Ppositive hepatic foci development initiated with diethyl- nitrosamine in the rat. Cancer Lett. 72: 59–64. https://doi.org/10.1016/0304-3835(93)90011-W.

Hussain, S., Arshad, M., Springael, D., Sørensen, S.R., Bending, G.D., Devers-Lamrani, M. and Martin-Laurent, F. 2015. Abiotic and biotic processes governing the fate of phenylurea herbicides in soils: A review. Crit. Rev. Env. Sci. Tech., 45(18): 1947–1998. https://doi.org/10.1080/10643389.2014.1001141.

Hussain, S., Devers-Lamrani, M., El-Azhari, N. and Martin-Laurent, F. 2011. Isolation and characterization of an isoproturon mineralizing Sphingomonas sp. strain SH from a French agricultural soil. Biodegradation 22: 637–650. DOI10.1007/s10532-010-9437-x.

Hussain, S., Devers-Lamrani, M., Spor, A., Rouard, N., Porcherot, M., Beguet, J. and Martin-Laurent, F. 2013. Mapping field spatial distribution patterns of isoproturon-mineralizing activity over a three-year winter wheat/rape seed/barley rotation. Chemosphere 90: 2499–2511. https://doi.org/10.1016/j.chemosphere.2012.10.080.

Hussain, S., Shahzad, T., Imran, M., Khalid, A. and Arshad, M. 2017. Bioremediation of isoproturon herbicide in agricultural soils. pp. 83–104. *In*: Microbe-Induced Degradation of Pesticides. Springer, Cham. https://doi.org/10.1007/978-3-319-45156-5_4.

Hussain, S., Sørensen, S.R., Devers-Lamrani, M., El-Sebai, T. and Martin-Laurent, F. 2009. Characterization of an isoproturon mineralizing bacterial culture enriched from a French agricultural soil. Chemosphere 77: 1052–1059. https://doi.org/10.1016/j.chemosphere.2009.09.020.

Issa, S. and Wood, M. 2005. Degradation of atrazine and isoproturon in surface and sub-surface soil materials undergoing different moisture and aeration conditions. Pest Manage. Sci. 61: 126–132. https://doi.org/10.1002/ps.951.

Jang, J., Khanom, S., Moon, Y., Shin, S. and Lee, O.R. 2020. PgCYP76B93 docks on phenylurea herbicides and its expression enhances chlorotoluron tolerance in Arabidopsis. App. Biol. Chem. 63(1): 1–11. https://doi.org/10.1186/s13765-020-00498-x.

Johannesen, H., Sørensen, S.R. and Aamand, J. 2003. Mineralization of soil-aged isoproturon and isoproturon metabolites by Sphingomonas sp. strain SRS2. J. Environ. Qual. 32: 1250–1257. https://doi.org/10.2134/jeq2003.1250.

Johnson, A.C., Besien, T.J., Bhardwaj, C.L., Dixon, A., Gooddy, D.C., Haria, A.H. and White, C. 2001. Penetration of herbicides to groundwater in an unconfined chalk aquifer following normal soil applications. J. Contam. Hydrol. 53: 101–117. https://doi.org/10.1016/S0169-7722(01)00139-5.

Johnson, A.C., Hughes, C.D., Williams, R.J. and Chilton, P.J. 1998. Potential for aerobic isoproturon biodegradation and sorption in the unsaturated and saturated zones of a chalk aquifer. J. Contam. Hydrol. 30: 281–297. https://doi.org/10.1016/S0169-7722(97)00048-X.

Juhler, R.K., Sørensen, S.R. and Larsen, L. 2001. Analysing transformation products of herbicide residues in environmental samples. Water Research 35: 1371–1378. https://doi.org/10.1016/S0043-1354(00)00409-7.

Kawahigashi, H., Hirose, S., Ohkawa, H. and Ohkawa, Y. 2007 Herbicide resistance of transgenic rice plants expressing human CYP1A1. Biotechnol. Adv. 25: 75–84. [CrossRef] https://doi.org/10.1016/j.biotechadv.2006.10.002.

Khadrani, A., Seigle-Murandi, F., Steiman, R. and Vroumsia, T. 1999. Degradation of three phenylurea herbicides (chlortoluron, isoproturon and diuron) by micromycetes isolated from soil. Chemosphere 38: 3041–3050. https://doi.org/10.1016/S0045-6535(98)00510-4.

Khan, M.S., Chaudhry, P., Wani, P.A. and Zaidi, A. 2006. Biotoxic effects of the herbicides on growth, seed yield, and grain protein of greengram. J. Appl. Sci. Environ. Mgt. 10: 141–6. https://doi.org/10.4314/jasem.v10i3.17333.

Khanom, S., Jang, J. and Lee, O.R. 2019. Overexpression of ginseng cytochrome P450 CYP736A12 alters plant growth and confers phenylurea herbicide tolerance in Arabidopsis. J. Ginseng Res. 43: 645–653. [CrossRef] [PubMed]. https://doi.org/10.1016/j.jgr.2019.04.005.

Khurana, J.L., Jackson, C.J., Scott, C., Pandey, G., Horne, I., Russell, R.J., Herlt, A., Easton, C.J. and Oakeshott, J.G. 2009. Characterization of the phenylurea hydrolases A and B: founding members of a novel amidohydrolase subgroup. Biochem. J. 418: 431–441. https://doi.org/10.1042/BJ20081488.

Kiesel, C.A. 2014. Enhanced Degradation of Isoproturon in Soils: Sustainability of Inoculated, Microbial Herbicide Degraders, and Adaptions of Mative Microbes (Doctorate Dissertation). Technische Universitat München, Munich, Germany. http://nbn-resolving.de/urn/resolver.pl?urn:nbn:de:bvb:91-diss-20141120-1188688-0-1.

Kleczkowski, L.A. 1994. Inhibitors of photosynthetic enzymes/carriers and metabolism. Annu. Rev. Plant Physiol. Plant Mol. Biol. 45: 339–367.

Knauert, S., Escher, B., Singer, H., Hollender, J. and Knauer, K. 2008. Mixture toxicity of three photosystem II inhibitors (atrazine, isoproturon, and diuron) towards photosynthesis of freshwater phytoplankton studied in outdoor mesocosms. Environ. Sci. Technol. 42(17): 6424–6430. https://doi.org/10.1021/es072037q.

Knauert, S., Singer H., Hollender J. and Knauer, K. 2010. Phytotoxicity of atrazine, isoproturon, and diuron to submersed macrophytes in outdoor mesocosms. Environ. Pollut. 158: 167–174. https://doi.org/10.1016/j.envpol.2009.07.023.

Kuiper, I., Lagendijk, E.L., Bloemberg, G.V. and Lugtenberg, B.J.J. 2004. Rhizoremediation: A beneficial plant-microbe interaction. Mol. Plant Microbe Interact. 17: 6–15. https://doi.org/10.1094/MPMI.2004.17.1.6.

Laviale, M., Morin, S. and Créach, A. 2011. Short term recovery of periphyton photosynthesis after pulse exposition to the photosystem II inhibitors atrazine and isoproturon. Chemosphere 84(5): 731–734.

Laviale, M., Prygiel, J. and Créach, A. 2010. Light modulated toxicity of isoproturon toward natural stream periphyton photosynthesis: A comparison between constant and dynamic light conditions. Aquat. Toxicol. 97: 334–342. https://doi.org/10.1016/j.aquatox.2010.01.004.

Lehr, S., Glassgen, W.E., Sandermann, H., Beese, F. and Scheunert, I. 1996. Metabolism of isoproturon in soils originating from different agricultural management systems and in cultures of isolated soil bacteria. Int. J. Environ. Anal. Chem. 65: 231–243. https://doi.org/10.1080/03067319608045558.

Liang, L., Lu, Y.L. and Yang, H. 2012. Toxicology of isoproturon to the food crop wheat as affected by salicylic acid. Environ. Sci. Pollut. Res. 19(2012): 2044–2054. https://doi.org/10.1007/s11356-011-0698-7.

Lopez-Munoz, M.J., Revilla, A. and Aguado, J. 2013. Heterogeneous photocatalytic degradation of isoproturon in aqueous solution: Experimental design and intermediate products analysis. Catal. Today 209: 99–107. https://doi.org/10.1016/j.cattod.2012.11.017.

Lu, Y.C., Zhang, S. and Yang, H. 2015. Acceleration of the herbicide isoproturon degradation in wheat by glycosyltransferases and salicylic acid. J. Hazard. Mater. 283: 806–814. https://doi.org/10.1016/j.jhazmat.2014.10.034.

Ma, L.Y., Zhang, S.H., Zhang, J.J., Zhang, A.P., Li, N., Wang, X.Q., Yu, Q.Q. and Yang, H. 2018. Jasmonic acids facilitate the degradation and detoxification of herbicide isoproturon residues in wheat crops (Triticum aestivum). Chem. Res. Toxicol. 31(8): 752–761. https://doi.org/10.1021/acs.chemrestox.8b00100.

Mamy, L., Vrignaud, P., Cheviron, N., Perreau, F.O., Belkacem, M., Brault, A., Breuil, S.B., Delarue, G., Petraud, J.P., Touton, I., Mougin, C. and Chaplain, V.R. 2011. No evidence for effect of soil compaction on the degradation and impact of isoproturon. Environ. Chem. Lett. 9: 145–150. DOI 10.1007/s10311-009-0273-3.

Mansour, M., Feicht, E.A., Behechti, A., Schramm, K.W. and Kettrup, A. 1999. Determination photostability of selected agrochemicals in water and soil. Chemosphere 39: 575–585. https://doi.org/10.1016/S0045-6535(99)00123-X.

Martin Laviale, Soizic Morin and Anne Créach. 2011. Short term recovery of periphyton photosynthesis after pulse exposition to the photosystem II inhibitors atrazine and isoproturon. Chemosphere 84: 731–734. https://doi.org/10.1016/j.chemosphere.2011.03.035.

Mittler, R. 2002. Oxidative stress, antioxidants and stress tolerance. Trends in Plant Science 7(9): 405–410.

Mordern, C.W. and Golden, S.S. 1989. PsbA gene indicates common ancestry of chlorophytes and chloroplasts. Nature 337: 382–385. https://doi.org/10.1038/337382a0.

Mostafa, F.I.Y. and Helling, C.S. 2001. Isoproturon degradation as affected by the growth of two algal species at different concentrations and pH values. J. Environ. Sci. Health (Part B) 36: 709–727. https://doi.org/10.1081/PFC-100107406.

Mudd, P.J., Hance, R.J. and Wright, S.J.L. 1983. The persistence and metabolism of isoproturon in soil. Weed Research 23: 239–246. https://doi.org/10.1111/j.1365-3180.1983.tb00545.x.

Mueen-ud-Din, A.L., Ahmad, S.B. and Ali, M. 2011. Effect of post emergence herbicides on narrow leaved weeds in wheat crop. J. Agric. Res. 49(2).

Mutlu, S., Atici, Ö. and Nalbantoglu, B. 2009. Effects of salicylic acid and salinity on apoplastic antioxidant enzymes in two wheat cultivars differing in salt tolerance. Biol. Plant. 53(2): 334–338.

Muud, P., Hance, R. and Wright, S. 1983. The persistence and metabolism of isoproturon in soil. Weed Res. 23: 239–246. https://doi.org/10.1111/j.1365 -3180.1983.tb00545.x.2.

Nadasy, E., Lehoczky, E., Lukacs, P. and Adam, P. 2000. Influence of different pre-emergent herbicides on the growth of soybean varieties. Zeitschrift fur Pflanzenkrankheiten und Pflanzenschutz 17: 635–639.

Navarro, S., Bermejo, S., Vela, N. and Hernandez, J. 2009. Rate of loss of Simazine, Terbuthylazine, Isoproturon and Methabenzthiazuron during soil solarisation. J. Agric. Food Chem. 57: 6375–6382. https://doi.org/10.1021/jf901102b.

Nemat Alla, M.M. and Younis, M.E. 1995. Herbicidal effect on phenolic metabolism in maize (Zea mays) and soybean (Glycine max) seedlings, J. Exp. Bot. 46: 1731–1736. https://doi.org/10.1093/jxb/46.11.1731.

Nemat Alla, M.M. and Hassan, N.M. 2007. Changes of antioxidants and GSH-associated enzymes in isoproturon-treated maize. Acta Physiol. Plant. 29: 247–258. DOI 10.1007/s11738-007-0031-8.

Nemat Alla, M.M. and Hassan, N.M. 2014. Alleviation of isoproturon toxicity to wheat by exogenous application of glutathione. Pestic. Biochem. Phisiol. 112: 56–62. http://doi.org/10.1016/j.pestbp.2014.04.012.

Nemat Alla, M.M., Badawi, A.M., Hassan, N.M., El-Bastawisy, Z.M. and Badran, E.G. 2008. Effect of metribuzin, butachlor and chlorimuron-ethyl on amino acid and protein formation in wheat and maize seedlings. Pestic. Biochem. Physiol. 90: 8–18. https://doi.org/10.1016/j.pestbp.2007.07.003.

Newman, L.A. and Reynolds, C.M. 2004. Phytodegradation of organic compounds. Curr. Opin. Biotechnol. 15: 225–230. https://doi.org/10.1016/j.copbio.2004.04.006.

Ostrofsky, E., Traina, S. and Tuovinen, O. 1997. Variation in atrazine mineralization rates in relation to agricultural management practice. J. Environ. Qual. 26: 647–657. https://doi.org/10.2134/jeq1997.00472425002600030009x.

Pandey, V.C. and Bajpai, O. 2019. Chapter 1—Phytoremediation: From Theory Toward Practice. pp. 1–49. *In*: Pandey, V.C., and Bauddh, K. (eds.). Phytomanagement of Polluted Sites. Elsevier: Amsterdam, The Netherlands. https://doi.org/10.1016/B978-0-12-813912-7.00001-6.

Paris-Palacios, S., Mosleh, Y.Y., Almohamad, M., Delahaut, L., Conrad, A., Arnoult, F. and Biagianti-Risbourg, S. 2010. Toxic effects and bioaccumulation of the herbicide isoproturon in Tubifex tubifex (Oligocheate, Tubificidae): A study of significance of autotomy and its utility as a biomarker. Aquat. Toxicol. 98: 8–14. https://doi.org/10.1016/j.aquatox.2010.01.006.

Parra, S., Sarria, V., Malato, S., Peringer, P. and Pulgarin, C. 2000. Photochemical versus coupled photochemical-biological flow system for the treatment of two biorecalcitrant herbicides: metobromuron and isoproturon. Appl. Catal. B: Environ. 27: 153–168. https://doi.org/10.1016/S0926-3373(00)00151-X.

Pascal-Lorber, S., Alsayeda, H., Jouanin, I., Debrauwer, L., Canlet, C. and Laurent, F.O. 2010. Metabolic fate of [14C] diuron and [14C] linuron in wheat (Triticum aestivum) and radish (Raphanus sativus). J. Agric. Food Chem. 58: 10935–10944. https://doi.org/10.1021/jf101937x.

Peng, R., Fu, X., Tian, Y., Zhao, W., Zhu, B., Xu, J., Wang, B., Wang, L. and Yao, Q. 2014a. Metabolic engineering of Arabidopsis for remediation of different polycyclic aromatic hydrocarbons using a hybrid bacterial dioxygenase complex. Metab. Eng. 26: 100–110. https://doi.org/10.1016/j.ymben.2014.09.005.

Peng, R.H., Fu, X.Y., Zhao, W., Tian, Y.S., Zhu, B., Han, H.J., Xu, J. and Yao, Q.H. 2014b. Phytoremediation of phenanthrene by transgenic plants transformed with a naphthalene dioxygenase system from Pseudomonas. Environ. Sci. Technol. 48: 12824–12832. https://doi.org/10.1021/es5015357.

Penning, H., Sørensen, S.R., Meyer, A.H., Aamand, J. and Elsner, M. 2010. C, N, and H isotope fractionation of the herbicide isoproturon reflects different microbial transformation pathways. Environ. Sci. Technol. 44: 2372–2378.
Pérés, F., Florin, D., Grollier, T., Feurtet-Mazef, A., Coste, M., Ribeyre, F., Ricard, M. and Boudou, A. 1996. Effect of the phenyl-urea herbicide isoproturon on periphytic diatom communities in fresh water indoor microcosms. Environ. Pollut. 94: 141–152. https://doi.org/10.1016/S0269-7491(96)00080-2.
Perrin-Ganier, C., Schiavon, F., Morel, J.L. and Schiavon, M. 2001. Effect of sludge amendment or nutrient addition on the biodegradation of the herbicide isoproturon in soil. Chemosphere 44(4): 887–892. https://doi.org/10.1016/S0045-6535(00)00283-6.
Pieuchot, M., PerrinGanier, C., Portal, J.M. and Schiavon, M. 1996. Study on the mineralization and degradation of isoproturon in three soils. Chemosphere 33: 467–478. https://doi.org/10.1016/0045-6535(96)00181-6.
Pot, V., Benoit, P., Etievant, V., Bernet, N., Labat, C., Coquet, Y. and Houot, S. 2011. Effects of tillage practice and repeated urban compost application on bromide and isoproturon transport in a loamy Albeluvisol. Eur. J. Soil Sci. 62: 797–810. https://doi.org/10.1111/j.1365-2389.2011.01402.x.
Prakash, O. and Lal, R. 2006. Description of Sphingobium fuliginis sp. nov., a phenanthrene-degrading bacterium from a fly ash dumping site, and reclassification of Sphingomonas cloacae as Sphingobium cloacae comb.nov. Int. J. Syst. Evol. Microbiol. 56: 2147–2152. doi:10.1099/ijs.0.64080-0.
Pussemier, L., Goux, S. and Vanderheyden, V. 1997. Rapid dissipation of atrazine in soils taken from various maize fields. Weed Res. 37: 171–179. doi:10.1046/j.1365-3180.1997.d01-18.x.
Reddy, P.A.K., Reddy, P.V.L., Sharma, V.M., Kumari, V.D. and Subrahmanyam, M. 2010. Photocatalytic degradation of isoproturon pesticide on C, N and S Doped TiO2. J. Water Resour. Protect. 2: 235–244. http://www.scirp.org/journal/PaperInformation.aspx?PaperID=1495.
Reddy, P.A.K., Srinivas, B., Kala, P., Kumari, V.D. and Subrahmanyam, M. 2011. Preparation and characterization of Bi-doped TiO2 and its solar photocatalytic activity for the degradation of isoproturon herbicide. Mater. Res. Bull. 46: 1766–1771. https://doi.org/10.1016/j.materresbull.2011.08.006.
Remde, A. and Traunspurger, W. 1994. A method to assess the toxicity of pollutants on anaerobic microbial degradation activity in sediments. Environ. Toxicol. Water Quality 9: 293–298. https://doi.org/10.1002/tox.2530090407.
Rigaud, J.P. and Lebreton, J.C. 2004. Points derepère Blédésherbage de postlevée. Chambre d' Agriculture de Mayenne. <http://www.mayenne.chambagri.Downloaded by fr/ services/documentation/ ble desherbage post levee.pdf>, denier acc`es octobre 2007.
Roberts, S.J., Walker, A., Cox, L. and Welch, S.J. 1998. Isolation of isoproturon degrading bacteria from treated soil via three different routes. J. Appl. Microbiol. 85: 309–316. https://doi.org/10.1046/j.1365-2672.1998.00507.x.
Robineau, T., Batard, Y., Nedelkina, S., Cabello-Hurtado, F., LeRet, M., Sorokine, O., Didierjean, L. and Werck-Reichhart, D. 1998. The chemically inducible plant cytochrome P450 CYP76B1 actively metabolizes phenylureas and other xenobiotics. Plant Physiol. 118(3): 1049–1056. DOI: https://doi.org/10.1104/pp.118.3.1049.
Ronhede, S., Jensen, B., Rosendahl, S., Kragelund, B.B., Juhler, R.K. and Aamand, J. 2005. Hydroxylation of the herbicide isoproturon by fungi isolated from agricultural soil. Appl. Environ. Microbiol. 71: 7927–7932. http://doi.org/10.1128/AEM.71.12.7927-7932.2005.
Ronhede, S., Sorensen, S.R., Jensen, B. and Aamand, J. (2007). Mineralization of hydroxylated isoproturon metabolites produced by fung. Soil Biol. Biochem. 39(7): 1751–1758. doi:10.1016/j.soilbio.2007.01.037.
Rubio, M.I.M., Gernjak, W., Alberola, O., Galvez, J.B., Fernandez-Ibanez, P. and Rodriguez, S. 2006. Photo-fenton degradation of alachlor, atrazine, chlorfenvinphos, diuron, isoproturon and pentachlorophenol at solar pilot plant. Int. J. Environ. Pollut. 27: 135–146. https://doi.org/10.1504/IJEP.2006.010459.
Rutherford, A.W. and Krieger-Liszkay, A. 2001. Herbicide-induced oxidative stress in photosystem II. Trends Biochem. Sci. 26: 648–653. https://doi.org/10.1016/S0968-0004(01)01953-3.
Rylott, E.L., Jackson, R.G., Edwards, J., Womack, G.L., Seth-Smith, H.M.B., Rathbone, D.A., Strand, S.E. and Bruce, N.C. 2006. An explosive-degrading cytochrome P450 activity and its targeted application for the phytoremediation of RDX. Nat. Biotechnol. 24: 216–219. https://doi.org/10.1038/ nbt118.
Samunder, S., Kirkwood, R.C. and Marshall, G. 1998. Effect of the monooxygenase inhibitor piperonyl butoxide on the herbicidal activity and metabolism of isoproturon in herbicide resistant and susceptible biotypes of phalaris minor and wheat. Pestic. Biochem. Physiol. 59: 143–153. https://doi.org/10.1006/pest.1998.2318.
Scarponi, L., Nemat Alla, M.M. and Martinetti, L. 1995. Consequences on nitrogen metabolism in soybean (Glycine max L.) as a result of imazethapyr action on acetohydroxy acid synthase. J. Agric. Food Chem. 43: 809–814.
Scarponi, L., Younis, M.E., Standardi, A., Martinetti, L. and Hassan, N.M. 1998. Changes in carbohydrate formation and stress symptoms in *vicia faba* L. treated with propachlor, chlorimuronethyl and imazethapyr. Agricoltura Mediterranea 128(2): 118–125.
Scheunert, I. and Reuter, S. 2000. Formation and release of residues of the 14C-labelled herbicide isoproturon and its metabolites bound in model polymers and in soil. Environ. Pollut. 108: 61–68. https://doi.org/10.1016/ S0269-7491(99)00202-X.
Schroll, R. and Kühn, S. 2004. Test system to establish mass balances for 14C-labeled substances in soil–plant–atmosphere systems under field conditions. Environ. Sci. Technol. 38: 1537–1544. https://doi.org/10.1021/es030088r.
Schuelein, J., Glaessgen, W.E., Hertkorn, N., Schroeder, P., Jr, H.S. and Kettrup, A. 1996. Detection and identification of the herbicide isoproturon and its metabolites in field samples after a heavy rainfall event. Int. J. Environ. Anal. Chem. 65: 193–202. https://doi.org/10.1080/03067319608045554.

Scow, K.M. and Hicks, K.A. 2005. Natural attenuation and enhanced bioremediation of organic contaminants in groundwater. Curr. Opinion Biotechnol. 16(3): 246–253. https://doi.org/10.1016/j.copbio.2005.03.009.

Sharma, M.V.P., Kumari, V.D. and Subrahmanyam, A. 2008b. Photocatalytic degradation of isoproturon herbicide over TiO2/Al-MCM-41 composite systems using solar light. Chemosphere 72: 644–651. https://doi.org/10.1016/j.chemosphere.2008.02.042.

Sharma, M.V.P., kumari, V.D. and Subrahmanyam, A. 2008a. Solar photocatalytic degradation of isoproturon over TiO2/H-MOR composite systems. J. Hazard. Mater. 160: 568–575. https://doi.org/10.1016/j.jhazmat.2008.03.042.

Sharma, P., Raina, V. and Kumari, R. 2006. Haloalkane dehalogenase linB is responsible for beta- and delta-hexachlorocyclohexane transformation in Sphingobium indicum B90A. Appl. Environ. Microbiol. 72: 5720–5727. doi:10.1128/AEM.00192-06.

Si, Y., Wang, M., Tian, C., Zhou, J. and Zhou, D. 2011. Effect of charcoal amendment on adsorption, leaching and degradation of isoproturon in soils. J. Contam. Hydrol. 123: 75–81. https://doi.org/10.1016/j.jconhyd.2010.12.008.

Singh, D.V., Adeppa, K. and Misra, K. 2012. Mechanism of isoproturon resistance in Phalaris minor: *in silico* design, synthesis and testing of some novel herbicides for regaining sensitivity. J. Mol. Model. 18: 1431–1445. https://doi.org/10.1007/s00894-011-1169-2.

Singh, D.V., Gaur, A.K. and Mishra, D.P. 2004. Biochemical and molecular mechanism of resistance against isoproturon in Phalaris minor biotypes: Variation in protein and RAPD profiles of isoproturon resistant and susceptible biotypes. Indian J. Weed Sci. 36: 256–259.

Singh, H., Singh, N.B., Singh, A., Hussain, I. and Yadav, V. 2016. Physiological and biochemical effects of salicylic acid on Pisum sativum exposed to isoproturon. Arch. Argon. Soil. Sci. 62(10): 1425–1436. https://doi.org/10.1080/03650340.2016.1144926.

Singh, S., Kirkwood, R.C. and Marshall, G. 1997. Effect of isoproturon on photosynthesis in susceptible and resistant biotypes of Phalaris minor. Weed Res. 37: 315. https://doi.org/10.1046/j.1365-3180.1997.d01-54.x.

Sørensen, S.R. and Aamand, J. 2001. Biodegradation of the phenylurea herbicide isoproturon and its metabolites in agricultural soils. Biodegradation 12: 69–77. https://doi.org/10.1023/A:1011902012131.

Sørensen, S.R. and Aamand, J. 2003. Rapid mineralisation of the herbicide isoproturon in soil from a previously treated Danish agricultural field. Pest Manage. Sci. 59: 1118–1124. https://doi.org/10.1002/ps.739.

Sørensen, S.R., Ronen, Z. and Aamand, J. 2002. Growth in co-culture stimulates metabolism of the phenylurea herbicide isoproturon by Sphingomonas sp strain SRS2. Appl. Environ. Microbiol., 68: 3478–3485. https://doi.org/10.1128/AEM.68.7.3478-3485.2002.

Steiman, R., Seigle-Murandi, F., Benoit-Guyod, J.L., Merlin, G. and Khadri, M. 1994. Assessment of isoproturon degradation by fungi. *In*: Proceedings of 5th International Workshop, Brussels (pp. 229–235). https://doi.org/10.1007/978-3-319-45156-5_4.

Su, X.N., Zhang, J.J., Liu, J.T., Zhang, N., Ma, L.Y., Lu, F.F., Chen, Z.J, Shi, Z., Si, W.J., Liu, C. and Yang, H. (2019). Biodegrading two pesticide residues in paddy plants and the environment by a genetically engineered approach. J. Agri. Food Chem. 67(17): 4947–4957. https://doi.org/10.1021/acs.jafc.8b07251.

Suhadolc, M., Schroll, R., Gattinger, A., Schloter, M., Munch, J.C. and Lestan, D. 2004. Effects of modified Pb, Zn, and Cd availability on the microbial communities and on the degradation of isoproturon in a heavy metal contaminated soil. Soil Biol. Biochem. 36: 1943–1954. https://doi.org/10.1016/j.soilbio.2004.05.015.

Sun, J., Huang, X. and He, J. 2006. Isolation identification of isoproturon degradation bacterium Y57 and its degradation characteristic. China Environ. Sci. 26: 315–319. In Chinese.

Sun, J.Q., Huang, X., Chen, Q.L., Liang, B., Qiu, J.G., Ali, S.W. and Li, S.P. 2009. Isolation and characterization of three Sphingobium sp strains capable of degrading isoproturon and cloning of the catechol 1,2-dioxygenase gene from these strains. World. J. Microbiol. Biotechnol. 25: 259–268. https://doi.org/ 10.1007/s11274-008-9888-y.

Sunohara, Y. and Matsumoto, H. 2004. Oxidative injury induced by the herbicide quinclorac on Echinochloa oryzicola Vasing and the involvement of anti-oxidative ability in its highly selective action in grass species. Plant Sci. 167: 597–606. https://doi.org/10.1016/j.plantsci.2004.05.005.

Tian, C., Wang, M.D. and Si, Y.B. 2010. Influences of charcoal amendment on adsorption-desorption of isoproturon in soils. Agric. Sci. China 9: 257–265.

Tixier, C., Sancelme, M., Ait-Aissa, S., Widehem, P., Bonnemoy, F., Cuer, A., Truffaut, N. and Veschambre, H. 2002. Biotransformation of phenylurea herbicides by a soil bacterial strain, Arthrobacter sp N2: structure, ecotoxicity and fate of diuron metabolite with soil fungi. Chemosphere 46: 519–526. https://doi.org/10.1016/S0045-6535(01)00193-X.

Trebst, A. 1991. The molecular basis of resistance of Photosystem II inhibitors Herbicide resistance in weeds and crops. Butterworth- Heinmann Ltd, Oxford UK.

Tripathi, M.K. 2003. Biochemical and molecular mechanism of isoproturon resistance in Phalaris minor (Doctoral dissertation, PhD Thesis GBPUA & T, Pantnagar, India).

Tripathi, M.K., Yadav, M.K., Gaur, A.K. and Mishra, D.P. 2005. PCR based isolation of psbA (herbicide binding protein encoding) gene using chloroplast and genomic DNA from Phalaris minor biotype(s). Physiol. Mol. Biol. Plants 11: 161–163.

Turnbull, G.A., Ousley, M. and Walker, A. 2001. Degradation of substituted phenylurea herbicides by Arthrobacter globiformis strain D47 and characterization of a plasmid-associated hydrolase gene, puhA. Appl. Environ. Microbiol. 67: 2270–2275. doi:10.1128/AEM.67.5.2270-2275.2001.

Uchida, E., Ouchi, T., Suzuki, Y., Yoshida, T., Habe, H., Yamaguchi, I., Omori, T. and Nojiri, H. 2005. Secretion of bacterial xenobiotic-degrading enzymes from transgenic plants by an apoplastic expressional system: an applicability for phytoremediation. Environ. Sci. Technol. 39: 7671–7677. https://doi.org/10.1021/es0506814.

van Dillewijn, P., Couselo, J.L., Corredoira, E., Delgado, A., Wittich, R.M., Ballester, A. and Ramos, J.L. 2008. Bioremediation of 2,4,6-trinitrotoluene by bacterial nitroreductase expressing transgenic aspen. Environ. Sci. Technol. 42: 7405–7410. https://doi.org/10.1021/es801231w.

Van Straalen, N. and Gestel, C.A.M. 1993. Soil invertebrates and microorganisms. pp. 251–277. *In*: Calow, P. (ed.). Handbook of Ecotoxicology, Vol. 1. Oxford, UK: Blackwell Sc. Publications.

Vieuble-Gonod, L., Benoit, P., Cohen, N. and Houot, S. 2009. Spatial and temporal heterogeneity of soil microorganisms and isoproturon degrading activity in a tilled soil amended with urban waste composts. Soil Biol. Biochem. 41: 2558–2567. https://doi.org/10.1016/j.soilbio.2009.09.017.

Vroumsia, T., Steiman, R., SeigleMurandi, F., BenoitGuyod, J.L. and Khadrani, A. 1996. Biodegradation of three substituted phenylurea herbicides (chlorotoluron, diuron, and isoproturon) by soil fungi. A comparative study. Chemosphere 33(10): 2045–2056. https://doi.org/10.1016/0045-6535(96)00318-9.

Walker, A., Jurado-Exposito, M., Bending, G.D. and Smith, V.J.R. 2001. Spatial variability in the degradation rate of isoproturon in soil. Environ. Pollut. 111: 407–415.

Walker, A., Turner, I.J., Cullington, J.E. and Welch, S.J. 1999. Aspects of the adsorption and degradation of isoproturon in a heavy clay soil. Soil Use Manage. 15: 9–13. https://doi.org/10.1016/S0269-7491(00)00092-0.

Wang, X., Aulenta, F., Puig, S., Esteve-Núñez, A., He, Y., Mu, Y. and Rabaey, K. 2020. Microbial electrochemistry for bioremediation. Environ. Sci. Ecotechnol. 1: 100013. https://doi.org/10.1016/j.ese.2020.100013.

Weyens, N., van der Lelie, D., Taghavi, S., Newman, L. and Vangronsveld, J. 2009. Exploiting plant-microbe partnerships to improve biomass production and remediation. Trends Biotechnol. 27: 591–598. https://doi.org/10 .1016/j.tibtech.2009.07.006.

Williams, S.L., Carranza, A., Kunzelman, J., Datta, S. and Kuivila, K.M. 2009. Effects of the herbicide diuron on cordgrass (*Spartina foliosa*) reflectance and photosynthetic parameters. Estuaries Coasts 32(1): 146–157. https://doi.org/10.1007/s12237-008-9114-z.

Yan, X., Huang, J., Xu, X., Chen, D., Xie, X., Tao, Q., He, J. and Jiang, J. 2018. Enhanced and complete removal of phenylurea herbicides by combinational transgenic plant-microbe remediation. Appl. Environ. Microbiol. 84(14). 10.1128/AEM.00273-18.

Yin, X.L., Jiang, L., Song, N.H. and Yang, H. 2008. Toxic reactivity of wheat (*Triticum aestivum*) plants to herbicide isoproturon, J. Agric. Food Chem. 56: 4825–4831. https://doi.org/10.1021/jf800795v.

Zehe, E. and Fluhler, H. 2001. Preferential transport of isoproturon at a plot scale and a field scale tile-drained site. J. Hydrol. 247: 100–115. https://doi.org/10.1016/S0022-1694(01)00370-5.

Zhang, J., Hong, Q., Li, Q., Li, C., Cao, L., Sun, J.Q., Yan, X. and Li, S.P. 2012. Characterization of isoproturon biodegradation pathway in Sphingobium sp. YBL2. Int. Biodeterior. Biodegrad. 70: 8–13. https://doi.org/10.1016/j.ibiod.2012.01.008.

CHAPTER 3

New Paradigm in Algae-based Wastewater Remediation

Pankaj Kumar Singh and *Archana Tiwari**

1. Introduction

Water gets contaminated from different sources, which fluctuate both in quality and volume. Wastewater is usually a combination of natural inorganic and organic materials and compounds which are created by man. Amino acids, fats, proteins, volatile acids and carbohydrates like organic carbon are present in sewage as 75% of its constitution (Gray, 1989). The large amount of sulphur, magnesium, bicarbonate, sodium, phosphate, potassium, ammonium salts, chlorine, heavy metals and bicarbonates are the main inorganic constituents of wastewater (Lim et al., 2010; Tebbutt, 1983; Horan, 1990). Different pollutant sources contain discharges from villages and towns which include treated or raw sewage, wastes released from industries or manufacturing plants, leachates from the site of solid waste disposal and agricultural land run off. These locations of toxic wastes are issues that need to be resolved (Horan, 1990). Shortage of water, their requirement for life and foodstuff is compelling one to investigate the purpose of reusing wastewater and resort to its recuperation (De la Noue and De Pauw, 1988). When wastewater is ready for discharge or reuse, it is mandatory to remove all toxic metals and other nutrients within acceptable limits from it (Moondra et al., 2020). Increased and harmful metals are taken out with the help of microalgae from wastewater (Neveux et al., 2016) as they work as decontaminators (Mishra and Mishra, 2017), due to their elevated activity of photosynthesis, adopting supplements, short-life expectancy combined with basic health necessities (Cai et al., 2013) and are additionally unaffected by ecological fluctuation (Batista et al., 2015; Goncalves et al., 2017). There are a number of problems which are based on pollutants and toxins from different formative activities faced by the world. The increase of contaminated water is the outcome of the explosion of the population of the world. The amount and the nature of squander created and released into normal waterway shave of late shown the requirement for various techniques to deal with the quality of water difficulties in areas. Bioremediation uses inhabiting microbes and different parts of the common habitat to include the nutrients of wastewater. Bioremediation can help in demonstrating less inexpensive advances that are used for cleaning unsafe waste (Validi, 2001). For the development of refining natural water, algae play a very important role universally (Han

Diatom Research Laboratory, Amity Institute of Biotechnology, Amity University, Noida, Uttar Pradesh, 201 313, India.
* Corresponding author: panarchana@gmail.com

et al., 2000; Olguin, 2003). Consequently, the use of micro algal species for elimination of nutrients from different wastelands has been shown by many researchers (Senegar and Sharma, 1987; Gupta and Rao, 1980; Williams, 1981; De la Noue, 1992; Kunikane et al., 1984; Benemann et al., 1977; Rao et al., 2011; DeBashan et al., 2002; Gantar et al., 1984; Queiroz et al., 2007; Tam and Wong, 1989). Furthermore, after harvesting algal biomass, it could be used to obtain the production of methane, generation of electricity and bioplastic material (Gao et al., 2009; Mulbry et al., 2010; Arbib et al., 2014).

In lotic systems, diatoms algae are excellent indicators for showing the quality of water. In milder areas, the impact of the substrate on diatom algae-based water appraisal in tropical streams is not completely understood. The capacity to use diatoms algae to assess the current and earlier periods on the importance of water and its conditions vary in a few amphibian conditions has been understood worldwide for a long time (Whitmore, 1989; Chessman, 1986; Van Dam 1974 and Patrick, 1973), however, it has been restricted to South Africa, mainly due to the difficult example of preparation and the tedious commitment needed to create key ordered aptitudes. State and territorial waterways projects in the United States for algae, generally depend on the assessment of diatom gatherings (Charles, 1996). Kociolek and Stoermer (2001) observed that reviews on precise scientific classification and environment of microalgae like diatoms in the 21st century will and should be connected. They investigated the world view that coordinates the diatom's scientific classification and environment to preserve science in which microbial networks (particularly diatoms) can be used to characterize regular living spaces requiring protection. Evaluation approaches dependent on the amount of diatoms to be created in the deltaic conditions and have since then been included the reverine frameworks (Eloranta and Soininen, 2002; Round, 1991a; Stevenson and Pan, 1999).

Elevated recurrence, a multi-constraint water feature checking a program is basically not able to comprehend the current circumstances and an option in favour of surveying varies after some period is very essential. The assessment of the living diatoms within a residue and lying onstones, with the means of invertebrates, gives a strategy to facilitate joining two free pointer frameworks on diverse trophic level (Hofmann, 1996; Smol, 1992). In the most recent decade, diatom-based records have gained increased extensive prominence throughout the world as a device to give a coordinated impression of water quality, and on the choices for rivers and water courses (Prygiel and Coste, 1993; Kelly, 2002; 1998b).

The greater part of the advanced work on the diatom algae has been done within the French seepage basin with test staking place for the size of a region, in large and typologically broadened areas, like France; empowering the overall use of these files on the continent of Europe (Prygiel and Coste, 1993). The effort on the functional use of microalgae diatoms as bio markers have continued with the end goal that diatom algae lists have supplanted those of invertebrates as the bio monitor technique for decisions in specific circumstances for example canalizing streams (Prygiel and Coste, 1993).

Pollution of freshwater resources has become one of the most important problems of humanity. The ecosystem of rivers is under danger from different human exercises over the globe prompting extensive changes in silt conveyance and stream designs, deteriorating the quality of water and reducing biodiversity (Vorosmarty et al., 2010). The oceanic life forms were used as bio pointers for contamination; likewise, as bio monitors to comprehend the collaboration between livings being's reactions to the natural change and their legitimate impact (Markert and Zechmeister, 2003).

Diatoms algae have been known as an excellent indicator of water quality (Stevenson, 2014), several diatoms index have been implemented in the regions of the earth, some examples are Trophic Diatom Index (TDI), IPS (Kelly and Whitton, 1995), also the Diatom Biological Index (IBD) (Coste et al., 2009), which is carried on a weighing standard equation. The benthic diatoms can be used as environmental pointers and viewed as significant in water quality estimation and inspection (Round, 1991).

Fresh water is an essential requirement for the advance and establishment of different human movements. Water resources provide in valuable foodstuff to farming establishments. However, liquefied and hard dispersal created by human beings which is discharged and involuntary movements contaminate the mainstream of the water resources around the globe. Due of large overall increment in the human populace, water will get perhaps become the most difficult asset in the 21st century (Day, 1996). India possesses just 3.29 million km^2 geological territory, which structures 2.4% to the ground region; it supports above 15% of the total population. Among the yearly exploitable water assets of the nation are 1086 km^3 which is just 4% of the earth's water assets (Kumar et al., 2005). The entire water necessity of the nation in 2050 is assessed to be 1450 km^3 which is elevated from the ebb and flow accessibility of 1086 km^3. A great part of the losses of development into water bodies are through the release of waterborne waste from households, industrialized and non-point resources transferring undesirable and unrecovered material (Welch, 1992). The amount of wastewater that go back to older periods, its management is a generally delayed development dating from the last part of the 1800s and mid 1900s (Chow et al., 1972). The most widely recognized wastewater treatment techniques in industrial nations are unified high-impact treatment plants of wastewater and tidal ponds intended for both household and industrialized wastewater. Current human advancement, equipped with rapidly compelling improvement and developing financial structure is less than the escalating hazard that comes from the contamination of water. In India uncontrolled development in metropolitan areas has led to the extension of water and sewage frameworks that are damaging and costly. In India, wastewater created by means of 299 class-1 urban communities is 16,625.5 MLD. Apart from this, approximately 59% is produced by 23 metro urban communities. About 23% of the total wastewater is generated only by the state Maharashtra and the supply of the basin of River Ganga is 31% of the entire wastewater produced in class-1 urban areas. From 16,662.5 MLD of wastewater created, just 4037.2 MLD (24%) can be treated before discharge; the rest is discarded untreated (Looker, 1998). Figure 3.1 elaborates the approaches towards algae-based wastewater treatment.

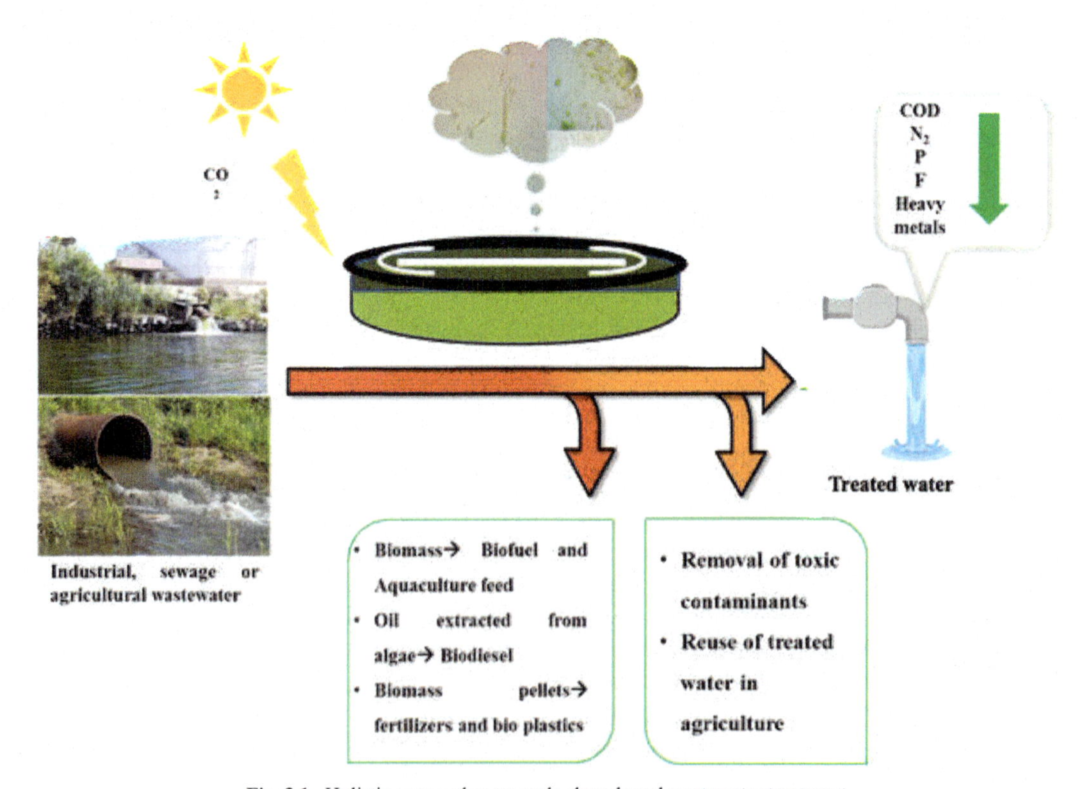

Fig. 3.1. Holistic approaches towards algae-based wastewater treatment.

2. Need for Treatment of Wastewater

Treatment for water and wastewater needs serious attention all over the world especially in countries with a dry climate. In wastewater, the most important pollutants are heavy metals. The compounds of heavy metals and the heavy metals themselves are used widely in many industrial applications resulting in their increase in aquatic system. The methods of elimination are precipitation by means of chemicals, membrane systems, coagulation with alum, ion exchange and adsorption (Bailey et al., 1999; Lazaridis et al., 2003; Walker et al., 1999). Among the above described methods, the adsorption systems are effective in diminishing the absorption of heavy metal. These systems are advantageous, as they are quick, suitable and simple for water and wastewater containing small or only a little amount of metal. The only disadvantage of these processes can be the high cost of some adsorbents (Ozverdi et al., 2006; Low et al., 2000). The utilization of any solid substance by means of a micro porous composition to be able to use anew adsorbent, e.g., charcoal, clays, iron oxides and bone, natural and synthetic zeolites, activated carbon and molecular sieve. Among the most significant adsorbents in the treatment of water and wastewater are activated carbon. However, because of the cost in the manufacturing expenditure of activated carbon, it is not appropriate for a country which is under developed (Pandey et al., 1985). The use of other alternatives like Diatomite ($Sio_2.nH_2o$), diatomaceous earth or diatoms can be used for the treatment of water and wastewater. Diatomite or diatomaceous earth is a pale-coloured, spongy, light-weight sedimentary stun made mainly of the silica of microfossils obtained from marine unicellular microalgae. They consist of a large diversity of shapes and sizes, on average 10 to 200 μm, in a composition containing up to 80 to 90% voids. They have a novel blend of chemical and physical properties, for example, elevated porosity, elevated penetrability, little molecule dimension, enormous shell zone and low thermal conductivity, which make diatomites appropriate for a wide range of industrialized application, for example, channel help or a channel. In aqueous solutions, diatomite particles have negative charges, so they possess a strong attractive force for particles with positive charges. There are many diatomaceous mines around the world, and it seems that diatomite will be used for many industrial applications in the future (Baojiao et al., 2005). Al-Degs et al. (2000) studied the possibility of using diatomite as a manganese-diatomite for the remediation of Cd^{2+}, Cu^{2+}& Pb^{2+} from wastewater and water. There are several diatomite mines in Iran, mostly in the Azerbaijan province, such as Tabriz diatomite in east Azerbaijan and the Mamaghan mine in Miyaneh. The diatomite can be used as an appropriate substitute for activated carbon (Mobbs, 1999).

The treatment of wastewater consists of the collapse of multifarious untreated compounds in wastewater to be able to easily strengthen those that are stable and the aggravation liberated, either physico-artificially and additionally by using miniature life forms (organic action). The antagonistic ecological effect of the accompanying wastewater needs to be unconstrained in groundwater, outside water bodies as well as land is as follow:

1. The decomposition of normal materials enclosed in wastewater can rapidly lead to developing large quantities of mouldy gases.
2. Unprocessed wastewater (sewage) restraining countless unusual matter, at any time unconfined into a watercourse/river, will increase the oxygen used for satisfying the Biochemical Oxygen Demand (BOD) of wastewater and in this way drain the collapsed oxygen of the watercourse, causing the death of fish and other unfortunate impacts.
3. Wastewater might similarly restrain accessories, which can cause the growth of sea-going shrubbery and algal vegetates, along with timely eutrophication of the ponds and watercourses.
4. Unprocessed wastewater usually includes a variety of pathogenic, or microorganisms that can create infections and toxicity intensifies, that are found in the region of human intestines or may be present inside established industrialized waste. These may possibly taint the soil or the stream bodies, where ever such manure is found.

However the motives for the management and removal of wastewater, are essential. There are many ways to treat wastewater and their utilization, but now a days, cultivation of microalgae like diatoms for the treatment of wastewater is increasing and these diatoms algae play a significant role for the process of treatment of wastewater. In the procedure of wastewater treatment with the help of diatoms algae, they consume a massive quantity of CO_2 in the process of photosynthesis and produce molecules of glucose and oxygen naturally. Diatoms algae-based wastewater treatment can eliminate the number of heavy metals, organic and inorganic materials, large amounts of nutrients and many other contaminants from wastewater (Dhote et al., 2012; Wang et al., 2019; 2010).

3. Wastewater Composition

The composition of wastewater depends on the area from where it is discharged or being released. There are many sources for wastewater. The composition of wastewater refers to the actual amount of different constituents present in wastewater like chemical, physical and biological. Generally, wastewater contains the following compositions given below

1. Physical constituent or properties
   i) Solids
   ii) Odour
   iii) Colour
   iv) Temperature

Solids: The solid contents in wastewater are defined as those substances which remain there even after the process of evaporation at temperatures of 103° to 105°. The source for the above-mentioned solids particles are domestic and industrial wastewaters.

Odour: In wastewater, odours are usually caused by gases formed by the disintegration of untreated material. Wastewater which is often fresh has some unpleasant odour. The source for these odours is industrial and decomposing wastewater.

Colour: The colour of wastewater depends on its source such as, the colour of clean wastewater is generally grey, septic wastewater reveals black appearing towards the intermediate. Besides this, the colour of industrial wastewater is not defined, which may contain many colouring substances. Here, the source for the mentioned characteristics is domestic and industrial wastewater.

Temperature: The temperature of wastewater also depends on the geographical location, which varies from 10–24°C or more. Wastewater temperature plays a significant role on aquatic life, different biological and chemical reactions rate and the solubility reactions of gases like depletion of oxygen. Here, the source for wastewater temperature is domestic and industrial wastewater.

2. Chemical Constituents: - There are two types of chemical constituents-
   a) Organic Constituents
   b) Inorganic Constituents

A) Organic
   i) Carbohydrates
   ii) Proteins
   iii) Pesticides
   iv) Phenols
   v) Surfactants
   vi) Fat, oils, grease.

B) Inorganic
   i) Nitrogen
   ii) Phosphorous

iii) Heavy metals
iv) Sulphur
v) Chlorides
vi) Alkalinity
vii) Toxic
viii) pH
ix) Hydrogen ions

The major group of untreated substance which come from wastewater are carbohydrates (25–50%), proteins (40–60%) and fat and oils (10%). Besides this, a mixture of hydrogen, oxygen and carbon with nitrogen is a common organic compound found in wastewaters. The most important element which serves like non-living metabolites are ammonia-nitrogen, carbon and phosphorous.

Besides organic and inorganic constituents, some gases like nitrogen, oxygen, Co_2, H_2So_4, ammonia and methane are usually found in raw wastewater.

3) Biological Constituents:
i) Animals
ii) Plants
iii) Protista
iv) Viruses

The sources of biological constituents are domestic wastes and open watercourses and treatment plants.

4. Sources and Compositions of Different Wastewaters

There are different sources of wastewaters, Fig. 3.2 elaborates the different types of wastewaters.

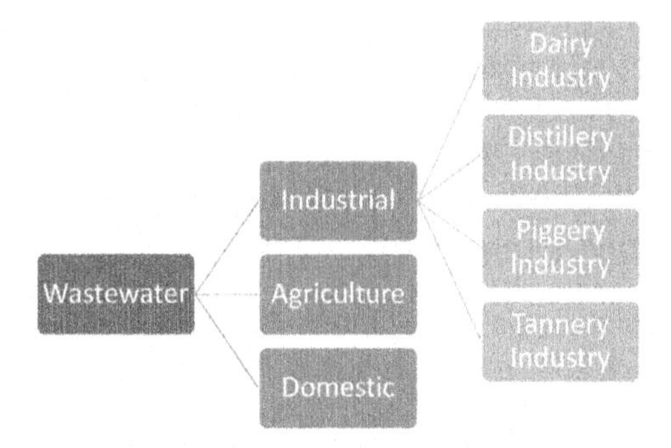

Fig. 3.2. Different types of wastewaters.

4.1 Dairy wastewater

In the sector of food processing, the dairy industry is the largest. The dairy industry needs an enormous amount of fresh water for the purpose of different types of hygienic work like washing, cleaning of floors, animals and for the purpose of sanitization and also used at the place of heat exchange (Tocchi et al., 2012). Wastewater released from the dairy industry contains various kinds of compounds like wasted milk with fats and lactose. They also include different types of detergent which are used for the purpose of washing, some sanitizing agents and nutrients are also a part of dairy wastewater. The pH of the dairy wastewater varies between 4.7 to 11 due to different seasons and milking processes. The characterization of dairy wastewater varies, examples include Biological Oxygen Demand or BOD of dairy wastewater between 40–48000 mgL^{-1}, Chemical Oxygen Demand or COD of dairy wastewater is between 80–95000 mgL^{-1}, Total Phosphorous or TP between 9–280 mgL^{-1} and Total Nitrogen or TN between 14–830 mgL^{-1} respectively. The number of different kinds of nutrients (macro and micro) like calcium, chlorine, sodium, potassium, magnesium, cobalt, manganese, iron and nickel may be more noticeable in dairy wastewater (Chokshi et al., 2016; Daneshvar et al., 2018). Wastewater which contains phosphorous and nitrogen can upset the balance of the ecosystem and also lead towards eutrophication in normal aquatic bodies (Nayak et al., 2016).

4.2 Textile wastewater

The processing of textile involves several diverse steps. And in the processing of these different steps, a huge amount of wastewater is produced. These are various factors which involve the composition and amount of wastewater, and this includes the type of the procedure and types of the processed fabric. The nature of the generated wastewater from the textile industry depends on the changes in machines, chemicals which are used and some other characteristics in the processes. In the textile industry, dyes are frequently used, and these are actually reactive dyes which are frequently used in the industry of textiles. In the dyeing of fibres, such as cellulosic ones, substantial amounts of dyes are not fixed on the fabric. The wastewaters of dye generally contain approximately 10–50 mgL^{-1} amount of dye in that solution. These types of concentration are large enough to make significant colour bodies where they are discharged (Alinsafi et al., 2007). Many toxic unmanageable organics are found in raw wastewater which is released from the textile industry.

4.3 Winery wastewater composition

Winery wastewaters are generally formed from the process of cleaning inside the plants of wineries, and as a result principally prepared with wine, some kinds of suspended solids, some amounts of juice of grapes and KoH, NaOH like agents which are generally used for the purpose of cleaning. But the exact component is not known. Wastewaters released from the majority of wineries are rich in organic contents, propionic acid, acetic acid, and tartaric acid such as, organic acids. Besides this, the study based on wineries wastewater indicates that these wastewaters also contain some sugars, polyphenolic compounds and esters (Malandra et al., 2003).

4.4 Olive mill wastewater composition

In the region of the Mediterranean, the production of olive oil and olive tree cultivation and its uses has been established and well-known during the last 7000 yr. In the areas of the Mediterranean, approximately 97% or more of olive oil production takes place. Worldwide, there are 750 million olive trees produced and approximately 98% of olive trees are from the Mediterranean region. Greece, Italy and Spain are three countries, which produce the maximum olive oil worldwide. There are generally three functioning steps which are used for the extraction of olive oil:

- The crushing of olives, in this process the cells of fruits are crushed and olive oil is extracted.
- Mixing, in this process left over crushed pastes are slowly blended to enhance the acquiesce of oil; and
- Separation of olive oil from the left-over waste.

When water containing olive vegetation and the water which is added in this process are mixed together, these mixed waters are called olive mill wastewater (OMWW) (El-Abbassi et al., 2012). The olive mill wastewater contains different types of compounds like phenolic compounds, some salts or heavy metals which are bad for the environment as well as for the soil where these wastes are dumped and also toxic for water sources such as rivers or ponds where olive mill wastewater is discarded. Heavy metals that are produced in manufacturing olive oil is a cause of a negative footprint from the ecological point of view (McNamara et al., 2008). The waste which is produced through the agro food industry contains a phenol like organic substance, low pH in nature which indicates its acidic nature, is very harmful for all ecosystem in which different organisms like bacteria, different aquatic species, plants and air are affected due to its extremely poisonous character and amount of harmful pollutants in them (Dermeche et al., 2013; El-Abbassi et al., 2012).

4.5 Piggery wastewater

Piggery wastewater appears mainly from the pig industry. As the human population is growing very fast, the demand for meat like pork is rising day by day. According to Niyiragira et al. (2018) the population of pigs was about 784.83 million globally. For the creation of a clean and hygiene internal environment, very large areas of fresh stream are necessary to support cleaning and appropriate washing of the excreta generated by pigs. Therefore, a very huge quantity of piggery wastewater is released from the pig industry. According to one calculation, 18 litres of piggery wastewater is generated per day by a pig in a piggery farm (Maraseni and Maroulis, 2008). The wastewater generated from piggery farms generally contain phosphorous, organic matter and nitrogen in large amounts. In these piggery wastewaters, phosphorous is present in the number of phosphates, while organic matters are present in the form of urea, while ammonium and nitrogen are present as nitrites or nitrate (Ganeshkumar et al., 2018).

4.6 Tannery wastewater

In the world, the tannery industry is known for being one of the most polluting industries. The main components of a tannery wastewater are chlorides, sulphides, sulphates, tannins, calcium, chrome and ammonium. Besides these components, wastewaters include dung, coats, hairs, blood and fat of animals, alkalinity, solutions of protein and their suspension. A higher concentration of trivalent chromium, organic and inorganic acids are usually present (Mannucci et al., 2010).

4.7 Pulp and paper mill wastewater

The producer of world's biggest plants-based wastewater is paper and pulp industries (Ried et al., 2008). For one ton of paper production, approximately 10 and 50 m^3 of fresh water required. However the quantity of water for every ton of paper production has been diminishing over time (Buyukkamaci and Koken, 2010; Pizzichini et al., 2005). The main composition of paper and pulp wastewater are enriched with carbon and also contain phosphorous and nitrogen in small quantities. A massive amount of wastewater is produced globally from the pulp and paper industries (Axelsson and Gentili, 2014). Besides this, the industry of pulp and paper is seen as a major user of water, energy and other natural assets. The main ecological effects of paper and pulp resulting from the pulping process and filtering measures include some contaminations produced into the environment and air, some of them are released into the wastewaters, and other wastes like solid wastes are created (Bhatti et al., 2021; Hubbe et al., 2016). In all, the effluents of a pulp and paper factory include a combination of different natural mixes, for example, some extractives, lignins and other products excreted from the degradation of the carbohydrates (Ugurlu et al., 2008). In the process of preparation of wood, processing of pulping, washing of pulp, operating paper machines, in the process of screening, in the process of coating and particularly in the bleaching process, different polluting waste matters are produced (Pokhrel and Viraraghavan, 2004; Bhatti et al., 2021). In Sweden, there is an undertaking in progress to build a procedure that adequately permits the microalgae to treat the supplements in factory wastewater and carbon dioxide fixation from gases which are generated through flue (Ekendahl et al., 2018).

4.8 Distillery wastewater

In distilleries, stillage (distillery wastewater) is the main accessory which is produced, and its volume is around multiple times more than that of ethanol. The use of stillage raises some significant issues, and numerous endeavours have been made everywhere in the world to resolve them. Wastewater from distilleries normally contains a high volume of incredibly acidic issues which presents numerous removal and treatment concerns. Squander streams by and large contain large levels of both inorganic materials and dissolved organic materials (Shivajirao, 2012). In Asia, the second largest producer of ethanol is India. Approximately, 319 distilleries are present in India with a set up capability of 3.25 billion litres of alcohol (Kalavathi et al., 2001; Uppal, 2004). Distillery industries are considered among the top 17 pollution-causing industries in India by the Central Pollution Control Board (C.P.C.B.). For each litre of liquor created 10 to 15 litres of profluent, are produced and there by a common refinery delivering ethanol from the molasses of cane produces almost a half million litres of waste matter every day (Tiwari et al., 2007; Ghosh et al., 2002). Roughly, 40 billion litres of liquid waste material are created every year only in India for the manufacturing of 2.3 billion litres of liquor. The distillery industry is quite probably the most pollution causing and development leaning industry (Kanimozhi and Vasudevan, 2010; Garg and Garg, 2016). The wastes from distillery industries mostly contain compounds known as melanoidin polymers which are dark brown in colour and produced during a reaction which occurs when carbonyl group and amino acid are present in wastes of the distillery (Wedzicha and Kaputo, 1992; Patel and Jamaluddin, 2018). Besides the high natural substance, wastewater from distilleries additionally contains supplements as potassium, nitrogen and phosphorus that can prompt the process of eutrophication of water bodies. Profluent removal even after regular treatment is precarious and has an elevated contamination prospective because of the aggregation of non-biodegradable intolerant composite, which are generally dark and colourful and in an extremely impenetrable condition, melanoidin have some of the properties like antioxidants, which are harmful to many microorganisms involved with wastewater treatment measures (Sirianuntapiboon et al., 2004a). The wastewater from distillery represents agenuine risk to the quality of water in a few areas of the countryside. Notwithstanding contamination, progressively severe ecological guidelines are constraining refineries to develop obtainable treatment and further investigate optional techniques for the management of effluents (Kanimozhi and Vasudevan, 2010). Wastewaters from distilleries are measured like a main resource of ecological contamination due to their elevated organic substance and acidic nature (Keyser et al., 2003; Borja et al., 1993; Patel and Jamaluddin, 2018). The pH estimations of wastewaters from wine distilleries vary from 3.5 to 5.0 (low pH) (Bustamante et al., 2005; Wolmarans and De Villiers, 2002; Goodwin et al., 2001; Genovesi et al., 2000; Rajeshwari et al., 2000; Badar et al., 2017; Patel and Jamaluddin, 2018), which is also poisonous for a number of living organisms.

4.9 Agricultural and municipal wastewater

Wastewater from agricultural activity is one of the fundamental issues that influence biodiversity in various areas around the world. The effluents generated from agriculture start with an overabundance of water used in water system frameworks. These crude waters contain remaining manures and pesticides which are unsafe and toxins, which ultimately pollute the resources of natural water like ponds or rivers which are generally used for the production of drinking water (Moss, 2008; Smith and Siciliano, 2015; Picos-Corrales et al., 2020). Some of pollutants like organochlorine pesticides have existed for a long time (Yadav et al., 2015; Chiesa et al., 2016). Subsequently, the chemical atmospheres of the water bodies are disturbed by agrochemicals left behind and simultaneously cause a disturbance of the environmental equilibrium and a possible hazardous effect on the health of human beings (Kim et al., 2017; Baqar et al., 2018; Choi et al., 2018). Municipal wastewater is a term which refers to the combination of household waste and the sewage of industries and all those runoffs which stream into the treatment plants of wastewater which are responsible for the process of treatment (Korzeniewska, 2011). In municipal wastewater, the bacteriological composition depends on various components, including, among others, food-preparing plants, slaughterhouses, are the principle source of microorganisms in flowing sewage etc. (Kim and Aga, 2007). For the development of microalgae, for the purpose of remediation, a high level of phosphorous and nitrogen is required and is generally found in agricultural and municipal wastewater. Usually, the concentration of Posphorous (P) and Nitrogen (N) may vary from > 1,000 mg L^{-1} in wastewater generated from farming fields and 10–100 mg L^{-1} in wastewater generated from metropolitan areas. These elevated level of phosphorous and nitrogen are vital supplements upholding strong development of a wide range of types of freshwater microalgae species and consequently lead to a successful remediation system simultaneously (Woertz et al., 2009; Bawiec et al., 2016; Mohsenpour et al., 2021).

Table 3.1 Elaborates different wastewater remediation mediated by algae.

Table 3.1. Use of various algal species in remediation of diverse wastewaters.

S. No.	Name of Algal Species	Type of Wastewater	References
1.	*Chlamydomonas, Scenedesmus* species, *Chlorella vulgaris*	Artificial wastewater	Voltolina et al., 1999 Feng et al., 2011
2.	*Chlorella* species, *Auxenochlorella protothecoides, Phaeodactylum tricornutum*	Municipal wastewater	Li et al., 2012 Wang et al., 2019 Zhou et al., 2012
3.	*Ankira* sp., *Phormidium* sp., *Cyanophyta cocal, Microspora* sp., *Chroococcus limneticus, Dactylococcopsis* sp., *Stigeoclonium* sp., *Chodatella* sp., *Scenedesmus* sp.,	Fish processing wastewater	Riano et al., 2011
4.	*Chlamydomonas incerta*	Agro-industrial wastewater	Kamyab et al., 2016
5.	*Chlorella pyrenoidosa*	Industrial wastewaters from a palm oil mill and pig farm and domestic sewage	Aziz and Ng, 1992
6.	*Nitzschia* species	Aquaculture wastewater	Xing et al., 2018
7.	*Cyanobacteria*	Secondarily treated domestic effluent and settled swine wastewater	Pouliot et al., 1989
8.	*Chlorella* sp., *Chlamydomonas* sp., *Protoderma* sp., *Oocystis* sp.	Digested swine manure	MolinuevoSalces et al., 2010
9.	*Chlorococcum humicola, Chlorella vulgaris, Selanastrum* sp., *Scenedesmous* sp.	Domestic wastewater	Thomas et al., 2016 Chandra et al., 2014 Xin et al., 2010
10.	*Phormidium* sp., *Euglena* sp.	Lake, sewage treatment plant	Mahapatra et al., 2013
11.	*Nitzschia* sp., *Oocystis* sp., *Chlamydomonas* sp., *Chlorella* sp., *Protoderma* sp., *Achnanthes* sp., *Microspora* sp., *Scenedesmus* sp.	Swine manure, fermented swine wastewater	de Godos et al., 2009 Kim et al., 2007
12.	Mixed culture of diatom and Chlorella species	Paper industry wastewater and wood-based pulp	Tarlan et al., 2002
13.	*Phormidium tergestinum Pinnularia* sp., *Chlamydomonas subcaudata, Nitzschia* sp., *Teilingia* sp.	Slaughterhouse wastewater	Hernandez et al., 2016
14.	*Chlorella* sp., *Nitzchia* sp., *Chlorella pyrenoidosa*	Settled domestic sewage	McGriff and McKinney, 1972 Tam and Wong, 1989
15.	*Scenedesmus* sp.	Chey Whey permeate, Artificial wastewater	Girard et al., 2014 Voltolina et al., 1999

Table 3.1 contd. ...

...*Table 3.1 contd.*

S. No.	Name of Algal Species	Type of Wastewater	References
16.	*Chlorella pyrenoidosa*	Piggery wastewater	Wang et al., 2013
17.	*Chlorella vulgaris*	Diluted pig slurry, Municipal wastewater, Synthetic wastewater	Fallowfield and Barret, 1985 Ji et al., 2013
18.	*Scenedesmus obliquos*	Secondary treated wastewater	Martínez et al., 2000
19.	*Chlorella pyrenoidosa*	Wastewater from soyabean process	Hongyang et al., 2011
20.	*Chlorella vulgaris*	Synthetic wastewater	Perez-Garcia et al., 2010
21.	Diatom consortium	Industrial & municipal wastewater	Marella et al., 2018
22.	*T. indica, spirulina, S. abundans, Nostoc muscorum, Botryococcus braunii*	Secondary treated wastewater	Amit and Ghosh, 2018 Oprez et al., 2009
23.	*Chlorella* sp., *Chlorococcum* sp.	Dairy effluents, dairy manure wastewater	Johnson and Wen, 2010; Ummalyma and Sukumaran, 2014

5. Role of Diatoms Algae in Wastewater Remediation

The development of microalgae in wastewater treatment is an advanced strategy for the purpose of decontamination that has led to a lot of concern in recent years for their noteworthy nutrient-elimination ability and energy-rich production of biomass of the microalgae (Sharma et al., 2020a; Nayak et al., 2019). The fundamental reasons for the management of wastewater are to get rid of the higher quantity of Phosphorus (P), Chemical Oxygen Demand (COD), Nitrogen (N) and other substance such as different organic materials (Nzayisenga et al., 2018). The treatment of natural wastewater with the help of microalgae is largely beneficial as compared with other treatment alternatives because of its operational costs, diminished slime arrangement and small amount of carbon (Udaiyappan et al., 2017; Ferro et al., 2019). The fundamental use of the microalgal-based innovation has led to a decrease of the amount of contaminants and allocation of carbon has been demonstrated as a useful method (Singh et al., 2020; Marella et al., 2019). The exploitation of microalgae diatoms in the field of wastewater is well suited because they are easily available and rapidly growing organisms. Furthermore, different species of algae or microalgae are available with their different characteristics (Su et al., 2012; Chaitanawisuti et al., 2011). The crucial class of algae which include benthic diatom can be found in natural environments by the process of separation and purification (Xing, 2007). Microalgae diatoms are environment friendly, relatively of low cost, most efficient for the purpose of wastewater treatments which are compared as an alternative technique of wastewater treatment (HII et al., 2011). In wastewater, micro algal cultivation helps in the biological treatment of wastewater because it consumes a large quantity of CO_2 and produces sugar molecules through photosynthesis and oxygen release as by-products (Al Darmaki et al., 2012). Various environmental challenges like climate change, global warming and ozone hole depletion can be solved as they sequester atmospheric CO_2 eliminating heavy metals, inorganic and organic materials, nutrients and toxic waste materials (Satpal and Khambete, 2016). Diatoms algae are a large component of benthic habitation (often 90 to 95%) and have developed into an imperative component of stream feature management. The community of benthic diatom is valuable and plays a significant role in the recognition of human beings motivated by the effects of the marine atmosphere. The main element is that they can be available in all surface waters, and tests gathered by them can be kept for a long period of time (Acs et al., 2004). The use of macro or microalgae for wastewater and waste management is known as phycoremediation (Phang et al., 2015). Microalgae diatom growths can probably bring about phycoremediation in different kinds of wastewater because of their effective cell components and their variation procedures. They can use different micro and macronutrients from wastewater and acclimatize them into their biomass (Olguin, 2012). A species of benthic diatom like *Nitzschia* sp.

can decompose organic material rapidly with better action in the layer of residue (Yamamoto et al., 2008). In the future, with the help of diatom algae there are a lot of possibilities to degrade pollutants biologically, and later in the future diatoms algae can be used for the purpose of bioremediation with a complete prospective, wide ranging and in a well-organized manner (Marella et al., 2020). For the purpose of pesticides, heavy metals and different waste stuff removal, a microalgae-based remediation process for wastewater has been suggested (Baghour, 2019; Marella et al., 2016, 2018; Tiwari and Marella, 2019). In the process of phycoremediation, the use of algae diatom is beneficial over conservative methods as diatoms algae are not harmful in nature, ubiquitous, can develop more rapidly, lucrative, environmentally good, eco-friendly, biodegradable, non-harmful, exceptionally versatile in nature, fit for heavy metal remediation with little power contribution due to autotrophy which has drawn the attention of researchers worldwide (Sunday et al., 2018). The large importance of diatoms algae is in phycoremediation of wastewater where heavy metals are found and focus on appropriation of heavy metals which advance accumulating lipid bodies for the production of biofuels (Yi et al., 2017). The cell wall of the diatom is made up of silica and known as frustules which are composed with a hard outer layer comprised of shapeless silica designed with nano-sized to micro sized pores in the environment measuring 20 to 200 nm besides the hyaline zone, spines and with different features (Marella et al., 2020). Microalgae are able to diversify the category of wastewater in a small phase and can be an unconventional medium for the elimination of different impurities and also can be an economical beneficial process (Moondra et al., 2020). For the elimination of contaminated heavy metals from wastewater, three-dimensional frustules of diatoms could be suppressed (Jamali et al., 2012). At a point when the diatom algae adsorb metals of heavy weight, they get exchanged and complexed as their frustules which give a large contact zone encourage substantial adsorption of the metal (Lin et al., 2020). Diatoms utilize mechanisms like abiotic and biotic responses to contaminants and are well adjusted to changing ecological conditions to gather pollutants from the environment (Congestri et al., 2005; Guzzon et al., 2008). Thus, a practical new methodology should be arranged which disinfect the harmful mixes as well as a viable supportable procedure to recover and reuse heavy metals. Of late, phycoremediation allows incorporating the use of macroalgae or microalgae for the remediation of weighty metal particle close to the watercourse (Azimi et al., 2017).

6. Nanoremediation of Wastewater—A New Paradigm

Nanoremediation techniques involve the use of receptive nanomaterials for change and detoxification of contamination. These nanomaterials have assets that empower both the compounds, decrease and catalysis to moderate the toxins (Karn et al., 2009). The development of various kinds of nanomaterials and their products are a newand advancing field which allows numerous chances for development. Nanotechnology has demonstrated its successful bio applications in different fields including drugs, medicine and also in farming (Hasnain et al., 2013). With the progress in industrialization, farming and metropolitan means, the degrees of groundwater contamination have expanded largely over the last couple of decades (Mehndiratta et al., 2013; Mura et al., 2013). In India, this has prompted an expansion in the centralization of numerous natural and inorganic pollutants far over the admissible reaches of drinking water principles. Most of the groundwater feature issues are brought about by pollution, over exploitation or a mix of the two. The importance of the top soil and groundwater are gradually declining all over. Direct unloading of unprocessed effluent into well sanderst while stream resources are significant reasons for the contamination of groundwater. In addition, broad utilization of pesticides has also prompted the expansion in poison focusing in groundwater. Largely destructive elements are intensifying their way into groundwaters through different illicit industrialized and farming actions incorporating different pesticides, organochlorines, heavy metals, other carcinogenic mixes and organophosphorous (Agarwal and Joshi, 2010). Groundwater remediation can be accomplished by using nanotechnology (Rajan, 2011). Comparative focuses have also been discussed of late (Kemp et al., 2013). As indicated by Rajan (2011), nanomaterials, for example, carbon nanotubes (CNTs) and zero-valent iron (nZVI) can be used in natural cleanups, for example, remediation of groundwater for the purpose of consumption and use again. Notwithstanding, there might be a few concerns left with respect to the expected dangers to nature, human wellbeing related with the use of nanomaterials (Kumar et al., 2014; Taghizadeh et al., 2013; Farrukh et al., 2013). There are different nanomaterials and nanoparticles used for the purpose of remediation of diverse kinds of contaminants and pollutants which is found in water or wastewater are explained below:

- Activated Carbon Fibres (ACFs) which are used against pollutants like xylene, ethylbenzene, toluene and benzene.
- Nanocrystalline zeolites which are used to remove pollutants like nitrogen dioxide and toluene.
- Single-walled CNTs (carbon nanotubes) with Fe can remove pollutants like Trihalomethanes (THMs).
- CeO_2 CNTs (carbon nanotubes) which removes pollutants like heavy metal ions.
- CNTs (carbon nanotubes) functionalized with polymers can remove pollutants such as benzene, p-Nitrophenols, heavy metal ions, dimethylbenzene and toluene.
- Multi-walled CNTs (carbon nanotubes) with Fe can removes pollutants, such as herbicides, microcystin toxins, heavy metal ions, chlorophenols, herbicides, microcystin toxins and THMs
- Self-Assembled Monolayer on Mesoporous Support (SAMMS) can remove pollutants like heavy metal ions and inorganic ions.
- Nanoparticles of biopolymers can remove pollutants such as heavy metal ions.
- Palladium(Pd)/Iron(Fe) based nanoparticles PCBs, chlorinated ethane, chlorinated methanes.
- Zero-valent iron nanoparticles (nZVI) can remove pollutants like chlorinated organic compounds, heavy metals, Polychlorinated biphenyls (PCBs) and inorganic ion.
- Supported TiO_2 nanoparticles can remove aromatic pollutants.
- Nanocrystalline TiO_2 nanoparticle can remove pollutants like heavy metal ions.
- Fe (III)-doped TiO_2 nanoparticle can remove pollutants like phenols.
- Nitrogen (N)-doped TiO_2 nanoparticle can remove pollutants such as Azo dyes from water or wastewater sample.
- Palladium(Pd)/Gold(Au) based nanoparticles and Nickel(Ni)/Iron(Fe) based nanoparticles can remediate chlorinated ethane, Trichloroethylene (TCE), Brominated Organic Compounds (BOCs), trichlorobenzene, and dichlorophenol like pollutants (Theron et al., 2008; Mansoori et al., 2008).

7. Challenges in Wastewater Treatment based on Algae

7.1 Harvesting of algal biomass

Biomass of harvested algae has been recognized as one of the main disadvantages for lucrative wastewater treatment and its downstream handling. The principle factors that convolute the harvested algae incorporates a little (equal to 0.05%) of green growth (algae) dry and the all-out suspension, minute range of a solitary cell, negative cell surface charge that keeps them from framing bigger and effectively harvestable particles just as fast as the development rate (Milledge and Heaven, 2013). These viewpoints fundamentally increment the all-out expenses of both green growth biomass gather and its purpose for the treatment of wastewater. It is assessed that expenditure of green growth biomass obtained can comprise up to 30% of the all-out creation costs (Grima et al., 2003), which is because of high vigour utilization, which, contingent on the picked strategy, changes somewhere in the range of 0.1 and 15 kWh m^{-3} (Singh et al., 2013). In addition, up to 90% of the all-out stock expenses are inferable to the gather and dewatering gadgets (Amer et al., 2011). Even though different algal biomass strategies have been created and are generally utilized, each of one of them has its recompenses and disadvantages, which empowers research on discovering all the more monetarily plausible, widespread and less difficult techniques. For the process of algae biomass harvest, there are some methods like physical, chemical and biological methods which are utilized. Sedimentation, which is the simplest technique used to harvest the biomass of algae and also recommend a modest solution to harvest the biomass of algae. Sedimentation is done through gravitational force, where biomass of algae is isolated from the fluid because of contrasts in their densities. In any case, the distinction between cell of algae and density of water is generally minute, making the sequence quite moderate. In addition, the rate of sedimentation is influenced by a progression of factors like biotic and abiotic. Distinctive sedimentation rates are recommended for different algal utilitarian gatherings, inward at 3.6 m d^{-1} for the wastewater lenient *Chlorella* sp. (Collet et al., 2011). As sedimentation is a generally moderate procedure, it is regularly combined with other techniques of harvesting or is adjusted for additional fast execution. In the process of centrifugation during sedimentation, the centrifugal force is used rather than the force of gravitational. Centrifugation provides an easy and fast separation of algae biomass from fluid and proficient for every species of the algae and cell size (Rawat et al., 2013). Nevertheless, because of the enormous venture and working costs, it becomes in apprioprate for large scope algal growth-based wastewater treatment frameworks (Christension and Sims, 2011; Udom et al., 2013), prompting a fourfold increment in costs of absolute treatment (Gouveia et al., 2016). Synthetic polymer and electrolytes are used for neutralization of the negative surface charge from the cells of algae and simultaneously endorse the development of large size particle which settle easily. Generally, the use of ferric chloride and aluminium sulphate occurs for the flocculation stimulates (Christension and Sims, 2011), and also 90% and above biomass of algae can recover with their use (Udom et al., 2013). Under environmental stress, some species of algae can naturally form flocs, for example, higher pH level, concentration of dissolved oxygen and nutrient change (Uduman et al., 2010). The use of electrophoresis is another technique for algal cell surface charge neutralization and promote flocculation (Christension and Sims, 2011). After flocculation, flotation may be the next stage. Dissolved air floatation is used most commonly which is a process like energy concentrated and effectual for the large-scale use (Zhang et al., 2012; Hanotu et al., 2012). Another effective and simple method to harvest algae is called filtration which provides algal biomass recovery and equals to 90% (Shen et al., 2009). Another procedure is microfiltration, which is used for species like *Spirulina*, which is generally a microscopic filamentous algal species requiring lower energy for the harvesting process (Shimamatsu, 2004). Membrane microfiltration is used for the harvesting of *Scenedesmus* and *Chlorella* like species because the diameter of their cell is between 5 to 20 μm only (Babel and Takizawa, 2010). Diverse methods are offered for harvesting by control of the algal cells into a polymeric grid, which is also helpful, economically a judicious elective for conservative algal biomass harvesting strategies (Moreno-Garrido, 2008).

7.2 Removal of heavy metals

In urban sewage, the most frequent pollutants are heavy metal ions which are the result of industrialization (Ranade and Bhandari, 2014), therefore, posing enormous threat to the natural aquatic system. Substantial metals are not biodegradable similar to natural impurities and will in general amass in living life forms. The organisms which contain these pollutants can lead to dysfunction because several heavy metal ions are generally carcinogenic or toxic in nature. Hence, heavy metal ions like copper, chromium, cadmium, zinc, mercury lead and nickel are the main causes of distress when they are treated for wastewater (Marella et al., 2020). There are different methods for reduction in the content of heavy metal in wastewater which includes absorption, chemical precipitation, coagulation and flocculation, membrane filtration, electrochemical treatment, ion exchange and floatation (Fu and Wang, 2011). For domestic wastewater treatment with *Chlorella vulgarish* in membrane photobioreactor, Gao et al. (2016) achieved complete decrease of Mn and Fe ions, while Al, Cu, and Zn ions were reduced by 93, 65 and 80% respectively (Sekomo et al., 2012). To remove the metals from stock solution with the capability of *Cladophora fracta*, the study by (Ji et al., 2012) showed that the removal of Cd, Cu, Hg and Zn were achieved 85 to 99% respectively. The studies based on removal of metals by algae from the wastewater have been broadly conducted. Algal cells are able to adapt the toxic material from the environment and can uptake heavy metals. The cells of algae in any condition, dead or alive can contribute to the sorption of metal in wastewater (Enduta et al., 2011). For the reduction of heavy metal content, sustainable and economical ways are still being searched, and utilization of algae seems to be a hopeful substitute (Ji et al., 2012). Table 3.2 describes the microalgae based heavy metal removal.

Table 3.2. Different mechanism for biosorption of heavy metals by microalgae.

S. No.	Name of Microalgae	Mechanism	References
1.	*Pseudomonas plecoglossicida*	Electrostatic interaction, Ion exchange	Guo et al., 2012
2.	*Cyclotella meninigiana*	Cellular intake by potassium channel, adsorption	Adam and Garnier-Laplace, 2003
3.	*Yarrowia lipolytica*	Ion exchange	Shinde et al., 2012
4.	*Cyanobacteria microcystis*	Complexation	Sun et al., 2011 Wu et al., 2012
5.	*Thalassiosira weissflogii; Minutissimum; Navicula minima*	Ionic strength, adsorbtion	Hernández-Ávila et al., 2017; Loix et al., 2017; Gélabert et al., 2007
6.	*Rhizopus arrhizus*	Electrostatic attraction	Shroff and Vaidya, 2011
7.	*Spirulina* sp.	Ion exchange, Physical adsorption	Chojnacka et al., 2005
8.	*Pheodactylum tricornutum*	Adsorption	Loix et al., 2017
9.	*Thalassiosira weissflogii*	Adsorption	Loix et al., 2017
10.	*Skeletonema costatum; Achnanthidium*	Ionic strength, adsorbtion	Loix et al., 2017; Gélabert et al., 2007; Hernández-Ávila et al., 2017
11.	*Skeletonema costatum*	Absorption	Loix et al., 2017; Jadoon and Malik, 2017
12.	*Thalassiosira pseudonana*	Absorption	Rijstenbil, 2002; Loix et al., 2017; Łukowski and Dec 2018
13.	*Nitzschia Closterium*	Absorption	Jadoon and Malik, 2017 Łukowski and Dec, 2018 Ma et al., 2018
14.	*Thalsssiosira oceanica*	Absorption	Desai et al., 2006 Lane et al., 2008 Loix et al., 2017

7.3 Removal of emerging contaminants

In wastewater, the content of Emerging Contaminants (ECs) has been a greater than ever apprehension over the more recent decades. These incorporate an enormous assortment of natural and inorganic micropollutants, for example, perfluorinated compound, micropollutants such as inorganic and organic, prescription and illicit drugs, pharmaceuticals, products related to personal care and many other substances (Richardson and Ternes, 2005). Emerging contaminants are also found in drinking water and natural water through wastewater effluents. Even though these pollutants are present mostly in trace concentration, there is some adverse outcome on human health, aquatic individuals as well as terrestrial organisms (Bolong et al., 2009; Prichard and Granek, 2016). For the removal of emerging contaminants, some highly developed processes of wastewater treatment have been applied. However, the accomplishments of the elimination of emerging contaminants mainly depend on the chemical assets of specific micro pollutants. The study conducted by (Miege et al., 2009) showed that a highly developed method of treatment similar to the flocculation, lime softening and coagulation are not able to decrease the total content of emerging contaminants, mostly due to the sorption surface pollutant competition. The methods which are used for grounding of drinking water with stimulated carbon powder and with the process of chlorination can be reduced by emerging contaminants, which is additionally more effective than other processes and the study based on it was confirmed by (Westerhoff et al., 2005). The technology, which is based on membrane filtration like nanofiltration, reverse osmosis is able to approximately remove emerging contaminants showed the most promising result (Yoon et al., 2006). But till now, more consumption of energy in membrane filtration is a major difficulty. For the reduction in emerging contaminants of wastewater a method such as energy-efficient called bioremediation has been proposed. In wastewater treatment system, which is based on algae, emerging contaminants elimination is also studied. Different studies based on elimination of emerging contaminants similar to pesticides, personal care products and pharmaceuticals have been completed in created wetlands (Lv et al., 2016; Chen et al., 2016), showing that the elimination rates for substances is highly variable.

8. Conclusion

Wastewater remediation is a largely important issue that needs to be envisaged with precision and utmost attention round the world. New paradigms in the algae mediation wastewater remediation can pose a sustainable solution to this worldwide problem. The exploration of novel technologies, state of art harvesting processes and cost economics can lead to better application of diverse algae in the removal of waste concomitant with valuable products from the algal biomass for a better tomorrow.

Acknowledgements

This work was supported by Department of Biotechnology (DBT), New Delhi, India Grant No: BT/PR/15650/AAQ/3/815/2016.

References

Ács, E., Szabó, K., Tóth, B. and Kiss, K.T. 2004. Investigation of benthic algal communities, especially diatoms of some Hungarian streams in connection with reference conditions of the water framework directives. Acta Bot. Hung. 46(3-4): 255–278. https://doi.org/10.1556/abot.46.2004.3-4.1.

Adam, C. and Garnier-Laplace, J. 2003. Bioaccumulation of silver-110m, cobalt-60, cesium-137, and manganese-54 by the freshwater algae *Scenedesmus obliquus and Cyclotellameneghiana* and by suspended matter collected during a summer bloom event. Limnol. Oceanogr. 48(6): 2303–2313. https://doi.org/10.4319/lo.2003.48.6.2303.

Agarwal, A. and Joshi, H. 2010. Application of nanotechnology in the remediation of contaminated groundwater: A short review. Recent Res. Sci. Technol. 2(6): 51–57. ISSN: 2076-5061.

Al-Darmaki, A., Govindrajan, L., Talebi, S., Al-Rajhi, S., Al-Barwani, T. and Al-Bulashi, Z. 2012. Cultivation and characterization of microalgae for wastewater treatment, Proceedings of the world congress on eng. Vol 1. ISBN: 978-988-19251-3-8.

Al-Degs, Y.S. and Tutunji, M.F. 2000. The feasibility of using diatomite and Mn-diatomite for remediation of Pb2+, Cu2+ and Cd2+ from water. Separation Sci. Tech. 35(14): 2299–2310. https://doi.org/10.1081/SS-100102103.

Alinsafi, A., Evenou, F., Abdulkarim, E.M., Pons, M.N., Zahraa, O., Benhammou, A., Yaacoubu, A. and Nejmeddine, A. 2007. Treatment of textile industry wastewater by supported photocatalysis. Dyes Pigm. 74(2): 439–445. https://doi.org/10.1016/j.dyepig.2006.02.024.

Amer, L., Adhikari, B. and Pellegrino, J. 2011. Technoeconomic analysis of five microalgae-to-biofuels processes of varying complexity. Bioresour. Technol. 102(20): 9350–9359. https://doi.org/10.1016/j.biortech.2011.08.010.

Amit, Ghosh, U.K. 2018. An approach for phycoremediation of different wastewaters and biodiesel production using microalgae. Environ. Sci. Pollut. Res. 25: 18673–18681. https://doi.org/10.1007/s11356-018-1967-5.

Arib, Z., Ruiz, J., Alvarez-Diaz, P., Garrido, C. and Perales, J.A. 2014. Capability of different microalgae species for phytoremediation processes: Wastewater tertiary treatment, CO2 bio-fixation and low cost biofuels production. Water Res. 49: 465–474. https://doi.org/10.1016/j.watres.2013.10.036.

Axelsson, M. and Gentili, F. 2014. A single-step method for rapid extraction of total lipids from green microalgae. PLoS ONE 9(2): e89643. https://doi.org/10.1371/journal.pone.0089643.

Azimi, A., Azari, A., Rezakazemi, M. and Ansarpour, M. 2017. Removal of heavy metals from industrial wastewaters: A review. Chem. Biol. Eng. Rev. 4: 37–59. https://doi.org/10.1002/cben.201600010.

Aziz, M.A. and Ng, W.J. 1992. Feasibility of wastewater treatment using the activated-algae process. Bioresour. Technol. 40: 205–208. https://doi.org/10.1016/0960-8524(92)90143-L.

Babel, S. and Takizawa, S. 2010. Microfiltration membrane fouling and cake behavior during algal filtration. Desalination 261(1-2): 46–51. https://doi.org/10.1016/j.desal.2010.05.038.

Badar, M., Qamar, M.K. and Akhtar, M.S. 2017. Recovery of chromium from the wastewater of leather industry. BFIJ 9(1): 52–55. ISSN No. (Online): 2249–3239.

Baghour, M. 2019. Algal degradation of organic pollutants. pp. 1–22. *In*: Martínez, L., Kharissova, O. and Kharisov, B. (eds.). Handbook of Ecomaterials. Springer, Cham. https://doi.org/10.1007/978-3-319-68255-6_86.

Bailey, S.E., Olin, T.J., Brika, R.M. and Adrian, D.D. 1999. A review of the potentially low cost sorbents for heavy metals. Water Res. 33(11): 2469–2479. https://doi.org/10.1016/S0043-1354(98)00475-8.

Baojiao, G., Pengfei, J. and Fuqiang, A. 2005. Studies on the surface modification of diatomite with polyethylenamine and trapping effect of the modified diatomite for phenol. Appl. Surf. Sci. 250(1): 273–279. https://doi.org/10.1016/j.apsusc.2005.02.119.

Baqar, M., Sadef, Y., Ahmad, S.R., Mahmood, A., Li, J. and Zhang, G. 2018. Organochlorine pesticides across the tributaries of river ravi, Pakistan: Human health risk assessment through dermal exposure, ecological risks, source fingerprints and spatio-temporal distribution. Sci. Total Environ. 618: 291–305. https://doi.org/10.1016/j.scitotenv.2017.10.234.

Batista, A.P., Ambrosano, L., Graça, S., Sousa, C., Marques, P.A.S.S., Ribeiro, B., Botrel, E.P., Castro Neto, P. and Gouveia, L. 2015. Combining urban wastewater treatment with biohydrogen production-integrated microalgae-based approach. Bioresour. Technol. 184: 230–235. https://doi.org/10.1016/j. biortech.2014.10.064.

Bawiec, A., Pawęska, K. and Jarząb, A. 2016. Changes in the microbial composition of municipal wastewater treated in biological processes. J. Ecol. Eng. 17(3): 41–46. https://doi.org/10.12911/22998993/63316.

Benemann, J.R., Weismann, J.C., Eisenberg, D.M., Koopman, B.L., Goebel R.P., Caskey, P.S., Thomson, R.D. and Oswald, W.J. 1977. A systems analysis of bioconversion with micro algae. Paper Presented at Symposium on Clean Fuels from Biomass and Wastes, Oriando, Florida, 101–126.

Bhatti, S., Richard, R. and McGinn, P. 2021. Screening of two freshwater green microalgae in pulp and paper mill wastewater effluents in Nova Scotia, Canada. Water Sci. Technol. In Press. https://doi.org/10.2166/wst.2021.001.

Bolong, N., Ismail, A.F., Salim, M.R. and Matsuura, T. 2009. A review of the effects of emerging contaminants in wastewater and options for their removal. Desalination 238(1-3): 229–246. https://doi.org/10.1016/j.desal.2008.03.020.

Borja, R., Martin, A., Luque, M. and Duran, M.M. 1993. Enhancement of the anaerobic digestion of wine distillery wastewater by the removal of phenolics inhibitors. Bioresour. Technol. 45: 99–104. https://doi.org/10.1016/0960-8524(93)90097-U.

Bustamante, M.A., Paredes, C., Moral, R., Moreno-Casalles, J., Perez-Espinosa, A. and Perez Murcia, M.D. 2005. Uses of winery and distillery effluents in agriculture: Characterization of nutrient and hazardous components. Water Sci. Technol. 51(1): 145–151. PMID: 15771110.

Buyukkamaci, N. and Koken, E. 2010. Economic evaluation of alternative wastewater treatment plant options for pulp and paper industry. Sci. Total Environ. 408(24): 6070–6078. https://doi.org/10.1016/j.scitotenv.2010.08.045.

Cai, T., Park, S.Y. and Li, Y. 2013. Nutrient recovery from wastewater streams by microalgae: Status and prospects. Renew. Sust. Energ. Rev. 19: 360–369. https://doi.org/10.1016/j.rser.2012.11.030.

Chaitanawisuti, N., Santhaweesuk, W. and Kritsanapuntu, S. 2011. Performance of the seaweeds Gracilaria salicornia and Caulerpa lentillifera as biofilters in a hatchery scale recirculating aquaculture system for juvenile spotted babylons (*Babylonia areolata*). Aquac. Int. 19: 1139–1150. https://doi.org/10.1007/s10499-011-9429-9.

Chandra, T.S., Suvidha, G., Mukherji, S., Chauhan, V.S., Vidyashankar, S., Krishnamurthi, K., Sarada, R. and Mudliar, S.N. 2014. Statistical optimization of thermal pretreatment conditions for enhanced biomethane production from defatted algal biomass. Bioresour. Technol. 162: 157–165. https://doi.org/10.1016/j.biortech.2014.03.080.

Charles, D.F. 1996. Use of algae for monitoring rivers in the United States. Some examples. pp. 109–118. *In*: Whitton, B.A. and Rott, E. (eds.). Use of Algae for Monitoring Rivers I. https://doi.org/10.1111/j.1442-9993.1995.tb00521.x.

Chen, Y., Vymazal, J., Brezinova, T., Koželuh, M., Kule, L., Huang, J. and Chen, Z. 2016. Occurrence, removal and environmental risk assessment of pharmaceuticals and personal care products in rural wastewater treatment wetlands. Sci. Total Environ. 566-567: 1660–1669. https://doi.org/10.1016/j.scitotenv.2016.06.069.

Chessman, B.C. 1986. Diatom flora of an Australian river system: Spatial patterns and environmental relationships. Freshwater Biol. 16: 805–819. https://doi.org/10.1111/j.1365-2427.1986.tb01018.x.

Chiesa, L.M., Labella, G.F., Giorgi, A., Panseri, S., Pavlovic, R., Bonacci, S. and Arioli, F. 2016. The occurrence of pesticides and persistent organic pollutants in Italian organic honeys from different productive areas in relation to potential environmental pollution. Chemosphere 154: 482–490. https:doi.org/10.1016/j.chemosphere.2016.04.004.

Choi, S., Kim, H.J., Kim, S., Choi, G., Kim, S., Park, J. et al. 2018. Current Status of Organochlorine Pesticides (OCPs) and Polychlorinated Biphenyls (PCBs) exposure among mothers and their babies of KoreaCHECK cohort study. Sci. Total Environ. 618: 674–681. https://doi.org/10.1016/j.scitotenv.2017.07.232.

Chojnacka, K., Chojnacki, A. and Gorecka, H. 2005. Biosorption of Cr3+, Cd2+ and Cu2+ ions by blue–green algae *Spirulina* sp.: Kinetics, equilibrium and the mechanism of the process. Chemosphere 59(1): 75–84. https://doi.org/10.1016/j.chemosphere.2004.10.005.

Chokshi, K., Pancha, I., Ghosh, A. and Mishra, S. 2016. Microalgal biomass generation by phycoremediation of dairy industry wastewater: An integrated approach towards sustainable biofuel production. Bioresour. Technol. 221: 455–460. https://doi.org/10.1016/j.biortech.2016.09.070.

Chow, V.T., Eliason, R. and Linsley, R.K. 1972. Development and trends in wastewater engineering. In Wastewater Engineering. McGraw-Hill Book Company. New York, St. Louis, Dusseldorf, Johannesburg, Kuala Lumpur, London, Mexico, Montreal, New Delhi, Panama, Rio de Janeiro, Singapore, Sydney, Toronto, 1–11.

Christenson, L. and Sims, R. 2011. Production and harvesting of microalgae for wastewater treatment, biofuels, and bioproducts. Biotechnol. Adv. 29(6): 686–702. https://doi.org/10.1016/j.biotechadv.2011.05.015.

Collet, P., Hélias, A., Lardon, L., Ras, M., Goy, R.-A. and Steyer, J.-P. 2011. Lifecycle assessment of microalgae culture coupled to biogas production. Bioresour. Technol. 102(1): 207–214. https://doi.org/10.1016/j.biortech.2010.06.154.

Congestri, R., Cox, E.J., Cavacini, P. and Albertano, P. 2005. Diatoms (Bacillariophyta) in phototrophic biofilms colonising an Italian wastewater treatment plant. Diatom Res. 20: 241–255. https://doi.org/10.1080/0269249X.2005.9705634.

Coste, M., Boutry, S., Tison-Rosebery, J. and Delmas, F. 2009. Improvements of the Biological Diatom Index (BDI): Description and efficiency of the new version (BDI-2006). Ecol. Indic. 9(4): 621–650. https://doi.org/10.1016/j.ecolind.2008.06.003.

Daneshvar, E., Zarrinmehr, M.J., Koutra, E., Kornaros, M., Farhadian, O. and Bhatnagar, A. 2018. Sequential cultivation of microalgae in raw and recycled dairy wastewater: Microalgal growth, wastewater treatment and biochemical composition. Bioresour. Technol. 273: 556–564. https://doi.org/10.1016/j.biortech.2018.11.059.

Day, D. 1996. How Australian social policy neglects environments. Aust. J. Soil Water Cons. 9: 3–9. ISSN: 1032-2426.

De-Bashan, L.E., Moreno, M., Hernandez, J.P. and Bashan, J. 2002. Ammonium and phosphorus removal from continuous and semi-continuous cultures by the microalgae *Chlorella vulgaris* coimmobilized in alginate beads with *Azospirillum brasilense*. Water Res. 36(12): 2941–2948. https://doi.org/10.1016/S0043-1354(01)00522-X.

De Godos, I., Blanco, S., García-Encina, P.A., Becares, E. and Muñoz, R. 2009. Long-term operation of high rate algal ponds for the bioremediation of piggery wastewaters at high loading rates. Bioresour. Technol. 100: 4332–9. https://doi.org/10.1016/j.biortech.2009.04.016.

De la Noue, J. and Proulx, D. 1988. Tertiary treatment of urban wastewater by chitosan-immobilized Phormidium sp. pp. 159–168. *In*: Stadler, T., Mollion, J., Verdus, M.C., Kamaranos, Y., Morvan, H. and Christaien, D. (eds.). Algal Biotechnol. Elsevier Applied Science, New York.

De La Noue, J., Laliberte, G. and Proulx, D. 1992. Algae and waste water. J. Appl. Phycol. 4: 247–254. https://doi.org/10.1007/BF02161210.

Dermeche, S., Nadour, M., Larroche, C., Moulti-Mati, F. and Michaud, P. 2013. Olive mill wastes: Biochemical characterizations and valorization strategies. Process Biochem. 48(10): 1532–1552. https://doi.org/10.1016/j.procbio.2013.07.010.

Desai, S.R., Verlecar, X.N. and Nagarajappa. 2006. Genotoxicity of cadmium in marine diatom *Chaetoceros tenuissimus* using the alkaline Comet assay. Ecotoxicol. 15: 359–363. https://doi.org/10.1007/s10646-006-0076-2.

Dhote, J., Ingole, S. and Chavhan, A. 2012. Review on wastewater treatment technology. IJERT 1(5): 01–10. ISSN: 2278-0181.

Ekendahl, S., Bark, M., Engelbrektsson, J., Karlsson, C., Niyitegeka, D. and Stromberg, N. 2018. Energy-efficient outdoor cultivation of oleaginous microalgae at northern latitudes using waste heat and flue gas from a pulp and paper mill. Algal Res. 31: 138–146. https://doi.org/10.1016/j.algal.2017.11.007.

El-Abbassi, A., Kiai, H. and Hafidi, A. 2012. Phenolic profile and antioxidant activities of olive mill wastewater. Food Chem. 132(1): 406–412. https://doi.org/10.1016/j.foodchem.2011.11.013.

Eloranta, P. and Soininen, J. 2002. Ecological studies of some Finnish rivers evaluated using benthic diatom communities. J. Appl. Phycol. 14: 1–7. https://doi.org/10.1023/A:1015275723489.

Enduta, A., Jusoh, A., Ali, N. and Wan Nik, W.B. 2011. Nutrient removal from aquaculture wastewater by vegetable production in aquaponics recirculation system. Desalin. Water Treat. 32(1-3): 422–430. https://doi.org/10.5004/dwt.2011.2761.

Fallowfield, H.D. and Barret, M.K. 1985. The photosynthetic treatment of pig slurry in temperate climatic conditions: A pilot plant study. Agricultural Wastes 12: 111–136. https://doi.org/10.1016/0141-4607(85)90003-4.

Farrukh, A., Akram, A., Ghaffar, A., Hanif, S., Hamid, A. and Duran, H. 2013. Design of polymer-brush-grafted magnetic nanoparticles for highly efficient water remediation. ACS Appl. Mater. Interfaces 5: 3784–3793. https://doi.org/10.1021/am400427n.

Feng, Y., Li, C. and Zhang, D. 2011. Lipid production of *Chlorella vulgaris* cultured in artificial wastewater medium. Bioresource. Technol. 102(1): 101–105. https://doi.org/10.1016/j.biortech.2010.06.016.

Ferro, L., Gojkovic, Z., Munoz, R. and Funk, C. 2019. Growth performance and nutrient removal of a Chlorella vulgaris-Rhizobium sp. co-culture during mixotrophicfeed-batch cultivation in synthetic wastewater. Algal. Res. 44: 101690. https://doi.org/10.1016/j.algal.2019.101690.

Fu, F. and Wang, Q. 2011. Removal of heavy metal ions from wastewaters: A review. J. Environ. Manage. 92(3): 407–418. https://doi.org/10.1016/j.jenvman.2010.11.011.

Ganeshkumar, V., Subashchandrabose, S.R., Dharmarajan, R., Venkateswarlu, K., Naidu, R. and Megharaj, M. 2018. Use of mixed wastewaters from piggery and winery for nutrient removal and lipid production by *Chlorella* sp. MM3, Bioresour. Technol. https://doi.org/10.1016/j.biortech.2018.02.025.

Gantar, M., Gajin, S. and Dalmacija, B. 1984. The possibility of phosphate elimination by the use of algae in the process of wastewater purification. Mikrobiologia 21(1): 63–73.

Gao, J., Xiong, Z., Zhang, J., Zhang, W. and Obono Mba, F. 2009. Phosphorus removal from water of eutrophic Lake Donghu by five submerged macrophytes. Desalination 242(1-3): 193–204. https://doi.org/10.1016/j.desal.2008.04.006.

Gao, F., Li, C., Yang, Z.H., Zeng, G.M., Feng, L.J., Liu, J. and Cai, H. 2016. Continuous microalgae cultivation in aquaculture wastewater by a membrane photobioreactor for biomass production and nutrients removal. Ecol. Eng. 92: 55–61. https://doi.org/10.1016/j.ecoleng.2016.03.046.

Garg, U.K. and Garg, H.K. 2016. Optimization of process parameters for metal ion remediation using agricultural materials. Int. J. Theor. Appl. Sci. 8(1): 17–24. ISSN No. (Online): 2249-3247.

Gelabert, A., Pokrovsky, O.S., Schott, J., Boudou, A. and Feurtet-Mazel, A. 2007. Cadmium and lead interaction with diatom surfaces: A combined thermodynamic and kinetic approach. Geochim. Cosmochim. Acta 71(15): 3698–3716. https://doi.org/10.1016/j.gca.2007.04.034.

Genovesi, A., Harmand, J. and Steyer, J.P. 2000. Integrated fault detection and isolation: Application to a winery's wastewater treatment plant. Applied. Intell. 13: 59–76. https://doi.org/10.1023/A:1008379329794.

Ghosh, M., Ganguli, A. and Tripathi, A.K. 2002. Treatment of anaerobically digested distillery spent wash in a two-stage bioreactor using Pseudomonas putida and Aeromonas sp. Process Biochem. 37(8): 857–862. https://doi.org/10.1016/s0032-9592(01)00281-3.

Girard, J.M., Roy, M.L., Hafsa, M.B., Gagnon, J., Faucheux, N., Heitz, M. and Deschênes, J.S. 2014. Mixotrophic cultivation of green microalgae Scenedesmus obliquus on cheese whey permeate for biodiesel production. Algal. Res. 5: 241–248. https://doi.org/10.1016/j.algal.2014.03.002.

Goodwin, J.A.S., Finlayson, J.M. and Low, E.W. 2001. A further study of the anaerobic biotreatment of malt whisky distillery pot ale using an UASB system. Bioresour. Technol. 78: 155–160. https://doi.org/10.1016/S0960-8524(01)00008-6.

Goncalves, A.L., Pires, J.C.M. and Simoes, M. 2017. A review on the use of microalgal consortia for wastewater treatment. Algal. Res. 24: 403–415. https://doi.org/10.1016/j.algal.2016.11.008.

Gouveia, L., Graça, S., Sousa, C., Ambrosano, L., Ribeiro, B., Botrel, E.P., Neto, P.C., Ferreira, A.F. and Silva, C.M. 2016. Microalgae biomass production using wastewater: Treatment and costs. Algal. Res. 16: 167–176. https://doi.org/10.1016/j.algal.2016.03.010.

Gray, N.F. 1989. Biology of Wastewater Treatment. Oxford Univ. Press, Oxford.

Grima, E.M., Acie, F.G., Medina, A.R. and Chisti, Y. 2003. Recovery of microalgal biomass and metabolites: Process options and economics. Biotechnol. Adv. 20(7-8): 491–515. https://doi.org/10.1016/s0734-9750(02)00050-2.

Guo, J., Zheng, X., Chen, Q.B., Zhang, L. and Xu, X.P. 2012. Biosorption of Cd(II) from Aqueous Solution by *Pseudomonas plecoglossicida*: Kinetics and Mechanism. Curr. Microbiol. 65: 350–355. https://doi.org/10.1007/s00284-012-0164-x.

Gupta, S.K. and Rao, P.V.S.S. 1980. Treatment of urea by algae, activated sludge and flocculation algal bacterial system—A comparative study. Indian J. Environ. Health 22: 103–112.

Guzzon, A., Bohn, A., Diociaiuti, M. and Albertano, P. 2008. Cultured phototrophic biofilms for phosphorus removal in wastewater treatment. Water Res. 42: 4357–4367. https://doi.org/10.1016/j.watres.2008.07.029.

Han, S.Q., Zhang, Z.H. and Yan, S.H. 2000. Present situation and developmental trend of wastewater treatment and eutrophication waters purification with alga technology. Agro. Environ. Development 63(1): 13–16.

Hanotu, J., Bandulasena, H.C.H. and Zimmerman, W.B. 2012. Microflotation performance for algal separation. Biotechnol. Bioeng. 109: 1663–1673. https://doi.org/10.1002/bit.24449.

Hasnain, S., Ali, A.S., Uddin, Z. and Zafar, R. 2013. Application of nanotechnology in health and environmental research: A review. Res. J. Environ. Earth. Sci. 5: 160–166. https://doi.org/10.19026/rjees.5.5653.

Hernandez, D., Riaño, B., Coca, M., Solana, M., Bertucco, A. and García-González, M.C. 2016. Microalgae cultivation in high rate algal ponds using slaughterhouse wastewater for biofuel applications. Chem. Eng. J. 285: 449–58. https://doi.org/10.1016/j.cej.2015.09.072.

Hernandez-Avila, J., Salinas-Rodriguez, E., Cerecedo-Saenz, E., Reyes-Valderrama, M., Arenas-Flores, A., Roman-Gutierrez, A. and Rodriguez-Lugo, V. 2017. Diatoms and their capability for heavy metal removal by cationic exchange. Metals 7(5): 169. https://doi.org/10.3390/met7050169.

HII, Y.S., SOO, C.L., Chuah, T.S., Mohd-Azmi, A. and Abol-Munafi, B. 2011. Interactive effect of ammonia and nitrate on the nitrogen uptake by *Nannochloropsis* sp. J. Sustain Sci. Manag. 6(1): 60–68. ISSN: 1823-8556.

Hofmann, G. 1996. Recent developments in the use of benthic diatoms for monitoring eutrophication and organic pollution in Germany and Austria. pp. 73–77. *In*: Whitton, B.A. and Rott, E. (eds.). Use of Algae for Monitoring Rivers I.

Hongyang, S., Yalei, Z., Chunmin, Z., Xuefei, Z. and Jinpeng, L. 2011. Cultivation of *Chlorella pyrenoidosa* in soybean processing wastewater. Bioresour. Technol. 102(21): 9884–9890. https://doi.org/10.1016/j.biortech.2011.08.016.

Horan, N.J. 1990. Biological Wastewater Treatment Systems. Theory and Operation. John Wiley and Sons Ltd. Baffins Lane, Chickester. West Sussex PO 191 UD, England. ISBN : 0471924253.

Hubbe, M.A., Metts, J.R., Hermosilla, D., Blanco, M.A., Yerushalmi, L., Haghighat, F. et al. 2016. Wastewater treatment and reclamation: A review of pulp and paper industry practices and opportunities. BioResources 11(3): 7953–8091. https://doi.org/10.15376/biores.11.3.Hubbe.

Jadoon, S. and Malik, A. 2017. DNA Damage by heavy metals in animals and human beings: An overview. Biochem. Pharmacol: Open Access. 06(03). https://doi.org/10.4172/2167-0501.1000235.

Jamali, A.A., Akbari, F., Ghorakhlu, M.M., de la Guardia, M. and Khosroushahi, A.Y. 2012. Applications of diatoms as potential microalgae in nanobiotechnology. Bioimpacts 2: 83–89. https://doi.org/10.5681/bi.2012.012.

Ji, L., Xie, S., Feng, J., Li, Y. and Chen, L. 2012. Heavy metal uptake capacities by the common freshwater green alga Cladophora fracta. J. Appl. Phycol. 24(4): 979–983. https://doi.org/10.1007/s10811-011-9721-0.

Ji, M.K., Abou-Shanab, R.A.I., Kim, S.H., Salama, E.S., Lee, S.H., Kabra, A.N., Lee, Y.S., Hong, S. and Jeon, B.H. 2013. Cultivation of microalgae species in tertiary municipal wastewater supplemented with CO_2 for nutrient removal and biomass production. Ecol. Eng. 58: 142–148. https://doi.org/10.1016/j.ecoleng.2013.06.020.

Johnson, M.B. and Wen, Z. 2010. Development of an attached microalgal growth system for biofuel production. Appl. Microbiol. Biotechnol. 85(3): 525–534. https://doi.org/10.1007/s00253-009-2133-2.

Kalavathi, D.F., Uma, L. and Subramanian, G. 2001. Degradation and metabolization of the pigment-melanoidin in distillery effluent by the marine Cyanobacterium Oscillatoriaboryana BDU 92181. Enzyme Microb. Technol. 29: 246–251. https://doi.org/10.1016/S0141-0229(01)00383-0.

Kamyab, H., Md Din, M.F., Ponraj, M., Keyvanfar, A., Rezania, S., Taib, S.M. and Abd Majid, M.Z. 2016. Isolation and screening of microalgae from agro-industrial wastewater (POME) for biomass and biodiesel sources. Desalin. Water Treat. 57(60): 29118–29125. https://doi.org/10.1080/19443994.2016.1139101.

Kanimozhi, R. and Vasudevan, N. 2010. An overview of wastewater treatment in distillery industry. Int. J. Environ. Eng. 2: 159–184. https://doi.org/10.1504/IJEE.2010.029826.

Karn, B., Kuiken, T. and Otto, M. 2009. Nanotechnology and *in situ* remediation: A review of the benefits and potential risks. Environ. Health Perspect. 117: 1823–1831. https://doi.org/10.1289/ehp.0900793.

Kelly, M.G. 1998b. Use of community-based indices to monitor eutrophication in European rivers. Envir. Cons. 25: 22–29. https://doi.org/10.1023/A:1003400910730.

Kelly, M.G. 2002. Role of benthic diatoms in the implementation of the Urban Wastewater Treatment Directive in the River Wear, North-East England. J. Appl. Phycol. 1: 9–18. https://doi.org/10.1023/A:1015236404305.

Kelly, M.G. and Whitton, B.A. 1995. The trophic Diatom Index: A new index for monitoring eutrophication in rivers. J. of Appl. Phyco. 7: 433–444. https://doi.org/10.1007/BF00003802.

Kemp, K.C., Seema, H., Saleh, M., Le, N.H., Mahesh, K. and Chandra, V. 2013. Environmental applications using graphene composites: Water remediation and gas adsorption. Nanoscale 5: 3149–3171. https://doi.org/10.1039/C3NR33708A.

Keyser, M., Witthuhn, R.C., Ronquest, L.C. and Britz, T.J. 2003. Treatment of winery effluent with upflow anaerobic sludge blanket (UASB)–granular sludges enriched with Enterobacter sakazakii. Biotechnol. Letters 25(22): 1893–1898. https://doi.org/10.1023/b:bile.0000003978.72266.96.

Kim, K.H., Kabir, E. and Jahan, S.A. 2017. Exposure to pesticides and the associated human health effects. Sci. Total Environ. 575: 525–535. https://doi.org/10.1016/j.scitotenv.2016.09.009.

Kim, M.K., Park, J.W., Park, C.S., Kim, S.J., Jeune, K.H., Chang, M.U. and Acreman, J. 2007. Enhanced production of Scenedesmus spp. (green microalgae) using a new medium containing fermented swine wastewater. Bioresour. Technol. 98(11): 2220–2228. https://doi.org/10.1016/j.biortech.2006.08.031.

Kim, S. and Aga, D.S. 2007. Potential ecological and human health impacts of antibiotics and antibiotic-resistant bacteria from wastewater treatment plants. J. Toxicol. Environ. Health B 10(8): 559–573. https://doi.org10.1080/15287390600975137.

Kociolek, J.P. and Stoemer, E.F. 2001. Opinion: Taxonomy and ecology: A marriage of necessity. Diatom Res. 16: 433–442. https://doi.org/10.1080/0269249X.2001.9705529.

Korzeniewska, E. 2011. Emission of bacteria and fungi in the air from wastewater treatment plants—A review. Front. Biosci. 1(3): 393–407. https://doi.org/10.2741/s159.

Kumar, R., Khan, M.A. and Haq, N. 2014. Application of carbon nanotubes in heavy metals remediation. Crit. Rev. Environ. Sci. Technol. 44(9): 1000–1035. https://doi.org/10.1080/10643389.2012.741314.

Kumar, R., Singh, R.D. and Sharma, K.D. 2005. Water resources of India. Curr. Sci. 89: 794–811. http://www.jstor.org/stable/24111024.

Kunikane, S., Kaneko, M. and Maehara, R. 1984. Growth and nutrient uptake of green alga, Scenedesmus dimorphus, under a wide range of nitrogen-phosphorus ratio (I). Experimental study. Water Res. 18(10): 1299–1311. https://doi.org/10.1016/0043-1354(84)90036-8.

Lane, E.S., Jang, K., Cullen, J.T. and Maldonado, M.T. 2008. The interaction between inorganic iron and cadmium uptake in the marine diatom Thalassiosira oceanica. Limnol. Oceanogr. 53(5): 1784–1789. doi:10.4319/lo.2008.53.5.1784.

Lazaridis, N.K., Karapantsios, T.D. and Georgantas, D. 2003. Kinetic analysis for removal of a reactive dye from aqueous solution on to hydrotalcite by adsorption. Water Res. 37: 3023–3033. https://doi.org/10.1016/S0043-1354(03)00121-0.

Li, Y., Zhou, W., Hu, B., Min, M., Chen, P. and Ruan, RR. 2012. Effect of light intensity on algal biomass accumulation and biodiesel production for mixotrophic strains Chlorella kessleri and Chlorella protothecoide cultivated in highly concentrated municipal wastewater. Biotechnol. Bioeng. 109(9): 2222–2229. https://doi.org/10.1002/bit.24491.

Lim, S., Chu, W. and Phang, S. 2010. Use of Chlorella vulgaris for bioremediation of textile wastewater. J. Bioresour. Technol. 101: 7314–7322. https://doi.org/10.1016/j.biortech.2010.04.092.

Lin, Z., Li, J., Luan, Y. and Dai, W. 2020. Application of algae for heavy metal adsorption: A 20-year meta-analysis. Ecotox. Environ. Safe. 190: 110089. https://doi.org/10.1016/j.ecoenv.2019.110089.

Loix, C., Huybrechts, M., Vangronsveld, J., Gielen, M., Keunen, E. and Cuypers, A. 2017. Reciprocal interactions between cadmium-induced cell wall responses and oxidative stress in plants. Front. Plant Sci. 8. https://doi.org/10.3389/fpls.2017.01867.

Looker, N. 1998. Municipal Wastewater Management in Latin America and the Caribbean, R.J. Burnside International Limited, Published for Roundtable on Municipal Water for the Canadian Environment Industry Association.

Low, K.S., Lee, C.K. and Liew, S.C. 2000. Sorption of cadmium and lead, from aqueous solutions by spent grain. Process Biochem. 36: 59–64. https://doi.org/10.1016/S0032-9592(00)00177-1.

Lukowski, A. and Dec, D. 2018. Influence of Zn, Cd, and Cu fractions on enzymatic activity of arable soils. Environ. Monit. Assess 190: 278. https://doi.org/10.1007/s10661-018-6651-1.
Lv, T., Zhang, Y., Zhang, L., Carvalho, P.N., Arias, C.A. and Brix, H. 2016. Removal of the pesticides imazalil and tebuconazole in saturated constructed wetland mesocosms. Water Res. 91: 126–136. https://doi.org/10.1016/j.watres.2016.01.007.
Ma, J., Zhou, B., Duan, D., Wei, Y. and Pan, K. 2018. Silicon limitation reduced the adsorption of cadmium in marine diatoms. Aquat. Toxicol. 202: 136–144. https://doi.org/10.1016/j.aquatox.2018.07.011.
Mahapatra, D.M., Chanakya, H.N. and Ramachandra, T.V. 2013. Euglena sp. as a suitable source of lipids for potential use as biofuel and sustainable wastewater treatment. J. Appl. Phycol. 25: 855–865. https://doi.org/10.1007/s10811-013-9979-5.
Malandra, L., Wolfaardt, G., Zietsman, A. and Viljoen-Bloom, M. 2003. Microbiology of a biological contactor for winery wastewater treatment. Water Res. 37(17): 4125–4134. https://doi.org/10.1016/s0043-1354(03)00339-7.
Mannucci, A., Munz, G., Mori, G. and Lubello, C. 2010. Anaerobic treatment of vegetable tannery wastewaters: A review. Desalination 264(1-2): 1–8. https://doi.org/10.1016/j.desal.2010.07.021.
Mansoori, G.A., Rohani, T., Bastami, A., Ahmadpour, Z. and Eshaghi. 2008. Environmental application of nanotechnology. Annu. Rev. Nano Res. 2(2): 1–73. DOI: http://dx.doi.org/10.1142/9789812790248_001.
Maraseni, T.N. and Maroulis, J. 2008. Piggery: From environmental pollution to a climate change solution. J. Environ. Sci. Health, Part B 43(4): 358–363. https://doi.org/10.1080/03601230801941717.
Marella, T.K., Bhaskar, M.V. and Tiwari, A. 2016. Phycoremediation of eutrophic lakes using diatom algae. Lake Sci. Climate Change. M. Nageeb Rashed, Intech Open, https://doi.org/10.5772/64111.
Marella, T.K., Parine, N.R. and Tiwari, A. 2018. Potential of diatom consortium developed by nutrient enrichment for biodiesel production and simultaneous nutrient removal from waste water. Saudi J. Biol. Sci. 25: 704–709. https://doi.org/10.1016/j.sjbs.2017.05.011.
Marella, T.K., Datta, A., Patil, M.D., Dixit, S. and Tiwari, A. 2019. Biodiesel production through algal cultivation in urban wastewater using algal floway. Bioresour. Technol. 280: 222–228. https://doi.org/10.1016/j.biortech.2019.02.031.
Marella, T.K., Saxena, A. and Tiwari, A. 2020. Diatom mediated heavy metal remediation: A review. Bioresour. Technol. 305: 123068. https://doi.org/10.1016/j. biortech.2020.123068.
Markert, B.A., Brier, A.M. and Zechmeister, H.G. 2003. Definition, strategies and priniciples for bioindicator/biomonitoring of the environment. pp. 285–327. *In*: Markert, B.A., Breure, A.M. and Zechmeister, H.G. (eds.). Elsevier, Oxford.
Martinez, M.E., Sánchez, S., Jiménez, J.M., El Yousfi, F. and Muñoz, L. 2000. Nitrogen and phosphorus removal from urban wastewater by the microalga Scenedesmus obliquus. Bioresour. Technol. 73: 263–272. https://doi.org/10.1016/S0960-8524(99)00121-2.
McGriff, E.C. and McKinney, R.C. 1972. The removal of nutrients and organics by activated algae. Water Res. 6(10): 1155. https://doi.org/10.1016/0043-1354(72)90015-2.
McNamara, C.J., Anastasiou, C.C., O'Flaherty, V. and Mitchell, R. 2008. Bioremediation of olive mill wastewater. Int. Bioderior. 61(2): 127–134. https://doi.org/10.1016/j.ibiod.2007.11.003.
Mehndiratta, P., Jain, A., Srivastava, S. and Gupta, N. 2013. Environmental pollution and nanotechnology. Environ. Pollut. 2: 49–58. https://doi.org/10.5539/ep.v2n2p49.
Miege, C., Choubert, J.M., Ribeiro, L., Eusèbe, M. and Coquery, M. 2009. Fate of pharmaceuticals and personal care products in wastewater treatment plants – Conception of a database and first results. Environ. Pollut. 157(5): 1721–6. https://doi.org/10.1016/j.envpol.2008.11.045.
Milledge, J.J. and Heaven, S. 2013. A review of the harvesting of micro-algae for biofuel production. Rev. Environ. Sci. Biotechnol. 12(2): 165–178. https://doi.org/10.1007/s11157-https://doi.org/10.1007/s11157-012-9301-z.
Mishra, N. and Mishra, N. 2017. Utilization of microalgae for integrated biomass production and phycoremediation of wastewater. J. Algal Biomass Utln. 8: 95–105.
Mobbs, P.M. 1999. Mineral Industry of Iran, Report of Ministry of Industries and Mines of Iran, 101–107.
Mohsenpour, S.F., Hennige, S., Willoughby, N., Adeloye, A. and Gutierrez, T. 2021. Integrating micro-algae into wastewater treatment: A review. Sci. Total Environ. 752: 142168. https://doi.org/10.1016/j.scitotenv.2020.142168.
Molinuevo-Salces, B., García-González, M.C. and González-Fernández, C. 2010. Performance comparison of two photobioreactors configurations (open and closed to the atmosphere) treating anaerobically degraded swine slurry. Bioresour. Technol. 101: 5144–9. https://doi.org/10.1016/j.biortech.2010.02.006.
Moondra, N., Jariwala, N.D. and Christian, R.A. 2020. Sustainable treatment of domestic wastewater through microalgae. Int. J. Phytoremediation 1–7. https://doi.org/10.1080/15226514.2020.1782829.
Moreno-Garrido, I. 2008. Microalgae immobilization: Current techniques and uses. Bioresour. Technol. 99(10): 3949–3964. https://doi.org/10.1016/j.biortech.2007.05.040.
Moss, B. 2008. Water pollution by agriculture. Philos. Trans. R. Soc. B 363: 659−666. https://doi.org/10.1098/rstb.2007.2176.
Mulbry, W., Kangas, P. and Kondrad, S. 2010. Toward scrubbing the bay: Nutrient removal using small algal turf scrubbers on Chesapeake Bay tributaries. Ecol. Eng. 36(4): 536–541. https://doi.org/10.1016/j.ecoleng.2009.11.026.
Mura, S., Seddaiu, G., Bacchini, F., Roggero, P.P. and Greppi, G.F. 2013. Advances of nanotechnology in agro-environmental studies. Ital. J. Agron. 8(3): e18. https://doi.org/10.4081/ija.2013.e18.
Nayak, M., Karemore, A. and Sen, R. 2016. Performance evaluation of microalgae for concomitant wastewater bioremediation, CO_2 biofixation and lipid biosynthesis for biodiesel application. Algal. Res. 16: 216–223. https://doi.org/10.1016/j.algal.2016.03.020.

Nayak, M., Swain, D.K. and Sen, R. 2019. Strategic valorization of de-oiled microalgal biomass waste as biofertilizer for sustainable and improved agriculture of rice (Oryza sativa L.) crop. Sci. Total Environ. 10(682): 475–484. https://doi.org/10.1016/j.scitotenv.2019.05.123.

Neveux, N., Magnusson, M., Mata, L., Whelan, A., de Nys, R. and Paul, N.A. 2016. The treatment of municipal wastewater by the Macroalga oedogonium sp. and its potential for the production of biocrude. Algal. Res. 13: 284–292. https://doi.org/10.1016/j.algal.2015.12.010.

Nzayisenga, J.C., Eriksson, K. and Sellstedt, A. 2018. Mixotrophic and heterotrophicproduction of lipids and carbohydrates by a locally isolated microalga using wastewater as a growth medium. Bioresour. Technol. 257: 260–265. https://doi.org/10.1016/j.biortech.2018.02.085.

Olguin, E.J. 2003. Phycoremediation: Key issues for cost-effective nutrient removal processes. Biotechnol. Adv. 22: 1–91. https://doi.org/10.1016/j.biotechadv.2003.08.009.

Olguin, E.J. 2012. Dual purpose microalgae-bacteria-based systems that treat wastewater and produce biodiesel and chemical products within a biorefinery. Biotechnol. Adv. 30: 1031–1046. https://doi.org/10.1016/j.biotechadv.2012.05.001.

Orpez, R., Martínez, M.E., Hodaifa, G., El Yousfi, F., Jbari, N. and Sánchez, S. 2009. Growth of the microalga Botryococcus braunii in secondarily treated sewage. Desalination 246: 625–630.

Ozverdi, A. and Erdem, M. 2006. Cu2+, Cd2+ and Pb2+ adsorption from aqueous solutions by pyrite and synthetic iron sulphide. J. Hazard. Mater. 137: 626–632. https://doi.org/10.1016/j.jhazmat.2006.02.05.

Pandey, K.K., Prasad, G. and Singh, V.N. 1985. Copper (II) removal from aqueous solution by fly ash. Water Res. 19: 869–873. https://doi.org/10.1016/0043-1354(85)90145-9.

Patel, S. and Jamaluddin. 2018. Treatment of distillery waste water: A review. Int. J. Theor. Appl. Sci. 10(1): 117–139. ISSN No. (Online): 2249–3247.

Patrick, R. 1973. Use of algae, especially diatoms in the assessment of water quality. pp. 76–95. *In*: Biological Methods for the Assessment of Water Quality. ASTM STP 528. American Society for Testing and Materials. https://doi.org/10.1520/STP34718S.

Perez-Garcia, O., deBashan, L.E., Hernandez, J.P. and Bashan, Y. 2010. Efficiency of growth and nutrient uptake from wastewater by heterotrophic, autotrophic and mixotrophic cultivation of Chlorella vulgaris immobilised with Azospirillum brasilense. J. Phycol. 46: 800–812. https://doi.org/10.1111/j.1529-8817.2010.00862.x.

Phang, S.M., Chu, W.L. and Rabiei, R. 2015. Phycoremediation, 357–389. https://doi.org/ 10.1007/978-94-017-7321-8_13.

Picos-Corrales, L.A., Sarmiento-Sánchez, J.I., Ruelas-Leyva, J.P., Crini, G., Hermosillo-Ochoa, E. and Gutierrez-Montes, J.A. 2020. Environment-friendly approach toward the treatment of raw agricultural wastewater and river water via flocculation using chitosan and bean straw flour as bioflocculants. ACS Omega. https://doi.org/10.1021/acsomega.9b03419.

Pizzichini, M., Russo, C. and Meo, C.D. 2005. Purification of pulp and paper wastewater, with membrane technology, for water reuse in a closed loop. Desalination 178(1-3): 351–359. https://doi.org/10.1016/j.desal.2004.11.045.

Pokhrel, D. and Viraraghavan, T. 2004. Treatment of pulp and paper mill wastewater—A review. Sci. Total Environ. 333(1-3): 37–58. https://doi.org/10.1016/j.scitotenv.2004.05.017.

Pouliot, Y., Buelna, G., Racine, C. and de la Noüe, J. 1989. Culture of cyanobacteria for tertiary wastewater treatment and biomass production. Biological Wastes 29: 81–91. https://doi.org/10.1016/0269-7483(89)90089-X.

Prichard, E. and Granek, E.F. 2016. Effects of pharmaceuticals and personal care products on marine organisms: From single-species studies to an ecosystem based approach. Environ. Sci. Pollut. R 23(22): 22365–22384. https://doi.org/10.1007/s11356-016-7282-0.

Prygiel, J. and Coste, M. 1993. The assessment of water quality in the Artois-Picardie water basin (France) by the use of diatom indices. Hydrobiol. 269/279: 343–349. 10.1007/BF00028033.

Queiroz, M.I., Lopes, E.J., Zepka, L.Q., Bastos, R.G. and Goldbeck, R. 2007. The kinetics of the removal of nitrogen and organic matter from parboiled rice effluent by cyanobacteria in a stirred batch reactor. Bioresour. Technol. 98: 2163–2169. https://doi.org/10.1016/j.biortech.2006.08.034.

Rajan, C.S. 2011. Nanotechnology in groundwater remediation. Int. J. Environ. Sci. Dev. 2: 182–187.

Rajeshwari, K.V., Balakrishnan, M., Kansal, A., Lata, K. and Kishore, V.V.N. 2000. State-of art of anaerobic digestion technology for industrial wastewater treatment. Renew. Sust. Energy. Rev. 4: 135–156.

Ranade, V.V. and Bhandari, V.M. 2014. Industrial wastewater treatment, recycling and reuse, industrial wastewater treatment. Recycling and Reuse 1–80. https://doi.org/10.1016/b978-0-08-099968-5.00001- 5.

Rao, H.P., Kumar, R.R., Raghavan, B.G., Subramanian, V.V. and Sivasubramanian, V. 2011. Application of phycoremediation technology in the treatment of wastewater from a leather-processing chemical manufacturing facility. Water Soil Air 37(1): 7–14. https://doi.org/10.4314/wsa.v37i1.64099.

Rawat, I., Ranjith Kumar, R., Mutanda, T. and Bux, F. 2013. Biodiesel from microalgae: A critical evaluation from laboratory to large scale production. Appl. Energy 103: 444–467. https://doi.org/10.1016/j.apenergy.2012.10.004.

Reid, N.M., Bowers, T.H. and Lloyd-Jones, G. 2008. Bacterial community composition of a wastewater treatment system reliant on N_2 fixation. Appl. Microbiol. Biotechnol. 79: 285–292. https://doi.org/10.1007/s00253-008-1413-6.

Riano, B., Molinuevo, B. and García-González, M.C. 2011. Treatment of fish processing wastewater with microalgae-containing microbiota. Bioresour. Technol. 102: 10829–33. doi: 10.1016/j.biortech.2011.09.022.

Richardson, S.D. and Ternes, T.A. 2005. Water analysis: Emerging contaminants and current issues. Anal. Chem. 77(12): 3807–3838. https://doi.org/10.1021/ac0301301.

Rijstenbil, J. 2002. Interactions of algal ligands, metal complexation and availability, and cell responses of the diatom Ditylum brightwellii with a gradual increase in copper. Aquat. Toxicol. 56(2): 115–131. https://doi.org/10.1016/s0166-445x(01)00188-6.

Round, F.E. 1991. Diatoms in river water-monitoring studies. J. Appl. Phycol. 3: 129–45. https://doi.org/10.1007/BF00003695.

Round, F.E. 1991a. Diatoms in river water-monitoring studies. J. Appl. Phycol. 3: 129–145. https://doi.org/10.1007/BF00003695.

Satpal, S. and Khambete, A.K. 2016. Treatment of municipal sewage with microalgae—A laboratory based study. IJSET 5(8): 415–417. https://doi.org/10.17950/ijset/v5s8/804.

Sekomo, C.B., Rousseau, D.P.L., Saleh, S.A. and Lens, P.N.L. 2012. Heavy metal removal in duckweed and algae ponds as a polishing step for textile wastewater treatment. Ecological Engineering 44: 102–110. https://doi.org/10.1016/j.ecoleng.2012.03.003.

Sharma, J., Kumar, V., Kumar, S.S., Malyan, S.K., Mathimani, T., Bishnoi, N.R. and Pugazhendhi, A. 2020a. Microalgal consortia for municipal wastewater treatment-lipid augmentation and fatty acid profiling for biodiesel production. J. Photochem. Photobiol. B 202: 111638. https://doi.org/10.1016/j.jphotobiol.2019.111638.

Shen, Y., Yuan, W., Pei, Z.P., Wu, Q. and Mao, E. 2009. Microalgae mass production methods. Transactions of the ASABE 52(4): 1275–1287. https://doi.org/10.13031/2013.27771.

Shimamatsu, H. 2004. Mass production of Spirulina, an edible microalgae. Hydrobiologia. 512(1): 39–44. https://doi.org/10.1023/b:hydr.0000020364.23796.04.

Shinde, N.R., Bankar, A.V., Kumar, A.R. and Zinjarde, S.S. 2012. Removal of Ni (II) ions from aqueous solutions by biosorption onto two strains of Yarrowia lipolytica. J. Environ. Manag. 102: 115–124. https://doi.org/10.1016/j.jenvman.2012.02.026.

Senegar, R.M.S. and Sharma, K.D. 1987. Tolerance of Phormidium corium and Chlamydomonas sp. against chemical elements present in polluted water of Yamuna river. Proceeding National Academic Science (India) 57: 56–64.

Shroff, K. and Vaidya, V.K. 2011. Effect of pre-treatments on the biosorption of Chromium (VI) ions by the dead biomass of Rhizopus arrhizus. J. Chem. Technol. Biotechnol. 87(2): 294–304. https://doi.org/10.1002/jctb.2715.

Singh, M., Shukla, R. and Das, K. 2013. Harvesting of microalgal biomass. Biotechnological Applications of Microalgae 77–88. https://doi.org/10.1201/b14920-7.

Singh, P.K., Bhattacharjya, R., Saxena, A., Mishra, B. and Tiwari, A. 2020. Utilization of wastewater as nutrient media and biomass valorization in marine Chrysophytes-Chaetoceros and Isochrysis. Energy Convers. Manag. X 100062. https://doi.org/10.1016/j.ecmx.2020.100062.

Sirianuntapiboon, S., Phothilangka, P. and Ohmomo, S. 2004a. Decolourization of molasses wastewater by a strain no. BP103 of acetogenic bacteria. Bioresour. Technol. 92: 31–39. https://doi.org/10.1016/j.biortech.2003.07.010.

Smith, L.E.D. and Siciliano, G.A. 2015. Comprehensive review of constraints to improved management of fertilizers in china and mitigation of diffuse water pollution from agriculture. Agric. Ecosyst. Environ. 209: 15–25. https://doi.org/10.1016/j.agee.2015.02.016.

Smol, J.P. 1992. Paleolimnology: An important tool for effective ecosystem management. J. Aquat. Ecosyst. Health 1: 49–58. https://doi.org/10.1007/BF00044408.

Stevenson, J. 2014. Ecological assessments with algae: A review and synthesis. J. Phycol. 50: 437–461. https://doi.org/10.1111/jpy.12189.

Stevenson, J. and Pan, Y. 1999. Assessing environmental conditions in rivers and streams with diatoms. *In*: Stoermer, E.F. and Smol, J.P. (eds.). The Diatoms: Applications for the Environmental and Earth Sciences. Cambridge University Press, Cambridge. https://doi.org/10.1017/CBO9780511763175.005.

Su, Y., Mennerich, A. and Urban, B. 2012. Synergistic cooperation between wastewater-born algae and activated sludge for wastewater treatment: Influence of algae and sludge inoculation ratios. Bioresour. Technol. 105: 67–73. https://doi.org/10.1016/j.biortech.2011.11.113.

Sun, F., Wu, F., Liao, H. and Xing, B. 2011. Biosorption of antimony(V) by freshwater cyanobacteria microcystis biomass: Chemical modification and biosorption mechanisms. Chem. Eng. 171(3): 1082–1090. https://doi.org/10.1016/j.cej.2011.05.004.

Sunday, E.R., Uyi, O.J. and Caleb, O.O. 2018. Phycoremediation: An eco-solution to environmental protection and sustainable remediation. J. Chem. Environ. Biol. Engr. 2–5. https://doi.org/10.11648/j.jcebe.20180201.12.

Taghizadeh, M., Kebria, D.Y., Darvishi, G. and Kootenaei, F.G. 2013. The use of nano zero valent iron in remediation of contaminated soil and groundwater. Int. J. Sci. Res. Environ. Sci. 1: 152–157. https://doi.org/10.12983/ijsres-2013-p152–157.

Tam, N.F.Y. and Wong, Y.S. 1989. Wastewater nutrient removal by Chlorella pyrenoidosa and Scenedesmus sp. Environ. Pollut. 58: 19–34. https://doi.org/10.1016/0269-7491(89)90234-0.

Tarlan, E., Dilek, F.B. and Yetis, U. 2002. Effectiveness of algae in the treatment of a wood-based pulp and paper industry wastewater. Bioresour. Technol. 84: 1–5. https://doi.org/10.1016/s0960-8524(02)00029-9.

Tebbutt, T.H.Y. 1983. Principles of Water Quality Control. Pergammon Press, Oxford. ISBN: 9781483285979.

Theron, J., Walker, J.A. and Cloete, T.E. 2008. Nanotechnology and water treatment: Applications and emerging opportunities. Crit. Rev. Microbiol. 34: 43–69. https://doi.org/10.1080/10408410701710442.

Thomas, D.G., Minj, N., Mohan, N. and Rao, P.H. 2016. Cultivation of microalgae in domestic wastewater for biofuel applications—an upstream approach. J. Algal Biomass Utln. 7(1): 62–70. ISSN 2229-6905.

Tiwari, A. and Marella, T.K. 2019. Potential and application of diatoms for industry-specific wastewater treatment. pp. 321–339. *In*: Gupta, S. and Bux, F. (eds.). Application of Microalgae in Wastewater Treatment. Springer, Cham. lopment. Renew. Sust. Energ. Rev. 30: 1035–1046. https://doi.org/10.1007/978-3-030-13913-1_15.

Tiwari, P.K., Batra, V.S. and Balakrishnan, M. 2007. Water management initiatives in sugarcane molasses based distilleries in India. Res. Conserv. Recycl. 52: 351–367. https://doi.org/10.1016/j.resconrec.2007.05.003.

Tocchi, C., Federici, E., Fidati, L., Manzi, R., Vincigurerra, V. and Petruccioli, M. 2012. Aerobic treatment of dairy wastewater in an industrial three-reactor plant: effect of aeration regime on performances and on protozoan and bacterial communities. Water Res. 46: 3334–3344. https://doi.org/10.1016/j.watres.2012.03.032.

Udaiyappan, A.F.M., Hasan, H.A., Takriff, M.S. and Abdullah, S.R.S. 2017. A review of the potentials, challenges and current status of microalgae biomass applications in industrial wastewater treatment. J. Water Process. Eng. 20: 8e21. https://doi.org/10.1016/J.JWPE.2017.09.006.

Udom, I., Zaribaf, B.H., Halfhide, T., Gillie, B., Dalrymple, O., Zhang, Q. and Ergas, S.J. 2013. Harvesting microalgae grown on wastewater. Bioresour. Technol. 139: 101–106. https://doi.org/10.1016/j.biortech.2013.04.002.

Uduman, N., Qi, Y., Danquah, M.K., Forde, G.M. and Hoadley, A. 2010. Dewatering of microalgal cultures: A major bottleneck to algae-based fuels. J. Renew. Sustain. Energy 2(1): 012701. https://doi.org/10.1063/1.3294480.

Ugurlu, M., Gurses, A., Dogar, C. and Yalçin, M. 2008. The removal of lignin and phenol form paper mill effluents by electrocoagulation. J. Environ. Manage. 87(3): 420–428. https://doi.org/10.1016/j.jenvman.2007.01.007.

Ummalyma, S.B. and Sukumaran, R.K. 2014. Cultivation of microalgae in dairy effluent for oil production and removal of organic pollution load. Bioresour. Technol. 165: 295–301. https://doi.org/10.1016/j.biortech.2014.03.028.

Uppal, J. 2004. Water utilization and effluent treatment in the Indian alcohol industry: An overview. pp. 13–19. *In*: Tewari, P.K. (ed.). Liquid Asset, Proceedings of the Indo-EU Workshop on Promoting Efficient Water Use in Agro-Based Industries. TERI Press, New Delhi, India.

Van Dam, H. 1974. The suitability of diatoms for biological water assessment. Hydrobiol. Bull. 8(3): 274–284. https://doi.org/10.1007/BF02257503.

Vidali, M. 2001. Bioremediation. An overview. Pure Appl. Chem. 73(7): 1163–1172. https://doi.org/10.1351/pac200173071163.

Voltolina, D., Cordero, B., Nieves, M. and Soto, L.P. 1999. Growth of Scenedesmus sp. in artificial wastewater. Bioresour. Technol. 68: 265–268. https://doi.org/10.1016/S0960-8524(98)00150-3.

Vorosmarty, C.J., McIntyre, P., Gessner, M.O., Dudgeon, D., Prusevich, A., Green, P., Glidden, S., Bunn, S.E., Sullivan, C.A. and Liermann, C.R. 2010. Global threats to human water security and river biodiversity. Nature 467: 555–561. https://doi.org/10.1038/nature09440.

Walker, G.M. and Weatherley, L.R. 1999. Kinetic of acid dye adsorption on GAC. Water Res. 33(8): 1895–1899. https://doi.org/10.1016/S0043-1354(98)00388-1.

Wang, L.C., Lee, T.Q., Chen, S.H. and Wu, J.T. 2010. Diatoms in Liyu Lake, Eastern Taiwan. Taiwania 55(3): 228–242. https://doi.org/10.6165/tai.2010.55(3).228.

Wang, M., Sahu Ashish, K., Rusten, B. and Park, C. 2013. Anaerobic co-digestion of microalgae Chlorella sp. and waste activated sludge. Bioresour. Technol. 142: 585–590. https://doi.org/10.1016/j.biortech.2013.05.096.

Wang, X.W., Huang, L., Ji, P.Y., Cheng, C.P., Li, X.S., Gao, Y.H. and liang, J.R. 2019. Using a mixture of wastewater and seawater as the growth medium for wastewater treatment and lipid production by the marine diatom Phaeodactylum tricornutum. Bioresour. Technol. 289: 121–681. https://doi.org/10.1016/j.biortech.2019.121681.

Wedzicha, B.L. and Kaputo, M.T. 1992. Melanoidins from glucose and glycine: composition, characteristics and reactivity towards sulphite-ion. Food Chem. 43(5): 359–367. https://doi.org/10.1016/0308-8146(92)90308-O.

Welch, E.B. 1992. Ecological Effects of Wastewater: Applied Limnology and Pollutant Effects. Chapman and Hall, New York. https://www.taylorfrancis.com/books/9.

Westerhoff, P., Snyder, S.A., Yoon, Y., Snyder, S.A. and Wert, E. 2005. Fate of endocrine-disruptor, pharmaceutical, and personal care product chemicals during simulated drinking water treatment processes. Environ. Sci. Technol. 39(17): 6649–6663. https://doi.org/10.1021/es0484799.

Whitmore, T.J. 1989. Florida diatom assemblages as indicators of trophic state and pH. Limnol. Oceanogr. 34: 882–895. https://doi.org/10.4319/lo.1989.34.5.0882.

Williams, W.D. 1981. Ecological use of sewage. Some approaches to saprobiological problems. Environ. Pollut. 58: 48–56.

Woertz, I., Feffer, A., Lundquist, T. and Nelson, Y. 2009. Algae grown on dairy and municipal wastewater for simultaneous nutrient removal and lipid production for biofuel feedstock. J. Environ. Eng. 135(11): 1115–1122. https://doi.org/10.1061/(ASCE)EE.1943-7870.0000129.

Wolmarans, B. and De Villiers, G. 2002. Start up of a UASB effluent treatment plant on distillery wastewater. Water SA 28(1): 63–68. https://doi.org10.4314/wsa.v28i1.4869.

Wu, F., Sun, F., Wu, S., Yan, Y. and Xing, B. 2012. Removal of antimony(III) from aqueous solution by freshwater Cyanobacteria microcystis biomass. Chem. Eng. 183: 172–179. https://doi.org/10.1016/j.cej.2011.12.050.

Xing, R.L. 2007. Isolation, Culture and Application of Benthic Diatoms (Doctoral Dissertation). Dalian, China: Dalian University of technology. (In Chinese with English abstract).

Xing, R., Ma, W., Shao, Y., Cao, X., Su, C., Song, H., Su, Q. and Zhou, G. 2018. Growth and potential purification ability of Nitzschia sp. benthic diatoms in sea cucumber aquaculture wastewater. Aquac. Res. 49(8): 2644–2652. https://doi.org/10.1111/are.13722.

Xin, L., Hong-ying, H., Ke, G. and Ying-xue, S, 2010. Effects of different nitrogen and phosphorus concentrations on the growth, nutrient uptake, and lipid accumulation of a freshwater microalga Scenedesmus sp. Bioresour. Technol. 101: 5494–5500. https://doi. org/10.1016/j.biortech.2010.02.016.

Yadav, I.C., Devi, N.L., Syed, J.H., Cheng, Z., Li, J., Zhang, G. et al. 2015. Current status of persistent organic pesticides residues in air, water, and soil, and their possible effect on neighboring countries: A comprehensive review of India. Sci. Total Environ. 511: 123–137. https://doi.org/10.1016/j.scitotenv.2014.12.041.

Yamamoto, T., Goto, I., Kawaguchi, O., Minagawa, K., Ariyoshi, E. and Matsuda, O. 2008. Phytoremediation of shallow organically enriched marine sediments using benthic microalgae. Mar. Pollut. Bull. 57(1-5): 108–115. https://doi. org/10.1016/j.marpolbul.2007.10.006.

Yi, Z., Xu, M., Di, X., Brynjólfsson, S. and Fu, W. 2017. Exploring valuable lipids in diatoms. Front. Mar. Sci. 4: 17. https:// doi.org/10.3389/fmars.2017.00017.

Yoon, Y., Westerhoff, P., Snyder, S.A. and Wert, C. 2006. Nanofiltration and ultrafiltration of endocrine disrupting compounds, pharmaceuticals and personal care products. J. Membr. Sci. 270(1-2): 88–100. https://doi.org/10.1016/j. memsci.2005.06.045.

Zhang, X., Amendola, P., Hewson, J.C., Sommerfeld, M. and Hu, Q. 2012. Influence of growth phase on harvesting of Chlorella zofingiensis by dissolved air flotation. Bioresour. Technol. 116: 477–484. https://doi.org/10.1016/j. biortech.2012.04.002.

Zhou, W., Li, Y., Min, M., Hu, B., Zhang, H., Ma, X., Li, L., Cheng, Y., Chen, P. and Ruan, R. 2012. Growing wastewater-born microalga Auxenochlorella protothecoides UMN280 on concentrated municipal wastewater for simultaneous nutrient removal and energy feedstock production. Appl. Energy. 98: 433–440. https://doi.org/10.1016/j.apenergy.2012.04.005.

CHAPTER 4

Microbial Communities and GHG Emissions from Polluted Agricultural Soils Remediated with Several Amendments

Gabriela Medina-Pérez,[1] *Fabián Fernández-Luqueño,*[2,*]
Fernando López-Valdez,[3] *Alfredo Madariaga-Navarrete,*[1]
Elizabeth Perez-Soto,[1] *Isaac Almaraz-Buendía,*[1]
Rafael Germán Campos-Montiel[1] and *Oscar Fernández-Fernández*[4]

1. Introduction

It is well known that microorganisms are the most abundant and diverse organisms on the planet. Species diversity is determined by two factors: species richness and species evenness. Species richness is a simple count of species in an environment, whereas species evenness quantifies how similar the abundances of the species are. Molecular techniques and advance knowledge have provided tools for analyzing the entire bacterial community. Microbial DNA analysis has revealed that soil environments contain about tens of thousands of bacterial taxa, but environmental stress, pollution and agricultural management reduce bacterial diversity (Fernández-Luqueño et al., 2011). Soil microbial communities are present in four physiological states: (i) active, (ii) potentially active, (iii) dormant, and (iv) dead (Blagodatskaya and Kuzyakov, 2013). Microorganisms from four physiological states could be involved as source and sink for greenhouse gases.

[1] ICAP – Instituto de Ciencias Agropecuarias, Universidad Autónoma del Estado de Hidalgo, Tulancingo, Hidalgo, 43000, México.
[2] Sustainability of Natural Resources and Energy Program, Cinvestav-Saltillo. Coahuila de Zaragoza. 25900. Mexico.
[3] Research Centre for Applied Biotechnology (CIBA - IPN), Instituto Politécnico Nacional, Tepetitla de Lardizábal, Tlaxcala. 90700. México.
[4] Dept. Soil. Sci. Universidad Autónoma Chapingo. Chapingo, Estado de México, 56230, Mexico.
* Corresponding author: cinves.cp.cha.luqueno@gmail.com

Light, air, water and soil are key factors to sustain life, preserve the ecological balance and harness natural resources worldwide. Large amounts of elements, minerals or particles are extracted, produced (synthesized), refined and introduced into the environment daily (Fernández-Luqueño et al., 2017a; 2017b).

There are several global soil pollution challenges that are intricately due to: (i) diverse pollution types and sources, (ii) large quantities of pollutants are involved, (iii) high concentration of contaminants, (iv) vast polluted areas, and (v) the number of people affected. Unfortunately, there are thousands of synthetic products (xenobiotics and recalcitrant compounds) which may potentially be soil pollutants and harm the soil ecosystem. In addition, there is no current comprehensive data on global soil contamination, but it is well known that there are millions of hectares of cities and farmland soils that are severely polluted, jeopardizing productivity. If this decline in productivity of worldwide soil resources continues, the outcome will be severe, not only for national or international economies, but also for the welfare of millions of rural households dependent on agriculture for meeting their livelihoods. Healthy and productive soils are essential to produce raw materials and safe food throughout the world. Reducing contaminant emissions and remediating polluted soils are enormous challenges faced by humanity today (Leon-Silva et al., 2016).

Bioremediation is the treatment of pollutants by using living organisms to eliminate, weaken, degrade, transform or break down undesirable substances to simpler components, such as CO_2, H_2O and NO_3. Pollution is the introduction of elements, compounds, materials or energy into the environment at concentrations that adversely alter its biological functioning or that pose a risk to humans or other targets that use or are linked to the environment (Fernández-Luqueño, 2017a). Recently, new strategies have been launched to dissipate pollutants from soils by bioremediation technologies such as *phytoremediation, phytobialremediation, bioaugmentation* or *biostimulation*, where *autochthonous, exogenous* or *genetically engineered organisms* are beneficial to increase the degradation rate. However, the remediation of pollutants increases greenhouse gases (GHG) production. Soils may also act as sources or sinks for GHG such as water vapor, CO_2, CH_4 and N_2O. Precise quantification of soil GHG storage and the emission capacities is currently essential to obtain reliable global budgets, which are required to find the best soil management practices to face global climate change and for performing climate research (Heil et al., 2016; Karna et al., 2017). It has been estimated that soil contain about 2,500 billion tons of carbon at one-meter depth, while the global carbon stocks in soil is approximately 3,700 billion tons. The soil organic carbon pool is the second largest carbon pool worldwide, only surpassed by oceans where carbon stocks are more than 40,000 billion tons. Global carbon stocks in the soil are formed directly by soil biota or by the organic matter that accumulates due to the activity of soil biota.

The objective of this chapter is to highlight the advance knowledge about the dissipation of pollutants by *bioremediation* (*phytoremediation, phytobialremediation, bioaugmentation* or *biostimulation*) and GHG sources and/or sinks by autochthonous, exogenous or genetically engineered organisms. The various microbial processes will not be analyzed in this chapter. However, only some of these processes if they are directly or indirectly useful for understanding microbial dissipation and the resulting GHG emissions when a contaminated agricultural soil is amended or not with organic or mineral substrates will be referred to.

2. Soil Biodiversity

Soil organisms are beneficial to agricultural ecosystems. More than a quarter of all living species on Earth inhabit the soil, and microorganisms such as algae, bacteria and fungi make up the majority of soil biomass (Joergensen and Emmerling, 2006). There are one billion bacterial cells and about 10,000 different bacterial genomes per 1 cm^3 of soil, but only about 1% of soil microorganism species have been identified by classical or molecular microbiology methods, and only 0.1% of the total microbial community is culturable on growth media in the laboratory (Mocali and Benedetti, 2010).

Bacteria are the most species-rich group on Earth, and the majority of these microorganisms live in soil. Bacterial biomass can amount to 1–2 t ha^{-1} in a temperate grassland soil. (Mocali and Benedetti, 2010). Fungal diversity has been conservatively estimated at around 1.5 million species (Hawksworth, 2001). Earthworms make up a significant part of fauna biomass in the soil, representing up to 60% in some soil ecosystems (Blouin et al., 2013).

Soil biodiversity can be defined as the variety of life in soil, from genetic variability to the range of communities and the array of soil habitats, from microaggregates to whole landscapes (Clermont-Dauphin et al., 2014). Soil ecosystems have the most species-rich habitats as they are home to significant biodiversity: most organisms spend a part of their life cycle in soil and more than a quarter of the terrestrial invertebrates and vertebrate's species live completely in soil or litter ecosystems. Soil organisms and their interrelationships manage many soil processes, which deliver essential services for human beings. Nonetheless, anthropogenic activities have led to drastic alteration to soil biodiversity, suggesting that biodiversity will be diminished or completely changed in a few decades (Nielsen et al., 2011).

According to Lilley et al. (2006), soil biodiversity is classified based on the type, organism size and trophic level. Microbiota includes organisms that range from 1–100 μm in size and comprise primary producers, herbivores, detritivores and predators such as nematodes (50–100 μm), protists (5–100 μm), fungi (5–50 μm) and bacteria (0.5–5 μm). Mesofauna includes organisms that range from > 100 μm^{-2} mm in size and comprise detritivores and shredders bacterial and fungal feeders such as tardigrades (0.1–1 mm), collembolan (0.2–6 mm) and mites (0.5–2 mm). Macrofauna includes organisms that range from > 2 mm–20 mm in size and comprise shredders and detritivores herbivores and predators such as worms, arthropods, mollusks and earthworms. Megafauna includes organisms that range from > 20 mm in size and comprise herbivores and predators such as arthropods, mollusks, mammals and birds (Fig. 4.1).

It is notable that soil microbial community diversity and activities or functions are affected by similar factors, including: electronic waste (e-waste) recycling activities (Wu et al., 2017), fumigation (Li et al., 2017), heavy metal pollution (Ding et al., 2017), crude oil or PAH pollution (Cury et al., 2015; Abbasian et al., 2016; Fernández-Luqueño et al., 2016), cypermethrin (an insecticide) contamination (Tejada et al., 2015), nanoparticles (Fernández-Luqueño et al., 2014; Samarajeewa et al., 2017), uranium (Yan et al., 2016), mineral fertilizer, organic manure or organic carbon (Zhong

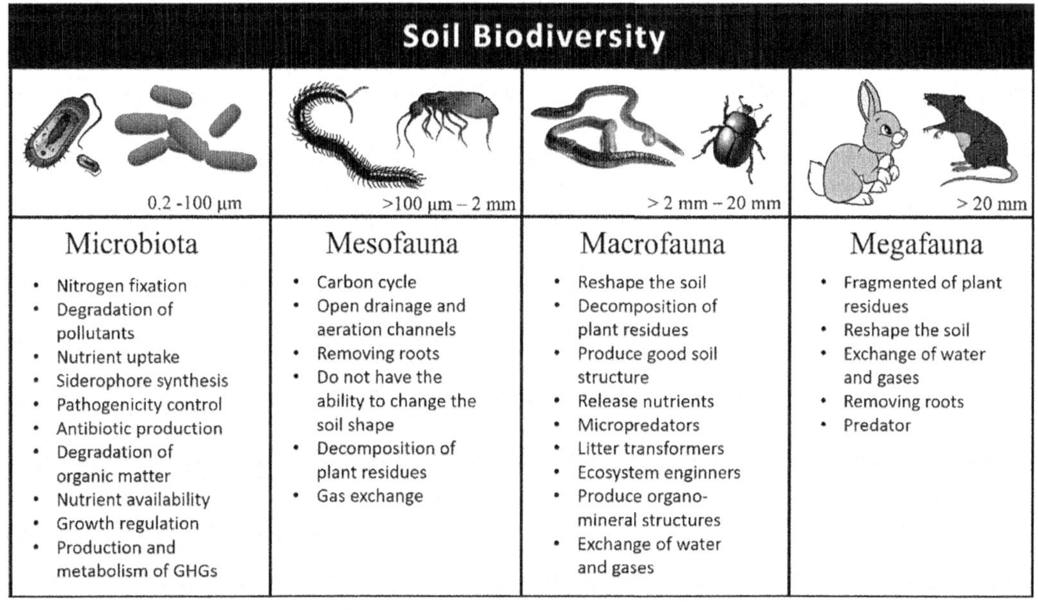

Fig. 4.1. Soil biodiversity effects on plant growth, soils development and the ecosystems.

et al., 2010), salinity (Zhao and Xu, 2016), UV-B radiation (Niu et al., 2014), land use, land management or agricultural practices (Bender et al., 2016; de Valenca et al., 2017; Waters et al., 2017; Wolinska et al., 2017), soil compaction (Devigne et al., 2016), erosion (Mabuhay et al., 2004), among many other natural or man-made factors.

Some essential functions performed by soil organisms are: (i) soil remediation, (ii) maintenance of soil structure, (iii) gas exchange and carbon sequestration, (iv) sources of raw materials, food and medicines, (v) symbiotic and non-symbiotic relationships with plants and their roots, (vi) plant growth promoting and control, (vii) suppression or control of pests, parasites and diseases, (viii) decomposition of organic matter, (ix) nutrient cycling, (x) regulation of soil hydrological processes, (xi) degradation and recycling of some materials, and (xii) purification of water and air (Barrios, 2007). However, 75 billion tons of agricultural soil are stripped each year by wind and water erosion worldwide, while millions of hectares are polluted every year worldwide, causing degradation and loss of soil quality (Pimentel and Burgess, 2013). This environmental degradation might lead to human catastrophes or famines because people would have to leave their homes in search of fertile cropland, which may be followed by loss of wealth and the welfare of humans.

2.1 Soil microorganisms

Studies concerning microbial communities are usually multidisciplinary, i.e., they not only focus on soil sciences but also on all related disciplines. Such studies require knowing which microorganisms are responsible for specific processes and, what percentage of the microbial biomass is responsible for the turnover of elements. According to Blagodatskaya and Kuzyakov (2013), soil microbial communities consist of an extensive range of organisms in four physiological states: (i) active, (ii) potentially active, (iii) dormant and (iv) dead.

Over 80,000 species of fungi that have been described live in soil, but many more remain undiscovered considering that the total fungal diversity is estimated at 1.5 million species (Hawksworth, 2001). One gram of soil may contain over a million individual fungi, while the fungi biomass can amount to 2.5–5 t ha^{-1} (De Vries et al., 2006). In agricultural soils, most of the biological activity occurs in the top 20 cm (the plow layer) while in non-cultivated soils, most biological activity occurs in the top 5 cm of soil. Soil biodiversity is essential for maintaining soil productivity since soil organisms perform different ecological functions and ecosystem services. Therefore, a significant loss of soil organisms that take part in these activities may result in severe effects including long-term soil degradation, modified landscape, decreased soil resilience and loss of agricultural productivity. It is important to highlight that soil health, soil quality and soil resilience are all fundamental to maintain productivity and viability of agricultural systems.

Microbial communities have a significant and relevant role in global GHG emissions. The GHG emissions from complex interactions between abiotic drivers and many microbial metabolic processes and mechanisms controlling CO_2, CH_4 and N_2O production have been well characterized in oxisols as well as in permafrost, i.e., in all types of soils.

3. GHG Emissions and Soil Pollution

It is increasingly recognized that plant-soil feedback could play an important role in improving the composition of the plant community, decreasing the rate of GHG emissions and improving the functioning of terrestrial ecosystems. On the other hand, soil contamination alters soil microbial biomass, diversity, structure, nutrient availability and GHG emissions, among other physical, chemical and biological characteristics.

Soil sources and sinks for methane have been well studied in some ecosystems such as rice fields, wetlands, forests, grasslands, polluted soils and cultivated soils. Land use and management significantly affect the net CH_4 exchange between soils and the atmosphere, but there are two different natural mitigation mechanisms. In the first mechanism, atmospheric methane is consumed by methanotrophs that have high affinity for the compound; this takes place primarily in aerobic soils. The second mechanism involves methanotrophs that act as biofilters at oxi-anoxic interfaces in wetlands, lake sediments and other environments with high CH_4 fluxes (Tate, 2015). The highest net CH_4 emissions occur in anaerobic environments, which include wetlands and rice paddies. Polluted soils produce CH_4 fluxes; nevertheless, wetlands and rice cultivations are the primary sources of methane worldwide, as a result of the following three processes. During the first process, methane is produced in the anoxic zones of flooded soils by methanogens through the anaerobic decomposition of plant material and soil organic matter, including manures. In the second phase, methane is transported to the atmosphere either indirectly via rice plants or directly in gas bubbles. Finally, during the third process CH_4 oxidation is carried out by methanotrophs in the oxic zone of soils, either in the rhizosphere or in the surface layer of rice paddy soils (Tate, 2015).

When compared with bulk or bare soil, most of the CH_4 produced in the rhizosphere is microbially oxidized (64–86% in dryland; 46–64% in flooded rice soil). Since methanogenic Archaea are globally ubiquitous, there is an observable interplay between methane production and consumption in soils with permanent or incipient anaerobic sites (Tate, 2015; Malyan et al., 2016). On a global scale, the role of soil as both a source and sink for CH_4 and the shifting balance between the two with changing land use or management, are not well understood yet.

In general, natural ecosystems exhibit the highest oxidation rates, particularly some forest soils. Disturbance or contamination of ecosystems often results in reduced methanotroph diversity, decreasing the soil CH_4 oxidation rates. However, some land-use changes, such as afforestation and reforestation, can reverse this occurrence; thus, increasing oxidation rates due to a marked shift in methanotrophic community compositions (Fig. 4.2; Tate, 2015; Finn et al., 2017).

Methane emission in rice soil is a biologically mediated phenomenon and depends on the rate of two opposite processes (production and oxidation), which are controlled by the production of methanogens and methanotrophs in the system. Both are conducted by many interplaying factors like soil organic matter content, soil pH, the texture of the soil, redox potential of soil, chemical or organic fertilization, soil temperature, among others. Methane emission processes are also affected by diurnal

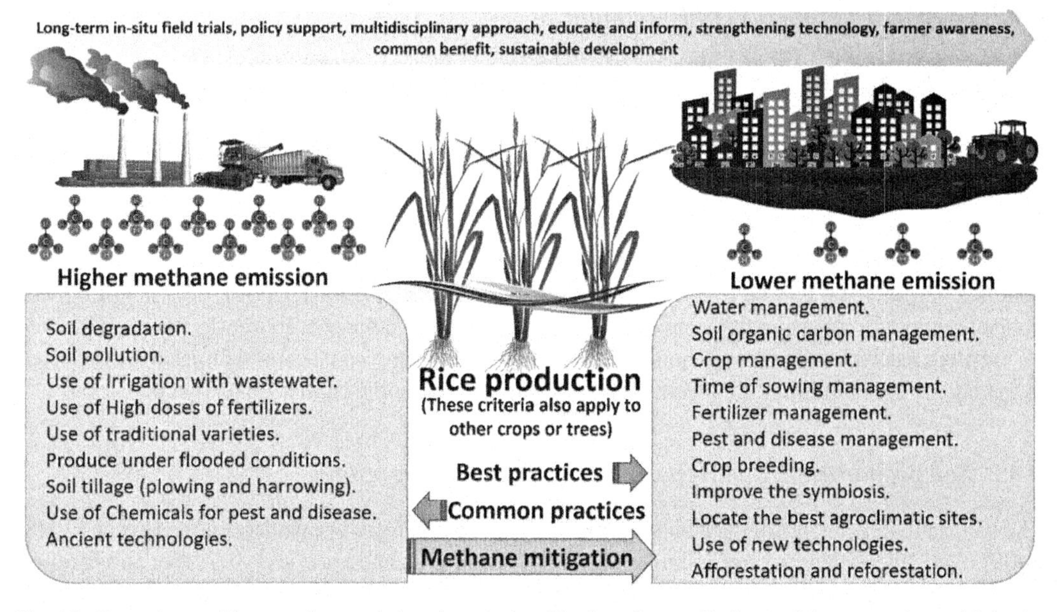

Fig. 4.2. Strategies to mitigate methane emissions in agricultural lands or forest soils. Some of these strategies could also be used to alleviate the emissions of others GHG.

and seasonal variation, elevated ozone or CO_2 levels, etc., along with management practices such as rice cultivar, nutrient applications, soil tillage, water management and pest and disease management (Fig. 4.2; Malyan et al., 2016).

4. Emission of GHG During the Remediation of Polluted Soils

Currently, there is uncertainty regarding how rates of soil carbon and nitrogen cycling change as CO_2, CH_4 and N_2O accumulate in the Earth's atmosphere. Zhou et al. (2014) carried out a 90-d laboratory incubation experiment which was conducted under the anaerobic condition in order to track changes in GHG evolution, i.e., CO_2, CH_4 and N_2O. The soil used was a heavy metal polluted paddy soil with polluted rice straw return, which was compared to unpolluted soil with unpolluted straw return. They found that total CO_2 and CH_4 evolution from contaminated soil amended with polluted straw decreased from 21 and 79%, respectively, compared to unpolluted straw modified in the unpolluted soil. They also found that the overall Global Warming Potential (GWP) increased by 31 and 58% with the polluted straw amendment in unpolluted and polluted soils, respectively. They further stated that the metal pollution in both soil and straw decreased the carbon mineralization, but the N_2O evolution increased and then enhanced the overall GWP in metal polluted rice paddies with rice straw return.

Our research team has carried out different experiments to determine the GHG emission during various processes of remediation and bioremediation involving Polycyclic Aromatic Hydrocarbons (PAH) or heavy metals. These experiments will be briefly described next.

4.1 Bioremediation of PAH polluted soils

An experiment was carried out in order to determine the dynamics of CO_2 and the removal of anthracene and phenanthrene from saline or agricultural soil. The objective was to investigate how wastewater sludge characteristics affect the removal of two PAH, phenanthrene and anthracene, from alkaline-saline soil of the former Lake Texcoco.

4.1.1 Area description and soil sampling

One sampling site is located at the former Lake of Texcoco in the valley of Mexico or Mexico city (Mexico) (N.L. 19° 30', W.L. 98° 53') at an altitude of 2255 m above sea level. Additional details about this soil could be found in Fernández-Luqueño et al. (2008; 2009). Soil sampling was done at random by augering 0–15 cm of the soil top-layer. The same procedure was repeated in three 0.7 ha plots. Each plot sample was pooled to obtain three soil samples. The other soil sampling site served as a control treatment. This site is located at Acolman, state of Mexico (N.L. 19°38' W.L. 98°55'). The first sampling site, the former Lake Texcoco, was also used as a control. Additional details about this soil can be found in Fernández-Luqueño et al. (2008; 2009). Soil sampling was done at random by augering 0–15 cm of the soil top-layer. The same procedure was repeated in three 0.7 ha plots. Each plot sample was pooled to obtain three soil samples. Six soil samples were collected at the end of the sampling, three from Acolman and three from the former Lake Texcoco.

4.1.2 Soil preparation, and PAH and sludge characterization

The soil was characterized at the laboratory, and treated according to Fernández-Luqueño et al. (2008; 2009). PAHs were obtained from Sigma (purity > 98% for phenanthrene, and > 97% for anthracene), while acetone was purchased from J.T. Baker (purity 99.7%). Sludge was obtained from Reciclagua (Lerma, state of Mexico, Mexico). Reciclagua treats wastewater from several sources, i.e., > 90 of the wastewater came from industries (alimentary industries) and under 10% from homes.

4.1.3 *Treatments and experimental setup*

One hundred and twenty-six sub-samples of 20 g soil of each of the six soil samples (three plots × two soils) were added to 120 mL glass flasks. Twenty-one flasks were used for each of the six treatments. Each treatment is described below:

i. In the first treatment, the soil was contaminated with 1,200 mg phenanthrene kg^{-1} dry soil plus 520 mg anthracene kg^{-1} dry soil and amended with 108 g dry sludge kg^{-1} dry soil (considered the S-SLUDGE treatment).
ii. In the second treatment, the soil was contaminated with 1,200 mg phenanthrene kg^{-1} dry soil plus 520 mg anthracene kg^{-1} dry soil and amended with 108 g sterilized dry sludge kg^{-1} dry soil (considered the S-STERILE treatment). Additional details regarding this treatment can be found in Fernández-Luqueño et al. (2008).
iii. In the third treatment, the soil was contaminated with 1,200 mg phenanthrene kg^{-1} dry soil plus 520 mg anthracene kg^{-1} dry soil, 108 g dry sludge kg^{-1} dry soil and 31 mL 0.1 M NaOH kg^{-1} dry Acolman soil and 5.8 mL 1 M NaOH kg^{-1} dry Texcoco soil (considered the S-BIO+NaOH treatment). Additional details regarding this treatment can be found in Fernández-Luqueño et al. (2008).
iv. In the fourth treatment, soil was contaminated with 1,200 mg phenanthrene kg^{-1} dry soil plus 520 mg anthracene kg^{-1} dry soil and spiked with 272 mg $(NH_4)_2SO_4$ kg^{-1} dry soil, 294.8 mg KH_2PO_4 kg^{-1} dry soil and 137.4 g glucose kg^{-1} dry soil for Acolman samples and 182 mg $(NH_4)_2SO_4$ kg^{-1} dry soil, 196.7 mg KH_2PO_4 kg^{-1} dry soil and 195.5 g glucose kg^{-1} dry soil for the Texcoco samples (considered the S-GLUCOSE treatment). Additional details regarding this treatment could be found in Fernández-Luqueño et al. (2008).
v. In the fifth treatment, soil was contaminated with 1,200 mg phenanthrene kg^{-1} dry soil plus 520 mg anthracene kg^{-1} dry soil (considered the S-PAH treatment).
vi. In the sixth treatment, an unamended uncontaminated soil served as a control (considered the S-CONTROL treatment).

The six treatments were designed such that (i) the effect of the microbial communities in the sludge on the removal of PAH could be estimated, i.e., S-STERILE vs. S-SLUDGE treatments, (ii) the effect of pH alteration, i.e., S-SLUDGE vs. S-BIO+NaOH treatments), (iii) the effect of components other than C, N and P in the sludge, i.e., S-STERILE vs. S-GLUCOSE treatments), and (iv) the effect of sludge, i.e., S-SLUDGE vs. S-PAH treatments). At the start of the experiment, three flasks were chosen randomly from each of the six soil treatments to determine inorganic N, PAH and extractable P. These were zero-time samples.

4.1.4 *Chemical analyses, PAH determination, and statistical analyses*

Soil pH was measured using a glass electrode (Thomas, 1996). Electrolytic Conductivity (EC) was determined as described by Rhoades et al. (1989). The WHC was measured were described by Fernández-Luqueño et al. (2008). Soil particle size distribution was determined as described by Gee and Bauder (1986). The organic or inorganic C in sludge and soil was measured by Fernández-Luqueño et al. (2008). Total N was measured by the Kjeldahl method (Bremner, 1996). The NH_4^+, NO_2^- and NO_3^- were determined and described by Mulvaney (1996). Total P was measured as described by Crosland et al. (1995) while extractable P was determined according to Murphy and Riley (1962). Concentrations of PAHs were extracted using a method developed by Song et al. (1995) and analyzed as described by Fernández-Luqueño et al. (2016).

Cumulative carbon dioxide (CO_2) production regressed on the elapsed time through a linear regression with different slopes (production rates). This was supported by the theoretical consideration that there was not CO_2 production at time zero. The concentration of ammonium, nitrite, nitrate and PAHs were subjected to one-way analysis of variance through PROC GLM (SAS Institute, 1989).

4.1.5 Some results about this experiment

Soils and sludge characteristics are described in Table 4.1. The soils and sludge used in this experiment have desirable characteristics such as pH, N and C content, textural classification, among others.

The activity rate of the microorganisms able to degrade organic or inorganic pollutants or materials depends on many factors, which are also interrelated, including contamination rate, microbial abundance and diversity, contaminant uptake, enzymatic activity, access to nutrients and bioavailability, concentration, toxicity and mobility of the pollutants. The rate at which CO_2 is emitted by microorganisms during the degradation and mineralization of organic carbon and CO_2 release from soils without plants might be a useful indicator of microbial activity and for quantitative measurements of biodegradation of PAH. Bench-top respirometry is an easy and effective method of measuring microbial respiration, i.e., the CO_2 production and the O_2 consumption (Fallgren et al., 2010). In the Acolman soil, there were no significant differences in the cumulative CO_2 production between the S-CONTROL and S-PAH treatments (Fig. 4.1a). However, in the Acolman soil, the addition of sludge or glucose significantly increased CO_2 emissions compared to the S-CONTROL and S-PAHs treatments ($P < 0.05$). In the Texcoco soil, the cumulative CO_2 production was significantly larger in the S-PAH treatment than in the S-CONTROL treatment (Fig. 4.3-b), while the addition of sludge or glucose to the Texcoco soil had the same effect as in the Acolman soil.

Contamination of the Acolman soil with phenanthrene or anthracene did not affect microbial activity as seen with CO_2 production, but increased in the Texcoco soil. Stimulation and inhibition of microbial activity have been found in PAH polluted soils (Fallgren et al., 2010), but long-term *in situ* field trials are also required. However, several short-term studies at laboratory scale have been published. Fernández-Luqueño et al. (2017c) stated that spiking soil with PAH or changing it with wastewater sludge increased the CO_2 emission rate, which decreased at higher EC. It was determined when they investigated how a different electrolytic conductivity affected the removal of phenanthrene and anthracene from wastewater sludge-amended soils, when soil with EC 6, 30, 80 or 146 dS m^{-1} were contaminated with PAH and modified or not with wastewater sludge. An additional experiment was carried out by Riveroll-Larios et al. (2015) to determine the natural attenuation potential of soils contaminated with hydrocarbons over two different periods (> 5 yr and < 1 mon). The basal respiratory rate of the recently contaminated soil (< 1 mon) was 1,538 mg C-CO_2 kg^{-1} h^{-1}, while

Table 4.1. Characteristics of the Texcoco and Acolman soils and the wastewater sludge.

Properties	Acolman Soil	Texcoco Soil	Sludge
pH_{H_2O}	6.0	9.3	6.4
Water holding capacity (g kg^{-1})[a]	674	659	ND[b]
Organic carbon (g kg^{-1})	8.1	58.2	509
Inorganic carbon (g kg^{-1})	0.2	0.8	ND
Total Kjeldahl nitrogen (g kg^{-1})	0.7	1.2	27.7
Total phosphorus (g kg^{-1})	0.2	0.3	1.7
Extractable phosphorus (mg kg^{-1})	2.0	0.3	600
Electrolytic conductivity (dS m^{-1})	2.4	7.3	5.7
Textural classification	Sandy loam	Loamy sand	ND

[a]On a dry base; [b]Not determined; [c]PAH were not detected ([c] is not in the table).

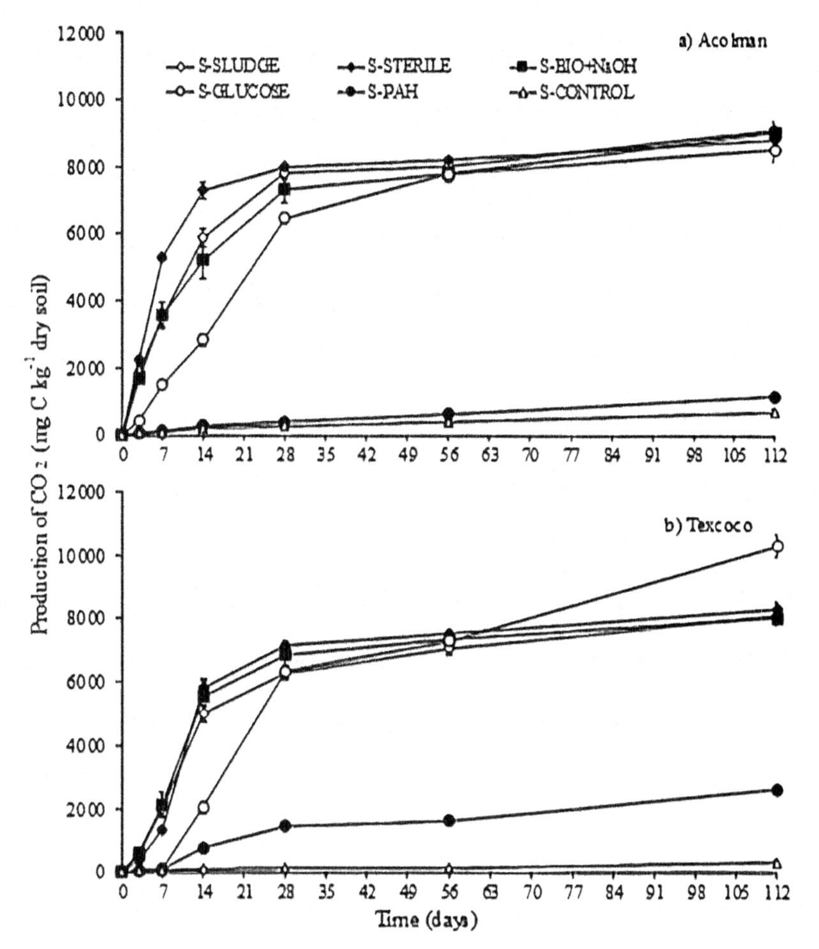

Fig. 4.3. Carbon dioxide emissions from a) Acolman and b) Texcoco soil. Soils were spiked with PAH and modified with different organic or inorganic substrates. The treatments are described later. Bars are standard errors of the estimates ($P < 0.05$) and each point on the graph is mean of n = 9.

the weathered soil had a greater basal mineralization capacity of 55–70 mg C-CO_2 kg^{-1} h^{-1}. Silva-Castro et al. (2015) suggested that the degree of stimulation in a bioremediation process is directly influenced by the initial biological activity of the treatment habitat. Thus, a soil with high biological activities, such as high levels of CO_2 production, would require a smaller amount of biostimulant compounds for the bioremediation process. However, Margesin et al. (2000) demonstrated that the CO_2 evolved from a contaminated habitat might be directly related to hydrocarbon concentration, biological activities and biodiversity of the microbial communities.

Phenanthrene was dissipated quickly in the Acolman soil, in spite of its high concentration at the start of the experiment (1,200 mg kg^{-1}). However, the dissipation of anthracene was slower as a consequence of its low solubility rate in aqueous systems like the one used in this experiment. The insolubility lowers PAH bioavailability, reduces microbial degradation and therefore anthracene dissipation decreases.

The microbial activity was stimulated when PAH was added to the alkaline soil (Texcoco soil), but not in the agricultural soil (Acolman soil) as witnessed by CO_2 emission. In addition, the degradation of PAH was regulated by the type of soil, the hydrocarbon and the sludge characteristics.

4.2 CO_2, CH_4, and N_2O emissions from soil irrigated with urban wastewater for maize cultivation

An experiment to determine how irrigation of *Zea mays* L. with wastewater from Mexico city at 120 kg N ha^{-1} affects the emission of CO_2, CH_4 and N_2O was carried out. Although urban wastewater has been used in agriculture for several centuries, but has gained recognition recently due to the increasing shortage of freshwater worldwide. In Mexico, more than 350,000 ha are irrigated with wastewater as it contains essential nutrients and organic matter for plants, though it also includes pollutants which might degrade, increase GHG emission from or contaminate the soil.

4.2.1 Area description and soil sampling

The soil sampling was carried out in the Mezquital Valley on agricultural land, which had been irrigated with wastewater for the past 120 yr. The soil was identified such as a loamy eutric Vertisol, which had been cultivated with alfalfa (*Medicago sativa* L.) or maize for more than 50 yr. The procedure for soil sampling was described earlier.

4.2.2 Soil preparation, treatments, and experimental set-up

The procedure for soil preparation was also described earlier. The experiment was carried out in a greenhouse using polyvinyl chloride (PVC) tubes filled with 6.5 kg dry soil. At the start of the experiment, five treatments were set-up: (i) S-WMAIZE (the crop, i.e., maize, was fertilized with wastewater), (ii) S-WASTE (soil was fertilized with wastewater, but maize was not grown), (iii) S-UMAIZE (the crop was fertilized with urea), (iv) S-UREA (soil was fertilized with urea but maize was not grown, and (v) CONTROL (soil was watered with tap water but maize was not grown and neither was the soil added with fertilizer). One hundred and twenty kg N ha^{-1} of inorganic N was included as urea or wastewater i.e., the suggested amount of N fertilizer for this crop. The treatment was irrigated with wastewater or tap water every 7 d , according to the treatment designs described above.

4.2.3 Chemical and statistical analyses

CO_2 and N_2O were determined with an Agilent 4890D gas chromatograph fitted with an electron capture detector coupled with a J&W Scientific GS-Q column to separate CO_2 and N_2O. The amount of CH_4 was determined with an Agilent 4890D gas chromatograph fitted with a flame ionization detector coupled with a Porapak Q column to separate CH_4 from the other gases. Concentrations of CO_2, N_2O and CH_4 were calculated by comparing peak areas against a standard curve prepared from known concentrations. Details regarding the determination of GHG can be found in Fernández-Luqueño et al. (2010). The procedures for chemical and statistical analyses are as described earlier. All data described below are the mean of three plants cultivated in agricultural soil in individual pots, i.e., PVC tubes, from three consecutive experiments which were carried out in a greenhouse, i.e., n = 27.

4.2.4 Some results from this experiment

The CO_2 emission rate ranged from 0.04 to 74 μg C kg^{-1} soil h^{-1} (Fig. 4.4). Urea did not have a significant ($p > 0.05$) effect on the CO_2 emission rate compared to the control treatment, while irrigation with wastewater significantly increased the CO_2 emission rate 2.5 times compared to the control treatment.

Lime mud, a by-product from the papermaking process, is conventionally used as a soil amendment to immobilize and remove pollutants in the liquid, solid and gas phase. Furthermore, it

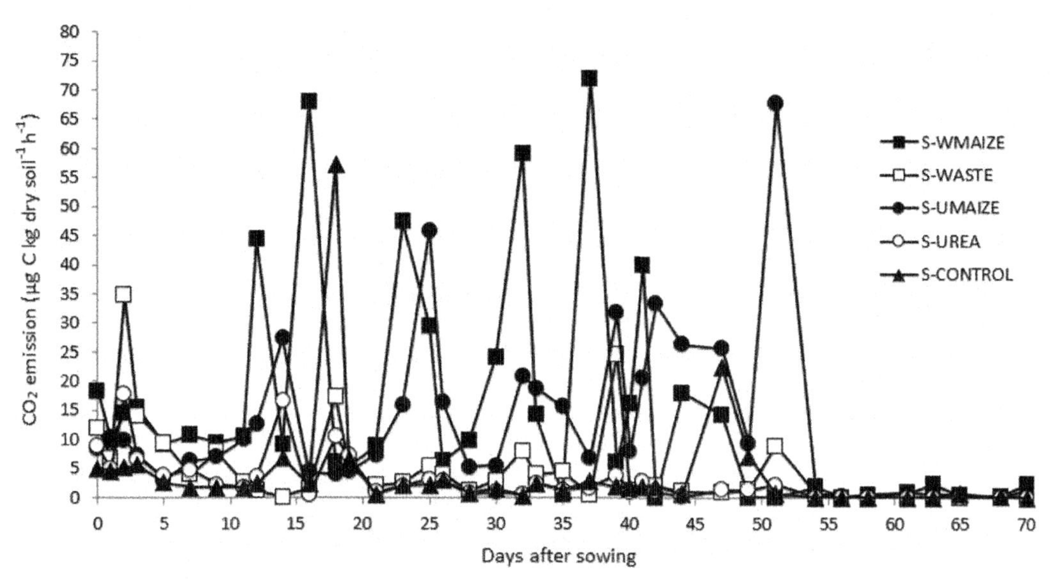

Fig. 4.4. CO_2 emissions from an agricultural soil cultivated with *Zea Mays* L. and irrigated with wastewater from Mexico city. Treatment descriptions can be seen earlier.

provides new opportunities for increasing the efficiency of the removal of dyes, organics pollutants, heavy metals and carbon dioxide (Zhang et al., 2015). Koptsik et al. (2015) carried out an experiment to determine if pines and microorganisms were affected during the remediation of a soil polluted with heavy metals. They found that pines were more sensitive to pollution compared to the relatively resistant microorganisms. They noted that soil respiration and the contribution of pine roots to the total respiration were positively correlated with the distance from the smelter and the content of carbon and nitrogen, but negatively correlated with the soil concentration and bioavailability of Cu and Ni.

The N_2O emission rate ranged from 0.0, i.e., undetectable amounts, to 0.056 μg N kg^{-1} soil h^{-1} (Fig. 4.5). The addition of urea increased the N_2O emission rate 2.4 times compared to the control treatment and irrigating with wastewater increased the mean N_2O emission rate 1.8 times, but maize irrigated with wastewater from Mexico city increased the N_2O emission rate 2.0 times.

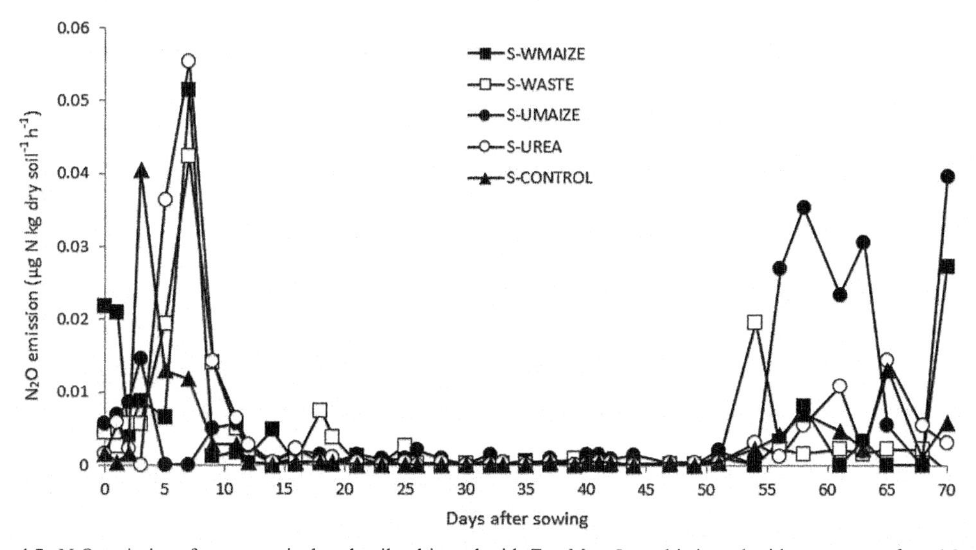

Fig. 4.5. N_2O emissions from an agricultural soil cultivated with *Zea Mays* L. and irrigated with wastewater from Mexico city. Treatment descriptions can be found earlier.

Agricultural soils are recognized as the primary source of NO and N_2O emissions worldwide, which have significant impacts on regional or global environments. Although N_2O is a growing risk to the environment, some methods such as biochar, activated carbon or others carbon materials might reduce N_2O emission from polluted or unpolluted agricultural soil. Sumaraj and Padhye (2016), published a review summarizing the current understanding of the interactions between inorganic nitrogen contaminants and carbon materials in agricultural soils.

Compost from organic materials is used to dissipate some pollutants during bioremediation technologies such as biostimulation, i.e., compost is a promising remediation strategy for soils. However, adding compost to the soils increases GHG emissions, primarily N_2O, which must be controlled and reduced during the production and use of compost (Yuan et al., 2017). Zhang et al. (2016) found that replacing ammonium with nitrate fertilizer and mixing with the nitrification inhibitor was an effective strategy to lower NO and N_2O emissions from arable soils.

The CH_4 emission rate ranged from –0.1 to 1.74 µg C kg^{-1} soil h^{-1} (Fig. 4.6). Urea addition did not affect the mean methane oxidation rate. The peaks observed in Fig. 4.6 for soil irrigated with wastewater from Mexico city occurred soon after the wastewater was added. It could be attributed to the sudden availability of organic matter and to the anaerobic conditions that were created after irrigation, thereby stimulating CH_4 production.

Microbial oxidation of CH_4 plays a significant role in reducing emissions to the atmosphere. Pariatamby et al. (2015) concluded that the addition of organic wastes at optimum ratio combined with compost as landfill cover material would have a significant effect on CH_4 emission reduction.

Wastewater from Mexico city increased the electrolytic conductivity of the soil top layer, i.e., 0–15 cm which might limit its use for several consecutive years. At present, hundreds of hectares of soil from the Mezquital Valley have already turned to saline soils as a consequence of excessive irrigation with wastewater for decades. Moreover, the addition of wastewater increased the emission of CO_2 and CH_4, but not the production of N_2O. It should be noted that irrigating with wastewater might be more environmentally friendly in the long-term, compared to irrigation with water from aquifers that take a long time to be replenished. It should be kept in mind that GHG emissions could be reduced by choosing the best management practices for soils and crops in terms of soil tillage, irrigation, fertilization or the control of pests and diseases.

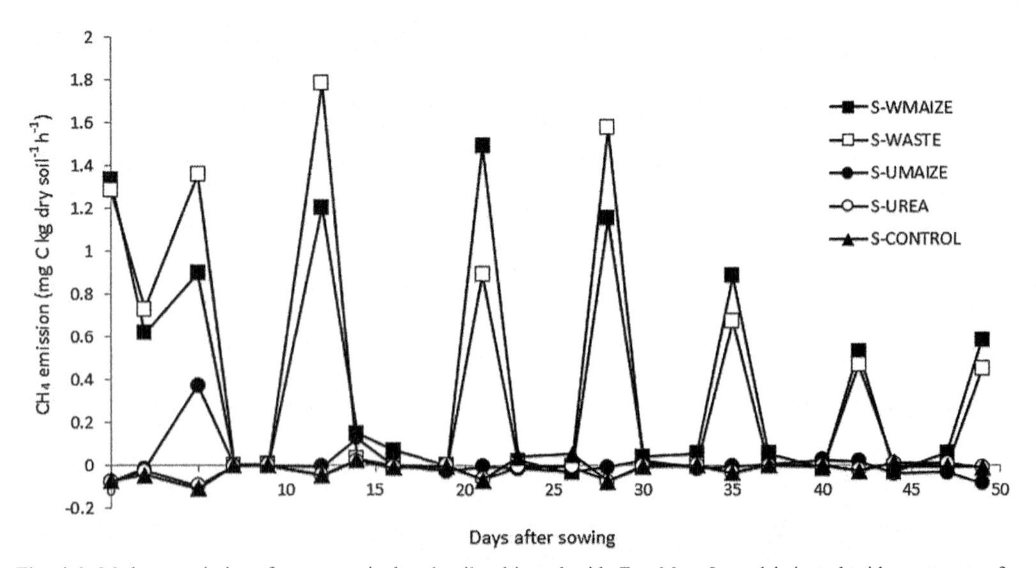

Fig. 4.6. Methane emissions from an agricultural soil cultivated with *Zea Mays* L. and irrigated with wastewater from Mexico city. Treatment descriptions can be found earlier.

5. Application of Carbon Materials or State of the Art Materials to Reduce GHG Emissions

Biochar, nanoparticles or smart materials have been studied in detail over the last years for their use in the reduction of GHG emission or for remediating the environment. For example, biochar is an emerging technology that has been well recognized for its potential role in CO_2 sequestration, GHG emissions reduction, soil quality improvement, crop productivity enhancement and environmental decontamination. Biochar is considered a carbon-negative strategy to mitigate GHG emissions; however, the impacts of its application for controlling long-term persistence and bioavailability of hazardous components is not clear. Moreover, the co-occurrence of low molecular weight VOC with PAH in biochar may lead to further phytotoxic effects (Dutta et al., 2017).

Biochar can reduce GHG and also has the ability to increase soil microbial community structure and enzymatic activity (Lone et al., 2015). Taxonomic studies revealed that soils modified with biochar contain higher numbers of operational taxonomic units compared to agricultural lands. Some arbuscular mycorrhizae have increased root colonization sites, enhanced the mycorrhizal plant association and improved the P availability in soils added with biochar (Qiao et al., 2015). Biochar has also provided a favorable habitat for microorganisms, thereby increasing microbial activity and basal respiration, and decreasing the efficiency of exchangeable fractions of heavy metals (Meier et al., 2017; Qiao et al., 2015). Bushnaf et al. (2011) stated that CO_2 evolution was similar in a soil polluted with BTEX compounds (benzene, toluene, ethylbenzene and xylene), amended or not with 2% of biochar. They also suggested that bulk concentrations of the BTEX compounds were not deleterious to microbes, thus there were no differences in CO_2 evolution in both treatments.

Soil fertilization is an important step in crop production, especially for high demand crops. Nitrogen (in the form of nitrate) is an essential macronutrient for crops that require additional supplements to achieve high productivity demanded by the global market. Over fertilization with this element could be a potential source of contamination, which might contribute to an increase in GHG emissions. To avoid this problem, the use of a 'smart fertilizer' is an alternative for intensive production systems. Souza et al. (2017) studied the behavior of chitosan-clay hybrid micro spheres (ferti-releaser) as a source of potassium nitrate in lettuce. They concluded that the micro spheres were more efficient for the controlled nitrogen release compared to conventional fertilization. Pereira et al. (2015) proposed that urea nanocomposites designed for slow- or controlled-release of nitrogen might increase the efficiency in the use of nitrogenous fertilizers. They described the impact of urea nanocomposites that were associated with exfoliated clay mineral, which they prepared with hydrophilic or hydrophobic polymers of varying concentrations, would have on the structure, urea release rate and how those nanocomposites would function in the field. Consequently, they concluded that nanocomposites made of polyacrylamide hydrogel or polycaprolactone considerably lowered nitrous oxide emissions in the field.

Greenhouse gas emissions are also significant in organic substrates such as manure. Gautam et al. (2017) stated that the application of bare nanoparticles could have adverse effects on human and environmental health. They experimented with nanoparticles entrapped in polymeric beads to treat livestock manure and reduce emissions of H_2S, CH_4 and CO_2. They found that ZnO nanoparticles entrapped in polymeric beads decreased H_2S, CH_4 and CO_2 emissions after swine manure was amended with zinc oxide nanoparticle alginate beads. Gautam et al. (2017) speculated that these GHG reductions were likely due to the inhibition of microbial communities induced by ZnO nanoparticles, as well as chemical conversion.

6. Conclusions

As part of the scientific and technological community, one needs to be able to deliver real technological developments to shape sustainable development without overlooking pollution . One should not shift the pollution from one place to another location, and also not change pollution from a solid to gas form or from soil to the biomass. Therefore, long-term *in situ* field trials are required. The aim to achieve global food security continuosly increases the demand for food production, but one should use less fertilizers, and more efficiently, to improve food distribution and waste less, and the need to decontaminate the environment through the dissipation or degradation of pollutants, but not through shifting pollution . The use of sustainability indicators for evaluating the performance of remediation technologies should also not be forgotten. The public needs to be educated and informed to avoid misperceptions regarding new technologies to dissipate pollutants or to decrease GHG such as genetic engineering or cloning. At the same time, implementing a review of regulatory guidelines based on several years of experience can help to highlight the benefits of bioremediation to provide affordable and safe food, human well-being and ecological sustainability.

Ideally, a multidisciplinary approach will result in a more profound understanding of harnessing natural resources, remediation of contaminated sites and the regulation of biogeochemical processes involved in GHG emissions. All these experiences and knowledge might subsequently lead to improving the sustainability of urban and agricultural systems.

This chapter also gives some insight into the current research gaps related to the studies involving GHG emissions, remediation of polluted soils and state of the art materials to reduce GHG emissions from agricultural lands under field conditions. Correct control procedures to degrade pollutants and reduce GHG emissions should be implemented and the associated risks of such methods must be improved. Unfortunately, pollution produced by emerging contaminants such as nanoparticles pose an additional threat to the soil system. Therefore, sustainable management of polluted soils is essential for maintaining ecosystem services worldwide.

Acknowledgements

This research was founded by the projects ‘Ciencia Básica SEP-CONACyT-151881’, ‘FONCYT-COAHUILA COAH-2019-C13-C006’, and ‘FONCYT-COAHUILA COAH-2021-C15-C095’, by the Sustainability of Natural Resources and Energy Program (Cinvestav-Saltillo), and by Cinvestav Zacatenco. To Instituto Politécnico Nacional for financial support (Project SIP-20161934, 20172065, and 20181453). G. M.-P., F. F.-L., and F. L.-V. received grant-aided support from ‘Sistema Nacional de Investigadores (SNI)’, Mexico. To Institute of Agricultural Sciences of the Autonomous University of the State of Hidalgo (ICAP-UAEH).

References

Abbasian, F., Lockington, R., Megharaj, M. and Naidu, R. 2016. The biodiversity changes in the microbial population of soils contaminated with crude oil. Curr. Microbiol. 72(6): 663–670. https://doi.org/10.1007/s00284-016-1001-4.

Barrios, E. 2007. Soil biota, ecosystem services and land productivity. Ecol. Econom. 64(2): 269–285. https://doi.org/10.1016/j.ecolecon.2007.03.004.

Bender, S.F., Wagg, C. and van der Heijden, M.G.A. 2016. An underground revolution: Biodiversity and soil ecological engineering for agricultural sustainability. Trends. Ecol. Evol. 31(6): 440–452. https://doi.org/10.1016/j.tree.2016.02.016.

Blagodatskaya, E. and Kuzyakov, Y. 2013. Active microorganisms in soil: Critical review of estimation criteria and approaches. Soil Biol. Biochem. 67: 192–211. https://doi.org/10.1016/j.soilbio.2013.08.024.

Blouin, M., Hodson, M.E., Delgado, E.A., Baker, G., Brussaard, L., Butt, K.R. and Brun, J.J. 2013. A review of earthworm impact on soil function and ecosystem services. Eur. J. Soil Sci. 64(2): 161–182. https://doi.org/10.1111/ejss.12025.

Bremner, J.M. 1996. Total nitrogen. pp. 1085–1122. *In*: Sparks, D.L. (ed.). Methods of Soil Analysis Chemical Methods, Part 3. Soil Science Society of America Inc., American Society of Agronomy, Madison, WI, USA. https://doi.org/10.2136/sssabookser5.3.c37.

Bushnaf, K.M., Puricelli, S., Saponaro, S. and Werner, D. 2011. Effect of biochar on the fate of volatile petroleum hydrocarbons in an aerobic sandy soil. J. Contam. Hydrol. 126(3-4): 208–215. https://doi.org/10.1016/j.jconhyd.2011.08.008.

Clermont-Dauphin, C., Blanchart, E., Loranger-Merciris, G. and Meynard, J.M. 2014. Cropping systems to improve soil biodiversity and ecosystem services: The outlook and lines of research. pp. 117–158. *In*: OzierLafontaine, H.

and Lesueur Jannoyer, M. (eds.). Sustainable Agricultural Reviews 14: Agroecology and Global Change. Springer. Switzerland. https://doi.org/10.1007/978-3-319-06016-3_5.

Crosland, A.R., Zhao, F.J., McGrath, S.P. and Lane, P.W. 1995. Comparison of aqua regia digestion with sodium carbonate fusion for the determination of total phosphorus in soil by inductively coupled plasma atomic emission spectroscopy (ICP). Commun Soil Sci. Plan. 26: 1357–1368. https://doi.org/10.1080/00103629509369377.

Cury, J.C., Jurelevicius, D.A., Villela, H.D.M., Jesus, H.E., Peixoto, R.S., Schaefer, C.E.G.R., Bicego, M.C., Seldin, L. and Rosado, A.S. 2015. Microbial diversity and hydrocarbon depletion in low and high diesel-polluted soil samples from Keller Peninsula, South Shetland Islands. Antarct. Sci. 27(3): 262–272. http://dx.doi.org/10.1017/S0954102014000728.

de Valenca, A.W., Vanek, S.J., Meza, K., Ccanto, R., Olivera, E., Scurrah, M., Lantinga, E.A. and Fonte, S.J. 2017. Land use as a driver of soil fertility and biodiversity across an agricultural landscape in the Central Peruvian Andes. Ecol. Appl. 27(4): 1138–1154. https://doi.org/10.1002/eap.1508.

Devigne, C., Mouchon, P. and Vanhee, B. 2016. Impact of soil compaction on soil biodiversity - does it matter in urban context? Uban. Ecosyst. 19(3): 1163–1178. https://doi.org/10.1007/s11252-016-0547-z.

De Vries, F.T., Hoffland, E., van Eekeren, N., Brussaard, L. and Bloem, J. 2006. Fungal/bacterial ratios in grasslands with contrasting nitrogen management. Soil Biol. Biochem. 38(8): 2092–2103. https://doi.org/10.1016/j.soilbio.2006.01.008.

Ding, Z.L., Wu, J.P., You, A.Q., Huang, B.Q. and Cao, C.G. 2017. Effects of heavy metals on soil microbial community structure and diversity in the rice (Oryza sativa L. subsp Japonica, Food Crops Institute of Jiangsu Academy of Agricultural Sciences) rhizosphere. Soil Sci. Plant. Nutr. 63(1): 75–83. https://doi.org/10.1080/00380768.2016.1247385.

Dutta, T., Kwon, E., Bhattacharya, S.S., Jeon, B.H., Deep, A., Uchimiya, M. and K.H. Kim. 2017. Polycyclic aromatic hydrocarbons and volatile organic compounds in biochar and biochar-amended soil: A review. GCB Bioenergy 9(6): 990–1004. https://doi.org/10.1111/gcbb.12363.

Fallgren, P.H., Jin, S., Zhang, R.D. and Stahl, P.D. 2010. Empirical models estimating carbon dioxide accumulation in two petroleum hydrocarbon-contaminated soils. Bioremediat. J. 14(2): 98–108. https://doi.org/10.1080/10889861003767084.

Fernández-Luqueño, F., Marcsh, R., Espinosa-Victoria, D., Thalasso, F., Lara, M.E.H., Munive, A., Luna-Guido, M.L. and Dendooven, L. 2008. Remediation of PAHs in a saline-alkaline soil amended with wastewater sludge and the effect on dynamics of C and N. Sci. Total Environ. 402(1): 18–28. https://doi.org/10.1016/j.scitotenv.2008.04.040.

Fernández-Luqueño, F., Thalasso, F., Luna-Guido, M.L., Ceballos-Ramírez, J.M., Ordoñez-Ruiz, I.M. and Dendooven, L. 2009. Flocculant in wastewater affects dynamics of inorganic N and accelerates removal of phenanthrene and anthracene in soil. J. Environ. Manage. 90(8): 2813–2818. https://doi.org/10.1016/j.jenvman.2009.03.010.

Fernández-Luqueño, F., Reyes-Varela, V., Cervantes-Santiago, F., Gómez-Juárez, C., Santillán-Arias, A. and Dendooven, L. 2010. Emissions of carbon dioxide, methane and nitrous oxide from soil receiving urban wastewater for maize (Zea maize L.) cultivation. Plant Soil 331: 203–215. https://doi.org/10.1007/s11104-009-0246-0.

Fernández-Luqueño, F., Valenzuela-Encinas, C., Marsch, R., Martínez-Suárez, C., Vazquez-Nuñez, E. and Dendooven, L. 2011. Microbial communities to mitigate contamination of PAHs in soil-possibilities and challenges: A review. Environ. Sci. Pollut. R 18(1): 12–30. https://doi.org/10.1007/s11104-009-0246-0.

Fernández-Luqueño, F., López-Valdez, F., Valerio-Rodríguez, M.F., Pariona, N., Hernández-López, J.L., García-Ortíz, I., López-Baltazar, J., Vega-Sánchez, M.C., Espinoza-Zapata, R. and Acosta-Gallegos, J.A. 2014. Effect of nanofertilizers on plant growth and development, and their interrelationship with the environment. pp. 211–224. *In*: López-Valdez, F. and Fernández-Luqueño, F. (eds.). Fertilizers: Components, Uses in Agriculture and Environmental Impacts. Nova, New York, USA.

Fernández-Luqueño, F., López-Valdez, F., Dendooven, L., Luna-Suarez, S. and Ceballos-Ramírez, J.M. 2016. Why wastewater sludge stimulates and accelerates removal of PAHs in polluted soils? Appl. Soil. Ecol. 101: 1–4. https://doi.org/10.1007/s11104-009-0246-0.

Fernández-Luqueño, F., López-Valdez, F., Sarabia-Castillo, C.R., García-Mayagoitia, S. and Pérez-Ríos, R. 2017a. Bioremediation of polycyclic aromatic hydrocarbons-polluted soils at laboratory and field scale: A review of the literature on plants and microorganisms. pp. 43–64. *In*: Anjun, N.A., Gill, S.S. and Tuteja, N. (eds.). Enhancing Cleanup of Environmental Pollutants. Volume 1: Biological approaches. Springer, Switzerland. https://doi.org/10.1007/978-3-319-55426-6_4.

Fernández-Luqueño, F., López-Valdez, F., Pérez-Morales, C., García-Mayagoitia, S., Sarabia-Castillo, C.R. and Pérez-Ríos, R. 2017b. Enhancing decontamination of PAHs-polluted soils: Role of organic and mineral amendments. pp. 339–368. *In*: Anjun, N.A., Gill, S.S. and Tuteja, N. (eds.). Enhancing Cleanup of Environmental Pollutants. Volume 2: Non-biological approaches. Springer, Switzerland. https://doi.org/10.1007/978-3-319-55423-5_11}.

Fernandez-Luqueño, F., Cabrera-Lazaro, G., Corlay-Chee, L., Lopez-Valdez, F. and Dendooven, L. 2017c. Dissipation of phenanthrene and anthracene from soil with increasing salt content amended with wastewater sludge. Pol. J. Environ. Stud. 26(1): 29–38. https://doi.org/10.15244/PJOES/64929.

Finn, D., Kopittke, P.M., Dennis, P.G. and Dalal, R.C. 2017. Microbial energy and matter transformation in agricultural soils. Soil. Biol. Biochem. 111: 176–192. https://doi.org/10.1016/j.soilbio.2017.04.010.

Gautam, D.P., Rahman, S., Fortuna, A.M., Borhan, M.S., Saini-Eidukat, B. and Bezbaruah, A.N. 2017. Characterization of zinc oxide nanoparticle (nZnO) alginate beads in reducing gaseous emission from swine manure. Environ. Technol. 38(9): 1061–1074. https://doi.org/10.1080/09593330.2016.1217056.

Gee, G.W. and Bauder, J.W. 1986. Particle size analysis. pp 383–411. *In*: Klute, A. (ed.). Methods of Soil Analysis, Part 1. Physical and Mineralogical Methods, Soil Science Society of America Inc., American Society of Agronomy, Madison, WI, USA. https://doi.org/10.2136/sssaj1979.03615995004300050038x.

Hawksworth, D.L. 2001. The magnitude of fungal diversity: the 1.5 million species estimate revisited. Mycol. Res. 105(12): 1422–1432. https://doi.org/10.1017/S0953756201004725.

Heil, J., Vereecken, H. and Brueggemann, N. 2016. A review of chemical reactions of nitrification intermediates and their role in nitrogen cycling and nitrogen trace gas formation in soil. Eur. J. Soil. Sci. 67(1): 23–37. https://doi.org/10.1111/ejss.12306.

Joergensen, R.G. and Emmerling, C. 2006. Methods for evaluating human impact on soil microorganisms based on their activity, biomass, and diversity in agricultural soils. J. Soil Sci. Plant. Nutr. 169(3): 295–309. https://doi.org/10.1002/jpln.200521941.

Karna, R.R., Luxton, T., Bronstein, K.E., Redmon, J.H. and Scheckel, K.G. 2017. State of the science review: Potential for beneficial use of waste by-products for in situ remediation of metal-contaminated soil and sediment. Crit. Rev. Env. Sci. Tec. 47(2): 65–129. https://doi.org/10.1080/10643389.2016.1275417.

Koptsik, G.N., Kadulin, M.S. and Zakharova, A.I. 2015. Impact of industrial pollution on emission of carbon dioxide by soils in the Kola Subarctic Region. Zhurnal. Obsh. Biol. 76(1): 48–62.

Leon-Silva, S., Fernández-Luqueño, F. and Lopez-Valdez, F. 2016. Silver nanoparticles (AgNP) in the environment: a review of potential risks on human and environmental health. Water Air. Soil. Poll. 227(9): Article number 306. https://doi.org/10.1007/s11270-016-3022-9.

Li, J., Huang, B., Wang, Q.X., Li, Y., Fang, W.S., Han, D.W., Yan, D.D., Guo, M.X. and Cao, A.C. 2017. Effects of fumigation with metam-sodium on soil microbial biomass, respiration, nitrogen transformation, bacterial community diversity and genes encoding key enzymes involved in nitrogen cycling. Sci. Total Environ. 598: 1027–1036. https://doi.org/10.1016/j.scitotenv.2017.02.058.

Lilley A.K., Bailey, M.J., Cartwright, C., Turner, S.L. and Hirsch, P.R. 2006. Life in earth: the impact of GM plants on soil ecology? Trends. Biotechnol. 24(1): 9–14. https://doi.org/10.1016/j.tibtech.2005.11.005.

Lone, A.H., Najar, G.R., Ganie, M.A., Sofi, J.A. and Ali, T. 2015. Biochar for sustainable soil health: A review of prospects and concerns. Pedosphere 25(5): 639–653. https://doi.org/10.1016/S1002-0160(15)30045-X.

Mabuhay, J.A., Nakagosh, N. and Isagi, Y. 2004. Influence of erosion on soil microbial biomass, abundance and community diversity. Land. Degrad. Dev. 15(2): 183–195. https://doi.org/10.1002/ldr.607.

Malyan, S.K., Bhatia, A., Kumar, A., Gupta, D.K., Singh, R., Kumar, S.S., Tomer, R., Kumar, O. and Jain, N. 2016. Methane production, oxidation and mitigation: A mechanistic understanding and comprehensive evaluation of influencing factors. Sci. Total Environ. 572: 874–896. https://doi.org/10.1016/j.scitotenv.2016.07.182.

Margesin, R., Walder, G. and Schinner, F. 2000. The impact of hydrocarbon remediation (diesel oil and plycyclic aromatic hydrocarbons) on enzyme activity and microbial properties of soil. Acta Biotechnol. 20: 313–333. https://doi.org/10.1002/abio.370200312.

Meier, S., Curaqueo, G., Khan, N., Bolan, N., Rilling, J., Vidal, C., Fernandez, N., Acuna, J., Gonzalez, M.E., Cornejo, P. and Borie, F. 2017. Effects of biochar on copper immobilization and soil microbial communities in a metal-contaminated soil. J. Soil. Sediment. 17(5): 1237–1250. https://doi.org/10.1007/s11368-015-1224-1.

Mocali, S. and Benedetti, A. 2010. Exploring research frontiers in microbiology: The challenge of metagenomics in soil microbiology. Res. Microbiol. 161(6): 497–505. https://doi.org/10.1016/j.resmic.2010.04.010.

Mulvaney, R.L. 1996. Nitrogen-Inorganic forms. pp. 1123–1184. *In*: Sparks, D.L. (ed.). Methods of Soils Analysis Chemical Methods Part 3. Soil Science Society of America Inc, American Society of Agronomy. Madison, WI, USA. https://doi.org/10.2136/sssabookser5.3.c38.

Murphy, J. and Riley, J.P. 1962. A modified single solution method for the determination of phosphate in natural water. Anal. Chim Acta 27: 31–36. https://doi.org/10.1016/S0003-2670(00)88444-5.

Nielsen, U.N., Ayres, E., Wall, D.H. and Bargett, R.D. 2011. Soil biodiversity and carbon cycling: A review and synthesis of studies examining diversity–function relationships. Eur. J. Soil. Sci. 62: 105–116. https://doi.org/10.1111/j.1365-2389.2010.01314.x.

Niu, F.J., He, J.X., Zhang, G.S., Liu, X.M., Liu, W., Dong, M.X., Wu, F.S., Liu, Y.J., Ma, X.J., An, L.Z. and Feng, H.Y. 2014. Effects of enhanced UV-B radiation on the diversity and activity of soil microorganism of alpine meadow ecosystem in Qinghai-Tibet Plateau. Ecotoxicology 23(10): 1833–1841. https://doi.org/10.1007/s10646-014-1314-7.

Pariatamby, A., Cheah, W.Y., Shrizal, R., Thamlarson, N., Lim, B.T. and Barasarathi, J. 2015. Enhancement of landfill methane oxidation using different types of organic wastes. Environ. Earth. Sci. 73(5): 2489–2496. https://doi.org/10.1007/s12665-014-3600-3.

Pereira, E.I., de Cruz, C.C.T., Solomon, A., Le, A., Cavigelli, M.A. and Ribeiro, C. 2015. Novel slow-release nanocomposite nitrogen fertilizers: The impact of polymers on nanocomposite properties and function. Ind. Eng. Chem. Res. 54(14): 3717–3725. https://doi.org/10.1021/acs.iecr.5b00176.

Pimentel, D. and Burgess, M. 2013. Soil erosion threatens food production. Agriculture 3(3): 443–463. https://doi.org/10.3390/agriculture3030443.

Qiao, Y.H., Croley, D., Wang, K., Zhang, H.Q. and Li, H.F. 2015. Effects of biochar and Arbuscular mycorrhizae on bioavailability of potentially toxic elements in an aged contaminated soil. Environ. Pollut. 206: 636–643. https://doi.org/10.1016/j.envpol.2015.08.029.

Rhoades, J.D., Mantghi, N.A., Shause, P.J. and Alves, W. 1989. Estimating soil salinity from saturated soil-paste electrical conductivity. Soil. Sci. Soc. Am. J. 53: 428–433. https://doi.org/10.2136/sssaj1989.03615995005300020019x.

Riveroll-Larios, J., Escalante-Espinosa, E., Focil-Monterrubio, R.L. and Diaz-Ramírez, I.J. 2015. Biological activity assessment in mexican tropical soils with different hydrocarbon contamination histories. Water Air. Soil. Poll. 226(10): Article number 353. https://doi.org/10.1007/s11270-015-2621-1.

Samarajeewa, A.D., Velicogna, J.R., Princz, J.I., Subasinghe, R.M., Scroggins, R.P. and Beaudette, L.A. 2017. Effect of silver nano-particles on soil microbial growth, activity and community diversity in a sandy loam soil. Environ. Pollut. 220: 504–513. https://doi: 10.1016/j.envpol.2016.09.094.

SAS Institute. 1989. Statistic guide for personal computers. Version 6.04, Edn. SAS Institute, Cary.

Silva-Castro, G.A., Uad, I., Rodríguez-Calvo, A. and Calvo, C. 2015. Response of autochthonous microbiota of diesel polluted soils to landfarming treatments. Environ. Res. 137: 49–58. https://doi: 10.1016/j.envres.2014.11.009.

Song, Y.F., Ou, Z.Q., Sun, T.H., Yediler, A., Lorinci, G. and Kettrup, A. 1995. Analytical method for polycyclic aromatic hydrocarbons (PAHs) in soil and plants samples. Chin. J. Appl. Ecol. 6: 92–96.

Souza, C.F., Faez, R., Bacalhau, F.B., Bacarin, M.F. and Pereira, T.S. 2017. *In situ* monitoring of a controlled release of fertilizers in lettuce crop. Engenharia Agricola 37: 656–664. https://doi.org/10.1590/1809-4430-eng.agric.v37n4p656-664/2017.

Sumaraj and Padhye, L.P. 2016. Influence of surface chemistry of carbon materials on their interactions with inorganic nitrogen contaminants in soil and water. Chemosphere 184: 532–547. https://doi.org/10.1016/j.chemosphere.2017.06.021.

Tate, K.R. 2015. Soil methane oxidation and land-use change—from process to mitigation. Soil Biol. Biochem. 80: 260–272. https://doi.org/10.1016/j.soilbio.2014.10.010.

Tejada, M., García, C., Hernández, T. and Gómez, I. 2015. Response of soil microbial activity and biodiversity in soils polluted with different concentrations of cypermethrin insecticide. Arch. Environ. Con. Tox. 69(1): 8–19. https://doi.org/10.1007/s00244-014-0124-5.

Thomas, G.W. 1996. Soil pH and soil acidity. pp. 475–490. *In*: Sparks, D.L. (ed.). Methods of Soil Analysis: Chemical Methods Part 3 Soil Science Society of America Inc., American Society of Agronomy Inc., Madison, Wisconsin, USA.

Waters, C.M., Orgill, S.E., Melville, G.J., Toole, I.D. and Smith, W.J. 2017. Management of grazing intensity in the semi-arid rangelands of Southern Australia: Effects on soil and biodiversity. Land. Degrad. Dev. 28(4): 1363–1375. https://doi.org/10.1002/ldr.2602.

Wolinska, A., Gorniak, D., Zielenkiewicz, U., Goryluk-Salmonowicz, A., Kuzniar, A., Stepniewska, Z. and Blaszczyk, M. 2017. Microbial biodiversity in arable soils is affected by agricultural practices. Int. Agrophys. 31(2): 259–271. https://doi:10.1515/intag-2016-0040.

Wu, W.C., Dong, C.X., Wu, J.H., Liu, X.W., Wu, Y.X., Chen, X.B. and Yu, S.X. 2017. Ecological effects of soil properties and metal concentrations on the composition and diversity of microbial communities associated with land use patterns in an electronic waste recycling region. Sci. Total Environ. 601: 57–65. https://doi.org/10.1016/j.scitotenv.2017.05.165.

Yan, X., Zhang, Y., Luo, X. and Yu, L. 2016. Effects of uranium on soil microbial biomass carbon, enzymes, plant biomass and microbial diversity in yellow soils. Radioprotection 51(3): 207–212. http://dx.doi.org/10.1051/radiopro/2016027.

Yuan, Y.H., Chen, H.H., Yuan, W.Q., Williams, D., Walker, J.T. and Shi, W. 2017. Is biochar-manure co-compost a better solution for soil health improvement and N_2O emissions mitigation? Soil. Biol. Biochem. 113: 14–25. http://doi:10.1016/j.soilbio.2017.05.025.

Zhang, J.S., Zheng, P.W. and Wang, Q.Q. 2015. Lime mud from papermaking process as a potential ameliorant for pollutants at ambient conditions: A review. J. Clean. Prod. 103: 828–836. http://doi:10.1016/j.jclepro.2014.06.052.

Zhang, Y.Y., Mu, Y.J., Zhou, Y.Z., Tian, D., Liu, J.F. and Zhang, C.L. 2016. NO and N_2O emissions from agricultural fields in the North China Plain: Origination and mitigation. Sci. Total. Environ. 551: 197–204. http://10.1016/j.scitotenv.2016.01.209.

Zhao, F. and Xu, K.D. 2016. Biodiversity patterns of soil ciliates along salinity gradients. Eur. J. Protistol. 53: 1–10. http://doi.10.1016/j.ejop.2015.12.006.

Zhong, W.H., Gu, T., Wang, W., Zhang, B., Lin, X.G., Huang, Q.R. and Shen, W.S. 2010. The effects of mineral fertilizer and organic manure on soil microbial community and diversity. Plant Soil 326(1-2): 511–522. http://doi:10.3389/fmicb.2018.01543.

Zhou, T., Pan, G.X., Li, L.Q., Zhang, X.H., Zheng, J.W., Zheng, J.F. and Chang, A. 2014. Changes in greenhouse gas evolution in heavy metal polluted paddy soils with rice straw return: A laboratory incubation study. Eur. J. Soil Biol. 63: 1–6. http://doi: 10.1016/j.ejsobi.2014.03.008.

CHAPTER 5

Agricultural Nanotechnologies to Improve the Crops' Quality, Increasing their Yields and Remediating Polluted Soils

Gabriela Medina-Pérez,[1] *Sergio R. Pérez-Ríos,*[2] *Fabián Fernández-Luqueño,*[3] *Rafael G. Campos-Montiel,*[2] *Alfredo Madariaga-Navarrete*[2] and *Fernando López-Valdez*[4,*]

1. Introduction

In the last decades, there has been increasing interest in the use of nanomaterials (NMs) in field of agriculture. However, literature published during the last years shows diverse effects from NM exposure on plants, from enhanced crop yield to acute cytotoxicity, including genetic alteration (Mukherjee et al., 2016a). These seemingly inconsistent research-outcomes present significant hurdles to the wide scale use of NM in agriculture worldwide. With the rise in global population, the necessity for increasing the supply of food has motivated scientists and engineers to look for methods in order to accelerate agricultural production (Baruah and Dutta, 2009). Nevertheless, increasing the crop production (tons per hectare) in agriculture may be achieved only by increasing productivity through Good Agricultural Practices (GAPs) and supporting agriculture with an efficient use of modern technology that take into account the limited availability of good quality soils and water resources. Advanced agronomic methods trying to boost agricultural yield through the use of more efficient fertilizers, water, pesticides and in the search of a way of hygienic storage of agricultural

[1] Instituto de Ciencias Agropecuarias, Universidad Autónoma del Estado de Hidalgo. Tulancingo, Hidalgo. 43600. Mexico.
[2] ICAP – Instituto de Ciencias Agropecuarias, Universidad Autónoma del Estado de Hidalgo. Tulancingo, Hidalgo. 43000. México.
[3] Sustainability of Natural Resources and Energy Program, Cinvestav-Saltillo, Coahuila de Zaragoza. 25900. Mexico.
[4] Research Centre in Applied Biotechnology (CIBA), Instituto Politécnico Nacional. Tepetitla de Lardizábal, Tlaxcala. 90700. Mexico.
* Corresponding author: flopez2072@yahoo.com

product that are harmless. However, the unfavorable effects of modern agricultural approaches on the environment have led to serious concerns among scientists, environmentalists, society and others related stakeholders (Baruah and Dutta, 2009).

The widespread use of chemical and persistent pesticides or synthetic fertilizers worldwide has contaminated water, air and soil resulting in diseases and collateral damages in non-target species such as soil microorganisms, wild plants, humans and animals. Nanotechnology can offer opportunities in the areas of pollution sensing and prevention, agriculture and environment, by exploiting novel properties of nanomaterials. In addition, it has also been stated that nanotechnology could augment the yield of crops, agricultural production, the breeding of livestock , boost the food processing industry and improve the human and environmental health through applications of these unique properties. However, inconsistent outcomes have been published by the best international reputable journals during the last years regarding benefits of NM on plants.

The aim of this chapter is to review and analyze the literature, studies of both positive and adverse effects of different NMs, mainly NPs. Some effects of NMs such as uptake of various contaminants and related materials will also be described. An attempt to highlight the gaps in knowledge or treat subjects more conventionally, while at the same time, the need for more soil-based investigations and efforts need to be focused on better understanding the underlying mechanisms of the NM-plant interactions.

2. Uptake, Movement and Accumulation of Nanosized Materials in Plants

The dimensions of nanoparticles (less than 100 nm) confer particular properties such as *increased reactivity*, a *lower surface* as well as *an unusual structure*, *greater stability*, *shape* and *aggregation* that differ from conventional bulk molecules (Nel et al., 2006). In addition to these characteristics of engineered nanomaterials, are known to be excellent *adsorbents*, *catalysts* and *sensors* (Carrillo et al., 2014). These properties have been exploited in a great number of applications: pharmaceutical, medicine, electronics, genetic, among others. Research in agriculture documented potentially benefic applications of nanotechnology in fertilization (Liu and Lal, 2014; Hossain et al., 2008), crop protection (Sahayaraj et al., 2016; Liu et al., 2006), germination and growth promoters, sensors of pollution and other substances and recovery and treatment of soil and water (Han, 2007; Stoimenov et al., 2002). The responses of plants are of interest, as they have both positive and negative effects.

According to Lin and Xing (2007), there are four categories of NPs such as: (a) ***Carbon-based materials***, that usually include *fullerene*, *single walled carbon nanotube* (SWCNT) and *multiwalled carbon nanotubes* (MWCNT); (b) ***Metal-based materials*** such as *quantum dots*, *nanogold*, *nanozinc*, *nanoaluminium* and *nanoscales metal oxides* such as TiO_2, ZnO and Al_2O_3; (c) ***Dendrimers***, which are nano-sized polymers built from branched units, capable of performing a specific chemical function, and (d) ***Composites***, which combine nanoparticles with other nanoparticles or with larger bulk-type materials (Lin and Xing, 2007), that present different morphologies such as: *spheres*, *tubes*, *rods* and *prisms* (Ju-Nam and Lead, 2008). Since plants are the first link in food chains, they strongly interact with the environment, and is a principal concern, because of the ability of NPs to move through membranes and walls of the cells (Nowack and Bucheli, 2007).

At present, it can be seen that the uptake of NP depends on the properties of nano-materials, for example: *size*, *aggregation* and *size-dependent sedimentation* or *diffusion* towards the cells, soil conditions, doses and the application method and plant conditions (Lin and Xing, 2008). For example, Raliya et al. (2016) reported that the uptake and translocation could depend on the shape of NP, for example if the NP are spherical, rhombic or cubic, etc. Lee et al. (2008) found that there is a linear relationship between high concentrations of CuNP in the growth media and higher uptake and accumulation of CuNP in plant tissues, so, this may be another factor influences the uptake and transport.

The uptake and accumulation of different NPs by plants is an incentive for conducting studies at present. Zhu et al. (2008) reported the first study about iron oxide nanoparticles (Fe_3O_4), where the NPs were taken up by roots of pumpkin (*Cucurbita maxima* Duchesne) and translocated in several plant tissues. When they were measured the quantity of NPs at the end of experiment, it was found that the amount of applied nanoparticles was 45.5% and were accumulated in roots, and approximately 0.6% of NPs were found in leaves. However, it should be noted that these results can vary for each species of plants. According to Watanabe et al. (2008) the NPs can form complexes associated with root exudates or membrane transporters, and then NPs could be transferred into the internal structures of plants. There are several ways that a vegetal cell can uptake the NPs, for example: through aquaporins, endocytosis, binding to carrier proteins, ion channels, by creating new channels with C nanoparticles (CNP) or binding to organic or polymeric chemicals, using natural or artificial compounds. The uptake depending on the application method of NPs, i.e., if this was on the root (via soil or water) or at the leaf level (foliar entry), so, the exposed tissue of the plant defines different transport and defense mechanism, and thus, the chemical and physiological responses correspond to the contact site.

It was also observed that the diameter of the cell structure pores or of the transport channels that determines the capability of NP to enter into cells. In the case of roots, according to Wang et al. (2016), it can find a size-exclusion barrier, (a) cell walls for the apoplastic transport pathway (5–20 nm), (b) symplastic transport (3–5 nm), or (c) the casparian strip transport (< 1 nm). Nevertheless, several reports have shown that when nanoparticle dimensions were more than 20 nm, led to the uptake and translocation. As stated by Wang et al. (2016), this could be explained for the entry of larger NPs than the exclusion barriers, rupture of membranes, formation of new large pores in the cell wall and some interactions between proteins, cations, viruses, among others, that may cause changes in the structure of the cell. In some cases as the study by Proseus and Boyer (2005), who evaluated the internalization of Au nanoparticles (as colloid solutions, no aggregated NPs) of different sizes into and across algal cell walls and found that AuNPs (~ 10 nm or above) were incapable of entering through the algal cell walls even under pressured conditions due to forming of trimeric and tetrameric clusters by a coalescence effect that were too large to pass through the cell walls, and found an accumulation of Au nanoparticles (spheres of 2–6 nm) that also bound to the internal walls. An interesting study by Birbaum et al. (2010), stated that the uptake of NPs did not depend on the opening or closing of the stomata, neither under dark and light exposure conditions and there was not translocation into newly grown leaves of maize plants by foliar NPs application. The results could indicate that natural entry barriers to some plants could be more resistant against NP translocation than mammalian barriers.

2.1 Foliar uptake and transport

One of the main routes for the internalization of nanoparticles into the plant system is aboveground organs and tissues by leaf spray, injection and atmospheric exposures (Corredor et al., 2009, Birbaum et al., 2010). There are two ways for foliar uptake of nanoparticles, (1) ***cuticular*** (Eichert and Goldbach, 2008) entry of NPs below 5 nm, and (2) ***stomatal*** (Eichert and Goldbach, 2008) entry of micrometer size range NPs. There are a few reports of foliar uptake of NPs (Eichert and Goldbach, 2008; Corredor et al., 2009; Birbaum et al., 2010), where it has been reported that magnetic ferrofluid of NPs enter into pumpkin plants cells via the vascular systems and stomata. Eichert and Goldbach (2008) found that the exclusion barrier in the stomatal foliar uptake for *Vicia faba* L. was above 10 nm and was important in enabling the transfer. To be able to confirm that true absorption of nanoparticles is being carried out, there should be some conditions, such as the particle size and NP concentration, the environment (light, temperature, water and gas), exposure time, plant species and NP methods of application (Wang et al., 2013). According to Eichert and Goldbach (2008) there are two possible ways for foliar uptake of NPs: cuticular and stomatal. The first, the cuticular pathway, is usually limited to NPs with sizes less than 5 nm due to cuticular pores sizes being extremely small (Eichert and Goldbach, 2008). The second, the stomatal pathways, allows the penetration of larger NP because the typical stomatal size is in a micrometer-size range (Willmer and Fricker, 1996; Eichert and Goldbach, 2008). Foliar uptake of atmospheric aerosols and particulate materials, including pollutants of diverse nature, on plants, has been well investigated by Uzu et al. (2010). It should be noted that these atmospheric aerosols or pollutants are found harmful for plant growth or development because they may cause alterations or damages in the photosynthetic activity, transpiration rates or thermal balance, as they obstruct the stomata due to their large sizes. However, there have been a few studies published that stated the foliar uptake of engineered NPs, such as Eichert and Goldbach (2008), Corredor et al. (2009) and Birbaum et al. (2010). Corredor et al. (2009) studied the entrance and transfer of magnetic NP using living pumpkin plants via injection bioferrofluid into the pith cavity of the leaf petiole and on leaf surface by also placing droplets of the ferrofluid. An Electron Microscopy Analysis showed that these particles get into the plant cells probably through the stomata and vascular systems. Additional studies were performed by Eichert and Goldbach (2008), where they found that in *Vicia faba* (L.) the size exclusion limited the stomatal foliar uptake for water-suspended NPs. These findings suggest that the stomatal pathway is largely capable due to its large size exclusion limit which was more than 10 nm, while an additional reason is its high transport velocity. Nevertheless, Birbaum et al. (2010) stated that cerium NP could not translocate in maize plants.

2.2 Root uptake and transport

Nanomaterials interact with an organism in three mechanisms: mechanical adhesion, electrostatic and biological interactions (Zhou et al., 2011). However, researchers are still investigating which of these mechanisms occur in the plant root. The uptake is easily confused with the adsorption, particularly the metal nanoparticles that are adsorbed at whole root level. Lin and Xing (2008) showed the aggregation of ZnO NPs on the root surface of cucumber plants using Scanning Electron Microscope (SEM) images, and the penetration of NP inside endodermal root cells. Lin and Xing (2007) found that ZnO NPs were attached to ryegrass root surfaces, but it was also found that a few individual particles were found in the protoplast and apoplast of the root endodermal and xylem cells and translocation from root to shoots of ryegrass were also reported. In another study, Martínez-Fernández and Komárek (2016) examined the agglomeration of iron NP on the root surface, causing blockage of water and nutrients uptake by plants, the lower dose treatment with 100 mg $nFe_2O_3 \cdot L^{-1}$, interfered with water transport, even in hydroponic conditions and it was seen that using nZVI (another type of nanomaterial) on the root in water and nutrients uptake seems to be lower than nFe_2O_3, and no effects in the exposed roots were found. Khataee et al. (2016) studied the uptake of CuO NPs, the fluorescent microscopic studies and agreed with the finding of CuO NPs into roots cells in *S. polyrrhiza*. However, in some species no uptake and transport were seen, for example in lima bean (*Phaseolus limensis* L.).

In rice (*Oryza sativa* L.) plants the uptake and translocation of carbon NMs was exhibited and fullerene (C70) was detected and easily taken up by roots and transported to shoots across to the vascular system, and could also be transported from leaves to roots (Lin et al., 2009). They also found similar results for MWCNT. Torney et al. (2007); Hussain et al. (2013), and Liu et al. (2009) reported similar results in MWCNTs and CNPs.

When NPs were assimilated by roots, they used the same routes of transport like nutrients and water. When they entered the cell wall, the pores restricted the entrance to large particles and aggregates (5–20 nm). Navarro et al. (2015) reported that particles can follow the symplastic route once internalized, in this pathway, the particles are translocated to the leaves. According to Aslani et al. (2014), the xylem is the most important way for translocation, accumulation and distribution of NPs. Anjum et al. (2016) reported that transport is facilitated if particles interact with other cellular components such as transport proteins, ion channels, etc. According to Etxeberria et al. (2012), it was proved that there is a mechanism of transport via endocytosis, this work also demonstrated that NP (nanospheres) with 40 nm enter vacuoles, while, nanospheres at 20 nm remain in the cytoplasm.

Another important pathway is via plasmodesmata for 20–50 nm, some research has also shown the potential for this route in the NP transport, that led endosomes or nano-protein complex to neighboring cells (Larue et al., 2012; Wang et al., 2011). According to Judy et al. (2012), the NPs could be integrated passively through the apoplast of the endodermis before reaching the stele. In addition, it was demonstrated that the xylem and phloem-mediated uptake, translocation and accumulation of nCuO (Wang et al., 2016). The casparian strip is the last barrier of translocation and it functions by keeping NPs free of transport, for which various plant species are reported (Anjum et al., 2016). The smart property of cell internalization of NPs of different sizes and compositions has been observed in different plant species (Lin et al., 2009; Liu et al., 2006; Torney et al., 2007) and it was proposed that some of them could be applied as carriers of DNA or other compounds, for example SWCNTs and mesoporous silica NPs.

2.3 Accumulation of NP

Lee et al. (2008) reported the accumulation of CuNPs in bean and wheat plant tissues. In *Medicago sativa* cells, bioaccumulation of quantum dots (CdSe/ZnS) appears specifically in the cytoplasm and nucleus (Santos et al., 2010). Leaves that have engineered NPs, accumulate first in the stomata, instead of the vascular bundle and later translocate to different parts via the phloem. When the NPs are absorbed by the root, they are transferred and accumulated in mature leaves due to their proximity to roots, on the other hand, mature leaves are usually more exposed than the young ones, therefore the exposure time is higher. The NPs translocation of grains, fruits and flowers has been reported only a little. Lin et al. (2009) showed that fullerene (C70) was capable of accumulating in *O. sativa* seeds. The translocation and bioaccumulation also seem to be species specific as the nano-CeO_2 case. Schwabe et al. (2013) found that there was a greater accumulation of Ce-ion in *Helianthus annuus* L. compared with an insignificant amount accumulated by *Cucurbita maxima* Duchesne and *Triticum aestivum* L. this may be attributed to the high ability to accumulate metals by *H. annuus*.

3. Interactions of NPs on Plants

The effect of NPs on plants could be both positive and negative. There are some important findings on the impact or engineered nanoparticles, and it depends on the *concentration*, *composition*, *size*, *physical* and *chemical properties* and the *plant species* (Ma et al., 2010; 2016).

3.1 Effects on germination

The NPs effect on the seed germination depends on NP concentration, the exposition time, the kind of NP and varies from plants to plants. Lee et al. (2010) found that SiO_2NP and Al_2O_2NP did not affect germination and growth of *Arabidopsis thaliana*, while ZnO NP hindered their germination. Nano-SiO_2 at low concentration improved tomato seed germination (Siddiqui et al., 2015). When exogenous nano-SiO_2 was applied, it enhanced seed germination of soybean by raising the activity of nitrate reductase enzyme, absorption potential and antioxidant system activity (Lu et al., 2002). The results were the same when nano-titanium dioxide (nano-TiO_2) was added.

In maize seeds, Suriyaprabha et al. (2012) found that nano-SiO_2 increased seed germination by expanding the availability of nutrients, by changing pH and conductivity to the growing medium. In tomato, nano-SiO_2 enhanced seed germination and the stimulation of antioxidant system under NaCl stress (Haghighi et al., 2012). De la Rosa et al. (2013) administered several concentrations of ZnO NPs on alfalfa, tomato and cucumber, only cucumber seed germination was enhanced. Other NPs, as silica, palladium, gold and copper were studied by Shah and Belozerova (2009), they found that all of them had a significant influence on lettuce seeds. Platinum nanoparticles exhibit a high germination index and had no negative effects in tomato and radish (Shiny et al., 2013). Several reports have shown that AuNP improve seed germination in cucumber, lettuce and *Brassica juncea* (Barrena et al., 2009; Arora et al., 2012). Savithramma et al. (2012), used biologically synthesized AgNP, they reported the improvement of seed germination of trees *Boswellia ovaliofoliolata.* Krishnaraj et al. (2012) studied the effect of the same nanoparticle AgNP on hydroponically grown *Bacopa monnier* and showed a significant effect on seed germination.

When carbon nanotubes are applied there were some positive effects. The MWCNT at the concentration of 10–40 mg L^{-1} enhanced seed germination and growth of tomato plants, the MWCNT can penetrate the thick seed coat of tomato (*Lycopersicon esculentum* Mill.) and tobacco (*Nicotiana tabacum* L.) (Khodakovskaya et al., 2009; 2012), the results showed an increased water uptake by seeds. Lin and Xing (2007) studied five types of NPs: MWCNT, nAl, nAl_2O_3, nZn and nZnO, on six plant species: lettuce (*Lactuca sativa* L.), ryegrass (*Lolium multiflorum* Lam.), radish (*Raphanus sativus* L.), oilseed rape (*Brassica napus* L.), maize (*Zea mays* L.) and cucumber (*Cucumis sativus* L.). They reported significant inhibition of germination of ryegrass by nZn and maize by nZnO or nAl_2O_3, but there were no inhibition for MWCNT.

3.2 Effects on roots

Under abiotic stress, nano-Si increases seed germination. Bao-shan et al. (2004) applied exogenous nano-SiO_2 on changbai larch (*Larix olgensis* A. Henry) seedlings and found that nano-SiO_2 enhanced seedling growth, by improving root length, mean height, root collar diameter, number of lateral roots and nano-SiO_2 generally induced the synthesis of chlorophyll.

Other NP such as nano-TiO_2 did not affect germination and root elongation of oilseed rape, wheat and *Arabidopsis* (Larue et al., 2011), cucumber, lettuce and radish (Wu et al., 2013). Different results were found by Asli and Neumann (2009), where colloidal suspensions of TiO_2 NPs (at 1 g · L^{-1}) can inhibit leaf (maize seedlings) growth and transpiration by damaging root water transport under hydroponic conditions; however, in another experiment, potted maize plants grown for 6 wk in a clay soil irrigated with nutrient solutions with TiO_2 at 1 g · L^{-1}, revealed minor inhibitory effects as compared with control treatments, where the effects were not statistically significant.

The MWCNT enhanced root growth and peroxidase and dehydrogenase enzyme activities (Smirnova et al., 2012; Tripathi and Sarkar, 2015). Aluminum oxide NPs were probed in five crop species, cabbage (*Brassica oleracea* L.), maize, cucumber, soybean and carrot (*Daucus carota* L.) the nanoparticles have effects on root elongation in hydroponic culture conditions (Yang and Watts, 2005). Wang et al. (2014) experimented with rice plant by applying Quantum Dots (QDs), and with silica coated with QDs, these promoted rice root growths.

3.3 Effects on plant growth

Nano-sized TiO_2 had a positive effect on the growth of spinach sprayed on leaves (Yang et al., 2006; Zheng et al., 2005; Hong et al., 2005). The nano-SiO_2 NP generally improves leaf biomass, proline content and chlorophyll, enhancing the plants tolerance to abiotic stress (Kalteh et al., 2014; Haghighi et al., 2012; Bo et al., 2012; Ramesh et al., 2014). Plants of tomato (*Lycopersicum esculentum*) when exposed to carbon nanotube, showed aquaporins regulation as a response to the stress caused by multi-walled CNTs (Khodakovskaya et al., 2009). Nano-SiO_2 enhances plant growth because it increases the gas exchange and the stomatal conductance, photosynthetic rate, transpiration rate, PSII potential activity, electron transport rate, effective photochemical efficiency, actual photochemical efficiency and photochemical quench (Ramesh et al., 2014; Yinfeng et al., 2011). In another study, exposure to SiO_2 NP did not cause any effects in zucchini (Stampoulis et al., 2009). Cu NP did not affect the plant system interactions in mung bean, but CuO and NiO NPs displayed harmful impacts on growth of lettuce, radish and cucumber showing increased lipid peroxidation, oxidized glutathione, Reactive Oxygen Species (ROS), peroxidase and catalase activities and decreased chlorophyll content.

Negative effects of nano-CuO were assessed in soybean (*Glycine max* [L.] Merr.) and chickpea (*Cicer arietinum* L.) (Adhikari et al., 2012), and in rice (Shaw and Hossain, 2013).

Hernandez-Viezcas et al. (2011) stated no significant toxic effect of ZnO NP in velvet mesquite plants (*Prosopis juliflora-velutina*). On ryegrass and Indian mustard (*Brassica juncea*) it was indicated that plant growth was inhibited by ZnO NP (Lin and Xing, 2008). Boonyanitipong et al. (2011) reported that TiO_2 and ZnO NP negatively affect rice and wheat (*Triticum aestivum*) growth. Ag NP were frequently reported as detrimental to plant growth, however several studies demonstrated the stimulatory effects of AgNP in plant growth (Syu et al., 2014; Yin et al., 2012; Salama, 2012). MWCNTs enhanced the growth of tobacco (*Nicotiana tabacum*) cell culture (Khodakovskaya et al., 2012). Dimkpa et al. (2012) found that Cu and Zn levels in shoots were the same when NPs or bulk materials were used. Additionally, they found that the oxidative stress in the NP-treated plants increased oxidized glutathione and lipid peroxidation in roots, but the chlorophyll content decreased, while higher catalase and peroxidase activities were reported in roots. Some authors reported the inhibitory effect of MWCNT on plants growth (Tiwari et al., 2014; Ikhtiari et al., 2013; Begum et al., 2014). Giraldo et al. (2014) reported that SWCNT augmented three times higher photosynthetic activity than controls by enhancing maximum electron transport rates, SWCNT also enabled plants to sense nitric oxide.

4. The Nanosized Materials Stimulate Some Physiological Changes in Plants

Currently, one of the objectives of agriculture is increased production and this includes different strategies that affect growth and the physiology of the plant. There is also an impact on the quality of the plant food. In this century it stands out as a strategy, while there are several studies for and against the use of nanoparticles, much more research needs to be conducted in the world of nanotechnology in agriculture. Agronanotechnology, which promises natural resource management through novel tools and technological platforms within limited resources of land and water (Mishra et al., 2014), can be a solution for food, sustainability in the production process and environmental improvement. Many works in areas on the uptake, translocation, accumulation and toxicity have been made in different spheres to explain the interaction between the relationship of the nanoparticle—plant in order to know the answer. Our research group found that some NPs at low doses can stimulate the plant growth in *H. annuus*, but more studies are needed in this plant and others to confirm these beneficial effects (data unpublished).

To date, investigations into the application of nanoparticles and its effect within the physiology of plants have only focused on the stages of seed germination and seedling growth. Information on other processes such as sprouting buds (vegetative and floral), flowering, fruit set and fruit need to be resolved. Additionally, some studies have reported various modifications in the physiology of the plant by the effect of the nanoparticle and the dose at which it was used.

Tomato seeds (*Solanum lycopersicum* L.) were exposed to doses of 125, 250, 500 and 1,000 mg L^{-1} $CoFe_2O_4$ NP, which clearly does not affect the germination and growth of tomato plants. A dose of 1,000 mg L^{-1} promoted root growth; translocation and Ca and Mg concentrations decreased to 250 mg L^{-1} or higher. Catalase activity decreased in roots and leaves. López-Moreno et al. (2016) proposed the need to determine whether the $CoFe_2O_4$ NP was transported to fruits. For AgNP applied to seedlings of Chinese cabbage (*Brassica rapa* L. subsp. *pekinensis*) at different concentrations to determine the impact on their growth and physiology, and it was found that low concentration acts as a growth stimulator and at high concentrations it retards it and there is an alteration of genes involved in glucosinolates, anthocyanins and antioxidants (Baskar et al., 2015).

At present it is not enough to find the effects and physiological changes in the early stages of the life cycle of a plant, answers need to be found on issues of absorption and translocation of nanoparticles, mainly in agricultural use food crops. For example, CeO_2NP and TiO_2NP influenced for

10 d the normal cycle of growth of barley (*Hordeum vulgare* L.). Specifically, the CeO_2NP reduced the number of tillers, leaf area and the number of ears per plant. Meanwhile, the TiO_2NP stimulated plant growth and offset the adverse effects of CeO_2NP. Crystalline TiO_2NP aggregates were detected in leaf tissues, while CeO_2NP was not present in the form of nanoclusters (Marchiol et al., 2016). The CeO_2NPs were evaluated in physiology, productivity and macromolecular composition in barley (*Hordeum vulgare* L.). A high dose (500 mg kg^{-1}) promoted the development of the plant with 331% increase in aboveground biomass compared with the control, but did not form grains. An average dose (250 mg kg^{-1}) enhanced the accumulation of Ce in the grain, thus 294% which was accompanied by increases in P, K, Ca, Mg, S, Fe, Zn, Cu and Al, as well as, increased methionine, aspartic acid, threonine, tyrosine, arginine and linolenic acid content in the grains up to 617, 31, 58, 141, 378 and 2.47%, respectively, compared with other treatments. In general, it modified stress levels in the leaves and there were no apparent signs of toxicity. The results illustrated the beneficial and harmful effects of this nanoparticle in barley (Rico et al., 2015). In some cases, titanium oxide application in barley (*Hordeum vulgare* L.) caused seed yield increment, while titanium dioxide nanoparticle (0.02%) was more effective in improving the yield. The highest seed yield was obtained when non-stressed plants under water deficit stress condition were treated with titanium dioxide nanoparticles (0.02%) during the stem elongation stage (Jaberzadeh et al., 2013). Not only did the nanoparticles cause concern to investigate its effect on the physiology of species of plant for agricultural use, but also, tests on its potential use in agroforestry species for improving the germination of those that contribute to the restoration of soil. *Quercus macdougallii* was threatened by habitat loss and is in criterion Vulnerable D2 (Nixon, 1998). Fe_3O_4 NP treatments help to improve the increase of germination up to 33% compared with the control, early growth and contribute to increasing the concentration of chlorophyll and dry biomass (Pariona et al., 2017). Another important research area are ornamentals, which do not provide food, only home-spaces, parks and landscaping visual comfort. Their diversity of flower phenotype and color, propagation, shades of foliage, flowering time, among others of phenological attributes that can be modified by the action of the nanoparticle.

Environmental damage due to particulate deposition is related to the competition pattern alteration among species that can result in a drastic effect in plant biodiversity: more sensitive species may be eliminated and growth, flowering and fructification of other species may be favored (Monica and Cremonini, 2009). The long-term impact of nano-particles on the environment and plants is an uncertain topic and research is gradually emerging that contributes to a response. In the *Brassica rapa* culture, CeO_2NPs were applied at concentrations ranging from 0–1,000 mg L^{-1} to evaluate the physiological and biochemical consequences of multi-generational exposure (3rd generation) and, among the results obtained, it was stated that second- and third-generation plants may have experienced greater oxidative stress than first-generation plants. This demonstrated the impact of CeO_2NP between generations (Ma et al., 2016). This chapter does not suggest a positive or negative opinion of the nanoparticles in the growth and development of the plant. The impact of nanoparticles on plants depends on the composition, concentration, size and physical and chemical properties of NP, and even the plant species characteristics (Ma et al., 2010), plant substrate (i.e., soil, hydroponics and culture medium) (Arruda et al., 2015) and the exposure duration of NP to crops (Rizwan et al., 2017). Nanomaterials have the potential for different agricultural applications, so further research to expand the application possibilities and methodologies in agriculture is necessary (Khot et al., 2012). In recent years, certain researchers have been working in order to obtain a better understanding the effects of different types of NPs on plants (Table 5.1) (Siddiqui et al., 2015).

5. Are Agricultural Nanotechnologies the Way to Feed a Hungry and Polluted World?

This could appear a simplistic question for a very complex problem. Such an issue may be addressed from different aspects that are involved from sociology, economy, politics and even theological points

Table 5.1. Main physiological changes in plants by use of nanoparticles (NPs). The positive or negative noteworthy effects are reported here.

NP	Size (nm)	Concentration	Plant	Observed Effect	References
Ag	< 10	1 and 10 mg L^{-1}	Wheat (*Triticum aestivum* L.)	NPs decreased the shoot and root length and fresh biomass. Treatment altered the expression of several proteins involved in primary metabolism and cell defense.	Vannini et al. (2014)
	< 20	0, 0.2, 0.5 and 1 µg mL^{-1}	Rice (*Oryza sativa* L.)	NPs significantly reduced the root elongation, shoot and root fresh weights, total chlorophyll and carotenoids contents. Additionally, production of Reactive Oxygen Species (ROS) increased in plants in a dose dependent manner.	Nair and Chung (2014a)
	10–15	0, 50, 100, 1000, 2500, 5000 µg mL^{-1}	Tomato (*Lycopersicon esculentum* L.)	Mature plants reduced root elongation, had lower chlorophyll contents, higher superoxide dismutase activity and less fruit productivity.	Song et al. (2013)
	5–25	0, 5, 10, 20, and 40 g mL^{-1} in agar. 0, 100, 300, 500, 1000 and 2000 mg kg^{-1} in artificial soil	Sorghum (*Sorghum bicolor* L.)	NPs inhibited the plant growth in agar media, but the plant growth was not affected in soil media.	Lee et al. (2008)
CeO_2	200	0, 125, 250 and 500 mg kg^{-1}	Barley (*Hordeum vulgare* L.)	NPs increased the plant height, chlorophyll contents, dry biomass, oxidative stress, K leakage and yield components.	Rico et al. (2015)
		0, 100, 500 and 2000 mg L^{-1}	Cotton (*Gossypium* sp.)	NPs decreased the plant height and shoot and root biomass. Additionally, content of Fe, Ca, Mg, Zn and Na decreased in roots exposed to NPs.	Li et al. (2014)
CuO	20–40	100 µg mL^{-1}	Maize (*Zea mays* L.)	Seed germination was not affected, but NPs inhibited the growth of maize seedlings.	Wang et al. (2012)
	< 50	0, 50, 100, 200, 400 and 500 µg mL^{-1}	Soybean (*Glycine max* [L.] Merr.)	Exposure to 500 mg L^{-1} of NPs reduced the shoot growth, weight and total chlorophyll content, while H_2O_2 contents increased.	Nair and Chung (2014b)
	< 50	0, 10, 50, 100, 500 and 1000 mg L^{-1}	Cucumber (*Cucumis sativus* L.)	Plant biomass decreased 75% compared to the control treatment, but NPs significantly increased the antioxidant enzyme activities.	Kim et al. (2012)
	< 50	0.5, 1.0 and 1.5 mM	Rice (*Oryza sativa* L.)	NPs exposure decreased the seed germination and seedling growth. In addition, NP caused a severe oxidative burst in plants.	Shaw and Hossain (2013)
	< 50	500 mg kg^{-1}	Wheat (*Triticum aestivum* L.)	NPs reduced the root and shoot length and biomass.	Dimkpa et al. (2012)
Fe_2O_3	20–30	0, 100, 500, 1000, 5000 and 10,000 ppm	Wheat (*Triticum aestivum* L.)	100 ppm NP increased the seed germination compared to the control, meanwhile NP reduced the seed germination with higher treatments.	Feizi et al. (2013)
TiO_2	< 100	10 g kg^{-1}	Wheat (*Triticum aestivum* L)	NPs decreased the growth and yield of plants.	Du et al. (2011)
	< 20	100–1000 µg mL^{-1}	Arabidopsis (*Arabidopsis thaliana* (L.) Heynh.)	TiO_2 NPs affects the antioxidant response of the plant due to changes in the expression of the vitamin E gene.	Szymańska et al. (2016)
ZnO	< 100	5 g kg^{-1}	Wheat (*Triticum aestivum* L)	NPs decreased the growth and yield of plants.	Du et al. (2011)
	< 50	0, 10, 50, 100, 500 and 1000 mg L^{-1}	Cucumber (*Cucumis sativus*)	Plant biomass decreased 35% compared to the control. NPs significantly increased antioxidant enzymes activities.	Kim et al. (2012)

of view. Nevertheless, one finds and there is a narrow relationship between hunger and science. This relationship is the most important one when others already mentioned attempts to solve hunger have failed. "God science" must represent the hope to solve hunger. "God scientist" is responsible to fight against hunger. "God's technological developments" are responsible to decrease the number of people who are victims of hunger. Science is also responsible for the development of technological tools with both objectives: fighting against hunger while taking into account the environment. This is quite a challenge since many technological developments have failed to do so (pesticides, modern plant breeding and transgenic, for instance). The speed in which science and technological developments are working, specifically towards the direction to feed a hungry world is extremely high, particularly in the last decade. As a result of this high development speed, nanotechnologies appear to make its contribution in the agrobiotechnology area. As an emerging technological option resulting from the experience of many other earlier developments, nanotechnology comes with the promise to efficiently combine a realistic option for food production and care of the environment.

5.1 The concept of hunger

While the term "hunger" appears to be the same for a number of people, the concept *per se* is not still clear. What does hunger mean for different people? What does hunger mean for different organizations? Does a charitable organization share the same concept of hunger than that of a bank, for instance? Some more ideas appear from this concept. What are the components of hunger? How are these components measured and by who? It is also important not to forget that hunger possesses several components, including quantitative, qualitative, psychological, marketing, politics and social ones.

A narrow concept that may be valid for the purpose of this discussion would be "the going without food for a specific period of time and the physical sensation of hunger" (Radimer et al., 2016). This definition represents, at least for the authors, the concept that makes the most sense . The physical sensation of hunger would be "the enemy" to be defeated. There is a "new concept" that is worth taking a closer look at, which are "*hunger policies*". This may be able to answer some of the questions already revealed. An important effort to solve the "conceptual issue" of hunger has been done by many organizations, countries, etc., through analysis and the approval of policies directed to "attend" to hunger issues. If these policies were effective, the question would be how the role of science and technological developments are involved in such policies. The most important objective of agriculture is the fight against the physical sensation of hunger. An interesting point of view is addressed in the book edited by Jean Dreze and the paradox there established: "*in a world of food surpluses and safety, hunger kill millions more people each year than wars or political repressions*". Since the intention of this chapter is to set a point of view, one sees hunger policies and scientifically and technological approaches working together for the fight against hunger, and will set the bases for this last statement in further analysis of this chapter.

5.2 Who is responsible for fainting for hunger?

As all the other issues have already been addressed, to answer the question about who is responsible for fainting from the hunger one will need to get closer to the main discussion of this chapter. Globally, people are involved in this concept as volunteers, political activists, global citizens and financial donors (Bocking-Welch, 2016). It is important to point out, the role of each of these participants and their responsibility in 'fainting' from hunger. In an analysis done by Breustedt and Qaim (2012), they eloquently exposed , why around one million people are starving. They concluded that, research in agriculture must be increased, public and private, in a developing country as in an industrialized one in order to reduce hunger. Research by Kaan and Liese (2011) compared public and private partnerships against hunger and concluded that there exists a link between business enrollment and output legitimacy. It was then suggested, that in order to give an answer to the proposed question, those responsible for fighting hunger are private and public organizations, especially public and private educational institutions and the research done by then as well as public and private research organizations and their research.

Areas of knowledge are supposed to be an outcome of the reflection of activities of human beings. They are the result of the use of intelligence to solve their own problems and, for the purpose of this discussion, it can be assumed that humans beings are the only ones to be aware of their situation and be capable of using science and technology to solve such problems. It could be, perhaps be a matter of economic issues as an excuse for not conducting research, however it is, instead, a matter of the lack of human resources that is the main factor that limits science and technology. It is then proposed that educational institutions and research centers, with the support of private and public funds be responsible to form new generations of researchers and technicians so they can develop tools to efficiently to fight against hunger.

5.3 To feed hunger without polluting the environment

History has shown that remarkable achievements have been made in the era of biotechnology and, moreover, in the post biotechnology area. Biotechnology and their developments have promised to be the way to fight against hunger and to solve the problem for now and permanently. The emergence of agrochemicals, the use of plant breeding and the soil mechanization to reach the genomics and proteomics time.

No doubt, all the efforts and the results from this period were positive and accomplished, that is, the promise to fight against hunger. The green revolution was on its way. How many green revolutions are there? How many green revolutions are needed to fight against hunger? And, if by boosting the development of green revolutions can one reduce the gap between the numbers of hungry human beings, then why, is it to date, that such an objective is still far from being accomplished? It theorize this by saying that research has forgotten a basic principle: all-natural systems are related and all are self-regulated, in a system with delicate equilibrium and robust at the same time. To feed the living organism on this planet, including human beings is part of the system. The environment is part of the system. Attempts to feed organisms without protecting the environment have been unsuccessful, as has been witnessed.

There is an urgent need to satisfy hunger. The only way is through science and technological developments and the political will, working together; where the key word is "will". These developments will be successful if it involves a strong environmental component. A new green revolution is necessary, but this one has to be definitive, has to take in to account all the experiences generated in the past revolutions (which had cost money, time and resources), accompanied with an awareness in the environment. There is no more time for experimentation. The second decade of the 21st century would be the last chance for science, technology and politics to solve hunger issues without polluting the environment.

5.4 Science, technology and sustainable development (nanotechnology)

Having made some statements about hunger, science and technology, it is important get to the main point of this chapter. As human beings if we appreciate the concept of sustainability, we have the right to use all the available resources (water, land, air, etc.), as long as we also respect this same right that future generations should have. Then, if we are to survive, we need to be intelligent enough to "force" science and technology to make a harmonic combination with sustainability. Only by making the concept of sustainability the base of the pyramid, will the proposed new green revolution be able to work.

Many components will be needed for sustainability (the concept that was missing in the combination, since science and technology were already involved in the fight against hunger). Some of these components are represented for the newest developments in the area of biotechnology such as proteomics, genomics, the use of the properties of molecules derived from natural products and nanotechnology.

Fight against hunger involves the production of food. This food production needs to be large, quicker, effective, nutritional and healthy. To be able to achieve all these characteristics, many processes should follow. Processes like planting (using the microcosm represented by the soil), delivering nutritional benefits to food, the protection of crops against pathogens and plague insects (which guarantee safer and healthier productions), crop fertilization (which ensure the effectiveness of crop production in a cleaner way) are efficiently supported by the so-called nanotechnology.

With support of nanotechnological techniques and processes to make a sustainable food production, implies a production of food to ensure future food without polluting the environment.

6. Agricultural Nanotechnologies to Decontaminate the Environment and Feeding Humans Worldwide

The relevance of understanding the diversity of the aspects involving NPs, plants and the environment if major advances in this field are to be made. For example, the interaction between plants, NPs, the environment and the consequences of this interactions have not been deciphered yet. However, there are several reviews published currently regarding the benefits or damages of NPs into the crop yield (Table 5.2).

Additionally, it should be stated that nanosensors and nanodevices are capable of detecting several variables such as microorganisms, humidity and toxic pollutants at high frequency rates (minutes). Organic pesticides or several types of pollutants can be degraded into harmless components by photocatalysis, using metal oxide semiconductor nanostructures. Despite the fact that a lot of information on specific NPs have been published, the toxicity level of many NP is still indefinable, so that NPs application is widely limited due to the lack of knowledge regarding the risk assessments, the regulatory framework and the effects on human and environmental health. Nanotechnology is gradually moving out from the experimental into the practical regime and is making its presence felt in agriculture and the food processing industry. However, there is insufficient information regarding the contributions of nanotechnology with respect to sensing and degradation of all pollutants for improved agricultural production with a conscious sustainable environmental protection.

It should be noted that NPs alter mineral nutrition, photosynthesis, oxidative stress and induce genotoxicity in crops. In addition, the activities of antioxidant enzymes increase at low NP toxicity, while decrease at higher NP toxicity in crops (Fernández-Luqueño et al., 2014; Leon-Silva et al., 2016; Rizwan et al., 2017). Additionally, Rizwan et al. (2017) found that due to exposure of crop plants to NPs, the concentration of these increases in several plant parts such as fruits and grains which may transfer to the food chain and pose a threat to human or environmental health (Leon-Silva et al., 2016). Most of the NP have both positive and negative effects on crops at physiological, morphological, biochemical and molecular levels, but the effects of NPs on crop plants vary greatly with plant species, growth stages, growth conditions, methods, dosage and duration of NPs exposure along with other factors (Fernández-Luqueño et al., 2014; 2016; Leon-Silva et al., 2016; Rizwan et al., 2017; López-Valdez et al., 2018).

The first stage for the removal of disease-causing microorganisms from food products or harmful contaminants from soil and groundwater is the effective detection of these deleterious elements. It is necessary to conduct analytical, physiological, biological and environmental studies, in order to gain further insight into the impact of the interaction NP-plant-environment. NMs are mostly being studied as tools for pest control (including defense against pests and diseases), improvement of growth and developmental processes and monitoring soil conditions. However, many of these studies are only at the laboratory or greenhouse level.

Table 5.2. Some benefits or damages of NPs into the crops that have been reported are highlighted here.

Main arguments and Findings	**References**
Although phytonanotechnology is in its infancy, it has the potential to generate: (i) new tools for smart delivery of agrochemicals, (ii) new ways to deliver particular bioactive molecules to manipulate plant breeding and genetic transformation, and (iii) new approaches for intracellular labeling and imaging.	Wang et al. (2016)
Metal and metal oxide NPs have both positive and negative effects on growth, yield and quality of important agricultural crops.	Rizwan et al. (2017)
The toxicity of AgNP is translocated from plants to other communities through the food chain and leads to the disruption of a balanced ecosystem.	Tripathi et al. (2017)
Nanoparticle toxicity occurs at multiple levels, for example, generation of oxidative stress, cytotoxicity, genotoxicity, germination rates, root and shoot growth and development.	Cox et al. (2017)
Given the lack of experimental standardization and the divergent responses, even within similar plant species, it is challenging to conclude what the effects of NMs are in plants.	Zuverza-Mena et al. (2017)
Metal NP might damage DNA and promote the cell cycle, induce the oxidative stress (promotes ROS) and lipid peroxidation. Phytotoxicity (leaf necrosis, lignin and callose development and DNA damage) and inhibition of nitrogen fixation.	Sadeghi et al. (2017)
The limited literature that does exist regarding NP into crops is mixed for most species, with both positive and negative effects being observed. The reasons for these mixed effects are numerous (different exposure scenarios, growth conditions, particle type/concentration and species, among others).	Mukherjee et al. (2016a)
Once NPs are in contact with plants, their physical and chemical properties (i.e., solubility, size, catalytic and binding properties and biotransformation) will dictate the mechanism of adsorption, uptake, transport and biotransformation, and in turn, phytotoxicity.	De la Rosa et al. (2017)
Growth of roselle (*Hibiscus sabdariffa*) under application of Fe_3O_4 nanoparticle treatment has appeared to be experimental-condition dependent.	bin Shuhaimi et al. (2019)
Spray application of B and Zn NPs at different concentrations were able to effect the qualitative and quantitative characteristics of Picual olive trees in two seasons, compared with the control treatment. Where, cultivar with nano-boron at 20 ppm + nano-zinc at 200 ppm was the best treatment to obtain maximum final fruit set, harvesting maximum fruits yield with high seed oil percentage and low acidity in both seasons.	Genaidy et al. (2020)
Foliar application of AgNPs (20, 40, and 60 mg L^{-1}) improved the growth parameters of the fenugreek plant (shoot length, number of leaves/plant and shoot dry weight) and photosynthetic pigment (chlorophyll a, chlorophyll b and carotenoids) and indole acetic acid contents, thus enhanced the yield quantity (number of pods/plant, number of seeds/pod, weight of seeds/plant and seed index) and quality (carbohydrate %, protein %, phenolics, flavonoids and tannins contents) of the yielded seeds as well as increasing antioxidant activity of the yielded seeds. The most effective treatment was 40 mg L^{-1}, where the highest increases were found.	Sadak (2019)

7. Remediation of Soils by Nanoparticles

Shooting activities are considered an important source of contamination by high concentrations of Pb in soils. The effectiveness of calcium phosphate nanoparticles (CaPNP) in the remediation of the small-arms firing range and trap shooting range soils were evaluated by Arenas-Lago et al. (2016). The extractable content of Pb, Cu and Zn was determined from soil samples treated with CaPNP, and after the treatment, it was found that the extractable contents of Cu, Pb and Zn decreased by retention of nanoparticles ($CaPO_3^+$, $CaPO_4^+$ and mainly, $CaPO_2^+$, as the most representative ions of CaPNP). The retention by nanoparticles was determined by TOF-SIMS (Time of Flight Secondary Ion Mass Spectrometry) and HR-TEM-EDS (High-Resolution Transmission Electron Microscopy with Energy Dispersive X-ray Spectroscopy). They suggested the formation of associations, compounds or aggregates of metals with CaPNP that cause the decrease in metal mobility.

According to Kumari and Singh (2016), nanotechnology offers nanoscale products with more efficient reactivity and larger surface area than its bulk phase. The potential of NPs is proposed for several applications such as to clean up petroleum hydrocarbons, pesticides and metals contaminated sites. The NPs also showed economical, eco-friendly and self-propelling attributes as major benefits compared with the conventional physicochemical methods of remediation. For removal of organic contaminants by adsorption or chemical modification, NPs can be applied directly. It was also observed that the microbial remediation of contaminants can be improved either by enhancing the microbial growth, by immobilizing the remediating agents or through an induced production of remediating microbial enzymes. NPs could also induce improved production of biosurfactants for microorganisms, contribute to enhance solubility of hydrophobic hydrocarbons and create a conducive environment for microbial degradation activity.

There are interesting studies regarding the degradation of organochlorines (OCs). These compounds are the most hazardous class of pesticides, restricted in several countries, due to their high persistence, toxicity and potential to bioaccumulation. The major sources of OCs are food industries, agriculture and sewage wastes. The advanced degradation techniques use nanomaterials from several kinds, for example, TiO_2 and Fe NPs as excellent adsorbents and efficient photocatalysts for degrading the OC, as well as, their toxic metabolites. These methods could be economic, fast and highly effective, which opens the opportunities for exploring various other nanoparticles as well (Rani et al., 2017).

The PAHs are hydrophobic organic groundwater and soil contaminants, are sorbed strongly to soils, and are difficult to remove as stated by Tungittiplakorn et al. (2004), who also reported the synthesis of engineered polymeric nanoparticles for soil remediation, using NPs that were of colloidal size of 17–97 nm, they stated that NPs (amphiphilic polyurethane) can be designed to have hydrophobic interior regions with a high affinity for phenanthrene and hydrophilic surfaces that can promote particle mobility into soil. Finally, they declared that the capability to control particle properties can offer the potential to produce different NPs optimized for several types of contaminants and soil conditions, that could be an interesting alternative for soil remediation.

According to Web of Science database, during the last few years, i.e., 2016 to 2020, some reviews have been published on nanoscale materials to remediate the environment. Additional information or discussions in this regard have not been included in this chapter, but the following reviews could be reviewed: Ezzatahmadi et al. (2017); Mahfoudhi and Boufi (2017); Lefevre et al. (2016); Li et al. (2016); Zou et al. (2016); Kamali et al., 2016; Mukherjee et al. (2016b); Malwal and Gopinath (2016); Wang et al. (2019) and Latif et al. (2020).

8. Conclusion

Agriculture is an important and stable economic sector because of its products and provides the main raw materials for food and feed, textile, pharmaceutical industries, among others, worldwide. It is well known that the findings on the consequences of environmental pollution, climate change, degradation of soils, land-use change and the decrease in biodiversity are dire, and humans beings are mainly responsible. Therefore, modern agriculture requires more knowledgeable practices that improve crop growth without negatively affecting the environment to shape the sustainable future, as resources such as water and arable land grow scarce, and at the same time, the human population continues to increase.

Nano-sciences and nanotechnologies are cutting-edge fields with the potential to revolutionize the most outstanding technological advances including industrial applications, agriculture and human and environmental care. As can be seen, nanotechnology has an important use currently. Nevertheless, these areas can lead to the degradation, destruction or imbalance in ecosystems with their unregulated release of nanomaterials posing toxic impacts on soils, plants, water, microorganisms or human beings. An additional important aspect is that results mainly come from laboratory or greenhouse experiments, are performed under controlled conditions, so it is a difficulty to predict if the responses will be the same at field conditions.

Nanoscale materials should be subjected to appropriate evaluations with regard to their safety before their widespread application. Additionally, future research should be focused on the molecular or genetic level at environmentally pertinent conditions. It should be noted that information gained from the whole genome, proteome or metabolome analyses of crops may be a powerful resource for assessing the risks of using NPs as nano fertilizers on cropping lands and shedding light on the main factors involved in the uptake, metabolism or storage of NPs. NPs could modify gene expression in plants totally, meanwhile, the potential effects in upper-level trophic communities are not well known yet. As one can see, there are a variety of NPs, plants and pollutants, that need to be researched in depth, in order to clarify the processes of absorption, translocation, accumulation and even degradation that occur under different complex scenarios. Comprehensive investigations of chronic exposure under environmentally realistic situations are also needed.

Acknowledgements

This research was founded by the projects 'Ciencia Básica SEP-CONACyT-151881', 'FONCYT-COAHUILA COAH-2019-C13-C006', and 'FONCYT-COAHUILA COAH-2021-C15-C095', by the Sustainability of Natural Resources and Energy Program (Cinvestav-Saltillo) and by Cinvestav Zacatenco. To Instituto Politécnico Nacional. G.M-P received grant-aided support from 'Becas Conacyt'. S.R. P.-R., F. F-L, R.G. C.-M., A.M-N and F. L-V, received grant-aided support from 'Sistema Nacional de Investigadores (SNI)', Mexico. To Institute of Agricultural Sciences of the Autonomous University of the State of Hidalgo (ICAP-UAEH) and Dr. Armando Peláez-Acero.

References

Adhikari, T., Kundu, S., Biswas, A.K., Tarafdar, J.C. and Rao, A.S. 2012. Effect of copper oxide nano particle on seed germination of selected crops. J. Agric. Sci. Technol. A 2: 815–823.

Anjum, N.A., Rodrigo, M.A.M., Moulick, A., Heger, Z., Kopel, P., Zítka, O. and Kizek, R. 2016. Transport phenomena of nanoparticles in plants and animals/humans. Environ. Res. 151: 233–243.

Arenas-Lago, D., Rodríguez-Seijo, A., Lago-Vila, M., Andrade Couce, L. and Vega, F.A. 2016. Using $Ca_3(PO4)_2$ nanoparticles to reduce metal mobility in shooting range soils. Sci. Total Environ. 571: 1136–1146.

Arora, S., Sharma, P., Kumar, S., Nayan, R., Khanna, P.K. and Zaidi, M.G.H. 2012. Gold-nanoparticle induced enhancement in growth and seed yield of *Brassica juncea*. Plant. Growth Regul. 66: 303–310.

Arruda, S.C.C., Silva, A.L.D., Galazzi, R.M., Azevedo, R.A. and Arruda, M.A.Z. 2015. Nanoparticles applied to plant science: A review. Talanta 131: 693–705.

Aslani, F., Bagheri, S., Julkapli, N.M., Juraimi, A.S., Hashemi, F.S.G. and Baghdadi, A. 2014. Effects of engineered nanomaterials on plants growth: an overview. Scie. World J. 2014: Article number 641759.

Asli, S. and Neumann, P.M. 2009. Colloidal suspensions of clay or titanium dioxide nanoparticles can inhibit leaf growth and transpiration via physical effects on root water transport. Plant, Cell Environ. 32: 577–584.

Bao-shan, L., Chun-Hui, L., Li-Jun, F., Shu-Chun, Q. and Min, Y. 2004. Effect of TMS (nanostructured silicon dioxide) on growth of *Changbai larch* seedlings. J. Forestry Res. 15: 138–140.

Barrena, R., Casals, E., Colón, J., Font, X., Sánchez, A. and Puntes, V. 2009. Evaluation of the ecotoxicity of model nanoparticles. Chemosphere 75: 850–857.

Baruah, S. and J. Dutta. 2009. Nanotechnology applications in pollution sensing and degradation in agriculture: A review. Environ. Chem. Lett. 7: 191–204.

Baskar, V., Venkatesh, J. and Park, S.W. 2015. Impact of biologically synthesized silver nanoparticles on the growth and physiological responses in *Brassica rapa* ssp. *pekinensis*. Environ. Sci. Pollut. Res. 22: 17672–17682.

Begum, P., Ikhtiari, R. and Fugetsu, B. 2014. Potential impact of multi-walled carbon nanotubes exposure to the seedling stage of selected plant species. Nanomaterials 4: 203–221.

bin Shuhaimi, S.AD.N., Kanakaraju, D. and Nori, H. 2019. Growth performance of roselle (*Hibiscus sabdariffa*) under application of food waste compost and Fe_3O_4 nanoparticle treatment. Int. J. Recycl. Org. Waste Agricult. 8: 299–309.

Birbaum, K., Brogioli, R., Schellenberg, M., Martinoia, E., Stark, W.J., Günther, D. and Limbach, L.K. 2010. No evidence for cerium dioxide nanoparticle translocation in maize plants. Environ. Science Technol. 44: 8718–8723.

Bo, L.I., Gongsheng, T.A.O., Yingfeng, X.I.E. and Xianlei, C.A.I. 2012. Physiological effects under the condition of spraying nano-SiO_2 onto the *Indocalamus barbatus* McClure leaves. J. Nanjing Forest. University (Nat. Sci. Edit.). 4: article number S795.

Bocking-Welch, A. 2016. Youth against hunger: service, activism and the mobilization of young humanitarians in 1960s Britain. Eur. Rev. Hist. 23: 154–170.

Boonyanitipong, P., Kumar, P., Kositsup, B., Baruah, S. and Dutta, J. 2011. Effects of zinc oxide nanoparticles on roots of rice *Oryza sativa* L. Int. Conf. Environ. Biosci. 21: 172–176.

Breustedt, G. and Qaim, M. 2012. Hunger in the world—facts, causes, recommendations. Ernährungs Umschau 59: 448–455.

Carrillo, G.R., Martínez, G.M.A. and González, C.M. 2014. Nanotecnología en la actividad agropecuaria y el ambiente. BBA. Mexico.

Corredor, E., Testillano, P.S., Coronado, M.J., González-Melendi, P., Fernández-Pacheco, R., Marquina, C. and Risueño, M.C. 2009. Nanoparticle penetration and transport in living pumpkin plants: *in situ* subcellular identification. BMC Plant Biology, 9: Article number 45.
Cox, A., Venkatachalam, P., Sahi, S. and Sharma, N. 2017. Reprint of: Silver and titanium dioxide nanoparticle toxicity in plants: A review of current research. Plant Physiol. Bioch. 110: 33–49.
De la Rosa, G., Garcia-Castaneda, C., Vazquez-Nunez, E., Alonso-Castro, A.J., Basurto-Islas, G., Mendoza, A., Cruz-Jimenez, G. and Molina, C. 2017. Physiological and biochemical response of plants to engineered NMs: Implications on future design. Plant Physiol. Bioch. 110: 226–235.
De la Rosa, G., Lopez-Moreno, M.L., de Haro, D., Botez, C.E., Peralta-Videa, J.R. and Gardea-Torresdey, J.L. 2013. Effects of ZnO nanoparticles in alfalfa, tomato, and cucumber at the germination stage: root development and X-ray absorption spectroscopy studies. Pure Appl. Chem. 85: 2161–2174.
Dimpka, C.O., McLean, J.E., Latta, D.E., Manangon, E., Britt, D.W., Johnson, W.P., Boyanov, M.I. and Anderson, A.J. 2012. CuO and ZnO nanoparticles: Phytotoxicity, metal speciation, and induction of oxidative stress in sand-grown wheat. J. Nanopart. Res. 14: 1125–1140.
Du, W., Sun, Y., Ji, R., Zhu, J., Wu, J. and Guo, H. 2011. TiO_2 and ZnO nanoparticles negatively affect wheat growth and soil enzyme activities in agricultural soil. J. Environ. Monitor. 13: 822–828.
Eichert, T. and Goldbach, H.E. 2008. Equivalent pore radii of hydrophilic foliar uptake routes in stomatous and astomatous leaf surfaces–further evidence for a stomatal pathway. Physiol. Plantarum 132: 491–502.
Etxeberria, E., Pozueta-Romero, J. and Fernández, E.B. 2012. Fluid-Phase endocytosis in plant cells pp. 107–122. *In*: Samaj J. [ed.]. Endocytosis in Plants. Springer, Berlin, Germany.
Ezzatahmadi, N., Ayoko, G.A., Millar, G.J., Speoght, R., Yan, C., Li, J.H., Li, S.Z., Zhu, J.X. and Xi, Y.F. 2017. Clay-supported nanoscale zero-valent iron composite materials for the remediation of contaminated aqueous solutions: A review. Chem. Eng. J. 312: 336–350.
Feizi, H., Moghaddam, P.R., Shahtahmassebi, N. and Fotovat, A. 2013. Assessment of concentrations of nano and bulk iron oxide particles on early growth of wheat (*Triticum aestivum* L.). Ann. Rev. Res. Biol. 3(4): 752–761.
Fernández-Luqueño, F., López-Valdez, F., Dendooven, L., Luna-Suarez, S. and Ceballos-Ramírez, J.M. 2016. Why wastewater sludge stimulates and accelerates removal of PAHs in polluted soils? Appl. Soil Ecol. 101: 1–4.
Fernández-Luqueño, F., López-Valdez, F., Valerio-Rodríguez, M.F., Pariona, N., Hernández-López, J.L., García-Ortíz, I., López-Baltazar, J., Vega-Sánchez, M.C., Espinoza-Zapata, R. and Acosta-Gallegos, J.A. 2014. Effect of nanofertilizers on plant growth and development, and their interrelationship with the environment. pp. 211–224. *In:* López-Valdez, F. and Fernández-Luqueño, F. [eds.]. Fertilizers: Components, uses in Agriculture and Environmental Impacts. Nova, New York, USA.
Genaidy, E.A.E., Abd-Alhamid, N., Hassan, H.S.A. et al. 2020. Effect of foliar application of boron trioxide and zinc oxide nanoparticles on leaves chemical composition, yield and fruit quality of *Olea europaea* L. cv. Picual. Bull. Natl. Res. Cent. 44: 106.
Giraldo, J.P., Landry, M.P., Faltermeier, S.M., McNicholas, T.P., Iverson, N.M., Boghossian, A.A. and Strano, M.S. 2014. Plant nanobionics approach to augment photosynthesis and biochemical sensing. Nat. Mater. 13(4): 400–408.
Haghighi, M., Afifipour, Z. and Mozafarian, M. 2012. The effect of N-Si on tomato seed germination under salinity levels. J. Biol. Environ. Sci. 6(16): 87–90.
Han, D. 2007. Arsenic Removal by Novel Nanoporous Adsorbents-kinetics, Equilibrium and Regenerability. Final Report, Texas A&M University, Texas, USA.
Hernandez-Viezcas, J.A., Castillo-Michel, H., Servin, A.D., Peralta-Videa, J.R. and Gardea-Torresdey, J.L. 2011. Spectroscopic verification of zinc absorption and distribution in the desert plant *Prosopis juliflora*-velutina (velvet mesquite) treated with ZnO nanoparticles. Chem. Eng. J. 170(2): 346–352.
Hong, F., Zhou, J., Liu, C., Yang, F., Wu, C., Zheng, L. and Yang, P. 2005. Effect of nano-TiO_2 on photochemical reaction of chloroplasts of spinach. Biol. Trace Elem. Res. 105(1-3): 269–279.
Hossain, K.Z., Monreal, C.M. and Sayari, A. 2008. Adsorption of urease on PE-MCM-41 and its catalytic effect on hydrolysis of urea. Colloid. Surface. B 62(1): 42–50.
Hussain, H.I., Yi, Z., Rookes, J.E., Kong, L.X. and Cahill, D.M. 2013. Mesoporous silica nanoparticles as a biomolecule delivery vehicle in plants. J. Nanopart. Res. 15(6): 1–15
Ikhtiari, R., Begum, P., Watari, F. and Fugetsu, B. 2013. Toxic effect of multiwalled carbon nanotubes on lettuce (*Lactuca sativa*). Nano Biomed. 5(1): 18–24.
Jaberzadeh, A., Moaveni, P., Moghadam, H.R.T. and Zahedi, H. 2013. Influence of bulk and nanoparticles titanium foliar application on some agronomic traits, seed gluten and starch contents of wheat subjected to water deficit stress. Not. Bot. Horti. Agrobo. 41(1): 201.
Judy, J.D., Unrine, J.M., Rao, W. and Bertsch, P.M. 2012. Bioaccumulation of gold nanomaterials by *Manduca sexta* through dietary uptake of surface contaminated plant tissue. Environ. Sci. Technol. 46(22): 12672–12678.
Ju-Nam, Y. and Lead, J.R. 2008. Manufactured nanoparticles: An overview of their chemistry, interactions and potential environmental implications. Sci. Total Environ. 400(1): 396–414.
Kaan, C. and Liese, A. 2011. Public private partnerships in global food governance: Business engagement and legitimacy in the global fight against hunger and malnutrition. Agr. Hum. Values 28(3): 385–399.
Kalteh, M., Alipour, Z.T., Ashraf, S., Aliabadi, M.M. and Nosratabadi, A.F. 2014. Effect of silica nanoparticles on basil (*Ocimum basilicum*) under salinity stress. J. Chem. Health Risks 4(3): 49–55.

Kamali, M., Gomes, A.P.D., Khodaparast, Z. and Seifi, T. 2016. Review in recents advances in environmental remediation and related toxicity of engineered nanoparticles. Environ. Eng. Manag. J. 15(4): 923–934.

Khataee, A., Movafeghi, A., Mojaver, N., Vafaei, F., Tarrahi, R. and Dadpour, M.R. 2017. Toxicity of copper oxide nanoparticles on Spirodelapolyrrhiza: assessing physiological parameters. Res. Chem. Intermediat. 43(2): 927–941.

Khodakovskaya, M., Dervishi, E., Mahmood, M., Xu, Y., Li, Z.R., Watanabe, F. and Biris, A.S. 2009. Carbon nanotubes are able to penetrate plant seed coat and dramatically affect seed germination and plant growth. ACS Nano. 3(10): 3221–3227.

Khodakovskaya, M.V., de Silva, K., Biris, A.S., Dervishi, E. and Villagarcia, H. 2012. Carbon nanotubes induce growth enhancement of tobacco cells. ACS Nano 6(3): 2128–2135.

Khot, L.R., Sankaran, S., Maja, J.M., Ehsani, R. and Schuster, E.W. 2012. Applications of nanomaterials in agricultural production and crop protection: A review. Crop Prot. 35: 64–70.

Kim, S., Lee, S. and Lee, I. 2012. Alteration of phytotoxicity and oxidant stress potential by metal oxide nanoparticles in *Cucumis sativus*, Water Air. Soil. Pollut. 223: 2799–2806.

Krishnaraj, C., Jagan, E.G., Ramachandran, R., Abirami, S.M., Mohan, N. and Kalaichelvan, P.T. 2012. Effect of biologically synthesized silver nanoparticles on *Bacopa monnieri* (Linn.) Wettst. plant growth metabolism. Process Biochem. 47(4): 651–658.

Kumari, B. and Singh, D.P. 2016. A review on multifaceted application of nanoparticles in the field of bioremediation of petroleum hydrocarbons. Ecol. Eng. 97: 98–105.

Larue, C., Khodja, H., Herlin-Boime, N., Brisset, F., Flank, A.M., Fayard, B. and Carrière, M. 2011. Investigation of titanium dioxide nanoparticles toxicity and uptake by plants. J. Phys. Conf. Ser. 304(1): Article number 012057.

Larue, C., Laurette, J., Herlin-Boime, N., Khodja, H., Fayard, B., Flank, A.M. and Carriere, M. 2012. Accumulation, translocation and impact of TiO_2 nanoparticles in wheat (*Triticum aestivum* spp.): Influence of diameter and crystal phase. Sci. Total Environ. 431: 197–208.

Latif, A., Sheng, D., Sun, K., Si, Y.B., Azeem, M., Abbas, A. and Bilal, M. 2020. Remediation of heavy metals polluted environment using Fe-based nanoparticles: Mechanisms, influencing factors, and environmental implications. Environ. Pollution 264.

Lee, C.W., Mahendra, S., Zodrow, K., Li, D., Tsai, Y.C., Braam, J. and Alvarez, P.J.J. 2010. Developmental phytotoxicity of metal oxide nanoparticles to *Arabidopsis thaliana*. Environ. Toxicol. Chem. 29(3): 669–675.

Lee, W.M., An, Y.J., Yoon, H. and Kweon, H.S. 2008. Toxicity and bioavailability of copper nanoparticles to the terrestrial plants mung bean (*Phaseolus radiatus*) and wheat (*Triticum aestivum*): Plant agar test for water-insoluble nanoparticles. Environ. Toxicol. Chem. 27(9): 1915–1921.

Lefevre, E., Bossa, N., Wiesner, M.R. and Gunsch, C.K. 2016. A review of the environmental implications of *in situ* remediation by nanoscale zero valent iron (nZVI): Behavior, transport and impacts on microbial communities. Sci. Total Environ. 565: 889–901.

Leon-Silva, S., Fernández-Luqueño, F. and Lopez-Valdez, F. 2016. Silver nanoparticles (AgNP) in the environment: A review of potential risks on human and environmental health. Water Air Soil Poll. 227(9): Article number 306.

Li, L.Y., Hu, J.W., Shi, X.D., Fan, M.Y., Luo, J. and Wei, X.H. 2016. Nanoscale zero-valent metals: A review of synthesis, characterization, and applications to environmental remediation. Environ. Sci. Pollut. Res. 23(18): 17880–17900.

Li, X., Gui, X., Rui, Y., Ji, W., Yu, Z. and Peng, S. 2014. Bt-transgenic cotton is more sensitive to CeO_2 nanoparticles than its parental non-transgenic cotton. J. Hazard. Mater. 274: 173–180.

Lin, D. and Xing, B. 2007. Phytotoxicity of nanoparticles: Inhibition of seed germination and root growth. Environ. Pollut. 150(2): 243–250.

Lin, D. and Xing, B. 2008. Root uptake and phytotoxicity of ZnO nanoparticles. Environ. Sci. Technol. 42(15): 5580–5585.

Lin, S., Reppert, J., Hu, Q., Hudson, J.S., Reid, M.L., Ratnikova, T.A. and Ke, P.C. 2009. Uptake, translocation, and transmission of carbon nanomaterials in rice plants. Small 5(10): 1128–1132.

Liu, F., Wen, L.X., Li, Z.Z., Yu, W., Sun, H.Y. and Chen, J.F. 2006. Porous hollow silica nanoparticles as controlled delivery system for water-soluble pesticide. Mater. Res. Bull. 41(12): 2268–2275.

Liu, Q., Chen, B., Wang, Q., Shi, X., Xiao, Z., Lin, J. and Fang, X. 2009. Carbon nanotubes as molecular transporters for walled plant cells. Nano Letters. 9(3): 1007–1010.

Liu, R. and Lal, R. 2014. Synthetic apatite nanoparticles as a phosphorus fertilizer for soybean (*Glycine max*). Sci. Rep. 4: Article number 5686.

Liu, X., Feng, Z., Zhang, S., Zhang, J., Xiao, Q. and Wang, Y. 2006. Preparation and testing of cementing nano-subnano composites of slow or controlled release of fertilizers. Sci. Agr. Sin. 39: 1598–1604.

López-Moreno, M.L., Lugo-Avilés, L., Guzmán-Pérez, N., Álamo-Irizarry, B., Perales, O., Cedeno-Mattei, Y. and Román, F. 2016. Effect of cobalt ferrite ($CoFe_2O_4$) nanoparticles on the growth and development of *Lycopersicon lycopersicum* (tomato plants). Sci. Total Environ. 550: 45–52.

López-Valdez, F., Miranda-Arámbula, M., Ríos-Cortés, A.M., Fernández-Luqueño, F. and de-la-Luz, V. 2018. Nanofertilizers and their controlled delivery of nutrients. pp. 35–48. *In*: López-Valdez, F. and Fernández-Luqueño, F. (eds.). Agricultural Nanobiotechnology, Modern Agriculture for a Sustainable Future. Springer Nature Switzerland AG. Cham, Switzerland.

Lu, C., Zhang, C., Wen, J., Wu, G. and Tao, M. 2002. Research of the effect of nanometer materials on germination and growth enhancement of *Glycine max* and its mechanism. Soybean Sci. 21(3): 168–171.

Ma, X., Geiser-Lee, J., Deng, Y. and Kolmakov, A. 2010. Interactions between engineered nanoparticles (NP) and plants: Phytotoxicity, uptake and accumulation. Sci. Total Environ. 408(16): 3053–3061.

Ma, X., Wang, Q., Rossi, L., Ebbs, S.D. and White, J.C. 2016. Multigenerational exposure to cerium oxide nanoparticles: Physiological and biochemical analysis reveals transmissible changes in rapid cycling *Brassica rapa*. NanoImpact. 1: 46–54.
Mahfoudhi, N. and Boufi, S. 2017. Nanocellulose as a novel nanostructured adsorbent for environmental remediation: A review. Cellilose 24(3): 1171–1197.
Malwal, D. and Gopinath, P. 2016. Fabrication and applications of ceramic nanofibers in water remediation: A review. Crit. Rev. Env. Sci. Tec. 46(5): 500–634.
Marchiol, L., Mattiello, A., Pošćić, F., Fellet, G., Zavalloni, C., Carlino, E. and Musetti, R. 2016. Changes in physiological and agronomical parameters of barley (*Hordeum vulgare*) exposed to cerium and titanium dioxide nanoparticles. Int. J. Env. Res. Pub. He. 13(3): Article number E332.
Martínez-Fernández, D. and Komárek, M. 2016. Comparative effects of nanoscale zero-valent iron (nZVI) and Fe_2O_3 nanoparticles on root hydraulic conductivity of *Solanum lycopersicum* L. Environ. Exp. Bot. 131: 128–136.
Mishra, V., Mishra, R.K., Dikshit, A. and Pandey, A.C. 2014. Interactions of nanoparticles with plants: An emerging prospective in the agriculture industry. pp. 159–180. *In*: Ahmad P. and Rasool, S. (eds.). Emerging Technologies and Management of Crop Stress Tolerance. Elsevier and Academic Press. CA, USA.
Monica, R.C. and Cremonini, R. 2009. Nanoparticles and higher plants. Caryologia 62(2): 161–165.
Mukherjee, A., Majumdar, S., Servin, A.D., Pagano, L., Dhankher, O.P. and White, J.C. 2016a. Carbon nanomaterials in agriculture: A critical review. Front. Plant. Sci. 7: Article number 172.
Mukherjee, R., Kumar, R., Sinha, A., Lama, Y. and Saha, A.K. 2016b. A review on synthesis, characterization, and applications of nano zero valent iron (nZVI) for environmental remediation. Crit. Rev. Env. Sci, Tec. 46(5): 443–466.
Nair, P.M.G. and Chung, I.M. 2014a. Physiological and molecular level effects of silver nanoparticles exposure in rice (*Oryza sativa* L.) seedlings. Chemosphere 112: 105–113.
Nair, P.M.G. and Chung, I.M. 2014b. A mechanistic study on the toxic effect of copper oxide nanoparticles in soybean (*Glycine max* L.) root development and lignification of root cells. Biol. Trace Elem. Res. 162: 342–352.
Navarro, E., Wagner, B., Odzak, N., Sigg, L. and Behra, R. 2015. Effects of differently coated silver nanoparticles on the photosynthesis of *Chlamydomonas reinhardtii*. Environ. Sci. Technol. 49(13): 8041–8047.
Nel, A., Xia, T., Mädler, L. and Li, N. 2006. Toxic potential of materials at the nano level. Science 311(5761): 622–627.
Nixon, K. 1998. *Quercus macdougallii*. The IUCN red list of threatened species. http://www.iucnredlist.org/.
Nowack, B. and Bucheli, T.D. 2007. Occurrence, behavior and effects of nanoparticles in the environment. Environ. Pollut. 150(1): 5–22.
Pariona, N., Martínez, A.I., Hernandez-Flores, H. and Clark-Tapia, R. 2017. Effect of magnetite nanoparticles on the germination and early growth of *Quercus macdougallii*. Sci. Total Environ. 575: 869–875.
Proseus, T.E. and Boyer, J.S. 2005. Turgor pressure moves polysaccharides into growing cell walls of *Chara corallina*. Ann. Bot-London 95(6): 967–979.
Radimer, K.L., Olson, C.M. and Campbell, C.C. 2016. Development of indicators to assess hunger. J. Nutr. 11: 1544–1548.
Raliya, R., Franke, C., Chavalmane, S., Nair, R., Reed, N. and Biswas, P. 2016. Quantitative understanding of nanoparticle uptake in watermelon plants. Front. Plant Sci. 7: Article number 1288.
Ramesh, M., Palanisamy, K., Babu, K. and Sharma, N.K. 2014. Effects of bulk & nano-titanium dioxide and zinc oxide on physio-morphological changes in *Triticum aestivum* Linn. J. Glob. Biosci. 3: 415–422.
Rani, M., Shanker, U. and Jassal, V. 2017. Recent strategies for removal and degradation of persistent & toxic organochlorine pesticides using nanoparticles: A review. J. Environ. Manage. 190: 208–222.
Rico, C.M., Barrios, A.C., Tan, W., Rubenecia, R., Lee, S.C., Varela-Ramirez, A.M. and Gardea-Torresdey, J.L. 2015. Physiological and biochemical response of soil-grown barley (*Hordeum vulgare* L.) to cerium oxide nanoparticles. Environ. Sci. Pollut. R 22(14): 10551–10558.
Rizwan, M., Ali, S., Qayyum, M.F., Ok, Y.S., Adrees, M., Ibrahim, M., Zia-ur-Rehmand, M., Faris, M. and F. Abbas. 2017. Effect of metal and metal oxide nanoparticles on growth and physiology of globally important food crops: A critical review. J. Hazard. Mater. 322: 2–16.
Sadak, M.S. 2019. Impact of silver nanoparticles on plant growth, some biochemical aspects, and yield of fenugreek plant (*Trigonella foenum-graecum*). Bull. Natl. Res. Cent. 43: 38.
Sadeghi, R., Rodriguez, R.J., Yao, Y., Kokini, J.L., Doyke, M.P. and Klaenhammer, T.R. 2017. Advances in nanotechnology as they pertain to food and agriculture: Benefits and risks. Annu. Rev. Food Sci. T. 8: 467–492.
Sahayaraj, K., Madasamy, M. and Radhika, S.A. 2016. Insecticidal activity of bio-silver and gold nanoparticles against *Pericallia ricini* Fab. (Lepidaptera: Archidae). J. Biopest. 9(1): 63–72.
Salama, H.M.H. 2012. Effects of silver nanoparticles in some crop plants, common bean (*Phaseolus vulgaris* L.) and corn (*Zea mays* L.). Int. Res. J. Biotech. 3(10): 190–197.
Santos, A.R., Miguel, A.S., Tomaz, L., Malhó, R., Maycock, C., Patto, M.C.V. and Oliva, A. 2010. The impact of CdSe/ZnS quantum dots in cells of *Medicago sativa* in suspension culture. J. Nanobiotechnol. 8: Article number 24.
Savithramma, N., Ankanna, S. and Bhumi, G. 2012. Effect of nanoparticles on seed germination and seedling growth of *Boswellia ovalifoliolata* an endemic and endangered medicinal tree taxon. Nano Vision 2: 61–68.
Schwabe, F., Schulin, R., Limbach, L.K., Stark, W., Bürge, D. and Nowack, B. 2013. Influence of two types of organic matter on interaction of CeO_2 nanoparticles with plants in hydroponic culture. Chemosphere 91: 512–520.
Shah, V. and Belozerova, I. 2009. Influence of metal nanoparticles on the soil microbial community and germination of lettuce seeds. Water Air Soil Poll. 197(1-4): 143–148.

Shaw, A.K. and Hossain, Z. 2013. Impact of nano-CuO stress on rice (*Oryza sativa* L.) seedlings. Chemosphere 93(6): 906–915.

Shiny, P.J., Mukerjee, A. and Chandrasekaran, N. 2013. Comparative assessment of the phytotoxicity of silver and platinum nanoparticles. Adv. Nano. Emerg. Eng. Technol. ICANMEET-2013: 391–393.

Siddiqui, M.H., Al-Whaibi, M.H., Firoz, M. and Al-Khaishany, M.Y. 2015. Role of nanoparticles in plants. pp. 19–35. *In*: Siddiqui, M.H., Al-Whaibi, M.H. and Firoz, M. (eds.). Nanotechnology and Plant Sciences. Springer. Switzerland.

Smirnova, E., Gusev, A., Zaytseva, O., Sheina, O., Tkachev, A., Kuznetsova, E. and Kirpichnikov, M. 2012. Uptake and accumulation of multiwalled carbon nanotubes change the morphometric and biochemical characteristics of *Onobrychis arenaria* seedlings. Front. Chem. Sci. Eng. 6(2): 132–138.

Song, U., Jun, H., Waldman, B., Roh, J., Kim, Y., Yi, J. and Lee, E.J. 2013. Functional analyses of nanoparticle toxicity: A comparative study of the effects of TiO_2 and Ag on tomatoes (*Lycopersicon esculentum*). Ecotox. Environ. Safe. 93: 60–67.

Stampoulis, D., Sinha, S.K. and White, J.C. 2009. Assay-dependent phytotoxicity of nanoparticles to plants. Environ. Sci. Technol. 43(24): 9473–9479.

Stoimenov, P.K., Klinger, R.L., Marchin, G.L. and Klabunde, K.J. 2002. Metal oxide nanoparticles as bactericidal agents. Langmuir 18: 6679–686.

Suriyaprabha, R., Karunakaran, G., Yuvakkumar, R., Prabu, P., Rajendran, V. and Kannan, N. 2012. Growth and physiological responses of maize (*Zea mays* L.) to porous silica nanoparticles in soil. J. Nanopart. Res. 14(12): Article number 1294.

Syu, Y.Y., Hung, J.H., Chen, J.C. and Chuang, H.W. 2014. Impacts of size and shape of silver nanoparticles on Arabidopsis plant growth and gene expression. Plant Physiol. Bioch. 83: 57–64.

Szymańska, R., Kołodziej, K., Ślesak, I., Zimak-Piekarczyk, P., Orzechowska, A., Gabruk, M. and Kruk, J. 2016. Titanium dioxide nanoparticles (100–1000 mg/l) can affect vitamin E response in *Arabidopsis thaliana*. Environ. Pollut. 213: 957–965.

Tiwari, D.K., Dasgupta-Schubert, N., Villaseñor-Cendejas, L.M., Villegas, J., Carreto-Montoya, L. and Borjas-García, S.E. 2014. Interfacing carbon nanotubes (CNT) with plants: Enhancement of growth, water and ionic nutrient uptake in maize (*Zea mays*) and implications for nanoagriculture. Appl. Nanosci. 4: 577–591.

Torney, F., Trewyn, B.G., Lin, V.S.Y. and Wang, K. 2007. Mesoporous silica nanoparticles deliver DNA and chemicals into plants. Nat. Nanotechnol. 2(5): 295–300.

Tripathi, D.K., Tripathi, A., Shweta, Singh, S., Singh, Y., Vishwakarma, K., Yadav, G., Sharma, S., Singh, V.K., Mishra, R.K., Upadhyay, R.G., Upadhyay, N.K., Lee, Y. and Chauhan, D.K. 2017. Uptake, accumulation and toxicity of silver nanoparticle in autotrophic plants, and heterotrophic microbes: A concentric review. Front. Microbiol. 8: Article number 7.

Tripathi, S. and Sarkar, S. 2015. Influence of water soluble carbon dots on the growth of wheat plant. Appl. Nanosci. 5(5): 609–616.

Tungittiplakorn, W., Lion, L.W., Cohen, C. and Kim, J.Y. 2004. Engineered polymeric nanoparticles for soil remediation. Environ. Sci. Technol. 38(5): 1605–1610.

Uzu, G., Sobanska, S., Sarret, G., Munoz, M. and Dumat, C. 2010. Foliar lead uptake by lettuce exposed to atmospheric fallouts. Environ. Sci. Technol. 44(3): 1036–1042.

Vannini, C., Domingo, G., Onelli, E., De Mattia, F., Bruni, I., Marsoni, M. and Bracale, M. 2014. Phytotoxic and genotoxic effects of silver nanoparticles exposure on germinating wheat seedlings. J. Plant Physiol. 171(13): 1142–1148.

Wang, A., Zheng, Y. and Peng, F. 2014. Thickness-controllable silica coating of CdTe QDs by reverse microemulsion method for the application in the growth of rice. J. Spectrosc. 2014: Article number 169245.

Wang, P., Lombi, E., Zhao, F.J. and Kopittke, P.M. 2016. Nanotechnology: A new opportunity in plant sciences. Trends Plant Sci. 21(6): 699–712.

Wang, P., Menzies, N.W., Lombi, E., McKenna, B.A., Johannessen, B., Glover, C.J. and Kopittke, P.M. 2013. Fate of ZnO nanoparticles in soils and cowpea (*Vigna unguiculata*). Environ. Sci. Technol. 47(23): 13822–13830.

Wang, S., Kurepa, J. and Smalle, J.A. 2011. Ultra-small TiO_2 nanoparticles disrupt microtubular networks in Arabidopsis thaliana. Plant. Cell Environ. 34(5): 811–820.

Wang, Y.N., O'Connor, D., Shen, Z.T., Lo, I.M.C., Tsang, D.C.W., Pehkonen, S., Pu, S.Y. and Hou, D.Y. 2019. Green synthesis of nanoparticles for the remediation of contaminated waters and soils: Constituents, synthesizing methods, and influencing factors. J. Clean. Prod. 226: 540–549.

Wang, Z., Xie, X., Zhao, J., Liu, X., Feng, W., White, J.C. and Xing, B. 2012. Xylem-and phloem-based transport of CuO nanoparticles in maize (*Zea mays* L.). Environ. Sci. Technol. 46(8): 4434–4441.

Watanabe, T., Misawa, S., Hiradate, S. and Osaki, M. 2008. Root mucilage enhances aluminum accumulation in *Melastoma malabathricum*, an aluminum accumulator. Plant Signal. Behave. 3(8): 603–605.

Willmer, C. and Fricker, M. 1996. Introduction. pp. 1–11. *In:* Willmer, C. and Fricker, M. [eds.]. Stomata. Springer, New York, USA.

Wu, S.G., Huang, L., Head, J., Chen, D., Kong, I.C. and Tang, Y.J. 2013. Phytotoxicity of metal oxide nanoparticles is related to both dissolved metals ions and adsorption of particles on seed surfaces. J. Petrol. Environ. Biotech. 3(4): Article number 1000126.

Yang, F., Hong, F., You, W., Liu, C., Gao, F., Wu, C. and Yang, P. 2006. Influence of nano-anatase TiO_2 on the nitrogen metabolism of growing spinach. Biol. Trace Elem. Res. 110(2): 179–190.

Yang, L. and Watts, D.J. 2005. Particle surface characteristics may play an important role in phytotoxicity of alumina nanoparticles. Toxicol. Lett. 158(2): 122–132.
Yin, L., Colman, B.P., McGill, B.M., Wright, J.P. and Bernhardt, E.S. 2012. Effects of silver nanoparticle exposure on germination and early growth of eleven wetland plants. Plos One 7: 1–7.
Yinfeng, X., Bo, L., Qianqian, Z., Chunxia, Z., Kouping, L. and Gongsheng, T. 2011. Effects of nano-TiO_2 on photosynthetic characteristics of Indocalamus barbatus. J. Northeast Forest. U. 3: Article number S795.
Zheng, L., Hong, F., Lu, S. and Liu, C. 2005. Effect of nano-TiO_2 on strength of naturally aged seeds and growth of spinach. Biol. Trace Elem. Res. 104(1): 83–91.
Zhou, D., Jin, S., Li, L., Wang, Y. and Weng, N. 2011. Quantifying the adsorption and uptake of CuO nanoparticles by wheat root based on chemical extractions. J. Environ. Sci-China. 23(11): 1852–1857.
Zhu, H., Han, J., Xiao, J.Q. and Jin, Y. 2008. Uptake, translocation, and accumulation of manufactured iron oxide nanoparticles by pumpkin plants. J. Environ. Monitor. 10(6): 713–717.
Zou, Y.D., Wang, X.X., Khan, A., Wang, P.Y., Liu, Y.H., Alsaedi, A., Hayat, T. and Wang, X.K. 2016. Environmental remediation and application of nanoscale zero-valent iron and its composites for the removal of heavy metal ions: A review. Environ. Sci. Technol. 50(14): 7290–7304.
Zuverza-Mena, N., Martinez-Fernandez, D., Du, W.C., Hernandez-Viezcas, J.A., Bonilla-Bird, N., Lopez-Moreno, M.L., Komarek, M., Peralta-Videa, J.R. and Gardea-Torresdey, J.L. 2017. Exposure of engineered nanomaterials to plants: Insights into the physiological and biochemical responses—A review. Plant. Physiol. Biochem. 110: 236–264.

SECTION 2

CHAPTER 6

Genetically Modified Plants for Phytoremediation Biotechnological Microbial Remediants

Enhanced Pedospheric Detoxification Accomplished via Transgenic Plants

Shaan Bibi Jaffri and *Khuram Shahzad Ahmad**

1. Introduction

Human beings have been trying to ease their lives by making many efforts. The access to food and energy are the integral factors for human survival. In fulfilling these demands, human beings have developed and adopted revolutionary mechanisms like agrochemicals for boosting agricultural crops. Additionally, the use of finite resources of fossil fuels is marked with non-sustainability. In this regard, the damage done to ecological sections by the introduction of synthetic chemicals have often been neglected until it began to appear significantly in the recent decades. Uncontrolled urbanized and industrialized approaches in the present era have introduced certain chemicals, e.g., pharmaceutical products, Polycyclic Aromatic Hydrocarbons (PAHs), explosives, petroleum hydrocarbons, pesticides, dyestuffs and halogenated hydrocarbons, etc. (Tahir et al., 2019a; Tahir et al., 2020; Ijaz et al., 2020) in different ecospheric zones. As a result of rushed development in the industrial sector, there has been increasing metal pollution since metallic components are largely needed in manufacturing jobs. Heavy metals possess large amount of toxicity even at lower concentrations (Naeem et al., 2020). Consequently, the environment as a whole becomes polluted due to the transboundary nature of such contaminants. Environmental deterioration is more pronounced because the contamination of one area is easily transferred to the other. For instance, pesticides released into the pedospheric zone do not remain there forever, but are transferred to atmospheric and hydrospheric zones through volatilization and leaching, respectively. In the past decades, the scientific community has paid special attention to

Department of Environmental Sciences, Fatima Jinnah Women University, The Mall, 46000, Rawalpindi, Pakistan.
Email: shaan.jaffri@outlook.com
*Corresponding author: chemist.phd33@yahoo.com; dr.k.s.ahmad@fjwu.edu.pk

the solution of this globally ecological pollution event (Tahir et al., 2019b; Ijaz et al., 2020). Efforts have been made for the quantification of such contamination and the pollution-related phenomenon which are inclusive of the development of certain technologies and novel materials tested for their efficiency in degradation of such pollutants (Jaffri et al., 2020; Jaffri and Ahmad, 2018).

The pedospheric zone has been especially exposed to the reception of various synthetic chemicals from anthropogenic activities in addition to the presence of inherently present contaminants received as a result of various implicit activities taking part on a geological scale under soils. Such chemicals are not only harmful for the health of the soil, but they have also been found to adversely impact life of biotic forms found in such pedospheric studies. Though a fixed quantity of such chemicals, e.g., chlorides, nitrates, sulfates and heavy metals are required for the normal growth and functioning of plants, but their presence beyond permissible limits can prove lethal to plant health in addition to contamination of the pedospheric area. A large amount of toxicity and persistence is associated with some organic pollutants which makes them difficult to remediate. Furthermore, they are also not easily soluble, as a result get accumulated in soils and sediments posing a serious threat (Jaffri and Ahmad, 2020a). Therefore, a very accurate approach is required for the release of such chemicals into soils because their remediation is marked by complex and unsatisfactory results. Remediation of soils has been primarily done through various physical and chemical approaches, e.g., filtration using membranous materials, reverse osmotic procedures, etc. (Shazia et al., 2018; Iram et al., 2018; Iftikhar et al., 2019). Nevertheless, the cost and complexity associated with such procedures in contaminated areas makes their adoptability a questionable matter in a practical sense. Therefore, such a case is indicative of the urgency in the development of cost effective, sustainable, cleaner and facile modes of remediation which not only detoxify soils but are also in conformity with the principles of green chemistry to retain soil quality. Utilization of biological agents for remediation of pedospheric zones is often broadly referred to as bioremediation. In a more specific manner, certain plants are used for detoxification of soils from heavy metals and other pollutants. Research is ongoing for exploration of plants which have greater potential of hyper-accumulation (Ishtiaq et al., 2020; Ahmad et al., 2020).

Phytoremediation is a well-known biotechnological method used for environmental decontamination and pollutants detoxification by the use of plants specialized for taking contaminants including heavy metals. Phytoremediation is operational in a dual mode, i.e., *in situ* and *ex situ*. The process of phytoremediation is preferably done at the location of the contamination in an *in situ* mode or sometimes the samples containing possible waste products are shifted to laboratory-based settings and then treated and tested for results in an *ex situ* mode. Phytoremediation is an emerging technology and further researches are being conducted it make it economically viable (Correa-García et al., 2021). Phyto-technological modes of remediation aim at the remediation, treatment, stabilization or controlling of substrates. Phytoremediation is specifically aimed at the elimination and extermination of contaminants. Phytoremediation is mainly based on the principle of exploitation of the natural process of plant physiology (Iram et al., 2020; Jaffri and Ahmad 2020). Other remediation technologies like microorganism-based bioremediation is known for effective results but the complexity associated with it needs to be explored further before adoption on a practical scale (Villanueva-Galindo and Moreno-Andrade, 2020). However, the already commendable inherent physiological mechanism of plants in taking up pollutants and heavy metals can be further improved by changing it by the introduction of effective genes from microbial species and other plants. Such plants, usually referred to as transgenic plants have been known for their effectiveness in remediation of contaminated sites by the virtue of their genetic makeup being transformed. A great deal of research has been done in this regard and is an emerging domain, where the scientific community is trying to develop plants with the most effective results for the detoxification of ecological areas, specifically soils from heavy metals and other types of contaminants (Antoniadis et al., 2021).

Environmental remediation with transgenic plants has received special attention from environmentalists and soil scientists, since this concept is in line with the goals set for sustainability.

To date, a number of research and review articles, book chapters and reports have been compiled on the use of transgenic plants in environmental remediation. However, this chapter is specifically aimed to describe the most recent advancements done in this field by elaboration of the basic concept of phytoremediation and its influential role in environmental decontamination; the greener aspect associated with transgenic plants-driven phytoremediation; introduction of microbial genes in plants for enhancement of their phytoremediation potential; and the future perspectives of this field have also been proposed.

2. Methods

2.1 Searching approach

The current work is explanatory of the phytoremediation potential of transgenic plants. Data collection was done in a meticulous manner by an assortment restricted to specific articles obtained from various electronic research focused databases, i.e., Google scholar, Embase, Web of Science, ProQuest, PubMed, Cochrane and Science direct, etc. The articles selected for compiling the present chapter included research articles, reviews, book chapters, perspectives, reports from various well reputed publishers like Springer, Springer Nature, The Institute of Electrical and Electronics Engineers, Elsevier, American Chemical Society, Jstore, Taylor and Francis, Royal Society of Chemistry etc. The pattern of investigation adopted is based on the analysis of the content in the selected research items. While publications produced during 2000–2020 were taken in consideration for the purpose of the analysis, those published between 2015–2020 were mainly emphasized. The approach for searching was adopted and varied selected search engines were chosen as: ["environmental pollution", "soil contamination", "phytoremediation", "transgenic plants", "cleaner remediation", "heavy metal tolerance", "hyper-accumulators", "heavy metal stress", "Arbuscular Mycorrhizae", "PGPR", "PGPB"].

3. Phytoremediation—An Eco-sustainable Detoxification Approach

Phytoremediation is representative of phyto-technology which aims at the remediation of contaminated lithospheric zones, sludges and sediments in addition to hydrospheric swatches from a myriad of contaminants inclusive of organic and inorganic ones through the use of specific plants (Terry and Banuelos, 2020). Since inception, plants have undergone an evolutionary process and acquired certain genetic adaptions and succeeded in handling the accumulated contaminants and pollutants occurring in their immediate ecological zones. Plants not only use such pollutants for their growth and maturity and can be harvested from such contaminated sites, but they can also be used as a passive remediation mode for cleaning up such sites having shallower or lower to moderate levels of environmental contamination (Jaffri and Ahmad, 2019). Specifically, phytoremediation is aimed at the cleaning up of metallic contaminations, i.e., metals and metalloids, leachates, pesticides and other agrochemical products, landfill materials, Poly-Aromatic Hydrocarbons (PAHs), solvents and explosives, etc. Due to the potential seen in phytoremediation, it has been largely investigated in different research-based dimensions and smaller scale demonstrations, however its adoption on a complete scale is as yet limited considering the optimization of parameters. Further investigation and development in this field will enhance the future prospects and acceptance of this technology (Moradi et al., 2021; Viana et al., 2021). An interesting factor associated with phytoremediation is that it is not only effective in the removal of the heavy metals from contaminated zones, but is also marked for its ability of removing the salt contents found in any polluted environment. Such a method is often referred to as salt phytoremediation. While plants have the potential to survive in metal-contaminated regions, only a few plants are naturally able to survive inside a region containing a higher extent of salinity. Halophytes, salt-loving plants can be used for phytoremediation and are often categorized in three classes of salt-excluders, salt-inducers and salt-accumulators (Tıpırdamaz et al., 2020; Turcios et al., 2021).

For normal growth and functioning of plants, various kinds of macronutrients and micronutrients are required. Such metals, e.g., copper (Cu), zinc (Zn), nickel (Ni) and iron (Fe) are present ii n soils and taken up by plants. These metals can enter plants by means of the plant root system with the use of various active mechanisms, i.e., conveyance of proteins inside the cellular membrane in addition to other passive mechanisms, e.g., transpiration. In a conventional sense, normal plants not having hyper- accumulator mechanisms are capable of taking up less 10 mg/L of different micronutrients, which is marked with some sufficiency in terms of fulfillment of their metabolic needs. Nevertheless, the fact of distorted growth and maturation of any plant inside soils having heavy metals exceeding the normal limit cannot be overlooked (Inobeme, 2021). The presence of different pollutants and heavy metals inside pedospheric matrices can be a source of phyto-toxicity if the permissible limits are being compromised. Plants having a phytoremediation potential, often referred to as metallophytes can be used for cleanup of such sites (Salmani- Ghabeshi et al., 2021). Different types of metallophytes also vary from one another in their functionality and thus, they are categorized as metal indicators, metal excluders and metal hyper-accumulators (Ievinsh et al., 2020). Metal indicators are usually those plants which take up higher concentrations of heavy metals from the pedospheric matrices and collect them inside their shoots and foliar regions. The extent of heavy metals taken up by such plants are quantified from shoots and leaves and in turn this is indicative of the concentration of heavy metals in soils. Plants not specialized in hyper-accumulation can be protected by such heavy metal stress by using metal indicator species. Metal excluders are known for the prevention of the locomotion of the heavy metals absorbed earlier from invading aboveground tissues, instead they tend to gather such up-take of heavy metals inside their roots and perform the function of phyto-stabilization (Chauhan and Kulshreshtha, 2021). Another class of metallophytes are metal hyper-accumulators which tend to take up and gather higher concentrations of the heavy metals inside their foliage regions (Romeroso et al., 2021). Strictly, only plants having a good potential of accumulation of heavy metals without signs of toxicity fall in this class. Accumulated heavy metals are used for the operation of various functions, e.g., ecological and physiological. Another interesting feature associated with metal hyper-accumulators is their defensive mechanism for themselves and other plants around them. They make the heavy metal laden leaves unpalatable for different consumers like herbivores and caterpillars. In this manner, metal hyper-accumulators stop the entrance of the heavy metals in higher profiles of the food chain (Calabrese and Agathokleous, 2021).

4. Categorization of Phytoremediation Processes

Phytoremediation has often been preferred due to its lower economical cost and is triggered by solar energy. Depending upon the mechanism of heavy metals and other contaminants uptake and cleanup of the environment, phytoremediation has been categorized in the various types as described below. Research in this regard is emerging daily in comparison to the traditional phytoremediation, i.e., heavy metals elimination efficiency and higher extent of the biomass production. Some of the recent advances accomplished in this field have been described in Table 6.1 and Fig. 6.1.

Phyto-stimulation: Phyto-stimulation is another phyto-technology based on the stimulation of different enzymatic substances inside the rhizospheric zone resulting in the biological remediation of the contaminants through the use of the microbial species in the form of bacteria, fungi, algae, etc., by the ejection of the exudates (Savani et al., 2021).

Phyto-extraction or phyto-accumulation: Phyto-extraction or phyto-accumulation is a type of phyto-technology used for the elimination of pollutants by the use of such plants which have the potential of accumulatining of certain pollutants inside pedospheric zones. Phyto-remediant plants store such pollutants inside their shoots and make them available on harvesting.

Table 6.1. Varied types of phytoremediation used environmental detoxification: an account of the recent studies in the relevant phytoremediation types.

Type	Plant Studied	Important Finding	References
Phytoextraction	*Bryophyllum laetivirens*	Pedsopheric remediation was successfully accomplished with *Bryophyllum laetivirens* by detoxifying the heavy metals obtained from municipal sludge. Since the sludge soils are abundant in heavy metal contents in comparison to common soils, plants grown in such contaminated soils exceeded 2–11 times in heavy metal content and expressed the promotion of the heavy metals remediation through phytoextraction.	Li et al., 2020a
	Zea mays	The synergistic impacts of the biological agents, i.e., maize (*Zea mays* L.) and chemical agents having an eco-friendly and effective biodegradation manner, i.e., methylglycinediacetic acid (MGDA) and N,N-Bis (carboxymethyl)-L-glutamic acid (GLDA) were studies in comparison to the synthetic ethylenediaminetetraacetic acid (EDTA). Results were indicative of the increased potential of the phytoextraction of Zn and Pb from contaminated calcareous pedospheric regions by the phyto-remediant *Zea mays* L. in association with the environmental friendly chelants.	Masoudi et al., 2020
	Arabidopsis	Experimental results of the wild-type *Arabidopsis* and the one inoculated with the abscisic acid (ABA)-catabolizing bacteria, i.e., *Rhodococcus qingshengii* expressed that microbial inoculation did not considerably enhanced wild-type *Arabidopsis*'s phytoextraction potential.	Lu et al., 2020
	Brassica napus and *Glycine max*	By adoption of the two variable phytoextraction methods, heavy metals dissolution was accomplished in waste biomass and an effective metal recovery was accomplished. The electrochemical removal efficiencies obtained for Cd, Pb and Zn were 80%, 94% and 68% in case of soybean biomass in a respective manner. While they were 97%, 99% and 46% for Cd, Pb and Zn in case of canola biomass, respectively.	Delil et al., 2020
	Corchorus capsularis	*Corchorus capsularis* expressed a profound level of tolerance in response to Cu-stress aided by the addition of the chelating agents which enhanced the growth and biomass production in Cu stressed *Corchorus capsularis* plants. In a comparative study, EDTA expressed an impressive performance over citric acid. Furthermore, transmission electron microscopy (TEM) was indicative of the destruction of the ultra-structure of the chloroplast due to Cu toxicity.	Saleem et al., 2020
	Lolium perenne	Results of a study done on a perennial ryegrass, *Lolium perenne* inoculated with Pb-resistant microbes expressed the enhancement of Pb extraction from the pedospheric region, leading to an effective and sustainable lithospheric remediation.	Sun et al., 2020
	Arabidopsis thaliana	Transgenic *Arabidopsis* plant generated from the genes of *Azospirillum brasilense* Sp245 was found to augment enzyme Target of Rapamycin (TOR) kinase expression in the target *Arabidopsis*. Furthermore, the development of the rhizobacterial-root association lead to the promotion of the S6K phosphorylation. In a particular manner, TOR signaling and phytostimulation were highly influenced by the auxins produced from bacteria.	Méndez-Gómez et al., 2020
	Arachis hypogaea	*Pseudomonas* spp. EGN 1 expressed a remarkable role in the bio-stimulation of stem rot present in *Arachis hypogaea* in addition to the acting as a biocontrol agent.	Archana et al., 2020

Table 6.1 contd. ...

...Table 6.1 contd.

Type	Plant Studied	Important Finding	References
Phytoextraction	*Scirpus grossus* and *Thypa angustifolia*	Transgenic *Scirpus grossus* and *Thypa angustifolia* showed an auspicious role in the hyper-accumulation and bioaccumulation of Al reaching up to 5.308 and 3.068 in a respective manner.	Purwanti et al., 2020
	Arabidopsis	Inoculation of *Arabidopsis* with bacterial enzymes modulated the genes responsive towards auxins in roots. Such enzymatic products are inclusive of the cyclodipeptides cyclo(L-Pro-L-Val), cyclo(L-Pro-L-Phe), and cyclo(L-Pro-L-Tyr) responsible for profound phytostimulation.	Ortiz-Castro et al., 2020
	Arundo donax L.	*Arundo donax* L. inoculated with bacterial species succeeded in taking up and volatilization of the As(III) and removed an efficiency of 50%. Such performance is attributable to the involvement of DNA methylation mechanism in response to As stress.	Guarino et al., 2020
	Eichhornia crassipes and *Lemna valdiviana*	The results of kinetics applied on the removal of As *Eichhornia crassipes* and *Lemna valdiviana* by towards the first-order reaction kinetics and there was a subsequent alleviation in the decay coefficient (k) with an augmentation in the nutrient's initial concentration.	de Souza et al., 2018
	Eucalyptus sideroxylon	Different parts of *Eucalyptus sideroxylon*, e.g.., leaves and trunk in addition to pedosphere was successfully remediated from Trichloroethylene (TCE) through phytovolatilization from a phytoremediation location.	Doucette et al., 2013
	Pinus massoniana	Masson pine plants with moderate or near maturation were found to store heavy metals inside woody tissues leading to the clean up of the contaminated pedospheric zones in the timber harvesting period.	Li et al., 2020b
	Swietenia macrophylla King., Anthocephalus macrophyllus (Roxb.) Havil., Maesopsis eminii Engl., and *Toona sureni*	Varied plant species tested for the phytovolatilization of 6 heavy metals, i.e., Cu, Fe, Hg, Mn, Pb and Zn expressed a variable response in terms of absorption. Nevertheless, the results confirmed that the selected plant species could not be classified as hyperaccumulator.	Siregar, 2019
	Enydra anagallis	Remarkable bio-concentration factor (BCF) and Plant Effective Number (PEN) were expressed by the *Enydra anagallis* in detoxification of the aquatic environment from different heavy metals, i.e., Al, As, Cr, Cu, Fe, Mn, Na, Ni, Pb, S, V, and Zn. Results pointed towards the candidacy of *Enydra* anagallis.	Demarco et al., 2020
	Pistia stratiotes	The results of this study declared *Pistia stratiots* to be an effective accumulator of Ni leading to the increment of levels of Malondialdehyde (MDA) to 8.214 nmol/g, when the concentration was	Leblebici et al., 2019
	Phragmites australis and *Kyllinga nemoralis*	The effective rhizofiltration system succeed in the removal of the pathogenic content present in the influent wastewater. This study was indicative of the future candidacy of macrophytes in the effective detoxification of both pedospheric and hydrospheric zone.	Odinga et al., 2019
	Pistia stratiotes	The results of this study declared *Pistia stratiots* to be an effective accumulator of Ni leading to the increment of levels of Malondialdehyde (MDA) to 8.214 nmol/g, when the concentration was kept at 20 mg l^{-1}, which can be associated with the concentration of the Ni and time span of experimental set up.	Leblebici et al., 2019

	Pistia stratiots	Environmental decontamination of Cu with *Pistia stratiots* expressed considerable rhizofiltration efficiency when the contact time was varied over a range of days from 5–20. The process of rhizofiltration was further aided by the utilization Sodium Lauryl Sulfate (SLS) which proved to be an effective chelating agent and increased the heavy metal availability for *Pistia stratiots*.	Niazy, 2020
	Eichhornia crassipes	Remarkable bioconcentration factor obtained for *Eichhornia crassipes* pointed towards its hyperaccumulation extent in case of Pb, while it also exhibited moderate accumulation of other heavy metals. Roots exceeded leaves in terms of heavy metals accumulation.	Peng et al., 2020
	Helianthus petiolaris	*Helianthus petiolaris* was effectively grown at a heavy metal contaminated site and assessment of its phytostabilization potential was done. Results were indicative of the successful growth of the *Helianthus petiolaris* in a pedospheric region thronged with 1000 mg/kg of Pb^{2+} and 50 of mg/kg Cd^{2+}. In fact, it succeeded in the accumulation of the pedospheric Cd in an impressive manner inside aerial parts in addition to the translocation of the remarkable quantities of Pb towards aerial zones when they were grown in a pedospheric zone having 500 mg/kg of Pb.	Saran et al., 2020
	Arundo donax, *Broussonetia papyrifera*, *Cryptomeria fortunei*, and *Robinia pseudoacacia*	Pedospheric zone contaminated with the different types of waste materials and heavy metals was phytostabilized using different plants and expressed the positive role played by the fractions of the particulate organic matter (POM) in the accurate estimation of the heavy metals distribution in the waste slag which was revegetated.	Luo et al., 2019
	Brassica juncea and *Dactylis glomerata*	As, Cd and Pb were effectively being remediated through phytostabilization experiment by utilization of organic amendments like compost and biochar. Utilization of *Brassica juncea* and *Dactylis glomerata* species and addition of the organic amendments effectively enhanced the pedospheric quality and immobilization of the heavy metals showing the synergestic impacts of the organic amendments and phytostabilization.	Visconti et al., 2020
	Arundo donax, *Broussonetia papyrifera*, *Robinia pseudoacacia*, and *Cryptomeria fortunei*	An extensive study spanning over the range of 5 years of revegetation was conclusive of the profound phytostabilization extent and expressed an advantageous vegetation cover. Higher heavy metal tolerance and lower accumulation was shown by these selected plants. The results were suggestive of the potential of these plants in terms of revegetation in addition to phytostabilization in the locations were Zn smelting is done.	Luo et al., 2019a
	Sapium sebiferum, *Salix matsudana*, *Hibiscus cannabinus*, *Corchorus capsularis*, *Ricinus communis*, and *Populus nigra*	The selected plant species possessed a considerable tolerance towards Pb and Zn with an impressive extent of the growth and bioaccumulation capacity for these heavy metals from mine tailings. Phytoremediation of the contaminated sites with the selected plants added with 10% of peat in tailing was accomplished successfully.	Tang et al., 2019

Fig. 6.1. Different types of plant species employed in varied kinds of phytoremediation for environmental detoxification.

Phyto-volatilization: Through the process of phyto-volatilization, different types of pollutants, contaminants and heavy metals having a volatile character drop on the foliar surficial regions due to different anthropogenic activities picked up by plants specialized in this process (Osama et al., 2020).

Phyto-degradation: Phyto-degradation is a comparatively advanced technique in which enzymatic substances are used for breaking of toxic pollutants marked with profound persistence, e.g., different types of agrochemicals inclusive of the herbicides and or others like trichloroethylene. Phyto-degradation can occur in the plant's internal or external regions since some plants are also capable of secretion of the external enzymes (Shrirangasami et al., 2020).

Phyto-rhizofiltration: By means of phyto-rhizofiltration, groundwater reservoirs and other hydrospheric bodies like streams etc., are protected from the release of organic pollutants released as a result of various anthropogenic activities. This is accomplished through the use of various plants having an extensive roots system which can filter such pollutants through absorption or adsorption mechanism (Yadav et al., 2020).

Phyto-transformation: By the process of phyto-transformation, the inherent physiological mechanisms of the selected plants trigger the conversion of the complex large-sized organic molecules into finely divided smaller and simpler molecules through degradation. The simpler molecules produced during such a reaction are then available for incorporation inside plant tissues (Abbas et al., 2020).

Phyto-stabilization: Phyto-stabilization effectively stabilizes the organic pollutants by means of complete deterrence of their movement inside pedospheric zones. Such timely prevention is associated with alleviation of the bioavailability of organic pollutants and consequent stoppage from entering food chains and destroying the health of consumers (Skorbiłowicz et al., 2020).

5. Dead Phyto-mass

In addition to living plants used for phytoremediation of the contaminated sites, biomass of the dead plants has also proved to be an effective media for heavy metals taking part and recovery of those metals from it. In this way, the dead plant biomass acts as a bio-sorbent material and provides attachment and absorbance sites for heavy metals and carry out the process of removal in an effective and sustainable way. Hence, this method has also achieved a certain level of attention by the scientific community in the recent era, which is attributable to its easier handling and economic viability (Sidek et al., 2018; Wilarso et al., 2020). For instance, dried roots of water hyacinth (Eichornia crassipes) have proved to be effective in the removal of the Cd and Pb from the samples of wastewater. Furthermore, the biomass of other plants, especially aquatic plants have also shown remarkable bio-sorbent materials for removing different heavy metals. For example, Eichhornia crassipes, Potamogetonlucens and Salvinia herzegoi have been known for remediating different heavy metals, e.g., Cr, Ni, Cd, Zn, Cu and Pb in an effective manner (Odjegba and Fasidi, 2007).

6. Phytoremediation with Transgenic Plants

In recent decades, the concept of phytoremediation has been further revolutionized by the use of transgenic plants adding to its effectiveness in environmental detoxification. Modification of plants having the potential of phytoremediation by incorporation of specific genes is associated with the augmentation of metabolism, heavy metals accumulation and also, enhances the uptake of different pollutants. Table 6.2 enlists the various types of plant species altered with different types of genes from variegated microbes and used for the detoxification of pedospheric matrices. In an ideal manner, plants chosen for genetic engineering by incorporation of different genes must possess a few significant characteristics, e.g., higher yielding biomass chosen for the local environment and a deep-rooted transformation procedure. Transgenic plants are known for the enhancement of ecological detoxification of various pollutants including organic ones and thus prevents the food chain from contamination in an eco-sustainable manner (Mishra et al., 2020; Dobrikova et al., 2021). Development of the transgenic plants was basically conducted with the special aim of detoxification of environmental areas laden with inorganic pollutants. However, with time, research succeeded in genetically engineered plants which could also effectively remediate organic contaminants. In the beginning of this development of these transgenic plants, *Nicotiana tabaccum* and *Arabidopsis thaliana* are representative of the first plants altered by genetic engineering and practiced for the elimination of different heavy metals like Cd, Hg, etc. However, through the years of large investigations and research, the scientific community concluded that plants aimed for phytoremediation need to be very carefully chosen (Kayıhan et al., 2021). In fact, in some cases, the wrong choices can lead to environmental deterioration instead of ecological detoxification.

More frequently , the sites of phyto-mining or phytoremediation are equipped with the hyper-accumulators or non hyper-accumulators which are capable of producing higher quantities of biomass. Nevertheless, such species might be the non-native ones. If invasive species are introduced, they can lead to disastrous consequences by the total disruption of the local ecosystem by severe competition with the already present endemic species through consumption of all available resources. For example, a study conducted by Che-Castaldo and Inouye (2015) was based on the introduction of a non-native hyper- accumulator specie, i.e., *Noccaea caerulescens* for remediation of Cd and Zn. During a field-based trial, results were indicative of the development of competitive interaction between *Noccaea caerulescens* and native species. In fact, the major inclination of this invasive hyper-accumulator was towards reproduction instead of vegetative growth (Mathakutha et al., 2019). Therefore, carrying out remediation with newer species needs careful monitoring for the prevention of such species becoming a source of nuisance in the future.

Different varieties of plants can be efficiently used for the degradation of a wide range of pollutants through genetic engineering. Transgenic plants are actually not associated with the up take or accumulation of organic target pollutants. However, various types of genes being incorporated in them are associated with the secretion of different enzymatic products specialized in the degradation of organic contaminants present inside the rhizosphere region. Such a resourceful approach is said to be a solution to the challenges associated with the harvesting of plants in addition to the toxic metals and chemicals-laden plants since the metallic pollution remediation and detoxification procedures mainly operate in the rhizosphere region through roots. An effective detoxification of pedospheric

Table 6.2. Utilization of transgenic plants modified with external genes for detoxification of cological compartments from heavy metals through specialized mechanisms.

Plant Species	Common Name	Microbial Species	Type	Heavy Metal	References
Triticum aestivum L.	Wheat	*Funneliformis mosseae* BGC HEB02 and *Rhizophagus intraradices* BGC HEB07D	Arbuscular mycorrhizal fungi (AMF)	Zn	Ma et al., 2019
Triticum aestivum and *Hordeum vulgare*	Wheat and barley	Rhizophagus irregularis	AMF	Zn	Coccina et al., 2019
Lactuca sativa	Lettuce	*Funneliformis mosseae, Rhizophagus intraradices, Claroideoglomus claroideum, Claroideoglomus etunicatum, Glomus microaggregatum* and *Funneliformis geosporum*	AMF	P, Zn	Konieczny and Kowalska, 2017
Tagetes patula	Marigold	*Glomus coronatum*	AMF	Cu	Zhou et al., 2017
Leucaena leucocephala	White leadtree	*Acaulospora morrowiae, Rhizophagus clarus, Gigaspora albida*	AMF	As	Schneider et al., 2017
Zea mays	Maize	*Claroideoglomus etunicatum*	AMF	La, Cd	Chang et al., 2018
Leucaena leucocephala	White leadtree	*Claroideoglomus etunicatum, Acaulospora scrobiculata*	AMF	Mn	Garcia et al., 2020
Prunus armeniaca L.	Apricot	*Glomus fasciculatum, G. mosseae, G. macrocarpum* and *Sclerocystis dussii*	AMF	Zn	Dutt Et al., 2013
Zea mays	Maize	*Glomus* sp.	AMF	Hg	Kodre et al., 2017
Apium graveolens Linn.	Celery	*Glomus mosseae*	AMF	Cd	Tanwar et al., 2015
Arabidopsis thaliana	Mouse-ear cress	*Saccharomyces cerevisiae*	AMF	As	Kumar et al., 2019
Populus deltoides	Cottonwood	*Glomus intraradices*	AMF	Cd	Chen et al., 2016
Hordeum vulgare	Barley	*Rhizophagus irregularis*	AMF	Zn	Watts-Williams and Cavagnaro, 2018
Triticum durum L.	Durum wheat	*Rhizophagus irregularis*	AMF	P, Zn, Fe	Tran et al., 2019
Pistacia vera	Pistachio nut	*Agaricus bisporus*	Ectomycorrhizal (ECM) fungi	Zn	Mohammadhasani et al., 2017
Zea mays	Maize	*Glomus mosseae*	AMF	Se	Yu et al., 2011
Solanum nigrum	Black nightshade	*Arthrobacter* sp. PGP41	Bacteria	Cd	Xu et al., 2018
Rosmarinus officinalis	Rosemary	*Glomus mosseae* and *Glomus intraradices*	AMF	Cu, Zn, Mn, Cd, Pb, Fe	Abbaslou et al., 2018
Triticum aestivum L.	Wheat	*Glomus* species	AMF	Zn	Kanwal et al., 2016

Helianthus annuus	Sunflower	*Pseudomonas libanensis* TR1 and *Claroideoglomus claroideum* BEG210	Bacterium and AMF	Ni and Na	Ma et al., 2019a
Helianthus annuus	Sunflower	*Glomus mosseae* and *Glomus intraradices*	AMF	Cr, Mn, Fe, Ni, Cu, Zn, Al, Cd, Pb, Si, Co, Mo	Sayın et al., 2019
Brachiaria mutica-a	Buffalo grass	*Rhizophagus irregularis*	AMF	Cr	Kullu et al., 2020
Solanum nigrum L.	Black nightshade	*Rhizophagus irregularis*	AMF	Cd	Wang et al., 2020
Zea mays and *Helianthus annuus*	Wheat and Sunflower	*Aspergillus niger, Aspergillus terreus, Aspergillus flavus, and Penicillium chrysogenum*	Trichocomaceae fungi and penicillium	Cr, Cu, Pb, Cd	Iram et al., 2019
Pinus massoniana	Horsetail pine	*Suillus luteus*	ECM	As, Cd, Cu, Cr, Mn, Ni, Pb, Zn, Se, Fe, P, K	Yu et al., 2020
Salix miyabeana	Seemen	*Sphaerosporella brunnea*	ECM	Pb, Sn, Zn	Dagher et al., 2020
Pinus thunbergii	Japanese black pine	*Pisolithus* sp. 1 and *Hebeloma vinosophyllum*	ECM	Cr	Shi et al., 2019
Pinus halepensis	Aleppo pine	*Rhizopogon* sp.	ECM	Pb, Zn, Cd	Hachani et al., 2020
Prosopis laevigata	Mesquite Trees	*Bacillus* sp. MH778713	Bacterium	Cr, Al, Fe, Ti, Zn	Ramírez et al., 2019
Phragmites communis	Reed	*Simplicillium chinense*	Fungi	Cd, Pb	Jin et al., 2019
Lolium perenne	Ryegrass	*Glomus* sp., *Acaulospora* sp., *Entrophospora* sp., and *Giaspora* sp., (MICO1), *Scutellospora* sp. (MICO2)	AMF	Hg	Leudo et al., 2020
Anadenanthera peregrina	Jopo	*AMF–Acaulospora scrobiculata and rhizobia (BH-ICB-A8)*	AMF and rhizobium	As	Gomes et al., 2020
Phragmites communis	Reed	*Bacillus* sp. PS-6	Bacterium	Fe, Cu, Zn, Cd, Mn, Ni, Pb, and As	Sharma et al., 2020
Suaeda vera	Shrubby sea-blite	*Pseudomonas fluorescens*	Bacteria	Cu, Zn, Cr	Gómez-Garrido et al., 2018
Helianthus annuus	Sunflower	*Pseudomonas fluorescens*	Bacteria	As	Shilev et al., 2006
Catharanthus roseus (L.)	Madagascar periwinkle	*Pseudomonas fluorescens RB4* and *Bacillus subtilis* 189	Bacteria	Cu, Pb	Khan et al., 2017

Table 6.2 contd....

...Table 6.2 contd.

Plant Species	Common Name	Microbial Species	Type	Heavy Metal	References
Sedum alfredii	Stonecrop	*Pseudomonas fluorescens*	Bacterium	Cd	Wu et al., 2020
Sedum alfredii Hance	Stonecrop	*Pseudomonas fluorescens* Sasm05 and IAA	Bacteria	Cd	Chen et al., 2017
Brassica napus	Canola	*Enterobacter cloacae* CAL2	PGPB	As	Nie et al., 2002
Brassica oxyrrhina	Radish	*Pseudomonas libanensis* TR1 and *Pseudomonas reactans* Ph3R3	Drought resistant ser-pentine rhizobacterial strains	Cu and Zn	Ma et al., 2016
Brassica napus	Rape	*Pseudomonas fluorescens* G10 *and Microbacterium sp.* G16	Endophytic bacteria	Pb	Sheng et al., 2008
Hordeum vulgare and *Avena sativa*	Barley and oats	*Pseudomonas* sp. and UW4 (*P*. sp.)	PGPB	Na	Chang et al., 2014
Zea Mays L.	Maize	*Priformospora indica* and *Pseudomonas fluorescens*	Fungus and bacteria	Pb	Asilian et al., 2018
Brassica juncea	Brown mustard	*Brevibacterium frigoritolerans* YSP40 and *Bacillus paralicheniformis* YSP151	Bacteria	Pb	Yahaghi et al., 2018
Eucalyptus camaldulensis L.	River red gum	*Glomus Mosseae*	PGPB	Cd	Moteshrezadeh et al., 2017
Mirabilis jalapa	Marvel of Peru	*Pseudomonas fluorescens*	Bacteria	Cr, Cu, Ni and Zn	Petriccione et al., 2013
Alnus firma	Siebold	*Bacillus thuringiensis* GDB-1	Endophytic bacillus	Pb, Zn, Cd, As, Cu, Ni	Babu et al., 2013
Alnus firma	Siebold	*Bacillus* sp. MN3-4	Endophytic bacillus	Pb	Shin et al., 2012
Vetiveria zizanioides L.	Bunchgrass	*Bacillus cereus* strain	Bacillus	Cr, Fe, Mn, Zn, Cd, Cu, Ni	Nayak et al., 2018
Vallisneria denseserrulata		*Bacillus* XZM	Bacillus	As	Irshad et al., 2020
Oryza sativa	Rice	*Ochrobactrum* sp. and *Bacillus* spp.	PGPR	Cd, Pb, As	Pandey et al., 2013
Oryza sativa	Rice	*Bacillus licheniformis*	Bacillus	Ni	Jamil et al., 2014
Aeschynomene fluminensis and *Polygonum acuminatum*	Joint Vetch and Smart weed	*Acinetobacter baumannii* BacI43, *Bacillus* sp. BacI34, *Enterobacter* sp. BacI14, *Klebsiella pneumoniae* BacI20, *Pantoea* sp.	Bacteria	Hg	Mello et al., 2020
		BacI23, *Pseudomonas* sp. BacI7, *Pseudomonas* sp. BacI38, and *Serratia marcescens* BacI56			

Spartina densiflora	Chilean cordgrass	*Aeromonas aquariorum* SDT13, *Pseudomonas composti* SDT3, and *Bacillus* sp. SDT14	Metal resistant PGPR	Cu	Andrades-Moreno et al., 2014
Brassica juncea L.	Indian mustard	*Bacillus* sp. JH 2-2	Bacillus	Cr	Shim et al., 2015
Raphanus sativus	Radish	*Bacillus* sp. CIK-512	Bacillus	Pb	Ahmad et al., 2018
Brassica juncea	Brown mustard	*Bacillus* sp. PZ-1	Bacillus	Pb	Yu et al., 2017
Broussonetia papyrifera	Paper mulberry	*Bacillus cereus* HM5 and *Bacillus thuringiensis* HM7	plant growth promoting (PGP) *Bacillus*	Mn	Huang et al., 2020
Saccharum spontaneum	Kans grass	*Bacillus anthracis strain* MHR2, *Staphylococcus* sp. *strain* MHR3 and *Bacillus* sp. *strain* MHR4	Phosphate solubilizing bacteria	P	Mukherjee et al., 2017

region from pollutants like 2,3- dihydroxybiphenyl (2,3-DHB) and 1-chlorobuatne has been reportedly accomplished with transgenic *Arabidopsis* and tobacco (Abhilash et al., 2009). The profound ability of transgenic plants in environmental remediation through modified phytoremediation is attributable to the diversity found in the microbial community, an augmented extent of the metabolic performance, and the discharge of substances from root exudates and enzymes in addition to the enhanced interaction developed between roots and target communities. In earlier periods when the concept of transgenic plants was introduced, such genetically engineered plants were used for the remediation of different types of organic pollutants and some halogenated compounds with organic composition. However with passage of time, advances have been made and presently such transgenic plants are being also used for the phytoremediation of diverse pollutants like phenolics, explosives, agrochemicals, etc. (RoyChowdhury et al., 2021) in addition to organic pollutants. Such engineered plants might either have transgenes which metabolize the organic pollutants or such transgenes can also play an influential role in increment of the resistance towards different contaminates. In the present era, an in-depth comprehension of the genetically modified plants has been achieved inclusive of their ability in taking up and degradation of the contaminants in an efficient manner.

Such an investigation-based comprehension of enzyme-related activities and xenobiotic metabolic substances led to the remarkable extent of the use of genetically engineered plants (Turcios et al., 2021). Tobacco plants engineered genetically with the human P4502 E1 gene led to an effective transformation of trichloroethylene exceeding the unmodified tobacco plants up to 640 times (Singh et al., 2011). There are problems associated with the use of transgenic microorganisms for modifying plants. These policy-concerns need to be addressed before using such microbes for altering plants and the use of different microbial strains marked by insignificant survival rates introduced inside contaminated pedospheric region. Such issues need to be explored and resolved through the adoption of advanced bioremediation approaches (Marapa et al., 2021). In response to the number of ecological stresses, several plants are known for developing exclusive features for the metabolization of contaminants having greater solubility extent (Mirhosseini et al., 2021).

7. Transgenic Plants: Aspects at Cellular and Molecular Levels

The occurrence of some serious symptoms is associated with the consistent exposure towards different types of pollutants and contaminants. Any pollutant coming in contact with the phyto-components can probably inhibit cellular activity through different mechanisms or can also lead to the disruption of the cellular structure, which can be attributed to the damage caused to the essential constituents of the plant cells (Ekanayake et al., 2021). In response to such disruption caused due to pollutants exposure, transgenic plants have also developed some well-defined mechanisms and approaches at both cellular and molecular levels. Such mechanisms and approaches are associated with the effective break down of the pollutants and contaminants of various kinds, e.g., pesticides, explosives, polyromantic hydrocarbons, heavy metals etc. In particular, these mechanisms are inclusive of the composition of the cell wall, environment of the root region, characteristics of the plasma membrane and veracity, enzymatic transfiguration, ligands driven complexation and compartmentalization of the vacuolar components (Punetha et al., 2021). In accordance with the type and characteristics of the pollutants in contact either organic or inorganic, plants can adopt any of these mechanisms and carry out the subsequent detoxification of the pedospheric zone (Lebrun et al., 2021). However, most research has highlighted the role of the hairy root cultures and modification of plants through genetic engineering as ideal approaches for the effective identification of the influential role of plants in environmental detoxification (Thakore and Srivastava, 2021; Han et al., 2021). Researchers can effectively carry out the monitoring and quantification of the contaminants and regularize the remediation approach with minute details. In the case of the hairy root culture, there is room for transformation through the adoption of several physiological assays. Such investigations prove effective in the selection of the most favorable plant species aimed at bioremediation of different pollutants (Galal et al.,

2021). A great deal of work has been reported with the *T. calerulescens* in the case of heavy metals transport due to its remarkable hyper-accumulator role (Luo et al., 2018). Molecular genetics-based investigations have compared varied ecotypes and expressed the role of certain genes belonging to families such as, ZIP, HMA, MATE, YSL and MTP in the transportation of the heavy metals beyond transmembrane region and hyper-accumulation (Dar et al., 2020).

8. Transgenics and Environmental Detoxification: Recent Advances

While a number of environmental detoxification mechanisms are attracting the attention of researchers for remediation of the pedospheric zone from different types of contaminants, transgenics are greatly appreciated due to their profound contribution towards the concept of sustainability and sustainable development (O'Connor et al., 2019). In this regard, the use of transgenic plants has a dual role of the transfiguration of the functional genes and also involves different types of some promoters which augment the functionality of the existing genes associated with the mechanisms, i.e., detoxification, accumulation and translocation of the heavy metals via introduction inside target plants. Table 6.3 shows such examples associated with environmental detoxification. In a work by Raldugina et al., (2018), transgenic canola varieties have been formed through modification of the *Brassica napus* L. cv. Westar, which is actually a wild-type by the use of the OsMyb4, a rice gene, consisting of the *Arabidopsis thaliana* COR15 a stress-inducible promoter. The results in this case were indicative of the significant role of OsMyb4 having an influential role in proline synthesis in addition to the positive regulation of the phenylpropanoid pathway leading to remarkable phytoremedial uses (Raldugina et al., 2018). In another study, a transgenic *A. thaliana* plants were generated through cloning and transferal of the VsCCoAOMT gene analog derived from *V. sativa* and used for the environmental detoxification and augmentation of lignin biosynthesis. The results obtained with this study also revealed the effective role of *V. sativa* driven VsCCoAOMT gene analog in beating Cd stress. In fact, a subsequent enhancement was obtained in the biomass production in addition to the increased mobility and loading of the Cd. In this regard, the phytoremedial role was especially highlighted by observing the successful piling up of the Cd inside the cell wall region of the transgenic plants (Xia et al., 2018).

Peptides usually referred to as 'PCs' generated from glutathione and ubiquitous cysteine-rich proteins known as 'MT' which are directly generated from genes and represent the main category of the substances which carry out the metallic chelation in plants (Gupta et al., 2013). Often these two enzymes are tailored for the enhancement of heavy metal tolerance in plants. In this regard, some species of tobacco hyper-accumulator e.g., *Nicotiana glauca* and *Nicotiana tabacum* are known for the proficient accumulation of the increased quantities of different heavy metals in accordance with the type of gene used for their transformation (Huang et al., 2012; Chen et al., 2015). For instance, when tobacco transgenic plant was generated with PCs gene from *Populus tomentosa*, it contributed towards the attenuation of Cd transportation towards shoots (Chen et al., 2015). However, when the tobacco transgenic-plant was generated with PCs genes acquired from an aquatic macrophyte known as *Ceratophyllum demersum*, the process of translocation was augmented in a remarkable manner (Shukla et al., 2013).

9. Conclusions and Future Perspectives

In the current period of an increased and irregular pattern of industrialization, urbanization and population boom, the need for food and other commodities has been increasing in a substantial manner. With such pressure, human beings are turning towards such mechanized and chemically governed approaches which might help them achieving short term goals of survival in terms of shelter and food. However, we cannot overlook the long-term adverse consequences associated with such short-sighted approaches (Jaffri and Ahmad, 2020a). The concept of sustainability and sustainable

Table 6.3. Examples of transgenic plants used for enhanced phytoremediation of varied pollutants at contaminated sites.

Gene Transferred	Origin	Target Plant	Type of Contaminant Remediated	Reference
CYP1A2	Bacteria	Arabidopsis thaliana	Herbicide Isoproturon phytoremediation	Azab et al., 2020
Oxygenase component (CndA) and bacterial acetochlor N- dealkylase system (CndABC)	Bacteria	Arabidopsis thaliana	Phytoremediation of acetochlor residue	Chu et al., 2020
Comamonas testosteroni bphC gene	Bacteria	Tobacco	PCB congener 2,3-dihydroxybiphenyl ring	Viktorová et al., 2014
Biphenyl 2,3-dioxygenase (BPDO) with BphAE, BphF, and BphG components	Bacteria	Tobacco	PCBs degradation	Mohammadi et al., 2007
Sinorhizobium meliloti genes	*Rhizosphere bacterium*	Medicago sativa	Dechlorination of 2′,3,4-trichlorobiphenyl	Chen et al., 2005
ACR2 gene	*Arabidopsis thaliana*	*Nicotiana tabacum*, var Sumsun	Arsenic phytoremediation	Nahar et al., 2017
N-demethylase PdmAB	Bacteria	Arabidopsis thaliana	Efficient removal of phenylurea herbicides (PHs)	Yan et al., 2018
StGCS-GS	*Streptococcus hermophilus*	Sugar beets	Heavy metal tolerance and accumulation of Cd, Zn and Cu	Liu et al., 2015
StSy	Bacteria	*Populus alba* L.	Analysis of Resistance to Cd, Co, Cu, Pb and Zn	Balestrazzi et al., 2009
tcu-1	*Neurospora crassa*	Nicotiana tabacum	Enhanced acquisition of Cu	Singh et al., 2011
MnPCS1 and MnPCS2	*Morus notabilis* PCS	*Arabidopsis* and tobacco	Zn and Cd tolerance and accumulation enhancement	Fan et al., 2018
tzn1	*Neurospora crassa* fungi	Nicotiana tabacum	Zn and Cd accumulation	Dixit et al., 2010
MerE	Tn21 mer operon bacteria	Arabidopsis thaliana	Methylmercury accumulation	Sone et al., 2013
xplA and xplB	Bacteria	Pascopyrum smithii	explosives hexahydro-1, 3, 5-trinitro-1, 3, 5-triazine (RDX), and 2, 4, 6-trinitrotoluene (TNT) remediation	Zhang et al.,2019
Caffeoyl-CoA *O*-methyltransferase (VsCCoAOMT)	Vicia sativa	Arabidopsis	Cd uptake and tolerance	Xia et al., 2018
PtABCC1	Populus trichocarpa	*Arabidopsis* and poplar	Hg tolerance and accumulation	Sun et al., 2018
CzcCBA	Pseudomonas putida	Tobacco	Cd uptake and accumulation	Nesler et al.,2017
AtStr5	Sulfurtransferase/ rhodanese family of *Arabidopsis thaliana*	*Arabidopsis* and *Nicotiana* tabacum	As phytoremediation	Parvin 2018
ScYCF1	Yeast cadmium *factor* 1	*Populus alba* X *P. tremula* var. glandulosa	Phytoremediation of mine tailing soil	Shim et al., 2013
PaGST, PaMT and PaFe-SOD	Yeast	Phytolacca americana	Adaptation to Cd toxicity	Zhao et al., 2019
bphC. B	*Agrobacterium tumefaciens* EHA105	Medicago sativa	Phytoremediation of polychlorinatedbiphenyls	Ren et al., 2018
SaNAS1	-	*Sedum alfredii* Hance	Cd/Zn tolerance and accumulation in plants	Chen et al., 2019
BrHMA3	Yeast and *Arabidopsis*	Brassica rapa	Cd accumulation	Zhan et al., 2019

NFSB	*Sulfurimonas denitrificans* DSM1251	Arabidopsis	Degradation of trinitrotoluene	Bo et al., 2018
NfsI	Bacteria	*Nicotiana tabacum* L	Detoxification of TNT	Zhang et al., 2017
OsMyb4	Rice	Canola	Cu and Zn tolerance	Raldugina et al., 2018
PtoHMA5	*Populus tomentosa* Carr. Via *Agrobacterium tumefaciens*	Tobacco	Higher Cd accumulation	Wang et al., 2018
--	Bacterial strains (*Massilia* sp. and *Pseudomonas* sp.) or saprophytic fungus (*Clitocybe* sp.)	Salix viminalis	Cd uptake increment	Złoch et al., 2015
CndABC	Bacteria	Arabidopsis thaliana	Degradation of acetochlor	Chu et al., 2020
merA and merB	Bacteria	Arabidopsis, tobacco, *tomato* and *rice*	Hg phytoremediation	Li et al., 2020e
CYP4502E1	--	*A. thaliana* and *S. grandiflora*	Heavy metal phytoremediation	Mouhamad et al., 2020
PeANN1 and AtANN1	Fungi	Populus euphratica	Cd phytoremediation	Zhang et al., 2020
IRT1	Arabidopsis thaliana	Solanum nigrum L.	Cd accumulation and tolerance	Ye et al., 2020
PsMT1 and PsMT2	Rhizobium leguminosarum	Pisum sativum (L.)	Cd phytoremediation	Tsyganov et al.,2020
BP-1	Thermosynechococcus elongatus	Arabidopsis thaliana	Thiocyanate phytoremediation	Gao et al., 2020
BM1	Serratia marcescens	Soyabean	Cd phytoremediation	El-Esawi et al., 2020
AtNHX1	--	Lemna turonifera	Cd phytoremediation	Yao et al., 2020
SlMYB14	Tomato	Tomato	Tolerance to 2,4,6- trichlorophenol	Li et al., 2020c
hrpZPsph and SP/hrpZPsph	*Pseudomonas syringae* pv.	Nicotiana benthamiana	Cd accumulation	Mitsopoulou et al., 2020
PvSR2	Agrobacterium	Nicotiana benthamiana	Removal of Cd contamination	Wu et al., 2020
SaPCR2	Yeast	*Sedum alfredii* Hance	Facilitation of Cd efflux in roots	Lin et al., 2020
PM21	Bacillus anthracis	Sesbania sesban	Metal accumulation	Ali et al., 2020
PgCYP76B93	Ginseng	Arabidopsis	Enhancement of chlorotoluron tolerance	Jang et al., 2020
BnMTP3	Brassica napus	Arabidopsis thaliana	Tolerance to Zn and Mn	Gu et al., 2020
SmCP	Salix matsudana	Arabidopsis thaliana	Resistance of salt stress	Li et al., 2020d
PvPht1;3, PvACR2 *and* PvACR3	--	Pteris vittata	As and V absorption and reduction	Wei et al., 2020
SaHsfA4c	Sedum alfredii Hance	Arabidopsis	Enhancement of Cd tolerance	Chen et al., 2020
CsMTP8.2	Camellia sinensis L.	Tobacco and onion	Enhanced of tolerance to Mn toxicity	Zhang et al., 2020a
BnMYB2	Boehmeria nivea	Arabidopsis thaliana	Cd tolerance	Zhu et al., 2020
SGECd[t]	*Variovorax paradoxus* 5C-2, *Rhizobium leguminosarum*	Pisum sativum	Cadmium tolerance and accumulation	Belimov et al., 2020

development is totally undermined when human beings challenge the environmental integrity through the use of harmful and persistent agrochemicals for boosting the production of crops, but actually leaving their residues inside the pedospheric region and translocation inside plants in the long term. For the detoxification of the pedosphere, varied approaches have been adopted such as physical and chemical methods, but the issues of economic unsuitability and operation complexity, such methods have been criticized. Restoration of the inherent health of soils by the use of nature's green factories in the form of plants signifies the green approach and has a role in the consolidation of the concept of the ecological sustainability. In recent times, the transformation of the inherent hyper-accumulation potential of some plants has further been augmented by the use of genes from other biological systems leading to the development of the efficient transgenic plants. The process of phytoremediation and utilization of transgenic plants in it is usually preferred over systems of remediation because it is an eco-friendly process which restores the health of the soil without causing any harm to the inherent pedospheric micro-floral or micro-faunal communities. However, a better understanding of the other interdisciplinary fields like environmental sciences, plant biology, soil chemistry and microbiology is needed for further improving the results obtained from the phytoremediation carried out with plants. Transgenic plants can be used for the detoxification of contaminated sites loaded with heavy metals, agrochemicals, industrial products and explosives, etc. In light of the results obtained from the research being conducted and those in progress, genetic modification of plants has been proved to be the most favorable biotechnological procedure which can be adopted by environmental scientists for environmental cleanup.

Conflict of Interest

None

References

Abbas, Q., Yousaf, B., Ullah, H., Ali, M.U., Ok, Y.S. and Rinklebe, J. 2020. Environmental transformation and nano-toxicity of engineered nano-particles (ENPs) in aquatic and terrestrial organisms. Critic. Rev. Envir. Sci. Tech. 50: 2523–2581. DOI: 10.1080/10643389.2019.1705721.

Abbaslou, H., Bakhtiari, S. and Hashemi, S.S. 2018. Rehabilitation of iron ore mine soil contaminated with heavy metals using rosemary phytoremediation-assisted mycorrhizal arbuscular fungi bioaugmentation and fibrous clay mineral immobilization. J. Iran. Sci. Tech. Trans. A 42: 431–441. DOI: 10.1007/s40995-018- 0543-7.

Abhilash, P.C., Jamil, S. and Singh, N. 2009. Transgenic plants for enhanced biodegradation and phytoremediation of organic xenobiotics. Biotech. Adv. 27: 474–488. DOI: 10.1016/j.biotechadv.2009.04.002.

Ahmad, I., Akhtar, M.J., Mehmood, S., Akhter, K., Tahir, M., Saeed, M.F., Hussain, M.B. and Hussain, S. 2018. Combined application of compost and *Bacillus* sp. CIK-512 ameliorated the lead toxicity in radish by regulating the homeostasis of antioxidants and lead. Ecotoxicol. Envir. Safety 148: 805–812. 10.1016/j.ecoenv.2017.11.054.

Ahmad, K.S., Nawaz, M. and Jaffri, S.B. 2020. Role of renewable energy and nanotechnology in sustainable desalination of water: Mini review. J. Int. Envir. Anal. Chem. 1: 1–20. DOI: 10.1080/03067319.2020.1837121.

Ali, J., Ali, F., Ahmad, I., Rafique, M., Munis, M.F., Hassan, S.W., Sultan, T., Iftikhar, M. and Chaudhary, H.J. 2020. Mechanistic elucidation of germination potential and growth of *Sesbania sesban* seedlings with *Bacillus anthracis* PM21 under heavy metals stress: An *in vitro* study. Ecotoxicol. Envir. Safety. 208: 111769. DOI: 10.1016/j. ecoenv.2020.111769.

Andrades-Moreno, L., Del Castillo, I., Parra, R., Doukkali, B., Redondo-Gómez, S., Pérez-Palacios, P., Caviedes, M.A., Pajuelo, E. and Rodríguez-Llorente, I.D. 2014. Prospecting metal-resistant plant-growth promoting rhizobacteria for rhizoremediation of metal contaminated estuaries using *Spartina densiflora*. Envir. Sci. Poll. Res. 21: 3713–721. DOI: 10.1007/s11356-013-2364-8.

Antoniadis, V., Shaheen, S.M., Stärk, H.J., Wennrich, R., Levizou, E., Merbach, I. and Rinklebe, J. 2021. Phytoremediation potential of twelve wild plant species for toxic elements in a contaminated soil. Envir. Int. 146: 106233. DOI: 10.1016/j. envint.2020.106233.

Archana, T., Rajendran, L., Manoranjitham, S.K., Krishnan, V.S., Paramasivan, M. and Karthikeyan, G. 2020. Culture-dependent analysis of seed bacterial endophyte, *Pseudomonas* spp. EGN 1 against the stem rot disease (*Sclerotium rolfsii* Sacc.) in groundnut. J. Egypt. Biol. Pest. Control. 30(1): 1–3. DOI: 10.1186/s41938-020- 00317-x.

Asilian, E., Ghasemi-Fasaei, R., Ronaghi, A., Sepehri, M. and Niazi, A. 2018. Effects of microbial inoculations and surfactant levels on biologically-and chemically-assisted phytoremediation of lead-contaminated soil by maize (*Zea Mays* L.). Chem. Ecol. 34: 964–977. DOI: 10.1080/02757540.2018.1520844.

Azab, E., Hegazy, A.K., Gobouri, A.A. and Elkelish, A. 2020. Impact of transgenic arabidopsis thaliana plants on herbicide isoproturon phytoremediation through expressing human cytochrome P450-1A2. Biol. 9: 362. DOI: 10.3390/biology9110362.

Babu, A.G., Kim, J.D. and Oh, B.T. 2013. Enhancement of heavy metal phytoremediation by Alnus firma with endophytic Bacillus thuringiensis GDB-1. J. Hazard. Mat. 250: 477–483. DOI: 10.1016/j.jhazmat.2013.02.014.

Baker, A.J.M. and Brooks, R. 1989. Terrestrial higher plants which hyperaccumulate metallic elements. A review of their distribution, ecology and phytochemistry. Biorecovery 1: 81–126.

Balestrazzi, A., Bonadei, M., Quattrini, E. and Carbonera, D. 2009. Occurrence of multiple metal-resistance in bacterial isolates associated with transgenic white poplars (*Populus alba* L.). Annals. Microbiol. 59: 17–23. DOI: 10.1007/BF03175593.

Belimov, A.A., Shaposhnikov, A.I., Azarova, T.S., Makarova, N.M., Safronova, V.I., Litvinskiy, V.A., Nosikov, V.V., Zavalin A.A. and Tikhonovich, I.A. 2020. Microbial consortium of PGPR, rhizobia and arbuscular mycorrhizal fungus makes pea mutant SGECdt comparable with indian mustard in cadmium tolerance and accumulation. Plants 9: 975. DOI: 10.3390/plants9080975.

Bo, Z., Hongjuan, H., Xiaoyan, F., Zhenjun, L., Jianjie, G. and Quanhong, Y. 2018. Degradation of trinitrotoluene by transgenic nitroreductase in *Arabidopsis* plants. Plant. Soil. Envir. 64: 379–385. DOI: 10.17221/655/2017-PSE.

Calabrese, E.J. and Agathokleous, E. 2021. Accumulator plants and hormesis. Envir. Poll. 1: 116526. DOI: 10.1016/j.envpol.2021.116526.

Chang, P., Gerhardt, K.E., Huang, X.D., Yu, X.M., Glick, B.R., Gerwing, P.D. and Greenberg, B.M. 2014. Plant growth-promoting bacteria facilitate the growth of barley and oats in salt-impacted soil: Implications for phytoremediation of saline soils. J. Int. Phytorem. 16: 1133–1147. DOI: 10.1080/15226514.2013.821447.

Chang, Q., Diao, F.W., Wang, Q.F., Pan, L., Dang, Z.H. and Guo, W. 2018. Effects of arbuscular mycorrhizal symbiosis on growth, nutrient and metal uptake by maize seedlings (*Zea mays* L.) grown in soils spiked with Lanthanum and Cadmium. Envir. Poll. 241: 607–615. DOI: 10.1016/j.envpol.2018.06.003.

Chauhan, S. and Kulshreshtha, S. 2021. Application of inorganic amendments to improve soil fertility. Phytomicrobiome Interactions and Sustainable Agriculture 1: 187–206. DOI: 10.1002/9781119644798.ch10.

Che-Castaldo, J.P. and Inouye, D.W. 2015. Interspecific competition between a nonnative metal-hyperaccumulating plant (*Noccaea caerulescens*, Brassicaceae) and a native congener across a soil-metal gradient. J. Aust. Bot. 63: 141–151. DOI: 10.1071/BT15045.

Chen, S., Yu, M., Li, H., Wang, Y., Lu, Z., Zhang, Y., Liu, M., Qiao, G., Wu, L., Han, X. and Zhuo, R. 2020. SaHsfA4c from *Sedum alfredii* hance enhances cadmium tolerance by regulating ROS-scavenger activities and heat shock proteins expression. Front. Plant. Sci. 11: 142. DOI: 10.3389/fpls.2020.00142.

Chen, S., Zhang, M., Feng, Y., Sahito, Z.A., Tian, S. and Yang, X. 2019. Nicotianamine Synthase Gene 1 from the hyperaccumulator *Sedum alfredii* Hance is associated with Cd/Zn tolerance and accumulation in plants. Plant. Soil. 443: 413–427. DOI: 10.1007/s11104-019-04233-4.

Chen, Y., Adam, A., Toure, O. and Dutta, S.K. 2005. Molecular evidence of genetic modification of *Sinorhizobium meliloti*: enhanced PCB bioremediation. J. Indust. Microbiol. Biotech. 32: 561–566. DOI: 10.1007/s10295- 005-0039-2.

Chen, Y., Liu, Y., Ding, Y., Wang, X. and Xu, J. 2015. Overexpression of PtPCS enhances cadmium tolerance and cadmium accumulation in tobacco. Plant. Cell. Tissue. Organ. Cult. 121: 389–396. DOI: 10.1007/s11240-015- 0710-x.

Chen, B., Luo, S., Wu, Y., Ye, J., Wang, Q., Xu, X. and Yang, X. 2017. The effects of the endophytic bacterium Pseudomonas fluorescens Sasm05 and IAA on the plant growth and cadmium uptake of *Sedum alfredii* Hance. Front. Microbiol. 8: 2538. DOI: 10.3389/fmicb.2017.02538.

Chen, L., Zhang, D., Yang, W., Liu, Y., Zhang, L. and Gao, S. 2016. Sex-specific responses of *Populus deltoides* to Glomus intraradices colonization and Cd pollution. Chemosphere 155: 196–206. DOI: 10.1016/j.chemosphere.2016.04.049.

Chu, C., Liu, B., Liu, J., He, J., Lv, L., Wang, H., Xie, X., Tao, Q. and Chen, Q. 2020. Phytoremediation of acetochlor residue by transgenic Arabidopsis expressing the acetochlor N-dealkylase from *Sphingomonas wittichii* DC-6. Sci. Total. Envir. 14: 138687. DOI: 10.1016/j.scitotenv.2020.138687.

Coccina, A., Cavagnaro, T.R., Pellegrino, E., Ercoli, L., McLaughlin, M.J. and Watts-Williams, S.J. 2019. The mycorrhizal pathway of zinc uptake contributes to zinc accumulation in barley and wheat grain. BMC Plant. Biol. 19: 1–14. DOI: 10.1186/s12870-019-1741-y.

Correa-García, S., Rheault, K., Tremblay, J., Séguin, A. and Yergeau, E. 2021. Soil characteristics constrain the response of microbial communities and associated hydrocarbon degradation genes during phytoremediation. Appl. Envir. Microbiol. 87: 1–10. DOI: 10.1128/AEM.02170-20.

Dagher, D.J., Pitre, F.E. and Hijri, M. 2020. Ectomycorrhizal fungal inoculation of *Sphaerosporella brunnea* significantly increased stem biomass of *Salix miyabeana* and decreased lead, tin, and zinc, soil concentrations during the phytoremediation of an industrial landfill. J. Fungi. 6: 87. DOI:10.3390/jof6020087.

Dar, F.A., Pirzadah, T.B. and Malik, B. 2020. Accumulation of heavy metals in medicinal and aromatic plants. pp. 113–127. *In*: Plant Micronutrients. Springer, Cham. DOI: 10.1007/978-3-030-49856-6_6.

de Souza, T.D., Borges, A.C., de Matos, A.T., Veloso, R.W. and Braga, A.F. 2018. Kinetics of arsenic absorption by the species *Eichhornia crassipes* and *Lemna valdiviana* under optimized conditions. Chemosphere 209: 866–874. DOI: 10.1016/j.chemosphere.2018.06.132.

Delil, A.D., Köleli, N., Dağhan, H. and Bahçeci, G. 2020. Recovery of heavy metals from canola (*Brassica napus*) and soybean (*Glycine max*) biomasses using electrochemical process. Envir. Technol. Innovat. 17: 100559. DOI: 10.1016/j.eti.2019.100559.

Demarco, C.F., Afonso, T.F., Pieniz, S., Quadro, M.S., de Oliveira Camargo, F.A. and Andreazza, R. 2020. Evaluation of Enydra anagallis remediation at a contaminated watercourse in south Brazil. J. Int. Phytorem. 1: 1–8. DOI: 10.1080/15226514.2020.1754759.

Dixit, P., Singh, S., Vancheeswaran, R., Patnala, K. and Eapen, S. 2010. Expression of a *Neurospora crassa* zinc transporter gene in transgenic *Nicotiana tabacum* enhances plant zinc accumulation without co-transport of cadmium. Plant Cell Envir. 33: 1697–1707. DOI: 10.1111/j.1365-3040.2010.02174.x.

Dobrikova, A.G., Apostolova, E.L., Hanć, A., Yotsova, E., Borisova, P., Sperdouli, I., Adamakis, I.D. and Moustakas, M. Cadmium toxicity in Salvia sclarea L. 2021. An integrative response of element uptake, oxidative stress markers, leaf structure and photosynthesis. Ecotoxicol. Envir. Safety. 209: 111851. DOI: 10.1016/j.ecoenv.2020.111851.

Doucette, W., Klein, H., Chard, J., Dupont, R., Plaehn, W. and Bugbee, B. 2013. Volatilization of trichloroethylene from trees and soil: Measurement and scaling approaches. Envir. Sci. Tech. 47: 5813–5820. DOI: 10.1021/es304115c.

Dutt, S., Sharma, S.D. and Kumar, P. 2013. Arbuscular mycorrhizas and Zn fertilization modify growth and physiological behavior of apricot (*Prunus armeniaca* L.). Scient. Hortic. 155: 97–104. DOI: 10.1016/j.scienta.2013.03.012.

Eevers, N., Hawthorne, J.R., White, J.C., Vangronsveld, J. and Weyens, N. 2018. Endophyte-enhanced phytoremediation of DDE-contaminated using Cucurbita pepo: A field trial. J. Int. Phytorem. 20: 301–310. DOI: 10.1080/15226514.2017.1377150.

Ekanayake, M.S., Udayanga, D., Wijesekara, I. and Manage, P. 2021. Phytoremediation of synthetic textile dyes: Biosorption and enzymatic degradation involved in efficient dye decolorization by *Eichhornia crassipes* (Mart.) Solms and *Pistia stratiotes* L. Envir. Sci. Poll. Res. 1: 1–1. DOI: 10.1007/s11356-020-11699-8.

El-Esawi, M.A., Elkelish, A., Soliman, M., Elansary, H.O., Zaid, A. and Wani, S.H. 2020. *Serratia marcescens* BM1 enhances cadmium stress tolerance and phytoremediation potential of soybean through modulation of osmolytes, leaf gas exchange, antioxidant machinery, and stress-responsive genes expression. Antioxidants 9: 43. DOI: 10.3390/antiox9010043.

Fan, W., Guo, Q., Liu, C., Liu, X., Zhang, M., Long, D., Xiang, Z. and Zhao, A. 2018. Two mulberry phytochelatin synthase genes confer zinc/cadmium tolerance and accumulation in transgenic *Arabidopsis* and tobacco. Gene. 645: 95–104. DOI: 10.1016/j.gene.2017.12.042.

Galal, T.M., Shedeed, Z.A., Gharib, F.A., Al-Yasi, H.M. and Mansour, K.H. 2021. The role of *Cyperus alopecuroides* Rottb. sedge in monitoring water pollution in contaminated wetlands in Egypt: A phytoremediation approach. Envir. Sci. Poll. Res. 1: 1–2. DOI: 10.1007/s11356-020-12308-4.

Gao, J.J., Zhang, L., Peng, R.H., Wang, B., Feng, H.J., Li, Z.J. and Yao, Q.H. 2020. Recombinant expression of *Thermosynechococcus elongatus* BP-1 glutathione S-transferase in Arabidopsis thaliana: An efficient tool for phytoremediation of thiocyanate. Biotechnol. Biotechnol. Equip. 34: 494–505. DOI: 10.1080/13102818.2020.1779127.

Gao, X., Ai, W.L., Gong, H., Cui, L.J., Chen, B.X., Luo, H.Y. and Qiu, B.S. 2016. Transgenic NfFeSOD *Sedum alfredii* plants exhibited profound growth impairments and better relative tolerance to longterm abiotic stresses. Plant. Biotechnol. Rep. 10: 117–128. DOI: 10.1007/s11816-016-0391-x.

Garcia, K.G.V., Mendes, Filho, P.F., Pinheiro, J.I., do Carmo, J.F., de Araújo Pereira, A.P., Martins, C.M. and de Souza Oliveira Filho, J. 2020. Attenuation of manganese-induced toxicity in *Leucaena leucocephala* colonized by arbuscular mycorrhizae. Water Air Soil Poll. 231: 22. DOI: 10.1007/s11270-019-4381-9.

Gomes, M.P., Marques, R.Z., Nascentes, C.C. and Scotti, M.R. 2020. Synergistic effects between arbuscular mycorrhizal fungi and rhizobium isolated from As-contaminated soils on the As-phytoremediation capacity of the tropical woody legume *Anadenanthera peregrina*. J. Int. Phytorem. 22: 1362–1371. DOI: 10.1080/15226514.2020.1775548.

Gómez-Garrido, M., Mora Navarro, J., Murcia Navarro, F.J. and Faz Cano, Á. 2018. The chelating effect of citric acid, oxalic acid, amino acids and *Pseudomonas fluorescens* bacteria on phytoremediation of Cu, Zn, and Cr from soil using *Suaeda vera*. J. Int. Phytorem. 20: 1033–1042. DOI: 10.1080/15226514.2018.1452189.

Gu, D., Zhou, X., Ma, Y., Xu, E., Yu, Y., Liu, Y., Chen, X. and Zhang, W. 2020. Expression of a *Brassica napus* metal transport protein (BnMTP3) in Arabidopsis thaliana confers tolerance to Zn and Mn. Plant Sci. 1: 110754. DOI: 10.1016/j.plantsci.2020.110754.

Guarino, F., Miranda, A., Castiglione, S. and Cicatelli, A. 2020. Arsenic phytovolatilization and epigenetic modifications in *Arundo donax* L. assisted by a PGPR consortium. Chemosphere 251: 126310. DOI: 10.1016/j.chemosphere.2020.126310.

Gupta, D.K., Vandenhove, H. and Inouhe, M. 2013. Role of phytochelatin in heavy metal stress and detoxification mechanisms in plants. pp. 73–94. *In*: Gupta, D.K., Corpas, F.J. and Palma, J.M. (eds.). Heavy Metal Stress in Plants. Springer, Berlin. DOI: 10.1007/978-3-642-38469-1_4.

Hachani, C., Lamhamedi, M.S., Cameselle, C., Gouveia, S., Zine El Abidine, A., Khasa, D.P. and Béjaoui, Z. 2020. Effects of Ectomycorrhizal Fungi and Heavy Metals (Pb, Zn, and Cd) on Growth and Mineral Nutrition of *Pinus halepensis* Seedlings in North Africa. Microorg. 8: 2033. DOI: 10.3390/microorganisms8122033.

Han, Y., Lee, J., Kim, C., Park, J., Lee, M. and Yang, M. 2021. Uranium rhizofiltration by *Lactuca sativa*, *Brassica campestris* L., *Raphanus sativus* L., *Oenanthe javanica* under different hydroponic conditions. Minerals 11: 41. DOI: 10.3390/min11010041.

Hirata, K., Tsuji, N. and Miyamoto, K. 2005. Biosynthetic regulation of phytochelatins, heavy metalbinding peptides. J. Biosci. Bioeng. 100: 593–599. DOI: 10.1263/jbb.100.593.

Huang, H., Zhao, Y., Fan, L., Jin, Q., Yang, G. and Xu, Z. 2020. Improvement of manganese phytoremediation by *Broussonetia papyrifera* with two plant growth promoting (PGP) *Bacillus* species. Chemosphere 260: 127614. DOI: 10.1016/j.chemosphere.2020.127614.

Ievinsh, G., Andersone-Ozola, U., Landorfa-Svalbe, Z., Karlsons, A. and Osvalde, A. 2020. Wild plants from coastal habitats as a potential resource for soil remediation. pp. 121–144. *In*: Soil Health. Springer, Cham. DOI: 10.1007/978-3-030-44364-1_8.

Iftikhar, S., Saleem, M., Ahmad, K.S. and Jaffri, S.B. 2019. Synergistic mycoflora–natural farming mediated biofertilization and heavy metals decontamination of lithospheric compartment in a sustainable mode via *Helianthus annuus*. J. Int. Envir. Sci. Tech. 16: 6735–6752. DOI: 10.1007/s13762-018-02180-8.

Ijaz, M., Zafar, M., Islam, A., Afsheen, S. and Iqbal, T. 2020. A review on antibacterial properties of biologically synthesized zinc oxide nanostructures. J. Inorg. Organomet. Polym. Mat. 30: 2815–2826. DOI: 10.1007/s10904-020-01603-9.

Inobeme, A. 2021. Effect of heavy metals on activities of soil microorganism. pp. 115–142. *In*: Microbial Rejuvenation of Polluted Environment. Springer, Singapore. DOI: 10.1007/978-981-15-7459-7_6.

Iram, S., Ahmad, K.S., Noureen, S. and Jaffri, S.B. 2018. Utilization of wheat (*Triticum aestivum*) and Berseem (*Trifolium alexandrinum*) dry biomass for heavy metals biosorption. Proceed. Pak. Acad. Sci. B 55: 61–70.

Iram, S., Iqbal, A., Ahmad, K.S. and Jaffri, S.B. 2020. Congruously designed eco-curative integrated farming model designing and employment for sustainable encompassments. Envir. Sci. Poll. Res. 1: 1–8. DOI: 10.1007/s11356-020-08499-5.

Iram, S., Basri, R., Ahmad, K.S. and Jaffri, S.B. 2019. Mycological assisted phytoremediation enhancement of bioenergy crops *Zea mays* and *Helianthus annuus* in heavy metal contaminated lithospheric zone. Soil Sed. Contam. 28: 411–430. DOI: 10.1080/15320383.2019.1597011.

Ishtiaq, M., Iram, S., Ahmad, K.S. and Jaffri, S.B. 2020. Multi-functional bio-sorbents triggered sustainable detoxification of eco-contaminants besmirched hydrospheric swatches. J. Int. Envir. Anal. Chem. 11: 1–6. DOI: 10.1080/03067319.2020.1776866.

Jaffri, S.B., Ahmad, K.S., Thebo, K.H. and Rehman, F. 2020. Sustainability consolidation via employment of biomimetic ecomaterials with an accentuated photo-catalytic potential: emerging progressions. Rev. Inorg. Chem. 1: 1–35. DOI: 10.1515/revic-2020-0018.

Jaffri, S.B. and Ahmad, K.S. 2020. Biomimetic detoxifier *Prunus cerasifera* Ehrh. silver nanoparticles: Innate green bullets for morbific pathogens and persistent pollutants. Envir. Sci. Poll. Res. 1: 1–7. DOI: 10.1007/s11356-020-07626-6.

Jaffri, S.B. and Ahmad, K.S. 2020a. Interfacial engineering revolutionizers: perovskite nanocrystals and quantum dots accentuated performance enhancement in perovskite solar cells. Critic. Rev. Solid. State. Mat. Sci. 1: 1–29. DOI: 10.1080/10408436.2020.1758627.

Jaffri, S.B. and Ahmad, K.S. 2019. Foliar-mediated Ag: ZnO nanophotocatalysts: Green synthesis, characterization, pollutants degradation, and *in vitro* biocidal activity. Green Proc. Synth. 8: 172–182. DOI: 10.1515/gps-2018-0058.

Jaffri, S.B., Nosheen, A., Iftikhar, S. and Ahmad, K.S. 2019. Pedospheric environmental forensics aspects. pp. 39–59. *In*: Trends of Environmental Forensics in Pakistan. Academic Press. DOI: 10.1016/B978-0-12-819436- 2.00003-0.

Jaffri, S.B. and Ahmad, K.S. 2018. Phytofunctionalized silver nanoparticles: Green biomaterial for biomedical and environmental applications. Rev. Inorg. Chem. 38: 127–149. DOI: 10.1515/revic-2018-0004.

Jamil, M., Zeb, S., Anees, M., Roohi, A., Ahmed , I., ur Rehman, S. and Rha, E.S. 2014. Role of *Bacillus licheniformis* in phytoremediation of nickel contaminated soil cultivated with rice. J. Int. Phyto. 16: 554–571. DOI: 10.1080/15226514.2013.798621.

Jang, J., Khanom, S., Moon, Y., Shin, S. and Lee, O.R. 2020. PgCYP76B93 docks on phenylurea herbicides and its expression enhances chlorotoluron tolerance in *Arabidopsis*. Appl. Biol. Chem. 63: 1–10. DOI: 10.1186/s13765-020-00498-x.

Jin, Z., Deng, S., Wen, Y., Jin, Y., Pan, L., Zhang, Y. and Zhang, D. 2019. Application of *Simplicillium chinense* for Cd and Pb biosorption and enhancing heavy metal phytoremediation of soils. Sci. Total. Envir. 697: 134148. DOI: 10.1016/j.scitotenv.2019.134148.

Kanwal, S., Bano, A. and Malik, R.N. 2016. Role of arbuscular mycorrhizal fungi in phytoremediation of heavy metals and effects on growth and biochemical activities of wheat (*Triticum aestivum* L.) plants in Zn contaminated soils. J. Afric. Biotech. 15: 872–883. DOI: 10.5897/AJB2016.15292.

Kayıhan, D.S., Kayıhan, C. and Özden Çiftçi, Y. 2021. Transgenic tobacco plants overexpressing a cold-adaptive nitroreductase gene exhibited enhanced 2, 4-dinitrotoluene detoxification rate at low temperature. J. Int. Phytorem. 23: 1–9. DOI: 10.1080/15226514.2020.1786795.

Khan, W.U., Ahmad, S.R., Yasin, N.A., Ali, A. and Ahmad, A. 2017. Effect of *Pseudomonas fluorescens* RB4 and Bacillus subtilis 189 on the phytoremediation potential of *Catharanthus roseus* (L.) in Cu and Pb-contaminated soils. J. Int. Phytorem. 19: 514–521. DOI: 10.1080/15226514.2016.1254154.

Kodre, A., Arčon, I., Debeljak, M., Potisek, M., Likar, M. and Vogel-Mikuš, K. 2017. Arbuscular mycorrhizal fungi alter Hg root uptake and ligand environment as studied by X-ray absorption fine structure. Envir. Experiment. Bot. 133: 12–23. DOI: 10.1016/j.envexpbot.2016.09.006.

Konieczny, A. and Kowalska, I. 2017. Effect of arbuscular mycorrhizal fungi on the content of zinc in lettuce grown at two phosphorus levels and an elevated zinc level in a nutrient solution. J. Element 22: 761–772. DOI: 10.5601/jelem.2016.21.4.1335.

Kullu, B., Patra, D.K., Acharya, S., Pradhan, C. and Patra, H.K. 2020. AM fungi mediated bioaccumulation of hexavalent chromium in *Brachiaria mutica*-a mycorrhizal phytoremediation approach. Chemosphere 258: 127337. DOI: 10.1016/j.chemosphere.2020.127337.

Kumar, S., Khare, R. and Trivedi, P.K. 2019. Arsenic-responsive high-affinity rice sulphate transporter, OsSultr1; 1, provides abiotic stress tolerance under limiting sulphur condition. J. Hazard. Mat. 373: 753–762. DOI: 10.1016/j.jhazmat.2019.04.011.

Leblebici, Z., Dalmiş, E. and Andeden, E.E. 2019. Determination of the potential of *Pistia stratiotes* L. in removing nickel from the environment by utilizing its rhizofiltration capacity. Brazil Archiv. Biol. Tech. 62: 1–10. DOI: 10.1590/1678-4324-2019180487.

Lebrun, M., Miard, F., Nandillon, R., Hattab-Hambli, N., Léger, J.C., Scippa, G.S., Morabito, D. and Bourgerie, S. 2021. Influence of biochar particle size and concentration on Pb and as availability in contaminated mining soil and phytoremediation potential of poplar assessed in a mesocosm experiment. Water. Air. Soil. Poll. 232: 1–21. DOI: 10.1007/s11270-020-04942-y.

Leudo, A.M., Cruz, Y., Montoya-Ruiz, C., Delgado, M.D.P. and Saldarriaga, J.F. 2020. Mercury phytoremediation with *Lolium perenne*-mycorrhizae in contaminated soils. Sustain. 12: 3795. DOI: 10.3390/su12093795.

Li, F., Yang, F., Chen, Y., Jin, H., Leng, Y. and Wang, J. 2020a. Chemical reagent-assisted phytoextraction of heavy metals by *Bryophyllum laetivirens* from garden soil made of sludge. Chemosphere 2020 Apr 2: 126574. DOI: 10.1016/j.chemosphere.2020.126574.

Li, H., Jiang, L., You, C., Tan, B. and Yang, W. 2020b. Dynamics of heavy metal uptake and soil heavy metal stocks across a series of Masson pine plantations. J. Clean. Prod. 1: 122395. DOI: 10.1016/j.jclepro.2020.122395.

Li, Z., Peng, R. and Yao, Q. 2020c. SlMYB14 promotes flavonoids accumulation and confers higher tolerance to 2, 4, 6-trichlorophenol in tomato. Plant Sci. 1: 110796. DOI: 10.1016/j.plantsci.2020.110796.

Li, H., Zheng, L., Wang, Y., Hu, X., Lu, Z., Fan, H., Zhang, Z., Li, Y., Zhuo, R. and Qiu, W. 2020d. Salt stress resistance in smcp-transgenic *Arabidopsis thaliana* as revealed by transcriptome analysis. Pak. J. Bot. 52: 735–52. DOI: 10.30848/PJB2020-3(8).

Li, R., Wu, H., Ding, J., Li, N., Fu, W., Gan, L. and Li, Y. 2020e. Transgenic merA and merB expression reduces mercury contamination in vegetables and grains grown in mercury-contaminated soil. Plant. Cell. Rep. 39: 1369–1380. DOI: 10.1007/s00299-020-02570-8.

Lin, J., Gao, X., Zhao, J., Zhang, J., Chen, S. and Lu, L. 2020. Plant cadmium resistance 2 (SaPCR2) facilitates cadmium efflux in the roots of hyperaccumulator *Sedum alfredii* Hance. Front. Plant. Sci. 1: 11. DOI: 10.3389/fpls.2020.568887.

Liu, D., An, Z., Mao, Z., Ma, L. and Lu, Z. 2015. Enhanced heavy metal tolerance and accumulation by transgenic sugar beets expressing *Streptococcus thermophilus* StGCS-GS in the presence of Cd, Zn and Cu alone or in combination. PLoS One 10: e0128824. DOI: 10.1371/journal.pone.0128824.

Lu, Q., Weng, Y., You, Y., Xu, Q., Li, H., Li, Y., Liu, H. and Du, S. 2020. Inoculation with abscisic acid (ABA)-catabolizing bacteria can improve phytoextraction of heavy metal in contaminated soil. Envir. Poll. 257: 113497. DOI: 10.1016/j.envpol.2019.113497.

Luo, J., Yin, D., Cheng, H., Davison, W. and Zhang, H. 2018. Plant induced changes to rhizosphere characteristics affecting supply of Cd to *Noccaea caerulescens* and Ni to *Thlaspi goesingense*. Envir. Sci. Tech. 52: 5085–5093. DOI: 10.1021/acs.est.7b04844.

Luo, Y., Wu, Y., Shu, J. and Wu, Z. 2019. Effect of particulate organic matter fractions on the distribution of heavy metals with aided phytostabilization at a zinc smelting waste slag site. Envir. Poll. 253: 330–341. DOI: 10.1016/j.envpol.2019.07.015.

Luo, Y., Wu, Y., Qiu, J., Wang, H. and Yang, L. 2019a. Suitability of four woody plant species for the phytostabilization of a zinc smelting slag site after 5 years of assisted revegetation. J. Soils. Sediment. 19: 702–715. DOI: 10.1007/s11368-018-2082-4.

Ma, X., Luo, W., Li, J. and Wu, F. 2019. Arbuscular mycorrhizal fungi increase both concentrations and bioavilability of Zn in wheat (*Triticum aestivum* L.) grain on Zn-spiked soils. Appl. Soil. Ecol. 135: 91–97. DOI: 10.1016/j.apsoil.2018.11.007.

Ma, Y., Rajkumar, M., Oliveira, R.S., Zhang, C. and Freitas, H. 2019a. Potential of plant beneficial bacteria and arbuscular mycorrhizal fungi in phytoremediation of metal-contaminated saline soils. J. Hazard. Mat. 379: 120813. DOI: 10.1016/j.jhazmat.2019.120813.

Ma, Y., Rajkumar, M., Zhang, C. and Freitas, H. 2016. Inoculation of *Brassica oxyrrhina* with plant growth promoting bacteria for the improvement of heavy metal phytoremediation under drought conditions. J. Hazard. Mat. 320: 36–44. DOI: 10.1016/j.jhazmat.2016.08.009.

Marappa, N., Dharumadurai, D. and Nooruddin, T. 2021. Bioremediation of noxious metals from e-waste printed circuit boards by Frankia. Microbiol. Res. 1: 126707. DOI: 10.1016/j.micres.2021.126707.

Masoudi, F., Shirvani, M., Shariatmadari, H. and Sabzalian, M.R. 2020. Performance of new biodegradable chelants in enhancing phytoextraction of heavy metals from a contaminated calcareous soil. J. Envir. Health Sci. Eng. 1: 1–0. DOI: 10.1007/s40201-020-00491-y.

Mathakutha, R., Steyn, C., Le Roux, P.C., Blom, I.J., Chown, S.L., Daru, B.H. and Greve, M. 2019. Invasive species differ in key functional traits from native and non-invasive alien plant species. J. Veget. Sci. 30: 994–1006. DOI: 10.1111/jvs.12772.

Mello, I.S., Targanski, S., Pietro-Souza, W., Stachack, F.F., Terezo, A.J. and Soares, M.A. 2020. Endophytic bacteria stimulate mercury phytoremediation by modulating its bioaccumulation and volatilization. Ecotoxicol. Envir. Safety 202: 110818. DOI: 10.1016/j.ecoenv.2020.110818.

Méndez-Gómez, M., Castro-Mercado, E., Peña-Uribe, C.A., Reyes-de la Cruz, H., López-Bucio, J. and García-Pineda, E. 2020. Target of rapamycin signaling plays a role in Arabidopsis growth promotion by *Azospirillum brasilense* Sp245. Plant. Sci. 293: 110416. DOI: 10.1016/j.plantsci.2020.110416.

Mirhosseini, M.S., Saeb, K., Rahnavard, A. and Kiadaliri, M. 2021. Phytoremediation of nickel and lead contaminated soils by *Hedera colchica*. Soil. Sed. Contam. 30: 122–133. DOI: 10.1080/15320383.2020.1832040.

Mishra, S.K., Kumar, P.R. and Singh, R.K. 2020. Transgenic plants in phytoremediation of organic pollutants. pp. 39–56. *In*: Bioremediation of Pollutants. Elsevier. DOI: 10.1016/B978-0-12-819025-8.00003-X.

Mitsopoulou, N., Lakiotis, K., Golia, E.E., Khah, E.M. and Pavli, O.I. 2020. Response of hrpZ Psph-transgenic *N. benthamiana* plants under cadmium stress. Envir. Sci. Poll. Res. 1: 1–0. DOI: 10.1007/s11356-020-09204-2.

Mohammadhasani, F., Ahmadimoghadam, A., Asrar, Z. and Mohammadi, S.Z. 2017. Effect of Zn toxicity on the level of lipid peroxidation and oxidative enzymes activity in Badami cultivar of pistachio (*Pistacia vera* L.) colonized by ectomycorrhizal fungus. J. Ind. Plant. Physiol. 22: 206–212. DOI: 10.1007/s40502-017-0300-5.

Mohammadi, M., Chalavi, V., Novakova-Sura, M., Laliberté, J.F. and Sylvestre, M. 2007. Expression of bacterial biphenyl-chlorobiphenyl dioxygenase genes in tobacco plants. Biotechnol. Bioeng. 97: 496–505. DOI: 10.1002/bit.21188.

Moradi, B., Maivan, H.Z., Hashtroudi, M.S., Sorahinobar, M. and Rohloff, J. 2021. Physiological responses and phytoremediation capability of *Avicennia marina* to oil contamination. Acta. Physiol. Plant. 43: 1–2. DOI: 10.1007/s11738-020-03177-y.

Motesharezadeh, B., Kamal-poor, S., Alikhani, H.A., Zariee, M. and Azimi, S. 2017. Investigating the effects of plant growth promoting bacteria and Glomus Mosseae on cadmium phytoremediation by *Eucalyptus camaldulensis* L. Poll. 3: 575–588. DOI: 10.22059/poll.2017.62774.

Mouhamad, R., Ibrahim, K.M. and Al-Daoude, A. 2020. Heavy metal phytoremediation potential of CYP4502E1 expressing *A. thaliana* and *S. grandiflora* plants. DYSONA-Life Sci. 1: 64–69. DOI: 10.30493/DLS.2020.222935.

Mukherjee, P., Roychowdhury, R. and Roy, M. 2017. Phytoremediation potential of rhizobacterial isolates from Kans grass (*Saccharum spontaneum*) of fly ash ponds. Clean Technol. Envir. Policy 19: 1373–1385. DOI: 10.1007/s10098-017-1336-y.

Naeem, H., Ahmad, K.S. and Jaffri, S.B. 2020. Biotechnological tools based lithospheric management of toxic Pyrethroid pesticides: A critical evaluation. J. Int. Envir. Anal. Chem. 1: 1–24. DOI: 10.1080/03067319.2020.1854240.

Nahar, N., Rahman, A., Nawani, N.N., Ghosh, S. and Mandal, A. 2017. Phytoremediation of arsenic from the contaminated soil using transgenic tobacco plants expressing ACR2 gene of *Arabidopsis thaliana*. J. Plant Physiol. 218: 121–126. DOI: 10.1016/j.jplph.2017.08.001.

Nayak, A.K., Panda, S.S., Basu, A. and Dhal, N.K. 2018. Enhancement of toxic Cr (VI), Fe, and other heavy metals hytoremediation by the synergistic combination of native *Bacillus cereus* strain and *Vetiveria zizanioides* L. J. Int. Phytorem. 20: 682–691. DOI: 10.1080/15226514.2017.1413332.

Nesler, A., DalCorso, G., Fasani, E., Manara, A., Di Sansebastiano, G.P., Argese, E. and Furini, A. 2017. Functional components of the bacterial CzcCBA efflux system reduce cadmium uptake and accumulation in transgenic tobacco plants. New Biotechnol. 35: 54–61.

Niazy, M.M. 2019. Rhizo-filtration of cu from water via water lettuce (*Pistia stratiots*) using na-lauryl sulphate surfactant. Egypt. J. Appl. Sci. 34: 317–334.

Nie, L., Shah, S., Rashid, A., Burd, G.I., Dixon, D.G. and Glick, B.R. 2002. Phytoremediation of arsenate contaminated soil by transgenic canola and the plant growth-promoting bacterium *Enterobacter cloacae* CAL2. Plant. Physiol. Biochem. 40: 355–361. DOI: 10.1016/S0981-9428(02)01375-X.

O'Connor, D., Zheng, X., Hou, D., Shen, Z., Li, G., Miao, G., O'Connell, S. and Guo, M. 2019. Phytoremediation: Climate change resilience and sustainability assessment at a coastal brown field redevelopment. Envir. Int. 130: 104945. DOI: 10.1016/j.envint.2019.104945.

Odinga, C.A., Kumar, A., Mthembu, M.S., Bux, F. and Swalaha, F.M. 2019. Rhizofiltration system consisting of *Phragmites australis* and *Kyllinga nemoralis*: Evaluation of efficient removal of metals and pathogenic microorganisms. Desal Water Treat. 169: 120–132. DOI: 10.5004/dwt.2019.24428.

Odjegba, V.J. and Fasidi, I.O. 2007. Phytoremediation of heavy metals by *Eichhornia crassipes*. Envir. 27: 349–355. DOI: 10.1007/s10669-007-9047-2.

Ortiz-Castro, R., Campos-García, J. and López-Bucio, J. 2020. *Pseudomonas putida* and *Pseudomonas fluorescens* influence *Arabidopsis* root system architecture through an auxin response mediated by bioactive cyclodipeptides. J. Plant. Growth Regul. 39: 254–265. DOI: 10.1007/s00344-019-09979-w.

Osama, R., Awad, H.M., Zha, S., Meng, F. and Tawfik, A. 2020. Greenhouse gases emissions from duckweed pond system treating polyester resin wastewater containing 1, 4-dioxane and heavy metals. Ecotoxicol. Envir. Safety 207: 111253. DOI: 10.1016/j.ecoenv.2020.111253.

Pandey, S., Ghosh, P.K., Ghosh, S., De, T.K. and Maiti, T.K. 2013. Role of heavy metal resistant *Ochrobactrum* sp. and *Bacillus* spp. strains in bioremediation of a rice cultivar and their PGPR like activities. J. Microbiol. 51: 11–17. DOI: 10.1007/s12275-013-2330-7.

Parvin, M.S. 2018. The functional characterization of AtStr5, a member of the sulfurtransferase/rhodanese family of *Arabidopsis thaliana*, and its arsenic phytoremediation studies. Hanover: Gottfried Wilhelm Leibniz Universität, Diss., IX, 147. DOI: 10.15488/3855.

Peng, H., Wang, Y., Tan, T.L. and Chen, Z. 2020. Exploring the phytoremediation potential of water hyacinth by FTIR Spectroscopy and ICP-OES for treatment of heavy metal contaminated water. J. Int. Phytorem. 1: 1–3. DOI: 10.1080/15226514.2020.1774499.

Petriccione, M., Di Patre, D., Ferrante, P., Papa, S., Bartoli, G., Fioretto, A. and Scortichini, M. 2013. Effects of *Pseudomonas fluorescens* seed bioinoculation on heavy metal accumulation for *Mirabilis jalapa* phytoextraction in smelter-contaminated soil. Water Air Soil Poll. 224: 1645. DOI: 10.1007/s11270-013- 1645-7.

Punetha, D., Tewari, G., Pande, C., Kharkwal, G. and Tripathi, S. 2021. Assessment of phytoremediation efficiency of *Coriandrum sativum* in metal polluted soil and sludge samples: A green approach. pp. 58–85. *In*: Handbook of Research on Waste Diversion and Minimization Technologies for the Industrial Sector. IGI Global. DOI: 10.4018/978-1-7998-4921-6.ch004.

Purwanti, I.F., Obenu, A., Tangahu, B.V., Kurniawan, S.B., Imron, M.F. and Abdullah, S.R. 2020. Bioaugmentation of *Vibrio alginolyticus* in phytoremediation of aluminium-contaminated soil using *Scirpus grossus* and *Thypa angustifolia*. Heliyon. 6: e05004. DOI: 10.1016/j.heliyon.2020.e05004.

Raldugina, G.N., Maree, M., Mattana, M., Shumkova, G., Mapelli, S., Kholodova, V.P., Karpichev, I.V. and Kuznetsov, V.V. 2018. Expression of rice OsMyb4 transcription factor improves tolerance to copper or zinc in canola plants. Biol. Plant 62: 511–520. DOI: 10.1007/s10535-018-0800-9.

Ramírez, V., Baez, A., López, P., Bustillos, M.D.R., Villalobos, M.A., Carreño, R. and Munive, J.A. 2019. Chromium hyper-tolerant *Bacillus* sp. MH778713 assists phytoremediation of heavy metals by mesquite trees (*Prosopis laevigata*). Front. Microbiol. 10: 1833. DOI: 10.3389/fmicb.2019.01833.

Ren, H., Wan, Y. and Zhao, Y. 2018. Phytoremediation of polychlorinated biphenyl-contaminated soil by transgenic alfalfa associated bioemulsifier AlnA. pp. 645–653. *In*: Twenty Years of Research and Development on Soil Pollution and Remediation in China. Springer, Singapore. DOI: 10.1007/978-981-10-6029- 8_39.

Romeroso, R., Tandang, D.N. and Navarrete, I.A. 2021. New distributional record of *Phyllanthus securinegoides* Merr. (Phyllanthaceae) and *Rinorea niccolifera* Fernando (Violaceae) of Homonhon Island, Philippines. J. Biodiver. Biol. Diver. 22: 1–10. DOI: 10.13057/biodiv/d220160.

RoyChowdhury, A., Mukherjee, P., Panja, S., Datta, R., Christodoulatos, C. and Sarkar, D. 2021. Evidence for phytoremediation and phytoexcretion of NTO from industrial wastewater by vetiver grass. Molecules 26: 74. DOI: 10.3390/molecules26010074.

Rui, H.Y., Chen, C., Zhang, X.X., Shen, Z.G. and Zhang, F.Q. 2016. Cd-induced oxidative stress and lignifcation in the roots of two *Vicia sativa* L. varieties with diferent Cd tolerances. J. Hazard. Mater. 301: 304–313. DOI: 10.1016/j.jhazmat.2015.08.052.

Saleem, M.H., Ali, S., Rehman, M., Rizwan, M., Kamran, M., Mohamed, I.A., Bamagoos, A.A., Alharby, H.F., Hakeem, K.R. and Liu, L. 2020. Individual and combined application of EDTA and citric acid assisted phytoextraction of copper using jute (*Corchorus capsularis* L.) seedlings. Envir. Tech. Innov. 1: 100895. DOI: 10.1016/j.eti.2020.100895.

Salmani-Ghabeshi, S., Fadic-Ruiz, X., Miró-Rodríguez, C., Pinilla-Gil, E. and Cereceda-Balic, F. 2021. Trace element levels in native plant species around the industrial site of *Puchuncaví-Ventanas* (Central Chile): Evaluation of the phytoremediation potential. Appl. Sci. 11: 713. DOI: 10.3390/app11020713.

Saran, A., Fernandez, L., Cora, F., Savio, M., Thijs, S., Vangronsveld, J. and Merini, L.J. 2020. Phytostabilization of Pb and Cd polluted soils using *Helianthus petiolaris* as pioneer aromatic plant species. J. Int. Phytorem. 22: 459–467. DOI: 10.1080/15226514.2019.1675140.

Savani, K.A., Bhattacharyya, A., Boro, R.C., Dinesh, K. and Jc, N.S. 2021. Exemplifying endophytes of banana (*Musa paradisiaca*) for their potential role in growth stimulation and management of *Fusarium oxysporum* f. sp cubense causing panama wilt. Folia. Microbiol. 1: 1–4. DOI: 10.1007/s12223-021-00853-5.

Sayın, F.E., Khalvati, M.A. and Erdinçler, A. 2019. Effects of sewage sludge application and arbuscular mycorrhizal fungi (*G. mosseae* and *G. intraradices*) interactions on the heavy metal phytoremediation in chrome mine tailings. 7th international conference on Sustainable waste management, Heraklion 2019, 26–29 June, 2019, 1–11.

Schneider, J., Bundschuh, J., de Melo Rangel, W. and Guilherme, L.R.G. 2017. Potential of different AM fungi (native from As-contaminated and uncontaminated soils) for supporting *Leucaena leucocephala* growth in As-contaminated soil. Envir. Poll. 224: 125–135. DOI: 10.1016/j.envpol.2017.01.071.

Sharma, P., Tripathi, S., Chaturvedi, P., Chaurasia, D. and Chandra, R. 2020. Newly isolated *Bacillus* sp. PS-6 assisted phytoremediation of heavy metals using *Phragmites communis*: Potential application in wastewater treatment. Bioresource Tech. 320: 124353. DOI: 10.1016/j.biortech.2020.124353.

Shazia, I., Ahmad, K.S. and Jaffri, S.B. 2018. Mycodriven enhancement and inherent phytoremediation potential exploration of plants for lithospheric remediation. Sydowia. 70: 141–153. DOI: 10.12905/0380.sydowia70- 2018-0141.

Sheng, X.F., Xia, J.J., Jiang, C.Y., He, L.Y. and Qian, M. 2008. Characterization of heavy metal-resistant endophytic bacteria from rape (*Brassica napus*) roots and their potential in promoting the growth and lead accumulation of rape. Envir. Poll. 156: 1164–1170. DOI: 10.1016/j.envpol.2008.04.007.

Shi, L., Deng, X., Yang, Y., Jia, Q., Wang, C., Shen, Z. and Chen Y. 2019. A Cr (VI)-tolerant strain, *Pisolithus* sp1, with a high accumulation capacity of Cr in mycelium and highly efficient assisting *Pinus thunbergii* for phytoremediation. Chemosphere 224: 862–872. DOI: 10.1016/j.chemosphere.2019.03.015.

Shilev, S., Fernández, A., Benlloch, M. and Sancho, E.D. 2006. Sunflower growth and tolerance to arsenic is increased by the rhizospheric bacteria *Pseudomonas fluorescens*. pp. 315–318. *In*: Phytoremediation of Metal-Contaminated Soils. Springer, Dordrecht. DOI: 10.1007/1-4020-4688-X_12.

Shim, D., Kim, S., Choi, Y.I., Song, W.Y., Park, J., Youk, E.S., Jeong, S.C., Martinoia, E., Noh, E.W. and Lee, Y. 2013. Transgenic poplar trees expressing yeast cadmium factor 1 exhibit the characteristics necessary for the phytoremediation of mine tailing soil. Chemosphere 90: 1478–1486. DOI: 10.1016/j.chemosphere.2012.09.044.

Shim, J., Kim, J.W., Shea, P.J. and Oh, B.T. 2015. IAA production by *Bacillus* sp. JH 2-2 promotes Indian mustard growth in the presence of hexavalent chromium. Basic Microbiol. 55: 652–658.

Shin, M.N., Shim, J., You, Y., Myung, H., Bang, K.S., Cho, M., Kamala-Kannan, S. and Oh, B.T. 2012. Characterization of lead resistant endophytic *Bacillus* sp. MN3-4 and its potential for promoting lead accumulation in metal hyperaccumulator *Alnus firma*. J. Hazard. Mat. 199: 314–320. DOI: 10.1016/j.jhazmat.2011.11.010.

Shrirangasami, S.R., Rakesh, S.S., Murugaragavan, R., Ramesh, P.T., Varadharaj, S., Elangovan, R. and Saravanakumar, S. 2020. Phytoremediation of contaminated soils—A review. Int. J. Curr. Microbiol. App. Sci. 9: 3269–3283.

Shukla, D., Kesari, R., Tiwari, M., Dwivedi, S., Tripathi, R.D., Nath, P. and Trivedi, P.K. 2013. Expression of *Ceratophyllum demersum* phytochelatin synthase, CdPCS1, in *Escherichia coli* and Arabidopsis enhances heavy metal(loid)s accumulation. Protoplasma 250: 1263–1272. DOI: 10.1007/s00709-013-0508-9.

Sidek, N.M., Abdullah, S.R., Draman, S.F., Rosli, M.M. and Sanusi, M.F. 2018. Phytoremediation of abandoned mining lake by water hyacinth and water lettuces in constructed wetlands. Jurnal Teknol. 80: 1–10. DOI: 10.11113/jt.v80.10992.

Singh, S., Korripally, P., Vancheeswaran, R. and Eapen, S. 2011. Transgenic *Nicotiana tabacum* plants expressing a fungal copper transporter gene show enhanced acquisition of copper. Plant. Cell. Rep. 30: 1929–1938. DOI: 10.1007/s00299-011-1101-3.

Singh, S., Sherkhane, P.D., Kale, S.P. and Eapen, S. 2011. Expression of a human cytochrome P4502E1 in *Nicotiana tabacum* enhances tolerance and remediation of γ-hexachlorocyclohexane. New Biotech. 28: 423–429. DOI: 10.1016/j.nbt.2011.03.010.

Siregar, U.J. 2019. Heavy metal absorption of four fast growing tree species on media containing tailing from Pongkor gold mining in Indonesia. *In*: IOP Conference Series: Earth and Environmental Science 394(1): 012070. IOP Publishing. DOI: 10.1088/1755-1315/394/1/012070.

Skorbiłowicz, E., Skorbiłowicz, M., Kisiel, A. and Klimowicz, J. 2020. Assessment of the condition of Biebrza river aquatic environment with the use of *Phragmites Australis* (Cav.) Trin. ex Steud (Poland). Desal. Water. Treat. 1: 186. DOI: 10.5004/dwt.2020.25329.

Sone, Y., Nakamura, R., Pan-Hou, H., Sato, M.H., Itoh, T. and Kiyono, M. 2013. Increase methylmercury accumulation in *Arabidopsis thaliana* expressing bacterial broad-spectrum mercury transporter MerE. AMB Express. 3: 52. DOI: 10.1186/2191-0855-3-52.

Sun, L., Ma, Y., Wang, H., Huang, W., Wang, X., Han, L., Sun, W., Han, E. and Wang, B. 2018. Overexpression of PtABCC1 contributes to mercury tolerance and accumulation in *Arabidopsis* and poplar. Biochem. Biophys. Res. Comm. 497: 997–1002. DOI: 10.1016/j.bbrc.2018.02.133.

Sun, X., Sun, M., Chao, Y., Wang, H., Pan, H., Yang, Q., Cui, X., Lou, Y. and Zhuge, Y. 2020. Alleviation of lead toxicity and phytostimulation in perennial ryegrass by the Pb-resistant fungus *Trichoderma asperellum* SD-5. Funct. Plant Biol. 1: 1–10. DOI: 10.1071/FP20237.

Tahir, M.B., Ahmad, A., Iqbal, T., Ijaz, M., Muhammad, S. and Siddeeg, S.M. 2019a. Advances in photo-catalysis approach for the removal of toxic personal care product in aqueous environment. Envir. Devel. Sust. 1: 1–24. DOI: 10.1007/s10668-019-00495-1.

Tahir, M.B., Tufail, S., Ahmad, A., Rafique, M., Iqbal, T., Abrar, M., Nawaz, T., Khan, M.Y. and Ijaz, M. 2019b. Semiconductor nanomaterials for the detoxification of dyes in real wastewater under visible-light photocatalysis. J. Int. Envir. Anal. Chem. 1: 1–5. DOI: 10.1080/03067319.2019.1686494.

Tahir, M.B., Malik, M.F., Ahmed, A., Nawaz, T., Ijaz, M., Min, H.S., Muhammad, S. and Siddeeg, S.M. 2020. Semiconductor based nanomaterials for harvesting green hydrogen energy under solar light irradiation. J. Int. Envir. Anal. Chem. 1: 1–7. DOI: 10.1080/03067319.2019.1700970.

Tang, C., Chen, Y., Zhang, Q., Li, J., Zhang, F. and Liu, Z. 2019. Effects of peat on plant growth and lead and zinc phytostabilization from lead-zinc mine tailing in southern China: Screening plant species resisting and accumulating metals. Ecotoxicol. Envir. Safety 176: 42–49. DOI: 10.1016/j.ecoenv.2019.03.078.

Tanwar, A., Aggarwal, A., Charaya, M.U. and Kumar, P. 2015. Cadmium remediation by arbuscular mycorrhizal fungus–colonized celery plants supplemented with ethylenediaminetetraacetic acid. J. Biorem. 19: 188–200. DOI: 10.1080/10889868.2014.995371.

Terry, N. and Banuelos, G.S. (eds.). 2020. Phytoremediation of Contaminated Soil and Water. CRC Press, 1–30.

Thakore, D. and Srivastava, A.K. 2021. Mass scale hairy root cultivation of *Catharanthus roseus* in bioreactor for indole alkaloid production. Plant Cell and Tissue Differentiation and Secondary Metabolites: Fundamentals and Applications 1: 487–502. DOI: 10.1007/978-3-030-30185-9_21.

Tıpırdamaz, R., Karakas, S. and Dikilitas, M. 2020. Halophytes and the future of agriculture. Handbook of Halophytes: From Molecules to Ecosystems towards Biosaline Agriculture 1: 1–5. DOI: 10.1007/978-3- 030-17854-3_91-1.

Tran, B.T., Cavagnaro, T.R. and Watts-Williams, S.J. 2019. Arbuscular mycorrhizal fungal inoculation and soil zinc fertilisation affect the productivity and the bioavailability of zinc and iron in durum wheat. Mycorrhiza 29: 445–457. DOI: 10.1007/s00572-019-00911-4.

Tsyganov, V.E., Tsyganova, A.V., Gorshkov, A.P., Seliverstova, E.V., Kim, V.E., Chizhevskaya, E.P., Belimov, A.A., Serova, T.A., Ivanova, K.A., Kulaeva, O.A. and Kusakin, P.G. 2002. Efficacy of a plant-microbe system: *Pisum sativum* (L.) cadmium-tolerant mutant and *Rhizobium leguminosarum* strains, expressing pea metallothionein genes PsMT1 and PsMT2, for cadmium phytoremediation. Front Microbiol. 11: 15. DOI: 10.3389/fmicb.2020.00015.

Turcios, A.E., Hielscher, M., Duarte, B., Fonseca, V.F., Caçador, I. and Papenbrock, J. 2021. Screening of emerging pollutants (EPs) in estuarine water and phytoremediation capacity of *Tripolium pannonicum* under controlled conditions. J. Int. Envir. Res. Public Health. 18: 943. DOI: 10.3390/ijerph18030943.

Viana, D.G., Pires, F.R., Ferreira, A.D., Egreja Filho, F.B., de Carvalho, C.F., Bonomo, R. and Martins, L.F.. 2021. Effect of planting density of the macrophyte consortium of *Typha domingensis* and *Eleocharis acutangula* on phytoremediation of barium from a flooded contaminated soil. Chemosphere. 262: 127869. DOI: 10.1016/j.chemosphere.2020.127869.

Viktorová, J., Novakova, M., Trbolova, L., Vrchotova, B., Lovecka, P., Mackova, M. and Macek, T. 2014. Characterization of transgenic tobacco plants containing bacterial bphc gene and study of their phytoremediation ability. J. Int. Phytorem. 16: 937–946. DOI: 10.1080/15226514.2013.810575.

Villanueva-Galindo, E. and Moreno-Andrade, I. 2020. Bioaugmentation on hydrogen production from food waste. J. Int. Hydrogen Energy 1: 1–10. DOI: 10.1016/j.ijhydene.2020.11.092.

Visconti, D., Álvarez-Robles, M.J., Fiorentino, N., Fagnano, M. and Clemente, R. 2020. Use of *Brassica juncea* and *Dactylis glomerata* for the phytostabilization of mine soils amended with compost or biochar. Chemosphere 260: 127661. DOI: 10.1016/j.chemosphere.2020.127661.

Wang, X., Zhi, J., Liu, X., Zhang, H., Liu, H., Xu, J. 2018. Transgenic tobacco plants expressing a P1B-ATPase gene from *Populus tomentosa* Carr. (PtoHMA5) demonstrate improved cadmium transport. J. Int. Biol. Macro. 113: 655–661. DOI: 10.1016/j.ijbiomac.2018.02.081

Wang, G., Wang, L., Ma, F., You, Y., Wang, Y. and Yang, D. 2020. Integration of earthworms and arbuscular mycorrhizal fungi into phytoremediation of cadmium-contaminated soil by *Solanum nigrum* L. J. Hazard. Mat. 389: 121873. DOI: 10.1016/j.jhazmat.2019.121873.

Watts-Williams, S.J. and Cavagnaro, T.R. 2018. Arbuscular mycorrhizal fungi increase grain zinc concentration and modify the expression of root ZIP transporter genes in a modern barley (*Hordeum vulgare*) cultivar. Plant Sci. 274: 163–170. DOI: 10.1016/j.plantsci.2018.05.015.

Wei, S., Kohda, Y.H., Inoue, C. and Chien, M.F. 2020. Expression of PvPht1; 3, PvACR2 and PvACR3 during arsenic processing in root of *Pteris vittata*. Envir. Exper. Bot. 182: 104312. DOI: 10.1016/j.envexpbot.2020.104312.

Wilarso, B.S., Wibowo, C., Sukendro, A. and Bekti, H.S. 2020. The Growth improvement of *Falcataria moluccana* inoculated with MycoSilvi grown in post-mining silica sand soil medium amended with soil ameliorants. Biodiver. J. Biol. Diver. 21: 1–10. DOI: 10.13057/biodiv/d210149.

Wu, J., Wang, Q. and Zhu, X. 2002. 2020. A new approach to protect tobacco plants from Cd contamination using the attenuated recombinant virus CMV△ 2b containing the PvSR2 gene. Biotech. Biotech. Equip. 34: 475–481. DOI: 10.1080/13102818.2020.1777898.

Wu, Y., Ma, L., Liu, Q., Sikder, M.M., Vestergård, M., Zhou, K. and Feng, Y. 2020. *Pseudomonas fluorescens* promote photosynthesis, carbon fixation and cadmium phytoremediation of hyperaccumulator *Sedum alfredii*. Sci. Total Envir. 1: 138554. DOI: 10.1016/j.scitotenv.2020.138554.

Xia, Y., Liu, J., Wang, Y., Zhang, X., Shen, Z. and Hu, Z. 2018. Ectopic expression of *Vicia sativa* Caffeoyl-CoA O-methyltransferase (VsCCoAOMT) increases the uptake and tolerance of cadmium in *Arabidopsis*. Envir. Exper. Bot. 145: 47–53. DOI: 10.1016/j.envexpbot.2017.10.019.

Xu, X., Xu, M., Zhao, Q., Xia, Y., Chen, C. and Shen, Z. 2018. Complete genome sequence of Cd (II)-resistant Arthrobacter sp. PGP41, a plant growth-promoting bacterium with potential in microbe-assisted phytoremediation. Current Microbiol. 75: 1231–1239. DOI: 10.1007/s00284-018-1515-z.

Yadav, A., Goyal, D., Prasad, M., Singh, T.B., Shrivastav, P., Ali, A. and Dantu, P.K. 2020. Bioremediation of toxic pollutants: features, strategies, and applications. pp. 361–383. *In*: Contaminants in Agriculture. Springer, Cham. DOI: 10.1007/978-3-030-41552-5_18.

Yahaghi, Z., Shirvani, M., Nourbakhsh, F., De La Pena, T.C., Pueyo, J.J. and Talebi, M. 2018. Isolation and characterization of Pb-solubilizing bacteria and their effects on Pb uptake by *Brassica juncea*: Implications for microbe-assisted phytoremediation. J. Microbiol. Biotechnol. 28: 1156–1167. DOI: 10.4014/jmb.1712.12038.

Yan, X., Huang, J., Xu, X., Chen, D., Xie, X., Tao, Q., He, J. and Jiang, J. 2018. Enhanced and complete removal of phenylurea herbicides by combinational transgenic plant-microbe remediation. Appl. Envir. Microbiol. 84: 1–10. DOI: 10.1128/AEM.00273-18.

Yao, J., Sun, J., Chen, Y., Shi, L., Yang, L. and Wang, Y. 2020. The molecular mechanism underlying cadmium resistance in NHX1 transgenic *Lemna turonifera* was studied by comparative transcriptome analysis. Plant Cell, Tissue and Organ Culture (PCTOC) 143: 189–200. DOI: 10.1007/s11240-020-01909-z.

Ye, P., Wang, M., Zhang, T., Liu, X., Jiang, H., Sun, Y., Cheng, X. and Yan, Q. 2020. Enhanced cadmium accumulation and tolerance in transgenic hairy roots of *Solanum nigrum* L. expressing iron-regulated transporter gene IRT1. Life 10: 324. DOI: 10.3390/life10120324.

Yu, S., Teng, C., Bai, X., Liang, J., Song, T., Dong, L., Jin, Y. and Qu, J. 2017. Optimization of siderophore production by Bacillus sp. PZ-1 and its potential enhancement of phytoextration of Pb from soil. J. Microbiol. Biotech. 27: 1500–1512. DOI: 10.4014/jmb.1705.05021.

Yu, P., Sun, Y., Huang, Z., Zhu, F., Sun, Y. and Jiang, L. 2020. The effects of ectomycorrhizal fungi on heavy metals' transport in *Pinus massoniana* and bacteria community in rhizosphere soil in mine tailing area. J. Hazard. Mat. 381: 121203. DOI: 10.1016/j.jhazmat.2019.121203.

Yu ,Y., Luo, L., Yang, K. and Zhang, S. 2011. Influence of mycorrhizal inoculation on the accumulation and speciation of selenium in maize growing in selenite and selenate spiked soils. Pedobiol. 54: 267–272. DOI: 10.1016/j.pedobi.2011.04.002.

Zhang, L., Rylott, E.L., Bruce, N.C. and Strand, S.E. 2019. Genetic modification of western wheatgrass (*Pascopyrum smithii*) for the phytoremediation of RDX and TNT. Planta 249: 1007–1015. DOI: 10.1007/s00425-018- 3057-9.

Zhang, L., Rylott, E.L., Bruce, N.C. and Strand, S.E. 2017. Phytodetoxification of TNT by transplastomic tobacco (*Nicotiana tabacum*) expressing a bacterial nitroreductase. Plant. Mol. Biol. 95: 99–109. DOI: 10.1007/s11103-017-0639-z.

Zhang, L., Wu, J., Tang, Z., Huang, X.Y., Wang, X., Salt, D.E. and Zhao, F.J. 2019. Variation in the BrHMA3 coding region controls natural variation in cadmium accumulation in *Brassica rapa* vegetables. J. Exp. Bot. 4: 5865–78. DOI: 10.1093/jxb/erz310.

Zhang, X., Li, Q., Xu, W., Zhao, H., Guo, F., Wang, P., Wang, Y., Ni, D., Wang, M. and Wei, C. 2020. Identification of MTP gene family in tea plant (*Camellia sinensis* L.) and characterization of CsMTP8. 2 in manganese toxicity. Ecotoxicol Envir. Safety 202: 110904. DOI: 10.1016/j.ecoenv.2020.110904.

Zhang, X., Rui, H., Zhang, F., Hu, Z., Xia, Y. and Shen, Z. 2018. Overexpression of a functional *Vicia sativa* PCS1 homolog increases cadmium tolerance and phytochelatins synthesis in *Arabidopsis*. Front. Plant. Sci. 1–10. DOI: 10.3389/fpls.2018.00107.

Zhang, Y., Sa, G., Zhang, Y., Hou, S., Wu, X., Zhao, N., Zhang, Y., Deng, S., Deng, C., Deng, J. and Zhang, H. 2020a. *Populus euphratica* annexin1 facilitates cadmium enrichment in transgenic Arabidopsis. J. Hazard. Mat. 124063. DOI: 10.1016/j.jhazmat.2020.124063.

Zhao, H., Wei, Y., Wang, J. and Chai, T. 2019. Isolation and expression analysis of cadmium-induced genes from Cd/Mn hyperaccumulator *Phytolacca americana* in response to high Cd exposure. Plant Biol. 21: 15–24. DOI: 10.1111/plb.12908.

Zhou, X., Fu, L., Xia, Y., Zheng, L., Chen, C., Shen, Z. and Chen, Y. 2017. Arbuscular mycorrhizal fungi enhance the copper tolerance of *Tagetes patula* through the sorption and barrier mechanisms of intraradical hyphae. Metal 9: 936–948. DOI: 10.1039/c7mt90028g.

Zhu, S., Shi, W., Jie, Y., Zhou, Q. and Song, C. 2020. A MYB transcription factor, BnMYB2, cloned from ramie (*Boehmeria nivea*) is involved in cadmium tolerance and accumulation. PloS One 15: e0233375. DOI: 10.1371/journal.pone.0233375.

Złoch, M., Tyburski, J. and Hrynkiewicz, K. 2015. Analysis of microbiologically stimulated biomass of *Salix viminalis* L. in the presence of Cd^{2+} under *in vitro* conditions–implications for phytoremediation. Acta. Biol. Cracs. Bot. 1–10. DOI: DOI: 10.1515/abcsb-2015-0024.

CHAPTER 7

Synthesis and Applications of Transition Metal Oxide Nanoparticles for Bacteria Disinfection and Virus Inactivation in the Environment

Yolanda G. Garcia-Huante,[1, 2] *Martha A. Gomez-Gallegos,*[3,*]
Oscar D. Máynez-Navarro,[3] *Erick R. Bandala*[4] and *Irwing M. Ramirez*[5,*]

1. Introduction

The rise in population has led to environmental pollution, the emerging of antibiotic-resistant bacteria and a faster spread of viruses. These issues have become one of the primary public concerns in recent times. One major apprehension is on surfaces, which can act as a reservoir of transmitting diseases to humans after contamination with body fluids, excretions and airborne pollutants (Benitha et al., 2019). Studies suggest that contaminated surfaces or fomites play a significant role in the rapid spread of diseases, particularly in indoor establishments such as educational, medical and catering (Barker et al., 2001; Vasickova et al., 2010). Therefore, active surfaces and technologies will be crucial to keep up essential public health standards for pathogen removal (Schijven et al., 2013; Yiu et al., 2013).

Many efforts have been made to develop surfaces with antiviral and antibacterial properties and decontamination technologies (Sportelli et al., 2020). Inorganic nanoparticles (NPs) have gained importance as active materials on covers because they can control infections by physic, catalytic or photocatalytic activity. An example of this, NPs based on transition metal oxides semiconductors

[1] Center for Systems and Synthetic Biology and Department of Molecular Biosciences, University of Texas at Austin, Texas 78712, USA.
[2] Tecnologico de Monterrey, School of Engineering and Sciences.
Email: yori_gh12@tec.mx
[3] Universidad de las Américas Puebla, Sta. Catarina Mártir, Cholula, Puebla, 72810, Mexico.
Email: oscar.maynezno@udlap.mx
[4] Division of Hydrologic Sciences. Desert Research Institute, 755 E. Flamingo Road, Las Vegas 89119-7363 Nevada, USA.
Email: erick.bandala@dri.edu
[5] Department of Civil, Architectural and Environmental Engineering, The University of Texas at Austin, Austin, TX, 78712, USA.
* Corresponding author: martha.gomezgs@udlap.mx; irwingmoises@gmail.com

represent an alternative to avoid spreading diseases because of their effectiveness as disinfectants and antiviral capabilities. These NPs based on transition metal oxides have also been applied in environmental problems, biomedical applications and industrial products (Kavitha et al., 2017; Anuradha and Raji, 2019a).

In particular, NPs based on cobalt (Co), nickel (Ni), copper (Cu) and zinc (Zn) oxide semiconductors have been recognized to have a significant activity against bacteria and viruses. For example, NPs based on these transition metal oxides can prevent the spread of disease and infections in the healthcare sector (Piedade et al., 2020), food packing (El Fawal et al., 2020), bacteriological concrete degradation (Dyshlyuk et al., 2020) and fruit preservation (Arroyo et al., 2020). Several physics, chemical and green methods have been reported, including chemical precipitation, chemical reduction, co-precipitation, electrospun, hydrothermal, microwave, pyrolysis, sol-gel, thermal decomposition, thermolysis, vacuum impregnation and wet chemical methods (Çolak et al., 2017; Abinaya et al., 2018; Dobrucka et al., 2018; Theophil et al., 2019).

For example, Gajendiran et al. (2020) found that Co_3O_4 antibiotic effectivity was comparable to amoxicillin. Derbalah et al. (2019) came across significant antiviral activity in NiO NPs . Murray and Laband (1979) found that CuO easily adsorbed and denatured virus proteins and suggested its activity was associated with strong Van der Waals interactions. Gad El-Rab et al. (2020) demonstrated that ZnO could be used against antibiotic-resistant bacteria Recently, Weiss et al. (2020) discovered CuO NPs and their composites could inactivate coronavirus, suggesting copper oxides could become promising antiviral agents to fight SARS-CoV-2.

In brief, it has been suggested that the main mechanisms explaining disinfection and virus inactivation capabilities of NPs based on Co, Ni, Cu and Zn oxides are based on the metal ion diffusion, physical interactions and the Reactive Oxygen Species (ROS) formation resulting from activation under radiation (Slavin et al., 2017). This chapter deals with the synthesis of oxide NPs based on cobalt, nickel, copper and zinc, as well as their catalytic and photocatalytic applications for bacteria disinfection and virus inactivation in the last 5 yr.

2. Synthesis

The synthetic route used for developing nanostructures plays an important role in their final properties because the shape, size and superficial properties of nanoparticulated materials can be carefully controlled during preparation. Thus, such nanostructure engineering involves chemical manipulation of nano-building blocks to control composition, morphology and size. In this regard, synthetic routes are divided into two main categories (Fig. 7.1): the top-down and the bottom-up approaches (Ghiuță et al., 2017).

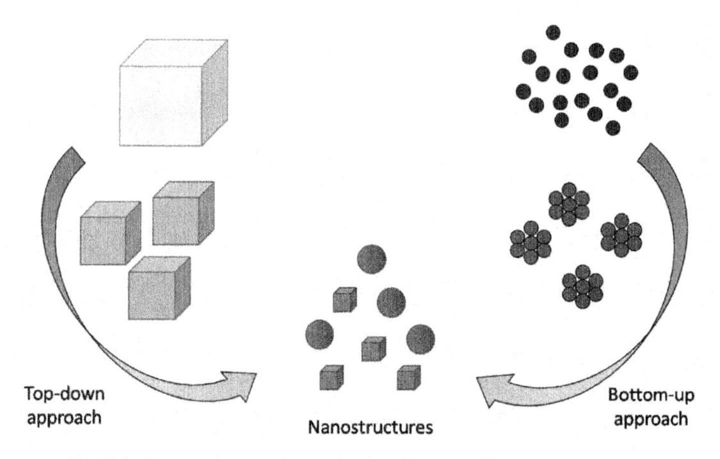

Fig. 7.1. Approaches on the synthesis of metal oxides nanostructures.

The top-down approach is a destructive method in which larger structures are decomposed into smaller units until they are suitable nanomaterials. The raw material reduction in nanoparticle size can be managed by physical or chemical means, such as mechanic grinding, physical vapor deposition and chemical vapor deposition (Khan et al., 2019). On the other hand, the bottom-up approach uses simple compounds as building blocks to build larger structures manipulating individual atoms and molecules through self-assembly processes, leading to larger nanoscale structures (Ghiuţă et al., 2017). Unlike the top-down approach, bottom-up involves atomic-level manipulation and has proved to be favorable in controlling the size, morphology and chemical composition of nanomaterials (Chaturvedi et al., 2012). Specifically, transition metal oxide nanoparticles based on Co, Ni, Cu and Zn have attracted significant attention because of their unique physicochemical properties and superficial structures (Cid and Simal-Gandara, 2020).

Nanostructures such as tricobalt tetraoxide (Co_3O_4), nickel oxide (NiO), cupric oxide (CuO) and zinc oxide (ZnO) display a wide variety of properties, which make their applications manifold. The electronic structure of these materials is particularly interesting because they can act as semiconductors, making them important materials for environmental applications in catalysis, photocatalysis, water splitting and disease control owing to a wide bandgap (E_g) and remarkable optical properties (Khan et al., 2019). Besides, these transition metal oxide nanostructures possess several advantages over their bulk counterparts because of higher surface area than those on the micro or macro scale and the improved biocompatibility by enhancing aqueous solubility. Synthesis approaches to obtain Co_3O_4, NiO, CuO and ZnO nanostructure used in bacteria disinfection and virus inactivation are described below.

2.1 Synthesis of tricobalt tetraoxide

Co_3O_4 is a spinel-structured material with E_g ranging from 1.48 to 2.19 eV, considered a p-type semiconducting material, where Co^{2+} ions and Co^{3+} occupy the tetragonal sites and octahedral sites, respectively, arranged in a cubic close-packed structure (Pagar et al., 2019). The Co_3O_4 electrochemical properties are attributed to such spatial arrays (Pagar et al., 2019). Various Co_3O_4 morphologies have been synthesized using different methodologies (see Table 7.1). Co_3O_4 was applied as an antimicrobial agent and demonstrated photocatalytic activity and biocompatibility (see Table 7.1).

Co_3O_4 NPs showed antibacterial activity against *Escherichia coli* (Gajendiran et al., 2020). Kavitha et al. (2017) studied the cytotoxicity of Co_3O_4 NPs against fibroblast cells and Gajendiran et al. (2020) tested the hemocompatibility. Additionally, Co_3O_4 presented photocatalytic activity in textile dyes degradation with good results (Jesudoss et al., 2017). Besides, Co_3O_4 NPs synthesized using the *Vitis vinifera* extract demonstrated the photocatalytic activity to remove contaminants present in textile dyeing wastewater in which the presence of nitroaromatic compounds was observed (Kombaiah et al., 2019).

2.2 Synthesis of nickel oxide

NiO possesses a wide E_g ranging (3.6 to 4.0 eV) and exhibits electrochemical stability. Furthermore, NiO has shown remarkable potential in disinfection, photocatalysis, electrodes for lithium-ion batteries, photocathodes, catalysts, magnetic material and electrochemical supercapacitors (see Table 7.2). The size and morphology of NiO nanostructures are two significant properties that depend on the synthesis route used for its preparation, among which the most common are sol-gel (Benitha et al., 2019), wet chemical synthesis (Paul and Neogi, 2019), hydrothermal (Derbalah et al., 2019) and biogenic approaches (Helan et al., 2016).

Chu et al. (2016) explored a bio template hierarchical synthesis of Ni/NiO nanorods using Tobacco Mosaic Virus (TMVs) as a template, managing to grow NiO uniformly. NiO has shown effective antibacterial activity, as proved by various groups (Table 7.2). For example, NiO nanoparticles can

Table 7.1. Synthesis and applications of Co_3O_4 nanostructures.

Metallic Oxide	Preparation	Characteristics	Application	Reference
Co_3O_4 NPs	Thermal decomposition	d: 20 nm, ζ: –7.30 mV@pH 6.5	Microbial fuel cell	(Bhowmick et al., 2019)
Co_3O_4 NPs	Thermal decomposition	d: 20 nm, ζ: –7.30 mV at pH 6.5	Antibacterial activity and biocompatibility test	(Kavitha et al., 2017)
Co_3O_4 NPs	Thermolysis	Cubic spinel structure, spherical NPs; Co_3O_4@ 450°C [ρ: 6.15 g cm^{-3}, d: 41–46 nm E_g: 1.56 eV]; Co_3O_4@600°C [ρ: 6.08 g cm^{-3}, d: 60–65 nm, E_g: 1.63 eV]	Hemo-compatibility and antibacterial activity	(Gajendiran et al., 2020)
Co_3O_4 NPs	Biosynthesis (*Hibiscus Rosa-Sinensis* flower extract)	Cubical spinel structure; Co_3O_4 at RT [D= 61.32 nm, E_g: 1.792 eV]; Co_3O_4 @ 600°C [d: 57.3 nm, E_g: 2.048 eV]; Co_3O_4 @ 700°C [d: 50.6 nm, E_g: 2.504eV]; Co_3O_4 @ 800°C [d: 40 nm, E_g: 2.91 eV]	Antimicrobial and antifungal activity	(Anuradha and Raji, 2019b)
Co_3O_4 NPs	Biosynthesis (*P. granatum* seed extract)	d: 1–5 nm, E_g: 1.89–2.40 eV	Photocatalytic, catalytic hydrogenation and antibacterial activities	(Jesudoss et al., 2017)
Co_3O_4 Nanorods	Hydrothermal method	Mesoporous nanorods; s_{BET}: 15.55 m^2/g	Microbial fuel cell	(Kumar et al., 2016)
Co_3O_4 Nanorods	Biosynthesis (*Vitis vinifera* leaf extract)	Spherical rods, single-crystal, cubic phase structure, antiferromagnetic, d: 10–20 nm, E_g: 2.1 and 2.4 eV	Catalytic, photocatalytic, and antibacterial activities	(Kombaiah et al., 2019)
Co_3O_4 Nanowires	Hydrothermal method	Needle-like NWs. l: 500 nm, R: 9.62 and 4.29 Ω for Co_3O_4 and SG-Co_3O_4	Microbial fuel cell	(Cheng et al., 2019)

ζ - Zeta potential; ρ - density; s_{BET} - specific surface area; D - crystallite size (calculated from Debye-Scherrer equation); d - particle size (measured from SEM); l - length; R= resistance; E_g - Bandgap; NP - Nanoparticles; RT - Room temperature.

Table 7.2. Synthesis and applications of NiO nanostructures.

Metallic Oxide	Preparation	Characteristics	Application	Reference
NiO NPs	Co-precipitation method	Cubic phase bunsenite, spherical structure d: 24 nm, E_g: 3.17 eV, ζ: 31.5 mV at pH 6.5, IEP: 11.8	Antibacterial activity	(Paul and Neogi, 2019)
NiO NPs (nanopigment)	Sol-gel method	Irregular and cubic structures, d: 30–200 nm, E_g: 2.64 eV, coating σ_{crit}: 15.13 mN/m	Antibacterial activity	(Benitha et al., 2019)
NiO NPs	Biosynthesis (Neem leaf extract) at two annealing temperatures.	Oblong-shaped NiO @ 350°C [E_g: 3.12 eV d: 32.1 nm, Hc: 25.169 G]; NiO @ 400°C [E_g: 3.01 eV, d: 33.24 nm, Hc: 27.160 G]	Antibacterial activity	(Helan et al., 2016)
NiO nanostructure	Hydrothermal method	Semi-spherical head (formed with nanotubes), connected to trunk-like arm particles d: 15–20 nm, S_{BET}: 30.1 m^2g^{-1}	Antiviral activity	(Derbalah et al., 2019)
NiO nanorods	Biotemplated hierarchical and thermal oxidation	Nanorod length: 130 nm	Electrochemical performance	(Chu et al., 2016)
NiO nanosticks	Green-fuel-mediated hot-plate combustion reaction	Face-centered cubic structure	Photocatalytic efficiency and antibacterial activity	(Rajan et al., 2017)

S_{BET} - specific surface area; ζ - Zeta potential; σ_{crit} - critical surface tension; d - particle size (measured from SEM); Eg - Bandgap; Hc - coercivity; NP - Nanoparticle; IEP - isoelectric point.

be used on antibacterial coatings (Benitha et al., 2019). NiO has also been harnessed on the direct control of Cucumber Mosaic Virus (CMV), reducing infected cucumber plants by CMV count and lowering disease severity, thus increasing the number of healthy plants (Derbalah et al., 2019). As a photocatalyst, NiO nanosticks improve the efficiency of Rose Bengal dye removal by varying pH values (Rajan et al., 2017).

2.3 Synthesis of copper oxide

CuO is a simple, cheap and stable copper p-type semiconductor possessing a high surface area and E_g 1.2 eV. A wide variety of morphologies, including nanoparticles, nanorods and nanoflowers, have been produced, as shown in Table 7.3. Different synthetic methodologies can be adopted to obtain CuO nanostructures. For instance, Chauhan et al. (2019) studied the effect on the morphology of the synthesis for CuO nanostructures; they observed that the synthetic route played an important role in the final properties of nanomaterials, resulting in potential and diverse applications such as catalysis, gas sensors, solar energy conversion, antimicrobial and antifungal activity and wastewater treatment.

CuO has been used in the biomedical sector because of its antimicrobial and biocide properties (Table 7.3). Bhadra et al. (2019) evaluated the *in vivo* use of CuO and CuO@SiO_2 nanoparticles for a potential anticancer drug, as it proved to be effective. Nishino et al. (2017) tested the Submerged Photo-Synthesis of Crystallites (SPSC) process to synthesize nanoflowers with several morphologies, which were dependent on synthetic conditions; either way, nanoflowers showed antibacterial activity on *E. coli* and *Staphylococcus aureus*. Additionally, CuO has been used as a catalyst in various reactions, such as the Azide-alkyne Cycloaddition Click reaction (ACC) (Kiani et al., 2020) and the alcoholysis-polyesterification process (A-PE) (Ong et al., 2016; Ruey et al., 2018), which produced shorter reaction times and created alternative green approaches.

2.4 Synthesis of zinc oxide

ZnO, one of the most used semiconductors, possesses a wide E_g of 3.37 eV and an excitation binding energy of 60 meV at room temperature (Talam et al., 2012; Žaharia et al., 2016). Many different applications emerge for ZnO, including disinfection, wastewater treatment, solar cells, electronic devices and photocatalysts. ZnO synthesis was carried out with relative simplicity, obtaining different morphologies as shown in Table 7.4 (Khan et al., 2016; Ashraf et al., 2017). Moreover, biogenic synthetic routes using various plant extracts such as Acalypha fruticose L. (Vijayakumar et al., 2020), Catharantus roseus (Gupta et al., 2018) and Aloe vera (Sharma et al., 2020) have also been explored. These resourceful and unique characteristics, have led to the formation of many nanostructured morphologies and attracted several researchers' attention.

ZnO has also demonstrated to be effective against Gram-negative and Gram-positive bacteria, making it a widely used material for disinfection processes (Table 7.4). Khan et al. (2016) synthesized thorn-like nanoparticles with effective antibacterial activity towards *Bacillus subtilis*, *E. coli*, and *Candida albicans*. Nanorods were grown on polyester support, aimed towards water disinfection using solar radiation (Danwittayakul et al., 2020). Moreover, Sathishkumar et al. (2017) synthesized ZnO nanoflakes with enhanced bactericidal efficacy that could indicate its potential use in medical applications.

Besides, ZnO nanorods and nanoflowers were synthesized using the hydrothermal method for viral detection, including the H1N1 influenza virus and Japanese encephalitis virus (Han et al., 2016; Ma et al., 2017). The catalytic activity was evaluated by synthesizing thiophene derivatives with good yields using this alternate synthetic route (Hamedani et al., 2020).

Table 7.3. Synthesis and applications of CuO nanostructures.

Metallic Oxide	Preparation	Characteristics	Application	Reference
CuO NPs	Sol-gel method	NPs were used as a doping agent in bioresin	Antimicrobial activity and A-PE process	(Ong et al., 2016)
CuO NPs	Sol-gel method	NPs were employed as a component in palm oil-based alkyd resin	Antimicrobial activity	(Ruey et al., 2018)
CuO NPs	Thermal decomposition method	Sphere-like NPs Monoclinic phase d: 39 nm	Catalytic and antimicrobial activity	(Kiani et al., 2020)
CuO NPs	Co-precipitation method	Monoclinic phase, cubic structure d: 16 nm, E_g: 3.78 eV, ζ: 10.5 mV at pH 6.5, IEP: 8.2, d: 50 ± 15 nm	Antibacterial activity	(Paul and Neogi, 2019)
CuO NPs	Hydrothermal method	Hexagonal, monoclinic and rhombohedral structures d: 26.1 nm	Antimicrobial and antifungal activity	(Bhadra et al., 2019)
CuO NPs	Biosynthesis (*S. tuberosum* starch)	Spherical NPs, d: 54.1 nm	Green synthesis and antibacterial activity	(Alishah et al., 2017)
CuO NPs	Chemical precipitation method	Spherical NPs, monoclinic phase d: 15–20 nm, S_{BET}: 40.3 m^2g^{-1}	Dye removal efficiency, kinetics studies, and antimicrobial activity	(Chauhan et al., 2019)
CuO NPs	Microwave irradiation	Spherical NPs, Monoclinic phase d: 25–30 nm, S_{BET}: 29.704 m^2g^{-1}	Dye removal and antimicrobial activity	(Chauhan et al., 2019)
CuO nanorods	Hydrothermal method	Monoclinic phase w: 100–150 nm, l: 200–300 nm, S_{BET}: 19.6 m^2g^{-1}	Dye removal and antimicrobial activity	(Chauhan et al., 2019)
CuO nanoflowers	SPSC process	Nanoflowers with varying ramifications depending on conditions	Antibacterial activity	(Nishino et al., 2017)

S_{BET} - specific surface area; d - particle size (measured from SEM); NP - Nanoparticle; E_g - Bandgap; IEP - isoelectric point; A-PE - alcoholysis-polyesterification process; SPSC - Submerged photo-synthesis of crystallites; w - width; l – length.

Table 7.4. Synthesis of ZnO nanostructures.

Metallic Oxide	**Preparation**	**Characteristics**	**Application**	**Reference**
ZnO NPs	Wet chemical route	d: 7–10 nm	Antimicrobial activity	(Sarwar et al., 2016)
ZnO NPs	Biosynthesis (*Prunus dulcis*)	Hexagonal wurtzite structure d: 35.4 nm; E_g: 5.17 eV	Antimicrobial activity	(Theophil et al., 2019)
ZnO NPs	Biosynthesis (*Acalypha fruticose L.* leaf extract)	Hexagonal wurtzite structure, d: 50–55 nm	Antimicrobial activity	(Vijayakumar et al., 2020)
ZnO NPs	Biosynthesis (aloe vera plant extract)	Hexagonal, spherical and cuboidal shapes d: 40–180 nm, E_g: 3.4 eV	Antibacterial activity and photocatalytic activity of MO	(Sharma et al., 2020)
ZnO NPs	Biosynthesis (Catharanthus roseus leaf extract)	NPs synthesized at various conditions d: 19.9–47.2 nm, d: 50–90 nm	Antibacterial activity	(Gupta et al., 2018)
ZnO NPs	Biosynthesis (*Aegle marmelos*)	Hexagonal wurtzite structure d: 20–30 nm; ζ: –36.9 mV	Antibacterial and photocatalytic activity of MB	(Anupama et al., 2018)
ZnO NPs	Sol-gel method	Thorn-like NPs d: 3–25 nm, the stirring conditions impacted morphology and size.	Antibacterial activity	(Khan et al., 2016)
ZnO NPs	Sol-gel method	d: 50–650 nm (dependent on NaOH concentration)	Antimicrobial activity and moisture management	(Ashraf et al., 2017)
ZnO Nanorods	Hydrothermal method	Nanorods were grown on a PDMS channel	Electrochemical immunosensor for viral detection	(Han et al., 2016)
ZnO Nanorods	Hydrothermal method	Nanorods were grown on cellulose and polyester d: 17–30 nm	Solar water disinfection	(Danwittayakul et al., 2020)
ZnO Nanorods	Solvothermal method	d: 30 nm	Catalytic activity	(Hamedani et al., 2020)
ZnO Nanofibers	Electrospun	Fiber d: 315–292 nm (annealed)	Antimicrobial activity	(Thakur et al., 2020)
ZnO Nanosheets	Biosynthesis (green tea leaves)	d: 30–40 nm, l: 1 μm	Antimicrobial activity	(Irshad et al., 2018)
ZnO Nanoflowers ZnO Nanorod and nanosheets assembled	Hydrothermal method	Nanorod (hexagonal) L: 1–3 μm Nanosheet l: 200–900 nm; d: 60 nm	Viral detection (*EG*)	(Ma et al., 2017)
Nanoflakes	Biosynthesis (*Couroupita guianensis Aubl.* Leaf extract)	Nanoflake clusters, Wurtzite structure ζ: –20.7 mV	Antibacterial activity and hemocompatibility	(Sathish kumar et al., 2017)

d - particle size (measured from SEM); E_g - Bandgap; MO - methyl orange; MB - Methylene blue; ζ - Zeta potential; *EG* - Japanese encephalitis virus; PDMS - Polydimethylsiloxane.

2.5 Composites

Electronic properties are one of the best features of metallic oxides. Their semiconducting nature makes them very sensitive and largely influenced by their chemical composition. Doping is a good strategy to improve photocatalytic activity and modify the electronic properties of the material. This approach can also help in avoiding the recombination process during photocatalysis (Ramírez-Sánchez et al., 2019).

For many metal oxides, doping can be achieved by exchanging oxygen or metal atom with another entity. This substitution can be performed by selecting a doping precursor and using it during the synthesis. Usually metals are added as doping agents to modify E_g values (Shi et al., 2016; Adhikary et al., 2016; Pugazhendhi et al., 2018; Andrade et al., 2019; Zare et al., 2019; Burlibaşa et al., 2020). In general, various percentages of doping will show different effects.

For example, when ZnO was paired with Ag, the E_g drop to ~ 3.2 e, enabling the material to absorb wavelengths in the visible spectrum range (AL-Jawad et al., 2018; Zare et al., 2019; Burlibaşa et al., 2020). Shi et al. (2016) reported that NiO doped with Ag displayed different inactivation results on Gram-positive and Gram-negative bacteria, which was dependent on the doping amount. Doping can also provide other effects on metal oxide, which could help other applications. In this regard, Adhikary et al. (2016) found the ferromagnetic behavior on CuO doped with Ag was useful on recyclable disinfection materials.

The incorporation of surface heterostructures on metal oxide leads to a different performance. This strategy aims to improve the photocatalytic process by reducing the recombination of the hole-electron pair or shifting the light absorption range by modifying the E_g. Granular Activated Carbon (GAC) is a commonly used material because of its excellent adsorption capacity and high surface area. GAC was doped with CuO to pair the antibacterial and antiviral properties of the metal oxide with GAC high adsorption capacity, resulting in a promising wastewater disinfectant (Shimabuku et al., 2017). TiO_2-Eu paired with CuO increased the BET surface and mineralization of phenol and *Enterococcus faecalis* inactivation (Michal et al., 2016). Improved photocatalytic efficiency towards brilliant green in the UV-vis range was achieved by coupling two metal oxides, NiO, and CuO, with reduced Graphene Oxide (rGO) (Sree et al., 2020).

Capping agents have various advantages such as stabilizing the nanostructure nucleation growth, avoiding the particle from overgrowing, keeping nucleated particles separated from each other through steric and electrostatic forces and preventing agglomeration (Ajitha et al., 2016). Singh et al. (2009) prepared ZnO with different organic capping agents, which had different particle sizes and presented different luminescence. Organic capping agents can also modify the nanoparticle solubility, which might be exploited for medical applications (Singh et al., 2009).

3. Morphological and Structural Characterization of Nanoparticles

Morphological and structural characterization techniques are used to assess nanostructure properties (Modena et al., 2019). Evaluating their properties allows a better understanding of the material nature and identifying further improvements. For instance, Scanning Electron Microscopy (SEM) creates magnified images that yield topographical and elemental information when it is coupled with Energy-Dispersive X-ray analysis (EDX) (Vernon-Parry, 2000). Elemental analysis using SEM-EDS helps obtaining qualitative and semi-quantitative elemental information and a correlation between microstructures and elemental composition (Newbury and Ritchie, 2013). Transmission Electron Microscopy (TEM) is ideal for obtaining morphological and crystalline information and nanostructure size (Su, 2017).

In addition to microscopic techniques, Ultraviolet-visible Spectroscopy (UV-vis), X-Ray Diffraction (XRD), EDX, thermogravimetric analysis (TG) and Fourier Transformed Infrared Spectroscopy (FTIR) are frequently used to characterize different moieties of capping agents,

Table 7.5. Synthesis of transition metal oxide nanocomposites.

Composite	Preparation	Characteristics	Application	Reference
ZnO-Ag	Chemical reduction	Hexagonal wurtzite; hexagonal prism-shaped NPs, D: 60.01–61.95 nm; d: 100 nm; l: 50–600 nm, E_g: 3.19–3.22 eV	Antibacterial activity	(Burlibaşa et al., 2020)
ZnO-Ag	Biosynthesis (*Thymus vulgaris* leaf extract)	Hexagonal structure with 2–5 Ag NPs on the ZnO surface E_g: 3.25 eV	Antibacterial, antioxidant, cytotoxic and photocatalytic activity	(Zare et al., 2019)
ZnO-Ag	Sol-gel method Spin coating process	Thin films with spherical NPs. Properties varied by the Ag% content. D: 9.7–18.3 nm; E_g: 3.55–3.23 eV	Antibacterial activity	(AL-Jawad et al., 2018)
ZnO-Se	Chemical method	Se-doped ZnO NPs, agglomeration is present, hexagonal wurtzite phase D: 16.3–18.0 nm; d: 50–150 nm	Antimicrobial activity and *in vivo* effect studies	(Kumar and Nesaraj 2018)
ZnO-Ga	Sol-gel method	Hexagonal wurtzite phase; Spherical-shape D: 25.5–29.9 nm; d: 25–30 nm	Photocatalytic and antibacterial activity	(Malhotra et al., 2019)
Zn-Fe, Zn-Pb, and Zn-Fe-Pb	Hydrothermal method	A series of ZnO nanoplates doped with Fe and Pb at various % hexagonal D: 31.08–19.75 nm; d: 20 nm E_g : 3.04–3.15 eV; ζ : -0.03 - +6. 62 mV (pH 5)	Photocatalytic and antibacterial activity	(Andrade et al., 2019)
Ag/ZnO (CMC-capped)	Chemical deposition (CD) or Mechanical deposition (MD)	d: 399 ± 11 nm (CD); d: 412 ± 12 nm (MD)	Antimicrobial activity	(Lungu et al., 2016)
Fe_3O_4/ZnO	Sol-gel method	Spherical NPs. As the Fe_3O_4: ZnO ratio increased, the morphology changed. d: 40–50 nm; D: 27 nm	Antibacterial activity	(Bahari et al., 2019)
Fe_2O_3/Ag-ZnO	Hydrothermal method	E_g: 3.17 eV	Photocatalytic and Antibacterial activity	(Karunakaran and Vinayagamoorthy, 2016)
Polyvinyl Alcohol/ Pluronic Blends /ZnO	Pyrolysis method	NPs were homogeneously dispersed into the polymer d: 34 nm	Antibacterial and antifungal activity	(Amin et al., 2020)
ZnO/GO	Vacuum impregnation and hydrothermal method	ZnO and GO incorporated the bamboo structure.	Antibacterial activity	(Zhang et al., 2017)
SC4- or SThC4-doped PPY-ZnO	Hydrothermal method	d: 20–25 nm	Antibacterial and electrocatalytic activity	(Waghmode et al., 2016)
The casein-based nanocomposite of ZnO	Hard-template method	Hollow nanospheres d: 5 nm	Antibacterial activity (*A. favlus*)	(Liu et al., 2017)
ZnO/FA	Chemical method	ZnO conjugated with FA; Wurtzite structure d: 16 nm; E_g: 3.32 eV	Antimicrobial and photocatalytic activity of MB	(El-Borady and El-Sayed 2020)

Table 7.5 contd. ...

...Table 7.5 contd.

Composite	Preparation	Characteristics	Application	Reference
PVC/ZnO-EDTA NCs	Sol-gel method	Spherical NPs d: 3 nm	Antibacterial activity and determination of physical properties	(Mallakpour and Javadpour 2017)
ZnO-CuO	Modified perfume spray pyrolysis method	Hexagonal and monoclinic structure (ZnO and CuO, respectively) D: 36 nm; E_g: 3.25 eV	Antibacterial activity	(Saravanakkumar et al., 2018)
NiO-Ag	Hydrothermal method	Ag/NiO nanosheets with different %Ag.	Antibacterial activity	(Shi et al., 2016)
NiO-Cu	Spray pyrolysis method	NiO-Cu porous thin films. D: 37.9–93.41 nm (at various ratios)	Antibacterial activity	(Aftab et al., 2020)
NiO-CuO-rGO	Hydrothermal method	NCS were synthesized for 5 and 10% of rGO D: 20.72–21.43 nm; d: 14 nm	Photocatalytic activity and antibacterial activity.	(Sree et al., 2020)
CuO-NiO	Co-precipitation method	Both CuO and NiO phases were found D: 25 nm; E_g: 3.34 eV; ζ: 25.5 mV at pH 6.5; IEP = 10.7; d = 8 ± 2 nm	Antibacterial activity	(Paul and Neogi 2019)
CuO-Ag	Pyrolytic decomposition	Hexagonal NPs d= 25 ± 5 nm; d_{mean} = 25 ± 5 nm; E_g = 2.88 eV; Hc= 30 Oe	Biological activity	(Adhikary et al., 2016)
CuO-Ag	Wet chemical process	Ag NPs onto CuO nanowires, with diameters of 200 nm at the bottom to 10 nm on the tip. Ag NPs are distributed uniformly	Antibacterial activity	(Yue et al., 2019)
CuO-Fe	Sol-gel method	Rectangular-shaped NPs with agglomerations d: 21 nm; E_g : 1.0 eV	Antibacterial activity	(Pugazhendhi et al., 2018)
CuO-Ag CuO-Mo CuO-Ag/Mo	Biosynthesis (*Azadirachta indica* leaf extract)	CuO-Ag [E_g = 2.39 eV; D= 17.20 nm; CuO-Mo [E_g = 2.59 eV; D= 14.57 nm]; CuO-Ag/Mo [E_g = 2.22 eV; D= 11.23 nm]	Photocatalytic and antibacterial activity	(Rajendaran et al., 2019)
GAC-CuO and GAC-Ag/CuO	Incipient wetness impregnation method.	D= 22–37 nm (CuO) d= 50 nm (GAC) ζ= –33.97 to –36.79 mV (GAC containing CuO)	Virus removal by filtration	(Shimabuku et al., 2017)
TiO_2-Eu /CuO	Complex precipitation method	S_{BET}= 72.6–46.0 m^2/g (varying %CuO) D= 31–35 nm (TiO_2-Eu/CuO); d= 2–5 nm (CuO)	Photocatalytic and antibacterial activity	(Michal et al., 2016)
Cu_2O/C_3N_4	Biosynthesis (*Citrus limon* leaves) and hydrothermal treatment	C_3N_4 nanosheet with Cu_2O NPs evenly dispersed d= 2–10 nm	Antimicrobial activity	(Meenakshisundaram et al., 2019)
Co_3O_4-Ag	Thermal decomposition	Polyindole doped with Co_3O_4-Ag D= 26 nm; d= 18.4–21.8 nm	Antibacterial and antifungal activity	(Elango et al., 2018)

D - crystallite size (calculated from Debye-Scherrer equation); d - particle size (measured from SEM); E_g - Band gap; NP - Nanoparticles; GO - graphene oxide; SC4 - 4-sulfato calix [4] arene hydrate; SThC4 - 4-sulfato tia calix [4] arene sodium salt; PPY= polypyrrole; PVC - poly(vinyl chloride); rGO - reduced graphene oxide; IEP - isoelectric point; ζ - Zeta potential; Hc - coercivity; GAC - granular activated carbon; AC - activated carbon; FA- folic acid; s_{BET} - specific surface area.

nanocomposites and doping agents (Mourdikoudis et al., 2018). Given their high sensitivity, the results obtained with these techniques are reliable and helpful in knowing the chemical identity of the sample (Modena et al., 2019). Dynamic Light Scattering (DLS) is also a technique that provides information about the size distribution of the sample and zeta potential (ζ) at different pH values to modify the electrical charge of nanomaterial surfaces. This information helps in understanding the material agglomeration and their behavior in aqueous solutions (Carvalho et al., 2018).

XRD is an important technique as it reveals the crystalline properties of the material by providing the diffraction pattern and particle size using the Debye-Scherer equation (Holder and Schaak, 2019). The method is effective on single and multiphase crystallographic identification, advantageous on nanocomposites, showing crystalline phases and identifying impurities (Whitfield and Mitchell, 2004). The characteristic peaks at 2θ observed for transition metal oxides were mentioned earlier and are in Table 7.6.

FTIR provides a spectrum of the chemical nature of the material that yields a particular spectrum, identifying substances, particularly those with functional groups visible on the infrared spectrum (Dutta, 2017). M-O stretching mode bands appear near the visible spectrum for transition metallic oxides as mentioned earlier, located around 700–400 cm^{-1}. Bands appearing above 700 cm^{-1} are mostly functional groups or organic traces, although they may have slight variations, as mentioned in Table 7.6.

UV-vis spectroscopy is a technique that records from 200 to 1000 nm and provides information on the number of incident photons that are reflected by the material. Absorption peaks can be found in the reflectance spectra, which depend on the valence electrons excitation energy of the metal oxide (Rastar et al., 2013). Bands appearing in the spectra also give one indirect information about the E_g using the Kubelka-Munk equation (Myrick et al., 2011).

Table 7.6. Structural characterization of transition metal oxide nanoparticles.

Nanostructure	XRD Patterns	Ref.	FTIR Bands	Ref.
Co_3O_4	19.0° (111), 31.3° (220), 36.9° (311), 38.6° (222), 44.8° (400), 55.7° (422), 59.3° (511), 65.2° (440)	(Jesudoss et al., 2017; Bhargava et al., 2018; Gajendiran et al., 2020)	Co-O stretching mode appears at ~ 573 cm^{-1} and O-Co-O stretching mode appears ~ 647 cm^{-1}	(Elango et al., 2018)
NiO	37.5° (111), 43.3° (200), 62.9° (220), 75.4° (311), 79.4° (222)	(Behera et al., 2019; Benitha et al., 2019).	Ni-O stretching modes appear at ~542 cm^{-1} and ~ 418 cm^{-1}	(Paul and Neogi, 2019)
CuO	32.6° (111), 35.5° (002), 39.0° (111), 48.8° (202), 53.4° (020), 58.2° (202), 61.6° (213), 66.0° (022), 68.0° (220)	(Bouazizi et al., 2015).	Cu-O stretching mode appear at ~600 cm^{-1}, ~ 510 cm^{-1} and ~ 435 cm^{-1}	(Bouazizi et al., 2015; Paul and Neogi, 2019)
ZnO	31.8° (100), 34.5° (002), 36.3° (101), 47.6° (102), 56.7° (110), 62.9° (103), 68.1° (002), 69.2° (101)	(Talam et al., 2012; Das et al., 2017)	Zn-O stretching mode appears at ~ 480 cm^{-1} and ~ 410 cm^{-1}	(Esparza-González et al., 2016; Gupta et al., 2018)

XRD - X-ray diffraction; FTIR- Fourier transformed infrared.

4. Applications for Disinfection

4.1 *Tricobalt tetraoxide*

Co_3O_4 has been recognized for its reactive activity and use as a disinfection agent. For instance, Co_3O_4 showed high antibacterial properties against *E. coli* (Gajendiran et al., 2020), obtaining 99% growth reduction (Kavitha et al., 2017). Co_3O_4 NPs have been used for counteracting Gram-positive bacteria (*S. aureus*, *Streptococcus mutans*), Gram-negative bacteria (*Klebsiella pneumonae*, *E. coli*), and fungi growth (*Aspergillus flavus* and *Aspergillus niger*) (Anuradha and Raji, 2019a).

The biogenic synthesis of $Co_3O_{4\,ao}$ nanostructures provides new options without affecting the environment (Hsu et al., 2018; Anuradha and Raji, 2019a; Kombaiah et al., 2019). For example, Co_3O_4 NPs were synthesized using the *Vitis vinifera* extract and it was demonstrated that these NPs inhibited the growth of Gram-positive-(*S. aureus* and *B. subtilis*) and Gram-negative bacteria (*Pseudomona aeruginosa* and *E. coli*) (Kombaiah et al., 2019). Co_3O_4 NPs synthesized with *Hibiscus Rosa-sinensis* aquatic flower extract showed antimicrobial activity against Gram-positive bacteria (*S. aureus*, *Streptococcus mutans*), Gram-negative (*K. pneumoniae*, *E. coli*) and fungi (*A. flavus*, *A. niger*) (Anuradha and Raji, 2019b). The synthesis of Co_3O_4 NPs induced precipitation using *B. pasteurii,* offering a nature-friendly technological option since successful biomineralization was obtained (Hsu et al., 2018).

4.2 Nickel oxide

NiO NPs have been widely used because of their electrochemical stability. High purity NiO NPs of optimal nanometric size demonstrated antimicrobial properties and effective removal of organic contaminants (Saleem et al., 2017; Ezhilarasi et al., 2018). Although NiO NPs showed good antibacterial properties against *E. coli*, such activity decreased when doped with 0.125% Cu (Aftab et al., 2020). Eucalyptus (*Eucalyptus globulus*) leaf extract was used to synthesize NiO NPs, which controlled microbial growth of ampicillin-resistant *E. coli* strain and against *P. aeruginosa*; however, NiO NP was not effective against *S. aureus* (Saleem et al., 2017). Besides, extracts from *Aegle marmelos* have been used for synthesizing NiO NPs, showing higher bactericidal activity against Gram-positive strains compared to Gram-negative bacteria and low cytotoxic activity (Ezhilarasi et al., 2018).

4.3 Cupric oxide

CuO NPs showed antimicrobial action against *E. coli*, *B. subtilis* and *Saccharomyces cerevisiae*, attributed to CuO NPs size (Salah et al., 2020). The antimicrobial properties were determined for porous CuO NPs, porous Ag NPs/CuO NPs, porous Ag NPs against Gram-negative (*E. coli*, *Salmonella*) and Gram-positive (*Listeria*) bacteria (Chen et al., 2017). In particular, CuO/Ag NPs showed the highest antimicrobial efficacy due to bimetallic synergy (Chen et al., 2017). Besides its powerful antimicrobial activity, CuO NPs have been used as an anticancer agent (Bhadra et al., 2019) and as an alternative to counteract antibiotic-resistant bacteria (Chen et al., 2017).

Biogenic synthesis strategies of CuO NPs showed a good relationship with the environment and, at the same time, inexpensive methods as compared to other synthesis methods (Radhakrishnan et al., 2018; Chauhan et al., 2019). The synthesis of CuO NPs using pomegranate juice generated nanoflowers with greater photocatalytic and antibacterial effects than the rest of the structures analyzed due to the large surface area (Radhakrishnan et al., 2018). CuO NPs, synthesized by the chemical precipitation method, were used for the removal of toxic chemicals such as Direct Red 81 (DR-81) and Coomasie Brilliant blue R 250 (BBR-250) from water, as well as the control of *S. aureus* (Chauhan et al., 2019).

4.4 Zinc oxide

ZnO NPs have shown antimicrobial activity against Gram-positive and Gram-negative bacteria and fungi (Shahmohammadi and Almasi, 2016; Tian et al., 2017; Amin et al., 2020). For example, ZnO NPs inactivated the growth of both Gram-negative (*E. coli* and *P. aeruginosa*) and Gram-positive bacteria (*Bacillus subtilis* and *S. aureus*) and avoided the growth of *Candida albicans* (Amin et al., 2020). The disinfection rate depended on the surface roughness, where the microbial activity increases with the concentration of NPs (Amin et al., 2020). Cellulose associated with ZnO NPs also performed antibacterial properties against Gram-positive and Gram-negative bacteria, *S. aureus*, *E. coli*, respectively (Shahmohammadi and Almasi, 2016).

Furthermore, several studies have been carried out based on the excellent antimicrobial properties of ZnO NPs against pathogenic bacteria in hospitals (Kumar and Nesaraj, 2017; Kaushik et al., 2019; Taghizadeh et al., 2020). ZnO NPs have been used in combination with blue light to counteract colistin- and imipenem-resistant *A. baumannii* and *K. pneumoniae* bacteria present in hospitals (Kaushik et al., 2019). ZnO films and the addition of copper and carbon nanocomposites were used to cover non-critical hospital surfaces and demonstrated antimicrobial activity against *P. aeruginosa* and *S. aureus* (Piedade et al., 2020).

In biomedicine, silver-doped ZnO NPs were used for the control of microorganisms and the prevention of infectious diseases in Swiss albino mice, using Gram-positive bacteria (*S. aureus* and *B. subtilis*) and Gram-negative bacteria (*E. coli* and *K. pneumoniae*) (Kumar and Nesaraj, 2017). In packing, ZnO NPs were found to reduce the risk of cross-contamination in packaged products since these NPs improved antimicrobial barriers and additionally prevented the migration of chemicals into food products (Amin et al., 2020).

The applications of these ZnO NPs have gained popularity for environmental treatments, especially in wastewater treatment (Liu et al., 2016). ZnO NPs allowed a reduction in Chemical Oxygen Demand (COD) concentration, nitrogen and phosphorus. Furthermore, the amount of bacteroidetes decreased about 20%, and the percentage of Proteobacteria increased almost 23% (Liu et al., 2016). However, *Chryseobacterium and Dechloromona,* and *Sediminibacterium*, and Blvii28 showed resistance and tolerance to ZnO NPs, respectively (Liu et al., 2016).

The development of NPs obtained by biogenic synthesis has been evaluated for their biocidal and antifungal activity (Dobrucka et al., 2018; Theophil et al., 2019). ZnO nanorods produced from *Chlorella vulgaris*-secretory carbohydrates showed a sustainable and cost-effective alternative to the existing techniques, and this played an essential role in directing both the NP morphology and NP size (Taghizadeh et al., 2020). The ZnO NPs synthesized with the extract of *Chelidonium majus* were used to investigate their cytotoxic and biocidal properties (Dobrucka et al., 2018). The use of a bacterial exopolymer from *Bacillus licheniformis* Dahb1 allowed an effective control against Gram-negative bacteria (*P. aeruginosa* and *Proteus vulgaris*), Gram-positive bacteria (*B. subtilis* and *B. pumilus*) and fungi (*C. albicans*) (Abinaya et al., 2018). ZnO NPs using horse chestnut (*Aesculus hippocastanum*) extract were used as bactericidal power against *Bacillus thuringiensis* (Duman and Karako, 2017). The ZnO NPs synthesis from almond gum showed great antimicrobial properties, inhibiting the growth of Gram-positive (*S. aureus*), Gram-negative bacteria (*E. coli, Salmonella paratyphi*) and fungal microorganisms (*C. albicans*). However, ZnO NPs did not show antimicrobial activity against *K. pneumoniae* and *Proteus mirabilis* (Theophil et al., 2019). The harness of Extracellular Polymeric Substances (EPS) was non-toxic, biodegradable and eliminated 100% of the population of *Anopheles stephensi* and *Aedes aegypti* larvae (Abinaya et al., 2018).

4.5 Nanocomposite

The use of NiO in association with other elements can provide considerable advantages for antibacterial applications. For example, Ag/NiO offered higher antimicrobial activity compared to NiO NPs when examined against Gram-positive bacteria (*S. aureus* and *B. subtilis*) and Gram-negative bacteria (*P. aeruginosa* and *E. coli*) (Shi et al., 2016). NiO, CdO and NiO-CdO showed a considerable effect on *K. pneumoniae* and *S. aureus* (Anitha et al., 2018). β-$CoMoO_4$-Co_3O_4 composite showed high antimicrobial activity on *S. aureus*, *P. aeruginosa,* and *E. coli* (Mobeen et al., 2018).

NPs composites as antimicrobial agents have an important use in food packaging for the inactivation of pathogenic microorganisms. ZnO NPs have been used in combination with other agents to verify the low toxicity shown in biological systems, providing free-microbial food (Shi et al., 2016; Zhang et al., 2017; Roy and Rhim, 2019). The use of ZnO NPs incorporated in the polylactic acid layer in antimicrobial packaging effectively inhibited the growth of *S. aureus* and *E. coli*, where the last was more susceptible to this agent(Zhang et al., 2017). Recent studies reported ZnO flake-shaped nanoparticles stabilized by melanin showed relevant microbial activity against Gram-negative bacteria (*E. coli*), but displayed slight activity against Gram-positive bacteria (*L. monocytogenes*) (Roy and Rhim, 2019). The combination of nanostructures traits has been applied in the biomedical area as an antimicrobial and anticancer agent(Demirel et al., 2018; Visnapuu et al., 2018; Bahari et al., 2019; Reyes-Torres et al., 2019). ZnO and Fe_3O_4 NPs showed that the antimicrobial benefits of these NPs increased when the particles were smaller and more spherical (Bahari et al., 2019).

The antimicrobial activity of ZnO NPs increased when they were doped with Ag and Cu, called MicNo ZnO (Ag-doped ZnO lamellae). The antimicrobial ZnO NPs effect grew when the Ag concentration doping increased, and were effective against *B. subtilis*, *E. coli*, *Salmonella typhimurium*, *S. aureus*, *C. albicans*, *Candida glabrata*, *Candida zeylanoides*, *A. flavus*, *A. flavus* and *Figpernarium fosslanum* (Demirel et al., 2018).

Surfaces covered with ZnO and ZnO/Ag NPs were evaluated in the dark and under UV-A radiation against Gram-negative bacteria (*Escherichia coli*), Gram-positive bacteria (*Staphylococcus aureus*) and yeast (*Candida albicans*). The antibacterial and photocatalytic activities were observed during several cycles, but surfaces at higher concentrations of Ag or the presence of Zn did not show antibacterial activity in the absence of radiation (Visnapuu et al., 2018).

The emergence of antibiotic-resistant microorganisms has encouraged the use of NPs associated with antibiotics to treat infections; for example, CuO nanobars with ZnO nanospheres inhibited the growth of *E. coli*, *S. aureus*, *E. faecalis* and *P. aeruginosa* (ATCC microorganisms) (Reyes-Torres et al., 2019). The antimicrobial activity of the combination of CuO with Fe against *S. aureus*, *S. epidermidis* and *C. albicans* was higher against *S. epidermis* and *C. albicans* than *S. aureus* (Pugazhendhi et al., 2018). When hospital wastewater is mixed with municipal wastewater, drug-resistant bacteria can transfer their resistance to non-resistant bacterial communities (Das et al., 2017). Therefore, the disinfection of this type of bacteria by conventional methods has become a challenge today. Fe-doped ZnO NPs were used to counteract the presence of antibiotic-resistant *E. coli* when it was used for disinfection as a solar catalyst, being more effective compared to ZnO and TiO_2 (Das et al., 2017).

The optimal conditions for the biogenic synthesis of NPs have been investigated using environmentally friendly materials (Anitha et al., 2018; Khatami et al., 2018; Hamedani et al., 2020). For example, Ag and ZnO, and Ag/ZnO NPs synthesized from *Prosophis fracta* and coffee were used to manufacture cotton dressings and showed antimicrobial effects that help the wound healing process (Khatami et al., 2018). The antimicrobial effect of Ag, ZnO, and Ag/ZnO composites was evaluated in *Acinetobacter baumannii* and *P. aeruginosa* cultures. ZnO nanochains were prepared to evaluate their therapeutic benefits using thiophenes and demonstrated effects against cancer cells and Gram-positive and Gram-negative bacteria (Hamedani et al., 2020).

When hospital wastewater is mixed with municipal wastewater, drug-resistant bacteria can transfer their resistance to non-resistant bacterial communities (Das et al., 2017). Therefore disinfection of this type of bacteria by conventional methods has become a challenge today. Fe-doped ZnO NPs were used to counteract the presence of antibiotic-resistant *E. coli* when it was used for disinfection as a solar catalyst, being more effective compared to ZnO and TiO_2 (Das et al., 2017).

5. Disinfection Mechanism

Literature has documented the disinfection mechanism of Co_3O_4, NiO, CuO and ZnO (Kombaiah et al., 2019; Kim et al., 2020; Salah et al., 2020; Kannan et al., 2020). It has been suggested that the disinfection process occurs when metal-transition oxides, such as Co_3O_4, NiO, CuO and ZnO, entails electrostatic interactions, cation diffusion, ROS production, as well as particle morphology and shape interactions (Slavin et al., 2017; Wang et al., 2017; Kumar et al., 2017).

Microorganism inhibition mechanisms of CuO have been attributed to membrane perturbation, oxidative stress by ROS and diffusions of Cu^{2+} ions (Salah et al., 2020; Alayande et al., 2020). Similarly, ZnO antimicrobial mechanisms may rely on particle size, Zn^{2+} diffusion, ROS formation and superficial charge of both cell membrane and particles (Kim et al., 2020). Furthermore, evidence of such mechanisms has been illustrated in SEM images in which *P. aeruginosa* leaked intracellular compounds after bacteria exposed to ZnO and visible and UV radiation (Piedade et al., 2020) and *E. coli* involved profound changes in the cell shape along with a ruptured membrane (Kavitha et al., 2017).

The electrostatic interaction depends on the characteristics of the outer cellular membrane, i.e., Gram-negative bacteria have a lipopolysaccharide and a thin peptidoglycan layer, which acts as a primary permeability barrier. Meanwhile, Gram-positive bacteria have a simple membrane formed of a peptidoglycan layer and teichoic and lipoteichoic acids (Khashan et al., 2017). Costa et al. (2020) suggested that differences in disinfection activity may be related to differences in the cell membrane; for example, the lipopolysaccharide layer may protect *E. coli* from oxidative stress . So Gram-negative bacteria with negative surface charge could interact with positively charge Zn^{2+} by electrostatic interaction, enhancing mechanical and oxidative stress (Kim et al., 2020). On the other hand, less negatively charged cells (Gram-positive) could resist ZnO NPs for electrostatic interactions (Kim et al., 2020). Studies have shown that Gram-positive bacteria were more sensitive to ZnO NPs than Gram-negative bacteria (Amin et al., 2020). However, a third group of studies pointed out that ZnO has high, moderate and absent antimicrobial activity against *Corynobacterium glutamicum* (Boura-Theodoridou et al., 2020). These results have led to a gap explaining the relationship between ZnO superficial charge, pH, cell membrane charge and disinfection activity.

Cation diffusion has been well-documented; for example, it was suggested Ni^{2+} could interact with the cell membrane, causing changes in cell functions and disintegrating the bacteria (Rajan et al., 2017; Kannan et al., 2020). NiO can also interfere with intracellular Ca^{2+} absorption and penetrate the cell, causing membrane damage by oxidative stress via ROS (Srihasam et al., 2020). At the same time, ZnO antibacterial activity may be a consequence Zn^{2+} diffusion and accumulation into cells (Taghizadeh et al., 2020), which inhibit respiratory enzyme activity (Kim et al., 2020), impairing substance transport (Sirelkhatim et al., 2015) and may increase oxidative stress owing to ROS generation (e.g., $^{\bullet}OH$) (Korshunov and Imlay, 2010).

When radiation is involved, the photocatalytic process has been documented for high resistant bacteria structures (*Bacillus subtilis* spores); in such processes, the amount of $^{\bullet}OH$ was found to contribute to disinfection (Huesca-Espitia et al., 2017). Specifically, NiO particles can produce ROS, such as $O_2^{\bullet-}$, H_2O_2 and $^{\bullet}OH$, which could break down proteins and lipid molecules present in the bacterial cell (Rajan et al., 2017). Co_3O_4 damaged the cell wall and cell membrane by intercellular penetration and oxidative stress, which could diminish the thickness membrane (Gajendiran et al., 2020). Alayande et al. (2020) confirmed using a probe of intracellular oxidative stress of CuO nanocomposite, finding a direct electron transfer and ROS generation such as $^{\bullet}OH$ and H_2O_2. In general, $^{\bullet}OH$ can damage lipids, DNA and proteins in the cell (El Fawal et al., 2020). Although ROS formation is relayed on a well-known photocatalytic activity of oxide semiconductors, ROS formation, i.e., $O_2^{\bullet-}$, in the absence of light on ZnO has been attributed to lattice defects (Hirota et al., 2010).

Membrane perturbation was due to piercing, influenced by CuO particle morphology (Alayande et al., 2020). It was suggested that the high surface ratio and shape of Co_3O_4 promote diffusion into cell membranes (Kombaiah et al., 2019). ZnO particle size also affected inhibitory activity against fungus, whose antimicrobial activity decreases with an increase in the particle size (Kaushik et al., 2019).

6. Virus Inactivation

Viruses are cellular parasites ubiquitous on Earth capable of infecting any cellular form, from eukaryotes (vertebrate and invertebrate animals, plants, fungi) to prokaryotes (bacteria and archaeas).

When viruses are outside from their host cells, known as virions, they act as a gene delivery system, protecting the virus genome and aiding in the host cell (Carter and Saunders, 2007).

Many viruses have a lipid membrane that is at the virion surface and is associated with one or more species of virus proteins. This lipid membrane is known as an envelope; however, other viruses lack this membrane (Fig. 7.2).

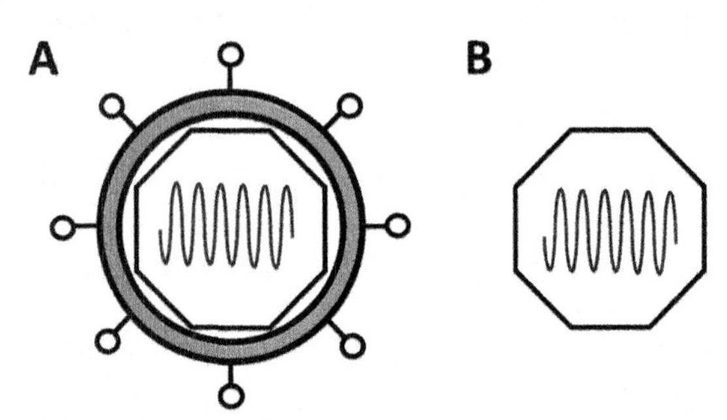

Fig. 7.2. (A) General structure of enveloped viruses, showing the lipid membrane surrounding the viral capside (hexagonal shape), which contains the viral genetic material (curved line). (B) General structure of nacked viruses.

The importance of virus studies stems from the significant challenge for the medical, pharmaceutical and biotechnological fields as it is estimated that 60% of human infections and deaths worldwide are caused by viruses (Vasickova et al., 2010; Rai et al., 2016). Respiratory and enteric viruses transmitted through several routes produce the most common viral illnesses (Vasickova et al., 2010).

Most respiratory tract diseases are produced by viruses, compared to any other microbial pathogens, which results in a significant toll on society and healthcare systems, giving rise to increased morbidity and mortality in both children and adults throughout the world (Hodinka, 2016). Infection sources could come from any activity, including coughing, sneezing, talking, bed-making or turning pages of books because they can generate microbial aerosols borne and disseminated by air movements. However, whether an infectious disease appears will depend on the viability and the infectivity of the inhaled microbes and their landing sites. Lipid-containing virus experiences desiccation, showing phase changes in their outer phospholipid membranes through water content changes and temperature changes. These changes, which could occur at mid to high Relativity Humidity (RH), could produce the crosslinking reactions of associated protein moieties. Otherwise, for surface protein moieties of lipid-free viruses, these reactions take place rapidly at low RH. Radiation, oxygen ozone and pollutants can also lead to the viability and infectivity reduction due to chemical, physical and biological modification of the different compounds that form a virus. The effect of the damage and the repair degree of the host's defense mechanisms controls it, if the disease is carried through air (Cox, 1989).

Thus, metal NPs-based technologies represent an alternative to avoid viral spread because of their high effectiveness as antiviral agents, their ability to target several types of viruses and to act either inside or outside of them, through the viral replication inhibition or by avoiding the entry of viral particles, respectively (Fig. 7.3).

A great deal of research has led to the evidence that the most effective virucides are those able to bind to the viral capsid, leading to their inactivation by NPs, which are more general and attractive than other physicochemical strategies (Cromeans et al., 2014; Broglie et al., 2015). The broad-spectrum of antimicrobial activity of metallic NPs is caused owing to particle size and superficial area (Ingle et al., 2008).

So far, it has been demonstrated that any type of metal might exert specific antiviral potential, silver NPs being the most effective metal-based antiviral agent (Rai et al., 2016). However, its high cost represents a main constraint. Therefore, cheaper metals than silver showing antiviral activity emerge as a cost-effective alternative to fight viral spreads and infections.

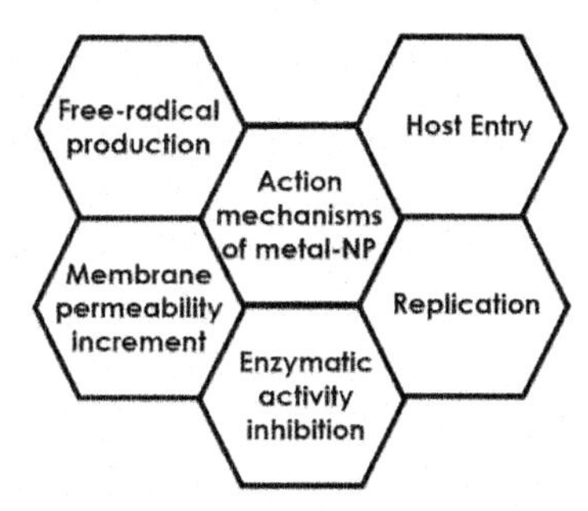

Fig. 7.3. Different levels at which metal-NPs show an effect on the viral activity.

In this regard, copper has shown the ability to destroy and inactivate viruses (Vincent et al., 2018). Cooper alloys were seen to be useful antiviral surfaces by inactivating murine virus (MNV-1) (Warnes and Keevil, 2013), which have a high rate of infection , causing viral gastroenteritis worldwide (Barker et al., 2004). The inactivation rate shown by the copper alloys was high and dependent on the content of copper . It was also determined that Cu^{2+} and specially Cu^{+} ions were the primary effectors of toxicity copper alloys by targeting the viral genome, thereby avoiding viral replication, reducing the production of important viral proteins that confers the viral infectivity (Warnes and Keevil, 2013). Copper nickels and copper in brasses also have a considerable influence on the alloying effect to inactivate norovirus through the massive breakdown of the viral capsid, thus leading to copper ion access to the viral genome (Warnes et al., 2015). In addition, a synergetic behavior with increased virucide efficacy is obtained with Cu and Zn alloys, which have low amounts of copper (Warnes et al., 2015).

On the other hand, evidence has shown that the antiviral activity from Cu^{2+} ions on touch surfaces produces morphological abnormalities in the viral particles (Noyce et al., 2007; Horie et al., 2008). Studies show that the sensitivity of enveloped viruses to Cu^{+2} ions is higher than non-enveloped ones, implying the viral lipid membrane turns out structurally by intracellular copper. For instance, nano copper surfaces could inactivate more than 95% of the influenza A virus triggered by copper ion diffusion through its lipid membrane (Sagripanti et al., 1993; Sundberg and Champagne, 2015). Additionally, copper surface hardness leads to copper ion diffusion from conventional copper surfaces, resulting in increased microbial destruction (Champagne and Helfritch, 2013).

Furthermore, the copper oxide in respiratory face masks was seen to bring biocidal properties on avian and human influenza A virus (Borkow et al., 2010). Likewise, it was demonstrated that copper oxide and copper oxide-impregnated fibers have antiviral activity against the Human Immunodeficiency Virus (HIV-1) (Borkow et al., 2008). It was inactivated in a copper dose-dependent manner, without cytotoxic effect or strain specificity, by targeting the viral genome and protease, essential for virus replication. Furthermore, other studies have shown that copper ions, Cu^{2+}, may inhibit the VIH-1 protease rapidly and irreversibly via interaction with its cysteine residues (Karlstrom and Levine 1991). Additionally, evidence pointing to the antiviral activity of Cu-NP against the Hepatitis C virus (HVC) was suggested resulting in its ability to inhibit viral infection by avoiding the entry of virus into the cells (Hang et al., 2015). Nano sized copper iodide (Cu^{+}) particles also showed an effective antiviral agent by generating $^{\cdot}OH$. In this work, the model study was the influenza A virus of swine origin, whose titers decreased in a dose-dependent manner upon incubation with the Cu^{+} particles (Fujimori et al., 2012).

Another proof-of-concept that metal NPs (metal-NP) showed effective and rapid viral inactivation was using Au-NP coated with CuS-nanoshell. These NP showed $\geq$ 75% inactivation (in 30 min) of norovirus-like particles by capside-protein degradation (Broglie et al., 2015). Thus, the harness of antimicrobial surfaces containing copper might help in reducing the spread of highly infectious pathogens in high-risk closed environments such as hospitals, schools and cruise ships, among others. (Warnes and Keevil, 2013).

The recent advent of zoonotic pathogens as avian influenza A viruses and coronaviruses causing Severe Acute Respiratory Syndrome (SARS), the Middle East Respiratory Syndrome coronavirus (MERS-CoV) and the cause of the virus for the current COVID-19 pandemic (SARS-CoV-2) has gained attention because of their unprecedented health and economic impacts owing to the ability to spread accurately among humans via the airborne route (Herfst et al., 2017). This situation has led to the importance of creating technology that can be harnessed to prevent the contagious spread of the infectious disease by sanitizing indoors and manufacturing clothes or masks to avoid the entry of the virus to the host.

Transmission through the air of SARS-CoV-2 via air-conditioning systems is a significant concern to the occurrence of the COVID-19 pandemic. Hence, considering that SARS-Cov-2 cannot resist temperatures above 70°C, investigators designed and fabricated efficient nickel (Ni) foam-based filters, which efficiently caught and killed more than 99% aerosolized SARS-CoV-2 by passing through it when the filter had been heated up to 200°C. Furthermore, this filter could catch and destroy 99.9% of spores from *Bacillus anthracis*, an airborne spore. Hence, this research demonstrates the application of a commercial Ni foam in an air-conditioner filter in closed and crowdy spaces such as airports, hospitals, schools, office buildings, cinemas and hotels entirely removed SARS-CoV-2 virus in circulating air, slowing the spread of COVID-19, as well as preventing transmission of other airborne infectious agents (Yu et al., 2020). The use of these brand-new nanotechnologies might slow the spread of COVID-19 and prevent the transmission of any airborne infectious agents.

Besides coronavirus, other viruses can represent a trait for public health, such as Avian Influenza Viruses (AIV). Therefore, to avoid the spread of AIV in areas of poultry, the implementation of biosecurity measures, clothes with antiviral effect were developed—as a novel and functional textile material-, in which zeolite was synthesized chemically on a 100% cotton textile through the structural components of zeolite and by replacing Na^+ by copper ions (Cu^{2+}). The modified zeolite was able to inactivate the H5N1 (highly pathogen) and H5N3 (lower pathogenic) viruses in a short time of contact (Imai et al., 2012). The antiviral Zeolite-Cu^{2+} effect was also sub type AIV-dependent.

Although zinc oxide NPs have shown antimicrobial activities against several pathogens, the studies addressing Zn NP-based antiviral activity are limited, and just a few instances have been seen on the inhibitory effect of ZnO-NP and polyethylene glycol (PEG)- coated ZnO-NP on Herpes Simplex Virus (HSV-1) (Tavakoli et al., 2018) and H1N1 influenza virus. In both cases, the antiviral activity was exerted only when NPs were added after viral infection cells, which was by the Real-Time quantitative PCR (RT-qPCR) test. In addition, the antiviral activity was in a dose-dependent manner, observing the highest inhibition rate (94.6%) at 200 μg/mL of ZnO-PEG-NP. Although the inhibitory potential of these NPs was effective post-infection, its antiviral activity before the exposure to cells should not be completely excluded (Ghaffari et al., 2019).

7. Conclusion

Significant progress has been achieved in the synthesis of NPs based on transition metal oxides such as tricobalt tetraoxide (Co_3O_4), Nickel Oxide (NiO), Cupric Oxide (CuO) and Zinc Oxide (ZnO), but the least studied semiconductors are Co_3O_4 and NiO. These results may indicate a gap in knowledge in the application of Co_3O_4 and NiO NPs in disinfection, virus inactivation, wastewater treatment and biomedical and food industry. Several methodologies have been reported, and the most notable ways have been the sol-gel process, hydrothermal method and co-precipitation techniques. One disadvantage of such synthetic routes is that they can be harmful to the environment due to the hazardous chemicals employed. Thus, green pathways must be optimized to increase the eco-friendliness and minimize costs. Therefore, green synthesis should become an alternative for the NPs production for disinfection.

Moreover, green synthesis for virus inactivation has hardly been reported. Consequently, virus inactivation by transition metal oxides currently has application and gaps in information that should be reduced to give alternatives to improve the population's health. The degradation of bacteria and virus inactivation by transition metal oxides have been demonstrated to be effective materials to limit the dissemination of bacteria and viruses in the environment.

Although literature has documented disinfection by transition metal oxide such as Co_3O_4, NiO, CuO and ZnO, the consensus in the mechanism is still discussed in literature. Consequently, there is a need to reduce the gap in disinfection and virus inactivation, for example, to clearly understand Fenton-like reaction in the interface of bacteria and virus membranes.

Transition metal oxides NPs could have promising particles for disinfection and sanitization purposes, which is currently required urgently. Acquired knowledge on these particles and their reaction mechanisms will help create surfaces, sanitizers, water disinfectants, face coverings that aid in controlling bacteria and viruses in the air and on surfaces. The quick action of NPs-based disinfectants would help prevent the spread of virus and its infections by reducing contaminated surfaces, which will allow one to repurpose and expand the applications of transition metal oxides NPs.

Acknowledgments

The authors contributed equally to this work. This chapter was supported by the University of Texas System and the Consejo Nacional de Ciencia y Tecnología (CONACYT) through the ConTex Postdoctoral Fellowship Program. The opinions expressed by the authors do not represent the views of these funding agencies. The authors thank the University of Texas at Austin for granting financial resources, access to facilities, training to conduct this work, and the bioprocess strategic research group (GIEE) at the Monterrey Institute of Technology and Higher Education (ITESM) for supporting this work.

References

Abinaya, M., Vaseeharan, B., Divya, M., Sharmili, A., Govindarajan, M., Alharbi, N.S., Kadaikunnan, S., Khaled, J.M. and Benelli, G. 2018. Bacterial exopolysaccharide (EPS)-coated ZnO nanoparticles showed high antibiofilm activity and larvicidal toxicity against malaria and Zika virus vectors. J. Trace. Elem. Med. Biol. 45: 93–103.

Adhikary, J., Das, B., Chatterjee, S., Dash, S.K., Chattopadhyay, S., Roy, S., Chen, J.W. and Chattopadhyay, T. 2016. Ag/CuO nanoparticles prepared from a novel trinuclear compound $[Cu(Imdz)_4(Ag(CN)_2)_2]$ (Imdz = imidazole) by a pyrolysis display excellent antimicrobial activity. J. Mol. Struct. 1113: 9–17.

Aftab, M., Butt, M.Z., Ali, D., Bashir, F. and Aftab, Z.H. 2020. Impact of copper doping in NiO thin films on their structure, morphology, and antibacterial activity against *Escherichia coli*. Ceram Int. 46: 5037–5049.

Ajitha, B., Kumar, R.Y.A., Reddy, P.S., Jeon, H.J. and Ahn, C.W. 2016. Role of capping agents in controlling silver nanoparticles size, antibacterial activity and potential application as optical hydrogen peroxide sensor. RSC Adv. 6: 36171–36179.

Alayande, A.B., Obaid, M. and Kim, I.S. 2020. Antimicrobial mechanism of reduced graphene oxide-copper oxide (rGO-CuO) nanocomposite films: The case of *Pseudomonas aeruginosa* PAO1. Mater. Sci. Eng. C 109: 110596.

Alishah, H., Pourseyedi, S., Ebrahimipour, S.Y., Mahani, S.E. and Rafiei, N. 2017. Green synthesis of starch-mediated CuO nanoparticles: Preparation, characterization, antimicrobial activities and *in vitro* MTT assay against MCF-7 cell line. Rend Lincei. 28: 65–71.

AL-Jawad, S.M.H., Sabeeh, S.H., Taha, A.A. and Jassim, H.A. 2018. Studying structural, morphological and optical properties of nanocrystalline ZnO:Ag films prepared by sol–gel method for antimicrobial activity. J. Sol-Gel. Sci. Technol. 87: 362–371.

Amin, K.M., Partila, A.M., Abd El-Rehim, H.A. and Deghiedy, N.M. 2020. Antimicrobial ZnO nanoparticle–doped polyvinyl alcohol/pluronic blends as active food packaging films. Part Part. Syst. Charact. 37: 2000006.

Andrade Neto, N.F., Matsui, K.N., Paskocimas, C.A., Bomio, M.R.D. and Motta, F.V. 2019. Study of the photocatalysis and increase of antimicrobial properties of Fe^{3+} and Pb^{2+} co-doped ZnO nanoparticles obtained by microwave-assisted hydrothermal method. Mater. Sci. Semicond Process. 93: 123–133.

Anitha, S., Suganya, M., Prabha, D., Srivind, J., Balamurugan, S. and Balu, A.R. 2018. Synthesis and characterization of NiO-CdO composite materials towards photoconductive and antibacterial applications. Mater. Chem. Phys. 211: 88–96.

Anupama, C., Kaphle, A., Udayabhanu and Nagaraju, G. 2018. *Aegle marmelos* assisted facile combustion synthesis of multifunctional ZnO nanoparticles: Study of their photoluminescence, photo catalytic and antimicrobial activities. J. Mater. Sci. Mater. Electron. 29: 4238–4249.

Anuradha, C.T. and Raji, P. 2019a. Synthesis, characterization and anti-microbial activity of oxalate-assisted CO_3O_4 nanoparticles derived from homogeneous co-precipitation method. Int. J. Nanosci. 18: 1950002.

Anuradha, C.T. and Raji, P. 2019b. Effect of annealing temperature on antibacterial, antifungal and structural properties of bio-synthesized Co_3O_4 nanoparticles using Hibiscus Rosa-sinensis. Mater. Res. Express. 6: 095063.

Arroyo, B.J., Bezerra, A.C., Oliveira, L.L., Arroyo, S.J., Melo, E.A. and De and Santos, A.M.P. 2020. Antimicrobial active edible coating of alginate and chitosan add ZnO nanoparticles applied in guavas (*Psidium guajava* L.). Food Chem. 309: 125566.

Ashraf, M., Siyal, M.I., Nazir, A. and Rehman, A. 2017. Single-step antimicrobial and moisture management finishing of Pc fabric using ZnO nanoparticles. Autex Res. J. 17: 259–262.

Bahari, A., Roeinfard, M., Ramzannezhad, A., Khodabakhshi, M. and Mohseni, M. 2019. Nanostructured features and antimicrobial properties of Fe_3O_4/ZnO Nanocomposites. Natl. Acad. Sci. Lett. 42: 9–12.

Barker, J., Stevens, D. and Bloomfield, S.F. 2001. Spread and prevention of some common viral infections in community facilities and domestic homes. J. Appl. Microbiol. 91: 7–21.

Barker, J., Vipond, I.B. and Bloomfield, S.F. 2004. Effects of cleaning and disinfection in reducing the spread of Norovirus contamination via environmental surfaces. J. Hosp. Infect. 58: 42–49.

Behera, N., Arakha, M., Priyadarshinee, M., Pattanayak, B.S., Soren, S., Jha, S. and Mallick, B.C. 2019. Oxidative stress generated at nickel oxide nanoparticle interface results in bacterial membrane damage leading to cell death. RSC Adv. 9: 24888–24894.

Benitha, V.S., Jeyasubramanian, K., Mala, R., Hikku, G.S. and Rajesh, K.R. 2019. New sol–gel synthesis of NiO antibacterial nano-pigment and its application as healthcare coating. J. Coatings. Technol. Res. 16: 59–70.

Bhadra, P., Dutta, B., Bhattyacharya, D. and Mukherjee, S. 2019. CuO and CuO@SiO_2 as a potential antimicrobial and anticancer drug. J. Microbiol. Biotechnol. Food Sci. 9: 63–69.

Bhargava, R., Khan, S., Ahmad, N. and Ansari, M.M.N. 2018. Investigation of structural, optical and electrical properties of Co_3O_4 nanoparticles. *In*: AIP Conf. Proc. 1953: 030034.

Bhowmick, G.D., Das, S., Verma, H.K., Neethu, B. and Ghangrekar, M.M. 2019. Improved performance of microbial fuel cell by using conductive ink printed cathode containing Co_3O_4 or Fe_3O_4. Electrochim. Acta 310: 173–183.

Borkow, G., Lara, H.H., Covington, C.Y., Nyamathi, A. and Gabbay, J. 2008. Deactivation of human immunodeficiency virus type 1 in medium by copper oxide-containing filters. Antimicrob Agents Chemother. 52: 518–525.

Borkow, G., Zhou, S.S., Page, T. and Gabbay, J. 2010. A novel anti-influenza copper oxide containing respiratory face mask. PLoS One 5: e11295.

Bouazizi, N., Bargougui, R., Oueslati, A. and Benslama, R. 2015. Effect of synthesis time on structural, optical and electrical properties of CuO nanoparticles synthesized by reflux condensation method. Adv. Mater. Lett. 6: 158–164.

Boura-Theodoridou, O., Giannakas, A., Katapodis, P., Stamatis, H., Ladavos, A. and Barkoula, N.-M. 2020. Performance of ZnO/chitosan nanocomposite films for antimicrobial packaging applications as a function of NaOH treatment and glycerol/PVOH blending. Food Packag. Shelf Life 23: 100456.

Broglie, J.J., Alston, B., Yang, C., Ma, L., Adcock, A.F., Chen, W. and Yang, L. 2015. Antiviral activity of gold/copper sulfide core/shell nanoparticles against human norovirus virus-like particles. PLoS One 10: e0141050.

Burlibaşa, L., Chifiriuc, M.C., Lungu, M.V., Lungulescu, E.M., Mitrea, S., Sbarcea, G., Popa, M., Măruţescu, L., Constantin, N., Bleotu, C. and Hermenean, A. 2020. Synthesis, physico-chemical characterization, antimicrobial activity and toxicological features of Ag ZnO nanoparticles. Arab. J. Chem. 13: 4180–4197.

Carter, J. and Saunders, V. 2007. Virology: Principles and Applications. Liverpool, UK: John Wiley and Sons, Ltd.

Carvalho, P.M., Felício, M.R., Santos, N.C., Gonçalves, S. and Domingues, M.M. 2018. Application of light scattering techniques to nanoparticle characterization and development. Front. Chem. 6: 237.

Champagne, V.K. and Helfritch, D.J. 2013. A demonstration of the antimicrobial effectiveness of various copper surfaces. J. Biol. Eng. 7: 8.

Chaturvedi, S., Dave, P.N. and Shah, N.K. 2012. Applications of nano-catalyst in new era. J. Saudi. Chem. Soc. 16: 307–325.

Chauhan, M., Sharma, B., Kumar, R., Chaudhary, G.R., Hassan, A.A. and Kumar, S. 2019. Green synthesis of CuO nanomaterials and their proficient use for organic waste removal and antimicrobial application. Environ. Res. 168: 85–95.

Chen, X., Ku, S., Weibel, J.A., Ximenes, E., Liu, X., Ladisch, M. and Garimella, S.V. 2017. Enhanced antimicrobial efficacy of bimetallic porous CuO microspheres decorated with Ag nanoparticles. ACS Appl. Mater. Interfaces 9: 39165–39173.

Cheng, C., Hu, Y., Shao, S., Yu, J., Zhou, W., Cheng, J., Chen, Y., Chen, S., Chen, J. and Zhang, L. 2019. Simultaneous Cr(VI) reduction and electricity generation in Plant-Sediment Microbial Fuel Cells (P-SMFCs): Synthesis of non-bonding Co_3O_4 nanowires onto cathodes. Environ. Pollut. 247: 647–657.

Chu, S., Gerasopoulos, K. and Ghodssi, R. 2016. Tobacco mosaic virus-templated hierarchical Ni/NiO with high electrochemical charge storage performances. Electrochim. Acta 220: 184–192.

Cid, A. and Simal-Gandara, J. 2020. Synthesis, characterization, and potential applications of transition metal nanoparticles. J. Inorg. Organomet. Polym. Mater. 30: 1011–1032.

Çolak, H., Karaköse, E. and Duman, F. 2017. High optoelectronic and antimicrobial performances of green synthesized ZnO nanoparticles using Aesculus hippocastanum. Environ. Chem. Lett. 15: 553–553.

Costa, S.I.G., Cauneto, V.D., Fiorentin-Ferrari, L.D., Almeida, P.B., Oliveira, R.C., Longo, E., Módenes, A.N. and Slusarski-Santana, V. 2020. Synthesis and characterization of $Nd(OH)_3$-ZnO composites for application in photocatalysis and disinfection. Chem. Eng. J. 392: 123737.

Cox, C. 1989. Airborne bacteria and viruses. Sci. Prog. 73: 469–499.

Cromeans, T., Park, G.W., Costantini, V., Lee, D., Wang, Q., Farkas, T., Lee, A. and Vinjé, J. 2014. Comprehensive comparison of cultivable norovirus surrogates in response to different inactivation and disinfection treatments. Appl. Environ. Microbiol. 80: 5743–5751.

Danwittayakul, S., Songngam, S. and Sukkasi, S. 2020. Enhanced solar water disinfection using ZnO supported photocatalysts. Environ. Technol. 41: 349–356.

Das, S., Sinha, S., Das, B., Jayabalan, R., Suar, M., Mishra, A., Tamhankar, A.J., Lundborg, C.S. and Tripathy, S.K. 2017. Disinfection of multidrug resistant Escherichia coli by solar-photocatalysis using Fe-doped ZnO nanoparticles. Sci. Rep. 7: 104.

Demirel, R., Suvacı, E., Şahin, İ., Dağ, S. and Kiliç, V. 2018. Antimicrobial activity of designed undoped and doped MicNo-ZnO particles. J. Drug. Deliv. Sci. Technol. 47: 309–321.

Derbalah, H., Soliman, A. and Elsharkawy, M.M. 2019. A new strategy to control Cucumber mosaic virus using fabricated NiO-nanostructures. J. Biotechnol. 306: 134–141.

Dobrucka, R., Dlugaszewska, J. and Kaczmarek, M. 2018. Cytotoxic and antimicrobial effects of biosynthesized ZnO nanoparticles using of *Chelidonium majus* extract. Biomed. Microdevices 20: 5.

Duman, F. and Karako, E. 2017. High optoelectronic and antimicrobial performances of green synthesized ZnO nanoparticles using Aesculus hippocastanum. 15: 547–552.

Dutta, A. 2017. Chapter 4—Fourier transform infrared spectroscopy. pp. 73–93. *In*: Thomas, S., Thomas, R. and Zachariah, A.K. (eds.). Mishra RKBT-SM for NC, editors. Micro. Nano. Technol.

Dyshlyuk, L., Babich, O., Ivanova, S., Vasilchenco, N., Atuchin, V., Korolkov, I., Russakov, D. and Prosekov, A. 2020. Antimicrobial potential of ZnO, TiO_2 and SiO_2 nanoparticles in protecting building materials from biodegradation. Int. Biodeterior. Biodegradation. 146: 104821.

Elango, M., Deepa, M., Subramanian, R. and Saraswathy, G. 2018. Synthesis, structural characterization and antimicrobial activities of polyindole stabilized Ag-Co_3O_4 nanocomposite by reflux condensation method. Mater. Chem. Phys. 216: 305–315.

El-Borady, O.M. and El-Sayed, A.F. 2020. Synthesis, morphological, spectral and thermal studies for folic acid conjugated ZnO nanoparticles: potency for multi-functional bio-nanocomposite as antimicrobial, antioxidant and photocatalytic agent. J. Mater. Res. Technol. 9: 1905–1917.

El Fawal, G., Hong, H., Song, X., Wu, J., Sun, M., He, C., Mo, X., Jiang, Y. and Wang, H. 2020. Fabrication of antimicrobial films based on hydroxyethylcellulose and ZnO for food packaging application. Food Packag. Shelf Life 23: 100462.

Esparza-González, S.C., Sánchez-Valdés, S., Ramírez-Barrón, S.N., Loera-Arias, M.J., Bernal, J., Meléndez-Ortiz, H.I. and Betancourt-Galindo, R. 2016. Effects of different surface modifying agents on the cytotoxic and antimicrobial properties of ZnO nanoparticles. Toxicol Vitr. 37: 134–141.

Ezhilarasi, A., Vijaya, J., Kaviyarasu, K., Kennedy, L.J., Ramalingam, R.J. and Al-Lohedan, H.A. 2018. Green synthesis of NiO nanoparticles using *Aegle marmelos* leaf extract for the evaluation of *in-vitro* cytotoxicity, antibacterial and photocatalytic properties. J. Photochem. Photobiol. B. Biol. 180: 39–50.

Fujimori, Y., Sato, T., Hayata, T., Nagao, T., Nakayama, M., Nakayama, T., Sugamata, R., Suzuki, K., Pure, W. and Industries, C. 2012. Novel antiviral characteristics of nanosized Copper(I) Iodide particles showing inactivation activity against 2009 Pandemic H1N1. 78: 951–955.

Gad El-Rab, S.M.F., Abo-Amer, A.E. and Asiri, A.M. 2020. Biogenic synthesis of ZnO nanoparticles and its potential use as antimicrobial agent against multidrug-resistant pathogens. Curr. Microbiol. 77: 1767–1779.

Gajendiran, J., Sivakumar, N., Reddy, C.P. and Ramya, J.R. 2020. The effect of calcination's temperature on the structural, morphological, optical behaviour, hemocompatibility and antibacterial activity of nanocrystalline Co_3O_4 powders. Ceram Int. 46: 5469–5476.

Ghaffari, H., Tavakoli, A., Moradi, A., Tabarraei, A., Bokharaei-Salim, F., Zahmatkeshan, M., Farahmand, M., Javanmard, D., Kiani, S.J. and Esghaei, M. 2019. Inhibition of H1N1 influenza virus infection by zinc oxide nanoparticles: Another emerging application of nanomedicine. 26: 70.

Ghiuță, I., Cristea, D. and Munteanu, D. 2017. Synthesis methods of metallic nanoparticles an overview. Bull. Transilv. Univ. Brașov. 10: 133–140.

Gupta, M., Tomar, R.S., Kaushik, S., Mishra, R.K. and Sharma, D. 2018. Effective antimicrobial activity of green ZnO nano particles of Catharanthus roseus. Front. Microbiol. 9: 2030.

Hamedani, N.F., Ghazvini, M., Sheikholeslami-Farahani, F. and Bagherian-Jamnani, M.T. 2020. ZnO nanorods as efficient catalyst for the green synthesis of thiophene derivatives: Investigation of antioxidant and antimicrobial activity. J. Heterocycl. Chem. 57: 1588–1598.

Han, J.H., Lee, D., Chew, C.H.C., Kim, T. and Pak, J.J. 2016. A multi-virus detectable microfluidic electrochemical immunosensor for simultaneous detection of H1N1, H5N1, and H7N9 virus using ZnO nanorods for sensitivity enhancement. Sensors Actuators, B. Chem. 228: 36–42.

Hang, X., Peng, H., Song, H., Qi, Z., Miao, X. and Xu, W. 2015. Antiviral activity of cuprous oxide nanoparticles against Hepatitis C virus *in vitro*. J. Virol. Methods 222: 150–157.

Helan, V., Prince, J.J., Al-Dhabi, N.A., Arasu, M.V., Ayeshamariam, A., Madhumitha, G., Roopan, S.M. and Jayachandran, M. 2016. Neem leaves mediated preparation of NiO nanoparticles and its magnetization, coercivity and antibacterial analysis. Results Phys. 6: 712–718.

Herfst, S., Böhringer, M., Karo, B., Lawrence, P., Lewis, N.S., Mina, M.J., Russell, C.J., Steel, J., de Swart, R.L. and Menge, C. 2017. Drivers of airborne human-to-human pathogen transmission. Curr. Opin. Virol. 22: 22–29.

Hirota, K., Sugimoto, M., Kato, M., Tsukagoshi, K., Tanigawa, T. and Sugimoto, H. 2010. Preparation of zinc oxide ceramics with a sustainable antibacterial activity under dark conditions. Ceram. Int. 36: 497–506.

Hodinka, R.L. 2016. Respiratory RNA viruses. Microbiol. Spectr. 4: 1–33.

Holder, C.F. and Schaak, R.E. 2019. Tutorial on Powder X-ray Diffraction for characterizing nanoscale materials. ACS Nano. 13: 7359–7365.

Horie, M., Ogawa, H., Yoshida, Y., Yamada, K., Hara, A., Ozawa, K., Matsuda, S., Mizota, C., Tani, M., Yamamoto, Y., Yamada, M., Nakamura, K. and Imai, K. 2008. Inactivation and morphological changes of avian influenza virus by copper ions. Arch. Virol. 153: 1467–1472.

Hsu, C.M., Huang, Y.H., Chen, H.J., Lee, W.C., Chiu, H.W., Maity, J.P., Chen, C.C., Kuo, Y.H. and Chen, C.Y. 2018. Green synthesis of nano-Co_3O_4 by Microbial Induced Precipitation (MIP) process using *Bacillus pasteurii* and its application as supercapacitor. Mater. Today Commun. 14: 302–311.

Huesca-Espitia, L.L., del, C., Aurioles-López, V., Ramírez, I., Sánchez-Salas, J.L. and Bandala, E.R. 2017. Photocatalytic inactivation of highly resistant microorganisms in water: A kinetic approach. J. Photochem. Photobiol. A. Chem. 337: 132–139.

Imai, K., Ogawa, H., Nghia, V., Inoue, H., Fukuda, J. and Ohba, M. 2012. Inactivation of high and low pathogenic avian influenza virus H5 subtypes by copper ions incorporated in zeolite-textile materials. Antiviral. Res. 93: 225–233.

Ingle, A., Gade, A., Pierrat, S., Sonnichsen, C. and Rai, M. 2008. Mycosynthesis of silver nanoparticles using the fungus Fusarium acuminatum and its activity against some human pathogenic bacteria. Curr. Nanosci. 4: 141–144.

Irshad, S., Salamat, A., Anjum, A.A., Sana, S., Saleem, R.S., Naheed, A. and Iqbal, A. 2018. Green tea leaves mediated ZnO nanoparticles and its antimicrobial activity. Cogent. Chem. 4: 1469207.

Jesudoss, S.K., Vijaya, J., Iyyappa Rajan, P., Kaviyarasu, K., Sivachidambaram, M., John Kennedy, L., Al-Lohedan, H.A., Jothiramalingam, R. and Munusamy, M.A. 2017. High performance multifunctional green Co_3O_4 spinel nanoparticles: photodegradation of textile dye effluents, catalytic hydrogenation of nitro-aromatics and antibacterial potential. Photochem. Photobiol. Sci. 16: 766–778.

Kannan, K., Radhika, D., Nikolova, M.P., Sadasivuni, K.K., Mahdizadeh, H. and Verma, U. 2020. Structural studies of bio-mediated NiO nanoparticles for photocatalytic and antibacterial activities. Inorg. Chem. Commun. 113: 107755.

Karlstrom, A.R. and Levine, R.L. 1991. Copper inhibits the protease from human immunodeficiency virus 1 by both cysteine-dependent and cysteine-independent mechanisms. Proc. Natl. Acad. Sci. 88: 5552–5556.

Karunakaran, C. and Vinayagamoorthy, P. 2016. Tri-functional Fe_2O_3-encased Ag-doped ZnO nanoframework: magnetically retrievable antimicrobial photocatalyst. Mater. Res. Express. 3: 115501.

Kaushik, M., Niranjan, R., Thangam, R., Madhan, B., Pandiyarasan, V., Ramachandran, C., Oh, D.H. and Venkatasubbu, G.D. 2019. Investigations on the antimicrobial activity and wound healing potential of ZnO nanoparticles. Appl. Surf. Sci. 479: 1169–1177.

Kavitha, T., Haider, S., Kamal, T. and Ul-islam, M. 2017. Thermal decomposition of metal complex precursor as route to the synthesis of Co_3O_4 nanoparticles: Antibacterial activity and mechanism. J. Alloys. Compd. 704: 296–302.

Khan, I., Saeed, K. and Idrees, K. 2019. Nanoparticles: Properties, applications and toxicities. Arab. J. Chem. 12: 908–931.

Khan, M.F., Ansari, A.H., Hameedullah, M., Ahmad, E., Husain, F.M., Zia, Q., Baig, U., Zaheer, M.R., Alam, M.M., Khan, A.M., AlOthman, Z.A., Ahmad, I., Ashraf, G.M.d and Alievd, G. 2016. Sol-gel synthesis of thorn-like ZnO nanoparticles endorsing mechanical stirring effect and their antimicrobial activities: Potential role as nano-Antibiotics. Sci. Rep. 6: 27689.

Khashan, K.S., Sulaiman, G.M., Hamad, A.H., Abdulameer, F.A. and Hadi, A. 2017. Generation of NiO nanoparticles via pulsed laser ablation in deionised water and their antibacterial activity. Appl. Phys. A 123: 190.

Khatami, M., Varma, R.S., Zafarnia, N., Yaghoobi, H., Sarani, M. and Kumar, V.G. 2018. Applications of green synthesized Ag, ZnO and Ag/ZnO nanoparticles for making clinical antimicrobial wound-healing bandages. Sustain. Chem. Pharm. 10: 9–15.

Kiani, M., Bagherzadeh, M., Meghdadi, S., Fadaei-Tirani, F., Babaie, M. and Schenk-Joß, K. 2020. Synthesis, characterisation and crystal structure of a new Cu(II)-carboxamide complex and CuO nanoparticles as new catalysts in the CuAAC reaction and investigation of their antibacterial activity. Inorganica. Chim. Acta 506: 119514.

Kim, I., Viswanathan, K., Kasi, G., Thanakkasaranee, S., Sadeghi, K. and Seo, J. 2020. ZnO Nanostructures in active antibacterial food packaging: Preparation methods, antimicrobial mechanisms, safety issues, future prospects, and challenges. Food. Rev. Int. 1–29.

Kombaiah, K., Vijaya, J.J., Kennedy, L.J., Kaviyarasu, K., Ramalingam, R.J. and Al-Lohedan, H.A. 2019. Green synthesis of Co_3O_4 nanorods for highly efficient catalytic, photocatalytic, and antibacterial activities. J. Nanosci. Nanotechnol. 19: 2590–2598.

Korshunov, S. and Imlay, J.A. 2010. Two sources of endogenous hydrogen peroxide in *Escherichia coli*. Mol. Microbiol. 75: 1389–1401.

Kumar, P.S. and Nesaraj, A.S. 2017. Soft chemical synthesis of Ag doped ZnO nanoparticles: Antimicrobial activity and in Vivo acute nanotoxicological impact on Swiss Albino. Mice 8: 2949–2959.

Kumar, P.S. and Nesaraj, A.S. 2018. Development of Se doped ZnO nanoparticles: Antimicrobial activity and in vivo acute nanotoxicological impact on Swiss Albino Mice. Int. J. Pharm. Sci. Res. 9: 2395–2404.

Kumar, R., Singh, L., Zularisam, A.W. and Hai, F.I. 2016. Potential of porous Co_3O_4 nanorods as cathode catalyst for oxygen reduction reaction in microbial fuel cells. Bioresour. Technol. 220: 537–542.

Kumar, R., Umar, A., Kumar, G. and Nalwa, H.S. 2017. Antimicrobial properties of ZnO nanomaterials: A review. Ceram. Int. 43: 3940–3961.
Liu, J., Hui, A., Ma, J., Chen, Z. and Peng, Y. 2017. Fabrication and application of hollow ZnO nanospheres in antimicrobial casein-based coatings. Int. J. Appl. Ceram. Technol. 14: 128–134.
Liu, Z., Zhou, H., Liu, J., Yin, X., Mao, Y., Liu, Z., Li, Z. and Xie, W. 2016. Microbiote shift in sequencing batch reactors in response to antimicrobial ZnO nanoparticles. RSC Adv. 6: 110108–110111.
Lungu, M.V., Vasile, E., Lucaci, M., Pătroi, D., Mihăilescu, N., Grigore, F., Marinescu, V., Brătulescu, A., Mitrea, S., Sobetkii, A., Sobetkii, A.A., Popa, M. and Chifiriuc, M.-C. 2016. Investigation of optical, structural, morphological and antimicrobial properties of carboxymethyl cellulose capped Ag-ZnO nanocomposites prepared by chemical and mechanical methods. Mater. Charact. 120: 69–81.
Ma, Y., Lu, Y., Guan, G., Luo, J., Niu, Q., Liu, J., Yin, H. and Liu, G. 2017. Flower-like ZnO nanostructure assisted loop-mediated isothermal amplification assay for detection of Japanese encephalitis virus. Virus Res. 232: 34–40.
Malhotra, J.S., Sharma, A., Singh, A.K., Kumar, S., Rana, B.S. and Kumar, S. 2019. Investigations on Photocatalytic, antimicrobial and magnetic properties of Sol–Gel-Synthesized Ga-Doped ZnO nanoparticles. Int. J. Nanosci. 18: 1850014.
Mallakpour, S. and Javadpour, M. 2017. Antimicrobial, mechanical, optical and thermal properties of PVC/ZnO-EDTA nanocomposite films. Polym. Adv. Technol. 28: 393–403.
Meenakshisundaram, I., Kalimuthu, S., Priya, P.G. and Karthikeyan, S. 2019. Facile green synthesis and antimicrobial performance of Cu_2O nanospheres decorated g-C_3N_4 nanocomposite. Mater. Res. Bull. 112: 331–335.
Michal, R., Dworniczek, E., Caplovicova, M., Monfort, O., Lianos, P., Caplovic, L. and Plesch, G. 2016. Photocatalytic properties and selective antimicrobial activity of TiO_2 (Eu)/CuO nanocomposite. Appl. Surf. Sci. 371: 538–546.
Mobeen, A.A., Shahina, S.K.J., Sundaram, R., Magdalane, C.M., Kaviyarasu, K., Letsholathebe, D., Mohamed, S.B., Kennedy, J. and Maaza, M. 2018. Antibacterial, magnetic, optical and humidity sensor studies of β-$CoMoO_4$-Co_3O_4 nanocomposites and its synthesis and characterization. J. Photochem. Photobiol. B. Biol. 183: 233–241.
Modena, M.M., Rühle, B., Burg, T.P. and Wuttke, S. 2019. Nanoparticle characterization: What to measure? Adv. Mater. 31: e1901556.
Mourdikoudis, S., Pallares, R.M. and Thanh, N.T.K. 2018. Characterization techniques for nanoparticles: Comparison and complementarity upon studying nanoparticle properties. Nanoscale 10: 12871–12934.
Murray, J.P. and Laband, S.J. 1979. Degradation of poliovirus by adsorption on inorganic surfaces. Appl. Environ. Microbiol. 37: 480–486.
Myrick, M.L., Simcock, M.N., Baranowski, M., Brooke, H., Morgan, S.L. and McCutcheon, J.N. 2011. The Kubelka-Munk diffuse reflectance formula revisited. Appl. Spectrosc. Rev. 46: 140–165.
Newbury, D.E. and Ritchie, N.W.M. 2013. Is scanning electron microscopy/energy dispersive X-ray spectrometry (SEM/EDS) quantitative? Scanning 35: 141–168.
Nishino, F., Jeem, M., Zhang, L., Okamoto, K., Okabe, S. and Watanabe, S. 2017. Formation of CuO nano-flowered surfaces via submerged photo- synthesis of crystallites and their antimicrobial activity. Sci. Rep. 7: 1063.
Noyce, J.O., Michels, H. and Keevil, C.W. 2007. Inactivation of influenza a Virus on Copper versus stainless steel surfaces. Appl. Environ. Microbiol. 73: 2748–2750.
Ong, H.R., Ramli, R., Khan, M.M.R. and Yunus, R.M. 2016. The influence of CuO nanoparticle on non-edible rubber seed oil based alkyd resin preparation and its antimicrobial activity. Prog. Org. Coatings 101: 245–252.
Pagar, T., Ghotekar, S., Pagar, K., Pansambal, S. and Oza, R. 2019. A review on bio-synthesized Co_3O_4 nanoparticles using plant extracts and their diverse applications. J. Chem. Rev. 1: 260–270.
Paul, D. and Neogi, S. 2019. Synthesis, characterization and a comparative antibacterial study of CuO, NiO and CuO-NiO mixed metal oxide. Mater. Res. Express 6: 055004.
Piedade, A.P., Pinho, A.C., Branco, R. and Morais, P.V. 2020. Evaluation of antimicrobial activity of ZnO based nanocomposites for the coating of non-critical equipment in medical-care facilities. Appl. Surf. Sci. 513: 145818.
Pugazhendhi, A., Kumar, S.S., Manikandan, M. and Saravanan, M. 2018. Photocatalytic properties and antimicrobial efficacy of Fe doped CuO nanoparticles against the pathogenic bacteria and fungi. Microb. Pathog. 122: 84–89.
Radhakrishnan, A., Rejani, P. and Beena, B. 2018. CuO nano structures as an ecofriendly nano photo catalyst and antimicrobial agent for environmental remediation. Int. J. Nano Dimens. 9: 145–157.
Rai, M., Deshmukh, S.D., Ingle, A.P., Gupta, I.R., Galdiero, M. and Galdero, S. 2016. Metal nanoparticles: The protective nanoshield against virus infection. Crit. Rev. Microbiol. 42: 46–56.
Rajan, P.I., Vijaya, J.J., Jesudoss, S.K., Kaviyarasu, K., Kennedy, L.J., Jothiramalingam, R., Al-Lohedan, H.A. and Vaali-Mohammed, M.-A. 2017. Green-fuel-mediated synthesis of self-assembled NiO nano-sticks for dual applications—photocatalytic activity on Rose Bengal dye and antimicrobial action on bacterial strains. Mater. Res. Express 4: 085030.
Rajendaran, K., Muthuramalingam, R. and Ayyadurai, S. 2019. Green synthesis of Ag-Mo/CuO nanoparticles using Azadirachta indica leaf extracts to study its solar photocatalytic and antimicrobial activities. Mater. Sci. Semicond. Process 91: 230–238.
Ramírez-Sánchez, I.M., Máynez-Navarro, O.D. and Bandala, E.R. 2019. Degradation of emerging contaminants using Fe-doped TiO_2 under UV and visible radiation BT—Advanced research in nanosciences for water technology. pp. 263–285. *In*: Prasad, R. and Karchiyappan, T. (eds.). Adv. Res. Nanosci. Water Technol. Cham: Springer International Publishing.

Rastar, A., Yazdanshenas, M.E., Rashidi, A. and Bidoki, S.M. 2013. Theoretical review of optical properties of nanoparticles. J. Eng. Fiber. Fabr. 8: 85–96.
Reyes-Torres, M.A., Mendoza-Mendoza, E., Miranda-Hernández, Á.M., Pérez-Díaz, M.A., López-Carrizales, M., Peralta-Rodríguez, R.D., Sánchez-Sánchez, R. and Martinez-Gutierrez, F. 2019. Synthesis of CuO and ZnO nanoparticles by a novel green route: Antimicrobial activity, cytotoxic effects and their synergism with ampicillin. Ceram. Int. 45: 24461–24468.
Roy, S. and Rhim, J.W. 2019. Carrageenan-based antimicrobial bionanocomposite films incorporated with ZnO nanoparticles stabilized by melanin. Food Hydrocoll. 90: 500–507.
Ruey, O.H., Khan, M.M.R., Ramli, R., Shein, H.C. and Mohd, Y.R. 2018. Influence of CuO nanoparticle on palm oil based alkyd resin preparation and its antimicrobial activity. IOP Conf. Ser. Mater. Sci. Eng. 324: 012027.
Sagripanti, J.L., Routson, L.B. and Lytle, C.D. 1993. Virus inactivation by copper or iron ions alone and in the presence of peroxide. Appl. Environ. Microbiol. 59: 4374–4376.
Salah, N., Alfawzan, A.M., Allafi, W., Baghdadi, N., Saeed, A., Alshahrie, A., Al-Shawafi, W.M. and Memic, A. 2020. Size-controlled, single-crystal CuO nanosheets and the resulting polyethylene–carbon nanotube nanocomposite as antimicrobial materials. Polym. Bull. 1–22.
Saleem, S., Ahmed, B., Saghir, M. and Al-shaeri, M. 2017. Microbial pathogenesis inhibition of growth and bio film formation of clinical bacterial isolates by NiO nanoparticles synthesized from Eucalyptus globulus plants. Microb. Pathog. 111: 375–387.
Saravanakkumar, D., Sivaranjani, S., Kaviyarasu, K., Ayeshamariam, A., Ravikumar, B., Pandiarajan, S., Veeralakshmi, C., Jayachandran, M. and Maaza, M. 2018. Synthesis and characterization of ZnO-CuO nanocomposites powder by modified perfume spray pyrolysis method and its antimicrobial investigation. J. Semicond. 39: 033001.
Sarwar, S., Chakraborti, S., Bera, S., Sheikh, I.A., Hoque, K.M. and Chakrabarti, P. 2016. The antimicrobial activity of ZnO nanoparticles against *Vibrio cholerae*: Variation in response depends on biotype. Nanomedicine Nanotechnology. Biol. Med. 12: 1499–1509.
Sathishkumar, G., Rajkuberan, C., Manikandan, K., Prabukumar, S. and Danieljohn, J. 2017. Facile biosynthesis of antimicrobial zinc oxide (ZnO) nano flakes using leaf extract of Couroupita guianensis. Aubl. Mater. Lett. 188: 383–386.
Schijven, J.F., Berg, H.H.J.L. Van Den, Colin, M., Dullemont, Y., Hijnen, W.A.M., Magic-Knezev, A., Oorthuizen, W.A. and Wubbels, G. 2013. A mathematical model for removal of human pathogenic viruses and bacteria by slow sand filtration under variable operational conditions. Water Res. 47: 2592–2602.
Shahmohammadi, J.F. and Almasi, H. 2016. Morphological, physical, antimicrobial and release properties of ZnO nanoparticles-loaded bacterial cellulose films. Carbohydr. Polym. 149: 8–19.
Sharma, S., Kumar, K., Thakur, N., Chauhan, S. and Chauhan, M.S. 2020. The effect of shape and size of ZnO nanoparticles on their antimicrobial and photocatalytic activities: A green approach. Bull. Mater. Sci. 43: 20.
Shi, C.E., Pan, L., Wang, C.R., He, Y., Wu, Y.F. and Xue, S.S. 2016. Facile preparation of Ag/NiO composite nanosheets and their antibacterial activity. Jom. 68: 324–329.
Shimabuku, Q.L., Arakawa, F.S., Fernandes, S.M., Ferri, C.P., Ueda-Nakamura, T., Fagundes-Klen, M.R. and Bergamasco, R. 2017. Water treatment with exceptional virus inactivation using activated carbon modified with silver (Ag) and copper oxide (CuO) nanoparticles. Environ. Technol. 38: 2058–2069.
Singh, A.K., Viswanath, V. and Janu, V.C. 2009. Synthesis, effect of capping agents, structural, optical and photoluminescence properties of ZnO nanoparticles. J. Lumin. 129: 874–878.
Sirelkhatim, A., Mahmud, S., Seeni, A., Kaus, N.H.M., Ann, L.C., Bakhori, S.K.M., Hasan, H. and Mohamad, D. 2015. Review on zinc oxide nanoparticles: Antibacterial activity and toxicity mechanism. Nano-Micro. Lett. 7: 219–242.
Slavin, Y.N., Asnis, J., Häfeli, U.O. and Bach, H. 2017. Metal nanoparticles: understanding the mechanisms behind antibacterial activity. J. Nanobiotechnology 15: 65.
Sportelli, M.C., Izzi, M., Kukushkina, E.A., Hossain, S.I., Picca, R.A., Ditaranto, N. and Cioffi, N. 2020. Can nanotechnology and materials science help the fight against SARS-CoV-2? Nanomaterials 10: 802.
Sree, G.S., Botsa, S.M., Reddy, B.J.M. and Ranjitha, K.V.B. 2020. Enhanced UV–Visible triggered photocatalytic degradation of brilliant green by reduced graphene oxide based NiO and CuO ternary nanocomposite and their antimicrobial activity. Arab. J. Chem. 13: 5137–5150.
Srihasam, S., Thyagarajan, K., Korivi, M., Lebaka, V.R. and Mallem, S.P.R. 2020. Phytogenic generation of NiO nanoparticles using stevia leaf extract and evaluation of their *in-vitro* antioxidant and antimicrobial properties. Biomolecules 10: 89.
Su, D. 2017. Advanced electron microscopy characterization of nanomaterials for catalysis. Green Energy Environ. 2: 70–83.
Sundberg, K. and Champagne, V.K. 2015. Effectiveness of nanomaterial copper cold spray surfaces on inactivation of Influenza A Virus. J. Biotechnol. Biomater. 5: 205.
Taghizadeh, S.M., Lal, N., Ebrahiminezhad, A., Moeini, F., Seifan, M., Ghasemi, Y. and Berenjian, A. 2020. Green and economic fabrication of Zinc Oxide (ZnO) nanorods as a broadband UV blocker and antimicrobial agent. Nanomaterials 10: 530.
Talam, S., Karumuri, S.R. and Gunnam, N. 2012. Synthesis, characterization, and spectroscopic properties of ZnO nanoparticles. ISRN Nanotechnol. 2012: 372505.
Tavakoli, A., Ataei-Pirkooh, A., Mm Sadeghi, G., Bokharaei-Salim, F., Sahrapour, P., Kiani, S.J., Moghoofei, M., Farahmand, M., Javanmard, D. and Monavari, S.H. 2018. Polyethylene glycol-coated zinc oxide nanoparticle: An efficient nanoweapon to fight against herpes simplex virus type 1. Nanomedicine 13: 2675–2690.

Thakur, S., Kaur, M., Lim, W.F. and Lal, M. 2020. Fabrication and characterization of electrospun ZnO nanofibers; antimicrobial assessment. Mater. Lett. 264: 127279.

Theophil, A.G., Renuka, D., Ramesh, R., Anandaraj, L., John Sundaram, S., Ramalingam, G., Magdalane, C.M., Bashir, A.K.H., Maaza, M. and Kaviyarasu, K. 2019. Green synthesis of ZnO nanoparticle using Prunus dulcis (Almond Gum) for antimicrobial and supercapacitor applications. Surfaces and Interfaces 17: 100376.

Tian, X., Li, Y., Wan, S., Wu, Z. and Wang, Z. 2017. Functional surface coating on cellulosic flexible substrates with improved water-resistant and antimicrobial properties by use of ZnO nanoparticles. J. Nanomater. 2017: 9689035.

Vasickova, P., Pavlik, I., Verani, M. and Carducci, A. 2010. Issues concerning survival of viruses on surfaces. Food Environ. Virol. 2: 24–34.

Vernon-Parry, K.D. 2000. Scanning electron microscopy: An introduction. III-Vs Rev. 13: 40–44.

Vijayakumar, S., Arulmozhi, P., Kumar, N., Sakthivel, B., Prathip Kumar, S. and Praseetha, P.K. 2020. *Acalypha fruticosa* L. leaf extract mediated synthesis of ZnO nanoparticles: Characterization and antimicrobial activities. Mater. Today Proc. 23: 73–80.

Vincent, M., Duval, R.E., Hartemann, P. and Engels-Deutsch, M. 2018. Contact killing and antimicrobial properties of copper. J. Appl. Microbiol. 124: 1032–1046.

Visnapuu, M., Rosenberg, M., Truska, E., Nõmmiste, E., Šutka, A., Kahru, A., Rähn, M., Vija, H., Orupõld, K., Kisand, V. and Ivask, A. 2018. UVA-induced antimicrobial activity of ZnO/Ag nanocomposite covered surfaces. Colloids Surfaces B Biointerfaces 169: 222–232.

Waghmode, B.J., Husain, Z., Joshi, M., Sathaye, S.D., Patil, K.R. and Malkhede, D.D. 2016. Synthesis and study of calixarene-doped polypyrrole-TiO_2/ZnO composites: Antimicrobial activity and electrochemical sensors. J. Polym. Res. 23: 35.

Wang, L., Hu, C. and Shao, L. 2017. The antimicrobial activity of nanoparticles: Present situation and prospects for the future. Int. J. Nanomedicine 12: 1227–1249.

Warnes, S.L. and Keevil, C.W. 2013. Inactivation of norovirus on dry copper alloy surfaces. PLoS One 8: e75017.

Warnes, S.L., Summersgill, E.N. and Keevil, C.W. 2015. Inactivation of murine norovirus on a range of copper alloy surfaces is accompanied by loss of capsid integrity. Appl. Environ. Microbiol. 81: 1085–1091.

Weiss, C., Carriere, M., Fusco, L., Capua, I., Regla-Nava, J.A., Pasquali, M., Scott, J.A., Vitale, F., Unal, M.A., Mattevi, C. et al. 2020. Toward nanotechnology-enabled approaches against the COVID-19 Pandemic. ACS Nano. 14: 6383–6406.

Whitfield, P. and Mitchell, L. 2004. X-Ray Diffraction analysis of nanoparticles: Recent developments, potential. Int. J. Nanosci. 3: 757–763.

Yiu, K., Wei, W., Zhang, Y., Dan, L., Leung, P.H.M., Li, Y. and Yang, M. 2013. Ultrasensitive detection of *E. coli* O157: H7 with biofunctional magnetic bead concentration via nanoporous membrane based electrochemical immunosensor. Biosensors and Bioelectronics 41: 532–537.

Yu, L., Peel, G.K., Cheema, F.H., Lawrence, W.S., Bukreyeva, N., Jinks, C.W., Peel, J.E., Peterson, J.W. and Paessler, S. 2020. Catching and killing of airborne SARS-CoV-2 to control spread of COVID-19 by a heated air disinfection system. Mater. Today Phys. 15: 100249.

Yue, L., Chen, S., Wang, S., Wang, C., Hao, X. and Cheng, Y.F. 2019. Water disinfection using Ag nanoparticle–CuO nanowire co-modified 3D copper foam nanocomposites in high flow under low voltages. Environ. Sci. Nano. 6: 2801–2809.

Zaharia, A., Muşat, V., Pleşcan Ghisman, V. and Baroiu, N. 2016. Antimicrobial hybrid biocompatible materials based on acrylic copolymers modified with (Ag)ZnO/chitosan composite nanoparticles. Eur. Polym. J. 84: 550–564.

Zare Mina, Namratha, K., Alghamdi, S., Mohammad, Y.H.E., Hezam, A., Zare Mohamad, Drmosh, Q.A., Byrappam, K., Chandrashekar, B.N., Ramakrishna, S. and Zhang, X. 2019. Novel green biomimetic approach for synthesis of ZnO-Ag nanocomposite; antimicrobial activity against food-borne pathogen. Biocompatibility and Solar Photocatalysis. Sci. Rep. 9: 8303.

Zhang, H., Hortal, M., Jordá-Beneyto, M., Rosa, E., Lara-Lledo, M. and Lorente, I. 2017. ZnO-PLA nanocomposite coated paper for antimicrobial packaging application. LWT. 78: 250–257.

Zhang, J., Zhang, B., Chen, X., Mi, B., Wei, P., Fei, B. and Mu, X. 2017. Antimicrobial bamboo materials functionalized with ZnO and graphene oxide nanocomposites. Materials (Basel) 10: 239.

CHAPTER 8

Synthesis of Carbon-based Nanomaterials and their use in Nanoremediation

Kien A. Vu and *Catherine N. Mulligan**

1. Introduction

1.1 Environmental problems

In recent years, the quality of the environment has been decreasing due to human activities and the uncontrolled use of natural resources. This situation may influence humans who use daily environmental resources, such as clean air, water and soil. The significant environmental impacts are air pollution, water pollution and soil pollution.

Air pollution is the discharge of particulates or chemicals to the atmosphere at such concentrations that it may be risky for people, animals, plants and surroundings. The use of fossils occurs in domestic and industrial activities, such as transportation, mining, power plants, buildings construction. Some essential air pollutants are carbon dioxide (CO_2) from fossil fuel combustion, carbon monoxide (CO) from industry and motor vehicles, ammonia (NH_3), sulfur oxide (SO_2), nitrogen oxide (NO_2), hydrogen sulfide (H_2S), methane (CH_4) or Volatile Organic Compounds (VOCs) like toluene (C_7H_8) from other activities. Their considerable risks include environmental effects (e.g., global warming, ozone depletion, acid rain) and human effects (e.g., lung cancer, heart disease, respiratory infections) (Singh and Singh, 2017).

Water pollution is any physical, chemical or biological change in waterbodies that negatively influences water quality or living organisms. Pollutants can be discharged into water bodies (e.g., groundwater, lakes, rivers, oceans), resulting from human activities, without suitable treatment. Some main pollutant supplies are natural (e.g., soil erosion, organic matter decomposition), industrial (e.g., detergents, heavy metals), agricultural (e.g., excess fertilizers and pesticides) and domestic (e.g., sewage water) source. A few significant effects of water pollution include diseases, ecosystem damage, food chain interruption and water deoxygenation (Singh and Singh, 2017).

Department of Building, Civil, and Environmental Engineering, Concordia University, Montreal, Quebec, Canada.
* Corresponding author: mulligan@civil.concordia.ca

Soil pollution is the degradation of the quality of soil that adversely affects the development of microbes or plants. It is caused by agricultural (e.g., pesticides, fertilizers, herbicides, insecticides, fungicides) and industrial (production process of pesticides or fertilizers) activities or the weather on rocks. Some soil pollutions are soil erosion, nutrient status reduction and structural stability decrease, making the soil unsuitable for growing crops (Singh and Singh, 2017). Moreover, soil pollutants' fate and transport to rivers or lakes may kill fish, plants or aquatic systems.

Hence, there is a need to research and develop some methods for remediating these contaminants from the environment.

1.2 Environmental remediation approaches

Environmental remediation is the technique used to clean up or remove contaminants from the environment (e.g., water, air, soil, sediments). There are two types of remediation: *in situ* and *ex situ* techniques, where contaminants are cleaned up in the ground and off site, respectively.

The general goal of environmental remediation is to reduce the adverse effects of contaminants by different methods. Choosing the best remediation method depends on factors such as site conditions, efficiency, cost and contaminant characteristics. Some remediation techniques are nanoremediation, bioremediation and nanobioremediation.

1.2.1 Nanoremediation

Nanoremediation is the technique for remediation of the pollutants using nanomaterials or nanoparticles. Depending on the nanoparticles' features, the process of nanoremediation may include reduction, oxidation, sorption or a combination (Lee et al., 2014). However, the main removal mechanisms are reduction and adsorption of contaminants on its surface.

In recent years, nanoparticles have been used as new adsorbents in remediation techniques. Compared with other materials, they have larger specific surface areas, extensive adsorption sites, and can be transported to a contaminated site's difficult target zones (Bains and Pal, 2019). Different types of nanoparticles, such as carbon-based nanomaterials, zero-valent iron nanoparticles, zinc oxide (ZnO), titanium dioxide (TiO_2) have been effectively investigated for remediation techniques over the past few years. They can be used in different ways to enhance their removal efficiency, such as catalysts, coatings, modifiers, composites. Some examples of the techniques of nanoremediation include the removal of heavy metals from soils, air pollution, inorganic and organic contaminants or pharmaceuticals and personal care products. In soil remediation, nanoparticles play a vital role in reducing, immobilization, conversion of heavy metals and degradation of organic contaminants (Araújo et al., 2015).

The advantages of nanoremediation are high removal rates with various contaminants and different environmental conditions, rapid degradation speed, fast clean up time at contaminated sites, low toxicity and potential capacity for broader application. The disadvantages of nanoremediation involve the high cost, aggregation or agglomeration of nanomaterials in living organisms, potential risks of nanomaterials to human health and the environment (Bardos et al., 2018).

1.2.2 Bioremediation

Bioremediation is the process that uses microorganisms, such as bacteria, fungi, plants or enzymes, for degradation of environmental contaminants (Mallikarjunaiah et al., 2020). The reactions induce contaminants' transformation (to harmless products) during different microorganisms' metabolism processes. The microbes used in bioremediation are nonpathogenic and originally from contaminated sites. Due to the high degradation efficiency and isolation simplicity, aerobic microbes are preferred in the bioremediation process.

Many factors influence the performance of bioremediation. It strongly depends on the type and properties of microorganisms. Each microorganism may degrade a particular contaminant and only survive under a specific condition. Hence, it is critical to choose the right microorganism under each condition. Besides, environmental factors may impact the biodegradation rate, which is due to the dependence of microorganisms' growth and activity on environmental conditions, such as pH, temperature, moisture, nutrients, oxygen content. Therefore, an ideal situation could improve the development of microorganisms and encourage them to break down contaminants.

The bioremediation method has some advantages. It does not impact natural ecosystems, thus is more eco-friendly than corresponding physical-chemical methods and is acceptable to the public. Due to the possibility of on-site degradation and the use of less energy , this method may save money (for excavation or transport of contaminants) and preserve the soil structure. However, bioremediation has some limitations. First, it only works well with biodegradable contaminants, which limits its complete application. Second, it will take a longer time than traditional treatment methods, such as incineration. Third, the products and by-products formed by this method may be more toxic or persistent than the original contaminants (Gouma et al., 2014).

1.2.3 Nanobioremediation

Nanobioremediation is the process of using both nanomaterials and bioremediation for the removal of contaminants. This method can effectively remove diverse contaminants, for instance, heavy metals, organic and inorganic pollutants in the environment (Mallikarjunaiah et al., 2020).

The removal mechanism of nanobioremediation is based on the adsorption by nanomaterials and degradation by microorganisms. Moreover, functional groups on the nanocomposite's surface may significantly improve their removal productivity and diversify the contaminant targets. This method can be applied for remediation of solid waste, groundwater and wastewater, heavy metal pollution or hydrocarbon. Some examples of nanobioremediation include the use of single-enzyme nanoparticles in bioremediation to improve the stability, longevity and reusability of enzymes that enhance their biocatalytic properties, and use of engineered polymeric nanoparticles for bioremediation of hydrophobic contaminants to increase the release of hydrophobic organic compounds, such as nonaqueous phase liquids (NAPLs), which leads to improving the *in situ* bioremediation rate. Use of nanoparticles prepared from *Noaea mucronata* vegetation may enhance the bioremediation of heavy metals (for example, copper (Cu), lead (Pb), nickel (Ni), zinc (Zn)) from a dried waste pool of a lead mine (Rizwan et al., 2014).

At present, different environmental remediation methods are available. Among various methods, the use of carbon-based nanomaterials (CNMs) is a new approach in the remediation technique. Therefore, this chapter will focus on this approach.

2. Some Carbon-based Nanomaterials

2.1 Carbon-based nanomaterials and applications

Nanomaterials are materials with at least one dimension less than 100 nm (1 nm = 10^{-9} m). Carbon-based nanomaterials (CNMs) are described as materials in which the main component is nanosize pure carbon. CNMs can be present in numerous forms with various dimensionalities (Fig. 8.1), such as zero-dimensional systems (fullerenes), one-dimensional systems (carbon nanotubes—CNTs) and two-dimensional systems (graphene). Some CNMs are present in nature, such as fullerenes in geological deposits and CNTs in ice cores. Pristine fullerenes and CNTs can also be formed from natural processes like volcanoes or forest fires, industrial combustion processes or engines (Kamrani and Nasr, 2010).

CNMs have attracted the public's interest due to the invention of C_{60} fullerene in 1985 (Kroto, et al., 1985).

Due to their structure, CNMs may show different features. SWCNTs could be metallic (armchair chirality) or semiconductor materials (zigzag tubes) depending on the way they are rolled up. Based on a specific purpose, a metallic or semiconductor SWCNT can be formed by unzipping a CNT.

The chemical property of CNMs is dependent on their morphology, such as Gaussian curvature and hybridization of carbon bonds (sp, sp^2, sp^3). Any surface modification in CNMs, such as doping atoms addition, defects generation or functionalization, may alter their interaction behavior with the surrounding environment and lead to changes in their chemical, electrical or magnetic features (Mauter and Elimelech, 2008).

In recent years, applying CNMs as new materials in remediation techniques has been developed (Table 8.1) due to their unique properties, low cost, low toxicity, high reactivity and simple synthesis process (Xu et al., 2016).

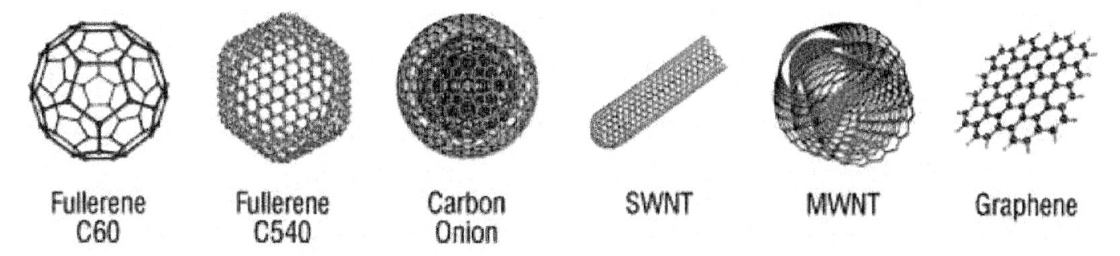

Fig. 8.1. Some carbon-based nanomaterials (Reprinted with permission from Mauter and Elimelech, 2008. Copyright (2008) America Chemical Society).

Table 8.1. Remediation application of carbon-based nanomaterials (Modified with permission from Mauter and Elimelech, 2008. Copyright (2008). America Chemical Society).

	Remediation Application				
Physical Properties	**Pollution Removal**	**Sorbents**	**Composite Filters**	**Antimicrobial Agents**	**Aligned CNT Membranes**
Size		X	X	X	X
Shape				X	X
Surface Area	X	X	X		
Molecular Specificity		X			
Hydrophobicity		X	X	X	X
Electric Conductivity					
Optical Activity		X			
Thermal Conductivity	X				

There are two ways of using CNMs: direct or indirect (i.e., combined with other functional groups through functionalization or oxidation), proactive (e.g., natural degradation prohibition, public health improvement, energy optimization) or retroactive (e.g., remediation, water treatment, pollutant removal) (Mauter and Elimelech, 2008). Due to their hollow or layered structure and interaction capacity with target contaminants, CNMs were considered suitable adsorbents for organic and inorganic pollutants. Their interaction with target compounds was caused by non-covalent forces, including hydrogen bonds, electrostatic forces or van der Waals forces (Xu et al., 2016). Different types of CNMs have been effectively studied for remediation techniques over the past few years. Some widely used CNMs are fullerenes, graphene, carbon nanotubes and carbon nanofibers.

2.2 Fullerenes

Fullerenes are the pentagonal or hexagonal rings of carbon molecules (e.g., C_{60}, C_{70}) with hollow shapes or ellipsoid, whereas covalent bonds connect these rings (Xu et al., 2016). The electronic structure influences fullerenes' properties. The lower the number of hexagonal rings, the more reactive the carbon sites. In some solutions, fullerenes might aggregate and create a steady crystalline nanoparticle form, which follows the Derjaguin-Landau-Verwey-Overbeek (DLVO) approach. Many factors could influence particle size, including pH, ionic strength, mixing time and performance and fullerene concentration. If humic acid is present in the solution, the aggregation rate could diminish due to macromolecules' adsorption on the fullerene nanoparticles (Mauter and Elimelech, 2008).

Fullerene has attracted interest due to their unique mechanical, chemical and physical characteristics. Fullerene's application started in the 1980s with the discovery of C_{60} buckminsterfullerene and the vaporization of graphite under laser irradiation. Since then, C_{60} material has been used in different fields, such as electronic sensors, water and wastewater treatment techniques and energy storage devices (Lofrano et al., 2017). The environmental applications of fullerenes are based on the covalent, supramolecular, endohedral transformations in their structure and their electric and conductive characteristics (Mauter and Elimelech, 2008).

The high adsorption of organic contaminants and micropollutants in wastewater by fullerenes is mostly due to their great active surface area and dispersion forces (Lofrano et al., 2017). Otherwise, doping different heteroatoms, such as Al, B, Si, N, P, S, may affect the interaction and adsorption activity of fullerenes with sulfur-containing compounds like H_2S, SO_2, C_4H_4S (Amiraslanzadeh, 2016). In particular, exothermic interactions occur in all sulfur adsorbates with heteroatoms-doped fullerenes. Besides, the adsorption capacity of SO_2 by fullerenes is higher than H_2S and C_4H_4S, whereas the Al-doped, B-doped and N-doped fullerenes showed the highest adsorption rate.

2.3 Graphene

Graphene involves a layer of pristine carbon atoms tightly organized in a hexagonal system (Lofrano et al., 2017). Graphene provides many exciting characteristics that have attracted the attention of scientists , such as strong mechanical characteristics, high elasticity, high thermal conductivity and a large surface area. It demonstrates high intrinsic strength, shear strength, fracture stress, Young's modulus and non-linear elastic properties. Due to the robust mechanical features, some defects (e.g., vacancies, grain beads, dislocations) are present on the graphene structure and may negatively influence its applications.

Like CNTs, the x-axes and y-axes parallel edges of graphene have a zigzag and armchair structure. This characteristic may impact on the electronic and electrochemical features of graphene. Moreover, two sides of the graphene sheet surface can be used for adsorption, which is more advantageous than CNTs and fullerenes.

Graphene and modified graphene materials have been widely used in environmental remediation. Pure graphene effectively adsorbs fluoride (F^-) from aqueous solution at different experimental conditions, and adsorption data fit well with the Langmuir isotherm equation. This adsorption activity can be described as a spontaneous, endothermic and irreversible process. Thus, graphene is recognized as one material with the potential for adsorption of fluoride in aqueous solutions. Due to the lower agglomeration of graphene layers on the surface, functionalized graphene has a higher specific surface area than pristine graphene, which leads to the greater adsorption capacity of environmental pollutants. The high adsorption efficiency of various environmental pollutants by functionalized graphene composite nanomaterials was shown in an earlier paper (Lofrano et al., 2017).

2.3.1 Graphene oxide

Graphene Oxide (GO) is a functionalized graphene, which is composed of oxygen-based functional groups on the graphene surface. GO can be produced by chemical oxidation or exfoliation of pristine graphite, such as oxidation of graphitizable carbons by a mixture of $KClO_4$ and fuming HNO_3 or heating of graphite with H_2SO_4, HNO_3 and $KClO_4$. Many oxygen-based functional groups, for instance, carboxylic and hydroxyl, exist on the GO's surface. The composition of GO depends on the synthesis method and condition. The GO sheet surface is relatively rough owing to the structure distortion and the existence of various covalent functional groups (Lofrano et al., 2017).

GO has exhibited unique properties of high surface area, intrinsic mobility, Young's modulus, thermal conductivity and optical transmittance. Moreover, it displays excellent handling properties and high chemical reactivity owing to the functional groups on its surface. Due to the layered structure and negatively charged surface, GO has shown its high adsorption capacity of different environmental contaminants, such as *Escherichia coli*, Rhodamine B, pesticides, pharmaceuticals. Furthermore, some functional groups containing oxygen on the surface of GO (e.g., carboxylic acids, hydroxyls) may react with primary gas pollutants, which leads to its effective adsorption of gaseous contaminants, such as Volatile Organic Compounds (VOCs) (Ashori et al., 2015).

GO may also interact with metal ions through oxygen atoms of hydroxyl and carboxyl groups in their composition, but their selectivity is low (de la Calle and Romero-Rivas, 2018). Hence, the GO surface functionalization by chemicals, such as chelating agents, can be used to enhance their metal ions selectivity (Sitko et al., 2015). For example, mercapto-functionalized GO performed high adsorption rates for metal ions like Cu (II), Cd (II), Co (II), Ni (II) and Pb (II) (Sitko et al., 2015).

2.3.2 Graphene-based composites

Graphene-Based Composites (GBCs) are the combined materials of graphene and other chemicals. They are formed by reducing GO under conditions like thermal, chemical, electrochemical, photothermal, photocatalytic, sonochemical, microwave, laser or plasmas (Lofrano et al., 2017).

GBCs represent strong mechanical properties, thermal stability and electrical conductivity. The characteristics of GBCs are strongly dependent on the reduction method. Compared with GO, GBCs have different crystal and detective structures, functional groups and optical features. More importantly, no oxygen-containing functional groups are present on the GBCs surface, contributing to their hydrophobicity and low dispersion in water (Lofrano et al., 2017).

GBCs have shown their high removal rate of organic and inorganic pollutants from aqueous solutions. The adsorption capacity of GBCs is dependent on their properties, such as texture and chemistry. For instance, the adsorption of NH_3 on the GBCs surface is due to the reaction with epoxide groups (to form amines). The adsorption of NH_3 on GBCs is usually weaker than on GO due to the absence of hydroxyl and epoxy functional groups (Lofrano et al., 2017).

2.4 Carbon nanotubes

Carbon nanotubes (CNTs) contain a nanometer dimension sheet of single-layer carbon atoms (graphene) rolled into a cylinder/tube (Wong et al., 2016). Some of their applications are in semiconductors, conductors or insulators due to their excellent physical, mechanical and electrical properties. For example, CNTs are rigid like diamonds, much more robust than steel and have the ability to store electrical energy, emit electron transmit and stabilize heat (Kamrani and Nasr, 2010).

CNTs provide a high adsorption capacity of organic and inorganic compounds, specifically the hydrophobic and low porosity contaminants (de la Calle and Romero-Rivas, 2018). The pollutant adsorption capacity of CNTs is dependent on their features, including adsorption site, surface area, cleanliness and existence of functional groups on their surface. The adsorption of positively charged metal ions is strongly enhanced by carboxyl and hydroxyl groups functionalized CNTs at high pH. Moreover, the more open and unblocked nanotubes, the more adsorption sites lead to higher adsorption capacity. Owing to van der Waals forces, CNTs tend to accumulate into bundles, which reduces the adsorption site areas and decreases their adsorption capacity. In particular, low-adsorption-energy contaminants may adsorb on external surfaces outside the CNTs, while contaminants with higher adsorption energy may localize between two adjoining tubes or inside a particular tube. Due to external sites' direct contact and adsorbate contaminants, the adsorption on external sites takes a shorter time to get equilibrium than internal sites. The process parameters of pH, temperature, contact time, ionic strength, Natural Organic Matter (NOM) content or initial solute concentration, also affect the adsorption capacity of CNTs because of their influence on pollutants (Jung et al., 2015). Oxygen content plays a vital role in the adsorption efficiency of CNTs. Functional groups containing oxygen on the CNTs surface, such as hydroxyl or carboxylic groups, may improve the adsorption capacity of CNTs via chemical bonding with contaminants (Jung et al., 2015).

Furthermore, the oxidation of CNTs by chemicals containing oxygen, such as HNO_3, NaOH and H_2O_2, enhances the active site on CNTs surface, leading to the higher adsorption capacity. Besides, the type of pores in aggregated MWCNTs may alter the adsorption performance, in which the effect of aggregated pores is higher than inner hollow cavities due to the stronger capillarity with the adsorbate. Moreover, pollutants' characteristics also play a crucial role in the adsorption rate of CNTs with these pollutants (Kamrani and Nasr, 2010).

The dispersion of CNTs in solvents is low due to the firm and attractive forces between nanotubes. Therefore, it is essential to disassemble nanotube covers by sonication, natural polyelectrolytes, aromatic polymers or DNA (Mauter and Elimelech, 2008). Besides, the surface functionalization may significantly improve the solubility of CNTs in aqueous solutions (de la Calle and Romero-Rivas, 2018).

CNTs are typically categorized into two types, single-walled (SWCNTs) and multi-walled (MWCNTs), whereas the MWCNTs demonstrated better interactions due to the high sheets rolled up (de la Calle and Romero-Rivas, 2018). For environmental applications, the application of bulk properties (e.g., high surface area) is more common than single properties (e.g., high conductivity).

2.4.1 Single-walled carbon nanotubes

The SWCNT is a CNT containing one layer of a rolled graphite held together by van der Waals bonds with a diameter of less than 1.5 nm. Due to low dispersion, it is easy to characterize and curve to form bundled structures. Its purity is dependent on the synthesis method, which is 30–50 wt% and 80 wt% by CVD and the arc discharge method, respectively. Moreover, its low solubility can be enhanced by adding functional groups, such as carboxylates or organic groups (Kamrani and Nasr, 2010).

SWCNTs may exist in three structures: armchair, zigzag and chiral shape (Notarianni et al., 2016). Depending on the armchair chirality or zigzag configuration, they can show metallic or semiconductor features. They demonstrated their excellent electrical properties as quantum wires. The production of SWCNTs typically requires a catalyst. Moreover, it is hard to synthesize the bulk of SWNTs due to the atmospheric condition control.

2.4.2 Multi-walled carbon nanotubes

MWCNT is a nest of several SWCNTs with various diameters, with a total diameter of more than 100 nm. Van der Waals bonds connect these SWCNTs in a MWCNT composition. The distance between layers of MWCNT is so small that no organic compounds can penetrate (Jung et al., 2015).

Due to homogeneous dispersion, MWCNTs may not form bundled structures. Their purity is higher than SWCNT (35–90 wt% by the CVD method), while their resistivity is much lower than

SWCNT. Due to the high electrical conductivity but low percolation, they can be used in epoxy composites, anti-fouling coatings or graphite electrode for Li-ion batteries (Jensen et al., 2015).

2.5 Carbon nanofibers

Carbon nanofibers (CNFs) are a nanometer dimension sheet of single-layer carbon atoms (graphene) arranged as stacked or herringbone, whereas graphene layers are stacked perpendicular to the fiber axis or at an angle between parallel, respectively (Poveda and Gupta, 2016).

The diameter of CNFs is generally in the range of 10–200 nm (Lee et al., 2018). Their structure depends on the properties of metallic catalyst particles and gases for the CNF process and synthesis methods. One of the most common types of CNFs is Vapor-Grown CNFs (VGCNFs). The structure of VGCNFs is formed by the convolution of graphene layers along the fiber axis with a hollow core (Poveda and Gupta, 2016), which makes them look like stacked cups.

The properties and structure of VGCNFs are dependent on the production approach. The Chemical Vapor Deposition (CVD) method is commonly used to fabricate VGCNFs. This technique is based on the combination of catalytic metal particles (e.g., Fe, Ni, Co, Cu) and carbon supply (e.g., CO) at a high temperature. Therefore, the CNFs diameter is greatly influenced by the catalyst particles size. Besides, the CNFs features are strongly impacted by the cups' orientation angle. A faceted catalyst particle was used to produce CNFs because it might form angled layers to deposit graphitic platelets at an angle (Zhang et al., 2014).

3. Properties of Carbon-based Nanomaterials

The properties of CNMs are greatly influenced by the configuration of carbon atoms , which relates to their hybridization state. Due to the small energy gap in the ground state of a carbon atom, one electron may move from the "s" orbital shell to the higher energy and empty the "p" orbital shell and hybridize into different configurations (such as sp, sp^2 or sp^3), which forms various carbon structures (e.g., fullerene, carbon nanotubes, graphene) (Mauter and Elimelech, 2008). The morphology of CNMs plays an essential role in their chemical properties. The modification of CNMs surface, such as doping atoms addition, functionalization, may alter their interaction behavior with contaminants or the surrounding environment and lead to a change in chemical properties (Lofrano et al., 2017).

Through the unique physical and chemical properties, CNMs (both single-molecule and bulk) have been widely used in environmental applications. Besides, some of their unique characteristics that attract more attention include size, shape, specific surface area, sorption, electronic, optical and thermal properties (Mauter and Elimelech, 2008).

3.1 Size, shape, and surface area

The structure and configuration of CNMs depend on their size, length and amount of fullerene cage layers. In particular, the quality of CNMs (e.g., structure, purity) are strongly affected by the synthesis method and conditions (e.g., temperature, pressure, catalyst, electron field, process gases). For example, the HiPCO synthesis method may produce SWNTs with an average diameter of 0.7–1 nm, while other techniques (e.g., laser ablation) form SWNTs with an average diameter of 1–2 nm (Mauter and Elimelech, 2008).

The size (or diameter) plays a crucial role in dictating the potential characteristics and applications of CNMs. The number of carbons in the spherical molecule may determine the size of fullerenes. For SWCNTs, the theoretical and practical diameters are 0.7 nm and 1.4 nm. The small diameter may modify the bond structure, which alters their mechanical, elastic, electronic, optical and thermal features. The size change also affects the environmental applications of CNTs. Owing to the small inner diameter, CNTs have been used in membrane filtration, nanowire synthesis, novel molding and separation. Nonetheless, for the small inner diameter CNTs, it is more difficult to purify the growth catalyst from bulk CNTs, limiting the applications of CNTs (Mauter and Elimelech, 2008).

The surface area to volume ratio of CNMs may be used to identify their size. If the $\Delta G_{surface}/\Delta G_{volume}$ ratio increases (ΔG is the free energy difference between the bulk material and structure), the CNMs will tend to be nanoscale. The greater the surface area is, the higher the adsorption capacity of CNMs will be (Mauter and Elimelech, 2008).

The size, shape and surface area of CNMs strongly rely on the aggregation (bundling) state and solvent properties (Mauter and Elimelech, 2008). Besides, the physical and chemical features of CNMs, such as thermal and electric properties, mechanical strength, aggregation behavior, are influenced by the adsorption of impurities (e.g., vapor, metals) on their surface. Nevertheless, these effects have not been studied well, which may limit the broad application of CNMs.

3.2 Molecular interactions and sorption characteristics

Conventional physical-chemical models and theories can describe molecular interactions and sorption processes of CNMs, such as electrostatics, adsorption, hydrophobicity and Hansen solubility parameters. Due to the nanosize, the physical and chemical properties of CNMs cannot be explained by experimental techniques. Fortunately, molecular models can be used for representing the features of CNMs.

Besides, the interaction of CNMs is mainly caused by the attractive van der Waals and repulsive Pauli forces between adjacent electron orbitals, which can be expressed by the Lennard-Jones continuum model. In addition, the DLVO theory can explain the electrostatic interactions of fullerene nanoparticles. During the functionalization process, the aggregation of nanoparticles will decrease, which is caused by the addition of polar functional groups on the surface of nanoparticles that improves their hydrophobicity (Mauter and Elimelech, 2008).

The high-adsorption sites of CNTs can customize their surface chemistry. CNTs are hydrophobic; therefore, they tend to be stabilized in an aqueous solution to protect their active surface from agglomeration (Lofrano et al., 2017). Energy supply may break down the water-water bonds, which reduce this hydrophobicity behavior. Besides, the dipole-induced dipole interactions and the attractive van der Waals forces play a vital role in attracting vapor or liquid molecules to nanotubes capillary.

The reduction of the medium's polarization ability leads to an increase of electrostatic interactions, which will supply the energy (entropy and enthalpy) for retaining water on nanotubes. The surface tension also influences the wettability of nanotubes. Liquids with low surface tension, such as water or organic solvents, may wet the nanotubes surface due to the cohesive forces and intermolecular interactions that cause capillary action. The surface tension limit liquids to moisten the nanotubes' surface to about 72 mN/m. In addition, the properties of fluid flow and capillarity can be described by the Laplace equation for the dimension of nanotubes (Mauter and Elimelech, 2008).

The adsorption properties of sorbates on CNMs are dependent on their hydrophobicity and capillarity. In general, their adsorption behavior can be described by traditional Langmuir, BET or Freundlich isotherms with high adsorption capacity, fast equilibrium rates and relatively independent on pH change. For unfunctionalized NMs, physisorption controls the sorption behavior. However, condensation might occur inside the nanotubes due to the dispersion energy by sorbent-sorbate and sorbent-sorbent interactions, which lead to filling of nanotubes. Furthermore, the adsorption process on their surface can be described by some models (Mauter and Elimelech, 2008).

3.3 Electronic, optical, and thermal characteristics

Due to the structure of the sp^2-bonded carbon atoms layer, the electrical and thermal conductivity and optical nonlinearity properties of CNMs are reasonably high. Therefore, they have promising potential applications in environmental sensing devices and productive solar cell or environmental remediation techniques.

Fullerenes have been widely used owing to their optical property under both UV and visible light. Due to the intense UV absorption, fullerenes may generate the Reactive Oxygen Species (ROS), such as single oxygen molecules, hydroxyl radicals and superoxide radical anions, by the photoexcitation process, which can degrade contaminants in an aqueous solution. ROS can be effectively produced by water-soluble fullerol with high quantum yields. Nevertheless, the self-quenching between fullerenes' molecules may limit their photochemical reactivity. Besides, the biocompatibility and aggregation state of CNMs in different environmental conditions are also concerns related to the production ROS that need further studies (Mauter and Elimelech, 2008).

SWCNTs have many interesting electronic properties that attract the attention of the public . Some notable characteristics are a great current-carrying capacity, small ionization potential, high field emission rate and efficiently tunable band gaps. These electronic features are dependent on chirality, diameter, length and a total of concentric tubules. However, the defect scattering and diffusive electron motion may restrict their applicability.

For nanotubes, the chirality and diameter may influence their band gaps properties. In particular, nanotubes' band gaps with zigzag or helical conformation carbon atom structures are small or large, respectively. If nanotubes' radius increases, their band gaps will decrease to zero, representing metallic properties. Besides, the conductivity of MWCNTs strongly depends on the electronic arrangement of the outermost tubules, which is similar to graphite. The metallic SWCNTs can apply as environmental sensors due to the sensitivity of their conductivity to the adsorption by neighboring liquid or gaseous molecules. Moreover, the low ionization potential of SWCNTs, which indicates the low energy required to excite an electron and emit from the molecule, contributes to their complete application of eco-friendly device designs in the electronics industry. This ionization potential can be lower with the presence of some adsorbates, such as water.

The thermal conductivity of CNTs, both individual tubules and bulk assemblies, is high. However, the heat loss at tube-tube junctions may limit this property. Fortunately, new nanotubes joining techniques have been developed to generate long fibers with high thermal conductivity.

The mechanical features of CNFs, for example, strength, stiffness, fracture behavior, toughness, are strongly dependent on their structure, diameter and production process. Remarkably, their tensile strength and stiffness measurement (Young's modulus) are 1.5–7 GPa and 228–724 GPa, respectively. The tensile strength of CNFs is lower than CNTs due to the polycrystalline and amorphous carbon in their structure. These mechanical properties can be enhanced by CNFs-based composites, such as polyacrylonitrile polymer (PAN) and MWCNTs composite CNFs may improve the tensile strength and elastic modulus 38 and 84%, respectively (Mohamed et al., 2017). At room temperature, the presence of water, solvents, acids or bases does not affect the mechanical properties of CNFs. Although their mechanical properties are not as impressive as CNTs, their cost is lower and easier for industrial manufacture than CNTs (Mauter and Elimelech, 2008).

The electrical conductivity of CNFs at room temperature is reasonably similar to the resistivity of CNTs. Moreover, CNFs can be used to improve the electrical and mechanical features of other materials. For example, by adding aligned CNFs, some nanocomposites and membrane's electrical conductivity improves about seven magnitudes (Ladani et al., 2015) and 10 times (Gu et al., 2015), respectively.

Besides, the CNFs size can strongly influence their electroactivity and strength. The smaller the size, the stronger conductivity and tensile strength, which is due to the geometric surface effect and the reduction of probability of including structure flaws, respectively.

4. Occurrence, Behavior, Fate and Transport of Carbon-based Nanomaterials in the Environment

The emission of CNMs into the environment may be due to natural, incidental or nanotechnological sources. Natural sources include combustion from volcanic eruptions, wildfires or CNMs aggregation on Earth, such as amorphous carbon NPs, MWCNT, fullerenes or CNTs (found in Greenland ice core). Unexpected sources are human activities that contribute to the emission of CNMs to the environment. For instance, the traffic activity may release many metal NPs or MWCNTs, which may travel through the atmosphere and deposit on the surface of soil and water bodies. Nanotechnological sources are manufacturing processes that may release engineered NMs to the environment. Among these three types, the amount of NMs emission from a nanotechnological source is the lowest. However, due to their diverse types and potential effects, they are considered emerging pollutants and need more study (Lofrano et al., 2017).

After release into the environment, CNMs may distribute among various media, such as water or porous means. The behavior, such as aggregation and adsorption of CNMs in the environment is strongly dependent on their primary properties and environmental conditions like pH, ionic strength or natural organic matter. Their transformation in the environment may influence their mobility and toxicity, which has not yet been understood due to the lack of analytical methods to quantify the number of CNMs in complex environmental samples. In recent years, some quantitative analytical methods for determining CNMs, mostly fullerenes in the environment, have been developed. In general, these methods include separation steps, such as liquid-liquid or solid-phase extraction and analysis by Liquid Chromatography coupled to a Mass Spectrometry instrument (LC-MS) (Lofrano et al., 2017).

4.1 Occurrence of carbon-based nanomaterials in the environment

The occurrence of CNMs in the environment has been studied for a long time. However, most studies were on the occurrence of fullerenes. Due to the difference in characteristics, such as lengths and structures, fullerenes' analytical method cannot be applied to other CNMs like CNTs. Therefore, some models were used to predict the fate, environmental behavior and potential risk of CNMs (Lofrano et al., 2017).

Besides, the lack of standardized methods for characterization of the toxicological properties of CNMs, such as the possible presence of impurities, suspension preparation, toxicity test procedures, also contributes to the limited amount of research on this topic. In particular, suspension preparations of CNMs may affect their surface features and their aggregation states that can change their potential toxicity or toxicity test procedures of bulk materials cannot be used for studying the toxicity of end-point and experimental design CNMs to avoid any artifacts (Lofrano et al., 2017).

4.2 Behavior, fate and transport of carbon-based nanomaterials in the environment

4.2.1 In an aquatic environment

In an aquatic environment, CNMs are generally insoluble in water. Hence, they are likely to attach to sediments and particulate matter or stay suspended in the wastewater effluents. Moreover, they may accumulate to form negatively charged clusters (Lofrano et al., 2017).

The aggregation, growth and deposition of CNMs in aqueous suspension depends on the equilibrium interactions of electrostatic and van der Waals forces, which can be explained by the Derjaguin-Landau-Verwey-Overbeek (DLVO) model (Mauter and Elimelech, 2008). In particular, the NPs surface is enclosed with a double layer of ions, whereas electrostatic repulsion forces balance the attractive van der Waals forces. The agglomeration of singular NPs is able to form large flocculated particles.

Under different ionic strength and pH conditions, the aggregation of CNMs suspension may be described by the DLVO model. For example, if the ionic strength is high, the energy barrier between colloids (caused by the electric double layer compression) will decrease, leading to the domination of attractive van der Waals interactions and the generation of larger agglomerates with reduced zeta potential or more NPs will be flocculated. Similarly, if the medium pH is low (acidic) or high (basic), the zeta potential value of NPs will be small or high, respectively. At neutral pH (a particle has zero net surface charge), the van der Waals interaction controls, leading to the production of larger agglomerates that may settle out of suspension or deposit on the NPs surface (Mauter and Elimelech, 2008).

The effect of Natural Organic Matter (NOM), sunlight, oxygen on aggregated fullerene flocculation stability has also been investigated in other papers. The distribution and aggregation properties of CNMs in the aquatic environment in recent papers are mostly C_{60} fullerene, while a few papers are on C_{70} and other functionalized fullerenes. Compared with C_{60} fullerene, C_{70} and other functionalized fullerenes tend to less agglomerated but form bigger and less stable agglomerates. Owing to the strong absorption of solar spectrum light, C_{60} clusters can be oxidized by a lamp or sunlight irradiation with the presence of oxygen, regardless of the presence of natural organic matter (fulvic acid), pH changes and the preparation approach of the clusters. In addition, its surface is able to be oxygenated and hydroxylated as pseudo-first-order under UVA irradiation and dissolved O_2 conditions. However, the NPs core is not influenced after 21 d of illumination. The aggregation and formation of water-stable C_{60} cluster (nC_{60}) were reduced, as seen in other papers due to the photochemical transformation under monochromatic UV light exposure at 254 nm. A new C_{60} cage product was formed containing various oxygen functional groups, such as epoxides and ethers (Lofrano et al., 2017).

The effect of environmental factors on CNMs transformation has been investigated. Under long-term UVA exposure and the presence of O_2, nC_{60} is likely to be dissolved and partially mineralized. Moreover, the addition of aquatic humic acid may scavenge ROS, which will reduce the transformation kinetics of nC_{60}. For MWCNTs, their physicochemical properties are strongly affected by the photochemical transformation in the environment. Under the UVA irradiation, the reactions of MWCNTs and photogenerated ROS may occur to produce single oxygen ($^1O_2$) and hydroxyl radical (·HO) on the MWCNTs surface. These hydroxyl radicals initially reacted and degraded the surface carboxylated carbonaceous fragments. Later on, they may react with the graphitic sidewall to generate functional groups and vacancies (Lofrano et al., 2017).

In summary, the elimination of carboxyl groups on the MWCNTs surface and the formation of other oxygen-based functional groups may lead to the reduction of total surface oxygen concentration. In other words, the surface potential and colloidal stability of MWCNTs will decrease or their mobility in aquatic systems shall diminish. Hence, the photochemical transformation of C_{60} in an aquatic environment should be carefully investigated for researching the fate and transport of CNMs in an aquatic environment.

4.2.2 *In soil*

The fate and transport of CNMs in soil have been investigated in many research studies. As a result of the possible commercial applications, CNMs may release to soil through deposition processes or application of treated sludge (biosolids) from wastewater treatment of plants for agricultural purposes. The mobility of aggregates in soil is strongly influenced by electrostatic repulsion of cations and sorption to clay (Lofrano et al., 2017).

The transport and retention properties of nanoscale fullerene agglomerates (nC_{60}) in water-saturated soil columns have been explored. The results showed that nC_{60} aggregates were largely retained in soil. In addition, the transfer of fullerenes in the soil column was improved by adding natural organic matter, for instance, humic acid. In other research, the transport of nC_{60} in saturated porous media, such as saturated sand and sandy soil, was investigated. The results demonstrated the little and significant restriction of flow velocity on the transport of nC_{60} in Ottawa sand containing mainly pure quartz and in Lula sandy soil containing mostly low organic matter soil, respectively, which was explained by the smaller particle size, more uneven and rougher shape and higher variety of Lula soil (Lofrano et al., 2017).

Moreover, the deposition of nC_{60} was improved with increasing ionic strength in both the sand and soil media. This outcome was demonstrated by the different responses of clay minerals (in Lula sandy soil) and quartz (in Ottawa sand) to the ionic strength and species change. Furthermore, ion species' presence might increase fullerene's deposition, whereas the effect of calcium ion (Ca^{2+}) was more substantial than sodium ion (Na^{+}). Additionally, fullerenes were not stable in superficial soils because of the O_3 oxidation effect, while their long-term durability might be explained by the retention in a reductive environment, such as sulfur-rich environments (Lofrano et al., 2017).

Many factors may influence the mobility of SWCNTs in soil. In particular, their mobility is affected by the presence of mesoporous and organic matter in municipal solid waste, such as paper, metal, food surrogates, plastic and glass. Moreover, due to the oxidization capacity of CNTs in the environment, the mobility of oxidized CNTs should be considered. In this research, the mobility and behavior of CNTs might be described by the DLVO theory, whereas ionic strength, pH, and organic material content strongly influenced the VNTs aggregates (Lofrano et al., 2017).

The mobility of graphene in soils is different. Due to the complete application in microelectronics, most graphene will be landfilled after their useful life. Despite this, the mobility of graphene sections in soils has not been fully understood. They were assumed to alter and leach in landfills, which may negatively affect the surrounding areas. However, graphene sheets and nano-flakes have low solubility in water, and therefore graphene is expected to float and aggregate in the water-air interface. In summary, the emissions, fate and transfer and toxicity of graphene in the environment need more study (Lofrano et al., 2017).

4.2.3 In the atmosphere

The fate and transport properties of CNMs in the atmosphere have not been fully recognized. CNMs can be released to the atmosphere from anthropogenic sources, such as incinerators, commercial products or natural sources, such as volcano, biomass burning. After emitting to the atmosphere, CNMs may grow in the atmosphere in many ways, such as clumping, accumulation with other particulates on their surfaces, and finally being lost through dry deposition by gravity or washout. Moreover, the oxidation of CNMs like C_{60} by UV radiation or ozone also contributes a crucial role to their mobility in the atmosphere (Lofrano et al., 2017).

5. Synthesis of Carbon-based Nanomaterials

5.1 Synthesis of fullerenes and modified fullerenes

The synthesis of fullerenes was developed in the 1990s. An AC arc discharge usually produced fullerenes between high purity graphite rods in a helium (He) or argon (Ar) gas at about 2000°C. Under this high temperature, carbon evaporation might form soot containing fullerenes, which condensed in a reactor. The soot consisted of about 15% fullerenes, 13% C_{60} and 2% C_{70}. One of the most common methods was the auto-loading version of an arc-discharge apparatus using 24 carbon strips. The auto-loaded anodes were consumed, while the cathode containing carbon wheel would clear away the accumulated carbon powder. Besides, C_{60} and C_{70} fullerenes may be produced by burning benzene due to the lack of oxygen. In particular, diluted benzene is injected at the central axis and a combustion chamber, while oxygen is supplied through a porous plate (Nakagawa et al., 2014).

After combustion of benzene inside the chamber, condensable combustion products containing Polycyclic Aromatic Hydrocarbons (PAH), fullerenes and soot were obtained on the pre-weighed filter system. Later, the filter was immersed in toluene and sonicated, then the final solution was filtered by a 0.45 μm nylon filter to collect the final product as fullerenes.

Fullerenes may be produced by chemical methods, such as pyrolysis hydrogenation of naphthalene and corannulene or dehydrohalogenation of a chloroaromatic compound. However, these methods are suitable for only research purposes because of the inefficiency of low production.

There are some methods to modify fullerenes. Dopants, such as transitional metal atoms or noble gas like helium, neon, argon and xenon, can be put inside the structure of fullerene to form endohedral fullerenes. If fullerene is linked with other chemical groups, exohedral fullerene or fullerene derivative will be obtained. For example, the reaction of C_{60} with diazoalcane may form [6,6] methanofullerene, which can be isomerized to obtain [5,6] fulleroid ester. This ester can be refluxed with the o-dichlorobenzene solution or trifluoroacetic acid to produce the [6,6]-phenyl-C_{61}-butyric acid methyl ester ($PC_{61}BM$) that is a fullerene derivative widely applied in organic solar cells (Nakagawa et al., 2014).

The synthesis of fullerenes may form a by-product called Fullerenes-Extracted Soot (FES), which can be used for water purification. In particular, the fullerenes-extracted soot adjusted with ethylenediamine (FES-ED) adsorbed Cr (VI) and methyl orange from water effectively. The adsorption rate of Cr (VI) depended highly on pH (pH optimal = 3.0) and the data significantly fitted with the Langmuir isotherm model. Besides, this functionalized FES composite can be reused with a desorption rate of 75% (Nakagawa et al., 2014).

5.2 Synthesis of graphene and graphene-based nanomaterials

Various methods have been developed to synthesize graphene. Some standard methods are mechanical or chemical exfoliation, chemical exfoliation by graphene oxide, CVD and production on silicon carbide (SiC). The mechanical exfoliation method was initially used to separate single graphene. The bulk of Highly Ordered Pyrolytic Graphite (HOPG) was peeled into single graphene sheets and transferred onto Si, SiO_2 or Ni substrate (Allen et al., 2010). This low-cost approach may be conducted at room temperature, and the obtained graphene has high charge carrier mobility. However, the graphene size is too large, which suggests using it only for research purposes.

The chemical exfoliation method involves the use of pencils or solid lubricants-based graphite in solvents to break the weak van der Waals interaction forces, which links the graphene sheets in graphite, due to the transfer of enthalpy and charge between graphene layers and solvent molecules. Moreover, sonication may enhance the exfoliation performance because of reduced enthalpy of graphite and solvents mixture. Solvents that fit with graphene surface energy, for instance, N, N-dimethylformamide (DMF), benzyl benzoate, γ-butyrolactone (GBL), 1-methyl-2-pyrrolidinone (NMP), N-vinyl-2-pyrrolidone (NVP) and N, N-dimethylacetamide (DMA), are preferred to other popular solvents, for example, ethanol, acetone and water (Zhang et al., 2010). However, a mixture of surfactants and polymers with water can be used to the graphite exfoliation because they limit the agglomeration. Among different types of chemical exfoliation, the electrochemical exfoliation method showed better quality graphene and shorter processing time (Notarianni et al., 2016).

Chemical exfoliation may produce a large quantity of graphene with small crystallite size, but the charge carrier mobility of products is low, limiting their applications in electronic and electrical fields.

The reduction (chemical or thermal) of Graphene Oxide (GO) involves the production of semiconducting GO material by chemical or thermal reduction. GO attracts the attention of the public due to its ability to be used as a precursor component in the fabrication and functionalization of graphene sheets with high productivity. The presence of C-O-C, C=O and –COOH functional groups in GO composition expresses its excellent mechanical, optical, thermal and electronic properties. Moreover, GO's chemical activity is more vigorous than pristine graphene due to a large number of functional groups containing oxygen.

GO can be synthesized by the oxidation of graphite. In particular, graphite flakes are oxidized by mixing sulfuric acid (H_2SO_4), sodium nitrate ($NaNO_3$) and potassium permanganate ($KMnO_4$) together to form the hydrophilic oxidized GO with excellent electrical conductivity. GO can be used to produce the reduced GO (rGO), a different type of graphene but not pristine graphene. The presence of defects, residual and functional groups on rGO make its charge carrier mobility, conductivity and concentration lower than pristine graphene (Notarianni et al., 2016).

Chemical reduction at room temperature is also an option to produce rGO. In particular, some reducing agents, such as hydrazine and its derivatives, ascorbic acid ($C_6H_8O_6$) or hydroiodic (HI) acid, are simply added to the aqueous GO solution to form rGO. Other rGO, such as graphene-based photocatalysts, can be fabricated by mixing GO with photocatalysts like TiO_2 particles in four methods (An and Yu, 2011).

In method I, photocatalysts are grown on the graphene sheets. It is essential to modify the graphene surfaces to improve their surface-active sites, which will enhance the interactions with photocatalysts and prevent the aggregation of graphene sheets. For example, electron beam pretreatment of the graphene surface might enhance the affinity of graphene to TiO_2 photocatalyst through the electrostatic forces, which contributed more sites for TiO_2 crystals and improved the growth of TiO_2 clusters on graphene.

Method II is the deposition of a well-defined structure photocatalyst on graphene oxide (GO) surface under vigorous stirring or ultrasound. In this method, TiO_2 photocatalyst-graphene oxide nanocomposites were synthesized by the reaction of GO and Ti $(SO_4)_2$ at a low temperature (80°C). Under the electrostatic attraction forces, Ti $(SO_4)_2$ may diffuse into the GO interlayer and form TiO_2 that will grow to produce TiO_2-graphene oxide nanocomposites with improved photocatalytic activity.

In method III, metal oxide-graphene nanocomposite was produced by a chemical deposition method with GO-precursor. In particular, GO was *in-situ* reduced to graphene in a consecutive reaction step. After the synthesis process, the metal oxide nanocrystals were produced and covered on graphene sheets. The graphene nanocomposite showed superparamagnetic features at room temperature.

Method IV is similar to method III, where graphene nanocomposite was obtained by a chemical deposition method using graphene oxide as a precursor. However, GO was reduced to graphene in an auto-redox reaction through a simple one-pot growth method.

The Chemical Vapor Deposition (CVD) approach is usually used to fabricate large-area graphene films. This method was developed in the 1960s to produce a single layer of graphite on Pt. Here graphene was synthesized from gases containing C on surfaces of catalytic metal or surface separation of dissolved carbon atoms in metals like Fe, Ni, Co, Pt and Pd. One of the main concerns is the diffusion of carbon atoms (in the gas source) into the metal, which is difficult to manage. Therefore, single crystal and atomically smooth metals are often used to develop the monolayer graphene with a high quality. Moreover, copper was usually preferred to be used in a pure CVD process due to the low diffusion of carbon atoms in copper. For example, roll-to-roll technique, which applies the CVD process on Cu foils, can synthesize of 76-centimeter graphene film for a transparent electrode (Notarianni et al., 2016).

The quality of graphene obtained by the CVD method is not good. In particular, aperiodic heptagon/pentagon pairs were formed on polycrystalline or bilayer regions were overlapped at the particle boundaries (Huang et al., 2011). The mechanical and electrical properties of graphene decreased with grain boundaries. The CVD method's other disadvantages include high cost (mostly energy cost), hard to transfer to dielectric substrates and require training to control the crystallographic orientation. For large scale fabrication, a low-temperature CVD process, such as plasma-enhanced CVD, is suggested to obtain graphene with a large area and high quality on different substrates (Notarianni et al., 2016).

Epitaxial growth on the SiC method can be used to achieve high-quality graphene. This involves the formation of graphene lattice on the SiC surface (graphitization) caused by the displacement of carbon atoms under high temperatures. Graphene may be developed on both C and Si faces, but the Si face is preferred due to the easier control over bulk graphene layers. Moreover, hexagonal polytypes of SiC with orientation are more commonly used owing to the fit of their lattice structure with the graphene lattice. This approach may control the thickness and specific crystallographic orientation of graphene. SiC substrate has been widely used in many electronic applications because it does not need to transfer to another substrate. Furthermore, it may connect with traditional silicon technologies (Notarianni et al., 2016).

To improve graphene quality during graphitization, SiC wafer pretreatment in ultrahigh vacuum (UHV) or other techniques, such as heating the sample under Si flux or hydrogen etching, is required. Temperature plays a vital role in graphene growth, which will affect the number of graphene layers grown and the capacity of Si diffusion. In UHV, graphene starts growing at the temperature of 1200–1350°C. However, in the inert gas atmosphere or excess of Si in the gas phase, graphene can develop at 1400–1600°C (Notarianni et al., 2016).

This approach has two significant limitations. First, the cost of SiC wafers is relatively high. This can be solved by growing thin SiC layers of about 100–300 nm on Si substrates. However, more studies are necessary to obtain high-quality and monolayer graphene. Second, the high temperature of this method is not appropriate for Si technology. This concern can be solved using plasma-enhanced CVD (PECVD) equipment, which will decrease the growth temperature. Despite this, more solutions are required before applying graphene to the electronic market (Notarianni et al., 2016).

Physical methods can be used for the fabrication of graphene-based NMs. Some examples of physical methods involve: peeling of Highly Ordered Pyrolytic Graphite (HOPG) films with a cellophane tape to collect a single layer of graphene, growth of nanopatterned epitaxial graphene films on a substrate, CVD of multilayered graphene or etching of poly (methyl methacrylate) PMMA-MWCNT film by an Argon (Ar) plasma (Notarianni et al., 2016).

Other methods were also used to produce graphene-based nanocomposites. For example, TiO_2-GO nanocomposite was fabricated by a self-assembly method or TiO_2-graphene nanocomposite was obtained by the solvothermal conditions (Notarianni et al., 2016).

5.3 Synthesis of carbon nanotubes

Different CNTs synthesis methods have been developed, such as chemical vapor deposition (CVD), laser ablation and arch discharge. The CVD method used hydrocarbon gases as the precursors at a temperature of 500–1000°C, while the two other methods used solid-state carbon as the precursors at a higher temperature. Therefore, CVD is suitable for MWCNTs synthesis and laser ablation and arc discharge are desirable for SWCNTs synthesis (de la Calle and Romero-Rivas, 2018).

In the arc discharge method, CNTs are formed through the arc vaporization of carbon rods by inert gas (e.g., helium, argon) under low pressure and high temperature. The efficiency of CNTs synthesis is dependent on the heat of the precipitate on the carbon electrode and the consistency of the plasma arc (Kamrani and Nasr, 2010). Besides, the inert gas type and pressure, metal concentration, current and system design also influence the properties of CNTs. The CNTs produced by this method are tight bundles and linked together by strong van der Waals forces (Notarianni et al., 2016).

The CVD method includes the Plasma-Enhanced Chemical Vapor Deposition (PECVD) method and Thermal Chemical Vapor Deposition (TCVD) method. In the PECVD method, the CNTs are generated by a glow discharge formation in a chamber under high-frequency power, in which CNTs will be developed on the nanoscale catalytic metal particles (e.g., Fe, Ni, Co) placed on the substrate of the electrode. The characteristics of CNTs (e.g., morphology, thickness, diameter, structure) are largely affected by catalysts. For MWCNTs synthesis, Ni is the best catalyst to obtain the nanotube diameter of about 15 nm at less than 330°C (Kamrani and Nasr, 2010).

In the TCVD method, catalytic metal particles (e.g., Fe, Ni, Co) are deposited on a substrate, etched in an HF solution and put into a CVD reaction furnace. This catalytic metal film will be etched once more by NH_3 gas at 750–1050°C to form the nanoscale catalytic metal particles. CNTs will gradually be developed on these nanosized catalytic metal particles. The thickness of catalytic film during the growing process of the CNTs on the catalytic film dictates the dimension of CNTs. For example, if the thicknesses of catalytic films are 13 nm and 27 nm, the diameters of MWCNTs will be between 30–40 nm and between 100–200 nm, respectively (Kamrani and Nasr, 2010).

Moreover, CNTs can be synthesized by high-pressure catalytic decomposition of carbon monoxide (HiPco). In this procedure, Fe $(CO)_5$ is decomposed in a high-pressure reactor at about 10 atm and temperature of 800–1200°C to form carbon monoxide and catalyst particles, in which CNTs will be gradually established. This method may provide the narrow CNTs with high production efficiency.

Finally, the laser ablation method was developed in 1995 by Smalley's group at Rice University in Houston, Texas. It is based on the vaporization of graphite in the oven under the irradiation of a continuous laser. At the temperature of 1200°C and pressure of 500 Torr inside the oven, small carbon molecules and atoms (in graphite) vaporize and expand. Then they are cooled quickly to form larger clusters. Moreover, the catalysts are also condensed and attached to the clusters, which will develop as the SWCNTs. The quality of CNTs produced by this method depends on the conditions of the reactor.

After the synthesis process, the CNTs may be contaminated by metal catalysts, amorphous carbon or fullerenes. Therefore, they need to be purified to obtain the highest purity production. Some common purification approaches are oxidation, annealing and filtration. However, due to the number of samples, it is still challenging to choose the best CNTs purification method, especially at commercial levels. The final CNTs will have excellent thermal and electrical conductivities, which can be used in various fields, such as catalysis, conductive coatings, water-purification systems (Lofrano et al., 2017).

5.4 Synthesis of carbon nanofibers

Various procedures have been used to synthesize carbon nanofibers (CNFs). Some common approaches are CVD, electrospinning, drawing, templating and self-assembly.

The catalytic chemical vapor deposition (CVD) method might produce CNFs (Poveda and Gupta, 2016). This technique is suitable for synthesizing relatively short and hard-to-use fibers, such as cup-stacked CNFs and platelet CNFs. In this method, several gases, such as C_2H_2, CH_4, CO, C_2H_6, were used as carbon sources, while different kinds of metal or alloys, for example, Fe, Co, Ni, Cr, V, were considered as catalysts. The decomposition of carbon-containing gases formed CNFs in the presence of metal catalyst powder in a quartz tube electric furnace reactor at high temperature (700–800°C).

The growth mechanism of CNFs is dependent on the properties of catalyst and gases. In particular, the catalyst particle size and manufacturing techniques may influence the outer diameter and structures of the CNFs, respectively. Moreover, the CNFs thickness depends on the size and category of metal particles and growth temperature (Mohamed, 2019).

The electrospinning method is an effective way to fabricate polymeric CNFs. This technique is simple, easy to control and able to synthesize many CNFs with various lengths and diameter ranges, stability and good electrospinnability. A typical electrospinning process was previously shown (Mohamed, 2019). In this system, a polymer precursor, such as cellulose, phenolic resin, polyacrylonitrile, polybenzimidazole, is continuously pumped through the metallic spinneret, while the solution is spurted out from the needle to a target by the high voltage power supply instrument. At a particular surface tension value, a fibrous form is developed and collected at the target. The nanofiber web is formed by the stretch and deposit of polymer solution on an electrode collector under the high electrical voltage effect during the electrospinning process. After the fast evaporation of the solvent and the jet stream's whipper and stretch, the solidified nanofibers will be obtained on the grounded collector (Mohamed, 2019).

The fiber produced by these collectors is not a homogeneous fiber mat due to the deposit of more fibers at the target center that may change the fiber thickness through the mat. Hence, a rotating drum collector was used to solve this problem and produce a consistent thickness fiber mat. Finally, the polymer nanofibers are carbonized by heat to form the final CNFs (Mohamed, 2019).

The characteristics of the final CNFs are dependent on many factors, such as a polymer solution and solvent type, capillary size, flow rate, solution concentration, viscosity and temperature. The increase of solution concentration and temperature will increase the diameter and formation of -phases in nanofibers' electrospun (Mohamed, 2019).

Other approaches to produce CNFs are template synthesis, drawing and phase separation. In these methods, CNFs are fabricated one-by-one through a template technique. Various raw material types, such as metals, semiconductors, carbons and different types of solvents, are used to produce CNFs. However, due to the time consumed , the electrospinning method is preferred to produce many continuous CNFs from different polymers (Mohamed, 2019).

5.5 Challenges and limitations

Due to the limitation of technology, most of the CNMs synthesis methods are in the laboratory. Hence, it is essential to research and develop new techniques and materials to synthesize CNTs at the industrial scale. Besides, the physical and chemical features of CNTs film formed (e.g., composition, structure) and their electronic and optical applications need more research. Among these three methods, laser ablation generates the best quality CNTs with the highest costs, while the arc discharge creates the lowest purity CNTs, and the CVD method needs more evaluation, especially on an industrial-scale (Kamrani and Nasr, 2010).

6. Use of Carbon-based Nanomaterials in Environmental Remediation

6.1 Carbon-based nanomaterials as adsorbents

Adsorption is the mass transfer of chemicals in the liquid phase onto a solid phase, which will remove that chemical from the liquid phase. Besides, it is also known as the adhesion in a skinny layer of gases, solutes and liquids molecules to the solid surfaces that contact them. Adsorption processes were widely used in many contaminant-removal processes. In water treatment, they were used to remove taste, odor, color-forming organics, Synthetic Organic Chemicals (SOCs) and Disinfection By-Product (DBPs) precursors (but trihalomethanes—THMs). They were also applied to remove inorganic constituents, for example, perchlorate, dechlorinate, arsenic or heavy metals. The adsorption efficiency of traditional adsorbents is dependent on many factors, such as their density, activation energy or mass transfer rate (Crittenden et al., 2012).

Traditional adsorbents have many disadvantages that limit their complete application. Some significant constraints are large dimensions, low specific surface area, unmanageable pore size distribution and uncontrolled surface chemistry. Fortunately, CNMs could solve these problems and have attracted more public attention. The excellent adsorption capacity, rapid equilibrium rates, high effectiveness for numerous contaminants under different environmental conditions and compatibility with various isotherms (e.g., BET, Langmuir, Freundlich) also make them suitable adsorbents for environmental applications.

The interactions between CNMs with organic and inorganic compounds may be categorized as chemisorption or adsorption. Chemisorption is the interactions associated with the formation of covalent bonds, while adsorption is the weak interactions that involve the non-covalent attachments, for example, hydrogen bonds, electrostatic, hydrophobic or π interactions (Lofrano et al., 2017). In this chapter, adsorption will be presented.

6.1.1 Adsorption of gas by carbon-based nanomaterials

The history of adsorption capabilities of CNMs started during the 90s. The first research was the gas adsorption of fullerene C_{60}. Later, the Inverse Gas Chromatography (IGC) technique was used to conduct gas adsorption experiments of fullerenes. Through this technique, the gas-solid partition coefficients of different gases on a mixture of C_{60} and C_{70} fullerenes were estimated, and the thermodynamic characteristics of adsorption heats and equilibrium constants of various organic compounds on C_{60} fullerene crystals and on graphitized black carbon were determined. In particular, the adsorption potential of fullerene surface was lower due to its weaker dispersive interactions, which led to the smaller adsorption equilibrium constants of alkanes and aromatic compounds on fuller. Meanwhile, the adsorption equilibrium constants of some electron-donor compounds like amines, ketones and nitriles on the fullerene surface are higher than on the graphitized black carbon surface. Besides, fullerenes' properties were comparable to alkenes rather than polyaromatic molecules due to the weak polarization (Lofrano et al., 2017).

The selectivity of C_{60} fullerene for organic molecules was also investigated. The main adsorption mechanisms of polar compounds and nonpolar compounds are physical adsorption and chemical adsorption, respectively. In particular, the adsorption rate of polar compounds decreases with the order of acids, aldehydes, amines, alcohols, ketones and the adsorption performance of nonpolar compounds decreases with the order of alkynes, alkanes, alkenes. By using infrared (IR) spectroscopy, the adsorption rates of a few gases such as CO, CO_2, N_2O and NO on C_{60} fullerene are more substantial than alternative carbon allotropes due to their higher interaction with the C_{60} fullerene surface. Furthermore, pristine fullerenes, modified fullerenes (Er et al., 2015) and SWCNTs virtually adsorbed hydrogen gas, which was applied as hydrogen storage materials.

Similarly, SWCNTs and MWCNTs demonstrated their high adsorption rate of different gases, such as NH_3, CO, NO_2, H_2, SF_6, Cl_2, which could be applied in gas-sensor fabrication (Meyyappan, 2016).

Graphene oxide (GO) and Graphene-Based Composites (GBCs) have shown different gas pollutants' effective adsorption. Both GO/polyoxometalate nanocomposites and GO showed a high adsorption rate of NH_3, SO_2, H_2S, whereas the adsorption rates by GBCs were higher than by GO. In particular, NH_3 might react with functional groups on the GBCs surface, which led to higher retention. Pre-humidification and biosurfactants' presence improved the adsorption efficiency due to the acid-base reactions of gases with water on GBCs surface and higher dispersion of graphene layers, respectively. Meanwhile, SO_2 and oxygen groups' interactions on graphene layers might form SO_3^{2-} and SO_4^{2-}, which led to the removal of SO_2. In summary, functional groups on layer-structured GO and GBCs play a vital role in the adsorption activity of gases. Moreover, adding some metals, such as Cu, Zn, into the GO and GBCs surface significantly improves gases' adsorption rate (Lofrano et al., 2017).

The adsorption rate of organic contaminants on CNMs is higher than activated carbon because of the polarization ability of electron or the interactions of Electron Donor-Acceptor (EDA) with organic compounds (adsorbates), lower adsorption energies and no pore diffusion during the adsorption process. Finally, the removal of H_2S depends on the physical adsorption of H_2S to the pore space formed between the functional groups and graphene layers, as well as the reactive adsorption by functional groups on GBCs surface (Lofrano et al., 2017).

For carbon-based nano adsorbents, the interaction potentials in curved geometry systems may improve the nonspecific van der Walls interaction forces, which increases the adsorption efficiency. Some models, such as Monte Carlo, can predict carbon nanostructures' influence on the adsorption of contaminants. For example, tetrafluoromethane's adsorption performance is highest with the CNTs' diameter of 1.05 nm (Lofrano et al., 2017). On the other hand, the specific complexation reactions may represent the mechanisms of hydrophobic adsorption of organic and inorganic compounds on carbon-based nano adsorbents. The effect of functional group density on inorganic adsorption efficiency is more than the total surface area. The presence of metal speciation or completing complexation reactions influences the adsorption capacity, especially with pH change. The adsorption capacity of oxyanion metal pollutants on the surface of carbon-based nano adsorbents is affected by the divalent metal cations concentration via the surface complex reaction.

The carbon-based nano adsorbents have been effectively used for the removal of organic contaminants (e.g., trihalomethanes, polycyclic aromatic hydrocarbons, naphthalene), residual taste or odor. Their adsorption productivity can be improved mainly by surface modification or functionalization. In particular, functionalized nano adsorbents have been used to target individual micro-pollutants, eliminate low concentration contaminants or enhance subsurface mobility. For example, the functionalized CNTs showed much higher adsorption of low molecular weight and polar compounds than activated carbon.

6.1.2 Adsorption of contaminants in an aqueous solution by carbon-based nanomaterials

CNMs have been effectively used to separate contaminants in aqueous solutions due to their simplicity and flexibility. Their adsorption rate is more significant than other common adsorbents in the wastewater treatment field, such as activated carbon, due to their higher specific surface. Besides, CNMs can be a support frame for the adsorption of other oxides or macromolecules. For instance, adsorption of arsenate and chromium from water by CeO_2-based CNTs nanoparticles, removal of fluoride by Al_2O_3-based CNTs nanoparticles, adsorption of lead ions from water by MnO_2-based CNTs, treatment of lead and copper ions from water by iron oxide-based CNTs nanocomposites.

CNTs are considered one of the best nano adsorbents to remove contaminants in aqueous solutions due to their unique characteristics and promising applications. The adsorption mechanisms of organic compounds on CNTs can be described as heterogeneous adsorption (Pan and Xing, 2008), which is different from a single model like Freundlich, Langmuir or BET. Two reasons cause heterogeneous adsorption. The first is high energy adsorption sites produced by functional groups, CNT defects and regions between CNT bundles. The second is the surface condensation process, such as surface and capillary condensation of adsorbates, caused by the multilayer adsorption of organic compounds on CNT surfaces that distributes adsorption energy. Adsorption on CNTs can also be represented by the hysteresis feature, which is the required potential to form a strong interaction of small molecules (e.g., methane, benzene) and polymers (e.g., poly-arylene ethynylene) with the surface of CNTs. These interactions may change the mechanisms of the adsorption process. Furthermore, multiple mechanisms acting simultaneously (e.g., electrostatic and π–π interactions, hydrogen bonds) also contribute to organic compounds' adsorption behavior on the CNT surface. These mechanisms can predict organic compounds' adsorption characteristics on CNTs under different environmental conditions (Pan and Xing, 2008).

CNTs have been widely used for the treatment of contaminants in water, such as dye (e.g., rhodamine B, Sudan) (Oyetade et al., 2015), metal ions (e.g., Bi^{3+}, Cd^{2+}, Cr^{6+}, Cu^{2+}, Pb^{2+}) (Al-Saidi et al., 2016), organic chemicals (e.g., sulfamethoxazole, ofloxacin, norfloxacin, bisphenol A, phenanthrene) (Li et al., 2016), TEX (e.g., toluene, ethylbenzene, xylene) (Yu et al., 2016), veterinary drug (e.g., nitrofurazone) (Ying-Ying and Zhen-Hu, 2016). At the same time, graphene also showed its high adsorption rate of contaminants in water, such as the removal of chlorophenols (Yan et al., 2016), dye (e.g., acridine orange, anionic orange IV, methylene blue) (Scalese et al., 2016), ions (Se^{4+}, Se^{6+}) (Fu et al., 2014), metal ions (e.g., Cu^{2+}) (Liu et al., 2016), PAHs (e.g., phenanthrene) (Zhao et al., 2014), plasticizer (e.g., bisphenol A) (Zhang et al., 2014), radioactive metal ions (e.g., U^{6+}) (Chen et al., 2016).

Moreover, they have shown a high adsorption rate of organic compounds from wastewater. In particular, both SWCNTs and MWCNTs may adsorb herbicide 4-chloro-2-methylphenoxyacetic acid (MCPA) from wastewater, whereas the removal efficiency of SWCNT is higher. Besides, the main adsorption mechanisms of herbicide and the CNTs surface are electron donor-acceptor interactions and hydrogen bonds. The desorption process in ethanol can reuse the SWCNT. Besides, CNTs activated by KOH dry etching can highly adsorb typical monoaromatic compounds (e.g., phenol, nitrobenzene) and pharmaceutical antibiotics (e.g., sulfamethoxazole, tetracycline, tylosin) in solution. The activation of KOH primarily increased the specific surface area of both SWCNTs and MWCNTs, significantly improving their adsorption capacities. In another research, MWCNT, activated carbon and carbon nanofibers were useful in the adsorption of pharmaceutical compounds and an endocrine disruptor from a municipal wastewater treatment plant, where the adsorption rate constant of MWCNT was higher than carbon nanofibers and activated carbon. Furthermore, pH, temperature, background Natural Organic Material (NOM), textural and surface chemistry of adsorbents could impact on the adsorption rate (Álvarez-Torrellas et al., 2016). The adsorption equilibrium data fit well with Freundlich and Langmuir isotherms.

The combination of carbonaceous-containing polymers and membrane can be used to remove trace organic compounds from wastewater. In particular, the incorporation of fullerene-containing polymers and membrane may form polymer-based membranes with enhanced properties. For example, fullerene-based poly 2,6-dimethyl-1,4-phenylene oxide (PPO) membrane displayed high adsorption (more than 96%) of estrogenic compounds from wastewater for up to 12 h (Jin et al., 2007).

Functionalized graphene materials have shown effective removal of organic and inorganic contaminants from aqueous solutions. Due to its simple operation, these graphene-based materials have been widely used for the removal of organic pollutants in wastewater, including glyphosate (Yamaguchi et al., 2016), chlorophenols (Yan et al., 2016), p-nitrophenol (Liu et al., 2016) and dyes (Wang et al., 2016).

Similar to CNTs, the interaction mechanisms of graphene-based materials consist of electrostatic forces, van der Waals forces, π–π bonds, anion-cation interaction and hydrogen bonds. Their adsorption mechanism is dependent on adsorbate type and properties. Graphene and graphene oxide may effectively adsorb the negatively charged and high surface area anionic and cationic dyes, whereas the interaction mechanism is caused by electrostatic and van der Waals forces. Besides, three-dimensional graphene oxide sponge (GO sponge) and graphene/GO-composites have displayed their high adsorption of contaminants. For instance, the GO sponge effectively removed methylene blue and methyl violet due to the π–π bonds and anion-cation interactions (Khan et al., 2017). Another example is to use methylene blue from an aqueous solution that was adsorbed on the high surface area and pore volume 3D graphene/Mg $(OH)_2$-composite that could be effectively reused by washing with ethanol (Liu et al. 2012). In another work, sulfonated GO nanosheets in Nafion membranes (Nafion-GO_{SULF}) productively adsorbed and photocatalytic degraded of methyl orange, cationic and anionic azo-dye from water (Scalese et al., 2016).

Additionally, flower-like TiO_2 microsphere-based graphene composite, formed by hydrothermal reduction of graphene oxide and TiO_2 microsphere with polyvinyl pyrrolidine, was used to adsorb organic dye from water (Wang et al., 2016). In another work, graphene oxide (GO) was used to remove organic pollutants, such as tetracycline antibiotics from aqueous solutions. Moreover, the strong adsorption of tetracycline on the surface of GO was caused by π–π interaction and cation-π bonding and could be described well by Langmuir and Temkin isotherm models (Khan et al., 2017). A graphene-like layered hexagonal boron nitride (g-BN) was used to adsorb more than 90% of fluoroquinolone antibiotic gatifloxacin (GTF) from an aqueous media, and the data results fitted the Langmuir isotherm model (Chao et al., 2014).

Other graphene-like composites, such as carbon nanotube-graphene hybrid aerogels (Sui et al., 2012), few-layered GO nanosheet (Zhao et al., 2011), magnetic graphene nanocomposites decorated with core@ double-shell nanoparticles (Zhu et al., 2012), amino-functionalized magnetic graphene oxide composite (Chen et al., 2016), triethylenetetramine-magnetic reduced graphene oxide composite (Chen et al., 2015), magnetic polyaniline/graphene oxide composites (Liu et al., 2016), polypyrole decorated reduced graphene oxide-Fe_3O_4 magnetic composites (Wang et al., 2015), showed their high adsorption of inorganic contaminants like heavy metals. The main mechanisms are adsorption, attraction and interaction between the metal ions and the functional groups on these material surfaces. For example, few-layered GO nanosheets strongly adsorbed of Cd^{2+} and Co^{2+}, Pb^{2+} and U(VI), Hg^{2+}, Au^{3+} and Pd^{2+}, As (III) and As (V), Cr^{6+} ions from aqueous solutions (Wang et al., 2013). Besides,

GO and graphene-based composites (GBCs) showed high adsorption capacities of some inorganic anions, such as F^-, ClO_4^-, PO_4^{3-} (Wang et al., 2013) in water. The primary mechanism was the surface-exchange between inorganic anions and functional groups.

Activated graphene with high specific surface area and a significant number of micropores enhanced some pharmaceutical antibiotics' adsorption capacity, such as ciprofloxacin (CIP) from an aqueous solution. The data fit with the Langmuir isotherm model. In addition, the adsorption kinetics performed well according to a pseudo-second-order model (Yu et al., 2015).

Similar to fullerenes and CNTs, the graphene-based materials also showed antibacterial properties. GO nanosheets, which consist of graphene and suspended hydroxyl, epoxy or carboxyl groups, are to some extent dispersed in water. When physically interacting with the cell membrane, they can form the aggregation of cell-graphene oxide as well as damage to the cell membrane. For instance, graphene oxide-chlorophyllin and graphene oxide-chlorophyllin-zinc showed high antimicrobial activity of *E. coli* by destroying the *E. coli* membrane (Azimi et al., 2014).

Nonetheless, physical adsorption, electrostatic attraction and chemical interaction are the primary mechanisms for the adsorption of inorganic contaminants, such as heavy metals on the surface of graphene-based nanocomposites. For instance, the adsorption of heavy metal ions from water on Few-layered Graphene Oxide (FGO) is caused by the Lewis acid-base interaction between metal ions and functional groups of FGO that contain oxygen (Wang et al., 2013).

Their high cost and potential toxicity limit the application of carbon-based nanoadsorbents in water treatment. However, they showed a much higher adsorption efficiency than traditional materials, such as activated carbon. Carbon-based nano adsorbents were also widely used in various environmental fields, such as hydrogen adsorption on SWCNTs in fuel cell design, virus adsorption on SWCNT-based filters in drinking water treatment plants or applying SWCNTs for environmental sensors.

6.2 Carbon-based nanomaterials for composite filters

Functionalized composite membranes have more advantages over regular membranes, such as high flowrate, great sorption capacity, vigorous antimicrobial activity or high thermal stability. For instance, poly (2,6-dimethyl-1,4-phenylene oxide) (PPO)-fullerene composite membranes show high permeate flux and adsorption rate of hydrophobic organics, e.g., dissolved estrogenic compounds. Furthermore, nanomaterials-based composite membranes also have specific features of nanomaterials. Therefore, the membrane lifespan in water and wastewater treatment plants may help to evaluate the immobilization of nanomaterials on the membrane surface.

The mechanical characteristic of carbon-based nanocomposite membranes is more substantial than regular reverse osmosis membranes due to the presence of CNTs-functionalized groups on their surface. Besides, membrane coated with a hydrophilic nanocomposite layer on the top, electrospun polyvinyl alcohol (PVA) layer in the middle and nanofibrous support layer shows a high-flux filtration, mechanical strength, durability, as well as improves the water permeability. Moreover, the higher concentration and oxidization of MWCNTs, the higher water permeability. This result implies that the presence of MWCNTs on membrane layers may break the polymer chain packing, form nanoscale holes on the hydrophilic top layer and lead to an increase in the water flow rate on the CNT-based membrane surface (Mauter and Elimelech, 2008). The smooth surface of MWCNT or small friction force of the MWCNT surface, also contributes to the CNT-based membrane.

The incorporation of MWCNTs and polymer may be used as a functionalized ultrafiltration membrane. The treatment efficiency of this membrane is dependent on the oxidized MWCNTs concentration. At the MWCNTs concentration of 1.5 wt% , the casting solution's viscosity increases, while its thermodynamic stability decreases. Therefore, the nanocomposite membrane's porosity is maximized, leading to improved treatment productivity of the MWCNT-based membrane. If the MWCNTs concentration in the casting solution is from 1.5 wt to 4 wt%, the functionalized membrane's flux and rejection rate increase, reducing its treatment performance (Choi et al., 2006).

Hybrid ceramic filters are another example of the SWCNTs-based composite membrane. Due to the large surface area, great thermal resistance, narrow diameter range and permeable aggregate structure of SWCNTs, the SWCNTs-functionalized filter effectively inactivates viral and bacterial pathogens in water (Mauter and Elimelech, 2008). In this inactivation process, bacteria and virus removal are evident on the surface and deep inside the filter. Therefore, a hybrid SWCNT-based ceramic filter is a promising technology for water treatment applications.

This filter's limitation is the immobilization of the nanotube on the surface, which prevents its wide use in the treatment of drinking water. One solution is to directly grow MWCNTs on the micron-sized metallic fibers surface, which significantly enhances the metal filter. This functionalized filter can be applied to remove some micropollutants in air streams, such as adsorption of nicotine and tar from cigarette smoke.

CNFs can be used as a filter to remove submicron particles. The electrospun CNF structured filter can be fabricated by the electrospinning method with high surface-area-to-volume ratio and high surface cohesion, which can effectively trap the tiny particles. In a US patent, the CNF filter can be used as a layer in the dust filter bag. Moreover, the electrostatically charged CNFs may enhance the filtration performance by changing the particles' electrostatic attraction. Likewise, some selective agents-based CNF filters may productively detect and filter chemical and biological weapon agents. Nanofiber infiltration will enhance the filtration efficiency and does not significantly influence the pressure drop. Due to the large specific surface area and porosity and reduced pore size, CNFs effectively adsorb or filter air and water pollutants, such as microorganisms, viruses and hazardous molecules (Schaefer et al., 2007).

6.3 Carbon-based nanomaterials as water treatment membranes

The membrane is a semi-permeable and thin film (typically less than 1 mm thickness) that allows some matters to pass through while leaving others behind (Crittenden et al., 2012). The membrane process has been widely used in the water treatment plant as a separation process. In the membrane, the driving force is supplied. The semi-permeable membrane properties mean it is highly permeable to some feed stream components and less permeable to others. In particular, permeable components will pass through the membrane, while impermeable components will be retained on the membrane's feed side.

The Reverse Osmosis (RO) membrane has been widely used in the treatment of drinking water plants among different membranes. The mechanism of conventional RO desalination membranes is due to solution-diffusion rates, where each component in the high-pressure side (feed) dissolves in the membrane and diffuses through the membrane polymer matrix follow the Fick's law of diffusion. In the membrane, selectivity is inversely proportional to water flux. Therefore, it is necessary to research and develop new RO membrane designs that may improve treatment efficiency.

The use of CNT as a novel membrane has shown many benefits, such as lower cost and energy. For instance, the insertion of gold nanotubes into one CNT-functionalized membrane may effectively separate small molecules under high permeate flux. The separation performance of the CNT-based membrane is dependent on the diameter of the nanotube pore and properties of the membrane surface.

High-flux CNT-based membrane has been effectively used in advanced water treatment. For example, MWCNTs-based polymer film can be used as an ultrafiltration membrane or double-walled CNTs-based membrane. However, the CNT-based membrane has not been able to separate salt well from seawater due to the high ionic strength and divalent cation concentration of seawater that declines the charge exclusion. Therefore, the CNT-based membrane is typically suitable for nanofiltration. Besides, the hydrated radius of sodium ions in seawater is affected by environmental conditions (e.g., pH, temperature, ionic strength). Hence, to obtain desalination purposes, the SWNT-based membrane must be produced with narrowed pore size distribution and sub nanometer pores. Moreover, membrane fouling, scale and cost should be taken into account to produce the CNT-based membrane (Lofrano et al., 2017).

At the same time, the CNT-based membranes can be used in nanofluidic and microfluidic systems requiring highly skilled techniques related to small samples, such as analytical separation and chip devices. Under a power discharge, the nanotube tips on CNT-based membrane become charged and hydrophilic, which change the behavioral interactions of the fluid and nanotubes and lead to the possible management of fluid wicking, flowrate and direction. The flow rate of water through CNT-based membranes is much higher than through traditional membranes. This difference is due to the small channel, hydrophobic surface and smooth internal walls of CNT-based membranes, which may decrease the friction forces between their surface and water molecules. Many factors could influence these flowrates, such as the functional groups of CNTs, physical barriers at the inlet and outlet of the nanotube channel. Some models have also be described (Lofrano et al., 2017).

The selectivity of CNT-based membranes is influenced by the properties of functional groups on the membrane surface. In particular, changing the length and structure of functional groups on nanotube tips may increase the membrane selectivity, but reduce the membrane permeability and flux. For instance, giant streptavidin molecules' functionalization at open tips of CNTs may significantly and reversibly reduce the flux through CNT-based membranes. Therefore, it is critical to balance selectivity forces and permeability forces when designing any high-flux CNT-based membrane.

6.4 Carbon-based nanomaterials as antimicrobial agent

CNMs can be used as antimicrobial agents due to their ability in cell membrane damage and oxidation properties. Although the mechanisms of antimicrobial activity have not been clearly understood, the antimicrobial properties of CNMs, especially in environmental and human health applications, have attracted the attention of scientists. In particular, CNMs can be used in water purification, medical therapy, antimicrobial activity or microbiology techniques in the laboratory. Some examples include the removal of antibiotics (Ncibi and Sillanpää, 2015), pharmaceuticals ((Shan et al., 2016), surfactants (Ncibi et al., 2015), removal of antibiotics and pharmaceuticals by graphene (Yu et al., 2015), removal of Pb and Cu ions in water by polyvinyl alcohol (PVA)/MWCNTs nanocomposite (Jing and Li, 2016).

Pristine fullerenes and functionalized fullerenes effectively removed various bacteria types, such as *E. coli*, *Salmonella* or *Streptococcus* spp. due to their high antibacterial activity. Among different types of fullerene derivatives, C_{60}-bis (7V,7V-dimethylpyrrolidinium iodide), a cationic fullerene derivative, displayed better antibacterial activity of negatively-charged *E. coli* and *Shewanella oneidensis* bacteria. Similarly, the cationic protonated amine (AF) derivative of fullerenes also demonstrated the antibacterial property of *E. coli* due to substantial and partially irreversible binding with *E. coli* cells. In contrast, the antibacterial activity of *E. coli* by anionic deprotonated carboxylic (CF) derivative of fullerene was not significant due to no binding with *E. coli* cells (Deryabin et al., 2014).

Moreover, fullerenes have also demonstrated the capacity for water disinfection due to their photocatalytic activity. In aqueous solutions, the suspension form of fullerene (nC_{60}) might contact bacterial cells directly and cause antibacterial effects. In particular, nC_{60} induce ROS-independent oxidative stress in *E. coli* K12 and *B. subtilis* 168 bacteria, such as protein oxidation, changes in cell membrane potential and cellular respiration interruption, that lead to the removal of these bacteria from water (Lofrano et al., 2017). This mechanism is different from earlier antibacterial mechanisms by NMs that relate to the ROS and ROS-mediated generation (by metal oxides) or leaching of toxic elements (by nanosilver).

The antibacterial effects of CNTs are influenced by their size (diameter), length and concentration. The removal rate of *E. coli* by SWCNTs was higher than MWCNTs due to their smaller diameter that led to faster penetration into the *E. coli* cell wall. Their larger surface area contributed to more contact and interaction with the *E. coli* cell surface, and their greater chemical reactivity. Moreover, the more potent antibacterial activity was observed in the longer SWCNTs due to their better aggregation with bacterial cells in suspensions. Besides, the antibacterial activities of SWCNTs on *Salmonella enteric* and *E. coli* dispersed in various surfactant solutions were enhanced with the increase of SWCNTs concentrations (Lofrano et al., 2017).

The disturbance of cell membranes is also a promising method for pathogen removal. In this procedure, the primary inactivation mechanism is membrane disruption. Antimicrobial coating on the CNTs surface is also a good option for treating bacterial colonization and biofilm development due to its antibacterial capacity. This technique can be applied to inactivate bacteria in drinking water treatment systems, medical implant devices or submerged surfaces (Mauter and Elimelech, 2008). Furthermore, CNTs can be combined with other antimicrobial/antiviral or photocatalytic nanoparticles for the inactivation of bacteria and viruses in water treatment and distribution systems, such as TiO_2-coated CNTs or Ag-based CNTs (Xie et al., 2014). At the same time, other CNMs, such as fullerol and stable fullerene water suspensions (nC_{60}), have proved their ability for pathogen removal in water or wastewater treatment systems. The pathogen inactivation mechanism of fullerol depends on the antiviral activity of reactive oxygen species, such as singlet oxygen or superoxide, generated in an aqueous solution under both visible and ultraviolet (UV) light sources. At the same time, nC_{60} shows their antibacterial activity to various bacteria under different environmental conditions.

Besides, CNTs can combine with other metal nanoparticles or membranes to form nanocomposites that have strong antibacterial activity (Shi et al., 2016) (Sharma et al., 2015). Significantly, the combination of Ag-NPs and CNTs showed many advantages over other nanocomposite materials (Chang et al., 2016; Karumuri et al., 2016).

Furthermore, the nanocomposites of CNMs and Ag NPs, such as SWCNTs-Ag or graphene oxide-Ag, displayed a high antibacterial property, both gram-negative and gram-positive bacteria, due to the antibacterial properties of Ag NPs and CNMs. The antibacterial property of SWCNTs-Ag is incredibly potent caused by the high dispersion of Ag NPs into the SWCNTs. Moreover, Ag-carbon nanocomplexes effectively inhibit the activity of various pathogens, for example *Burkholderia cepacia* and multidrug-resistant *Acinetobacter baumannii* (Leid et al., 2012).

Similarly, GO nanosheets have shown their antibacterial activity. In particular, they may physically interact with the cell membrane, form the cell-graphene oxide agglomeration and damage the cell membrane. For instance, some functionalized graphene oxides, such as GO-chlorophyllin and graphene oxide-chlorophyllin-Zn, showed potent antibacterial activity of *E. coli* by damaging the membrane (Lofrano et al., 2017).

The toxicity of CNMs, such as CNTs, on various microbial communities is one of the main concerns that need more research. Besides, the immobilization and separation of nanomaterials from treated water are also challenging, requiring further study.

6.5 Carbon-based nanomaterials for photocatalytic approaches

CNMs can be used for contaminant remediation through photocatalytic approaches. The photocatalytic approach mechanism is based on the photocatalytic activity of carbon-based semiconductor catalysts, such as CNTs, fullerenes, nanocomposites. A carbon-based semiconductor catalyst consists of two energy bands, a valence band (with lower energy) and a conduction band (with higher energy) and a band gap with the energy of about 1.1 eV between the two energy bands. When it is illuminated by UV light with photons of energy equal or greater than its band gap energy, a negatively charged electron at the valence band is excited and crosses over to the conduction band, leaving behind a positively charged hole. These charged electrons and holes will react with O_2 and H_2O in the atmosphere to form reactive oxygen species (ROS), such as $O_2^{\cdot -}$ and ·HO (hydroxyl radical). These ROS will react with organic contaminants to form CO_2 and H_2O as the final products. Graphene has been used to produce the photocatalytic nanocomposites in earlier papers, whereas TiO_2 NPs-based graphene composites display more significant photocatalytic activity than only TiO_2 NPs due to the higher conductivity properties.

TiO_2-graphene nanocomposites have been effectively used to remove benzene. Like other TiO_2-based nanocomposites, such as TiO_2-CNTs, TiO_2-fullerenes, the photocatalytic activity of

TiO_2 in TiO_2-graphene nanocomposite was improved due to the higher light absorption intensity and longer electron-hole pairs lifetime, which led to higher treatment of pollutants. In this research, benzene was degraded by the photocatalytic activity of nanocomposite and was adsorbed on the graphene surface. Therefore, the degradation rate of benzene by TiO_2-graphene photocatalyst was higher than bare TiO_2. Moreover, the removal rate of benzene was maintained after 28 hr of experiment, demonstrating the stable activity of the TiO_2-graphene photocatalyst. Besides, benzene treatment efficiency was dependent on factors such as the ratio of graphene in nanocomposite composition (Lofrano et al., 2017).

At the same time, other carbon-based nanocomposites, such as ZnO-graphene and CdS-graphene, have been applied to remove contaminants in water systems. In particular, ZnO-graphene nanocomposite showed a higher degradation rate of Cr^{6+} suspension than pure ZnO due to its better photocatalytic activity under UV light illumination. The degradation rate is influenced by the ratio of the graphene amount in the nanocomposite composition. Similarly, CdS-graphene nanocomposite also displayed high removal efficiency of Cr^{6+} ions in water, whereas its photolytic activity relied on the graphene ratio in the composition of CdS-graphene nanocomposite (Zhang et al., 2013). In summary, the removal rate of contaminants by functionalized carbon-based nanocomposites was much higher than pure material.

7. Toxicity of Carbon-based Nanomaterials

Concerns on the toxicity of CNMs have been raised for a long time. CNMs may be dangerous for human health, animals, the environment, aquatic communities or some sensitive bacterial strains. The toxicity of CNMs is influenced by their physicochemical and morphological properties, for instance, surface component, density, length and size. Due to the nano size, it is easy for them to penetrate human, animal and plant cells, which may induce adverse effects. The CNMs toxicity includes their toxic nature and the toxicity of substances adsorbed on their surface.

In most cytotoxicity study of fullerene, nC_{60} fullerenes are typically used due to their solubility in water. In particular, the nC_{60} colloidal suspension may break the cellular function through lipid peroxidation, while ROS produced on their surface can damage the membrane. Otherwise, fullerenes also play a vital role in the repairing mechanisms of DNA (Dinesh et al., 2012).

The toxicity mechanism of CNTs includes passing through cell membranes, producing ROS species, accumulating in tissues and causing oxidative damage or physical disruption. Their toxicity behavior is influenced by the properties of CNTs, such as type or length, aggregate size, structural imperfections and the presence of functional groups or metal impurities (Dinesh et al., 2012).

Besides, toxicity tests are controversial due to their unsuitability. In particular, toxicity tests for general CNMs are not suitable for raw materials like carbon. For instance, the possible water suspension preparation methods for fullerenes' aquatic toxicity tests are the long-term stirring method, solvent transfer method or ultrasound-assisted method. However, the suspensions prepared by the two latter methods showed higher toxicity than the first method. Therefore, the long-term stirring method tends to be accepted due to its more ecologically representative.

7.1 Toxicity tests in aqueous solutions

The effect of C_{60} agglomerates in water communities has been researched (Tao et al., 2011). Their toxicity values to the digestive tract proved to be dependent on the preparation process. The toxicity of suspensions obtained by vigorous shaking and ultra-sonicated of SWCNTs on the behavior of fish was investigated. After 10 d of exposure with a 0.5 mg/L SWCNTs suspension prepared by the ultrasound method, abnormal respiration rates and increased aggressive behaviors, such as attack to each other, were recognized in the rainbow trout fish. In other words, the observation of many physiological abnormalities in fish confirmed the harmful toxicity of CNTs on fish behavior. In other papers, toxicity measurements, LC_{50} concentrations, of suspension of fullerene aggregates obtained by the solvent transfer and ultrasound method are 460 μg/L and 7.9 μg/L, respectively, or the LC_{50} concentrations to *Daphnia magna* by SWCNT and MWCNT are 2.425 mg/L and 22.751 mg/L, respectively.

The antimicrobial activity of CNTs on the bacterial cell membrane has been studied earlier. However, some bacteria, such as *Cupriavidus metallidurans*, may have the resistance capacity to the MWCNT aggregates due to the detoxification behavior caused by their particular efflux systems (Simon-Deckers et al., 2009).

Graphene and GO are potentially toxic to aquatic organisms. However, their toxicity is different according to the bacteria type. GO showed potent antimicrobial activity to *E. coli* or nematode *Caenorhabditis elegans*, but low toxicity to *Amphibalanus amphitrite nauplii* (Mesarič et al., 2013). At the same time, pristine graphene nanoparticles are toxic to marine organisms, such as *Vibrio fischeri* and *Dunaliella tertiolecta*, whereas toxicity increased with the smaller particle size (Pretti et al., 2014). In other works, SWCNTs displayed high toxicity to freshwater organisms, such as microalgae (*Raphidocelis subcapitata* and *Chlorella vulgaris*), micro-crustacean (*D. magna*) and fish (*Oryzias latipes*). In particular, SWCNTs prevented the growth of the algae *R. subcapitata* and *C. vulgaris* (Sohn et al., 2015).

7.2 Toxicity tests based on sediment and soil organisms

CNMs may be toxic to sediment and soil organisms. Earlier papers showed the toxicity of C_{60} and CNTs to different types of sediment and soil organisms. The high concentration of CNTs may influence benthic populations. For instance, 2 mg/kg of MWCNTs affected the biodiversity of a benthic community after 15 months (Velzeboer et al., 2013).

However, fullerenes showed low toxicity to the soil microbial communities, particularly to the respiration, biomass, number, diversity of some bacteria and protozoans after 14 d. For example, C_{60} fullerene expressed low toxicity to *Lumbricus rubellus* earthworm adults but displayed significant harm to their cocoon production, juvenile growth rate and mortality.

On the other hand, the ecotoxicity assessment of CNMs has some challenges that need further research. One concern is the need for analytical techniques to monitor the presence of CNMs in the environment. In recent years, the quantitative methods for fullerenes have been developed, but not for CNTs or graphene. For example, only the analytical technique for analyzing radio-labeled graphene in the laboratory was available, which raised concerns about non-labeled analysis procedures. The toxicity of CNM suspensions in aqueous solutions is also influenced by their low stability. Besides, stabilizing agents can prevent aggregation and settle out of solution in an aqueous test media, but it may cause some public concerns. Hence, it should be approved by the research community, which would lead to its revision to adapt to the new scenarios (Lofrano et al., 2017).

The lack of published knowledge about the test methods' properties, such as accuracy, robustness, reproducibility, variability within and among laboratories, may cause some concerns. The correct use of ecotoxicological tests of pure CNMs standards for evaluating potential risks of CNMs have raised concerns due to the various forms of CNMs after releasing them into the environment. Lastly, the combined presence of CNMs and other contaminants in the environment needs more study. In particular, the influence of adsorption capabilities and hysteresis on the bioavailability, transport and toxicity of contaminants in the environment should be considered (Lofrano et al., 2017).

8. Discussion and future directions

CNMs have been mostly used as adsorbents for the removal of contaminants. Despite high removal efficiency, the adsorption method is not considered a sustainable treatment approach for the future. Other treatment processes are required for the contaminant-containing adsorbents after the adsorption procedure. Moreover, the adsorption process may form Disinfection By-Products (DBPs), which can be more harmful than the original contaminants. Hence, more research about the potential DBPs, especially their form and toxicity, is required before using CNMs in the adsorption process.

Based on the treatment plant and purposes, CNMs can be used in a pre-treatment step or a stand-alone system to remove contaminants from the environment. Therefore, more study on the application range of CNMs is needed to use them in the treatment systems widely.

A combination of some CNM adsorbents in one treatment system is a potential option that needs more research. Some possible ways include using graphene with carbon nanotubes or graphene with carbon nanofibers. Graphene is cheap and easy to apply, it may be used as the primary treatment to reduce the contaminant concentration. Carbon nanotubes and carbon nanofibers may be used next to remove contaminants from the environment altogether.

Besides, more adsorption modeling software should be developed and applied for future study. The right mode will save time and money to research the adsorption activity of any contaminant and predict the potential risks caused by CNMs.

Although the synthesis methods of CNMs have been widely used, understanding these processes is required to acquire better control. Furthermore, more investigation of these industrial manufacturing approaches is necessary to fill the gap between scientific research and possible use.

The retention and reuse of CNMs should be considered when designing and applying them to the industrial scale. In water treatment, using a membrane to immobilize the NMs is an excellent way to lower the operation and maintenance cost, while keeping high treatment efficiency. However, membrane filtration is not a suitable option to filter suspended matter in a wastewater treatment plant due to clogging. In this case, pre-treatment of raw water is required to decrease the turbidity and suspended solids in the wastewater source, which will improve the treatment performance. Some other materials can be used to collect the NMs after-treatment process. For example, CNTs can be integrated into micrometer-sized colloidal particles to form CNTs ponytails (CNTPs) and used as an enhanced post-treatment separation of NMs.

CNMs can move together with a water source during the treatment process, which will reduce removal productivity. Therefore, the immobilization of CNMs on resins or membranes will be an excellent way to solve this issue. Furthermore, 3D graphene sponge material can be used as a support for NMs. However, any immobilization technique that will lead to the reduction of treatment efficiency should be considered in any water/wastewater treatment plant design and operation. Additionally, magnetic NPs-nanocomposites or low-field magnetic separation are also some energy-efficient ways to solve the problem.

In general, the release of NMs during the treatment process has not been completely understood yet. Further research in this field is mandatory, especially with CNMs used in devices. The potential release of each CNM has a strong relationship with its type, immobilization method or separation technique. Due to their possible toxicity and behavior, more attention should be considered to release metal ions from CNMs during or after the treatment process. More studies for the detection and analytical methods of released NMs in the environment should also be developed .

The toxicity of CNMs to human health and the environment also needs more study. Due to the large number of CNMs produced and used, concerns about their potential risks when released to the environment are arising, which relates to the CNMs-based treatment system. According to earlier research, CNMs may be toxic in some specific conditions. Therefore, more investigations on the toxicity of CNMs are required to study the potential application of CNMs in a large-scale treatment process, which will affect the complete application of these materials.

9. Conclusions

In conclusion, carbon-based nanomaterials have shown their practical use in environmental remediation. Due to the small pore size, high specific surface area, low cost and high removal efficiency, carbon-based nanomaterials have been widely used to remove various contaminants under different conditions, both in the laboratory and in the environment. Their adsorption capacity is dependent on many factors, such as properties of contaminants and CNMs, type and characteristic of functional groups, environmental conditions.

Moreover, the functionalization of CNMs with metal/oxide and organics to form functionalized CNMs or nanocomposites may improve their removal performance and useful lifetime. There are many types of functionalized CNMs, depending on the functional groups. Each functionalized CNM will be suitable for a specific purpose.

Many approaches to synthesize of CNMs and functionalized CNMs for environmental remediation were studied. Each method may produce CNMs with different properties. Some standard methods are physical, Chemical Vapor Deposition (CVD), thermal CVD, arc discharge, laser vaporization, mechanical or chemical exfoliation. Overall, the synthesis and use of carbon-based nanomaterials in environmental remediation have both advantages and disadvantages. Choosing the best method and material depend on many factors, such as site-specific conditions, cost or contaminant properties. Besides, their potential risks, such as toxicity, aggregation or accumulation in the environment, should be considered when applying CNMs for environmental remediation.

References

Abhilash, P.C., Tripathi, V., Edrisi, S.A., Dubey, R.K., Bakshi, M., Dubey, P.K., Singh, H.B. and Ebbs, S.D. 2016. Sustainability of crop production from polluted lands. Energy, Ecology and Environment 1: 54–65. DOI: 10.1007/s40974-016-0007-x.

Afreen, S., Omar, R.A., Talreja, N., Chauhan, D. and Ashfaq, M. 2018. Carbon-based nanostructured materials for energy and environmental remediation applications. pp. 369–392. *In*: Prasad, R. and Aranda, E. (eds.). Approaches in Bioremediation. Switzerland: Springer; DOI: 10.1007/978-3-030-02369-0_17.

Allen, M.J., Tung, V.C. and Kaner, R.B. 2010. Honeycomb carbon: A review of graphene. Chemical Reviews 110: 132–145. DOI: 10.1021/cr900070d.

Al-Saidi, H.M., Abdel-Fadeel, M.A., El-Sonbati, A.Z. and El-Bindary, A.A. 2016. Multi-walled carbon nanotubes as an adsorbent material for the solid phase extraction of bismuth from aqueous media: Kinetic and thermodynamic studies and analytical applications. Journal of Molecular Liquids 216: 693–698. DOI: 10.1016/j.molliq.2016.01.086.

Álvarez-Torrellas, S., Rodríguez, A., Ovejero, G. and García, J. 2016. Comparative adsorption performance of ibuprofen and tetracycline from aqueous solution by carbonaceous materials. Chemical Engineering Journal 283: 936–947. DOI: 10.1016/j.cej.2015.08.023.

Amiraslanzadeh, S. 2016. The effect of doping different heteroatoms on the interaction and adsorption abilities of fullerene. Heteroatom Chemistry 27: 23–31. DOI: 10.1002/hc.21284.

An, X. and Yu, J.C. 2011. Graphene-based photocatalytic composites. RSC Advances 8: 1426–1434. DOI: 10.1039/C1RA00382H.

Araújo, R., Castro, A.C.M. and Fiúza, A. 2015. The use of nanoparticles in soil and water remediation processes. Materials Today: Proceedings 2: 315–320. DOI: 10.1016/j.matpr.2015.04.055.

Ashori, A., Rahmani, H. and Bahrami, R. 2015. Preparation and characterization of functionalized graphene oxide/carbon fiber/epoxy nanocomposites. Polymer Testing 48: 82–88. DOI: 10.1016/j.polymertesting.2015.09.010.

Azimi, S., Behin, J., Abiri, R., Rajabi, L., Derakhshan, A.A. and Karimnezhad, H. 2014. Synthesis, characterization and antibacterial activity of chlorophyllin functionalized graphene oxide nanostructures. Science of Advanced Materials 6: 771–781. DOI: 10.1166/sam.2014.1767.

Bains, U. and Pal, R. 2019. *In-situ* continuous monitoring of the viscosity of surfactant-stabilized and nanoparticles-stabilized pickering emulsions. Applied Sciences 19: 4044–4053. DOI: 10.3390/app9194044.

Bardos, P., Merly, C., Kvapil, P. and Koschitzky, H. 2018. Status of nanoremediation and its potential for future deployment: Risk-benefit and benchmarking appraisals. Remediation Journal 28: 43–56. DOI: 10.1002/rem.21559.

Chang, Y., Gong, J., Zeng, G., Ou, X., Song, B., Guo, M., Zhang, J. and Liu, H. 2016. Antimicrobial behavior comparison and antimicrobial mechanism of silver coated carbon nanocomposites. Process Safety and Environmental Protection 102: 596–605. DOI: 10.1016/j.psep.2016.05.023.

Chao, Y., Zhu, W., Chen, J., Wu, P., Wu, X., Li, H., Han, C. and Yan, S. 2014. Development of novel graphene-like layered hexagonal boron nitride for adsorptive removal of antibiotic gatifloxacin from aqueous solution. Green Chemistry Letters and Reviews 7: 330–336. DOI: 10.1080/17518253.2014.944941.

Chen, J.H., Xing, H.T., Sun, X., Su, Z.B., Huang, Y.H., Weng, W., Hu, S.R., Guo, H.X., Wu, W.B. and He, Y.S. 2015. Highly effective removal of Cu (II) by triethylenetetramine-magnetic reduced graphene oxide composite. Applied Surface Science 356: 355–363. DOI: 10.1016/j.apsusc.2015.08.076.

Chen, L., Zhao, D., Chen, S., Wang, X. and Chen, C. 2016. One-step fabrication of amino functionalized magnetic graphene oxide composite for uranium (VI) removal. Journal of Colloid and Interface Science 472: 99–107. DOI: 10.1016/j.jcis.2016.03.044.

Choi, J., Jegal, J. and Kim, W. 2006. Fabrication and characterization of multi-walled carbon nanotubes/polymer blend membranes. Journal of Membrane Science 284: 406–415. DOI: 10.1016/j.memsci.2006.08.013.

Crittenden, J.C., Trussell, R.R., Hand, D.W., Howe, K.J. and Tchobanoglous, G. 2012. MWH's Water Treatment: Principles and Design [third edition]. United States: John Wiley & Sons; pp. 1–1901. DOI: 10.1002/9781118131473.

Cui, X., Zhang, C., Hao, R. and Hou, Y. 2011. Liquid-phase exfoliation, functionalization and applications of graphene. Nanoscale 3: 2118–2126. DOI: 10.1039/C1NR10127G.

De La Calle, I. and Romero-Rivas, V. 2018. The role of nanomaterials in analytical chemistry: Trace metal analysis. pp. 251–301. *In*: Bhagyaraj, S.M., Oluwafemi, O.S., Kalarikkal, N. and Thomas, S. (eds.). Applications of Nanomaterials. Duxford (United Kingdom): Elsevier. DOI: 10.1016/B978-0-08-101971–9.00010-7.

Deryabin, D.G., Davydova, O.K., Yankina, Z.Z., Vasilchenko, A.S., Miroshnikov, S.A., Kornev, A.B., Ivanchikhina, A.V. and Troshin, P.A. 2014. The activity of [60] fullerene derivatives bearing amine and carboxylic solubilizing groups against escherichia coli: A comparative study. Journal of Nanomaterials 2014: 1–9. DOI: 10.1155/2014/907435.

Dinesh, R., Anandaraj, M., Srinivasan, V. and Hamza, S. 2012. Engineered nanoparticles in the soil and their potential implications to microbial activity. Geoderma 173: 19–27. DOI: 10.1016/j.geoderma.2011.12.018.

Du, W., Jiang, X. and Zhu, L. 2013. From graphite to graphene: Direct liquid-phase exfoliation of graphite to produce single- and few-layered pristine graphene. Journal of Materials Chemistry A 36: 10592–10606. DOI: 10.1039/C3TA12212C.

Ealias, A.M. and Saravanakumar, M.P. 2017. A review on the classification, characterisation, synthesis of nanoparticles and their application. IOP Conf. Ser. Mater. Sci. Eng. 263: 2–16. DOI: 10.1088/1757–899X/263/3/032019.

Er, S., de Wijs, G.A. and Brocks, G. 2015. Improved hydrogen storage in ca-decorated boron heterofullerenes: A theoretical study. Journal of Materials Chemistry A 15: 7710–7714. DOI: 10.1039/C4TA06818A.

Farre, M., Sanchis, J. and Barcelo, D. 2017. Adsorption and desorption properties of carbon nanomaterials, the potential for water treatments and associated risks. pp. 137–183. *In*: Lofrano, G., Libralato, G. and Brown, J. (eds.). Nanotechnologies for Environmental Remediation. Cham (Switzerland): Springer. DOI: 10.1007/978-3-319-53162-5.

Fu, Y., Wang, J., Liu, Q. and Zeng, H. 2014. Water-dispersible magnetic nanoparticle–graphene oxide composites for selenium removal. Carbon 77: 710–721. DOI: 10.1016/j.carbon.2014.05.076.

Gouma, S., Fragoeiro, S., Bastos, A.C. and Magan, N. 2014. Bacterial and fungal bioremediation strategies. pp. 301–323. *In*: Das, S. (ed.). Microbial Biodegradation and Bioremediation. London (United Kingdom): Elsevier. DOI: 10.1016/B978-0-12-800021-2.00013-3.

Gu, H., Li, Y. and Li, N. 2015. Electrical conductive and structural characterization of electrospun aligned carbon nanofibers membrane. Fibers and Polymers 16: 2601–2608. DOI: 10.1007/s12221-015-5543-z.

Guadagno, L., Raimondo, M, .Vittoria, V., Vertuccio, L., Lafdi, K., De Vivo, B., Lamberti, P., Spinelli, G. and Tucci, V. 2013. The role of carbon nanofiber defects on the electrical and mechanical properties of CNF-based resins. Nanotechnology 24: 1–10. DOI: 10.1088/0957-4484/24/30/305704.

Guerra, F.D., Attia, M.F., Whitehead, D.C. and Alexis, F. 2018. Nanotechnology for environmental remediation: Materials and applications. Molecules 23: 1760–1783. DOI: 10.3390/molecules23071760.

Huang, P.Y., Ruiz-Vargas, C.S., Van Der Zande, Arend, M., Whitney, W.S., Levendorf, M.P., Kevek, J.W., Garg, S., Alden, J.S., Hustedt, C.J. and Zhu, Y. 2011. Grains and grain boundaries in single-layer graphene atomic patchwork quilts. Nature 469: 389–392. DOI: 10.1038/nature09718.

Jain, R., Jordan, N., Schild, D., Van Hullebusch, E.D., Weiss, S., Franzen, C., Farges, F., Hübner, R. and Lens, P.N. 2015. Adsorption of zinc by biogenic elemental selenium nanoparticles. Chemical Engineering Journal 260: 855–863. DOI: 10.1016/j.cej.2014.09.057.

Jensen, K.A., Bøgelund, J., Jackson, P., Jacobsen, N.R., Birkedal, R., Clausen, P.A., Saber, A.T., Wallin, H. and Vogel, U.B. 2015. Carbon Nanotubes–Types, products, market and provisional assessment of the associated risks to man and the environment. Environmental Project 1805: 80004–80002. ISBN: 978–8793352988.

Jing, L. and Li, X. 2016. Facile synthesis of PVA/CNTs for enhanced adsorption of Pb^{2+} and Cu^{2+} in single and binary system. Desalination and Water Treatment 57: 21391–21404. DOI: 10.1080/19443994.2015.1119739.

Jung, C., Son, A., Her, N., Zoh, K., Cho, J. and Yoon, Y. 2015. Removal of endocrine disrupting compounds, pharmaceuticals, and personal care products in water using carbon nanotubes: A review. Journal of Industrial and Engineering Chemistry 27: 1–11. DOI: 10.1016/j.jiec.2014.12.035.

Kamrani, A.K. and Nasr, E.A. 2010. Design for manufacture and assembly. pp. 141–183. *In*: Engineering Design and Rapid Prototyping. Boston (United States): Springer. DOI: 10.1007/978-0-387-95863-7.

Karumuri, A.K., Oswal, D.P., Hostetler, H.A. and Mukhopadhyay, S.M. 2016. Silver nanoparticles supported on carbon nanotube carpets: Influence of surface functionalization. Nanotechnology 27: 1–14. DOI: 10.1088/0957-4484/27/14/145603.

Khan, A., Wang, J., Li, J., Wang, X., Chen, Z., Alsaedi, A., Hayat, T., Chen, Y. and Wang, X. 2017. The role of graphene oxide and graphene oxide-based nanomaterials in the removal of pharmaceuticals from aqueous media: A review. Environmental Science and Pollution Research 24: 7938–7958. DOI: 10.1007/s11356-017-8388-8.

Kim, H.H., Kang, B., Suk, J.W., Li, N., Kim, K.S., Ruoff, R.S., Lee, W.H. and Cho, K. 2015. Clean transfer of wafer-scale graphene via liquid phase removal of polycyclic aromatic hydrocarbons. ACS Nano. 9: 4726–4733. DOI: 10.1021/nn5066556.

Kroto, H.W., Heath, J.R., O'Brien, S.C., Curl, R.F. and Smalley, R.E. 1985. C60: Buckminsterfullerene. Nature 318: 162–163. DOI: 10.1038/318162a0.

Ladani, R.B., Wu, S., Kinloch, A.J., Ghorbani, K., Zhang, J., Mouritz, A.P. and Wang, C.H. 2015. Improving the toughness and electrical conductivity of epoxy nanocomposites by using aligned carbon nanofibres. Composites Science and Technology 117: 146–158. DOI: 10.1016/j.compscitech.2015.06.006.

Mishakov, I.V., Vedyagin, A.A., Bauman, Y.I., Buyanov, R.A. and Shubin, Y.V. 2018. Synthesis of carbon nanofibers via catalytic chemical vapor deposition of halogenated hydrocarbons. pp. 77–183. *In*: Lee, C.S. (ed.). Carbon Nanofibers: Synthesis, Applications and Performance. New York (United States): Nova Science. ISBN: 978-1536134339.

Lee, C., Lien, H., Wu, S., Doong, R. and Chao, C. 2014. Reduction of priority pollutants by nanoscale zerovalent iron in subsurface environments. pp. 63–97. *In*: Reisner, D.E. and Pradeep, T. (eds.). Aquananotechnology Global Prospects. Florida (United States): Taylor & Francis Group. ISBN: 978-1466512245.

Lee, C., Kim, Y.J., Yang, K.S., Kim, D.W., Kim, Y.A., Mishakov, I.V., Vedyagin, A.A., Bauman, Y.I., Buyanov, R.A. and Shubin, Y.V. 2018. Carbon Nanofibers: Synthesis, Applications and Performance [first edition]. United States: Nova Science, pp. 1–327. ISBN: 978-1-53613-433-9.

Leid, J.G., Ditto, A.J., Knapp, A., Shah, P.N., Wright, B.D., Blust, R., Christensen, L., Clemons, C.B., Wilber, J.P. and Young, G.W. 2012. *In vitro* antimicrobial studies of silver carbene complexes: Activity of free and nanoparticle carbene formulations against clinical isolates of pathogenic bacteria. Journal of Antimicrobial Chemotherapy 67: 138–148. DOI: 10.1093/jac/dkr408.

Li, H., Zheng, N., Liang, N., Zhang, D., Wu, M. and Pan, B. 2016. Adsorption mechanism of different organic chemicals on fluorinated carbon nanotubes. Chemosphere 154: 258–265. DOI: 10.1016/j.chemosphere.2016.03.099.

Li, S., Niu, Z., Zhong, X., Yang, H., Lei, Y., Zhang, F., Hu, W., Dong, Z., Jin, J. and Ma, J. 2012. Fabrication of magnetic ni nanoparticles functionalized water-soluble graphene sheets nanocomposites as sorbent for aromatic compounds removal. Journal of Hazardous Materials 229: 42–47. DOI: 10.1016/j.jhazmat.2012.05.053.

Li, Y., Du, Q., Liu, T., Sun, J., Jiao, Y., Xia, Y., Xia, L., Wang, Z., Zhang, W. and Wang, K. 2012. Equilibrium, kinetic and thermodynamic studies on the adsorption of phenol onto graphene. Materials Research Bulletin 47: 1898–1904. DOI: 10.1016/j.materresbull.2012.04.021.

Li, Z., Wang, R., Young, R.J., Deng, L., Yang, F., Hao, L., Jiao, W. and Liu, W. 2013. Control of the functionality of graphene oxide for its application in epoxy nanocomposites. Polymer 54: 6437–6446. DOI: 10.1016/j.polymer.2013.09.054.

Liu, F., Wu, Z., Wang, D., Yu, J., Jiang, X. and Chen, X. 2016. Magnetic porous silica–graphene oxide hybrid composite as a potential adsorbent for aqueous removal of p-nitrophenol. Colloids and Surfaces A: Physicochemical and Engineering Aspects 490: 207–214. DOI: 10.1016/j.colsurfa.2015.11.053.

Liu, T., Li, Y., Du, Q., Sun, J., Jiao, Y., Yang, G., Wang, Z., Xia, Y., Zhang, W. and Wang, K. 2012. Adsorption of methylene blue from aqueous solution by graphene. Colloids and Surfaces B: Biointerfaces 90: 197–203. DOI: 10.1016/j.colsurfb.2011.10.019.

Liu, Y., Chen, L., Li, Y., Wang, P. and Dong, Y. 2016. Synthesis of magnetic polyaniline/graphene oxide composites and their application in the efficient removal of Cu (II) from aqueous solutions. Journal of Environmental Chemical Engineering 4: 825–834. DOI: 10.1016/j.jece.2015.12.023.

Lofrano, G., Libralato, G. and Brown, J. 2017. Nanotechnologies for Environmental Remediation [first edition]. United States: Springer; pp. 1–332. DOI: 10.1007/978-3-319-53162-5_1.

Mallikarjunaiah, S., Pattabhiramaiah, M. and Metikurki, B. 2020. Application of nanotechnology in the bioremediation of heavy metals and wastewater management. pp. 297–321. *In*: Thangadurai, D., Sangeetha, J. and Prasad, R. (eds.). Nanotechnology for Food, Agriculture, and Environment. Cham (Switzerland): Springer. DOI: 10.1007/978-3-030-31938-0.

Mauter, M.S. and Elimelech, M. 2008. Environmental applications of carbon-based nanomaterials. Environmental Science and Technology 42: 5843–5859. DOI: 10.1021/es8006904.

Mesarič, T., Sepčič, K., Piazza, V., Gambardella, C., Garaventa, F., Drobne, D. and Faimali, M. 2013. Effects of nano carbon black and single-layer graphene oxide on settlement, survival and swimming behaviour of amphibalanus amphitrite larvae. Chemistry and Ecology 29: 643–652. DOI: 10.1080/02757540.2013.817563.

Meyyappan, M. 2016. Carbon nanotube-based chemical sensors. Small 12: 2118–2129. DOI: 10.1002/smll.201502555.

Mohamed, A. 2019. Synthesis, characterization, and applications carbon nanofibers. pp. 243–257. *In*: Yaragalla, S., Mishra, R., Thomas, S., Kalarikkal, N. and Maria, H.J. (eds.). Carbon-based Nanofillers and their Rubber Nanocomposites. Amsterdam (The Netherlands): Elsevier. DOI: 10.1016/B978-0-12–813248–7.00008-0.

Mohamed, A., Yousef, S., Abdelnaby, M.A., Osman, T.A., Hamawandi, B., Toprak, M.S., Muhammed, M. and Uheida A. 2017. Photocatalytic degradation of organic dyes and enhanced mechanical properties of PAN/CNTs composite nanofibers. Separation and Purification Technology 182: 219–223. DOI: 10.1016/j.seppur.2017.03.051.

Mohan, V.B., Lau, K., Hui, D. and Bhattacharyya, D. 2018. Graphene-based materials and their composites: A review on production, applications and product limitations. Composites Part B: Engineering 142: 200–220. DOI: 10.1016/j.compositesb.2018.01.013.

Nakagawa, T., Kokubo, K. and Moriwaki, H. 2014. Application of fullerenes-extracted soot modified with ethylenediamine as a novel adsorbent of hexavalent chromium in water. Journal of Environmental Chemical Engineering 2: 1191–1198. DOI: 10.1016/j.jece.2014.05.003.

Ncibi, M.C., Gaspard, S. and Sillanpää, M. 2015. As-synthesized multi-walled carbon nanotubes for the removal of ionic and non-ionic surfactants. Journal of Hazardous Materials 286: 195–203. DOI: 10.1016/j.jhazmat.2014.12.039.

Ncibi, M.C. and Sillanpää, M. 2015. Optimized removal of antibiotic drugs from aqueous solutions using single, double and multi-walled carbon nanotubes. Journal of Hazardous Materials 298: 102–110. DOI: 10.1016/j.jhazmat.2015.05.025.

Notarianni, M., Liu, J., Vernon, K. and Motta, N. 2016. Synthesis and applications of carbon nanomaterials for energy generation and storage. Beilstein Journal of Nanotechnology 7: 149–196. DOI: 10.3762/bjnano.7.17.

Oyetade, O.A., Nyamori, V.O., Martincigh, B.S. and Jonnalagadda, S.B. 2015. Effectiveness of carbon nanotube–cobalt ferrite nanocomposites for the adsorption of rhodamine B from aqueous solutions. RSC Advances 29: 22724–22739. DOI: 10.1039/C4RA15446K.

Palmeri, M.J., Putz, K.W., Ramanathan, T. and Brinson, L.C. 2011. Multi-scale reinforcement of CFRPs using carbon nanofibers. Composites Science and Technology 71: 79–86. DOI: 10.1016/j.compscitech.2010.10.006.

Pan, B. and Xing, B. 2008. Adsorption mechanisms of organic chemicals on carbon nanotubes. Environmental Science & Technology 42: 9005–9013. DOI: 10.1021/es801777n.

Poveda, R.L. and Gupta, N. 2016. Carbon nanofibers: Structure and fabrication. pp. 11–26. *In*: Carbon Nanofiber Reinforced Polymer Composites. Cham (Switzerland): Springer. DOI: 10.1007/978-3–319–23787-9_2.

Pretti, C., Oliva, M., Di Pietro, R., Monni, G., Cevasco, G., Chiellini, F., Pomelli, C. and Chiappe, C. 2014. Ecotoxicity of pristine graphene to marine organisms. Ecotoxicology and Environmental Safety 101: 138–145. DOI: 10.1016/j.ecoenv.2013.11.008.

Pumera, M. 2010. Graphene-based nanomaterials and their electrochemistry. Chemical Society Reviews 39: 4146–4157. DOI: 10.1039/C002690P.

Rizwan M., Singh, M., Mitra, C.K. and Morve, R.K. 2014. Ecofriendly application of nanomaterials: Nanobioremediation. Journal of Nanoparticles 2014: 1–7. DOI: 10.1155/2014/431787.

Russier, J., Ménard-Moyon, C., Venturelli, E., Gravel, E., Marcolongo, G., Meneghetti, M., Doris, E. and Bianco A. 2011. Oxidative biodegradation of single-and multi-walled carbon nanotubes. Nanoscale 3: 893–896. DOI: 10.1039/c0nr00779j.

Scalese, S., Nicotera, I., D'Angelo, D., Filice, S., Libertino, S., Simari, C., Dimos, K. and Privitera, V. 2016. Cationic and anionic azo-dye removal from water by sulfonated graphene oxide nanosheets in nafion membranes. New Journal of Chemistry 40: 3654–3663. DOI: 10.1039/C5NJ03096J.

Schaefer, K., Thomas, H., Dalton, P. and Moeller, M. 2007. Nano-fibres for filter materials. pp. 125–138. *In*: Duquesne, S., Magniez, C. and Camino, G. (eds.). Multifunctional Barriers for Flexible Structure. Berlin (Germany): Springer. DOI: 10.1007/978-3-540-71920-5_7.

Seredych, M. and Bandosz, T.J. 2012. Manganese oxide and graphite oxide/MnO_2 composites as reactive adsorbents of ammonia at ambient conditions. Microporous and Mesoporous Materials 150: 55–63. DOI: 10.1016/j.micromeso.2011.09.010.

Shan, D., Deng, S., Zhao, T., Yu, G., Winglee, J. and Wiesner, M.R. 2016. Preparation of regenerable granular carbon nanotubes by a simple heating-filtration method for efficient removal of typical pharmaceuticals. Chemical Engineering Journal 294: 353–361. DOI: 10.1016/j.cej.2016.02.118.

Sharma, M., Madras, G. and Bose, S. 2015. Unique nanoporous antibacterial membranes derived through crystallization induced phase separation in PVDF/PMMA blends. Journal of Materials Chemistry A 11: 5991–6003. DOI: 10.1039/C5TA00237K.

Shi, H., Liu, H., Luan, S., Shi, D., Yan, S., Liu, C., Li, R.K. and Yin, J. 2016. Effect of polyethylene glycol on the antibacterial properties of polyurethane/carbon nanotube electrospun nanofibers. RSC Advances 23: 19238–19244. DOI: 10.1039/C6RA00363J.

Simon-Deckers, A., Loo, S., Mayne-L'hermite, M., Herlin-Boime, N., Menguy, N., Reynaud, C., Gouget, B. and Carriere, M. 2009. Size-, composition-and shape-dependent toxicological impact of metal oxide nanoparticles and carbon nanotubes toward bacteria. Environmental Science & Technology 43: 8423–8429. DOI: 10.1021/es9016975.

Singh, R.L. and Singh, P.K. 2017. Global environmental problems. pp. 13–41. *In*: Sing, R.L. (ed.). Principles and Applications of Environmental Biotechnology for a Sustainable Future. Singapore (Singapore): Springer. DOI: 10.1007/978-981-10-1866-4_2.

Sitko, R., Janik, P., Zawisza, B., Talik, E., Margui, E. and Queralt, I. 2015. Green approach for ultratrace determination of divalent metal ions and arsenic species using total-reflection X-ray fluorescence spectrometry and mercapto-modified graphene oxide nanosheets as a novel adsorbent. Analytical Chemistry 87: 3535–3542. DOI: 10.1021/acs.analchem.5b00283.

Sohn, E.K., Chung, Y.S., Johari, S.A., Kim, T.G., Kim, J.K., Lee, J.H., Lee, Y.H., Kang, S.W. and Yu, I.J. 2015. Acute toxicity comparison of single-walled carbon nanotubes in various freshwater organisms. BioMed. Research International 2015: 1–7. DOI: 10.1155/2015/323090.

Sui, Z., Meng, Q., Zhang, X., Ma, R. and Cao, B. 2012. Green synthesis of carbon nanotube–graphene hybrid aerogels and their use as versatile agents for water purification. Journal of Materials Chemistry 22: 8767–8771. DOI: 10.1039/C2JM00055E.

Taha, M.R. and Mobasser, S. 2015. Adsorption of DDT and PCB by nanomaterials from residual soil. PLoS One 10: 1–16. DOI: 10.1371/journal.pone.0144071.

Tang, Y., Guo, H., Xiao, L., Yu, S., Gao, N. and Wang, Y. 2013. Synthesis of reduced graphene oxide/magnetite composites and investigation of their adsorption performance of fluoroquinolone antibiotics. Colloids and Surfaces A: Physicochemical and Engineering Aspects 424: 74–80. DOI: 10.1016/j.colsurfa.2013.02.030.

Tao, X., He, Y., Zhang, B., Chen, Y. and Hughes, J.B. 2011. Effects of stable aqueous fullerene nanocrystal (nC60) on daphnia magna: Evaluation of hop frequency and accumulations under different conditions. Journal of Environmental Sciences 23: 322–329. DOI: 10.1016/s1001-0742(10)60409-3.

Velzeboer, I., Peeters, E. and Koelmans, A.A. 2013. Multiwalled carbon nanotubes at environmentally relevant concentrations affect the composition of benthic communities. Environmental Science & Technology 47: 7475–7482. DOI: 10.1021/es400777j.

Vidu, R., Rahman, M., Mahmoudi, M., Enachescu, M., Poteca, T.D. and Opris, I. 2014. Nanostructures: A platform for brain repair and augmentation. Frontiers in Systems Neuroscience 8: 91–115. DOI: 10.3389/fnsys.2014.00091.

Wang, H., Yuan, X., Wu, Y., Chen, X., Leng, L., Wang, H., Li, H. and Zeng, G. 2015. Facile synthesis of polypyrrole decorated reduced graphene oxide–Fe_3O_4 magnetic composites and its application for the Cr (VI) removal. Chemical Engineering Journal 262: 597–606. DOI: 10.1016/j.cej.2014.10.020.

Wang, H., Chen, Y. and Wei, Y. 2016. A novel magnetic calcium silicate/graphene oxide composite material for selective adsorption of acridine orange from aqueous solutions. RSC Advances 41: 34770–34781. DOI: 10.1039/C6RA07625D.

Wang, S., Sun, H., Ang, H. and Tadé, M.O. 2013. Adsorptive remediation of environmental pollutants using novel graphene-based nanomaterials. Chemical Engineering Journal 226: 336–347. DOI: 10.1016/j.cej.2013.04.070.

Wang, Y., Mo, Z., Zhang, P., Zhang, C., Han, L., Guo, R., Gou, H., Wei, X. and Hu, R. 2016. Synthesis of flower-like TiO_2 microsphere/graphene composite for removal of organic dye from water. Materials and Design 99: 378–388. DOI: 10.1016/j.matdes.2016.03.066.

Wong, K.T., Yoon, Y., Snyder, S.A. and Jang, M. 2016. Phenyl-functionalized magnetic palm-based powdered activated carbon for the effective removal of selected pharmaceutical and endocrine-disruptive compounds. Chemosphere 152: 71–80. DOI: 10.1016/j.chemosphere.2016.02.090.

Xie, W., Vu, K., Yang, G., Tawfiq, K. and Chen, G. 2014. Escherichia coli growth and transport in the presence of nanosilver under variable growth conditions. Environmental Technology 35: 2306–2313. DOI: 10.1080/09593330.2014.902112.

Xu, L., Qi, X., Li, X., Bai, Y. and Liu, H. 2016. Recent advances in applications of nanomaterials for sample preparation. Talanta 146: 714–726. DOI: 10.1016/j.talanta.2015.06.036.

Yadav, K.K., Singh, J.K., Gupta, N. and Kumar, V. 2017. A review of nanobioremediation technologies for environmental cleanup: A novel biological approach. J. Mater. Environ. Sci. 8: 740–757. ISSN: 2028–2508.

Yamaguchi, N.U., Bergamasco, R. and Hamoudi, S. 2016. Magnetic $MnFe_2O_4$-graphene hybrid composite for efficient removal of glyphosate from water. Chemical Engineering Journal 295: 391–402. DOI: 10.1016/j.cej.2016.03.051.

Yan, H., Du, Q., Yang, H., Li, A. and Cheng, R. 2016. Efficient removal of chlorophenols from water with a magnetic reduced graphene oxide composite. Science China Chemistry 59: 350–359. DOI: 10.1007/s11426-015-5482-y.

Ying-Ying, W. and Zhen-Hu, X. 2016. Multi-walled carbon nanotubes and powder-activated carbon adsorbents for the removal of nitrofurazone from aqueous solution. Journal of Dispersion Science and Technology 37: 613–624. DOI: 10.1080/01932691.2014.981337.

Yu, F., Ma, J. and Bi, D. 2015. Enhanced adsorptive removal of selected pharmaceutical antibiotics from aqueous solution by activated graphene. Environmental Science and Pollution Research 22: 4715–4724. DOI: 10.1007/s11356-014-3723-9.

Yu, F., Ma, J., Wang, J., Zhang, M. and Zheng, J. 2016. Magnetic iron oxide nanoparticles functionalized multi-walled carbon nanotubes for toluene, ethylbenzene and xylene removal from aqueous solution. Chemosphere 146: 162–172. DOI: 10.1016/j.chemosphere.2015.12.018.

Zhang, L., Aboagye, A., Kelkar, A., Lai, C. and Fong, H. 2014. A review: Carbon nanofibers from electrospun polyacrylonitrile and their applications. Journal of Materials Science 49: 463–480. DOI: 10.1007/s10853-013-7705-y.

Zhang, N., Yang, M., Tang, Z. and Xu, Y. 2013. CdS-graphene nanocomposites as visible light photocatalyst for redox reactions in water: A green route for selective transformation and environmental remediation. Journal of Catalysis 303: 60–69. DOI: 10.1016/j.jcat.2013.02.026.

Zhang, X., Coleman, A.C., Katsonis, N., Browne, W.R., Van Wees, B.J. and Feringa, B.L. 2010. Dispersion of graphene in ethanol using a simple solvent exchange method. Chemical Communications 46: 7539–7541. DOI: 10.1039/C0CC02688C.

Zhang, Y., Cheng, Y., Chen, N., Zhou, Y., Li, B., Gu, W., Shi, X. and Xian, Y. 2014. Recyclable removal of bisphenol A from aqueous solution by reduced graphene oxide–magnetic nanoparticles: Adsorption and desorption. Journal of Colloid and Interface Science 421: 85–92. DOI: 10.1016/j.jcis.2014.01.022.

Zhao, G., Ren, X., Gao, X., Tan, X., Li, J., Chen, C., Huang, Y. and Wang, X. 2011. Removal of Pb (II) ions from aqueous solutions on few-layered graphene oxide nanosheets. Dalton Transactions 40: 10945–10952. DOI: 10.1039/C1DT11005E.

Zhao, J., Wang, Z., Zhao, Q. and Xing, B. 2014. Adsorption of phenanthrene on multilayer graphene as affected by surfactant and exfoliation. Environmental Science and Technology 48: 331–339. DOI: 10.1021/es403873r.

Zhu, J., Wei, S., Gu, H., Rapole, S.B., Wang, Q., Luo, Z., Haldolaarachchige, N., Young, D.P. and Guo, Z. 2012. One-pot synthesis of magnetic graphene nanocomposites decorated with core@ double-shell nanoparticles for fast chromium removal. Environmental Science and Technology 46: 977–985. DOI: 10.1021/es2014133.

CHAPTER 9

Green Synthesis of Nanomaterials and their use in Bio-and Nanoremediation

Manuel Palencia[1,*] and *Angélica García–Quintero*[2]

1. Introduction

Since the 19th century, humanity has witnessed constant demographic, industrial and urban growth, which has led to an intensification in the extraction of natural resources, as well as in the use of anthropogenic substances for the generation of efficient production mechanisms (Cecchin et al., 2017; Haider et al., 2015; Livingston et al., 2020). These processes have led to economic growth and contributed to the access of food and essential products for human communities (Haider et al., 2015). However, the lack of a specific strategy, policy and regulations, regarding the final dispositions and environmental considerations of the reagents, products and by-products, have led to annually more than three million metric tons of toxic substances, associated with the generation of public and environmental health problems, being released into the environment (Cecchin et al., 2017). In this way, a remarkable proportion of contemporary soils and aquifers have a wide variety of polluting substances, both natural and anthropogenic, ranging from heavy metals, hydrocarbons and agrochemicals to pharmaceuticals and personal care products (Mehdinia and Mehrabi, 2019). It is now widely recognized that these substances have generated a devastating impact on the aquatic and terrestrial ecosystems, leading to the degradation of their physical, chemical, biological and ecosystem properties, to the point that the new developments resulting from all this, become an imminent risk to human health in many parts of the world, since these substances can not only bioaccumulate, biomagnify and generate various diseases, but also their negative effects can be transferred to future generations (Bhandari 2018; Hasan et al., 2020; Iavicoli et al., 2017; Melendez and Polanski 2020).

[1] Research Group in Science with Technological Applications (GI-CAT), Department of Chemistry, Faculty of Natural and Exact Sciences, Universidad del Valle, Cali – Colombia.

[2] Research group in Science with Technological Applications (GI-CAT), Department of Chemistry, Faculty of Natural and Exact Sciences, Universidad del Valle, Cali - Colombia. Mindtech Research Group (Mindtech-RG), Mindtech s.a.s., Cali - Colombia.

Email: angelica.garcia.quintero@correounivalle.edu.co

* Corresponding author: manuel.palencia@correounivalle.edu.co

There is a need to develop efficient remediation practices for natural systems, which should be carried out preferably in shorter periods, by using environmentally friendly technologies and characterized by reducing the potential adverse environmental impacts of the remediation processes (Bhandari, 2018; Hou and O'Connor, 2020). In this context, the use of green nanomaterials for remediation processes has emerged as a promising alternative because these substances can be manufactured in the framework of sustainability and the Life Cycle Analysis defining by green chemistry philosophy (Pandey, 2018). At the same time, due to their nanoscale physicochemical properties, they can decrease the concentration of a wide variety of pollutants, through different mechanisms including adsorption, oxidative and catalytic processes, among others (Shah et al., 2015). That is why, in recent years, research related to the use of these nanotechnological systems in environmental remediation has increased, both individually and in a hybrid way, i.e., through processes including bioremediation and nanoremediation, sequential configurations and/or in metabolic bio-stimulation (Cecchin et al., 2017; Iavicoli et al., 2017; Singh et al., 2020).

With the purpose of gaining wider knowledge in this emerging area in environmental remediation field, this chapter is structured around three main areas, the first one is related to general aspects of contamination of soils and water and environmental remediation processes, emphasizing in bioremediation, nanoremediation and bionanoremediation techniques. The second is related to green synthesis of nanomaterials to be used in sustainable remediation. In such a way that the following thematic axis highlights research carried out on the use of green nanomaterials in reducing the concentration of pollutants, both in individual (nanoremediation) and hybrid processes based on biological agents (nanobioremediation), and in turn, highlights the limitations and perspectives related to this field.

2. Pollution and Environmental Remediation

2.1 General context of environmental pollution in soil and groundwater

The increasing industrialization of human societies has been intrinsically linked with higher production and use of dangerous substances with potential toxicity, as well as the massive use and dumping of substances that are in principle innocuous when they are in low concentrations. Nevertheless, in many countries, a few production systems lack specific regulations and appropriate supervision, making it difficult to protect natural systems and establishing sustainable operations (Saxena and Bharagava, 2017; Yadav et al., 2018). This situation can be seen in the report on the development of water in the world since 2017, which estimated that around 75% of industrial discharges are carried out without do any treatment against potential pollutants. As a consequence, the presence of polluting substances in natural environments has become one of the most significant public and environmental health problems, because the concentrations and complexity of contamination have increased over time due to the development of new industrial technologies and processes, the increase in the population and the incorporation of excessive consumption habits (Bhandari, 2018; Das et al., 2015; Saxena and Bharagava, 2017; Yadav et al., 2018). Figure 9.1 illustrates some of the different types of pollutants present in natural water and soil systems. These dangerous substances can be classified according to their structural characteristics and chemical nature as organic and inorganic pollutants.

- Organic pollutants: this category is made up of chemical substances whose structure is mainly composed of carbon and hydrogen atoms. These chemical compounds can present a wide variety of functional groups with the presence of aromatic segments, unsaturations and/or heteroatoms (Isac-García et al., 2016). In addition, their structural diversity allows notable differences in their physicochemical properties, toxicity and stability, as they are used in a wide range of applications such as dyes, industrial reagents, pharmaceutical and agrochemical products. Nevertheless, these compounds have been associated with harmful effects in humans, such as neurological, hepatic and renal diseases, as well as environmental effects, ranging from the

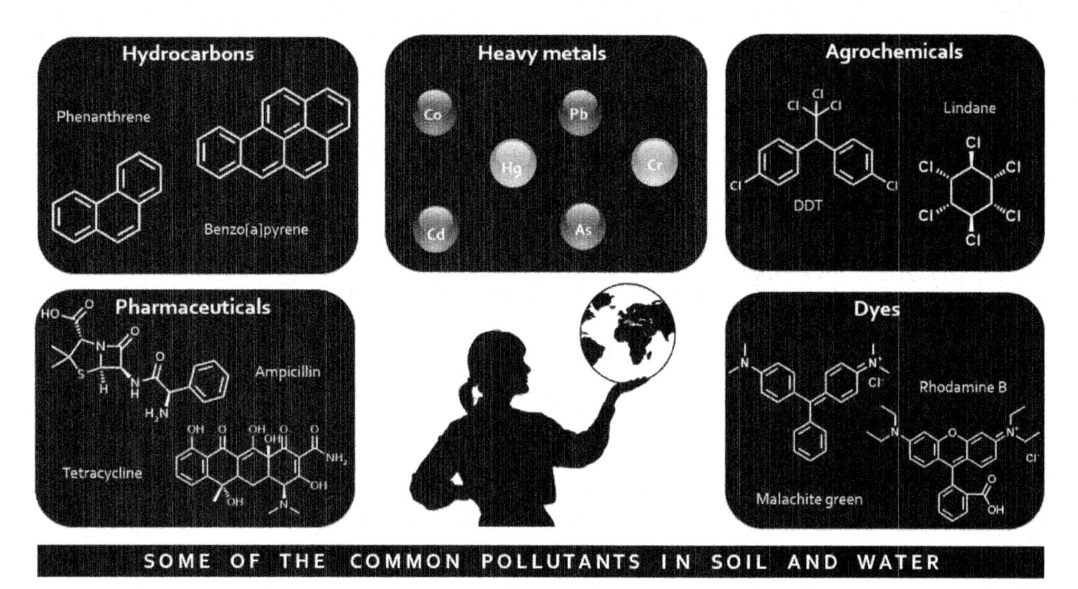

Fig. 9.1. Some types of pollutants present in soil and groundwater: Hydrocarbons, heavy metals, agrochemicals, pharmaceuticals and dyes.

alteration of microbial and animal populations to damages in the photosynthetic process, leading to a detriment in the physiological development of plants and the ecosystem functionality of natural systems (Haider et al., 2015; Saxena and Bharagava, 2017).

- Inorganic pollutants: these substances include inorganic salts and heavy metals, mainly from mining activities, electronic products and agrochemicals (Fajardo et al., 2020). However, heavy metals are the most critical inorganic pollutants in the environment, due to their high toxicity at low concentrations and the fact that, in many cases, are present as elemental entities that cannot undergo degradation processes (Haider et al., 2015; Singh and Singh, 2017). In some cases, high-toxicity chemical speciation is related with the formation of organometallic compounds, e.g., methyl mercury. On the other hand, heavy metals can interact with cellular components and generate Reactive Oxygen Species (ROS) producing the peroxidation of the unsaturated segments of the phospholipids of the cell membrane, the displacement of active centers of biomolecules and the alteration of genetic and proteinic material (Haider et al., 2015; Kanwar et al., 2017).

A wide range of pollutants is currently present in water bodies and soils; however, since soils are not a fluid phase, conventional remediation techniques used for aqueous systems are not applicable or easily adapted to them. In particular, in soils the behavior of pollutants is markedly more complex because soils are open, dynamic and complex systems, with high spatial and temporal variability (Brevik et al., 2020; Dror et al., 2017). As a result for remediation purposes, soils represent a challenge yet to be resolved, therefore, some inherent aspects associated with the dynamics of pollutants in soils will be examined. The first point to highlight is that soils can act as reservoirs of high levels of contamination as a consequence of their disturbance by both anthropogenic and natural factors (Combatt et al., 2017; Hulisz et al., 2017), as is the case of sulfate acid soils that have been of the most documented (Hulisz et al., 2017). Thus, it is well-known that soils are natural reservoirs for inactivating polluting species as a result of their chemical speciation. In soils, for example, organic components can interact with pollutants by electrostatic interactions with ionic groups, by dispersion interactions with hydrophobic segments, establish ion-dipole interactions with polar groups or generate chemical complexes between metals and ligands (Dror et al., 2017). While inorganic components, primarily clays can present electrostatic interactions and adsorption processes on the surface and interlaminar, altering their microstructure (Dror et al., 2017). Chemical and structural changes in soils affect pollutant's bioavailability, leading to variations in cation exchange properties, the polarity of components, retention of solutes, affect the microbial population and transport of water and nutrients, among others (Dror et al., 2017; Kanwar et al., 2017). In addition, pollutants can also be mobilized through the soil horizons hydrologically by the percolation of the soil solution, runoff and erosion, generating water bodies' contamination (Garcés et al., 2018; Hasan et al., 2020; Lerma et al., 2018; Sohail et al., 2019; Tripathi et al., 2017).

2.2 Environmental remediation

Environmental remediation refers to the processes and techniques used to reduce the concentration of a pollutant to safer levels for planetary health (Speight, 2020). A wide variety of technological processes have been designed using, individually, sequentially or simultaneously, three main mechanisms: (i) extraction of the contaminated phase, (ii) chemical transformation of the contaminant, and (iii) contaminant immobilization using adsorbents (Zhang et al., 2019). The optimal choice of the remediation strategy depends on the hydrological, geological, climatic and geographical conditions of the affected area; of economic, governmental and operational factors; as well as the concentration, nature and availability of the pollutant (Azubuike et al., 2016; Bharagava et al., 2017; Singh and Singh, 2017; Vasudevan and Odukkathil, 2017). The previously mentioned factors were analyzed to determine the place of application of the remediation strategy (i.e., by *in situ* or *ex situ* approach) (Azubuike et al., 2016). In the *in situ* processes, it is not necessary to extract the area affected by contamination, in the case of soil, this avoid the disturbance of the structure and porosity; however, this remediation requires a more significant amount of time, implying the risk of mobility of pollutants and monitoring the remediation process is complicated with limited veracity (Bharagava et al., 2017). While in *ex situ* processes the extraction of the affected area is required, causing disturbances in the contaminated zone; nevertheless, it allows in carrying out more significant control over the variables of the procedure and its monitoring (Bharagava et al., 2017).

In addition, environmental remediation can also be classified according to the nature of the treatment as physical, chemical and biological remediation (Azubuike et al., 2016; Speight, 2020). However, it is important to note that this classification does not include mixed remediation processes resulting from sequential or mechanistically combined configurations. The physical remediation is based on the use of the pollutant's physical properties to carry out its extraction from a natural matrix. Some examples are vapor-phase extraction of volatile pollutants, pollutant extraction by soil washing using solvents, adsorption processes based on the affinity of the adsorbent and adsorbate or pollutant, among others (Azubuike et al., 2016; Speight, 2020). On the other hand, chemical remediation is based on the use of agents to transform pollutants into innocuous or easily removable substances. Typical chemical reactions include oxidation processes, chemical precipitation mediated by pH or by interaction with chelating agents, photodegradation reactions, among others (Azubuike et al., 2016; Speight, 2020). Table 9.1 lists some of the most common physical and chemical remediations and their target contaminants.

The disadvantages of physical and chemical remediation are high operating costs, energy intensity, contamination transfer, use and generation (in some cases) of dangerous substances with potential toxicity (Bharadwaj, 2018; Khan, 2020). It should be noted that, if chemical remediation cannot efficiently reduce the concentration of pollution in a targeted area, i.e., for example, without generating secondary pollutant substances, this means that the remediation is not, in an integrated way, environmentally friendly, and on the contrary produces an anthropogenic disturbance on ecosystems (Bhandari, 2018; Hou and O'Connor, 2020). On the other hand, bioremediation is the use of biological mechanisms present in microorganisms and plants to reduce the pollutant concentrations, thus, this strategy is recognized to be most in concordance with sustainability principles, because it allows the use of substances of natural origin, with low danger, and in most cases, high selectivity, low economic impact, efficient use of resources and a positive complete impact, making it the remediation process with the highest public acceptance (Hou and O'Connor, 2020; Koduru et al., 2017).

Table 9.1, Typical physical and chemical remediations, as well as their target pollutants.

Process	Target Pollutants	References
	Physical Remediation	
Soil vapor extraction	Volatile hydrocarbons like toluene, benzene, and ethylbenzene.	Yang et al., 2017
Surfactant flushing	Halogenated hydrocarbons like dichlorobenzenes and hexachlorocyclohexanes.	Dominguez et al., 2019; Pei et al., 2017
Pump and Treat	Metals, chemical compounds dissolved in the soil's solution, volatile and semi-volatile organic compounds like petroleum oil derivates.	Ossai et al., 2020
Adsorption	Heavy metals, metalloids, organic contaminants like tetracycline, ethylbenzene, Methyl orange and Bisphenol A.	Awad et al., 2020; Zhang et al., 2019
	Chemical Remediation	
Electrochemical and redox reactions	Heavy metals ions, organic pollutants like hexachlorobenzene, pyrene, and atrazine.	Andrade and dos Santos, 2020; Gulati et al., 2018; Zhang et al., 2019
Catalytic degradation	Heavy metal ions, chlorinated organic compounds like pesticides and polychlorinated biphenyls.	Andrade and dos Santos, 2020; Bhandari 2018
Precipitation reactions	Heavy metals.	Gulati et al., 2018; Speight 2020

However, bioremediation shows notable disadvantages against persistent and xenobiotic pollutants; since these tend to present structures that are not recognized for biological mechanisms responsible of biodegradation and bioaccumulation; for this bioremediation is not an efficient alternative to a wide variety of emerging pollutants (Bharadwaj, 2018). In this context the use of nanomaterials synthesized within the framework of green chemistry emerges as a promising alternative to carry out sustainable and efficient processes of environmental remediation against persistent pollutants (Zhang et al., 2019; Ortega-Calvo et al., 2016).

2.2.1 Bioremediation

Bioremediation consists of reducing the pollutant concentration to safe levels by biological pathways, which can be carried out through mechanisms such as biodegradation, biomineralization, biotransformation, bioadsorption, bioleaching and bioaccumulation (Bharagava et al., 2017; Kanwar et al., 2017; Mehdinia and Mehrabi, 2019; Tripathi et al., 2018). These processes can be carried out by various unicellular and multicellular organisms, like plants (phytoremediation), algae (phycoremediation), fungi (mycoremediation) and bacteria (bacterial bioremediation) (Bharagava et al., 2017; Dror et al., 2017; Tripathi et al., 2018).

On the other hand, bioremediation techniques are based primarily on two approaches: bioaugmentation and biostimulation (Azubuike et al., 2016). Biostimulation is related to the promotion of environmental and nutritional conditions allowing an adequate growth and development of the autochthonous organisms responsible for bioremediation, while bioaugmentation refers to the incorporation of remedial organisms in the contaminated matrix (Azubuike et al., 2016). Nevertheless, the addition of exogenous organisms can generate competition processes with autochthonous species in *in situ* remediation (Tripathi et al., 2018). In this way, the most common *in situ* bioremediation technologies use the controlled stimulation of autochthonous microorganisms through oxygen flow and the addition of nutrients, as in bioventing and bioslurping processes; however, these require the use of aeration streams, which generate operational costs and intensive energy use (Azubuike et al., 2016). In the case of *ex situ* bioremediations, the control of operational parameters of remediation is possible allowing to verify the efficiency of the remediation in real-time; the most used *ex situ* processes are windrow, biopiles and bioreactors (Azubuike et al., 2016; Umadevi et al., 2017).

In the case of bioremediations conducted out in soils and water, the most widely used biological systems are plants and microorganisms by strategies such as bioaccumulation, bioabsorption and

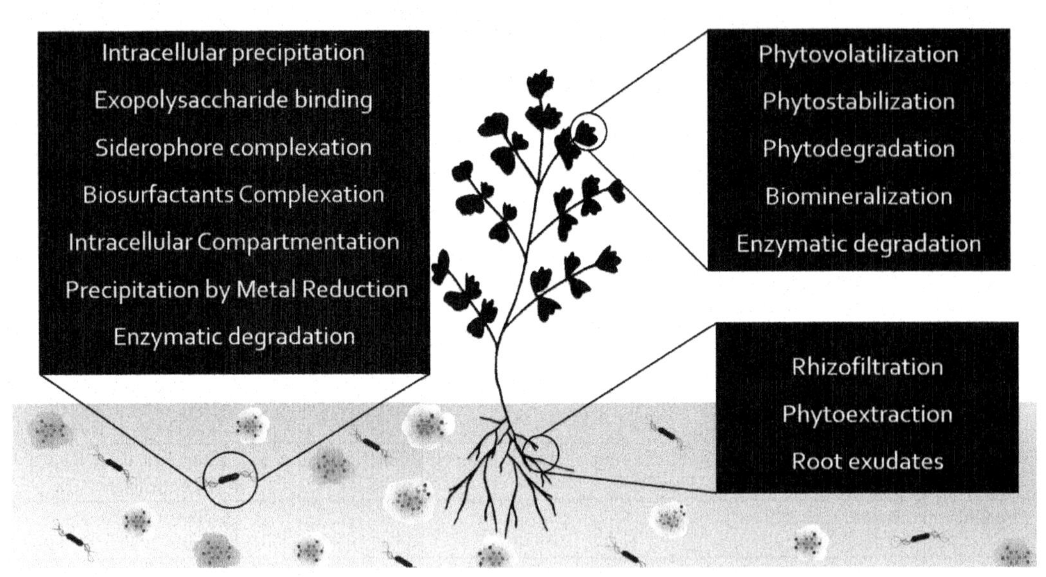

Fig. 9.2. Primary strategies for bioremediation of plants and microorganisms.

biodegradation through enzymatic processes, oxidative reactions by laccases present in plants such as *Toxicodendron vernicifluum* and soil microorganisms such as *Azospirillum lipoferum* are some examples. The laccase enzymes allow the generation of free radicals through monoelectronic transfers, generating the formation of less toxic substances from transformation of various aromatic compounds (Arregui et al., 2019; Madhava et al., 2018). However, due to their genotypic differences, microorganisms and plants allow different bioremediation strategies to be carried out, as illustrated in Fig. 9.2.

The strategies used by plants to carry out environmental remediation processes are based on the interactions between the rhizosphere, pollutants, root exudates and organic and inorganic components of soil and water determining the bioavailability of polluting substances; physical or chemical adsorption, transfer and the compartmentalization being the most common mechanisms (Bharagava et al., 2017; Tripathi et al., 2018). It is notable that phytoremediation must take into account plant-rhizosphere-microorganism interactions, since these can promote synergistic effects between plants and remediation microorganisms through the segregation of exudates that favor the microbial colonization and development processes like phytohormone production and nitrogen fixing (Singh and Singh, 2017). In recent years, the plant-microorganism synergistic relationship has been widely studied for developing phytoremediation with plants and crops. Thus, by the generation of biodegradable growth substrates for nitrogen-fixing bacteria, such as *Azotobacter crococum*, and matrices with the ability to release bioactive substances for plants as auxin-type phytohormones (Garcés et al., 2017; Palencia et al., 2017; Palencia et al., 2020; Singh and Singh, 2017).

The wide variety of bioremediation strategies allows efficient control of remediation, against various pollutants in a considerable amount of time (weeks or months). However, this process cannot be useful when the accumulation rate is higher than the organism's remediation rate, the structure of the pollutant makes it recalcitrant, and when the concentration of the pollutant is higher than live system tolerance (Azubuike et al., 2016; Cecchin et al., 2017; Davies et al., 2017; Kanwar et al., 2017). On the other hand, process efficiency is limited by the pollutant's bioavailability, the selectivity of the biological system, the physicochemical conditions of the matrix, the interactions between the pollutant and the bioremediation agent, the rate of mass transfer, as well as the depth and accessibility of the remediation agent to the contaminant (Davies et al., 2017; Kanwar et al., 2017; Umadevi et al., 2017).

In some cases, the metabolic degradation of pollutants results in the formation of dangerous metabolites; for example, the partial degradation of DDT produce dichlorodiphenyldichloroethane and dichlorodiphenyldichloroethylene, which are toxic and persistent substances (Tarannum and Khan, 2020; Vasudevan and Odukkathil, 2017). In consequence, it is clear that bioremediation is not a viable or efficient alternative for some kinds of pollutants (Zhang et al., 2019).

2.2.2 Nanoremediation

Nanoremediation consists of the use of nanomaterials to reduce the pollutant concentration to safe levels (Corsi et al., 2018; Iavicoli et al., 2017). Nanoremediation is directly associated with the technological potential of nanomaterials (Corsi et al., 2018). At this point, it is relevant to develop the concept of nanomaterial, their classes and properties, because these are directly responsible for remediation based on nanomaterials being a technological innovation that allows carrying out more economical decontamination processes, in less time, more sustainable and efficient compared with traditional remediation strategies (Corsi et al., 2018; Khan, 2020).

Nanomaterials are the result of manipulating matter at the nanoscale by nanoscience, which was a term introduced, in conjunction with the concept of nanotechnology, in 1974 by a professor at University of Tokyo, Norio Taniguchi. However, whereas nanoscience is focused on the study, design, synthesis, manipulation and characterization of materials that have their dimensions or structure, external or internal, at the nanoscale, nanotechnology is the use of nanometric properties in some technological applications, understanding as a nanometric property those properties that emerge as a result of size in the nanoscale (1–100 nm in at least one dimension), and through which the nanomaterial can be differentiated from its counterpart on a larger scale (Halada and Orlov, 2018; Khan, 2020; Kolahalam et al., 2019; Mohammadi et al., 2019; Rahman et al., 2020; Sohail et al., 2019; Villalobos et al., 2019; Zelić et al., 2018). Thus, the wide variety of nanomaterials applications is directly related to their size due to their high surface area triggering a greater preponderance of the surface phenomenon compared to the surface bulk phenomenon, the generation of quantum electronic confinement determining their electrical and optical properties, the diffusion capacity in structures that are not easily accessible to macroscopic materials and potential intimate interaction with biomolecules (Davies et al., 2017; González-Ballesteros et al., 2020; Shah et al., 2015; Sohail et al., 2019; Villalobos et al., 2019).

Nanomaterials are usually classified according to their chemical nature (i.e., organic, inorganic, polymeric or hybrid nanomaterials), shape (i.e., nanoparticles (NPs), nanorods, nanowires, quantum dots, among others) or the number of dimensions at the nanoscale (i.e., nano-objects and nanostructures) (Kolahalam et al., 2019; Livingston et al., 2020; López-Villareal et al., 2019; Rahman et al., 2020). In particular, nano-objects present some of their external dimensions on nanometric level and are sub-classified to be 0D when they present three dimensions on the nanoscale (e.g., nanoparticles); 1D when they present two dimensions at the nanoscale (e.g., nanotubes and nanowires); 2D when they have one dimension at the nanoscale (e.g., exfoliated clay layers and nanometric thin films) (Afreen et al., 2018; Kolahalam et al., 2019; López-Villareal et al., 2019; Rahman et al., 2020). On other hand, 3D nanomaterials are macroscopic materials composed of nano-objects, that is, their internal structure contains nanomaterials (i.e., nanocomposites) (Afreen et al., 2018; Devatha and Thala, 2018; Lerma et al., 2018; Sohail et al., 2019; Villalobos et al., 2019).

Remediation processes through the use of nanomaterials can be carried out through different mechanisms, including:

- Adsorption: which is a physical process allows the extraction of pollutants from natural systems by the use of interfacial properties due to the large surface area of nanomaterials, their high reactivity and adsorption kinetics. However, the effectiveness of this process is conditioned by the nature of the adsorbent and the adsorbate, the availability of the pollutant and the interactions between both substances; in addition, it can be favored by nanomaterial surface modification by chemical treatments, insertion of polymer chains or ligands (Yadav et al., 2018). An example of one of the many nanomaterials with pollutant adsorption capacity is found in the research performed by Ramanayaka et al. (2020) who studied the pollutant removal from aqueous solutions using adsorptive processes. Contaminant models were the antibiotic oxytetracycline, the pesticide Glyphosate, and the heavy metals Cr (VI) and Cd (II), which were adsorbed on the surface of a graphitic nanobiochar with a maximum adsorption capacity at equilibrium time of 520.0, 83.0, 7.5 and 922.0 mg/g, respectively. According to the authors, the nanomaterial showed the multifunctionality for adsorption processes because the removal of organic and inorganic polluting substances can be carried out using the same material. However, it is important to note that often pollutants are included in complex matrixes, and therefore, multifunctionality should not confused with lack of selectivity since a low selectivity is associated with a greater facility for the saturation of the adsorbent surface, which can have repercussions in low efficiency in real systems. For contaminants of the same type, pharmaceuticals, pesticides and heavy metals, several nano-adsorbents have been studied (Fiorati et al., 2020; Stan et al., 2017).

- Catalytic degradation: the catalytic capacity of nanomaterials allows reducing the activation energy of a reaction, through the surface immobilization of a reagent and polarization of its bond due to their high surface area, interfacial energy and surface properties (Bhandari 2018; Kolahalam et al., 2019; Livingston et al., 2020). Oxidation-reduction processes are the most common degradation reactions. Usually, the oxidation-reduction process occurs in metallic nanomaterials depending on the reduction potential of the pollutant and the nanomaterial component, consequently, electronic transfers can be carried out to promote the change in the speciation or reduction of organic pollutants (Yadav et al., 2018). On the other hand, redox-type reactions in which nanomaterials are present are also influenced by their adsorption capacity since their surface properties favor the nanomaterial-contaminant interaction through molecular forces (Yadav et al., 2018). Degradation of potentially dangerous dyes, recalcitrant contaminants of agrochemical origin such as DDT, the reduction of Cr (VI), among others, have been carried out by the catalytic strategy (El-Aziz et al., 2018; Jyoti and Singh, 2016; Wei et al., 2019).
- Photocatalysis: is the increase in the reaction rate of a photoreaction by the use of visible electromagnetic spectrum in the presence of a photocatalyst, which is a semiconductor nanomaterial. When light energy incising is equal to the energy gap between the valence and conduction bands of the photoactive component of the nanomaterial, an electronic transfer is produced towards the conduction band. Thus, generated holes (h^+) in the valence band produce the breakdown of water molecules to form hydronium cations (H^+), hydroxyl radicals (•OH) and hydrogen peroxide (H_2O_2); the latter can undergo a dissociation reaction to produce two hydroxyl radicals, that can react with organic pollutants and lead to their degradation. On the other hand, conduction band electrons can be transferred to oxygen molecules adsorbed on the surface of the nanomaterial, generating superoxide anion radicals ($\bullet O_2^-$), which are also involved in the degradation process of organic pollutants since they can react with a H^+ to form H_2O_2 (Bethi and Sonawane, 2018; Bolade et al., 2020; Das et al., 2018; Sohail et al., 2019). By using photocatalysis reactions it has been possible to carry out the degradation of different classes of polluting substances, for example, photoreduction using sunlight of Cr (VI) to Cr (III) and polycyclic aromatic hydrocarbons such as anthracene, phenanthrene, benzo[a] pyrene, fluorene and chrysene (Ramar et al., 2019; Shanker et al., 2017). The reactions involved in the photocatalytic process are illustrated in Fig. 9.3.
- Fenton reaction: this mechanism is specific for nanomaterials containing transition metals in their composition, where iron is the most common metal used due to its environmental availability and its capacity to generate the degradation of a wide range of pollutants by oxidation processes with •OH (Das et al., 2018). In the case of iron, the Fenton reaction is an advanced oxidative process in which hydroxyl radicals are generated by the reaction of Fe (II) with H_2O_2 (see Fig. 9.4). The Fe (III) cations generated can react with H_2O_2 to form the potent oxidizing agents •OH and •OOH carrying out the oxidation of organic pollutants (Das et al., 2018). As illustrated in Fig. 9.4, the reactions between Fe, O_2 and H_2O lead to oxidative processes generating

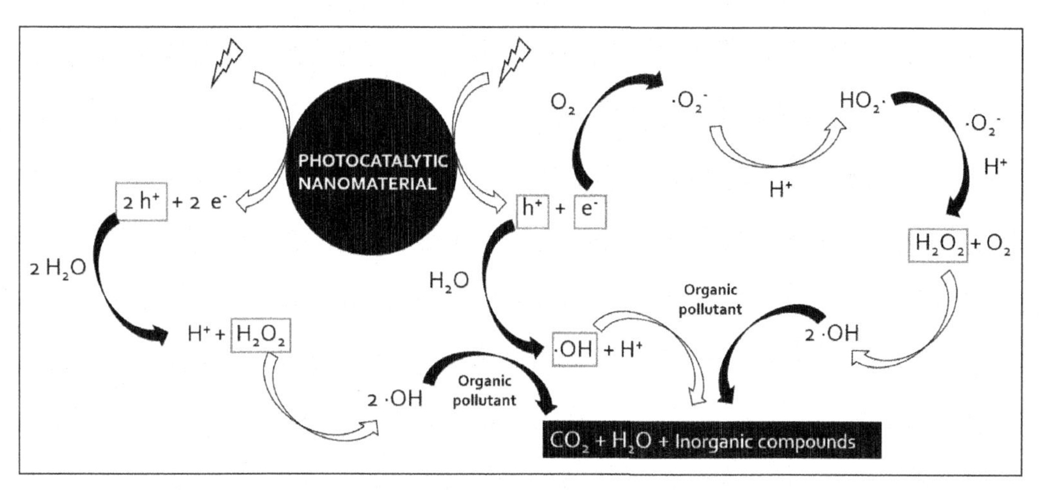

Fig. 9.3. Mechanism of photocatalysis by nanomaterials.

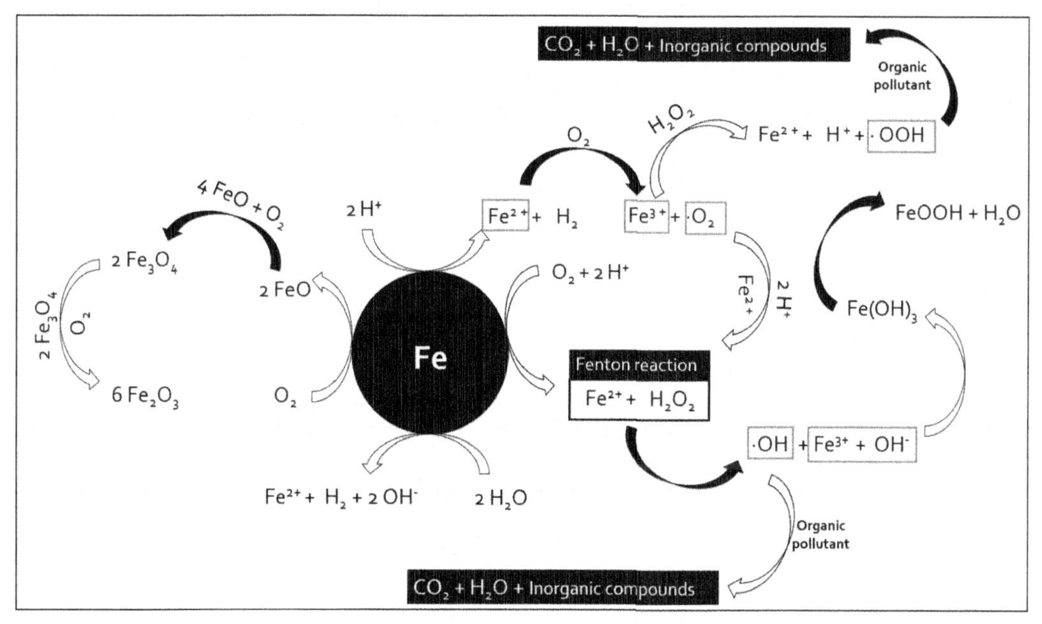

Fig. 9.4. Fenton and oxidation reactions of iron nanomaterial.

H_2 which acts as a reducing agent, different species of iron oxides and hydroxides (Lu et al., 2016; Mu et al., 2017). In addition, iron derivatives act as a favorable surface for the coprecipitation of organic pollutants and surface adsorption processes, easing the interaction of the pollutant with the active iron core, but also participate in the electronic transfer between them due to their behavior as a n-type class semiconductor (Das et al., 2018; Lu et al., 2016; Mu et al., 2017). By Fenton-type reactions the degradation of emerging contaminants such as ciprofloxacin using bi-metallic composition of Fe/Cu nanoparticles has been carried out (Chen et al., 2020); the concentration of heavy metals such as Pb (II) has also been reduced to non-detectable levels using iron nanoparticles (Kumar et al., 2015).

Despite the excellent results that can be seen in controlled laboratory conditions for Fenton-type reactions, in the case of *in situ* remediation, biogeochemical features can significantly influence the remediation capacity of nanomaterials because they interact not only with pollutants but also with biotic and abiotic components from soils and groundwater. In addition, depending on the nature of the contamination, its remediation using Fenton-type reactions assisted by nanomaterials may lead to the generation of toxic degradation products and in consequence, the generation of potential emerging contamination is a risk that must always be considered and evaluated (Cecchin et al., 2017; Iavicoli et al., 2017; Koduru et al., 2017; Singh et al., 2020; Tripathi et al., 2018; Zhang et al., 2019).

2.2.3 Nanobioremediation

Nanobioremediation is a term used to refer to the hybrid remediation processes based on the combined use of nanotechnology and biotechnology in order to achieve a more efficient, sustainable and multifunctional decontamination strategy in comparison with individual remediation technologies (Sebastian et al., 2020; Singh et al., 2020). Due to these characteristics, the nanobioremediation field has undergone a notable development in recent years. Many researches focusing on nanobioremediation are based on four approaches: (i) sequential nanobioremediation, (ii) simultaneous nanobioremediation, (iii) nanobioestimulating and (iv) biogenic nanoremediation.

The sequential nanobioremediation is based on the use of bioremediation and nanoremediation by an independent-stage configuration. In this case, at the first stage, the nanomaterial-mediated transformation of the pollutant to less toxic substances or the degradation to tolerance levels of the bioremediating organism is performed; later, biodegradation by biological systems is carried out. (Cecchin et al., 2017; Davies et al., 2017; Pandey, 2018; Singh et al., 2020; Tripathi et al., 2018; Yadav et al., 2018). On the other hand, in simultaneous nanobioremediation, nano- and bioremediation coexist in the same system, carrying out the transformations in parallel or acting synergistically (Cecchin et al., 2017; Iavicoli et al., 2017; Tripathi et al., 2018; Zhang et al., 2019). Studies have shown that nanobioremediation is a more efficient alternative compared with their individual strategies. For example, the degradation of 2,2',4,4'-tetrabromodiphenyl ether (BDE47) in an aqueous solution by nanoremediation and bioremediation in individual and simultaneous configurations has been described. Thus, the bioremediation of BDE47 by the strain *Pseudomonas putida* allowed a 34% degradation after 10 days of exposure, while the degradation by advanced oxidative processes, using bimetallic Fe/Pd nanoparticles in a solution at pH 3, allowed the degradation of 11% of the contaminant after 90 min. However, by applying sequential nanobioremediation, the total degradation of BDE47 to CO_2 and H_2O was achieved, through the initial dehalogenation reaction in a nitrogen atmosphere, generated by a reductive mechanism mediated by the nanomaterial, to carry out the transformation of BDE47 to biphenyl ether. Later, the biphenyl ether in the presence of the nanomaterial and oxygen underwent an advanced Fenton-type oxidation process that, after 90 min, allowed the degradation of BDE47 to form the phenolic compounds used as substrates in the tricarboxylic acid cycle of the *P. putida* allowing the total degradation (Lv et al., 2016).

Nanobioestimulating is defined as the use of nanomaterials to stimulate the organisms responsible for bioremediation; in this case, nanomaterials can be used indirectly in the remediation process in order to favor the nutritional conditions and requirements of the biological systems (Bharagava et al., 2017; Davies et al., 2017; Thyagarajan et al., 2017; Zhang et al., 2019). Among the most widely used alternatives for this process is the use of nanomaterials capable of releasing oxygen molecules in an anoxic media; in addition, nanomaterials are used to allow the release of substrates for the metabolic degradation pathways of the remedial agent, stimulate a response in the biological system permitting the release of extracellular substances able to modify the solubility and mobility of the pollutant and promote the growth and development of bioremediation organisms, as well as allow the controlled release of agrochemicals for their efficient use by phytoremediation plants. The biostimulation process can be carried out by adding the nanomaterial to the growth medium of the bioremediation agent or to vegetative organs such as leaves or fruits in the case of plants, but nanomaterials can also be synthesized inter or extracellularly by the biological agent with the presence of their precursors in the growth medium (Bharagava et al., 2017; Davies et al., 2017; Qian et al., 2020; Thyagarajan et al., 2017; Zhang et al., 2019).

Biogenic nanoremediation is defined as the remediation by using of biogenically-synthesized nanomaterials (Davies et al., 2017; Pandey 2018; Singh and Khan, 2018; Thyagarajan et al., 2017; Yadav et al., 2018). Thus, biogenic nanoremediation is related to sustainable remediation as it is based on the synthesis of nanomaterials using biological agents (Sebastian et al., 2020). Nanomaterials resulting from using biological nano-factories or biomolecules are assumed to promote their assimilation in the environment (Bandala and Berli, 2018), consequently, nanobiotechnology emerges as an area of great interest for remediation since conceptually it is a more sustainable, economical and an efficient alternative (Pandey, 2018). Biogenic synthesis of nanomaterials for the eco-friendly production of remediation agents and their application in model systems and environmental matrices are described below.

3. Green Nanotechnology

3.1 Principles of green chemistry in nanotechnology

Since the 1960s ethical responsibility of human beings with respect to planetary health has been unquestionable, in consequence, this has become evident with political and social interests for preserving and remedying the environment; in this context, the environmental movement and the philosophy of sustainable development appeared (Albini and Protti, 2016). With time, environmental sustainability focused on the development of increasingly eco-friendly technologies and processes, including the green chemistry principles (see Fig. 9.5) (Albini and Protti, 2016; Dicks and Hents, 2015; Upreti et al., 2015). This line of thinking promoted and stimulated a change in the paradigm of science and technology, in such a way that in 1996 the ISO 14001 emerged as an international standard for environmental management; which was proposed with a holistic and integrated approach for the Life Cycle Analysis and environmental impact of the processes, as well as their monitoring, energy and material optimization (Dicks and Hents, 2015; Iavicoli et al., 2017; Nasrollahzadeh et al., 2019; Upreti et al., 2015).

The 12 principles of green chemistry can be synthesized as: (i) avoid the production of waste, (ii) atom economy, (iii) safer synthetic processes, (iv) safer products, (v) avoid the use of auxiliaries, (vi) energy efficiency, (vii) use of reagents from renewable sources, (viii) avoiding derivatization as well as minimizing reaction steps, (ix) use substances with a high surface area to favor the efficiency and

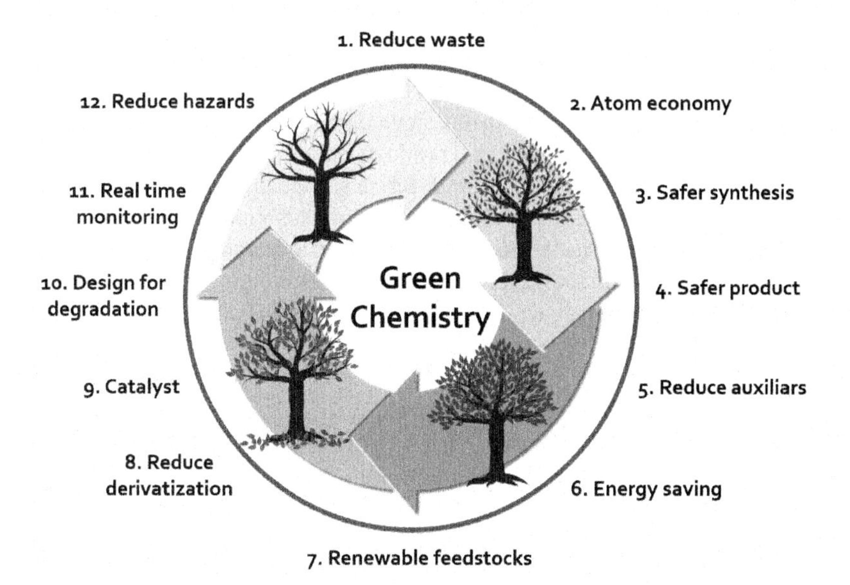

Fig. 9.5. The 12 principles of green chemistry.

selectivity of chemical reactions, (x) design for final disposal, (xi) real-time monitoring and prevention and (xii) implementing safer processes. Ample descriptions and examples of green chemistry principles are available in specialized literature (Anastas et al., 2018; Bhandari, 2018; Ciambelli et al., 2019; Dicks and Hents, 2015; Khan and Lee, 2015; Li et al., 2018; Livingston et al., 2020; Pandey, 2018; Sanjay, 2019; Silva et al., 2015; Upreti et al., 2015; Villalobos et al., 2019).

In this way, green chemistry allows advancing the development of eco-friendly and competitive technologies, at present many processes show a certain degree of approach to the principles of green chemistry, while advancing in the optimization of the unsustainable parameters of the process using eco-innovation and creativity to redesign or replace traditional processes (Albini and Protti, 2016; Iavicoli et al., 2017; Livingston et al., 2020). Thus, while many synthetic procedures are not entirely designed under the principles of green chemistry, it is possible that they are eco-friendlier than the earlier processes, and therefore, "greener" processes. In consequence, it can be concluded that the development of fully sustainable synthesis or processes is an iterative methodology using the precepts of green chemistry as a purpose of innovation (Albini and Protti, 2016). Although the principles of green chemistry represent a design philosophy, they do not allow a quantifiable comparison between different processes; and therefore, it is necessary to implement other types of metrics based on the 12 principles of green chemistry, which establish quantitative values on the eco-friendliness of the synthesis and its sustainability (Iavicoli et al., 2017; Leng et al., 2015; Livingston et al., 2020). Examples are the eco-scale and the Life Cycle Analysis with different perspectives such as cradle-to-grave, cradle-to-gate, cradle-to-cradle, gate-to-grave, among others (Dicks and Hents, 2015; Iavicoli et al., 2017; Leng et al., 2015; Livingston et al., 2020). Nevertheless, it is evident that there is no universal green metric for processes and quantitative approaches usually involve some subjective criteria, associated with a specific scientific overview, a market tendency or cultural premises. That is why there is no regulation or minimum criteria to define a process as green, so some research studies may make an incorrect use of the term considering it as only related to the use of substances of natural origin (Albini and Protti, 2016; Dicks and Hents, 2015).

On the other hand, from implementing green chemistry in nanotechnology, the concept of green nanotechnology appeared at the American Chemical Society, Green Chemistry Institute and the Oregon Nanoscience and Microtechnologies, focusing the development and implementation of the synthesis of nanomaterials based on the principles of green chemistry, in such a way that these products can contribute to sustainability and remediation, avoiding the use of dangerous agents or carrying out decontamination processes (Ciambelli et al., 2019; Khan, 2020; Pandey, 2018; Nasrollahzadeh et al., 2019; Sanjay, 2019). The above indicated that research and application of environmental technologies, such as nanoremediation or nanosensors, using eco-friendly nanomaterials, holistically involves the fundamentals of green nanotechnology. Therefore, the application of the philosophy of sustainable design in nanomaterials has contributed to the development of nanotechnology, notably in the public acceptance of potential applications and products based on these strategies, and at the same time, nanomaterials themselves represent an approach to sustainability. However, as established by James Hutchison, Director of the Safer Nanomaterials and Nanomanufacturing Initiative (SNNI), it is necessary to advance in the investigation of the implications and applications of the life cycle of nanomaterials in order to avoid the generation of potential emerging pollutants (Ciambelli et al., 2019; Khan, 2020; Nasrollahzadeh et al., 2019; Upreti et al., 2015).

3.2 Green synthesis of nanomaterials

Green synthesis is one that has been developed using the principles of green chemistry. This implies that green synthesis consists of synthetic processes minimizing the use of dangerous reagents, using renewable raw materials, implementing processes of low energy intensity, among other aspects described earlier. In this manner, obtaining nanomaterials through biological systems and biomolecules or nanobiotechnology, has emerged as the approach most in line with these parameters for the synthesis of nanomaterials, since it involves lower costs, a simple operation, easy purification, mild conditions, potential industrial scalability and safer processes (Khan and Lee, 2015; Livingston et al., 2020; Nasrollahzadeh et al., 2019; Sasidharan et al., 2019; Yadav et al., 2018). However, it should be kept in mind that not all synthetic processes using natural substances are by-definition green processes due to the sustainable design philosophy that proposes the generation of eco-friendlier processes than the ones used earlier, and the use of renewable raw materials does not guarantee that the objectives of green chemistry are completely achieved through this approach. This fact was evident in the analysis of E-factor for the synthesis of nanomaterials carried out by Eckelman et al. (2008). The E-factor is the quotient between the mass of total waste in respect to the mass of the product, and allows to evaluate, in a forthright manner, a synthesis through the atom economy perspective. The E-factor has been used in research to evaluate different synthesis of nanomaterials: Carbone nanotubes through thermal chemical vapor deposition (E = 170), fullerenes through the benzene-oxygen flame (E = 950), TiO_2 nanoparticles through hydrolysis and calcination (E = 17800), AuNPs stabilized with phosphines (E = 7200), Au nanoparticles functionalized with thiols-ionic liquids (E = 99400), Au nanoparticles synthesized using alfalfa plants as bioreactors (E = 163), AuNPs produced with starch-glucose (E = 29600). In consequence, it was demonstrated that in the biogenic synthesis, the use of alfalfa as a bioreactor allowed obtaining the lowest E-factor. In contrast, the use of reducing biomolecules presented the second highest value, meaning that the atom economy of this process is low compared with physical or chemical synthesis, by this simple approach to one of the metrics of green chemistry, synthesis of nanomaterials based on biomolecules are not necessarily completely sustainable. However, by Life Cycle Analysis, biogenic processes usually present greater sustainability compared with classical synthesis, following a complete way of the precepts of green chemistry (Bharagava et al., 2017; Dicks and Hents, 2015; Pandey, 2018; Rahman et al., 2020; Sasidharan et al., 2019; Singh and Khan, 2018; Shah et al., 2015). The research performed by Pourzahedi and Eckelman (2015) on the synthesis of silver nanoparticles gave evidence to the earlier situation. In their investigation, they carried out the Life Cycle Analysis "cradle-to-gate", i.e., from resource extraction to the factory gate of product, corresponding the above to the stage before its transfer to the market, in addition, the use and disposal stages are not taken in consideration. They analyzed the effects on the production of CO_2, ozone depletion, generation of smog, acidification, eutrophication, respiratory effects, ecotoxicity and impacts on planetary health associated with toxicity and carcinogenicity. Based on this, seven different representative approaches were analyzed using various reducing agents (sodium borohydride, sodium citrate, ethylene glycol and starch) and techniques (flame spray pyrolysis, arc plasma and reactive magnetron sputtering in Ar-N_2 atmosphere). They established that the physical methods had the most significant environmental impacts due to energy use, in addition, concluded that the least eco-friendly method was flame spray pyrolysis, while chemical methods presented higher yields, but larger impacts on planetary health; meanwhile the biological method exhibited a notable decrease in eutrophication and human health carcinogenicity. Nevertheless, the data available in most of the investigations do not carry out Life Cycle Analysis of the processes to evaluate their sustainability holistically. Consequently, it is assumed that the biogenic synthesis processes and those that use alternative sources of energy are procedures that frame their design in the principles of green chemistry and the objective of sustainability, these, therefore are good strategies to achieve the green synthesis of nanomaterials.

Considering alternative energy sources, the most used are microwaves and ultrasound, because these allow efficient synthetic processes to be carried out in reduced times and with low energy intensity (Rahman et al., 2020). In the case of microwave radiation, it allows heating processes to be carried out by generating rotation of the polar molecules according to their dipole moment, which leads to the transformation of kinetic energy into thermal energy through molecular friction; additionally, it favors dispersion and nucleation in the formation of nanoparticles (Bolade et al., 2020; Khan and Lee, 2015; Rahman et al., 2020). While ultrasound uses high-energy ultrasonic waves that generate cavitation bubbles resulting from pressure differences, which warms up the system, favoring synthesis efficiency (Sasidharan et al., 2019). However, the combined use of technologies should always be looked at very carefully when conducting an objective analysis, for example, though microwave technology can be identified as a mature and well-established technology from the energy-efficiency perspective, for which electrical consumption during its use is the main contributor to environmental impacts; however, it is important taking into account that the resources used for making of microwaves produces a strong reduction of abiotic elements (67%), a relevant contribution of photochemical oxidants (34%), ozone layer depletion (33%) and acidification (31%) (Gallego-Schmid et al., 2018).

On the other hand, biogenic synthesis is a process framed in the Bottom-Up approach and, can be carried out using a wide variety of biological systems and biomolecules, ranging from higher plants to microorganisms, such as fungi and bacteria and some of their molecular components, like proteins, carbohydrates and vitamins, are illustrated in Fig. 9.6.

The diversity of possible bioreactors and biomolecules permits synthesis to be carried out both *ex situ* and *in situ* of the organism, allowing to obtain nanomaterials in an aqueous media under slight conditions in the *ex situ* case, and in the *in-situ* case monodisperse nanomaterials, and in some processes, physicochemical properties of interest are presented to microorganisms. For example, magnetization properties have been generated in bacterial cells through the *in situ* biosynthesis of metal oxides, which permit the separation of these organisms using a magnetic field; this application is of notable interest in the use of bioremediation agents in bioreactors (Stan et al., 2017). In this manner, it is possible to obtain a great diversity of nanostructures with a wide variety of sizes, shapes, coating agents, and therefore, physicochemical properties (Khan and Lee, 2015; Shah et al., 2015).

Biogenic synthesis usually involves redox-type reactions in which biomolecules such as amino acids, flavonoids, peptides, proteins, secondary metabolites, terpenes, polysaccharides, electron transport molecules and vitamins, are involved; but, at the molecular level, these are biomolecules with functional groups in their composition capable of undergoing reduction-oxidation reactions, such as thiols, alcohols, aldehydes, unsaturated segments and amines (Nagajyothi and Tvm, 2015; Kolahalam et al., 2019). However, the chemical composition and concentration of active molecules differs depending on the nature of the organism and their growing conditions, so the features of each biological system determines the synthesis procedure and the morphological characteristics, size and properties of the synthesized nanomaterial, however, not all biological entities have the necessary components to obtain nanomaterials (Pandey, 2018). Furthermore, biogenic synthesis can

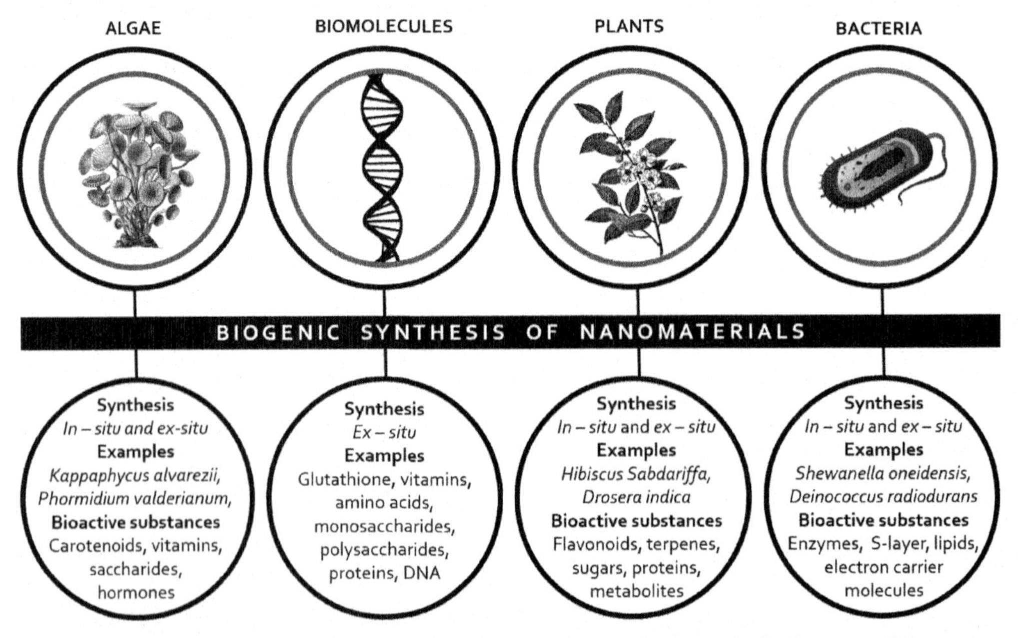

Fig. 9.6. Some of the biological systems and biomolecules used for the synthesis of nanomaterials.

be broadly differentiated into two groups: those carried out by microorganisms and those carried out by plants.

3.2.1 Green synthesis of nanomaterials based on microorganisms and algae

Microorganisms are biological entities that cannot be directly observed; therefore, their classification presents a high diversity, including biological entities such as bacteria, unicellular algae and fungi. Consequently, the biosynthetic processes and mechanisms for formation of nanomaterials from each species are different; however, in general, it can be established that the generation of nanomaterials by microorganisms is directly associated with defense mechanisms allowing the biotransformation of potentially harmful substances for the organism in safer products for them (Devatha and Thala, 2018; Nagajyothi and Tvm, 2015; Shah et al., 2015). Some examples of the use of these microorganisms in the biogenic synthesis of nanomaterials are shown in Table 9.2. Nanomaterial biosynthesis can be intracellular or extracellular; in the case of intracellular synthesis, nanomaterials are produced through metabolic processes, primarily by enzymatic mechanisms, involving oxidation-reduction mechanisms, through reductases and electron carriers and biomineralization (Kanwar et al., 2017; Kolahalam et al., 2019; Shah et al., 2015; Wei et al., 2018). On the other hand, extracellular synthesis is based on molecular interactions between the material precursors and biomolecules of the microorganism's coating such as the surface proteins, lipids and polysaccharides, by chemical reduction through the transfer of electrons, which leads to nucleation and subsequent growth of the nano-objects; extracellular synthesis can also be carried out through interactions between the precursors and extracellular compounds able to reduce and stabilize nanomaterials, such as polysaccharides and proteins (Bolade et al., 2020; Pandey, 2018; Shah et al., 2015). Usually, the manufacturing processes of green nanomaterials through microorganisms are carried out using microbial cultures and under appropriate conditions to carry out an efficient synthesis, most often these parameters are the amount of light, concentration of nutrients, pH, temperature, concentration of nanomaterial precursors and purity of the culture (Pandey, 2018; Shah et al., 2015).

In the case of the synthesis of nanomaterials using bacteria, is one of the most widely used approaches due to the high growth rate of these organisms, their potential industrial scalability and their ease of developing resistance mechanisms against adverse conditions (Devatha and Thala, 2018; Nagajyoti and Tvm, 2015). Therefore, many bacteria have been used to obtain nanomaterials, examples of this are *Escherichia coli*, *Pseudomonas aeruginosa*, *Deinococcus radiodurans*, *Acidithiobacillus*, *Bacillus megaterium*, among others (Ali et al., 2017; Shah et al., 2015; Thyagarajan et al., 2017; Ulloa et al., 2016; Zhou et al., 2016). Despite the wide use of this biotechnological approach, purification processes in intracellular synthesis involve dangerous solvents; furthermore, the efficiency of the synthesis in terms of time and quantity is an unfavorable condition and presents a restriction in morphology and dimensions (Devatha and Thala, 2018; Silva et al., 2015). On the other hand, it is common for bacterial cultures to present contamination, so maintaining optimal culture conditions can be arduous from an operational point of view (Devatha and Thala, 2018; Silva et al., 2015). It should be noted that one of the most complex situations in the synthesis of nanomaterials by bacterial organisms depends on their concentration, nanomaterials can constitute a harmful agent for the bioreactor (Devatha and Thala, 2018; Leng et al., 2015). This condition can be seen for example, in the research carried out by Thyagarajan et al. (2017) who evaluated the capacity of *Bacillus* bacteria strains to synthesize AgNPs with a decolorization activity against the emerging contaminant Congo Red. For this, the researchers used the strains *Bacillus megaterium*, *Bacillus subtilis* and *Bacillus licheniformis*. The synthesis was carried out through individual inoculation in a broth-like medium, and subsequent extraction of the bacterial pellet, which was incubated for 11 days with $AgNO_3$, obtaining AgNPs that exhibited decolorization properties against Congo Red, and antimicrobial properties against the strains used as a synthesizer organism.

In the case of nanomaterial synthesis mediated by fungi, this represents a notable area of interest due to their high internalization capacity of heavy metals and their ability to produce a greater quantity

Table 9.2. Some examples of microorganism-based synthesis of nanomaterials.

Type of Nanomaterial	Microorganism	Remarks	References
Bacteria			
Spherical AgNPs (5–45 nm)	*Pseudomonas aeruginosa* JP1	Extracellular synthesis of AgNPs via enzymatic pathway by nitrate reductase.	Ali et al., 2017
CdS NPs (5–400 nm)	*Acidithiobacillus: A. ferrooxidans, A. thiooxidans*, and *A. caldus*	Intracellular synthesis of CdS assisted with sulfured molecules (cysteine and glutathione).	Ulloa et al., 2016
CuS NPs (~ 5 nm)	*Shewanella oneidensis* MR-1	Anaerobic extracellular synthesis with $Na_2S_2O_3$ as an electron acceptor; obtaining CuS NPs with potential as a photothermal agent.	Zhou et al., 2016
Spherical, irregular, and triangular AuNPs (~ 43.8 nm)	*Deinococcus radiodurans*	The strain's cell pellets were incubated for 8 hours at 32°C with $HAuCl_4$ to obtain the NPs.	Li et al., 2016
Spherical AgNPs (~ 20 nm)	*Escherichia coli*	The *Candida albicans* metallothionein gene was introduced into *Escherichia coli* strains favoring their tolerance to Ag^+ through the formation of NPs.	Yuan et al., 2019
Fungi			
AgNPs	*Trametes trogii*	Extracellular synthesis of AgNPs by an extract of the strain at pH 13, and an incubation period of 72 hr.	Kobashigawa et al., 2019
Spherical AgNPs (~ 18.7 nm)	*Phomopsis liquidambaris*	Extracellular synthesis using an aqueous extract of the strain obtained from the plant *Salacia chinensis*, which was incubated with $AgNO_3$ for 24 hr, obtaining AgNPs with non-hemolytic properties.	Seetharaman et al., 2018
Quasi-spherical ZnO NPs (16–78 nm)	*Periconium* sp.	Extracellular synthesis using strain extract, which acts as a chelating and gelling agent for Zn(II), to obtain wurtzite-type particles.	Ganesan et al., 2020
AgNPs (15–35 nm)	*Cryptococcus laurentii, Rhodotorula glutinis*	Synthesis using the culture supernatant of the strain extracted from apple peel, cultivated for 48 hr at 28°C with $AgNO_3$, obtaining AgNPs with antifungal properties.	Fernández et al., 2016
Algae			
Cubic, triangular, and spherical AgNPs (40–90 nm)	*Botryococcus braunii*	The nanoparticles were synthesized using algae aqueous extract and $AgNO_3$, at ambient conditions. The synthesized NPs catalyzed the reduction of 2-nitroaniline to *o*-phenylenediamine, which in the presence of aldehydes helped to obtain 2-aryl benzimidazoles.	Arya et al., 2019
Spherical AgNPs (70–75 nm)	*Portieria hornemannii*	The synthesis was carried out using the aqueous extract of the seaweed acquired from the Gulf of Mannar and a 1 mM $AgNO_3$ solution; obtaining NPs with antibacterial activity against pathogenic fish bacteria.	Fatima et al., 2020
Spherical AgNPs (45–57 nm)	*Saccharina cichorioide, Fucus evanescens*	The synthesis was carried out using as reducers and stabilizers, individually, polysaccharides extracted from algae (alginate, fucoidan, laminaran); obtaining AgNPs with cytotoxic activity against tumor cells of the rat C6 glioma type.	Yugay et al., 2020

of oxidoreductase enzymes in comparison with bacterial systems, and therefore, by using fungi it is possible to obtain a higher proportion of nanoparticles in shorter periods through both extracellular and intracellular bioreductions. In addition, quinone-type reducing agents from fungi allow the secretion of reducing biomolecules through the cell wall and consequently, fungi have a greater capacity to effect the reduction of metal precursors by simultaneous synthesis through extracellular and intracellular routes (Khan and Lee, 2015; Shah et al., 2015; Wei et al., 2018). The fungi used for this nanobiotechnological process are yeast strain such as MKY3, *Trichoderma asperellum*, *Aspergillus clavatus*, *Penicillium* sp., among others (Jacob et al., 2018; Wei et al., 2018).

On the other hand, nanomaterial synthesis using algae, mainly multicellular eukaryotes, is an area of significant interest in recent years due to its molecular composition and its promising use in efficient nanobiotechnological processes. One of the attractions of algae-based bionanosynthesis is associated with the presence of a greater variety of reducing bioactive components compared with bacteria-based bionanosynthesis, as algae have phenolic compounds, pigments, vitamins, nucleic acids, exopolysaccharides, polyols, proteins, among other reductive biomolecules (Dahoumane et al., 2017; González-Ballesteros et al., 2020; Wei et al., 2018). Biogenic synthesis through the use of aqueous extracts from algae depends on concentration and nature of extracellular substances, polarity of the solvent and co-solutes present in the reaction medium (González-Ballesteros et al., 2020). In the case of intracellular synthesis, nanomaterial production is carried out through biological mechanisms mediated by enzymes and processes like cellular respiration and photosynthesis (Khanna et al., 2019). The algae used to carry out the nanomaterial biogenic synthesis are *Phormidium valderianum*, *Tetraselmis kochinensis*, *Spirulina subsalsa*, *Kappaphycus alvarezii*, among others (Khanna et al., 2019; Pachiyappan et al., 2020).

3.2.2 Plant-based green synthesis of nanomaterials

Synthesis of nanomaterials using plants and their extracts is the biogenic synthesis of most research. The popularity of this nanobiotechnological practice is directly associated with the ease of performing nanomaterial synthesis in one step, under ambient conditions, with high efficiency, speed and material conversion. The evident operational simplicity is a fundamental aspect that gives it a significant advantage over methods that use bacteria and other types of microorganisms (Kolahalam et al., 2019; Mohammadi et al., 2019; Nagajyothi and Tvm, 2015; Pandey 2018; Prakash et al., 2017; Salouti and Derakhshan, 2019; Shah et al., 2015). Plants represent in the context of green chemistry an affordable and easily propagated renewable resource, with a wide variety of species and phytochemicals capable of synthesizing and modulating nanomaterial properties. In this way, it is possible to use components such as leaves, roots, fruits, flowers, seeds, peels or stems for bionanosynthesis; these vegetal organs present an extensive diversity of phytochemicals acting as electrostatic reducers and steric stabilizers. Among numerous structural richness of chemical species obtained from plants are compounds with functional groups such as amines, aldehydes, carboxylic acids, ketones and unsaturations; these phytochemicals comprise biomolecules such as tannins, steroids, flavonoids, stilbenes, terpenoids, peptides, saponins, vitamins, polyphenols, pigments, proteins, aglycones, amino acids, alkaloids, carbohydrates and resins; the majority of these biomolecules are soluble in aqueous systems, which permits carrying out the synthesis in the green solvent par excellence: water (Bolade et al., 2020; Mohammadi et al., 2019; Nagajyothi and Tvm, 2015; Nasrollahzadeh et al., 2019; Salouti and Derakhshan, 2019; Singh et al., 2019; Tarannum and Khan, 2020).

The most widely used plant-based synthesis of nanomaterials is *ex situ* since an efficient aqueous extraction of biomolecules can be easily carried out, and the synthesis can be executed with this extract in a reduced time (Salouti and Derakhshan, 2019) (see Table 9.3). The general *ex situ* synthesis procedure consists, in the first case, of determining the plant to be used and selecting the tissue according to the molecular distribution and concentration of the efficient species toward the nanobiosynthesis of the target material. This biomass is collected and sanitized, subsequently subjected to drying and maceration, then the particles obtained are placed in aqueous or alcoholic solutions

Table 9.3. Some examples of plant-based synthesis of nanomaterials.

Type of Nanomaterial	Plant	Remarks	Reference
Spherical CeO_2 NPs (~ 3.9 nm)	*Hibiscus Sabdariffa*	The flowers were placed in water, at room temperature for 2 hr, to extract the anthocyanins responsible for the red color (cyanidin-3-glucoside, cyanidin-3-sambubioside, delphinidin-3-sambubioside), which acted as reducing agents and stabilizers of cerium hydroxides, which were used to obtain CeO_2 thermally.	Thovhogi et al., 2015
Quasi-spherical Fe/Pd NPs (2–20 nm)	*Vitis*	A grape extract was obtained in an aqueous medium by heating at 100ºC for 1 hr; this extract with oximes, polyphenols, quinones and natural aldehydes, carried out the reduction and stabilization of polydisperse bimetallic iron and palladium nanoparticles.	Luo et al., 2016
Quasi-spherical AgNPs (~ 50 nm)	*Elaphantopus scaber* *Azadirachta indica*	Leaves extract was obtained in an aqueous medium for 10 min at 95ºC; this solution allowed the reduction of Ag^+, obtaining NPs with biological activity against the *Culex quinquefasciatus* mosquito.	Hajra et al., 2016
AgNPs (5–300 nm)	*Drosera indica* *Drosera binata, Drosera spatulata, Dionaea muscipula*	The plant extract was obtained in an aqueous medium using microwave energy, these solutions were added, individually, to systems with the presence of polyvinylpyrrolidone and $AgNO_3$; and they were microwaved for 2 min , obtaining AgNPs with biological properties against *S. aureus*, *P. aeruginosa*, *C. Albicans*, *P. atrosepticum*, *P. parmentieri*, and *D. dadantii*.	Banasiuk et al., 2020
Spherical AgNPs (~ 20 nm)	*Jasminoides ellis*	The synthesis of AgNPs was carried out using the plant extract in an aqueous medium, obtained by boiling the seeds for 10 min. This process achieved the synthesis of NPs with cytotoxic activity against HeLa cells and photocatalytic activity in the degradation of the Coomassie blue dye.	Saravanakumar et al., 2018
Spherical and oval AgNPs (10–50 nm)	*Aloe barbadensis*	The aqueous extract of the *Aloe vera* pulp was obtained in an aqueous medium by boiling, and used in the reduction of Ag^+ ions; obtaining AgNPs that inhibit the aggregation of α-chymotrypsinogen A as a function of concentration.	Alam et al., 2018
Dendritic silver nanostructures composed by AgNPs	*Vitis* pomace	Bio-dendritic silver structures were obtained through the aqueous extract of the agricultural residue white grape pomace and $AgNO_3$ with microwave irradiation at 700 W for 40 sec, obtaining structures with antifungal activity against *F. graminearum*, and potential electroactive activity in applications as nanosensor.	Carbone et al., 2019
Hexagonal and spherical ZnO NPs (26.57 nm)	*Cinnamomun tamala*	The leaf extract, obtained in an aqueous medium by boiling for 20 min, was added to a zinc oxide solution, allowing its reduction and stabilization to obtain NPs with antimicrobial activity against *Staphylococcus aureus*.	Agarwal et al., 2019
Cubic phase and flower like form MgO NPs (4 – 8 nm)	*Limonia accidisima*	An aqueous extract of the fruit pulp was obtained by contact for 5 hr, this solution was used for the reduction of Mg (II), and with a thermal method was able to achieve obtaining of MgO NPs, these presented different morphologies according to the concentration of the extract, managing to obtain a flower morphology associated with isoleucine-MgO complexes.	Nijalingappa et al., 2019
Spheric Multiphase $Fe_X(O)_Y$ NPs	*Syzygium aromaticum*	The plant extract was obtained in an aqueous medium by heating at 100ºC for 1 hr; this allowed to obtain $Fe_X(O)_Y$ and super-paramagnetic NPs, through the presence of Fe (II) and Fe (III) salts in the system.	Kumar et al., 2020

and, usually, subjected to heating to carry out the extraction of the biomolecules. After which, the solution is filtered and the extract obtained is used to carry out the synthesis of the nanomaterial, usually under room pressure and temperature (Abbas et al., 2020; Liu et al., 2017; Sharma et al., 2017). At this stage evaluating structural characteristics and the optimization of synthesis parameters in order to warrant its reproducibility, is an aspect associated not only with the synthesis procedure but also with phylogenetic diversity. This effect of reproducibility resulting from the variability in the molecular composition from plants of the same species, is understood as a direct consequence of differences in the nutritional conditions of plants depending on its place of growth, the temperature and exposure to the sun, that vary with the seasons and geographical conditions, as well as the microbial population, the plant-rhizosphere interaction and the presence of stressors (Bolade et al., 2020; Davies et al., 2017; López-Valdez et al., 2018).

On the other hand, the biogenic synthesis of nanomaterials by plants is a versatile process, because by varying the synthesis conditions, nanomaterials of the same nature, but different sizes and morphologies, can be obtained. For example, during the synthesis of AuNPs using an ethanolic extract from leaves of *Artemisia dracunculus*, it was seen that, by increasing the extract concentration, nano-triangles with a smaller dimension were obtained (80–130 nm for 3%, 40–75 nm for 4% and 35–50 nm for 5%). Regarding the concentration of the metallic precursor, it was determined that in a concentration between 0.05–0.28 mM, spherical nanoparticles were obtained; while at 0.50 mM the nanoparticles showed triangular, hexagonal and spherical morphologies and at 1.00 mM, the morphologies observed were triangular, hexagonal, and spherical, but with a greater dimension than those obtained with 0.50 mM. Concerning pH, in the values of pH between 2.8–4.0, nano-triangles were obtained, which lost their morphological definition as pH was increased till pH 5.0, where it was indicated that the particles obtained were spherical (Wacławek et al., 2018).

It is relevant to note that the use of plants in synthesis of nanomaterials represents an approach to the principles of green synthesis not only by using renewable raw materials but also by the opportunity to use residual biomass from agro-industrial processes and food consumption (López-Valdez et al., 2018). Two examples of these are the use of a pomegranate peel extract residue to synthesize silver nanoparticles (Kotp et al., 2019) and the stem of the *Musa Paradisiaca Linn* (banana) to carry out the synthesis of Fe_2O_3 nanorods (Ramar et al., 2019).

3.2.3 Biomolecules-based green synthesis of nanomaterials

In the processes of intra- and extracellular biogenic synthesis, not all the molecular components of the organism contribute to the synthesis and stabilization of nanomaterials; however, specifically defining biomolecules participating in the synthesis process is an intricate analysis in most cases, due to the molecular complexity of biological systems and the structural similarity present in some families of biomolecules; for example, more than 8000 chemical compounds are classified as polyphenols in plants due to their structural characteristics, many of these antioxidants are capable of reducing metal salts to nanoparticles (Bolade et al., 2020; Nasrollahzadeh et al., 2019). Inspite of this, nanobiosynthesis processes have been primarily associated with biomolecules able to experience oxidation-reduction reactions, which can be extracted though in many cases, their isolation is a complex and low-yield procedure, however, these can be commercially obtained, and therefore, the production of the materials will have additional costs, not only economical but ambiently associated with extraction, purification, commercialization and transfer, among others. In consequence, it is evident that to carry out bionanosynthesis reactions, under mild conditions, with greater control over the concentration of the bio-reductive agent, and in the case of nanoparticles, with adequate control of nature of the capping agent, is an unsolved problem. The latter is a major limitation when bionanosynthesis is analyzed for specific applications, since, for example, the surface properties of nanoparticles is controlled by the nature and composition of their interface, i.e., nanostructured surface includes both the main precursor defining the nature of particle as the capping agent, therefore it is very important to know whether the interface is a better correspondence between the properties of the material and its surface characteristics (Palencia et al., 2017, Shah et al., 2015). Consequently, to investigate the research line of bionanosynthesis is to use extracts or biological entities, and is focused for making of nanomaterials using biomolecules with a known nature and clearly identified such as amino acids, proteins, polysaccharides and pigments (Palencia et al., 2017; Salouti and Derakhshan, 2019; Srivastava et al., 2015).

In the case of amino acids, these are biomolecules containing amino and carboxylic acid functional groups in their fundamental structure, therefore, they are capable of coordinating metal ions through non-binding electrons; but in turn, the functionality of the side chain favors the generation of redox reactions through functional groups on the side chain like hydroxyl, thiol, amine and carboxylic acid (Mansi et al., 2020; Palencia et al., 2016; Palencia et al., 2017; Salouti and Derakhshan, 2019; Srivastava et al., 2015). Accordingly, biomolecules such as peptides and proteins also have functional groups that are able to carry out reduction reactions, however, because of greater molecular weight compared with amino acids, stabilization of nanomaterials is also associated with their size and macromolecular behavior (Davaeifar et al., 2019; Marimón-Bolívar and González, 2018).

4. Green Synthesized Nanomaterials in Bio- and Nanoremediation Processes

Nanomaterials synthesized by green chemistry principles are a promising alternative for developing t efficient, eco-friendly and sustainable novel remediation methods to eliminate emerging pollutants and mixed contaminations. Nevertheless, green-synthesized nanomaterials are, from a broad view, resourceful products from a still-emerging area; they have been mostly evaluated for reducing the pollutant concentration at the laboratory scale or in simplified models of natural systems, therefore, their synergy or biostimulation along bioremediation processes has been carried out primarily with microorganisms, and consequently, its analysis at the field level or in real conditions of use is still developing. The use of bionanomaterials for the remediation of natural systems requires advancing key aspects to warrant the reproducibility of nanobiotechnological synthesis and decrease the potential risk that entails the release of manufactured materials to the environment without adequately knowing their structure-ecotoxic activity relation, as well as its mobility, environmental fate and Life Cycle Analysis (Cecchin et al., 2017; Srivastava et al., 2015). A large amount of research has been conducted in order to develop and evaluate "green" nanomaterials as a potential technological solution, and this way, to carry out "green" remediation processes. In this context, the most investigated nanomaterials are related to metals, their oxides and nanocomposites manufactured with agro-industrial waste, the last nanomaterials of increasing interest due to their mechanical stability and direct relationship with the implementation of circular economy's concepts (Khan, 2020, Ortega-Calvo et al., 2016).

4.1 Metallic nanomaterials

Metallic nanoparticles have been widely studied due to their catalytic, optical and biological properties, and also, due to their redox capacity and the ease in their manufacture. Thus, obtaining nanomaterials with different sizes, morphologies and functional characteristics allow them to participate in several pollutant remediation mechanisms, which can occur for one or more molecular targets, such as adsorptive, catalytic and oxidative processes (Bethi and Sonawane, 2018; Das et al., 2018; Livingston et al., 2020; López-Valdez et al., 2018). Metallic nanomaterials that are used often, include Fe, Cu, Ag, and Au can be synthesized by a green method using microorganisms and plants and have shown the ability to reduce the concentration, in model systems and aqueous solutions, of a wide variety of contaminants such as colorants, heavy metals and pharmaceuticals (see Table 9.4).

On the other hand, among various metallic nanomaterials developed to carry out remediation processes in soils and groundwater, the most relevant are those based on iron, it is estimated that more than 80% of the investigations and 60% of the environmental remediation techniques are related to

Table 9.4. Green synthesized metal nanomaterials and their use in remediation.

Nanomaterial	Synthesis	Pollutant	Mechanism	Ref.
Spherical AgNPs (5–20 nm)	The NPs were obtained in 1 hr, under ambient conditions, using $AgNO_3$ and aqueous extract of the leaves of *Trigonella foenum-graecum*.	Reactive blue 19 (RB19) Reactive yellow 186 (RY186)	Photocatalysis (88% for RB19 and 86% for RY186, in 3 hr)	Singh et al., 2019
Amorphous FeNPs (15–100 nm)	Aqueous extract of *Ilex paraguariensis* leaves, composed primarily of catechins that reduced Fe (III) to nano zero valent iron (nZVI).	Cr (VI)	Oxidation-reduction reactions (up to 85% in 20 min)	García et al., 2019
Quasi-spherical AuNPs (~ 72.32 nm)	The *Cladosporium oxysporum* AJP03 fungus was isolated from a soil sample from the district of Goa, India. The fungus was cultivated and extracellular metabolites were extracted; these biomolecules allowed the reduction of the metal of $AuHCl_4$ salt.	Rhodamine B	Catalysis (100% in 7 min, in the presence of $NaBH_4$)	Bhargava et al., 2016
Spherical FeNPs (~ 22.6 nm)	The aqueous extract of the leaves of *Emblica officinalis* was used for the reduction of Fe (III), at ambient conditions during 10 min of stirring.	Pb(II)	Catalysis of Fenton, redox reactions, and adsorption (up to 100 ppm in 24 hr)	Kumar et al., 2015
Spherical, triangular, and hexagonal AuNPs (20 nm)	The *Trichoderma sp.* WL-Go was isolated from a sludge sample from Dalian, China. The fungus was cultivated and incubated with $HAuCl_4$ in the optimal conditions: pH 7–11, 1.0 mM of $HAuCl_4$, and 0.5 g of the biomass; conditions that allowed to obtain AuNPs.	Acid Brilliant Scarlet GR (ABSGR) Acid Red B (ARB) Acid Orange G (AOG) Acid Black 1 (AB1) Reactive Red X-3B (RRX-3B) Reactive Black (RB) Reactive Red (RR) Cation Red (CR)	Catalysis (94.7% for ABSGR in 120 min, 84.8% for ARB in 120 min, 79.7% for AOG in 100 min, 82.9% for AB1 in 100 min, 73.3% for RRX-3B in 120 min, 56.2% for RB in 100 min, 46.1% for RR in 180 min and, 41.7% for CR in 180 min)	Qu et al., 2017
Spherical FeNPs (4.6 –30.6 nm)	The NPs were synthesized using the extract of the leaves of *Plantago major* and $FeCl_3$, under ambient conditions for 24 hr.	Methyl orange	Catalysis (83.3% in 6 hr, in the presence of H_2O_2)	Lohrasbi et al., 2019
Spherical AgNPs (60 –80 nm)	An aqueous cell extract of the cyanobacterium *Microchaete* NCCU-342 and $AgNO_3$ was used to obtain the NPs after 1 hr at 60°C and exposure to sunlight.	Methyl red	Catalysis (84.60% in 2 hr)	Husain et al., 2019
Spherical FeNPs (~ 70 nm)	The NPs were synthesized using the aqueous extract of *Eucalyptus* leaves and $FeSO_4$ as a source of Fe, which was carried out at room temperature with an N_2 atmosphere for 30 min.	Cr (VI)	Adsorption (98.6% in 8 hr)	Jin et al., 2018
Monoclinic and tetragonal Zr NPs (~15 nm)	The NPs were obtained by cell-free supernatant from *Pseudomonas aeruginosa* and $Cl_2H_{18}O_9Zr$.	Tetracycline	Adsorption (up to to 98.73% in 30 min)	Debnath et al., 2020
Amorphous FeNPs (2–120 nm)	The extract of *Eucalyptus urophylla*, *Eucalyptus grandis* leaves and $FeNO_3 \cdot 6H_2O$ were used to synthesize FeNPs, under ambient temperature, and N_2 atmosphere, in 8 min.	As(V)	Adsorption (55.9% in 96 hr)	Wu et al., 2019

zero valent iron type materials (Bandala and Berli, 2018; Bhandari 2018; Cecchin et al., 2017; Fajardo et al., 2020; Ortega-Calvo et al., 2016). The suitability of iron-based materials is related to their ability to carry out remediation processes through oxidative mechanisms (associated with their reduction potential of −0.44 V allowing a superficial covering of iron oxide on the surface of zero valent iron particles), catalytic dehydrohalogenation reactions, precipitation reactions and surface adsorption processes (Azubuike et al., 2016; Davies et al., 2017; Halada and Orlov, 2018; Lin et al., 2020; Lu et al., 2016; Mu et al., 2017).

In addition, compared with other metals, precursors of iron-based nanomaterials present a greater abundance, a lower cost, reduced handling risks and a greater ease in their operation; besides, a lesser potential risk of ecotoxicity. Accordingly, in the case of soils, iron is a micronutrient and promoter for the growth of phytoremediation plants (e.g., *Lolium perenne*), as well as a stimulating of metabolic activity of soil microorganisms (Cecchin et al., 2017; Iavicoli et al., 2017; Lu et al., 2016; Mu et al., 2017; Zelić et al., 2018). An example of the cost-benefit advantage of iron-based nanomaterials, compared with analogous nanomaterials obtained from noble metals such as silver, can be found in research directed toward the study of degradation efficiency of pollutants by the use of nanobioremediation in aqueous systems, simultaneously using plants and nanomaterials supported on activated carbon. In this way, iron nanoparticles bionanosynthesized using the aqueous extract of *Ipomoea carnea* leaves and $FeCl_3$ and silver nanoparticles bionanosynthesized using the aqueous extract of *Brassica alba seeds* and $AgNO_3$ were compared in a same remediation process. For bionanoremediation, the nanomaterials were used separately, but also in conjunction with *Plantago major* (which is a land plant species from class Angiospermae or flowering plants and from the family Plantaginaceae). Results showed that after 24 hr, the iron nanoparticles allowed the reduction of about 94% of the pollutant, corresponding to an effectivity higher than 23% than those obtained by individual bioremediation processes, and about 2% higher compared with the use of silver nanoparticles by the combined system. It should be noted that though efficiency obtained between combined systems is not significant, other aspects it should be taken in consideration such as the cost of $AgNO_3$ in respect to the cost of $FeCl_3$, the frequent presence, and larger concentrations of iron in respect to the presence and concentrations of silver in the environment (Romeh and Saber, 2020).

At present, iron-based nanomaterials have shown an ability to operate in soil nanoremediation under variable conditions of pH, temperature, geography and nutritional conditions, making these materials versatile remediation agents with industrial and commercial scalability (Bardos et al., 2018). Their commercialization is substantive in the market, through products such as Carbon-Iron®, NANOFER 25S®, NANOFER STAR® and the use of Permeable Reactive Barriers (PRB) (Bardos et al., 2018; Cecchin et al., 2017; Halada and Orlov, 2018; Kim et al., 2019; Zelić et al., 2018).

4.2 Metal oxide nanomaterials

Nanomaterials based on metal oxides are usually synthesized using transition metals due to their low cost, redox activity and catalytic capacity (Manav et al., 2018). This preference is also manifested in the development of green synthesis of nanomaterials and their evaluation in sustainable remediation processes. Table 9.5 demonstrates the use of Cu, Ni, Ti, Zn and Fe oxides for decrease in the concentration of polluting substances such as heavy metals and emerging pollutants such as colorants and pharmaceutical products and by-products.

The most widely used transition metal in research on green synthesis of oxide-based nanomaterials is iron; as mentioned earlier, the nano zero valent iron nanoparticles are iron cores with a superficial covering of iron oxides that favors adsorption and precipitation of pollutants, in consequence, nanomaterials made up only of iron oxides exhibit these properties, in addition to magnetization characteristics facilitating their separation in heterogeneous systems (Iavicoli et al., 2017; Lu et al., 2016; Mu et al., 2017; Yadav et al., 2018). In addition, iron oxides present a wide variety of compounds such as oxides, hydroxides and oxyhydroxides, among which magnetite (Fe_3O_4), maghemite (γ- Fe_2O_4) and hematite (α-Fe_2O_3) are the most common in natural matrices, due to this, it is feasible to assume

Table 9.5. Green synthesized metal oxide nanomaterials and their use in remediation.

Nanomaterial	Synthesis	Pollutant	Mechanism	References
α-Fe_2O_3 NPs (20–40 nm)	The culture supernatant of the *Bacillus cereus* SVK1 bacteria was used for the synthesis of hematite NPs, in alkaline medium with Fe (III), at 37°C for 48 hr.	Carbamazepine	Adsorption (~ 100% in 150 min)	Rajendran and Sen, 2016; Rajendran and Sen, 2018
FeOOH NPs (9–15 nm)	The exopolysaccharides of the strain *Klebsiella oxytoca* DSM 29614 were used to synthesize FeOOH NPs through the citric fermentation process and the formation of a ferric exopolymeric hydrogel.	As (III) As(V)	Adsorption (87–95% of As(V) in 5 min, and 45–61% of As (III) in 5 min)	Casentini et al., 2019
NiO nanorod (~ 32 nm)	The nanorods were synthesized using the aqueous extract of the *Phoenix dactylifera* plant and Ni $(NO_3)_2$, at ambient conditions for 30 min, and subsequent dehydration at 250°C for 20 min.	4-chlorophenol	Photocatalytic degradation (up to 92% in 210 min, under conditions of sunlight)	Ezhilarasi et al., 2020
Fe_2O_3 and Fe_3O_4 NPs (10–80 nm)	The NPs were obtained by reducing the $FeCl_3$ and $FeSO_4$ salts with the *Parkia speciosa Hassk* pod extract under ambient conditions and subsequent calcination at 400 °C for 2 hr.	Bromophenol blue	Photocatalytic degradation (98% under visible light)	Fatimah et al., 2020
Circular and Hexagonal CuO NPs (10– 50 nm)	The leaves extract of *Azadirachta indica* was used for the reduction of Cu (II) of the Cu_2O_3S salt, at 75 – 80 °C for 6 hr.	Reactive red 120	Catalysis (~ 80% in 4 hr, in presence of $NaBH_4$)	Thirumurugan et al., 2017
Mesoporous Fe_3O_4 NPs (pore diameter ~ 7.66 nm)	*Peltophorum pterocarpum* pod extract and a ferrous sulfate heptahydrate solution were used to obtain NPs in alkaline pH, at 90°C for 10 min.	Methylene Blue	Adsorption (89% in 45 min)	Dash et al., 2019
Irregular ZnO (20–40 nm)	The NPs were synthesized using an extract of *Eucalyptus spp.* leaves and zinc nitrate solution at 110°C for 2 hr; subsequently, the solution was sonicated for 45 min and alkalized to pH 11.	Malachite green Congo Red	Adsorption (89.56% of Malachite green and 82.97% of Congo red)	Chauhan et al., 2020
Spherical CuO NPs (5–20 nm)	*Citrus limon* juice, polyvinylpyrrolidone, and Cu $(CH_3COO)_2 \cdot H_2O$, were subjected to heating at 110°C for 2 hr obtaining an orange dispersion of CuO NPs.	Cr (VI)	Adsorption (98.8%)	Mohan et al., 2015
Spherical TiO_2 (~ 10 nm)	The NPs were synthesized using ovalbumin from expired eggs and titanium (IV) butoxide, which was subjected to ultrasound for 5 min obtaining a gel, subjected to drying at 100°C for 24 hr.	Rhodamine B	Photocatalytic degradation (96.6% in 210 min)	Kadam et al., 2020

that the synthetic nanomaterials of these oxides represent a lower risk of ecotoxicity compared to other metallic oxides (Davies et al., 2017; Lu et al., 2016).

An example of the above is the synthesis of Fe_3O_4 nanoparticles from iron (III) minerals (ferrihydrite and goethite) or iron oxides obtained for the sheaths of the *Leptothrix* sp. bacterium, by *Geobacter sulfurreducens* with acetate as an electron donor. Nanoparticles obtained by this procedure were described as carriers of Fe (II) and showed an ability to reduce Cr (VI) to Cr (III) by up to 85% (Joshi et al., 2018); in addition, by using an analogous procedure, it was possible to obtain Fe_3O_4 nanoparticles that showed the ability to remove Tc (VII) and reduce it to Tc (IV) (Newsome et al., 2019). On the other hand, magnetite nanomaterials, with a spherical shape and a diameter between 8–12 nm, have been used in eliminating emerging organic contaminants from pharmaceutical products (e.g., erythromycin, piperacillin, ampicillin, tetracycline and tazobactam); these nanomaterials have been obtained using extracts of agroindustrial residues (e.g., *Citrus limon*, *Vitis vinifera* and *Cucumis sativus*) and the salts $FeCl_2 \cdot 4H_2O$ and $FeCl_3 \cdot 6H_2O$ in an alkaline medium at 80°C (Stan et al., 2017).

4.3 Nanocomposites

A disadvantage of nano-objects is that these tend to agglomerate and destabilize, leading to the decrease or loss of their functional properties; in turn, in the field of environmental remediation, the release of nanometric systems to the environment is associated with low control over their mobility and, therefore, a limited knowledge about their environmental fate, with their recovery being a non-viable alternative in many situations during *ex situ* operations (Lu et al., 2016). Therefore, a promising alternative to solve these drawbacks is based on obtaining nanocomposites, which can be defined as materials made up of two or more components, in which at least one of their phases is on the nanometric scale (Córdoba and Palencia, 2017). This is the case of nanomaterials composed of different types of nanometric phases that provide functional properties to the system, such as magnetization or synergy in their adsorptive characteristics; as well as the inclusion of nano-objects in macroscopic matrices, which entails greater ease in handling and favors the control of the release of nanoparticles to the environment (Córdoba and Palencia, 2017; Lu et al., 2016; Villalobos et al., 2019). In environmental remediation, both strategies have been used to evaluate the decrease in the concentration of polluting substances, in model systems or aqueous solutions (see Table 9.6). However, the use of phases with some of their dimensions on a macroscopic scale has been one of the most investigated alternatives because the phase tends to have notable mechanical and thermal properties, in addition to the fact that it can be obtained through agro-industrial waste or the use of natural sources such as nanoclays (Zhao et al., 2020).

Some examples of nanocomposites in bionanoremediation are: Bimetallic Fe-Cu nanoparticles supported on bentonite that are synthesized using ethanolic extract of the shell of *Punica granatum* for the reduction of $FeCl_3$ and $CuCl_2$ in the presence of carboxymethylcellulose as a stabilizing agent, and bentonite as macroscopic phase; in addition, these nanocomposites have been used for the removal of tetracycline showing an efficiency of 95%, but they also showed greater efficiency than that obtained from nanoparticles without the presence of bentonite (Gopal et al., 2020). Montmorillonite have been chemically modified with functional polymers and biomolecules such as sorbitol and citric acid to be used as growth substrates for the inoculation of nitrogen-fixing bacteria, e.g., *Azotobacter chroococcum*. This approach called geomimicry has a remarkable potential in environmental remediation in degraded soils which can be adapted for the application of phytoremediation strategies (Lerma et al., 2018). Porous nanocomposites are able of removing anthracene and aromatic hydrocarbons from an aqueous solution after 60 min (~ 96%) have been synthesized from palm residues and MgO nanoparticles. In general, palm residues were activated by alkaline KOH solution, their size was reduced by maceration until obtaining particles with a diameter lower than 50 mm and subjected to carbonization. Finally, the particles were modified by adding of MgO nanoparticles synthesized from Mg $(NO_3)_2$ and aqueous extract of *Azadirachta indica* leaves in alkaline medium (Jagadeesan et al., 2019).

Table 9.6. Green synthesized nanocomposites and their use in remediation.

Nanomaterial	**Synthesis**	**Pollutant**	**Mechanism**	**References**
Spherical Fe_2O_3/Ag (15–25 nm)	The nanocomposite was obtained by *Aloe vera* gel extract, Fe $(NO_3)_3$ and $AgNO_3$, using microwave radiation equivalent to 120°C for 1 hr.	Methylene Blue	Photocatalysis (88.2% in 140 min)	Saranya et al., 2020
Irregular Fe_2O_3/Ag (50–90 nm)	*Psidium guajava* leaf extract was added to an aqueous solution of $AgNO_3$ and Fe $(NO_3)_3$, in ambient conditions, that allowing obtained the precipitate of nanoparticles in only 15 min.	Cr (VI)	Adsorption (up to 97% in 15 min)	Biswal et al., 2020
CuNPs NPs/eggshell, Fe_3O_4NPs/eggshell, and CuNPs/Fe_3O_4NPs/eggshell nanocomposite	The Leaves extract of *Orchis mascula* L. was used for the synthesis of the composite in the presence of eggshell, $FeCl_3$, and/or $CuCl_2$, with heating at 70°C for 3 hr.	Congo red 4-nitrophenol Methyl orange Methylene blue Rhodamine B	Catalytic reduction (~ 100%)	Nasrollahzadeh et al., 2016
Reduced graphene oxide (rGO)/FeNPs	The nanocomposite was obtained by green tea extract, $FeCl_3$, CH_3COONa, and graphene oxide, these were kept in contact for 8 hr at 80°C.	Mitoxantrone	Adsorption (95% in 5 min)	Wu et al., 2020
AgNPs/Seashell (20–50 nm)	The nanocomposite was obtained through an aqueous *Bunium persicum* extract, with the presence of $AgNO_3$ and seashell, at a temperature of 70°C for 1 hr.	Congo Red 4-nitrophenol Methyl Orange Methylene Blue	Catalytic reduction (100% with $NaBH_4$)	Rostami-Vartooni et al., 2016
rGO/FeNPs (20–40 nm)	*Eucalyptus* leaf extract carried out the reduction of graphene oxide and $FeCl_3$ by ultrasound for 30 min.	Pb(II)	Adsorption (72.7% in 10 min)	Xiao et al., 2019
Spherical Ag/FeO_2 NPs (~ 92 nm)	The aqueous extract of the *Amaranthus blitum* leaf was used as a reducing agent for Fe $(NO_3)_3$ and $AgNO_3$ at basic pH at 50°C for 3 hr.	Caffeine	Photocatalysis (up to 99.9%, under a fluorescent lamp)	Muthukumar et al., 2020

5. Limitations and Prospects of Green Nanomaterials in Bio or Nanoremediation

As stated earlier, it is clear that the use of green-based nanomaterials in processes for bio- or nanoremediation is a promising technology for sustainable environmental remediation, representing a field of interest not only for the scientific community but also for government organizations that aspire implementing this technology, to tackle the technological problem that involves the remediation of natural systems. In this way, projects such as NanoRem (Nanotechnology for the Remediation of Contaminated Soils) (2013–2017) have emerged, which allowed the development of research on the application of iron-based nanomaterials for environmental remediation in the European Union, as well as the commercialization of nanoproducts for the *in situ* remediation of soils and groundwater, oriented towards an ecological design. These models permitted progress in the development of the structure-activity and risk-benefit relationships of nanoremediation, their market and Life Cycle Analysis (Bardos et al., 2018; Corsi et al., 2018). From the perspective of sustainability, the CReO FESR NANOBOND project (2014–2020) focused on establishing a sustainable remediation of sludge and sediment through hybrid technologies, using dehydration tubes and nanostructured materials generated from renewable resources, with adsorption capacity organic and inorganic pollutants, under an environmentally safe guidance (Bardos et al., 2018). Similarly, the TANIA project (Pollution Treatment through Nanoremediation) (2017–2021) seeks to advance in the development of standardized methodologies, pilot applications and regulations of nanomaterials in soil and water nanoremediation processes through a sustainable and safe approach (Bardos et al., 2018). These projects demonstrate the relevance and market potential of synthesized nanomaterials through holistic sustainability considerations in environmental remediation.

However, green nanomaterials for environmental remediation present notable limitations, primarily associated with intrinsic nature of nanostructures, the inherent molecular variability of biosynthetic agents and the complexity in the nanomaterial-natural system interactions that hinder their Life Cycle Analysis (Khan, 2020). More specifically, nanomaterials synthesized through biological molecules or organisms present difficulties in their reproducibility because their molecular composition is not static but dependent on nutritional conditions, stressors and interactions with the environment (Joshi et al., 2018; Srivastava et al., 2015).

In addition, due to the size and reactivity of nanomaterials, they can interact with molecular components of biological systems, leading to disruption in cell membranes, oxidation of enzymatic and lipid components, affecting the development of biological organisms (Iavicoli et al., 2017; Ortega-Calvo et al., 2016; Sohail et al., 2019). The above acquires particular relevance for the environmental applications of nanomaterials, since inevitably the risk of release and interaction of them with natural systems emerges, for instance, in the case of soil, its complex composition including living organisms, humified organic substances, inorganic components, allow numerous interaction routes with the nanoparticles, making it practically impossible to establish the effect resulting from the interaction with each individual component, with multiple components and with the entire system (Dror et al., 2017; Qian et al., 2020). Besides, nanomaterials can be assumed to be anthropogenic substances with high potential for being dangerous to the environment because they can undergo agglomeration, oxidation, ligand exchanges and affecting plants and soil microbial communities, through an alteration in their metabolic development, as well as bioaccumulation in the trophic chain through the apoplastic and symplastic pathways of plants, allowing the accumulation of the nanomaterial in organs such as leaves and fruits, in consequence, nanomaterials are typified as emergent pollutants at the same level than that of pharmaceutical products and by-products (Bandala and Berli, 2018; Bhandari 2018; Livingston et al., 2020; López-Valdez et al., 2018; Poynton and Robinson, 2018; Qian et al., 2020; Sohail et al., 2019). However, to avoid in generalizations that are grave, it is clear that nanomaterials based on the principles of green chemistry, considering the Life Cycle Analysis and the sustainability philosophy from their design, are expected to show planetary health risks lower than many toxic substances currently released into the environment (León-Silva et al., 2018).

On the other hand, the scientific community and governments must advance in the development of international parameters allowing to define standards, protocols and universal regulations for the evaluation and implementation of nanotechnological materials for environmental applications. It is also necessary to advance in the development of nanomaterial libraries in order to achieve generalized and available information about surface properties, interactions and the risk of the nanomaterials (Iavicoli et al., 2017; Li et al., 2018; Palencia et al., 2015; Sánchez et al., 2018; Sohail et al., 2019).

6. Conclusions

The continuous release of manufactured residues, products and by-products represents a constant danger for the sustainability of life on the planet. Therefore, environmental pollution, primarily related to soils and water, is a reality including social, public, environmental and economic aspects; and consequently, mitigation and remediation strategies able to improve or restore the ecosystems without producing a direct or indirect negative impact are largely required. Between efficient, fast and eco-friendly alternatives for environmental remediation processes are the use of nanomaterials, which should be ideally synthesized under the design precepts of green chemistry and environmental sustainability. In general, numerous nanostructured systems can be designed and manufactured by the use of microorganisms and plants, besides, these can be applied as active components in remediation systems; depending on the nature, synthesis, composition, morphological features and using the environment, nanomaterials can interact, reduce, immobilize and remove a wide variety of substances, such as hydrocarbon-type organic compounds, halogenated, pesticides, pharmaceutical products, dyes, among others. However, due to this, environmental bionanotechnology is an area that is still emerging, several aspects must be evaluated and carefully analyzed before their application at a commercial level and under real conditions.

References

Abbas, S., Nasreen, S., Haroon, A. and Ashraf, M.A. 2020. Synthesis of silver and copper nanoparticles from plants and application as adsorbents for naphthalene decontamination. Saudi. J. Biol. Sci. 27: 1016–1023. DOI: 10.1016/j.sjbs.2020.02.011.

Afreen, S., Omar, R.A., Talreja, N., Chauhan, D. and Ashfaq, M. 2018. Carbon-based nanostructured materials for energy and environmental applications. pp. 369–392. *In*: Prasad, R. and Aranda, E. (eds.). Approaches in Bioremediation: The New Era of Environmental Microbiology and Nanobiotechnology. Cham. (Switzerland): Springer International Publishing. DOI: 10.1007/978-3-030-02369-0_17.

Agarwal, H., Nakara, A., Menon, S. and Shanmugam, V.K. 2019. Eco-friendly synthesis of zinc oxide nanoparticles using *Cinnamomum tamala* leaf extract and its promising effect towards the antibacterial activity. J. Drug. Deliv. Sci. Tec. 53: 101212. DOI: 10.1016/j.jddst.2019.101212.

Alam, M.T., Rauf, M.A., Siddiqui, G.A., Owais, M. and Naeem, A. 2018. Green synthesis of silver nanoparticles, its characterization, and chaperone-like activity in the aggregation inhibition of α-chymotrypsinogen A. Int. J. Biol. Macromol. 120(B): 2381–2389. DOI: 10.1016/j.ijbiomac.2018.09.006.

Albini, A. and Protti, S. 2016. Introduction. pp. 1–10. *In*: Albini, A. and Protti, S. (eds.). Paradigms in Green Chemistry and Technology. Cham (Switzerland): Springer International Publishing. DOI: 10.1007/978-3-319-25895-9_1.

Albini, A. and Protti, S. 2016. Green metrics, an abridged glossary. pp. 11–24. *In*: Albini, A. and Protti, S. (eds.). Paradigms in Green Chemistry and Technology. Cham (Switzerland): Springer International Publishing. DOI: 10.1007/978-3-319-25895-9_2.

Ali, J., Ali, N., Jamil, S.U.U., Waseem, H., Khan, K. and Pan, G. 2017. Insight into eco-friendly fabrication of silver nanoparticles by *Pseudomonas aeruginosa* and its potential impacts. J. Environ. Chem. Eng. 5(4): 3266–3272. DOI: 10.1016/j.jece.2017.06.038.

Anastas, N.D., Leazer, J., Gonzalez, M.A. and DeVito, S.C. 2018. Expanding rational molecular design beyond pharma. pp. 29–48. *In*: Constable, D.J.C. and Jimenéz-González, C. (eds.). Handbook of Green Chemistry, Volume 11: Green Metrics. Weinheim (Germany): Wiley BCH. DOI: 10.1002/9783527628698.hgc125.

Andrade, D.C. and dos Santos, E.V. 2020. Combination of electrokinetic remediation with permeable reactive barriers to remove organic compounds from soils. Curr. Opin. Electrochem. 22: 136–144. DOI: 10.1016/j.coelec.2020.06.002.

Arregui, L., Ayala, M., Goméz-Gil, X., Gutiérrez-Soto, G., Hernández-Luna, C.E., Herrera, M., Levin, L., Rojo-Domínguez, A., Romero-Martínez, D., Saparrat, M.C.N., Trujillo-Roldán, M.A. and Valdez-Cruz, N.A. 2019. Laccases: Structure, function, and potential application in water bioremediation. Microb. Cell. Fact. 18: 200. DOI: 10.1186/s12934-019-1248-0.

Arya, A., Mishra, V. and Chundawat, T.S. 2019. Green synthesis of silver nanoparticles from green algae (*Botryococcus braunii*) and its catalytic behavior for the synthesis of benzimidazoles. Chem. Data Collect. 20: 100190. DOI: 10.1016/j.cdc.2019.100190.

Awad, A.M., Jalab, R., Benamor, A., Nasser, M.S., Ba-Abbad, M.M., El-Naas, M. and Mohammad, A.W. 2020. Adsorption of organic pollutants by nanomaterial-based adsorbents. J. Mol. Liq. 301(1): 112335. DOI: 10.1016/j.molliq.2019.112335.

Azubuike, C.C., Chikere, C.B. and Okpokwasili, G.C. 2016. Bioremediation techniques-classification based on site of application. World J. Microbiol. Biotechnol. 32: 180. DOI: 10.1007/s11274-016-2137-x.

Banasiuk, R., Krychowiak, M., Swigon, D., Tomaszewicz, W., Michalak, A., Chylewska, A., Ziabka, M., Lapinski, M., Koscielska, B., Narajczyk, M. and Krolicka, A. 2020. Carnivorous plants used for green synthesis of silver nanoparticles with broad-spectrum antimicrobial activity. Arab. J. Chem. 13(1): 1415–1428. DOI: 10.1016/j.arabjc.2017.11.013.

Bandala, E.R. and Berli, M. 2018. Nanomaterials: New agrotechnology tools to improve soil quality? pp. 127–140. *In*: López-Valdez, F. and Fernández-Luqueño, F. (eds.). Agricultural Nanobiotechnology. Cham (Switzerland): Springer International Publishing. DOI: 10.1007/978-3-319-96719-6_7.

Bardos, P., Merely, C., Kvapil, P. and Koschitzky, H.P. 2018. Status of nanoremediation and its potential for future deployment: Risk-benefit and benchmarking appraisals. Remediation: The Journal of Environmental Cleanup Costs, Technologies & Techniques 28(3): 43–56. DOI: 10.1002/rem.21559.

Bethi, B. and Sonawane, S.H. 2018. Nanomaterials and its application for clean environment. pp. 385–409. *In*: Bhanvase, B.A., Pawade, V.J., Dhoble, S.J., Sonawane, S.H. and Ashokkumar, M. (eds.). Nanomaterials for Green Energy. United States: Elsevier. DOI: 10.1016/B978-0-12-813731-4.00012-6.

Bhandari, G. 2018. Environmental nanotechnology: Applications of nanoparticles for bioremediation. pp. 301–315. *In*: Prasad, R. and Aranda, E. (eds.). Approaches in Bioremediation: The New Era of environmental Microbiology and Nanobiotechnology. Cham (Switzerland): Springer International Publishing. DOI: 10.1007/978-3-030-02369-0_13.

Bharadwaj, A. 2018. Bioremediation of xenobiotics. pp. 1–13. *In*: Parmar, V.s., Malhotra, P. and Mathur, D. (eds.). Green Chemistry in Environmental Sustainability and Chemical Education. Singapore: Springer Singapore. DOI: 10.1007/978-981-10-8390-7_1.

Bharagava, R.N., Chowdary, P. and Saxena, G. 2017. Bioremediation: An eco-sustainable green technology, its applications and limitations. pp. 1–22. *In*: Bharagava, R.N. (ed.). Environmental Pollutants and their Bioremediation Approaches. Boca Raton (United States): CRC Press. DOI: 10.1201/9781315173351-2.

Bhargava, A., Jain, N., Khan, M.A., Pareek, V., Dilip, R.V. and Panwar, J. 2016. Utilizing metal tolerance potential of soil fungus for efficient synthesis of gold nanoparticles with superior catalytic activity for degradation of rhodamine B. J. Environ. Manag. 183(1): 22–32. DOI: 10.1016/j.jenvman.2016.08.021.

Biswal, S.K., Panigrahi, G.K. and Sahoo, S.K. 2020. Green synthesis of Fe_2O_3-Ag nanocomposite using *Psidium guajava* leaf extract: An eco-friendly and recyclable adsorbent for remediation of Cr(VI) from aqueous media. Biophys. Chem. 263: 106392. DOI: 10.1016/j.bpc.2020.106392.

Bolade, O.P., Williams, A.B. and Benson, N.U. 2020. Green synthesis of iron-based nanomaterials for environmental remediation: a review. Environ. Nanotech Monit. & Manage. 13: 100279. DOI: 10.1016/j.enmm.2019.100279.

Brevik, E.C., Slaughter, L., Singh, B.R., Steffan, J.J., Collier, D., Barnhart, P. and Pereira, P. 2020. Soil and human health: Current status and future needs. Air, Soil and Water Research 13: 1–23. DOI: 10.1177/1178622120934441.

Carbone, K., Paliotta, M., Micheli, L., Mazzuca, C., Cacciotti, I., Nocente, F., Ciampa, A. and Dell'Abate, M.T. 2019. A completely green approach to the synthesis of dendritic silver nanostructures starting from white grape pomace as a potential nanofactory. J. Arab. Chem. 12: 597–609. DOI: 10.1016/j.arabjc.2018.08.001.

Casentini, B., Gallo, M. and Baldi, F. 2019. Arsenate and arsenite removal from contaminated water by iron oxides nanoparticles formed inside a bacterial exopolysaccharide. J. Environ. Chem. Eng. 7: 102908. DOI: 10.1016/j.jece.2019.102908.

Cecchin, I., Reddy, K.R., Thomé, A., Tessaro, E.F. and Shnaid, F. 2017. Nanobioremediation: Integration of nanoparticles and bioremediation for sustainable remediation of chlorinated organic contaminants in soils. Int. Biodeterior. Biodegrad. 119: 419–428. DOI: 10.1016/j.ibiod.2016.09.027.

Chauhan, A.K., Nataria, N. and Garg, V.K. 2020. Green fabrication of ZnO nanoparticles using *Eucalyptus* spp. Leaves extract and their application in wastewater remediation. Chemosphere 247: 125803. DOI: 10.1016/j.chemosphere.2019.125803.

Chen, L., Ni, R., Yuan, T., Gao, Y., Kong, W., Yue, Q. and Gao, B. 2020. Effects of green synthesis, magnetization and regeneration on ciprofloxacin removal by bimetallic nZVI/Cu composites and insights of degradation mechanism. J. Hazard Mater. 382: 121008. DOI: 10.1016/j.jhazmat.2019.121008.

Ciambelli, P., La Guardia, G. and Vitale, L. 2019. Nanotechnology for green materials and processes. pp. 97–116. *In*: Basile, A., Centi, G., De Falco, M. and Iaguaniello, G. (eds.). Catalysis, Green Chemistry and Sustainable Energy. Oxford (United Kingdom): Elsevier. DOI: 10.1016/B978-0-444-64337-7.00007-0.

Combatt, E., Polo, V., Racero, L., Gómez, L. and Cantero, M. 2017. Process of soil degradation as a result of agricultural activities of different zones in the Montería municipality, department of Córdoba-Colombia. J. Sci. Technol. Appl. 3: 5–14. DOI: 10.34294/j.jsta.17.3.20.

Córdoba, A. and Palencia, M. 2017. Development of nanostructured crosslinkers with antimicrobial properties for free-radical polymerization. J. Sci. Technol. Appl. 2: 54–64. DOI: 10.34294/j.jsta.17.2.14.

Corsi, I., Winther-Nielsen, M., Sethi, R., Punta, C., Torre, C.D., Libralato, G., Lofrano, G., Sabatini, L., Fiordi, L., Cinuzzi, F., Caneschi, A., Pellegrini, D. and Buttino, I. 2018. Ecofriendly nanotechnologies and nanomaterials for environmental applications. Ecotoxicol. Environ. Saf. 154(15): 237–244. DOI: 10.1016/j.ecoenv.2018.02.037.

Dahoumane, S.A., Mechouet, M., Wijesekera, K., Filipe, C.D.M., Sicard, C., Bazylinski, D.A. and Jeffryes, C. 2017. Algae-mediated biosynthesis of inorganic nanomaterials as a promising route in nanobiotechnology—A review. Green Chem. 19: 552–587. DOI: 10.1039/C6GC02346K.

Das, S., Chakraborty, J., Chatterjee, S. and Kumar, H. 2018. Prospects of biosynthesized nanomaterials for the remediation of organic and inorganic environmental contaminants. Environ. Sci. Nano. 5: 2784–2808. DOI: 10.1039/C8EN00799C.

Das, S., Sen, B. and Debnath, N. 2015. Recent trends in nanomaterials applications in environmental monitoring and remediation. Environ. Sci. Pollut. Res. 22: 18333–18344. DOI: 10.1007/s11356-015-5491-6.

Dash, A., Ahmed, M.T. and Selvaraj, R. 2019. Mesoporous magnetite nanoparticles synthesis using the *Peltophorum pterocarpum* pod extract, their antibacterial efficacy against pathogens and ability to remove a pollutant dye. J. Mol. Struc. 1178: 268–273. DOI: 10.1016/j.molstruc.2018.10.042.

Davaeifar, S., Modarresi, M.H., Mohammadi, M., Hashemi, E., Shafiei, M., Maleki, H., Vali, H., Zahiri, H.S., Noghabi, K.A. 2019. Synthesizing, characterizing, and toxicity evaluating of Phycocyanin-ZnO nanorod composites: A back to nature approaches. Colloids Surf. B 175: 221–230. DOI: 10.1016/j.colsurfb.2018.12.002.

Davies, A.S., Prakash, P. and Thamaraiselvi, K. 2017. Nanobioremediation technologies for sustainable environment. pp. 581–293. *In*: Prashanthi, M., Sundaram, R., Jeyaseelan, A. and Kaliannan, T. (eds.). Bioremediation and Sustainable Technologies for Cleaner Environment. Cham. (Switzerland): Springer International Publishing. DOI: 10.1007/978-3-319-48439-6_2.

Debnath, B., Majumdar, M., Bhowmik, M., Bhowmik, K.L., Debnath, A. and Roy, D.N. 2020. The effective adsorption of tetracycline onto zirconia nanoparticles synthesized by novel microbial green Technology. 261: 110235. DOI: 10.1016/j.jenvman.2020.110235.

Devatha, C.P. and Thala, A.K. 2018. Green synthesis of nanomaterials. pp. 169–148. *In*: Bhagyaraj, S.M., Oluwafemi, O.S., Kalarikkal, N. and Thomas, S. (eds.). Synthesis of Inorganic Nanomaterials. Cambridge (United Kingdom): Elsevier. DOI: 10.1016/B978-0-08-101975-7.00007-5.

Dicks, A.P. and Hent, A. 2015. An introduction to life cycle assessment. pp. 81–95. *In*: Dicks, A.P. and Hent, A. (eds.). Green Chemistry Metrics. Cham. (Switzerland): Springer International Publishing. DOI: 10.1007/978-3-319-10500-0_5.

Dicks, A.P. and Hent, A. 2015. Atom economy and reaction mass efficiency. pp. 17–44. *In*: Dicks, A.P. and Hent, A. (eds.). Green Chemistry Metrics. Cham. (Switzerland): Springer International Publishing. DOI: 10.1007/978-3-319-10500-0_2.

Dicks, A.P. and Hent, A. 2015. Green chemistry and associated metrics. pp. 1–16. *In*: Dicks, A.P. and Hent, A. (eds.). Green Chemistry Metrics. Cham. (Switzerland): Springer International Publishing. DOI: 10.1007/978-3-319-10500-0_0.

Dicks, A.P. and Hent, A. 2015. Selected qualitative green metrics. pp. 69–80. *In*: Dicks, A.P. and Hent, A. (eds.). Green Chemistry Metrics. Cham (Switzerland): Springer International Publishing. DOI: 10.1007/978-3-319-10500-0_4.

Dominguez, C.M., Romero, A. and Santos, A. 2019. Selective removal of chlorinated organic compounds from lindane wastes by combination of nonionic surfactant soil flushing and Fenton oxidation. Chem. Eng J. 376: 120009. DOI: 10.1016/j.cej.2018.09.170.

Dror, I., Yaron, B. and Berkowitz, B. 2017. Microchemical contaminants as forming agents of anthropogenic soils. Ambio. 46: 109–120. DOI: 10.1007/s13280-016-0804-7.

Eckelman, M.J., Zimmerman, J.B. and Anastas, P.T. 2008. Toward green nano. J. Ind. Ecol. 12(3): 316–328. DOI: 10.1111/j.1530-9290.2008.00043.x.

El-Aziz, A.R.A., Al-Othman, M.R. and Mahmoud, M.A. 2018. Degradation of DDT by gold nanoparticles synthesized using *Lawsonia inermis* for environmental safety. Biotechnol. Biotechnol. Equip. 32(5): 1174–1182. DOI: 10.1080/13102818.2018.1502051.

Ezhilarasi, A.A., Vijaya, J.J., Kennedy, L.J. and Kaviyarasu. 2020. Green mediated NiO nano-rods using *Phoenix dactylifera* (Dates) extract for biomedical and environmental applications. Mater. Chem. Phys. 241: 122419. DOI: 10.1016/j.matchemphys.2019.122419.

Fajardo, C., Martín, M., Nande, M., Botías, P., García-Cantalejo, J., Mengs, G. and Costa, G. 2020. Ecotoxicogenomic analisis of stress induced on *Carnorhabditis elegans* in heavy metal contaminated soil after nZVI treatment. Chemosphere 254: 126909. DOI: 10.1016/j.chemosphere.2020.126909.

Fatima, R., Priya, M., Indurthi, L. and Radhakrishnan, Sudhakaran, R. 2020. Biosynthesis of silver nanoparticles using red algae *Portieria hornemannii* and its antibacterial activity against fish pathogens. Microb. Pathog. 138: 103780. DOI: 10.1016/j.micpath.2019.103780.

Fatimah, I., Pratiwi, E.Z. and Wicaksono, W.P. 2020. Synthesis of magnetic nanoparticles using *Parkia speciosa hassk* pod extract and photocatalytic activity for bromophenol blue degradation. Egypt. J. Aquat. Res. 46: 35–40. DOI: 10.1016/j.ejar.2020.01.001.

Fernández, J.G., Fernández-Baldo, M.A., Berni, E., Camí, G., Durán, N., Raba, J. and Sanz, M.I. 2016. Production of silver nanoparticles using yeasts and evaluation of their antifungal activity against phytopathogenic fungi. Process Biochem. 51(9): 1306–1313. DOI: 10.1016/j.procbio.2016.05.021.

Fiorati, A., Grassi, G., Graziano, A., Liberatori, G., Pastori, N., Melone, L., Bonciani, L., Pontorno, L., Punta, C. and Corsi, I. 2020. Eco-design of nanostructured cellulose sponges for sea-water decontamination from heavy metal ions. J. Cleaner. Prod. 246(10): 119009. DOI: 10.1016/j.jclepro.2019.119009.

Gallego-Schmid, A., Mendoza, J.M. and Azapagic, A. 2018. Environmental assessment of microwaves and the effect of European energy efficiency and waste management legislation. Sci. Total Environ. 618: 487–499. DOI: /10.1016/j.scitotenv.2017.11.064.

Ganesan, V., Hariram, M., Vivekanandhan, S. and Muthuramkumar, S. 2020. *Periconium* sp. (endophytic fungi) extract mediated sol-gel synthesis of ZnO nanoparticles for antimicrobial and antioxidant applications. Mater. Sci. Semicond. Process. 105: 104739. DOI: 10.1016/j.mssp.2019.104739.

Garcés, V., Lerma, T.A., Palencia, M. and Combatt, E.M. 2018. Building a probe-type passive sampler to study the contents of heavy metals in the aqueous phase in the soil. J. Sci. Technol. App. 4: 4–16. DOI: 10.34294/j.jsta.18.4.26.

Garcés, V., Palencia, M. and Combatt, E.M. 2017 Development of bacterial inoculums based on biodegradable hydrogels for agricultural applications. J. Sci. Technol. App. 2: 13–23. DOI: 10.34294/j.jsta.17.2.10.

García, F.E., Senn, A.M., Meichtry, J.M., Scott, T.B., Pullin, H., Leyva, A.G., Halac, E.B., Ramos, C.P., Sacanell, J., Mizrahi, M., Requejo, F.G. and Litter, M.I. 2019. Iron-based nanoparticles prepared from yerba mate extract: Synthesis, characterization and use on chromium removal. 235: 1–8. DOI: 10.1016/j.jenvman.2019.01.002.

González-Ballesteros, N. and Rodríguez-Argüelles, M.C. 2020. Seaweeds: A promising bionanofactory for ecofriendly synthesis of gold and silver nanoparticles. pp. 507–541. *In*: Torres, M.D., Kraan, S. and Dominguez, H. (eds.). Sustainable Seaweed Technologies. United States: Elsevier. DOI: 10.1016/B978-0-12-817943-7.00018-4.

Gopal, G., Sankar, H., Natarajan, C. and Mukherjee, A. 2020. Tectracycline removal using green synthesized bimetallic nZVI-Cu and bentonite supported green nZVI-Cu nanocomposite: A comparative study. J. Environ. Manage. 254: 109812. DOI: 10.1016/j.jenvman.2019.109812.

Gulati, K., Jeyaseelan, C. and Rattan, S. 2018. Pine needles as green material for removal of metal ions and dyes from waste water. pp. 61–72. *In*: Parmar, V.S., Malhotra, P. and Mathur, D. (eds.). Green Chemistry in Environmental Sustainability and Chemical Education. Singapore: Springer Singapore. DOI: 10.1007/978-981-10-8390-7_6.

Haider, S., Haider, A., Ahmad, A., Khan, S.U., Almasry, W.A. and Sarfarz, M. 2015. Electrospun nanofibers affinity membranes for water hazards remediation. Nanotechnol. Res. J. 8(4): 511–541. DOI: 10.1080/15583724.2017.1309664.

Hajra, A. and Mondal, N.K. 2016. Phytofabrication of silver nanoparticles using *Elephantopus scaber* and *Azadirachta indica* leaf extract and its effect on larval and pupal mortality of *Culex quinquefasciatus.* Asian Pac. J. Trop. Dis. 6(12): 979–986. DOI: 10.1016/S2222-1808(16)61168-4.

Halada, G.P. and Orlov, A. 2018. Environmental degradation of engineered nanomaterials. pp. 225–239. *In*: Kutz, M. (ed.). Handbook of Environmental Degradation of Materials. Norwich (United States): William Andrew Publishing. DOI: 10.1016/B978-0-323-52472-8.00011-3.

Hasan, M.K., Shopan, J. and Ahammed, G.J. 2020. Nanomaterials and soil health for agricultural crop production: Current status and future prospects. pp. 289–312. *In*: Husen, A. and Jawaid, M. (eds.). Nanomaterials for Agriculture and Forestry Applications. United States: Elsevier. DOI: 10.1016/B978-0-12-817852-2.00012-3.

Hou, D. and O'Connor, D. 2020. Green and sustainable remediation: Concepts, principles, and pertaining research. pp. 1–17. *In*: Hou, D. (ed.). Sustainable Remediation of Contaminated Soil and Groundwater. Woburn (United States): Elsevier. DOI: 10.1016/B978-0-12-817982-6.00001-X.

Hulisz, P., Kwasowski, W., Pracz, J. and Malinowski, R. 2017. Coastal acid sulphate soils in Poland: A review. Soil Sci. Ann. 68: 46–54. DOI: 10.1515/ssa-2017-0006.

Husain, S., Afreen, S., Hemlata, G.S., Yasin, D., Afzal, B. and Fatma, T. 2019. Cyanobacteria as a bioreactor for synthesis of silver nanoparticles-an effect of different reaction conditions on the size of nanoparticles and their dye decolorization ability. J. Microbiol. Methods 162: 77–82. DOI: 10.1016/j.mimet.2019.05.011.

Iavicoli, I., Leso, V., Beezhold, D.H. and Shvedova, A.A. 2017. Nanotechnology in agriculture: Opportunities, toxicological implications, and occupational risks. Toxicol. Appl. Pharmacol. 329: 96–111. DOI: 10.1016/j.taap.2017.05.025.

Ingle, A., Seabra, A.B., Duran, N. and Rai, M. 2014. Nanoremediation. pp. 233–250. *In*: Das, S. (ed.). Microbial Biodegradation and Bioremediation. United States: Elsevier. DOI: B978-0-12-800021-2.00009-1.

Isac-García, J., Dobado, J.A., Calvo-Flores, F.G. and Martínez, H. 2016. Functional-group analysis. pp. 177–206. *In*: Isac-García, J., Dobado, J.A., Calvo-Flores, F.G. and Martínez, H. (eds.). Experimental Organic Chemistry. United States: Academic Press; DOI: 10.1016/b978-0-12-803893-2.50006-1.

Jacob, J.M., Rajan, R., Aji, M., Kurup, G.G. and Pugazhendhi, A. 2018. Bio-inspired ZnS quantum dots as efficient photo catalysts for the degradation of methylene blue in aqueous phase. Ceram. Int. 45(4): 4857–4862. DOI: 10.1016/j.ceramint.2018.11.182.

Jagadeesan, K.A., Duvuru, J.A., Jabasingh, A., Ponnusamy, S.K., Kabali, V.A., Gopakumaran, N., Selvaraj, K.R., Thangavelu, K., Sunny, S., Somasundaram, P.P. and Devaerajan. 2019. One pot green synthesis of nano magnesium oxide-carbon composite: Preparation, characterization and application towards anthracene adsorption. J. Clean. Prod. 237: 117691. DOI: 0.1016/j.jclepro.2019.117691.

Jin, X., Liu, Y., Tan, J., Owens, G. and Chen, Z. 2018. Removal of Cr (VI) from aqueous solutions via reduction and absorption by green synthesized iron nanoparticles. J. Clean. Prod. 176: 929–936. DOI: 10.1016/j.jclepro.2017.12.026.

Joshi, N., Filip, J., Coker, V.S., Sadhukhan, J., Safarik, I., Bagshaw, H. and Lloyd, J.R. 2018. Microbial reduction off natural Fe (III) minerals; toward the sustainable production of functional magnetic nanoparticles. Front. Environ. Sci. 6: 127. DOI: 10.3389/fenvs.2018.00127.

Jyoti, K. and Singh, A. 2016. Green synthesis of nanostructures silver particles and their catalytic application in dye degradation. J. Genet. Eng. Biotechnol. 14(2): 311–317. DOI: 10.1016/j.jgeb.2016.09.005.

Kadam, A.N., Salunkhe, T.T., Kim, H. and Lee, SW. 2020. Biogenic synthesis of mesoporous N–S–C tri-doped TiO_2 photocatalyst via ultrasonic-assisted derivatization of biotemplate from expired egg white protein. Appl. Surf. Sci. 518: 146194. DOI: 10.1016/j.apsusc.2020.146194.

Kanwar, P., Mishra, T. and Mukherjee, G. 2017. Microbial remediation of hazardous heavy metals. pp. 581–293. *In*: Prashanthi, M., Sundaram, R., Jeyaseelan, A. and Kaliannan, T. (eds.). Bioremediation and Sustainable Technologies for Cleaner Environment. Cham (Switzerland): Springer International Publishing. DOI: 10.1007/978-3-319-48439-6_21.

Khan, S.A. and Lee, C. 2015. Green biological synthesis of nanoparticles and their biomedical applications. pp. 207–235. *In*: Basiuk, V.A. and Basiuk, E.V. (eds.). Green Processes for Nanotechnology. Cham. (Switzerland): Springer International Publishing. DOI: 10.1007/978-3-030-44176-0_10.

Khan, S.H. 2020. Green nanotechnology for the environment and sustainable development. pp. 13–46. *In*: Naushad, M. and Lichtfouse, E. (eds.). Green Materials for Wastewater Treatment. Cham. (Switzerland): Springer International Publishing. DOI: 10.1007/978-3-030-17724-9_2.

Khanna, P., Kaur, A. and Goyal, D. 2019. Algae-based metallic nanoparticles: Synthesis, characterization and applications. J. Microbiol. Methods 163: 105656. DOI: 10.1016/j.mimet.2019.105656.

Kim, H.K., Jeong, S.W., Yang, J.E. and Choi, Y.J. 2019. Highly efficient and stable removal of arsenic by live cell fabricated magnetic nanoparticles. Int. J. Mol. Sci. 20(14): 3566. DOI: 10.3390/ijms20143566.

Kobashigawa, J.M., Robles, C.A., Martínez, M.L. and Carmarán, C.C. 2019. Influence of strong bases on the synthesis of silver nanoparticles using the ligninolytic fungi Trametes Troggi. Saudi. J. Biol. Sci. 26(7): 1331–1337. DOI: 10.1016/j.sjbs.2018.09.006.

Koduru, J.R., Shankar, S., More, N.S., Shikha, Lingamdinne, L.P. and Singh, J. 2017. Toxic metals contamination in the environment. pp. 209–240. *In*: Bharagava, R.N. (ed.). Environmental Pollutants and their Bioremediation Approaches. Boca. Raton. (United States): CRC Press. DOI: 10.1201/9781315173351-8.

Kolahalam, L.A., Kasi, I.V., Diwakar, B.S., Govindh, B., Reddy, V. and Murthy, Y.L.N. 2019. Review on nanomaterials. Mater Today: Proc. 18: 2182–2190. DOI: 10.1016/j.matpr.2019.07.371.

Kotp, A.A., Farghali, A.A., Amin, R.M., Moaty, S.A., El-Deen, A.G., Gadelhak, Y.M., El-Ela, F.A., Youns, H.A., Syame, S.M. and Mahmoud, R.K. 2019. Green-synthesis of Ag nanoparticles and its composite with PVA nanofiber as a promising Cd^{2+} adsorbent and antimicrobial agent. J. Environ. Chem. Eng. 7: 102977. DOI: 10.1016/j.jece.2019.102977.

Kumar, R., Singh, N. and Pandey, S.N. 2015. Potential of green synthesized zero-valent iron nanoparticles for remediation of lead-contaminated water. Int. J. Environ. Sci. Technol. 12: 3943–3950. DOI: 10.1007/s13762-015-0751-z.

Kumar, S.J., Akshayb, V.R., Vasundharab, M. and Arumugama, M. 2020. Biosynthesis of multiphase iron nanoparticles using *Syzygium aromaticum*. Colloids Surf. A 603: 125241. DOI: 10.1016/j.colsurfa.2020.125241.

Leng, W., Pati, P. and Vikesland, P. 2015. Room temperature seed mediated growth of gold nanoparticles. Environ. Sci. Nano. 2: 440–453. DOI: 10.1039/c5en00026b.

León-Silva, S., Arrieta-Cortés, R., Fernández-Luqueño, F. and López-Valdez, F. 2018. Design and production of nanofertilizers. pp. 17–34. *In*: López-Valdez, F. and Fernández-Luqueño, F. (eds.). Agricultural Nanobiotechnology. Cham. (Switzerland): Springer International Publishing. DOI: 10.1007/978-3-319-96719-6_2.

Lerma, T.A., Combatt, E.M. and Palencia, M. 2018. Soil-mimicking hybrid composites based on clay, polymers and nitrogen-fixing bacteria for the development of remediation systems of degraded soil. J. Sci. Technol. App. 4: 17–27. DOI: 10.34294/j.jsta.18.4.27.

Li, J., Li, Q., Ma, X., Tian, B., Li, T., Yu, J., Dai, S., Weng, Y. and Hua, Y. 2016. Biosynthesis of gold nanoparticles by the extreme bacterium *Deinococcus radiodurans* and an evaluation of their antibacterial properties. Int. J. Nanomedicine. 9(11): 5931–5944. DOI: 10.2147/IJN.S119618.

Li, Y., Wang, J., Zhao, F., Bai, B., Nie, G., Nel, A.E. and Zhao, Y. 2018. Nanomaterial libraries and model organisms for rapid high-content analysis of nanosafety. Natl. Sci. Rev. 5(3): 365–388. DOI: 10.1093/nsr/nwx120.

Lin, Z., Weng, X., Owens, G. and Chen, Z. 2020. Simultaneous removal of Pb(II) and rifampicin from wastewater by iron nanoparticles synthesized by a tea extract. J. Clean. Prod. 242: 118476. DOI: 10.1016/j.jclepro.2019.118476.

Liu, H., Ren, M., Qu, J., Feng, Y., Song, X., Zhang, Q., Cong, Q. and Yuan, X. 2017. A cost-effective method for recycling carbon and metals in plants: synthesizing nanomaterials. Environ. Sci. Nano. 4: 461–469. DOI: 10.1039/C6EN00287K.

Livingston, A., Trout, B.L., Horvath, I.T., Johnson, M.D., Vaccaro, L., Coronas, J., Babbitt, C.W., Zhang, X., Pradeep, T., Drioli, E., Hayler, J.D., Tam, K.C., Kappe, O., Fane, A.G. and Szekely, G. 2020. Challenges and directions for green chemical engineering—Role of nanoscale materials. pp. 1–18. *In*: Livingston, A. and Szekely, G. (eds.). Sustainable Nanoscale Engineering: from Materials Design to Chemical Processing. Cambridge (United Kingdom): Elsevier. DOI: 10.1016/B978-0-12-814681-1.00001-1.

Lohrasbi, S., Kouhbanani, M.A.J., Beheshtkhoo, N., Ghasemi, Y., Amani, A.M. and Taghizadeh, S. 2019. Green synthesis of iron nanoparticles using *Plantago major* leaf extract and their application as a catalyst for the decolorization of azo dye. BioNanoScience 9(2): 317–322. DOI: 10.1007/s12668-019-0596-x.

López-Valdez, F., Miranda-Arámbula, M., Ríos-Cortés, A.M., Fernández-Luqueño, F. and de-la-Luz, V. 2018. Nanofertilizers and their controlled delivery of nutrients. p. 35–48. *In*: López-Valdez, F. and Fernández-Luqueño, F. (eds.). Agricultural Nanobiotechnology. Cham. (Switzerland): Springer International Publishing; DOI: 10.1007/978-3-319-96719-6_3.

López-Villareal, J., Castro-Peña, V.C., Solis-Pomar, F., Gutierrez-Lazos, C.D., Meléndrez, M.F., Martínez-Guerra, E., Pérez-Tijerina, E. and Fundora, A. 2019. Thermal treatment of ZnO seed layer into atomic layer deposition to improve crystal quality of ZnO nanorod arrays. J. Sci. Technol. Appl. 6: 108–117. DOI: 10.34294/j.jsta.19.6.45.

Lu, H., Wang, J., Stoller, M., Wang, T., Bao, Y. and Hao, H. 2016. An overview of nanomaterials for water and wastewater treatment. Adv. Mater. Sci. Eng. 4964828: 1–10. DOI: 10.1155/2016/4964828.

Luo, F., Yang, D., Chen, Z., Megharaj, M. and Naidu, R. 2016. Characterization of bimetallic Fe/Pd nanoparticles by grape leaf aqueous extract and identification of active biomolecules involved in the synthesis. Sci. Total Environ. 562: 526–532. DOI: 10.1016/j.scitotenv.2016.04.060.

Lv, Y., Zhang, Z., Chen, Y. and Hu, Y. 2016. A novel three-stages nano bimetallic reduction/oxidation/biodegradation treatment for remediation of 2,2',4,4'-tetrabromodiphenyl ether. Chem. Eng. J. 289: 382–390. DOI: 10.1016/j.cej.2015.12.097.

Madhava, A.K., Baskaralingam, P., Asthika, A.R.S. and Sivanesan, S. 2018. Role of bacterial consortia in bioremediation of textile recalcitrant compounds. pp. 165–184. *In*: Varjani, S.J., Gnansounou, E., Gurunathan, B., Pant, D. and Zakaria, Z.A. (eds.). Waste Bioremediation. Singapore: Springer. DOI: 10.1007/978-981-10-7413-4_8.

Manav, N., Dwivedi, V. and Bhagi, A.K. 2018. Degradation of DDT, a pesticide by mixed metal oxides nanoparticles. pp. 93–99. *In*: Parmar, V.s., Malhotra, P. and Mathur, D. (eds.). Green Chemistry in Environmental Sustainability and Chemical Education. Singapore: Springer Singapore; DOI: 10.1007/978-981-10-8390-7_9.

Mansi, K., Kumar, R., Kaur, J., Mehta, S.K., Pandey, S.K., Kumar, D., Dash, A.K. and Gupt, N. 2020. DL-Valine assisted fabrication of quercetin loaded CuO nanoleaves through microwave irradiation method: augmentation in its catalytic and antimicrobial efficiencies. Environ. Nanotechnol. Monit. Manag. 14: 100306. DOI: 10.1016/j.enmm.2020.100306.

Marimón-Bolívar, W. and González, E.E. 2018. Green synthesis with enhanced magnetization and life cycle assessment of Fe_3O_4 nanoparticles. Environ. Nanotechnol. Monit. Manag. 9: 58–66. DOI: 10.1016/j.enmm.2017.12.003.

Mehdinia, A. and Mehrabi, H. 2019. Application of nanomaterials for removal environmental pollution. pp. 365–402. *In*: Thomas, S., Grohens, Y. and Pottathara, Y.B. (eds.). Industrial Applications of Nanomaterials. United States: Elsevier. DOI: 10.1016/B978-0-12-815749-7.00013-X.

Melendez, M.A. and Polanski, A. 2020. Dirty neighbors—Pollution in an interlinked world. Energy Economy 86: 104636. DOI: 10.1016/j.eneco.2019.104636.

Mohammadi, P., Hesari, M., Samadian, H., Hajialyani, M., Bayrami, Z., Farzaei, M.H. and Abdollahi, M. 2019. Recent advancements and new perspective of phytonanotecnhology. pp. 1–22. *In*: Verma, S.K. and Das, A.K. (eds.). Analysis, Fate, and Toxicity of Engineered Nanomaterials in Plants. United States: Elsevier. DOI: 10.1016/bs.coac.2019.04.011.

Mohan, S., Singh, Y., Verma, D.K. and Hasan, S.H. 2015. Synthesis of CuO nanoparticles through green route using Citrus limon juice and its application as nanosorbent for Cr(VI) remediation. 96: 156–166. DOI: 10.1016/j.psep.2015.05.005.

Mu, Y., Jia, F., Ai, Z. and Zhang, L. 2017. Iron oxide shell mediated environmental remediation properties of nano zero-valent iron. Environ. Sci. Nano. 4: 27–45. DOI: 10.1039/C6EN00398B.

Muthukumar, H., Shanmuga, M.K. and Gummadi, S.N. 2020. Caffeine degradation in synthetic coffee wastewater using silver ferrite nanoparticles fabricated via green route using *Amaranthus blitum* leaf aqueous extract. J. Water Process. Eng. 36: 101382. DOI: 10.1016/j.jwpe.2020.101382.

Nagajyothi, P.C. and Tvm, S. 2015. Green synthesis of metallic and metal oxide nanoparticles and their antibacterial activities. pp. 99–117. *In*: Basiuk, V.A. and Basiuk, E.V. (eds.). Green Processes for Nanotechnology. Cham. (Switzerland): Springer International Publishing. DOI: 10.1007/978-3-319-15461-9_4.

Nasrollahzadeh, M., Atarod, M., Sajjadi, M., Sajadi, S.M. and Issaabadi, Z. 2019. Green nanotechnology. pp. 145–198. *In*: Nasrollahzadeh, M., Sajadi, M.S., Atarod, M., Sajjadi, M. and Isaabadi, Z. (eds.). An Introduction to Green Nanotechnology. San Diego (United States): Elsevier. DOI: 10.1016/B978-0-12-813586-0.00005-5.

Nasrollahzadeh, M., Atarod, M., Sajjadi, M., Sajadi, S.M. and Issaabadi, Z. 2019. Plant-mediated green synthesis of nanostructures. pp. 199–322. *In*: Nasrollahzadeh, M., Sajadi, M.S., Atarod, M., Sajjadi, M. and Isaabadi, Z. (eds.). An Introduction to Green Nanotechnology. San Diego (United States): Elsevier. DOI: 10.1016/B978-0-12-813586-0.00006-7.

Nasrollahzadeh, M., Sajadi, M.S. and Hatamifard, A. 2016. Waste chicken eggshell as a natural valuable resource and environmentally benign support for biosynthesis of catalytically active Cu/eggshell, Fe_3O_4/eggshell and Cu/Fe_3O_4/eggshell nanocomposites. Appl. Catal. B 191(15): 209–227. DOI: 10.1016/j.apcatb.2016.02.042.

Newsome, L., Morris, K., Cleary, A., Masters-Waage, N.K., Boothman, C., Joshi, N., Atherton, N. and Lloyd, J.R. 2019. The impact of iron nanoparticles on technetium-contaminated groundwater and sediment microbial communities. J. Hazar. Matter. 364(15): 134–142. DOI: 10.1016/j.jhazmat.2018.10.008.

Nijalingappa, T.B., Veeraiahb, M.K., Basavarajc, R.B., Darshand, G.P., Sharmae, S.C. and Nagabhushanac, H. 2019. Antimicrobial properties of green synthesis of MgO micro architectures via *Limonia acidissima* fruit extract. Biocatal. Agric. Biotechnol. 18: 100991. DOI: 10.1016/j.bcab.2019.01.029.

Ortega-Calvo, J.J., Jimenez-Sanchez, C., Pratarolo, P., Pullin, H., Scott, T.B. and Thompson, I.P. 2016 Tactic response of bacteria to zero-valent iron nanoparticles. Environ. Pollut. 213: 438–445. DOI: 10.1016/j.envpol.2016.01.093.

Ossai, I.C., Ahmed, A., Hassan, A. and Hamid, F.S. 2020. Remediation of soil and water contaminated with petroleum hydrocarbon. Env. Technol. Innovation. 17: 100526. DOI: 10.1016/j.eti.2019.100526.

Pachiyapan, J., Gnanasundaram, N. and Rao, G.L. 2020. Preparation and characterization of ZnO, MgO and ZnO–MgO hybrid nanomaterials using green chemistry approach. Results in Materials 7: 100104. DOI: 10.1016/j.rinma.2020.100104.

Palencia, M., Berrío, M.E. and Melendrez, M.F. 2016. Nanostructured polymer composites with potential applications into the storage of blood and hemoderivates. J. Sci. Technol. Appl. 1: 4–14. DOI: 10.34294/j.jsta.16.1.1.

Palencia, M., Berrío, M.E. and Palencia, S.L. 2017. Effect of capping agent and diffusivity of different silver nanoparticles on their antibacterial properties. J. Nanosci. Nanotechnol. 17: 5197–5204. DOI: 10.1166/jnn.2017.13850.

Palencia, S.L., Medina, A. and Palencia, M. 2015. Interaction mechanisms of inorganic nanoparticles and biomolecular systems of microorganisms. Curr. Chem. Biol. 9: 10–22. DOI: 10.2174/2212796809666151022201811.

Palencia, M., Mora, M.A. and Palencia, S.L. 2017. Biodegradable polymer hydrogels based in sorbitol and citric acid for controlled release of bioactive substances from plants. Curr. Chem. Biol. 11(1): 36–43. DOI: 10.2174/221279681066 6161028114432.

Palencia, M., Mora, M.A. and Lerma, T.A. 2020. Environment-friendly stimulus-sensitive polyurethanes based on cationic aminoglycosides for the controlled release of phytohormones. Smart Sustainable Built. Environ. Ahead-of-print. DOI: 10.1108/SASBE-09-2019-0126.

Pandey, G. 2018. Prospects of nanobioremediation in environmental cleanup. Orient. J. Chem. 34(6): 2828–2840. DOI: 10.13005/ojc/340622.

Pei, G., Zhu, Y., Cai, X., Shi, W. and Li, H. 2017. Surfactant flushing remediation of *o*-dichlorobenzene and *p*-dichlorobenzene contaminated soil. Chemosphere 185: 1112–1121. DOI: 10.1016/j.chemosphere.2017.07.098.

Prakash, S., Selvaraju, M., Ravikumar, K. and Punnagaiarasi, A. 2017. The role of decomposer animals in bioremediation of soils. pp. 57–64. *In*: Prashanthi, M., Sundaram, R., Jeyaseelan, A. and Kaliannan, T. (eds.). Bioremediation and Sustainable Technologies for Cleaner Environment. Cham. (Switzerland): Springer International Publishing. DOI: 10.1007/978-3-319-48439-6_6.

Pourzahedi, L. and Eckelman, M.J. 2015. Comparative life cycle assessment of silver nanoparticle synthesis routes. Environ. Sci. Nano. 2: 361–369. DOI: 10.1039/c5en00075k.

Poynton, H. and Robinson, W.E. 2018. Contaminants of emerging concern, with an emphasis on nanomaterials and pharmaceuticals. pp. 291–315. *In*: Török, B. and Dransfield, T. (eds.). Green Chemistry. United States: Elsevier. DOI: 10.1016/B978-0-12-809270-5.00012-1.

Qian, Y., Qin, C., Chen, M. and Lin, S. 2020. Nanotechnology in soil remediation—applications vs. implications. Ecotox. Environ. Safe. 201: 110815. DOI: 10.1016/j.ecoenv.2020.110815.

Qu, Y., Shen, W., Pei, X., Ma, F., You, S., Li, S., Wang, J. and Zhou, J. 2017. Biosynthesis of gold nanoparticles by *Trichoderma* sp. WL-Go for azo dyes decolorization. J. Environ. Sci. 56: 79–86. DOI: 10.1016/j.jes.2016.09.007.

Rahman, A., Kumar, S. and Nawaz, T. 2020. Biosynthesis of nanomaterials using algae. pp. 265–279. *In*: Yousuf, A. (ed.). Microalgae Cultivation for Biofuels Production. San Diego (United States): Academic Press. DOI: 10.1016/B978-0-12-817536-1.00017-5.

Rajendran, K. and Sen, S. 2016. Optimization of process parameters for the rapid biosynthesis of hematite nanoparticles. J. Photochem. Photobiol. B 159: 82–87. DOI: 10.1016/j.jphotobiol.2016.03.023.

Rajendran, K. and Sen, S. 2018. Adsorptive removal of carbamazepine using biosynthesized hematite nanoparticles. Environ. Nanotechnol. Monit. Manag. 19: 122–127. DOI: 10.1016/j.enmm.2018.01.001.

Ramanayaka, S., Tsang, D.C.W., Hou, D., Sik, O.k.Y. and Vithanage, M. 2020. Green synthesis of graphitic nanobiochar for the removal of emerging contaminants in aqueous media. Sci. Total Environ. 1(706): 135725. DOI: 10.1016/j.scitotenv.2019.135725.

Ramar, K., Ahamed, A.J. and Muralidharan, K. 2019. Robust green synthetic approach for the production of iron oxide nanorods and its potential environmental and cytotoxicity applications. Adv. Powder Technol. 30(11): 2636–2648. DOI: 10.1016/j.apt.2019.08.011.

Romeh, A.A. and Saber, R.A.I. 2020. Green nano-phytoremediation and solubility improving agents for the remediation of chlorfenapyr contaminated soil and water. J. Environ. Manage. 260: 110104. DOI: 10.1016/j.jenvman.2020.110104.

Rostami-Vartooni, A., Nasrollahzadeh, M. and Alizadeh, M. 2016. Green synthesis of seashell supported silver nanoparticles using *Bunium persicum* seeds extract: Application of the particles for catalytic reduction of organic dyes. J. Colloid. Inter. Sci. 470: 268–275. DOI: 10.1016/j.jcis.2016.02.060.

Salouti, M. and Derakhshan, F.K. 2019. Phytosynthesis of nanoscale materials. pp. 45–121. *In*: Ghorbanpour, M. and Wani, S.H. (eds.). Advances in Phytonanotechnology. San Diego (United States): Elsevier. DOI: 10.1016/B978-0-12-815322-2.00003-1.

Sanchez, Y., García-Quintero, A. and Palencia, M. 2018. Determination and distribution of size of inorganic particles by spectral deconvolution of surface plasmon resonance. J. Sci. Technol. Appl. 5: 45–54. DOI: 10.34294/j.jsta.18.5.34.

Sanjay, S.S. 2019. Safe nano is green nano. pp. 27–356. *In*: Shukla, A.K. and Iravani, S. (eds.). Green Synthesis, Characterization and Applications of Nanoparticles. London (United Kingdom): Elsevier. DOI: 10.1016/B978-0-08-102579-6.00002-2.

Saranya, A., Alomayri, T., Ramar, K., Priyadharsan, A., Raj, V., Murugan, K., Alsawalha, M. and Maheshwaran, P. 2020. Facile one pot microwave-assisted green synthesis of Fe_2O_3/Ag nanocomposites by phytoreduction. J. Photochem. Photobiol. B 207: 111885. DOI: 10.1016/j.jphotobiol.2020.111885.

Saravanakumar, K., Chelliah, R., Shanmugam, S., Varukattu, N.B., Oh, D.H., Kathiresan, K. and Wang, M.H. 2018. Green synthesis and characterization of biologically active nanosilver from seed extract of Gardenia *Jasminoides ellis*. J. Photochem. Photobiol. B 185: 126–135. DOI: 10.1016/j.jphotobiol.2018.05.032.

Sasidharan, S., Raj, S., Sonawane, S., Sonawane, S., Pinjari, D., Pandit, A.B. and Saudagar, P. 2019. Nanomaterial synthesis: chemical and biological route and applications. pp. 27–51. *In*: Pottathara, Y.B., Thomas, S., Kalarikkal, N., Grohens, Y. and Kokol, V. (eds.). Nanomaterials Synthesis: Design, Fabrication and Applications. United States: Elsevier. DOI: 10.1016/B978-0-12-815751-0.00002-X.

Saxena, G. and Bharagava, R.N. 2017. Organic and inorganic pollutants in industrial wastes, their ecotoxicological effects, health hazards and bioremediation approaches. pp. 23–56. *In*: Bharagava, R.N. (ed.). Environmental Pollutants and their Bioremediation Approaches. Boca Raton (United States): CRC Press. DOI: 10.1201/9781315173351-3.

Sebastian, A., Nangia, A. and Majeti, P. 2020. Advances in agrochemical remediation using nanoparticles. pp. 465–485. *In*: Prasad, M.N.V. (ed.). Agrochemicals: Detection, Treatment and Remediation. Oxford (United Kingdom): Butterworth-Heinemann. DOI: 10.1016/B978-0-08-103017-2.00018-0.

Seetharaman, P.K., Chandrasekaran, R., Gnanasekar, S., Chandrakasan, G., Gupta, M., Manikandan, D.B. and Sivaperumal, S. 2018. Antimicrobial and larvicidal activity of eco-friendly silver nanoparticles synthesized from endophytic fungi *Phomopsis liquidambaris*. Biocatal. Agric. Biotechnol. 16: 22–30. DOI: 10.1016/j.bcab.2018.07.006.

Shah, M., Fawcett, D., Sharma, S., Tripathy, S.K. and Ponern, G.E.J. 2015. Green synthesis of metallic nanoparticles via biological entities. Materials 8(11): 7278–7308. DOI: 10.3390/ma8115377.

Shanker, U., Jassal, V. and Rani, M. 2017. Green synthesis of iron hexacyanoferrate nanoparticles. J. Environ. Chem. Eng. 5(4): 4108–4020. DOI: 10.1016/j.jece.2017.07.042.

Sharma, J.K., Srivastava, P., Ameen, S., Akhtar, M.S., Sengupta, S.K. and Singh, G. 2017. Phytoconstituents assisted green synthesis of cerium oxide nanoparticles for thermal decomposition and dye remediation. Mater. Res. Bull. 91: 98–107. DOI: 10.1016/j.materresbull.2017.03.034.

Silva, L.P., Garcez, I. and Bonatto, C.C. 2015. Green synthesis of metal nanoparticles by plants. pp. 259–275. *In*: Basiuk, V.A. and Basiuk, E.V. (eds.). Green Processes for Nanotechnology. Cham (Switzerland): Springer International Publishing. DOI: 10.1007/978-3-319-15461-9_9.

Singh, J. and Singh, A.V. 2017. Microbial strategies for enhanced phytoremediation of heavy metal-contaminated soils. pp. 257–272. *In*: Bharagava, R.N. (ed.). Environmental Pollutants and their Bioremediation Approaches. Boca Raton (United States): CRC Press. DOI: 10.1201/9781315173351-4.

Singh, J., Dutta, T., Kim, K., Rawat, M., Samddar, P. and Kumar, P. 2018. 'Green' synthesis of metals and their oxide nanoparticles: Applications for environmental remediation. J. Nanobiotechnol. 16: 84. DOI: 10.1186/s12951-018-0408-4.

Singh, J., Kumar, V., Jolly, S.S., Kum, H., Rawat, M., Kukkar, D. and Tsang, Y.F. 2019. Biogenic synthesis of silver nanoparticles and its photocatalytic applications for removal of organic pollutants in water. J. Ind. Eng. Chem. 80: 247–257. DOI: 10.1016/j.jiec.2019.08.002.

Singh, N. and Khan, S. 2018. Nanobioremediation: An innovative approach to fluoride (F^-) contamination. pp. 343–353. *In*: Prasad, R. and Aranda, E. (eds.). Approaches in Bioremediation: The New Era of Environmental Microbiology and Nanobiotechnology. Cham (Switzerland): Springer International Publishing. DOI: 10.1007/978-3-030-02369-0_15.

Singh, R., Behera, M. and Kumar, S. 2020. Nano-bioremediation: an innovative remediation technology for treatment and management of contaminated sites. pp. 165–182. *In*: Saxena, G. and Bharagava, R.N. (eds.). Bioremediation of Industrial Waste for Environmental Safety. Singapore: Springer Singapore. DOI: 10.1007/978-981-13-3426-9_7.

Sohail, M.I., Waris, A.A., Ayub, M.A., Usman, M., Rehman, M.Z., Sabir, M. and Faiz, T. 2019. Environmental application of nanomaterials. Compr. Anal. Chem. 87: 1–54. DOI: 10.1016/bs.coac.2019.10.002.

Speight, J.G. 2020. Remediation technologies. pp. 263–303. *In*: Speight, J.G. (ed.). Natural Water Remediation. Woburn (United States): Butterworth-Heinemann. DOI: 10.1016/B978-0-12-803810-9.00008-5.

Srivastava, S.K., Ogino, C. and Kondo, A. 2015. Nanoparticle synthesis by biogenic approach. pp. 237–257. *In*: Basiuk, V.A. and Basiuk, E.V. (eds.). Green Processes for Nanotechnology. Cham. (Switzerland): Springer International Publishing. DOI: 10.1007/978-3-319-15461-9_8.

Stan, M., Lung, I., Soran, M.L., Leostean, C., Popa, A., Stefan, M., Lazar, M.D., Opris, O., Silipas, T.D. and Porav, A.S. 2017. Removal of antibiotics from aqueous solutions by green synthesized magnetite nanoparticles with selected agro-waste extracts. Process. Saf. Environ. Prot. 107: 357–372. DOI: 10.1016/j.psep.2017.03.003.

Tarannum, N. and Khan, R. 2020. Cost-effective green materials for the removal of pesticides from aqueous medium. pp. 100–118. *In*: Naushad, M. and Lichtfouse, E. (eds.). Green Materials for Wastewater Treatment. Cham. (Switzerland): Springer International Publishing. DOI: 10.1007/978-3-030-17724-9_5.

Thirumurugan, A., Harshini, E., Marakathanandhini, B.D., Kannan, S.R. and Muthukumaran, P. 2017. Catalytic degradation of Reactive red 120 by copper oxide nanoparticles synthesized by *Azadirachta indica*. pp. 95–102. *In*: Prashanthi, M., Sundaram, R., Jeyaseelan, A. and Kaliannan, T. (eds.). Bioremediation and Sustainable Technologies for Cleaner Environment. Cham. (Switzerland): Springer International Publishing. DOI: 10.1007/978-3-319-48439-6_9.

Thovhogi, N., Diallo, A., Gurib-Fakim, A. and Maaza, M. 2015. Nanoparticles green synthesis by *Hibiscus Sabdariffa* flower extract: main physical properties. J. Alloys Compd. 647(25): 392–396. DOI: 10.1016/j.jallcom.2015.06.076.

Thyagarajan, L.P., Sudhakar, S. and Meenambal, T. 2017. Bioremediation of Congo-Red dye by using silver nanoparticles synthesized from *Bacillus* sps. pp. 119–132. *In*: Prashanthi, M., Sundaram, R., Jeyaseelan, A. and Kaliannan, T. (eds.). Bioremediation and Sustainable Technologies for Cleaner Environment. Cham. (Switzerland): Springer International Publishing. DOI: 10.1007/978-3-319-48439-6_11.

Tripathi, V., Edrisi, S.A., Chen, B., Gupta, V.K., Vilu, R., Gathergood, N. and Abhilash, P.C. 2017. Biotechnological advances for restoring degraded land for sustainable development. Trends Biotechnol. 35(9): 847–859. DOI: 10.1016/j.tibtech.2017.05.001.

Ulloa, G., Collao, B., Araneda, M., Escobar, B., Álvarez, S., Bravo, D. and Pérez-Donoso, J.M. 2016. Use of acidophilic bacteria of the genus *Acidithiobacillus* to biosynthesize CdS fluorescent nanoparticles (quatum dots) with high tolerance to acidic pH. Enzyme Microb. Technol. 95: 217–224. DOI: 10.1016/j.enzmictec.2016.09.005.

Umadevi, S., Ayyasamy, P.M. and Rajakumar, S. 2017. Biological perspective and role of bacteria in pesticide degradation. pp. 57–64. *In*: Prashanthi, M., Sundaram, R., Jeyaseelan, A. and Kaliannan, T. (eds.). Bioremediation and Sustainable

Technologies for Cleaner Environment. Cham. (Switzerland): Springer International Publishing. DOI: 10.1007/978-3-319-48439-6_1.

Upreti, G., Dhingra, R., Naidu, S., Atuahene, I. and Sawhney, R. 2015. Life cycle assessment of nanomaterials. pp. 393–408. *In*: Basiuk, V.A. and Basiuk, E.V. (eds.). Green Processes for Nanotechnology. Cham (Switzerland): Springer International Publishing. DOI: 10.1007/978-3-319-15461-9_14.

Vasudevan, N. and Odukkathil, G. 2017. Pesticides contamination in the environment. pp. 57–102. *In*: Bharagava, R.N. (ed.). Environmental Pollutants and their Bioremediation Approaches. Boca Raton (United States): CRC Press. DOI: 10.1201/9781315173351-4.

Villalobos, R.J., Medina, C., Canales, C., Meléndrez, M. and Flores, P. 2019. Development of nanocomposite materials with thermostable matrix from nanoreinforcement of titanium dioxide. J. Sci. Technol. Appl. 6: 65–78. DOI: 10.34294/j.jsta.19.6.42.

Villalobos, R.J., Rojas, D., Díaz, A., Ramírez, J., Meléndrez, M. and Flores, P. 2019. Synthesis and characterization of 1D metal oxide nano-structures: titanium dioxide nanotubes (TNTs) and zinc oxide nanobars (ZNBs). J. Sci. Technol. Appl. 6: 79–95. DOI: 10.34294/j.jsta.19.6.43.

Wacławek, S., Gončuková, Z., Adach, K., Fijałkowski, M. and Černík, M. 2018. Green synthesis of gold nanoparticles using *Artemisia dracunculus* extract. Environ. Sci. Pollut. Res. 25(24): 24210–24219. DOI: 10.1007/s11356-018-2510-4.

Wei, J., Tu, C., Yuan, G., Bi, D., Xiao, L., Theng, B.K.G., Wang, H. and Sik, Ok.Y. 2019. Carbon-coated montmorillonite nanocomposite for the removal of chromium (VI) from aqueous solutions. J. Hazard Mater. 368: 541–549. DOI: 10.1016/j.jhazmat.2019.01.080.

Wei, Y., Chen, L., Zhang, Z., Zhu, C. and Zhang, S. 2018. Fungal nanoparticles formed in saline environments are conducive to soil health and remediation. pp. 317–341. *In*: Approaches in Bioremediation: The New Era of Environmental Microbiology and Nanobiotechnology. Cham (Switzerland): Springer International Publishing. DOI: 10.1007/978-3-030-02369-0_14.

Wu, J., Lin, Z., Weng, X., Owens, G. and Chen, Z. 2020. Removal mechanism of mitoxantrone by a green synthesized hybrid reduced graphene oxide @ iron nanoparticles. Chemosphere 246: 125700. DOI: 10.1016/j.chemosphere.2019.125700.

Wu, Z., Su, X., Lin, Z., Owens, G. and Chen, Z. 2019. Mechanism of As (V) removal by green synthesized iron nanoparticles. J. Hazardous Mater. 379: 120811. DOI: 10.1016/j.jhazmat.2019.120811.

Xiao, X., Wang, Q., Owens, G., Chiellini, F. and Chen, Z. 2019. Reduced graphene oxide/iron nanoparticles used for the removal of Pb (II) by one step green synthesis. J. Coll. Int. Sci. 557: 598–607. DOI: 10.1016/j.jcis.2019.09.058.

Yadav, M. and Khan, S. 2018. Nanotechnology: A new scientific outlook for bioremediation of dye effluents. pp. 301–315. *In*: Prasad, R. and Aranda, E. (eds.). Approaches in Bioremediation: The New Era of Environmental Microbiology and Nanobiotechnology. Cham. (Switzerland): Springer International Publishing. DOI: 10.1007/978-3-030-02369-0_16.

Yang, Y., Li, J., Xi, B., Wang, Y., Tang, J., Wang, J., Wang, Y. and Zhao, C. 2017. Modeling BTEX migration with soil vapor extraction remediation under low-temperature conditions. J. Environ. Manage. 203(1): 114–122. DOI: 10.1016/j.jenvman.2017.07.068.

Yuan, Q., Bomma, M. and Xiao, Z. 2019. Enhanced silver nanoparticle synthesis by *Escherichia coli* transformed with *Candida Albicans* metallothionein gene. Materials 12(24): 4180. DOI: 10.3390/ma12244180.

Yugay, Y.A., Usoltseva, R.V., Silantev, V.E., Egorova, A.E., Karabtsov, A.A., Kumeiko, V.V., Ermakova, S.P., Bulgakov, V.P. and Shkryl, Y.N. 2020. Synthesis of bioactive silver nanoparticles using alginate, fucoidan and laminaran from brown algae as a reducing and stabilizing agent. Carbohydr. Polym. 245: 116547. DOI: 10.1016/j.carbpol.2020.116547.

Zelić, E., Vuković, Ž. and Halkijević, I. 2018. Application of nanotechnology in wastewater treatment. Građevinar. 70(4): 315–323. DOI: 10.14256/JCE.2165.2017.

Zhang, T., Lowry, G.V., Capiro, N.L., Chen, J., Chen, W., Chen, Y., Dionysiou, D.D., Elliott, D.W., Ghoshal, S., Hofmann, T., Hsu-Kim, H., Hughes, J., Jiang, C., Jiang, G., Jing, C., Kavanaugh, M., Li, Q., Liu, S., Ma, J., Pan, B., Phenrat, T., Qu, X., Quan, X., Saleh, N., Vikesland, P.J., Wang, Q., Westerhoff, P., Wong, M.S., Xia, T., Xing, B., Yan, B., Zhang, L., Zhou, D. and Alvarez, P.J.J. 2019. *In situ* remediation of subsurface contamination. Environ. Sci. Nano. 6: 1283–1302. DOI: 10.1039/C9EN00143C.

Zhao, Q., Zhao, X. and Cao, J. 2020. Advanced nanomaterials for degrading persistent organic pollutants. pp. 249–305. *In*: Zhao, Q. (ed.). Advanced Nanomaterials for Pollutant Sensing and Environmental Catalysis. United States: Elsevier Science Publishing. DOI: 10.1016/b978-0-12-814796-2.00007-1.

Zhou, N.Q., Tian, L.J., Wang, Y.C., Li, D.B., Li, P.P., Zhang, X. and Yu, H.Q. 2016. Extracellular biosynthesis of copper sulfide nanoparticles by *Shewanella oneidensis* MR-1 as a photothermal agent. Enzyme Microb. Technol. 95: 230–235. DOI: 10.1016/j.enzmictec.2016.04.002.

CHAPTER 10

Nanoremediation of Hazardous Environmental Pollutants

Anita Grozdanov, Perica Paunović, Katerina Burevska Atkovska* and *Mirko Marinkovski*

1. Introduction

The increased environmental concern as well as the need for successful sustainable development has forced people around the world, to proceed with faster implementation of many advanced products and new technologies (Bardos et al., 2015; Nagar et al., 2020; Shukla et al., 2020). Active application of nanotechnology products is also included here. Many of the new developed nanostructures are designed for solving some of the environmental problems, especially related to the remediation of air, waters and soil contaminants. A wide range of new technologies, including nanotechnologies are available. A variety of novel technologies is continuously being developed for more efficient removal of most of the pollutants (organic compounds, pesticides, heavy metals, toxic gases, herbicides) (Ali et al., 2019; Kumar et al., 2020).

Successful and effective management of environmental pollutants consists of capture and degradation technologies (adsorption, solvent extraction, membrane filtration, ion exchange, chemical precipitation, reverse osmosis, electrolysis), which is rather complex due to the particular composition of individual materials as well as the high volatility of many toxic pollutants and compounds (Wang et al., 2005; Xing et al., 2007; Bakalar et al., 2009; Rahul, 2013; Ersahin et al., 2012; Wanees et al., 2013; Vindoh et al., 2011; Al-Degs et al., 2006). While conventional remediation technologies are mainly time-consuming and expensive (pump and treat methods), in the last years, "nano materials and "nanotechnology" have become one of the most effective tools for the management of environmental pollutants owing to the unique properties of nanostructures. In fact, the process of remediation of air, soils or groundwaters by using nanoparticles, known as "nanoremediation" can open a new route in remediation technologies based on experiment-scale findings, which provide a solid base for increased degradation speed of wider lists of hazardous pollutants or stabilization could be substanstially increased versus the conventional remediation (Bardos et al., 2015; Shukla et al., 2020). By definition, nanoremediation presents a process of *in situ* application of nanoparticles for cleaning polluted soils and waters. Based on the type of the applied nanoparticles, nanoremediation processes usually include oxidation, reduction, sortion or their combination. The final aim of nanoremediation

Faculty of Technology and Metallurgy, University Ss Cyril and Methodius in Skopje, 1000 Skopje, North Macedonia.
* Corresponding author: anita.grozdanov@yahoo.com

is to decrease the overall costs and to reduce the time for cleaning of contaminated sites, and parallel eliminating the need for treatment of toxic contaminant concentrations almost close to minimal or even zero. Largely prominent characteristics of nanomaterials are their huge specific surface area and reactivity, which provide their higher effectiveness in comparison with their bulk counterparts.

What is more promising, regarding their surface, nanomaterials provided more possibilities for proper modification in order to improve and increase their efficiency in nanoremediation (Gupta et al., 2015; Pyrzynska and Bystrzejewski, 2010; Di et al., 2006; Xu et al., 2008). The main criteria which should be met, is that the final properties of the modified nanosurfaces should not have negative effects on the environment.

On the other hand, the list of different contaminants released to the air, soils and groundwaters is wide and extensive, and usually contain various organic contaminants (bacteria and viruses) and many toxic heavy metal ions. Due to the fact that heavy metal ions are persistent and highly stable, they belong to the group of extremely hazardous pollutants, and their toxicity is a significant problem for the safe evolution of biodiversity and the human population (Nagajyoti et al., 2010). Many heavy metal ions, among them ions of As, Pb, Cd, Cr, Co, Fe, Hg, Mn, Ni, Zn, are highly toxic contaminants in all areas and can seriously destroy all living organisms and humans in a polluted environment. Usually, through industrial activities or accidents, the metal ions are released in rivers and lakes waters, and in concentrations that are higher than the regulative limits, which could create a lot of damage and environmental degradation. Additionally, food and water consumption, polluted with heavy metals ions with higher concentrations, much higher that the regulated limits, can cause different metabolic problems due to the fact that heavy metals can accumulate in the organism and are hardly processed and disposed of it. Higher pressure was also given by the world-established regulations such as, the maximum copper concentration of 2.0 mg/dm^3 in drinking water that was determined by the World Health Organization.

In this chapter some of the published research data as well as some of our research activities and results were reviewed concerning the application of carbon-based nanostructures (CNSs) - carbon nanotubes (CNTs) and graphene) and TiO_2 based nanomaterials in nanoremediation, where CNSs were applied as nanosorbents for removing heavy metal ions from contaminated waters and TiO_2 nanostructures for photocatalytic degradation of organic pollutants.

2. Application of Nanomaterials for Nanoremediation

Besides conventional materials used as adsorbents (clays minerals, zeolites, activated carbon, metal oxides) for remediation of contaminated waters and soils, for removing heavy metals from polluted waters, during the last years, nanomaterials based on carbon, such as carbon nanotubes and graphene have found a wide application (Zhao et al., 2011; Wang et al., 2012). As opposed to the conventional sorbent materials, nanostructured adsorbents based on carbon nanomaterials (graphene and carbon nanotubes), demonstrated much higher efficiency and faster rates in water treatment (Atkovska et al., 2018). In order to be used as adsorbents for heavy metal ions elimination, it is necessary that nanomaterials meet some of the listed criteria: (1) to be a non-toxic; (2) to exhibit a relatively higher adsorption capacity and selectivity at low concentrations of contaminants; (3) the adsorbed pollutants be able to easily be removed from the nanosorbent's surface; (4) adsorbents should be recyclable and recover easily (Gupta et al., 2015; Pillay et al., 2009). Till now, for removing heavy metal ions from contaminated waters, different nanostructures and their nanocomposites based on carbon, graphene, nanometals or metal oxides, polymeric sorbents were studied (Zhao et al., 2011; Mauter et al., 2008). It was found out that all these tested nanomaterials exhibited very high adsorption efficiency and capacity. During the last few years, especially nanomaterials based on carbon, due to their non-toxicity and high adsorption capacity, have been widely applied and tested for removing hazardous heavy metals from wastewaters (Atkovska et al., 2018; Mauter et al., 2008). Continuous improvement and new designs of the nanotechnologies provide production of better as well as cheaper carbon nanomaterials (tubes and graphene), nanometals and silicates and contribute to their higher application as nanosorbents for nanoremediation.

2.1 Carbon based nanostructures in nanoremediation

2.1.1 Carbon nanotubes

Since they were discovered by the Japanese researcher Sumio Iijima in 1991, carbon nanotubes attracted a great deal of scientific and industrial interest (Iijima, 1991). CNT were widely tested, due to their unique properties and application possibilities. Basically, CNTs were designed and developed into two types, as a single walled carbon nanotubes (SWCNTs) and multi walled carbon nanotubes (MWCNTs), as shown in Fig. 10.1. Due to their unique behavior and structural characteristics, CNTs are considered as relatively new adsorbents for removing heavy metals in wastewater treatment. Most of the analysis show that in order to increase the adsorption activity and suitability of carbon-based nanomaterials for nanoremediation applications, it is very important to first perform surface treatment and proper functionalization of the raw carbon nanomaterials (Chaudhery, 2020).

Usually, SWCNTs are aligned in a hexagonal configuration (i.e., one nanotube surrounded by six others), creating bundles of aligned tubes with a heterogeneous, porous structure. Ren et al. (2011) studied the adsorption capacity of SWCNTs. His team reported that for an open-ended SWCNT bundle, adsorption takes place in four different available sites. They also found out two kinds of active sites: first those with lower adsorption energy, mainly determined on the outer surfaces of the external SWCNTs and second -characterized with higher adsorption energy found between two neighboring tubes or within an individual tube. The adsorption process performed on the external sites can reach a stable configuration more rapidly than adsorption realized in internal sites as a result of primary (direct) contact of the external sites with the adsorbing compounds. For MWCNTs, it was seen that they do not exist in bundle configurations. MWCNT with bundle configuration were produced only with specials methods which were strictly targeted for such configurations. Yang et al. (2011) performed nitrogen adsorption studies and reported that the inner and aggregated pores can be used to create a multi-stage adsorption process. It was confirmed that the aggregated pores vs inner pores are responsible for the sorption properties of MWCNTs. It was found that the adsorption efficiency of CNTs is also affected by its oxygen content. Namely, functionalized CNTs contain different functional groups such as –OH, –COO and –COOH, depending on the surface modifications and

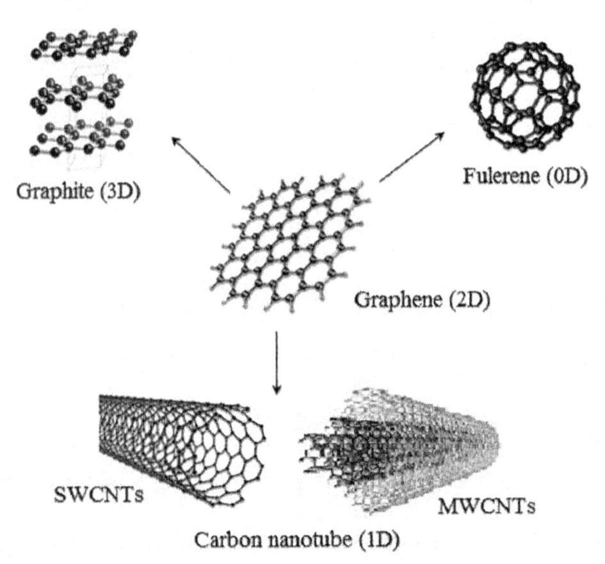

Fig. 10.1. Carbon based nanostructures.

purification methods. These groups can increase the adsorption capacity significantly and different chemicals (pure acids or their mixtures, such as HNO_3, H_2SO_4, $KMnO_4$) can be used to oxidize these CNTs. CNTs oxidized with HNO_3 have shown increased adsorption capacities against Pb, Cu, Ni and Cd ions (Khin et al., 2012).

In order to improve the adsorption capacity of CNTs, several methods were developed and reported, such as oxidation, oxygen plasma, ozone treatment, UV-radiation (Castro et al., 2019; Gupta et al., 2011; Kosa et al., 2012; Rao et al., 2007; Zhao et al., 2010). Usually, untreated CNTs exhibited relatively small adsorption capacity for metal ions, and it significantly increased after the oxidation with HNO_3, H_2O_2 and $KMnO_4$. Li et al. (2003) performed oxidation of CNTs with HNO_3, H_2O_2 and $KMnO_4$. Oxidized CNT exhibited increased adsorption capacities for removing of Cd (II) ions, compared to adsorption capacity of the untreated CNTs that was 1,1 mg/g (Li et al., 2003). After oxidation, the obtained adsorption capacities were 2,6; 5,1 and 11,0 $mg \cdot g^{-1}$ for CNTs oxidized by H_2O_2, HNO_3 and $KMnO_4$, respectively. CNTs acidic treatment was also tested by Akbar and Parviz (2012). They performed acid activation of MWCNTs with HNO_3, H_2O_2 and $KMnO_4$, and examined their adsorption characteristics for removing lead and copper ions from aqueous solutions. The obtained results confirmed that the adsorption capacity of oxidized MWCNTs significantly improved due to the increased number of functional groups containing oxygen (–COOH, –OH or –C=O) on the surface of the CNTs. These functional groups were potential reactive sites which can respond with Pb^{2+} and Cu^{2+} and form salts or complexes (Akbar and Parviz, 2012). Farghali et al. (2017) reported results for CVD produced MWCNTs that were purified and modified by the acid mixture of H_2O_2/HNO_3 (acids ratio of 1:3 (v/v) at 25°C) which led to functionalized CNT's surface. These acid treatments resulted in numerous chemical sorption sites on MWCNTs surface which provided increased adsorption of Pb^{2+}, Ni^{2+}, Cu^{2+} and Cd^{2+} in both single and quaternary solutions. The adsorption enlarged with the increase of the solution pH. They confirmed that the large adsorption capacity of acidified MWCNTs for metal ions was mainly due to the oxygen functional groups on CNTs surfaces which can react with metals to form salt or complex deposits on the surface of MWCNTs. Farghali et al. (2017) also studied the adsorption kinetics and their experimental results were in agreement with the Langmuir and Freundlich isotherms.

Rao et al. (2007) published a review of several publications that studied the removal of divalent metal ions (Cd^{2+}, Cu^{2+}, Ni^{2+}, Pb^{2+}, Zn^{2+}) from aqueous solutions, by the application of different types of CNTs. The effects and variation of several factors on the adsorption efficiency was studied: pH of the solution, initial concentration of the metal ions, the amount of adsorbent, contact time. It was found that the adsorption of metal ions to different CNTs, approximately follow the order: $Pb^{2+} > Ni^{2+} > Zn^{2+} > Cu^{2+} > Cd^{2+}$. The process parameters such as surface total acidity, pH and temperature, also played a key role in determining sorption rate of metal ion on CNTs.

The present authors (Grozdanov et al., 2018), tested MWCNTs (China cheap, purity > 95%, length = 50 nm), raw and modified ones. The adsorption ability of MWCNTs, in removing of Ni (II), Pb (II) and Fe (II) ions from aqueous solutions was investigated. Modified MWCNTs were activated by acidic treatment. Acidic oxidation treatment was performed using a mixture of concentrated H_2SO_4 (96%) and HNO_3 (62%) with the volume ratio of H_2SO_4/HNO_3 = 3:1. The centrifuged and dried MWCNTs were collected and stored in a desiccator. The metal solution is a system of three components: Cu^{2+}, Ni^{2+} and Fe^{2+} ions and the adsorption experiments were observed for different initial concentrations of 0.3; 0,6 and 1.2 $mg \cdot dm^{-3}$ of metal ions, at pH = 5. The concentration of MWCNT used for adsorption was 0.5 $g \cdot dm^{-3}$, and the adsorption was performed for 2 hr on a magnetic stirrer with 300 rotations per minute.

Characteristic SEM photographs of the tested MWCNTs nanosorbents are shown in Fig. 10.2. Due to the acidic treatment, carbon nanoparticles were shortened in length (Fig. 10.2b).

However, the comparison of the adsorption efficiency of pristine and modified MWCNTs in removing the metal ions suggested that the applied acid activation of CNTs was not justified (see in Fig. 10.3). Non-modified MWCNTs proved to be more efficient. It was found that 30 min of adsorption contact time was sufficient to reach the equilibrium, for all analyzed systems. The results

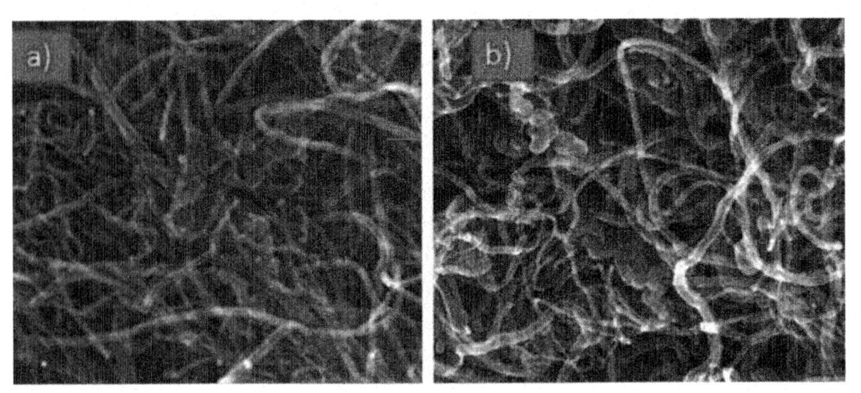

Fig. 10.2. SEM morphology of MWCNT a) pristine MWCNT and b) acid treated MWCNTs.

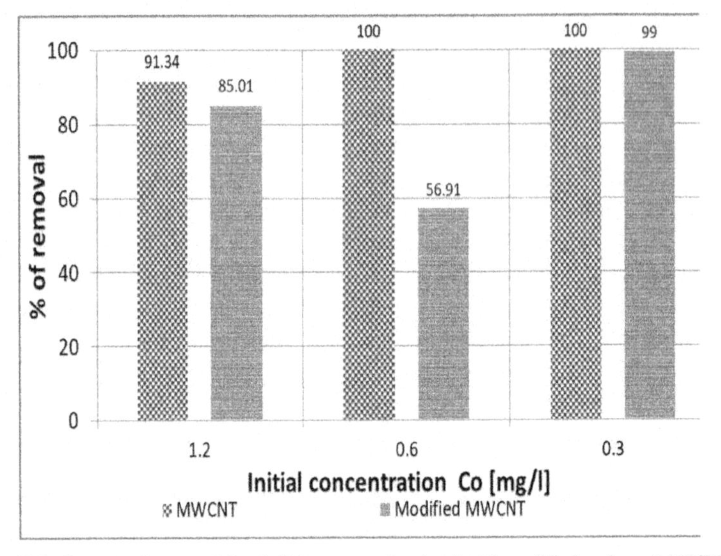

Fig. 10.3. Percent of removal for Cu^{2+} by nanosorbents of acid-modified and raw MWCNTs.

have shown that the effectiveness of MWCNT in the metal ion adsorption was in the following order: $Fe^{2+} > Cu^{2+} > Ni^{2+}$. The highest percentage removal was noticed for copper and iron with initial concentration of 0.6 $mg.dm^{-3}$ and 0.3 $mg.dm^{-3}$. It was interesting that higher adsorption efficiency was obtained in the solution with lower initial concentration of the Cu^{2+} metal ions.

2.1.2 Graphene

The second tested carbon-based nanosorbent is graphene. Graphene is a two-dimensional allotrope of carbon with a single atom-thick layer of hexagonal arrangement of sp^2-hybridized carbon atoms that has excellent electrical, thermal and mechanical properties (Ciriminna et al., 2015). In 1962, Hanns-Peter Boehm was the first who reported the concept of single layer carbon foil based on the SEM-photos, however researchers Geim and Novoselov were the pioneers who discovered graphene as the monolayer material (Novoselov et al., 2004). With a specific surface area of about 2300–2630 $m^2 \cdot g^{-1}$ (by the theory), graphene exhibited an exceptional adsorption capacity and great potential to be exploited as an adsorbent for heavy metal ions removal from aqueous solutions (Ciriminna et al., 2015; Wang et al., 2013; Zhao et al., 2011; Chandra et al., 2010). During the last 10 yrs, many scientists reported that graphene and its modified forms are effective sorbents for heavy metal removal, due to their unique structure and surface characteristics. Namely, for the applications in water purification, graphene and its derivatives have indicated several structural advantages. First, graphene has a typical mono-layered structure, which suggests that all atoms are mainly surface atoms. The active surface of graphene is even higher than that of CNTs, as the inner surface of CNTs is not easily accessed for pollutants. Second, comparison of adsorption kinetics of pollutants have shown that adsorption rate with graphene was faster compared to traditional adsorbents. The porous structure of graphene adsorbents allows much faster diffusion of pollutants, which in turn allows quicker adsorption. Third, it was demonstrated that the production costs of graphene adsorbents are lower than the production costs of other high-performance adsorbents. Fourth, the graphene adsorbents can be used for removing various pollutants simultaneously (Wang et al., 2013; Zhao et al., 2011; Chandra et al., 2010). Wang et al. (2012) worked on synthesis of a few-layered graphene oxide nanosheets using the modified Hummers method in the laboratory. The produced graphene nanostructures were used as sorbents for the removal of Cd^{2+} and Co^{2+} ions from an aqueous solution. The results confirmed that heavy metal ions adsorption on nano layers was dependent on pH and ionic strength. It was also found that the oxygen-containing functional groups on the surfaces of graphene oxide nano layers played an important role on sorption. Chandra et al. (2010) reported that modified magnetite-graphene adsorbents with dimensions of nanoparticles of ~ 10 nm exhibited a high binding capacity for As^{3+} and As^{5+}. The results suggested that the higher adsorption capacity was registered due to the increased adsorption sites in the graphene composite. This was supported by the theory that adsorption escalates with increasing pH, which is consistent with a mechanism that requires the release of protons for metal adsorption. Released protons should come from functional groups as the carbon backbone does not contain hydrogen.

Besides the pristine, monolayer carbon form of graphene as a nanoadsorbent also used its oxide, i.e., Graphene Oxide (GO) which is cheaper and easier to produce (Peng et al., 2017; Jayakaran et al., 2019). Graphene oxide actually represents an extension of carbon material and it has 2D structure obtained by the chemical oxidation of graphite layers. According to research studies, it was shown that GO contains oxygen functional groups of about 20 to 30% in the basal plane. At first, GO was used as an usual adsorbent for removing and reducing the metal ions (Zn, Cu, Pb, Cd, Co) since its adsorption affinity to different metal ions varied from the types of metal ions. It was shown that metal ions with higher electronegativity were strongly attracted to the negatively electrified wide GO surface. Accordingly, it was explained that better adsorption of nano heavy metals on the GO-surface was obtained due to the presence of hydroxyl as well as carboxyl functional group in GO (Jaroniec et al., 2017).

Jayakaran et al. (2019) reported qualitative and quantitative analysis of graphene-based adsorbents in wastewater treatment.

Najafi (2015) published a study on adsorption of Zn (II) ions from an aqueous solution at 298 K and pH 6, using grapheme oxide and functionalized grapheme oxide with glycine as the adsorbent surface. In order to confirm the functionalized graphene oxide, the glycine amino group was added to the surface of graphene oxide. He studied the effects of the initial concentration of Zn (II) ions and contact time, and the obtained results showed that with increasing initial concentration of Zn (II) ions, the adsorption capacity increased. Optimal time for the adsorption process was chosen at 50 min , as a substantial increase in the adsorption capacity of Zn (II) ions on graphene oxide and GO–G after 50 min was not found.

Mi et al. (2012) synthesized and used Graphene Oxide (GO) aerogel that was able to absorb Cu^{2+}. Their results showed that the adsorption kinetic followed the Langmuir model. They obtained the maximum capacity of 19.65 $mg \cdot g^{-1}$. It was also found that GO aerogel had smaller adsorption capacity than GO suspension which was due to the fact that some of the GO nano layers were folded during the drying phase, making some sites unavailable for Cu^{2+}.

It was found that graphene was also a good adsorbent for nonmetal ions. Luo et al. (2012) studied the adsorption of As (III) and As (V) ions from polluted water with the application of magnetite Fe_3O_4-RGO-MnO_2. They found that the adsorption of As (III) and As (V) on Fe_3O_4-RGO-MnO_2 from polluted water reached the equilibrium in about 120 min. The adsorption process could be fitted by the Langmuir model, which resulted in maximum adsorption capacity of 14.04 mg·g^{-1} for As (III) and 12.22 mg·g^{-1} for As (V).

Vilela et al. (2019) reported highly efficient removal and recovery of toxic heavy metal (Pb) from contaminated water. They used new attractive forms of graphene oxide (GOx) such as tubular micromotors and dubbed microbots. A procedure for the recovery of hazardous Pb after its removal from polluted water was also demonstrated by them. Namely, the procedure of lead recovery was based on treatment of GOx-microbots in different chemical conditions in order to induce the desorption of Pb (II) from their surfaces. This method resulted in the subsequent reusability of GOx-microbots too.

Dimitrov et al. (2004); Atkovska et al. (2020) tested graphene produced at the laboratory of Faculty of Technology and Metallurgy in Skopje by molten salt electrolysis. Pristine and acid-activated graphene were compared as nanosorbents. The acid-activation was done using 65% HNO_3 at room temperature under magnetic stirring for 3 hr at 600 rpm. The adsorption experiments were carried out for a three-component system – an aqueous solution of Cu^{2+}, Ni^{2+} and Fe^{2+} ions for different initial concentrations of 0.3, 0.6 and 1.2 mg·dm^{-3}. 0.5 g·dm^{-3} sorbent was used, and the adsorption was performed for 2 hr on a magnetic stirrer with 300 rotations per minute (Atkovska et al., 2020).

Typical SEM photos of the morphology of the tested graphene-based nanosorbents are shown in Fig. 10.4. The photographs clearly present the obtained exfoliated graphene layers from the graphite electrode (4a) and the morphology of more intensively divided lists of the activated graphene (4b).

The obtained results for adsorption efficiency of the graphene and its activated form for the removal of heavy metal ions are shown on the Figs. 10.5 and 10.6. However, it was noted that there

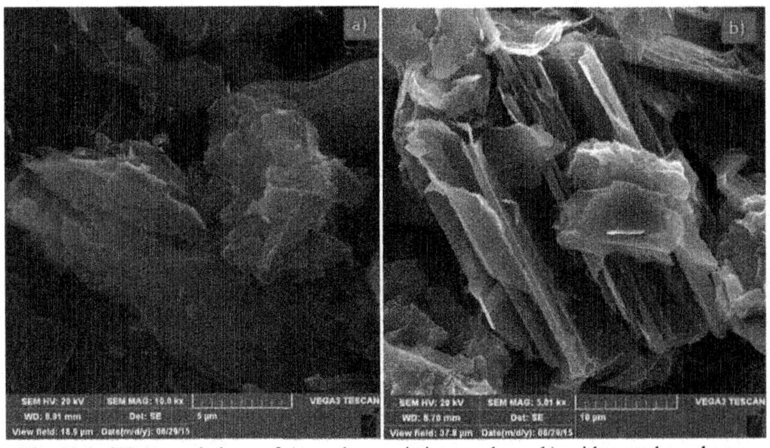

Fig. 10.4. SEM morphology of a) graphene pristine graphene, b) acid treated grapheme.

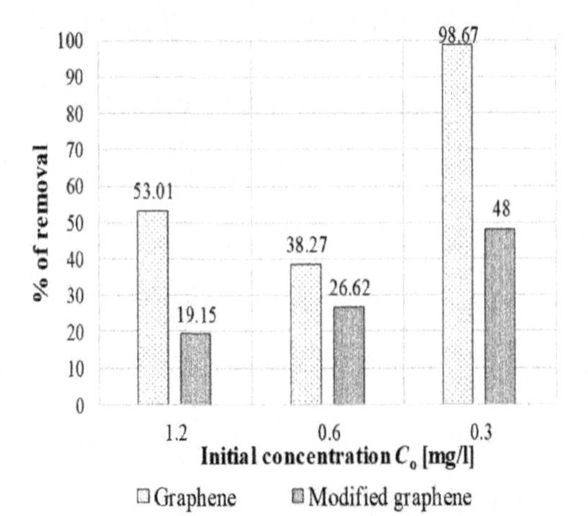

Fig. 10.5. Comparison of the percent of removal for Cu^{2+} among raw and modified grapheme.

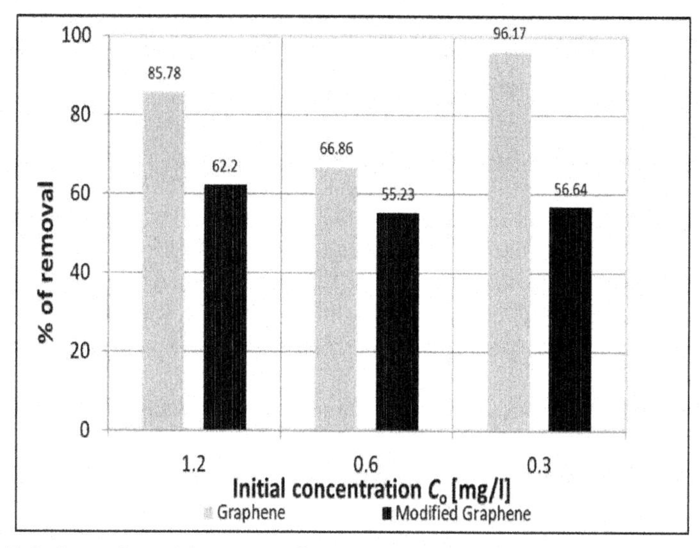

Fig. 10.6. Comparison of the percent of removal for Fe^{2+} among raw and modified graphene.

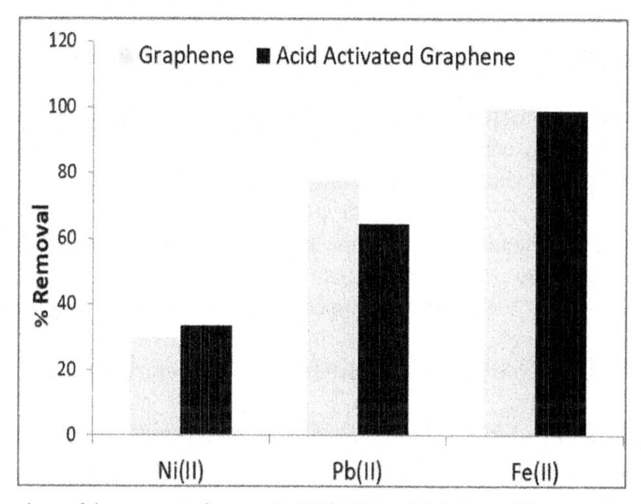

Fig. 10.7. Comparison of the percent of removal of Ni^{2+}, Pb^{+} and Fe^{2+} ions with raw and modified graphene.

is no positive effect of the acid activation of the graphene structure and raw graphene proved to be more effective compared to modified graphene. The results have shown the following order for the effectiveness of graphene in the metal ion adsorption: $Fe^{2+} > Cu^{2+} > Ni^{2+}$ (see Fig. 10.7). The highest percentage removal was noticed for copper with initial concentration of 0,3 mg/l. The largest amount adsorbed at 2 hr was measured for iron with initial concentration of 1.2 mg· dm^{-3}.

The maximum percentage of removal of Ni (II) ions with graphene and activated graphene was 30.0 and 33.3%, respectively. In the other two metals, graphene exhibited a better percentage of elimination compared to activated graphene, and for lead ions it was 77.8 and 64.4%, respectively. The removal of Fe (II) ions by applying graphene was complete (at 100%). The acid-activated graphene demonstrated slightly less efficiency and removed iron ions at 99.1%. Higher adsorption activity of the starting, i.e., inactivated graphene in relation to activated—modified graphene has been registered by other authors. Namely, the slightly higher adsorption capacity of the starting (unmodified) graphene is due to the strong coordination capacity and the high specific surface area of the starting graphene in an aqueous solution (Huang et al., 2003). Based on these results, it was concluded that the acid activation of graphene did not meet the expectations for improving the absorption capacity of graphene and thus achieving better efficiency for the elimination of heavy metal ions. Considering that activated graphene had an efficiency improvement in relation to graphene only for nickel ions of only 3.3%, this activation was practically and economically not justified.

2.2 TiO_2 nanostructures in bio and nanoremediation

The achievements of environmental nanotechnology include invention of numerous materials, processes and technologies for detecting and remediation of the environment from a number of pollutants. Among many materials used in this field, TiO_2 was recognized as one of the most significant materials in the environmental technology as a result of its non-toxicity, biocompatibility, abundant resources, affordable cost, high chemical stability and appropriate physical properties with a highlight on the semiconductivity.

These appropriate characteristics of TiO_2 are mainly due to its electronic and crystalline structure.

2.2.1 Structure and properties

Titanium oxide exists in three natural polymorphic forms such as rutile, anatase and brookite, where rutile is thermodynamically the most stable one. Rutile and anatase are the most abundant and and a diversity of applications have been found. As the synthesis of brookite involves by many difficulties, its application is very limited. The basic crystalline unit for all polymorph is an octahedron (TiO_6) in which one titanium atom is surrounded by six oxygen atoms (Diebold, 2003). Depending on the connection of these octahedra, TiO_2 obtained a corresponding polymorph form (Fig. 10.8). The connection of the octahedra by edges led to the formation of rutile, by vertices to anatase, while the connection by both edges and vertices showed the formation of brookite.

This structure involves various appropriate properties of TiO_2, which makes it suitable for a number of environmentally friendly applications. The most important properties are summarized in Table 10.1. Both the abounded polymorphs (anatase and rutile) have a very high melting point of 1870°C and boiling point of 2972°C, making it appear as a solid. Their particles are insoluble in water. As an oxide, TiO_2 shows high chemical stability in a wide range of temperatures. TiO_2 absorbs the UV light, becoming white in color and can be used as skin protector from the UV irradiation caused by sun-light.

TiO_2 is a semiconductor of n-type with $3d^2 4s^2$ outer electron configuration (Tang et al., 2017) with a wide band-gap (3 eV for rutile and 3.2 eV for anatase) (Zhang et al., 2014). This electronic

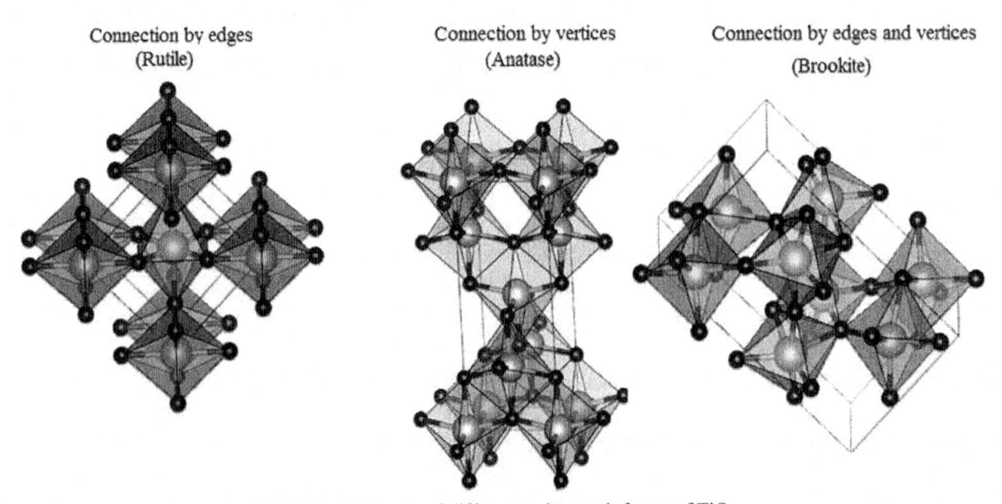

Fig. 10.8. Unit cells of different polymorph forms of TiO_2.

Table 10.1. List of the TiO_2 properties.

Property	Anatase	Rutile
Appearance	White, odorless	White, odorless
Crystal lattice	Tetragonal (con. by vertices)	Tetragonal (con. by edges)
Lattice parameters	$a = 3.733$ Å $c = 9.370$ Å	$a = 4.584$ Å $c = 2.953$ Å
Melting point (°C)	1870	
Boiling point (°C)	2927	
Bulk density ($g \times cm^{-3}$)	3.8	4.2
Mohs hardness	5.5–6.0	6.0–6.5
Specific heat capacity ($J \times kg^{-1} \cdot K^{-1}$)	710	
Thermal conductivity ($W \times m^{-1} \times K^{-1}$)	12	
Electroconductivity	semiconductors	
Energy of band gap (eV)	3.2	3.0
Chemical resistivity	High	

structure of semiconducting TiO_2 is suitable for promoting photocatalytic processes and the sensing application. Therefore, the semiconductivity and corresponding energy of the band gap is one of the most important from the environmental point of view.

2.2.2 Applications

All of the above mentioned characteristics of TiO_2 have enabled the majority of its applications to be directed to environmental monitoring and protection. These researches are the cornerstone of the TiO_2 application in environmental protection. Fujishima and Honda (1972) first performed water splitting, using TiO_2 (rutile) semiconductor photoelectrode and Pt counter electrode. This marked the beginning of a new era in photocatalysis applied in the field of energy production and environmental protection. Furthermore, Frank and Bard (1977) for the first time applied TiO_2 in water purification (removal of CN^- ions by reduction in water). Pruden and Ollis (1983) showed complete degradation of trichloroethylene (Cl_2CCClH) to HCl and CO_2 in water, using illuminated TiO_2. Wang et al. (1997) prepared a thin TiO_2 polycrystalline film from anatase sol on a glass substrate, showing excellent anti-fogging and self-cleaning properties. Until now , numerous research studies were conducted (Bai et al., 2014; Ahmad et al., 2015; Yan et al., 2015; Abdullah and Kamarudin, 2015; Paunovic,2018; Lee and Park, 2013; Ameta et al., 2013; Zhang et al., 2013; Maziarz et al., 2016; Wang et al., 2017) in the field of TiO_2 application in environmental protection, which can be distinguished in three main groups, as shown in Fig. 10.9.

There are a number of highly toxic and hardly degradable organic pollutants which can be successfully degraded using TiO_2 as a photocatalysts. Ameta et al. (2013) showed a list of them, such as phenols, halogenated hydrocarbons, pesticides, cationic, anionic and non-ionic surfactants, nitrogen containing, dyes, cellulose, formaldehyde, polymers (polyethylene PE, polystyrene PS and polyvinyl chloride PVC). Some kind of microorganisms (e.g., *Escherichia coli* and *Lactobacillus helveticus*) can be inactivated by the photocatalytic action of TiO_2 (Liu et al., 2003). Among the different methods for degradation of halogenated organic compounds, such as chemical treatment, incineration, extraction, sonochemical and electrochemical treatment, photocatalytic degradation has been recognized as the most effective one (Ameta et al., 2013). In their review, Lazar et al. (2012) emphasized the efficiency of the photocatalytic remediation of water using TiO_2 as a photocatalysts, which is applicable for wastewaters coming from different sources such as the cosmetics and pharmaceutical industry, paper mill and municipal wastewater. They highlighted the efficiency of cheaper bulk TiO_2 materials such as pigment and iron-containing industrial TiO_2 by-products in photocatalytic remediation of wastewaters

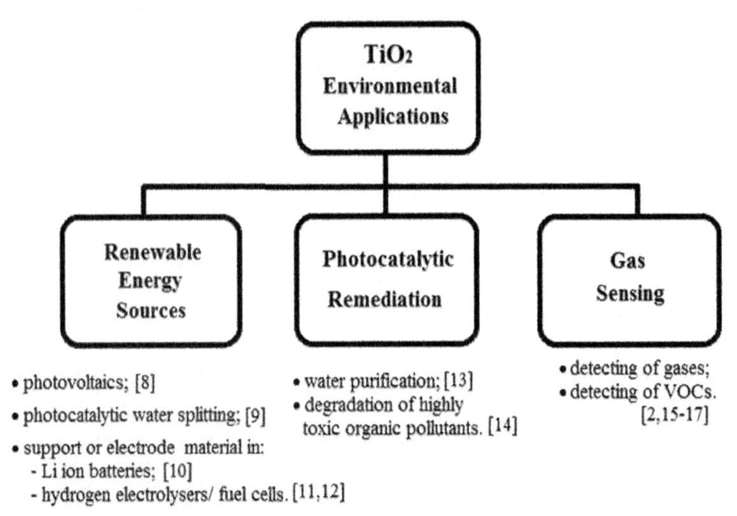

Fig. 10.9. The main fields of TiO_2 applications in environmental protection.

containing humic acid and phenol. The photocatalytic effectiveness is comparable with that of more expensive commercial TiO_2 – Degussa P 25.

2.3.3 Improvement of the photocatalytic activity

Material science in the recent few decades has focused on the development of nano-sized materials and nanotechnology. Nanomaterials have shown unique and superior properties over those of corresponding bulk materials. The performance of titanium dioxide was improved by developing methods for its synthesis in nano-dimensions. Nano-dimensioned TiO_2 can be obtained in a variety of morphological forms (Fig. 10.10). such as: spherical (Paunovic et al., 2015) or polygonal (rhombic-shaped or octahedrons) (Calatayud et al., 2015) particles, nanotubes (Tang et al., 2017), nanorods (Huang et al., 2016), nanowires (Hadia, 2014), nanosheets (Sajan Ponnappa et al., 2015), nanobelts (Zhou et al., 2011) and nanoflowers (Zhou et al., 2011; Mali et al., 2013). The main advantage of the nano-sized TiO_2 for photocatalytic purposes is the extensive higher specific surface area than bulk material, i.e., a larger place on which phototocatalytic reaction occurs. For example, the commercial TiO_2-Hombikat UV-100 (anatase nanoparticles smaller than 10 nm), has specific surface area of 250 $m^2 \cdot g^{-1}$.

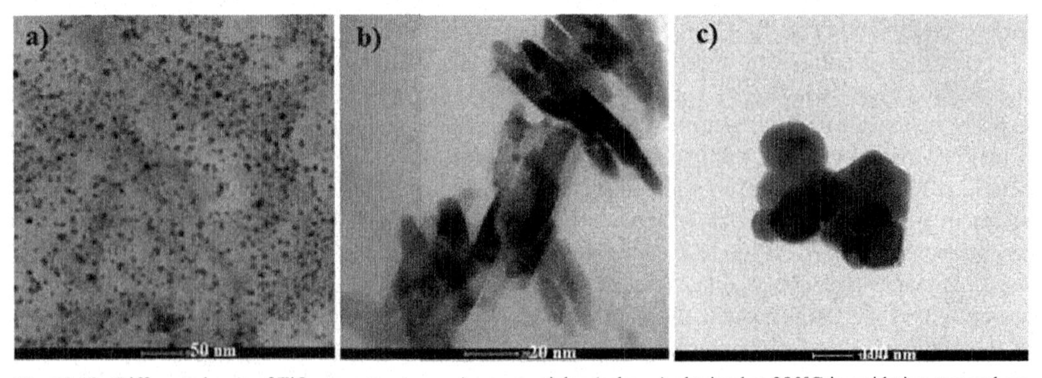

Fig. 10.10. Different shapes of TiO_2 nanostructures: a) nanoparticles (spheres) obtained at 380°C in oxidative atmosphere, b) nanorods obtained at 800°C in oxidative atmosphere and c) polygonal nanoparticles obtained at 800°C in reductive atmosphere (unpublished results of the present authors).

Highly developed surface area and a large number of active places for photocatalytic reaction is a physical aspect of the overall photocatalytic activity of the material. The other is the intrinsic photocatalytic activity, which depends on the width and the energy of the band gap. Although anatase has a higher value of the band gap energy (3.2 eV) than rutile (3.0 eV), it was recognized as a better photocatalyst. This can be explained by the following points (Zhang et al., 2014; Luttrell et al., 2014): (i) higher adsorption affinity of anatase to the hydroxyl groups and lower affinity to the oxygen adsorption, as a result of the higher Fermi level for 0.1 eV compared with rutile; (ii) smaller particle size of anatase as a result of the lower synthesis temperatures and (iii) the indirect band gap of anatase opposed to the direct band gap of rutile. Within the indirect band gap, the excited electron can be stabilized at the lower level in the conduction band, resulting in its higher mobility and longer life. The band gap energy of 3.2 eV means that anatase could be photocatalytic active under the UV region of light. Therefore, the next approach for improving the performances of nano TiO_2 is changing the energy of the band gap in the direction to shift the photocatalytic activity in the region of visible light.

This is a chemical approach, and generally, it means adding of some ions (doping) within the TiO_2 crystalline lattice causing a decrease of the band gap. This event was discovered by Sato (1986), during the synthesis of TiO_2 using addition of NH_4OH in titania sol. The produced TiO_2 showed improved photoactivity in the region of visible light, as a result of incorporation of N atoms within the TiO_2 crystalline lattice. Later, numerous researches were made including doping of anions (N, C, F, S and B), cations of transition metals (Fe, Co, Ni, V, Mo, Nb, Cr etc.) or cations of noble metals (Pt, Ag, Au and Pd) (Pelaez et al., 2012; Zaleska, 2008). Yan et al. (2014) pointed out that mono-doping into TiO_2 is associated with problems such as instability of photo-excited electron–hole pairs, reducing the photo-excited current or the carrier mobility. These problems can be overcome by anion-cation (acceptor-donor: N-Zr or N-Ta) co-doping of TiO_2, raising the valence band edge and increasing the conductive band edge and consequently considerable narrowing of the band gap. Doping of TiO_2 either with anions or cations also leads to formation of oxygen vacancies within the crystalline lattice, which are promoters of photocatalytic activity.

Similar effects of ion-doping could be induced by ionizing irradiation of TiO_2 such as e-beam or x-ray (Jun et al., 2006; Paunovic et al., 2020). Namely, during the passage of ionizing radiation through solids, there is an interaction between them. At the atomic level, this interaction is manifested through various phenomena, such as: (i) electron excitation, causing ionizing of the atoms and formation of electron/ positron pairs; (ii) Compton effect; (iii) phonon formation and consequently heating of the material; (iv) collective oscillation of electrons, i.e., plasmons; (v) photons emission and (vi) bonds breaking or cross-linking (Paunović et al., 2020). As result of these events, TiO_2 undergoes changes at the crystal level.

Research by the present authors has shown that interplanar distance and lattice parameters of anatase titania (obtained by simplified sol-gel method) increased as a result of its irradiation by e-beam and x-ray. Reduction of the particle size was also detected. This means improving the photcatalytic activity by the physical approach. The changes in the crystal lattice cause changes in the properties. At the crystal level, some events could be observed such as formation of oxygen vacancies and change of the Ti valence state (increasing of Ti^{3+}/Ti^{4+} ratio), causing a decrease of band gap energy and consequently, the increase of the photocatalytic activity. This is an improvement of the intrinsic photocatalytic activity, i.e., an advance by the chemical approach. Therefore, the ionizing irradiation cause changes in the structure which improves both the surface and intrinsic photocatalytic activity of TiO_2. All these effects have been more pronounced in the case of x-ray irradiation.

The other way to promote photocatalytic activity of TiO_2 is formation of oxygen vacancies by a reduction process, where the reaction products are non-stoichiometric titanium oxides. They were discovered in 1950 by the Swedish chemist and crystallographist Arne Magneli, after whom they were named—Magneli phases. Magneli phases have a homologous order with a general formula Ti_nO_{2n-1}. Non-stoichiometry of the non-stoichiometric titanium oxides (Magneli phases with common formula Ti_nO_{2n-1}), is a consequence of the lack of oxygen in the crystal lattice, created during the reductive rutile transformation (Paunovic et al., 2015). The first homolog of Magneli phases Ti_4O_7

is built of three TiO_2 octahedra and one TiO octahedron, where the oxygen vacancies are created at the edges rather than vertices (Smith et al., 1998). Thus, the basic crystalline unit of Ti_4O_7 and the whole homologous order of Magneli phases is triclinic. (Xu et al., 2017; Malik et al., 2020). Magneli phases are oxides with mixed valence. For example, the first homolog Ti_4O_7 has two Ti^{4+} ($3d^0$) and Ti^{3+} ($3d^1$) configurations. Therefore, it is considered as an electron doped TiO_2 containing oxygen deficiency (Xu et al., 2016). Physical and chemical properties of the Magneli phase are similar to those of TiO_2, except the color of appearance and electrical conductivity. Instead of a clear white appearance of anatase and rutile, the appearance of the Magneli phases is largely dark-blue. Ti_4O_7 is the best conductor within the homologous order of Magneli phases, showing electrical conductivity of 1030 to 1995 $S \cdot cm^{-1}$ (Xu et al., 2016), which is comparable with the electrical conductivity of carbon nanostructures. As the number of Ti atoms n increases, the electrical conductivity decreases. The energy of the band gap (E_g) of the homologs with $n < 9$ is at least 10 times smaller than the corresponding value of E_g for TiO_2 (E_g = 3.2 eV for anatase and E_g = 3.0 eV for rutile) (Radecka et al., 2007). The presence of lattice defects, i.e., oxygen vacancies, in either an ordered or disordered manner, seems to be an important factor to promote photocatalytic activity under visible light. Toyoda et al. (2008) obtained Magneli phases by coating of rutile with polyvinyl alcohol (PVA) and heated within the temperature range of 700 to 1100°C. They confirmed that such carbon-coated Magneli phases are photocatalytic active under visible light through the study of degradation of iminoctadine triacetate (IT, $C_{24}H_{53}N_7O_6$) and phenol. One of the problems associated with the Magneli phases is the inability to obtain fine nanoparticles due to high rutile reduction temperatures in the range of 800 to 1100°C. Among the most prominent achievements of the bottom-up based synthesis methods is that of (Nguyen et al., 2012), who produced Magneli phases with specific surface area of 45 $m^2 \cdot g^{-1}$, corresponding to approximately of 32 nm size of the particles. The synthesis was performed by heating the commercial anatase (255 $m^2 \cdot g^{-1}$) with a hydrogen flow at 1050°C for 6 hr. The best achievements of the top-down based procedures are considerably lower. Paunovic et al. (2018) obtained Magneli phases with specific surface area of 4.2 $m^2 \cdot g^{-1}$ (particle size of about 200 nm), using a mechanical treatment on the micro-sized commercial Magneli phases by planetary mill, for 20 hr.

2.3.4 Synthesis of TiO_2 nanostructures

The two main approaches for formation of TiO_2 nanostructures are: top-down and bottom-up (Fig. 10.11). The top-down approach includes mechanical reduction of the particle size of the initial bulk material, using milling, attrition, etc. However, due to the thermodynamic limitations, its practical application is insignificant. Namely, the reduction of the particles size is accompanied with a high increase of its surface energy, and after reaching the critical values, the particles start to agglomerate. The processes are accompanied with contamination and formation of a lot of defects within the structure of the treated material. Therefore, the procedures based on the bottom-up approach implies formation of the nanostructures atom by atom, are of great importance for nano-sized TiO_2 synthesis. As the driving force of the nanostructure formation is reduction of the free Gibbs energy, the obtained

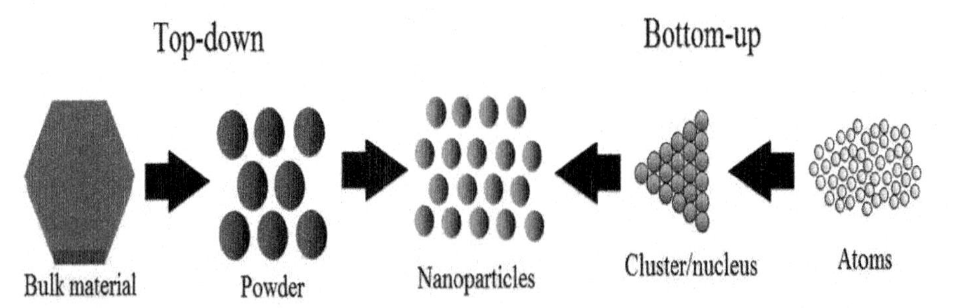

Fig. 10.11. Schematic view of the top-down and bottom-up approaches for obtaining TiO_2 nanostructures.

nanostructure is in a state very close to equilibrium (Cao, 2004). That is why such nanomaterials have less defects and a more stable structure.

The bottom-up procedures for synthesis of TiO_2 nanostructures (pure or co- doped with some ions) could be divided in two main groups (Carp et al., 2004): solution routes and gas phase methods. Solution routes include the following methods: (1) sol-gel, (2) solvothermal, (3) microemulsion, (4) precipitation of hydroxides, (5) electrochemical deposition and (6) combustion, while the gas phase methods include: (1) Chemical Vapor Deposition (CVD), (2) Physical Vapor Deposition (PVD), (3) Spray Pyrolysis Deposition (SPD), (4) Direct Current (DC) or Radio Frequency (RF) sputtering, (5) dynamic ion beam mixing and (6) Ion implantation.

Sol-gel is very suitable and a common method for production of different types of TiO_2 nanostructures, such as nanofilms or coatings, spherical nanoparticles, nanorods or nanotubes, etc. It includes hydrolysis and condensation of some titanium compound as a precursor to form sol and further, after evaporation of the solvent and formation of gel. Most of TiO_2 precursors are very reactive in water, especially the titanium alkoxide, known as titanium tetraisopropoxide (TTIP). One of the ways to provide better control of the hydrolysis process is using complexing agents which improve the stability of the precursor (Marien et al., 2017). For example, acetic acid acts as an chelating agent and increases the coordination number of TiO_2 precursors and improves its stability. The other way is using a non-aqueous solvent, such as isopropyl alcohol or anhydrous ethanol, while as a hydrolysis agent HNO_3 is used with ratio of precursor:HNO3 = 10:1 (Paunović et al., 2015). As a result of the series of hydrolysis and condensation steps, a colloidal suspension—sol is formed (Ullatil and Periyat, 2017). With a different treatment of this sol, several nanoproducts can be obtained such as dense films, nanoparticles (spheres, rods), fibers and powders with different morphology (Fig. 10.12). TiO_2 structure (anatase or rutile) can be adjusted by the appropriate temperature selection in the further thermal treatment of the produced nanostructure.

A simplified sol-gel method for TiO_2 synthesis used by (Paunović et al., 2020) are presented here. All processes within the procedure occur at an ambient pressure in laboratory glass (Fig. 10.13). As a precursor titanium tetraisopropoxide (TTIP) was used, while as a solvent anhydrous ethanol in ratio TTIP: EtOH = 1:8. 1M HNO_3 has role of a hydrolyzer and to improve the precursor's stability. It was dosed in ratio HNO_3: TTIP = 1:10. The entire process including dissolving of TTIP, hydrolysis, formation of gel, evaporation of the solvent to formation of fine light-yellow $Ti(OH)_4$ powder, took

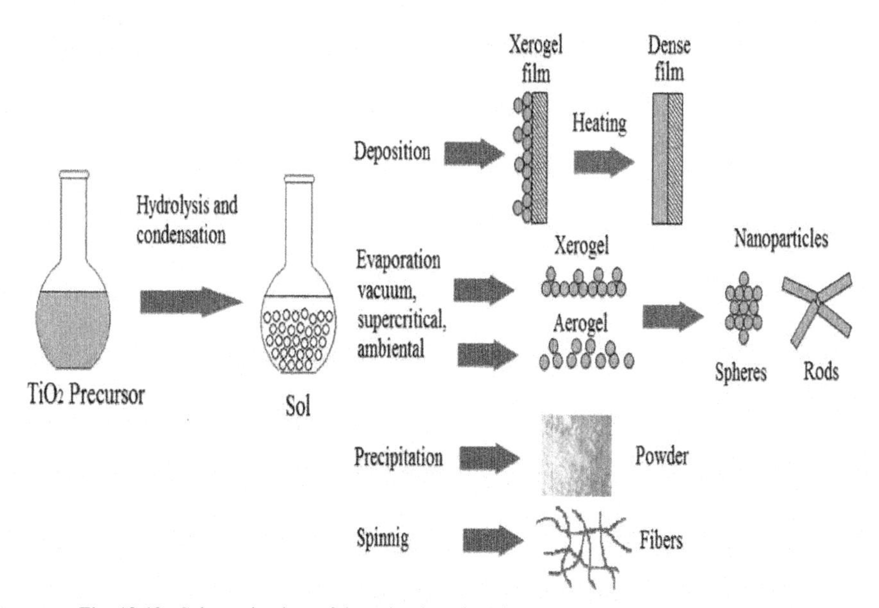

Fig. 10.12. Schematic view of the sol-gel method for synthesis of TiO_2 nanostructures.

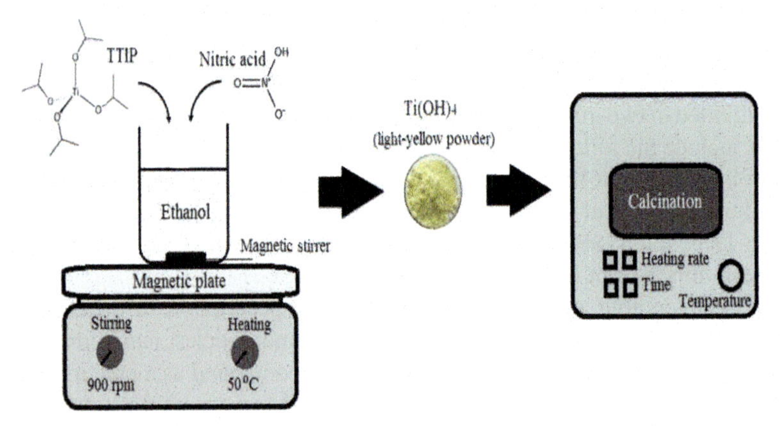

Fig. 10.13. Schematic view of the simplified sol-gel method.

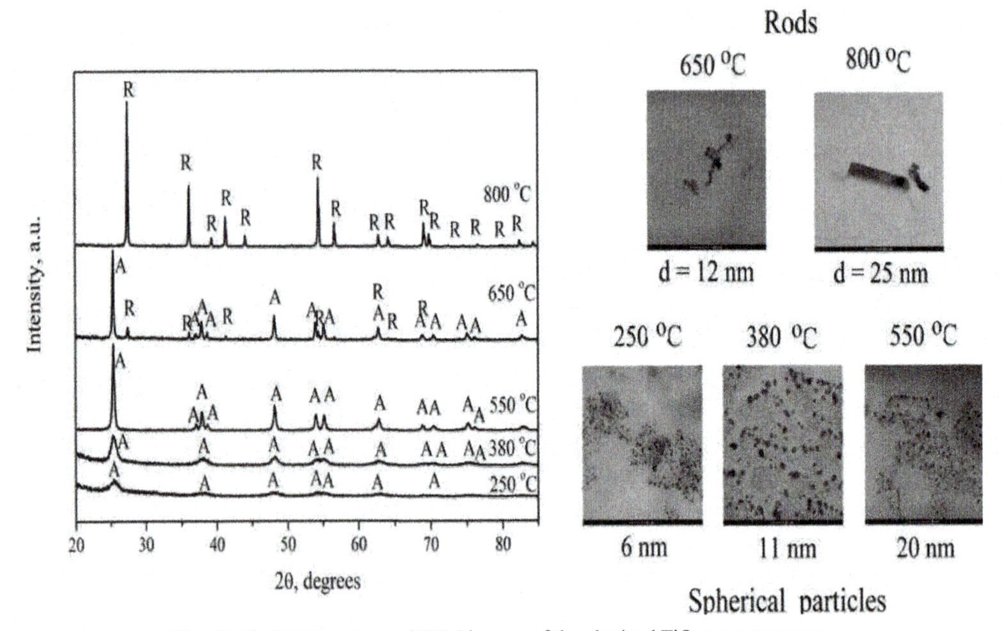

Fig. 10.14. XRD spectra and TEM images of the obtained TiO_2 nanostructures.

place in 8 hr, under stirring of 900 rpm at temperature of 50°C. The hydrolysis of TTIP occurs after the following chemical reaction:

$$Ti(OCH(CH_3)_2)_4 + 4H_2O \longrightarrow Ti(OH)_4 + 4(CH_3)_2CH(OH)$$

Transformation of the Ti $(OH)_4$ powder (calcination) was performed in a chamber furnace for 2 hr at different temperatures (250, 380, 550, 650 and 800°C), in order to produce different TiO_2 nanostructures. In Fig. 10.14 the XRD spectra and TEM images of the obtained TiO_2 nanostructures are shown. At 250°C anatase TiO_2 was formed with spherical particles of average size of 6 nm. The broad XRD spectrum points out the presence of amorphous TiO_2. At 380 and 550°C was detected in only an anatase crystalline structure with h spherical particles with average size of 11 and 20 nm, respectively. At 650°C, changes of the structure and morphology could be observed. Instead of spherical particles, nanorods were formed with average diameter of 12 nm and length of 35 to 45 nm. Except for anatase, the rutile phase was detected. At 800°C nanorods of rutile were obtained with average diameter of 25 nm and length of 50 to 100 nm.

2.3.5 A case study of naphthalene degradation

As an example of application of titanium oxides for air remediation from toxic and hardly degradable aromatic organic pollutants, a study on photocatalytic degradation of naphthalene was made by (Paunovic et al., 2015; Marinkovski et al., 2012). Naphthalene can be found in ambient and indoor air, as a result of the emissions from chemical industries, extractive metallurgy plants, oil refineries, vehicles, tobacco smoking, fumigants and deodorizers, etc. (Jia and Batterman, 2010). As an aromatic hydrocarbon consists of two fused benzene rings, it is considered a stable and barely degradable pollutant. People exposed to naphthalene through inhalation of ambient and indoor air, and consequently to the risk of non-cancer (hyperplasia and metaplasia in respiratory system) and cancer endpoint (nasal tumors) (Jia and Batterman, 2010).

Experimental oxidation of naphthalene was performed using pure oxygen. Solid-state naphthalene (Sigma-Aldrich, 99%) was evaporated in a Pyrex glass vacuum line (pressure of 10–15 Pa) presented elsewhere (Paunovic et al., 2015). The mixture of oxygen and naphthalene was collected via the vacuum line in the Pyrex glass reactor and was exposed under UV irradiation. As photocatalytic material on which naphthalene degradation occurs was top-down prepared Magneli phases and commercial TiO_2 HOMBIKAT UV-100. Micro-sized Magneli phases (Ebonex®, Altraverda, UK) were mechanically treated by Fritsch planetary mill (Pulverisisette 5) without a binder for 20 hr. The obtained size of the particles was 215 nm (Paunovic et al., 2010), while the specific surface area 4.2 $m^2 \cdot g^{-1}$ (Paunovic et al., 2018).

Magneli phases (Fig. 10.15a) form aggregates in micron dimensions. The presence of a hole can be appropriate for better inter-particle porosity and favorable for catalytic purposes. TiO_2 HOMBIKAT UV-100 (Fig. 10.15b) forms smaller aggregates (50 ÷ 200 nm) and shows considerably higher surface area (declared BET surface area is 250 $m^2 \cdot g^{-1}$). Raman analysis (Fig. 10.16a) has shown that Magneli phases are derived from the rutile phase, as all Raman modes correspond to the rutile phase, but shift

Fig. 10.15. SEM images of a) Magneli phases, and b) TiO2 HOMBIKATUV-100.

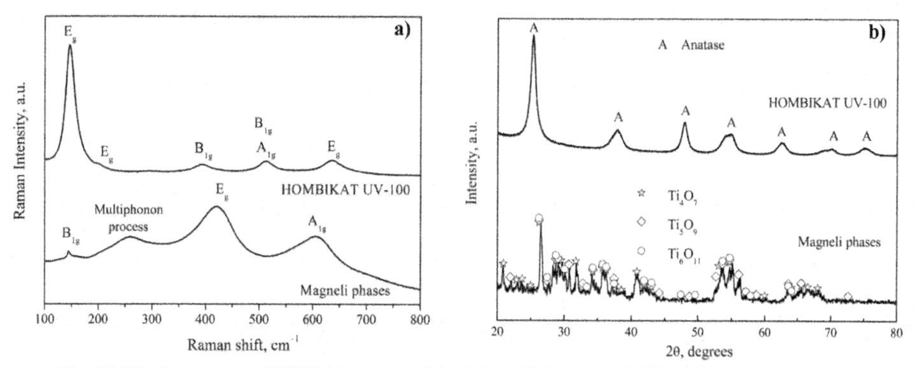

Fig. 10.16. Raman a) and XRD b) spectra of the Magneli phases and TiO_2 HOMBIKATUV-100.

to higher or lower values (E_g and A_{1g} to lower, multiphonon process mode and B_{1g} to higher values). Raman vibrational modes of TiO_2 HOMBIKAT UV-100 point out that it consisted only of the anatase crystalline form. The same can be concluded from the XRD spectrum (Fig. 10.16b). Characteristic XRD peaks of Magneli phases are on completely different positions than those of rutile or anatase and correspond to the three first homologs : Ti_4O_7, Ti_5O_9 and Ti_6O_{11} and show uniform distribution. This does not mean that the latter homologous oxides are not present, but due to the very small amounts, their characteristic peaks cannot be seen.

The oxidation of naphthalene takes place after the reaction:

$$C_{10}H_8 + O_2 + H\nu + \text{photocatalyst} \xrightarrow{\text{intermediaries}} CO_2 + H_2O + \text{photocatalyst}$$

The level of photocatalytic degradation of naphthalene was investigated by *in situ* FTIR spectroscopy measurements. In Fig. 10.17, FTIR spectra of non-treated naphthalene and treated by

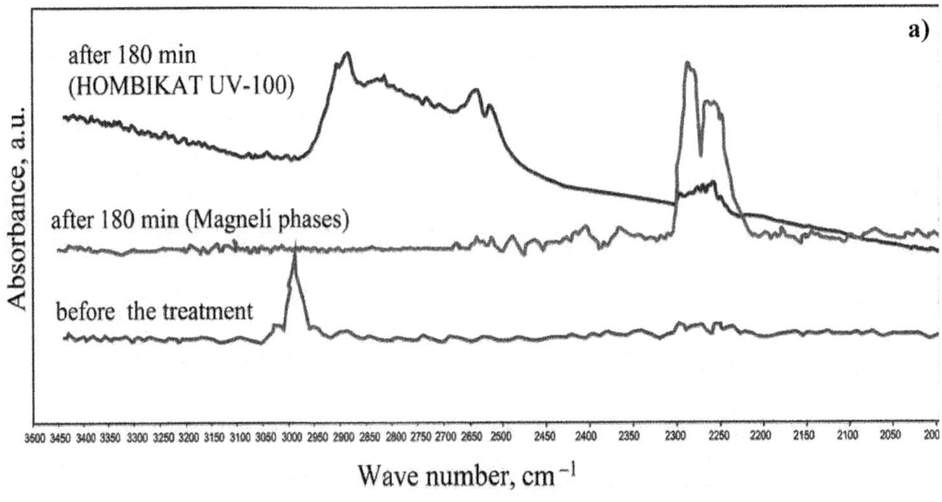

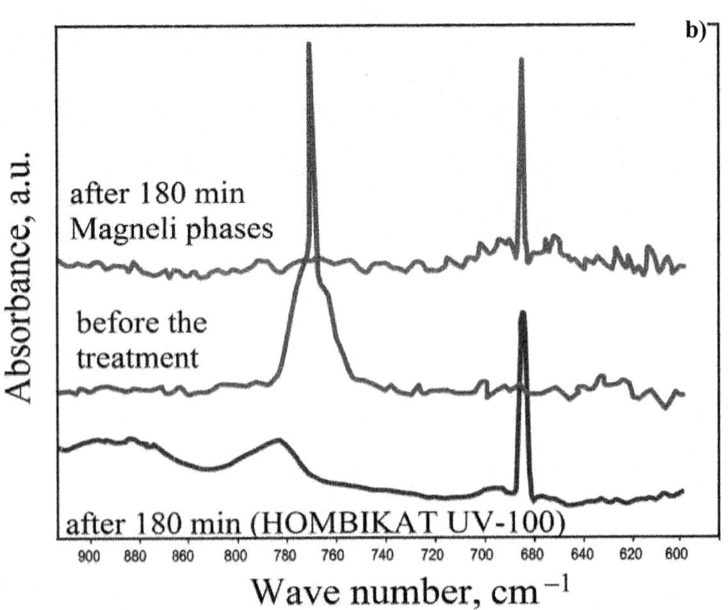

Fig. 10.17. FTIR spectra of naphthalene before the photocatalytic treatment under UV-irradiation and after 180 min photocatalytic treatment on the Magneli phases and TiO_2 HOMBIKATUV-100 surface, in the region of a) higher and b) lower wave numbers.

UV irradiation using different types of photocatalysts for 180 min. Two intensive peaks characteristic for naphthalene can be seen in the spectrum of non-treated naphthalene, at position of 3060.3 cm^{-1} and 781.8 cm^{-1}.

Analyzing the spectrum of UV-irradiated naphthalene using TiO_2 HOMBIKAT UV-100, it can be seen that the naphthalene peak at 3060.3 cm^{-1} almost disappeared, but as result of that it appeared as a broad peak in the region from 2660 to 3040 cm^{-1} with a several maximums at position 2705, 2733, 2900 and 2970 cm^{-1}. The first two maximums correspond to aliphatic aldehyde CH_2-CHO (Byun et al., 2000). The next one at 2900 cm^{-1} corresponds to CH_3 stretching from aliphatic hydrocarbons. The last one (2970 cm^{-1}) appeared as result of C–H stretching from aliphatic hydrocarbons (Byun et al., 2000). The peak in the region of 2300 to 2400 cm^{-1} corresponds to a surface CO_2 (Jogi et al., 2006). The intensive naphthalene peak positioned at 781,8 cm^{-1} is reduced by about 60%, and as a result of that a new peak at 668 cm^{-1} appeared. It corresponds to CO_2.

In the case of Magneli phases, it is obvious 3066 cm^{-1} totally disappeared as a result of degradation of aromatic rings. In the region of 2300 to 2400 cm^{-1} a very pronounced peak appeared as a result of surface CO_2 (Jogi et al., 2006), pointing out the high level of naphthalene degradation. The intensive peak at 781,8 cm^{-1} reduced more than 95%, and consequently, a new peak characteristic for CO_2 appears at 668 cm^{-1} (Jogi et al., 2006).

The degree of naphthalene degradation was determined by the ratio of the most intensive naphthalene peak positioned a 781,8 cm^{-1} from the spectra of the irradiated vs. spectrum of non-irradiated naphthalene. In presence of Magneli phases the level of degradation was determined to be 95%, while in the presence of TiO_2 HOMBIKAT UV the corresponding degradation was nearly 60%. Table 10.2 summarizes the achievements of the level of degradation of naphthalene in the presence of the two materials and another type of TiO_2 studied by the present authors.

Table 10.2. Degree of naphthalene degradation using different TiO_2 based photocatalysts.

Material	Synthesis Method	Degree of Degradation, %
HOMBIKAT UV-100	(commercial)	55
Amorphous TiO_2	Sol-gel method using TTIP as a precursor, thermally treated at 250°C in N_2 atmosphere	70
Cr doped TiO_2	Chemical vapor deposition ($TiCl_4$ and CrO_2Cl_2 precursors)	80
TiO_2 anatase	Sol-gel method using TTIP as a precursor, thermally treated at 480°C in N_2 atmosphere	85
Magneli phases	Mechanical milling of commercial Ebonex®, Altraverda, UK	95

2.4 Silica nanomaterials

In the last few decades, silica nanomaterials have attracted a special interest because of their highly efficient application in environmental remediation. The main characteristics of silica nanostructures consist of a high surface area, large pore volume, appropriate pore size and surface that can be easily modified. These materials exhibited very high efficiency when they were used as adsorbents especially for removal of pollutants in a gaseous phase (Huang et al., 2003; Guerra et al., 2018; Tsai et al., 2016). Especially for the higher potential for use as adsorbents for environmental hazardous pollutants obtained from mesoporous silica materials, which have a well-developed porous structure and corresponding specific surface area, as well as the possibility for modification of their surface using organic functional groups (Tsai et al., 2016). Functionalized mesoporous silicas were considered as highly efficient type of adsorbents for the removal of Volatile Organic Compounds (VOCs). Besides, removal of the gaseous pollutants, silicates based nanosorbents were applied for nanoremediation of many metal ions (Cd, Co, Cr, Cu, Zn, Pb) (Vunain et al., 2016; Guerra et al., 2018). These silicates-based sorbents were mainly functionalized with amino, aminopropyl and thiol functional compounds (Nakanishi et al., 2015; Wang et al., 2015).

The most important role for the remediation activity and for further surface modification of hydroxyl groups is their location on the surface of silica materials. Grafting and increasing the quantity of the functional groups onto the walls of the silica pore was one well-known strategy to design new adsorbents and catalysts (Jarmolinska et al., 2020). Diagboya and Dikio Ezekiel (2018) reviewed the pristine and specifically functionalized mesoporous silicates of the so called 'designer silicates' that were used as adsorbents for removal of aqueous pollutants and water treatment. The synthesis of ordered mesoporous silicates and their properties, surfactants removal and preparation of the designer silicates were briefly introduced and their applications in the removal of inorganic and organic pollutants were described. Various types of designer silicates, which contain the nitrogen/thiol components, magnetics and the composites, were presented with their applications for the removal of toxic metal cations, anionic species, dyes, pesticides, industrial organics, pharmaceuticals and other emerging pollutants. Chen et al. (2020) developed for the first time hollow-structured Porous Silica Nanocapsules (PSNs) functionalized with phenyl and n-octyl groups, which exhibited highly mesopores structure and a functional hydrophobic surface. Toluene was selected for testing the adsorption of the VOC molecule. They established the continuous-flow adsorption method in order to analyze and study the adsorption performances. The dynamic adsorption–desorption behaviors for PSNs were systematically evaluated and compared with those of other typical adsorbents (silicalite-1, KIT-6, SBA-15 and AC). The hydrophobicity of PSN-based materials was investigated under different relative humidity (11–50% RH) in the test conditions. The obtained results have shown that under dry conditions, the dynamic adsorption capacity and the desorption efficiency of PSNs were much better than those of the traditional silicates. Under wet conditions, the functionalized PSN materials made up the defects of pure PSNs, holding an excellent adsorption capacity and reusability.

From a family of silicates, the adsorption behavior of expanded perlite was studied (Burevska -Atkovska, 2017). It was reported that both types of perlites (obsidian and pitchstone) are volcanic glasses with different water content. Namely, the obsidian contains less than 2% H_2O; perlite, 2–5%; and pitchstone, more than 5%. Generally, the range of composition is 70–75% SiO_2 and about 10–15% Al_2O_3. The water content is necessary for expansion, but obsidians with as little as 0.2% H_2O are reported to have expanded satisfactorily. Originally, perlite means a glassy rock characterized by concentric cracks, but commercial use includes any glassy rock (except pumice) that can be expanded by heating to about 1100°C. Numerous studies have shown that perlite with its high porosity and specific surface area, as well as its low cost compared to other adsorbents, has the potential to remove heavy metals. Mohammed et al. (2011) investigated the effectiveness of perlite in removing lead and cadmium ions from industrial wastewater. They determined the pH values of the medium, the amount of adsorbent and the contact time at which they obtained the maximum removal of the ions of the examined heavy metals. They found that at pH = 7 the elimination efficiency was almost 100% for lead and 97.7% for cadmium, with the application of 10 g/dm^3 perlite and at a contact time of 1.5 hr. The adsorption properties of expanded perlite in the removal of cadmium and nickel ions from aqueous systems were analyzed by Torab-Mostaedi et al. (20110). They determined that the optimum pH for adsorption of both metals was 6 with the removal efficiency of Cd (II) being 88.8% when using 10 g/dm^3 expanded perlite, while for Ni (II) with 8 g/dm^3 expanded perlite removal rate was 93.3. The maximum capacity of expanded perlite was determined to be 1.79 mg/g and 2.24 mg/g for Cd (II) and Ni (II), respectively. Experimental data for both metals matched well with the Freundlich isotherm, and the adsorption kinetics corresponded to the pseudo II order of reaction (Torab-Mostaedi et al., 2010). Ahmet et al. (2007) by the use of the Langmuir's isothermal model, calculated the maximum adsorption capacities of the examined expanded perlite to remove Cu (II) and Pb (II) ions. They obtained 8.62 mg/g and 13.39 mg/g for both metals, respectively. At an initial ion concentration of 10 g/dm^3, at pH 5, a contact time of 90 min, at a temperature of 20°C and an amount of expanded perlite of 20 g/dm^3, the maximum adsorption percentages of 95% for Pb were determined (II) and 80% for Cu (II) ions. The adsorption kinetics were also investigated and it was found that the adsorption process for both metals closely follow the pseudo II order kinetics (Ahmet et al., 2007). Under similar conditions, the adsorption properties of expanded perlite in the adsorption of cobalt and lead ions by Hamid et al. (2010) were investigated. Absorption equilibrium was reached in 150 min for cobalt ions and for 90 min for lead where the maximum removal rates were 46% and 99% for Co (II) and Pb (II), respectively. Langmuir-'s and Dubinin-Radushkevich's isothermal models showed good agreement with the experimental values for both metals. The maximum capacity of expanded perlite was calculated using the Langmuir equation and was 1.05 mg/g for Co (II) and 6.27 mg/ g for Pb (II). In this case, also, the pseudo-second order reaction best describes the adsorption kinetics of the process for the two metal ions. The elimination of Cu (II) ions from industrial waste effluent using expanded perlite from Turkey was investigated by Yuksel et al. (2014). They determined the most efficient ion removal of over 90% at pH 7, using 15 g/dm^3 adsorbent, and the adsorption equilibrium was achieved in 30 min. Second-order pseudo-kinetic models, the Elovic model and interparticle diffusion were used, and the results showed the highest value of the correlation coefficient for second-order pseudo-kinetics. In two papers, (Serpil, 2015; Serpil, 2016) investigated the adsorption capacity of alternative perlite-based composite nanosorbents to remove Cr (VI) ions. These nanosorbents were obtained by modifying perlite with α-MnO_2 (RAM) and S-Fe_2O_3 (PGI) nanoparticles. The maximum adsorption capacities for hexavalent chromium removal were determined to be 8.64 mg/g and 7.60 mg/g for PGI and PAM, respectively, and compared with the adsorption capacity of pure perlite as a Cr (VI) adsorbent, these values were higher. Experimental data for PGI followed the Dubinin-Radushkevich isotherm well, while Temkin's isotherm model showed the best agreement with the experimental values of RAM. The pseudo II order kinetic model was found to be suitable for explaining the adsorption kinetics of both nanocomposite materials. The adsorption potential of Expanded Perlite (EP) and modified manganese oxide expanded perlite (Mn-MEP) to remove Sb (III) ions was investigated by Ahmet et al., ER and Mn-MEP were characterized before and after the adsorption process using FTIR and SEM analysis techniques.

Characteristic morphology of the tested perlite sample in our research is shown in Fig. 10.18. The effects of pH on adsorption of Ni (II), Pb (II) and Fe (II) on expanded perlite are presented in Table 10.3.

The optimal pH for the removal of Ni (II) and Pb (II) ions for expanded perlite was found to be pH = 7. This value was in correlation with the values of the zero charge point. which for perlite pHPZC had a value of 6.7. For removing iron ions, the best results were achieved at pH 4.5 and 5 respectively, for expanded perlite. This was obtained as a result of the oxidation of Fe (II) into Fe (III) ions and the increase of the concentration of $FeOH^+$ and $Fe(OH)_2$ forms at higher pH values. Examination of the effect of the initial concentration of Ni (II) and Pb (II) ions (0.3; 0.5; 0.6 and 0.7 g/dm^3) as well as of Fe (II) ions (0.8 ; 1.0; 1.5 and 2.0 g/dm^3) showed that with increasing initial concentration there is a decrease in the percentage of removal and in expanded perlite. This occurs

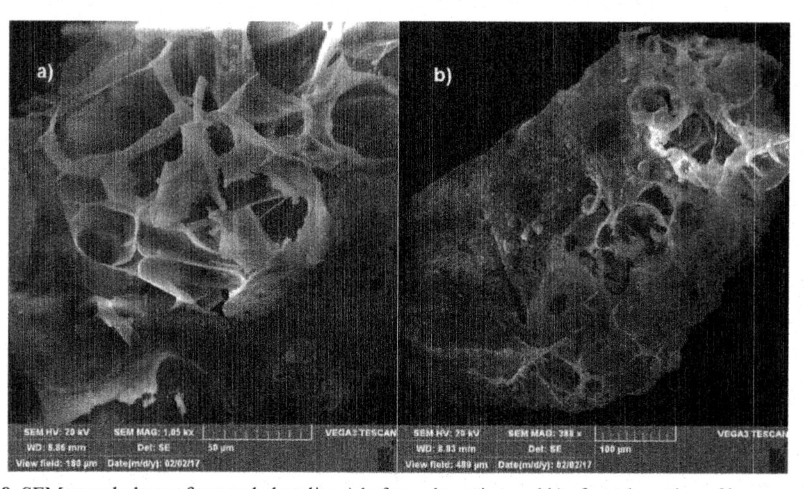

Fig. 10.18. SEM morphology of expanded perlite a) before adsorption and b) after adsorption of heavy metal ions.

Table 10.3. Effects of pH on adsorption of Ni (II), Pb (II) and Fe (II) on expanded perlite.

	% Removal		
pH	**Ni (II)**	**Pb (II)**	**Fe (II)**
3	/	/	50,2
4	19,2	51,9	69,4
5	56,4	72,3	73,1
6	60,1	85,5	31,2
7	72,0	91,6	19,0
8	64,0	89,2	/

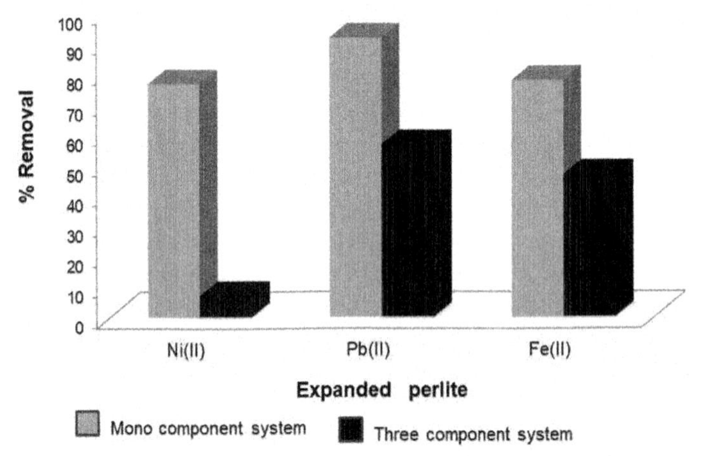

Fig. 10.19. Adsorption efficiency of expanded perlite in mono and three component systems.

due to the fact that at higher initial concentrations of metal ions the adsorption centers of sorbents become saturated and therefore more metal ions remain unabsorbed in the solution values.

Comparison of adsorption efficency of expanded perlite in mono and three component system are presented in Fig. 10.19, has shown that remarkably higher efficiency was obtained in mono component system.

3. Challenges and the Future of Nanoremediation

Literature review has shown some of the gaps in knowledge as well as the practical experience on the use of nanotechnology products for the remediation of soil and groundwater contaminations are limited. The narrow accessibility (variety and quantity) of different NPs for various contaminants, the relatively high costs for using NPs for remediation as well as the concerns about health and safety led to a rather restricted practical use of nanoremediation. These facts indicate that an "operating windows" of nano products in nanoremediation are still not clear and they face different challenges.

For example, concerning the graphene-based nanosorbents, inspite of many achievements, several questions are still left to be solved. First, the reported adsorption capacities of the same types of graphene-based adsorbents for different hazardous pollutants varied significantly. These differences could be the result of starting raw materials, manufacturing protocols and also the parameters of batch experiments. Secondly, the influence of the physicochemical properties of graphene on its adsorption abilities is not very well explained and described completely. These questions seriously hinder the design of high-performance graphene adsorbents. Third, the treatment of real wastewater samples has been rarely studied and most of the published reports were based on the model pollutants. Therefore, it is worth analyzing and evaluating graphene adsorbents in real applications and situations.

The design of new types of high-performance adsorbents based on carbon nanostructure is always a real challenge. Several issues should be taken into account when designing a new nanoadsorbent.

First, electrostatic attractions are preferred, but electrostatic repulsion should be avoided. It is recommended to perform the following transformations such as conversion of the charge of GO, the reduction or functionalization of GO. Second, specificity from certain functional groups should be developed in the future. The selectivity of graphene adsorbents has not received enough attention as yet. In particular, such an attempt might also benefit the environmental analysis using graphene to extract pollutants. Third, the dispersion state of graphene adsorbents should be optimized. Well-dispersed graphene has faster adsorption kinetics, but suffers the separation after treating pollutants. Probably, the development of magnetic and dispersible graphene adsorbents will solve the problem. Fourth, the production cost of graphene adsorbents needs to be reduced. The economic route for the preparation of graphene adsorbents should be developed. Moreover, the parameters of the adsorption experiments should be carefully optimized. Avoiding very harsh conditions is also necessary.

Another issue is to highlight the recycling of graphene adsorbents. Although many studies have demonstrated the recycling of graphene adsorbents, these studies only evaluated several cycles (Xu et al., 2013; Wu et al., 2011). For practical applications of the graphene-based adsorbents more cycles are important, which would definitely reduce the cost of water treatment. The process of the optimization of the binding strength is crucial to achieve the recycling of the nanosorbent. It is recommended to avoid very strong interaction between pollutants and graphene-based nanoadsorbent because they hinder the process of desorption. An attractive proposal for recycling is burning on fire for recycling graphene sponge. The new concern is how to control the burning parameters. It was reported that direct burning would more or less lead to the loss of adsorbents.

For better and the more successful future of nanoremediation, it important that the economical impact of technologies that use nanoparticles ensure that their full potential is achieved. The economic expansion of nanotechnology will provide industrial stakeholders and governmental policy-makers with the means to design and assess more stable economic forecasts and benefits of nanoremediation (Vijaya et al., 2018).

The sustainability in clean water is one of the main global environmental challenges (Nagar and Pradeep, 2017) of the world and it is considered that sustainable nanotechnology has provided a great contribution to solve this issue. What is very important for the future, is to increase and the focus all governments on the sustainable solution to provide clean waters by wider application of new designed and developed sustainable nanotechnologies and to increase their financial supports and investments in these research activities.

References

Abdullah, N. and Kamarudin, S.K. 2015. Titanium dioxide in fuel cell technology: An overview. J. Power Sources 278: 109–118. DOI: 10.1016/j.jpowsour.2014.12.014.

Ahmad, H., Kamarudin, S.K., Minggu, L.J. and Kassim, M. 2015. Hydrogen from photo-catalytic water splitting process: A review. Renewable Sustainable Energy Rev. 43: 599–610. DOI: 10.1016/j.rser.2014.10.101.

Ahmet, S., Mustafa, T., Demirhan, C. and Mustafa, S. 2007. Adsorption characteristics of Cu(II) and Pb(II) onto expanded perlite from aqueous solution. Journal of Hazardous Materials 148: 387–394. DOI: 10.1016/j.jhazmat.2007.02.052.

Ali, I., Al Arsh, B., Mbianda, X.Y., Burakove, A., Galunine, E., Burakovae, I., Mkrtchyane, E., Tkacheve, A. and Grachev, V. 2019. Graphene based adsorbents for remediation of noxious pollutants from wastewater. Environment International 127: 160–180. https://doi.org/10.1016/j.envint.2019.03.029.

Akbar, S.F. and Parviz, A.A. 2012. Removal of Cu^{2+} and Pb^{2+} from aqueous solutions by oxidized multiwalled carbon nanotubes. Proceedings of the 4th International Conference on Nanostructures, Kish Island, I. R. Iran.

Al-Degs, Y.S., El-Barghouthi, M.I, Issa, A.A., Khraisheh, M.A. and Walker, G.M. 2006. Sorption of Zn(II), Pb(II) and Co(II) using natural sorbents: equilibrium and kinetic studies. Water Res. 40: 2645–2658. DOI: 10.1016/j.watres.2006.05.018.

Ameta, R., Benjamin, S., Ameta, A. and Ameta, S.C. 2013. Photocatalytic degradation of organic pollutants: A review. Mater. Sci. Forum. 34: 247–272. DOI: 10.4028/www.scientific.net/MSF.734.247.

Atkovska, K., Lisickov, K., Ruseska, G., Dimitrov, A.T. and Grozdanov, A. 2018. Removal of heavy metal ions from wastewater using conventional and nanosorbents: A Review. J. Chem. Technol. Metall. 53: 202–219.

Atkovska, K., Paunovic, P., Dimitrov, A.T., Lisichkov, K., Alghuthaymi, M. and Grozdanov, A. 2020. Graphene and activated graphene as adsorbents for removal of heavy metals from water resources. pp. 177–191. *In*: Kamel, A., Abd-Elsalam, K.A. (eds.). Elsevier, Edited by Kamel A. Abd-Elsalam. DOI: 10.1016/B978-0-12-819786-8.00009-8.

Bai, Y., Mora-Sero, I., De Angelis, F., Bisquert, J. and Wang, P. 2014. Titanium dioxide nanomaterials for photovoltaic applications. Chem. Rev. 114: 10095–10130. DOI: 10.1021/cr400606n.
Bakalar, T., Bugel, M. and Gajdosova, L. 2009. Heavy metal removal using reverse osmosis. Acta Montan. Slovaca. 14: 250–253.
Bardos, P., Bone, B., Černík, M., Daniel, Elliott W. and Jones Merly, C. 2015. Nanoremediation and international environmental restoration markets. Rem. J. 25: 83–94. DOI: 10.1002/rem.21426.
Burevska Atkovska, K. 2017. Ph.D. Thesis University Ss Cyril and Methodius in Skopje, Skopje.
Byun, D., Jin, Y., Kim, B., Lee, J.K. and Park, D. 2000. Photocatalytic TiO_2 deposition by chemical vapor deposition. J. Hazard. Mater. 73: 199–206. DOI: 10.1016/S0304-3894(99)00179-X.
Calatayud, D.G., Rodríguez, M. and Jardiel, T. 2015. Controlling the morphology of TiO_2 nanocrystals with different capping agents. Bol. Soc. Esp. Ceram. Vidrio. 54: 159–195. DOI: 10.1016/j.bsecv.2015.07.001.
Cao, G. 2004. Nanostructures and Nanomaterials, Synthesis, Properties & Applications. Imperial College Press, London.
Carp, O., Huisman, C.L. and Reller, A. 2004. Photoinduced reactivity of titanium dioxide. Prog. Solid. State Chem. 32: 33–177. DOI: 10.1016/j.progsolidstchem.2004.08.001.
Castro, V.G., Costa, J.B., Medeiros, F.S., Siqueira, E.J., Kasama, A.H., Fiqueiredo, K.C.S., Lavall, R.L. and Silva, G.G. 2019. Improved functionalization of MWCNT in ultra-low acid volume: Effect pf solid/liquid interface. J. Braz. Chem. Soc. 30(11): 2477–2487. DOI:10.21577/0103-5053.20190166.
Chaudhery, M.H. 2020. Handbook of Functionalized Nanomaterials for Industrial Applications, Elsevier, p. 299. ISBN 978-0-12-816787-8.
Chandra, V., Park, J., Chun, Y., Lee, J.W., Hwang, I.C. and Kim, K.S. 2010. Water-dispersible magnetite-reduced graphene oxide composites for arsenic removal. ACS Nano. 4: 3979–3986. DOI: 10.1021/nn1008897.
Chen, J., Sun, C., Huang, Z., Qin, F., Xu, H. and Shen, W. 2020. Fabrication of functionalized porous silica nanocapsules with a hollow structure for high performance of toluene adsorption−desorption. ACS Omega 5(11): 5805–5814. DOI: 10.1021/acsomega.9b03982.
Ciriminna, R., Zhang, N., Yang, M.Q., Meneguzzo, F., Xu, Y.J. and Pagliaro, M. 2015. Commercialization of graphene-based technologies: A critical insight. Chem. Commun. 51: 7090–7095. DOI: 10.1039/c5cc01411e.
Di, Z.C., Ding, J., Peng, X.J., Li, H.Y., Luan, K.Z. and Liang, J. 2006. Chromium adsorption by aligned carbon nanotubes supported ceria nanoparticles. Chemosphere 62: 861–865. DOI: 10.1016/j.chemosphere. 2004.06.044.
Diagboya Ezekiel, P. and DikioEzekiel, D. 2018, Silica-based mesoporous materials; emerging designer adsorbents for aqueous pollutants removal and water treatment. Microporous and Mesoporous Materials 266: 252–267. DOI: 10.1016/j.micromeso.2018.03.008.
Diebold, U. 2003. The surface science of TiO_2. Surf. Sci. Rep. 48: 53–229. DOI: 10.1016/S0167-5729(02)00100-0.
Dimitrov, A.T., Fray, D.J. and Schwandt, C. 2004. GB Patent Application Number 0421869.9, 2004. Title "Electrolytic Method, Apparatus and Product".
Ersahin, M.E., Ozgun, H., Dereli, R.K., Ozturk, I. and Roest, K. 2012. A review on dynamic membrane filtration: Materials, applications and future perspectives. Bioresour. Technol. 122: 196–206. DOI: 10.1016/j.biortech.2012.03.086.
Farghali, A.A., Abdel Tawab, H.A., Abdel Moaty, S.A. and Khaled, R. 2017. Functionalization of acidified multi-walled carbon nanotubes for removal of heavy metals in aqueous solutions. J. Nanostruct. Chem. 7: 101–11. DOI: 10.1007/s40097-017-0227-4.
Frank, S.N. and Bard, A.J. 1977. Heterogeneous photocatalytic oxidation of cyanide ion in aqueous solutions at titanium dioxide powder. J. Am. Chem. Soc. 99: 303–304. DOI: 10.1002/chin.197714134.
Fujishima, A. and Honda, K. 1972. Electrochemical photolysis of water at a semiconductor electrode. Nature 238: 37–38. DOI: 10.1038/238037a0.
Grozdanov, A., Atkovska, J., Lisickov, K., Ruseska, G. and Dimitrov, A.T. 2018. Removal of heavy metal ions from wastewater using bio- and nanosorbents. Springer International Publishing AG. pp. 239–244. *In*: Cocca, M. et al. (eds.). Springer Water. DOI: 10.1007/978-3-319-71279-6_33.
Guerra, F.D., Attia, M.F., Whitehead, D.C. and Alexis, F. 2018. Nanotechnology for environmental remediation: Materials and applications. Molecules 23: 1760. DOI: 10.3390/molecules23071760.
Gupta, V.K., Agarawal, S. and Saleh, T.A. 2011. Synthesis and characterization of alumina-coated nanotubes and their application for lead removal. J. Hazard. Mater. 185: 17–23. DOI: 10.1016/j.jhazmat.2010.08.053.
Gupta, V.K., Tyagi, I., Sadegh, H., Ghoshekandi, R.S., Makhlouf, A.S.H. and Maazinejad, B. 2015. Nanoparticles as adsorbent; a positive approach for removal of noxious metal ions: A review. Sci. Tech. and Devel. 34(1): 195–214.
Hadia, N.M.A. 2014. Annealing effect on structural and optical properties of hydrothermally synthesized TiO_2 nanowires. J. Nanosci. Nanotechnol. 14: 5574–5580. DOI: 10.1166/jnn.2014.9241.
Hamid, G., Torab-Mostaedi, M., Ahmad, M., Ghannadi, M.M., Ahmadi, S.J. and Zaheri, P. 2010. Characterizations of Co(II) and Pb(II) removal process from aqueous solutions using expanded perlite. Desalination 261: 73–79.
Huang, H.Y., Yang, R.T., Chinn, D. and Munson, C.L. 2003. Amine-grafted MCM-48 and silica xerogel as superior sorbents for acidic gas removal from natural gas. Ind. Eng. Chem. Res. 42: 2427–2433. DOI: 10.1021/ie020440u.
Huang, X., Meng, L., Du, M. and Li, Y. 2016. TiO_2 nanorods: hydrothermal fabrication and photocatalytic activities. J. Mater. Sci. Mater. Electron. 27: 7222–7226. DOI 10.1007/s10854-016-4687-y.
Iijima, S. 1991. Helical microtubules of graphitic carbon. Nature 354: 56–58. DOI: 10.1038/354056a0.
Jarmolińska, S., Feliczak-Guzik, A. and Nowak, I. 2020, Synthesis, Characterization and use of mesoporous silicas of the following types SBA-1, SBA-2, HMM-1 and HMM-2. Materials 13: 4385. DOI:10.3390/ma13194385.

Jaroniec, V., Kim, K.H., Park, J.W., Hong, J. and Kumar, S. 2017. Graphene and its nano composites as a platform for environmental applications. Chem. Eng. J. 315. 8: 1–79.
Jayakaran, P., Nirmala, G.S. and Govindarajan, L. 2019. Qualitative and quantitative analysis of graphene-based adsorbents in wastewater treatment. Int. J. of Chem. Eng. DOI:10.1155/2019/9872502.
Jia, C. and Batterman, S. 2010. A critical review of naphthalene sources and exposures relevant to indoor and outdoor air. Int. J. Environ. Res. Public. Health. 7: 2903–2939; DOI:10.3390/ijerph7072903.
Jogi, I., Kukli, K., Aarik, J., Aidla, A. and Lu, J. 2006. Precursor-dependent structural and electrical characteristics of atomic layer deposited films: Case study on titanium oxide. Mater. Sci. Semicond. Process. 9: 1084–1089. DOI: 10.1016/j.mssp.2006.10.027.
Jun, J., Dhayal, M., Shin, J., Kim, J. and Getoff, N. 2006. Surface properties and photoactivity of TiO_2 treated with electron beam. Radiat. Phys. Chem. 75: 583–589. DOI: 10.1016/j.radphyschem.2005.10.015.
Khin, M.M., Nair, A.S., Babu, V.J., Murugan, R. and Ramakrishna, S. 2012. A review on nanomaterials for environmental remediation. Energy Environ. Sci. 5: 8075–8109. DOI: 10.1039/c2ee21818f.
Kosa, S.A., Al-Zhrani, G. and Salam, M.A. 2012. Removal of heavy metals from aqueous solution by multi-walled carbon nanotubes modified with 8-hydroxyquinoline. Chem. Eng. J. 181–182: 159–168. DOI: 10.1016/j.cej.2011.11.044.
Kumar, V., Lee, Y., Shin, J.W., Ki-Hyun Kim, K.H., Kukkar, D. and Tsang, Y.F. 2020. Potential applications of graphene-based nanomaterials as adsorbent for removal of volatile organic compounds. Env. Inter. 135: 105356. DOI: 10.1016/j.envint.2019.105356.
Lazar, M.A., Varghese, S. and Nair, S.S. 2012. Photocatalytic water treatment by titanium dioxide: Recent updates. Catalysts 2: 572–601. DOI: 10.3390/catal2040572.
Lee, S.Y. and Park, S.J. 2013. TiO_2 photocatalyst for water treatment applications. J. Ind. Eng. Chem. 19: 1761–1769. DOI: 10.1016/j.jiec.2013.07.012.
Li, Y.H., Wang, S., Wei, J., Zhang, X., Xu, C., Luan, Z., Wu, D. and Wei, B. 2003. Adsorption of cadmium (II) from aqueous solution by surface oxidized carbon nanotubes. Carbon 41: 1057–1062. DOI: 10.1016/S0008-6223(02)00440-2.
Liu, H.L. and Yang, T.C.K. 2003. Photocatalytic inactivation of Escherichia coli and Lactobacillus helveticus by ZnO and TiO_2 activated with ultraviolet light. Process Biochem. 39: 475–481. DOI: 10.1016/S0032-9592(03)00084-0.
Luo, X., Wang, C., Luo, S., Dong, R., Tu, X. and Zeng, G. 2012. Adsorption of As (III) and As (V) from water using magnetite Fe_3O_4-reduced graphite oxide-MnO_2 nanocomposites. Chem. Eng. J. 187: 45–52. DOI: 10.1016/j.cej.2012.01.073.
Luttrell, T., Halpegamage, S., Tao, J., Kramer, A., Sutter, E. and Batzill, M. 2014. Why is anatase a better photocatalyst than rutile? - Model studies on epitaxial TiO_2 films. Sci. Rep. 4: 4043. DOI:10.1038/srep04043.
Mali, S.S., Kim, H., Shim, C.S., Patil, P.S., Kim, J.H. and Hong, C.K. 2013. Surfactant free most probable TiO_2 nanostructures via hydrothermal and its dye sensitized solar cell properties. Sci. Rep. 3: 3004. DOI: 10.1038/srep03004.
Malik, H., Sarkar, S., Mohanty, S. and Carlson, K. 2020. Modelling and synthesis of Magnéli phases in ordered titanium oxide nanotubes with preserved morphology. Sci. Rep. 10: 8050. DOI: 10.1038/s41598-020-64918-0.
Marien, C.B.D., Marchal, C., Koch, A., Robert, D. and Drogui, P. 2017. Sol-gel synthesis of TiO_2 nanoparticles: Effect of pluronic P123 on particle's morphology and photocatalytic degradation of paraquat. Environ. Sci. Pollut. Res. 24: 12582–12588. DOI: 10.1007/s11356-016-7681-2.
Marinkovski, M., Paunović, P., Blaževska Gilev, J. and Načevski, G. 2012. Photodegradation of naphthalene by non-stoichiometric titanium oxides Magneli phases. Adv. Nat. Sci. Theory & Appl. 1: 215–224.
Mauter, M.S. and Elimelech, M. 2008. Environmental applications of carbon-based nanomaterials. Environ. Sci. Technol. 42: 5843–5859. DOI: 10.1021/es8006904.
Maziarz, W. Kusior and Trenczek-Zajac. 2016. Nanostructured TiO_2-based gas sensors with enhanced sensitivity to reducing gases. Beilstein. J. Nanotechnol. 7: 1718–1726. DOI: 10.3762/bjnano.7.164.
Mi, X., Huang, G., Xie, W., Wang, W., Liu, Y. and Gao, J. 2012. Preparation of graphene oxide aerogel and its adsorption for Cu^{2+} ions. Carbon 50: 4856–4864. DOI: 10.1016/j.carbon.2012.06.013.
Mohammad, M., Nemat-allah, J. and Hiwa, H. 2011. Efficiency of perlite as a low cost adsorbent applied to removal of Pb and Cd from paint industry effluent. Deaslination and Water Treatment 26(1-3): 243–249.
Nakanishi, K., Tomita, M. and Kato, K. 2015. Synthesis of amino-functionalized mesoporous silica sheets and their application for metal ion capture. J. Asian. Ceram. Soc. 3: 70–76.
Nagajyoti, P.C., Lee, K.D. and Sreekanth, T.V.M. 2010. Heavy metals, occurrence and toxicity for plants: A review. Environ. Chem Lett. 8: 199–216. DOI: 10.1007/s10311-010-0297-8.
Nagar, A. and Pradeep, T. 2020. Clean water through nanotechnology: needs, gaps, and fulfillment. ACS Nano. 14: 6420–6435.
Najafi, F. 2015. Removal of zinc(II) ion by graphene oxide (GO) and functionalized graphene oxide–glycine (GO–G) as adsorbents from aqueous solution: kinetics studies. Int. Nano. Lett. DOI:10.1007/ s40089-015-0151-x.
Nguyen, S.T., Lee, J.M., Yang, Y. and Wang, X. 2012. Exellent durability of substoichiometric titanium oxide as a catalyst support of Pd in alkaline direct ethanol fuel cell. Ind. Eng. Chem. Res. 51: 9966–9972. DOI: 10.1021/ie202696z.
Novoselov, K.S., Geim, A.K., Morozov, S.V., Jiang, D., Zhang, Y., Dubonos, S.V., Grigorieva, I.V. and Firsov, A.A. 2004. Electric Field Effect in Atomically Thin Carbon Films. Science 306(5696): 666–669. DOI: 10.1126/science.1102896.
Paunović, P. 2018. Enhancing the activity of electrode materials in hydrogen economy. LAP Lambert Academic Publishing. ISBN: 978-613-9-96188-7.
Paunović, P., Grozdanov, A., Makreski, P., Gentile, G. and Dimitrov, A.T. 2020. Application of ionizing irradiation for structure modification of nanomaterials. pp. 23–43. *In*: Petkov, P. et al. (eds.). Springer Nature B.V. DOI: 10.1007/978-94-024-2018-0_2.

Paunović, P., Grozdanov, A., Češnovar, A., Ranguelov, B., Makreski, P., Gentile, G. and Fidančevska, E. 2015. Characterization of nano-scaled TiO_2 produced by simplified sol-gel method using organometallic precursor. J. Eng. Mater. Technol. 137:021003. DOI: 10.1115/1.4029112.

Paunović, P., Grozdanov, A., Makreski, P. and Gentile, G. 2020. Structural changes of TiO_2 as a result of Irradiation by e-beam and x-rays. J. Eng. Mater. Technol. 142: 041003. DOI: 10.1115/1.4046944.

Paunović, P., Petrovski, A., Načevski, G., Grozdanov, A., Marinkovski, M., Andonović, B., Makreski, P., Popovski, O. and Dimitrov, A.T. 2015. Pathways for the production of non-stoichiometric titanium oxides. pp. 239–253. *In*: Petkov, P. et al. (eds.). Springer Science+Business Media B.V. DOI: 10.1007/978-94-017-9697-2_24.

Paunović, P., Popovski, O., Fidančevska, E., Ranguelov, B., Stoevska Gogovska, D., Dimitrov, A.T. and Hadži Jordanov, S. 2010. Co-Magneli phases electrocatalysts for hydrogen/oxygen evolution. Int. J. Hydrogen Energy 35: 10073–10080. DOI: 10.1016/j.ijhydene.2010.07.143.

Paunović, P., Popovski, O., Načevski, G., Lefterova, E., Grozdanov, A. and Dimitrov, A.T. 2018. Electrocatalysts with reduced noble metals aimed for hydrogen/oxygen evolution supported on Magneli phases. Part I: Physical. Characterization. Bulg. Chem. Commun. 50 A: 82–88.

Pelaez, M., Nolan, N., Pillai, S., Seery, M., Falaras, P., Kontos, A.G., Dunlop, P.S.M., Hamiltone, J.W.J., Byrne, J.A., O'Shea, K., Entezari, M.H. and Dionysiou, D.D. 2012. A review on the visible light active titanium dioxide photocatalysts for environmental applications. Appl. Catal. B 125: 331–349. DOI: 10.1016/j.apcatb.2012.05.036.

Peng, W., Li, H., Liu, Y. and Song, S. 2017. A review on heavy metal ions adsorption from water by graphene oxide and its composites. Journal of Molecular Liquids 230(1): 496–504.

Pillay, K., Cukrowska, E.M. and Coville, N.J. 2009. Multi-walled carbon nanotubes as adsorbents for the removal of parts per billion levels of hexavalent chromium from aqueous solution. J. Hazard. Mater. 166: 1067–1075. DOI: 10.1016/j.jhazmat.2008.12.011.

Pruden, A.L. and Ollis, D.F. 1983. Photoassisted heterogeneous catalysis: The degradation of trichloroethylene in water. J. Catal. 82: 404–417. DOI: 10.1016/0021-9517(83)90207-5.

Pyrzyńska, K. and Bystrzejewski, M. 2010. Comparative study of heavy metal ions sorption onto activated carbon, carbon nanotubes and carbon-encapsulated magnetic nanaoparticles. Colloids. Surf. A 362: 102–109. DOI: 10.1016/j.colsurfa.2010.03.047.

Radecka, M., Trenczek-Zajac, A., Zakrzewska, K. and Rekas, M. 2007. Effect of oxygen nonstoichiometry on photo-electrochemical properties of TiO_{2-x}. J. Power Sources 173: 816–821. DOI: 10.1016/j.jpowsour.2007.05.065.

Rahul, K.J. 2013. Application of electro-dialysis (ED) to remove divalent metals ions from wastewater. Int. J. Chem. Sci. Appl. 4: 68–72.

Rao, G.P., Lu, C. and Su, F. 2007. Sorption of divalent metal ions from aqueous solution by carbon nanotubes: A review. Sep. Purif. Technol. 58: 224–231. DOI: 10.1016/j.seppur.2006.12.006.

Ren, X., Chen, C., Nagatsu, M. and Wang, X. 2011. Carbon nanotubes as adsorbents in environmental pollution management: A review. Chem. Eng. J. 170: 395–410. DOI: 10.1016/j.cej.2010.08.045.

Sajan Ponnappa, C., Wageh, S., Al-Ghamdi, A.A., Yu, J. and Cao, S. 2015. TiO_2 nanosheets with exposed {001} facets for photocatalytic applications. Nano. Res. 9: 1–26. DOI: 10.1007/s12274-015-0919-3.

Sato, S. 1986. Photocatalytic activity of NO_x-doped TiO_2 in the visible light region. Chem. Phys. Lett. 123: 126–128. DOI: 10.1016/0009-2614(86)87026-9.

Serpil, E. 2015. Alternative composites nanosorbents based on Turkish perlite for the removal of Cr(VI) from aqueous solution. Journal of Nanomaterials (3): 1–7.

Serpil, E. 2016. Kinetics investigation of Cr(VI) removal by modified perlite with Fe_2O_3 and MnO_2 nanomaterials. International Journal of Chemical Engineering and Applications 7(3): 165–168.

Shukla, S., Khan, R. and Hussain, C.M. 2020. Nanoremediation. pp. 443–467. *In*: Hussain, C.M. (ed.). The Royal Society of Chemistry. DOI: 10.1039/9781788016261-00443.

Smith, J.R., Walsh, F.C. and Clarke, R.L. 1998. Electrodes based on Magnli phase titanium oxides: The properties and applications of Ebonex materials. J. Appl. Electrochem. 28: 1021–1033. DOI: 10.1023/A:1003469427858.

Tang, J., Zhang, X., Xiao, S. and Zeng, F. 2017. Application of TiO_2 nanotubes gas sensors in online monitoring of SF_6 insulated equipment. Intech. Open Science. DOI: 10.5772/intechopen.68328.

Torab-Mostaedi, M., Ghassabzadeh, H., Ghannadi, M.M., Ahmadi, S.J. and Taheri, H. 2010. Removal of cadmium and nickel from aqueous solution using expanded perlite. Brazilian Journal of Chemical Engineering 27(2): 299–308.

Toyoda, M., Yano, T., Tryba, B., Mozia, S., Tsumura, T. and Inagaki, M. 2008. Preparation of carbon-coated Magneli phases Ti_nO_{2n-1} and their photocatalytic activity under visible light. Appl. Catal. B. Environment 88: 160–164. DOI:10.1016/j.apcatb.2008.09.009.

Tsai, C.H., Chang, W.C., Saikia, D., Wu, C.E. and Kao, H.M. 2016. Functionalization of cubic mesoporous silica SBA-16 with carboxylic acid via one-pot synthesis route for effective removal of cationic dyes. J. Hazard. Mater. 309: 236–248.

Ullattil, S.G. and Periyat, P. 2017. Sol-gel synthesis of titanium dioxide. pp. 271–283. *In*: Pillai, S.C. and Hehir, S. (eds.). Springer International Publishing AG. DOI: 10.1007/978-3-319-50144-4_9.

Vindoh, R., Padmavathi, R. and Sangeetha, D. 2011. Separation of heavy metals from water samples using anion exchange polymers by adsorption process. Desalination 267: 267–276. DOI: 10.1016/j.desal.2010.09.039.

Vijaya, J.J., Adinaveen, T. and Bououdina, M. 2018. Economic aspects of functionalized nanomaterials for environment, in nanotechnology in environmental science. *In*: Hussain, C.M. and Mishra, A.K. (eds.). Wiley-VCH Verlag GmbH & Co. KGaA, DOI: 10.1002/9783527808854.

Vilela, D., Parmar, J., Zeng, Y., Zhao, Y. and Sanchezs, S. 2016. Graphene-based microbots for toxic heavy metal removal and recovery from water. Nano. Letters 16: 2860–2866. doi: 10.1021/acs.nanolett.6b00768.

Vunain, E., Mishra, A.K., Mamba, B.B. 2016. Dendrimers, mesoporous silicas and chitosan-based nanosorbents for the removal of heavy-metal ions: A review. Int. J. Biol. Macromol. 86: 570–586.

Wanees, S.A., Ahmed, A.M.M., Adam, M.S. and Mohamed, M.A. 2013. Adsorption studies on the removal of hexavalent chromium—contaminated wastewater using activated carbon and bentonite. Asian J. Chem. 25: 8245–8252. DOI: 10.14233/ajchem.2013.13559.

Wang, H., Yuan, X., Wu, Y., Huang, H., Peng, X., Zeng, G., Zhong, H., Liang, J. and Ren, M. 2013. Graphene-based materials: fabrication, characterization and application for the decontamination of wastewater and wastegas and the hydrogen storage/generation. Adv. Colloid. Interface. Sci. 195–196: 19–40. DOI: 10.1016/j.cis.2013.03.009.

Wang, L.K., Vaccari, D.A., Li, Y. and Shammas, N.K. 2005. Chemical precipitation. pp. 3141–3197. *In*: Wang, L.K., Hung, Y.T. and Shammas, N.K. (eds.). Humana Press. DOI: 10.1385/1-59259-820-x:141.

Wang, R., Hashimoto, K., Fujishima, A., Chikuni, M., Kojima, E., Kitamura, A., Shimohigoshi, M. and Watanabe, T. 1997. Light-induced amphiphilic surfaces. Nature 388: 431–432. DOI: 10.1038/41233.

Wang, X., Guo, Y., Yang, L., Han, M., Zhao, J. and Cheng, X. 2012. Nanomaterials as sorbents to remove heavy metal ions in wastewater treatment. J. Environ. Anal. Toxicol. 2: 1000154 (7 pages). DOI: 10.4172/2161-0525.1000154.

Wang, Y., Wu, T., Zhou, Y., Meng, C., Zhu, W. and Liu, L. 2017. TiO_2-based nanoheterostructures for promoting gas sensitivity performance: Designs, developments, and prospects. Sensors 17: 1971. DOI:10.3390/s17091971.

Wang, S., Wang, K., Dai, C., Shi, H. and Li, J. 2015. Adsorption of Pb2+ on amino-functionalized core–shell magnetic mesoporous SBA-15 silica composite. Chem. Eng. J. 262: 897–903.

Wu, T., Cai, X., Tan, S., Li, H., Liu, J. and Yang, W. 2011. Adsorption characteristics of acrylonitrile, p-toluenesulfonic acid, 1-naphthalenesulfonic acid and methyl blue on graphene in aqueous solutions. Chem. Eng. J. 173: 144–149. DOI: 10.1016/j.cej.2011.07.050.

Xing, Y., Chen, X. and Wang, D. 2007. Electrically regenerated ion exchange for removal and recovery of Cr(VI) from wastewater. Environ. Sci. Technol. 41: 1439–1443. DOI: 10.1021/es0614991.

Xu, B., Sohn, H.Y., Mohassab, Y. and Lan, Y. 2016. Structures, preparation and applications of titanium suboxides. RCS Adv. 6: 79706–79722. DOI: 10.1039/c6ra14507h.

Xu, D., Tan, X., Chen, C. and Wang, X. 2008. Removal of Pb(II) from aqueous solution by oxidized multiwalled carbon nanotubes. J. Hazard. Mater. 154: 407–416. DOI: 10.1016/j.jhazmat.2007.10.059.

Xu, J., Lv, H., Yang, S.T. and Luo, J. 2013. Preparation of graphene adsorbents and their applications in water purification. Rev. Inorg. Chem. 33: 139–160. DOI: 10.1515/revic-2013-0007.

Yan, H., Wang, X., Yaon, M. and Yao, X. 2014. Band structure design of semiconductors for enhanced photocatalytic activity: The case of TiO_2. Prog. Nat. Sci. Mater. Int. 23: 402–407. DOI: 10.1016/j.pnsc.2013.06.002.

Yan, X., Wang, Z., He, M., Hou, Z., Xia, T., Liu, G. and Chen, X. 2015. TiO_2 nanomaterials as anode materials for lithium-ion rechargeable batteries. Energy Technol. 3: 801–814. DOI: 10.1002/ente.201500039.

Yang, Q.H., Hou, P.X., Bai, S., Wang, M.Z. and Cheng, H.M. 2011, Adsorption and capillarity of nitrogen in inside channel of carbon nanotubes. Chem. Phys. Lett. 221: 18–24. DOI: 10.1016/S0009-2614(01)00848-X.

Yuksel, A., Gamze, N.T. and Fulya, A.T. 2014. Cu(II) removal from industrial waste leachate by adsorption using expanded perlite. Journal of Institute of Natural & Applied Sciences 19(1-2): 54–61.

Zaleska, A. 2008. Doped-TiO_2: A review. Recent. Pat. Eng. 2: 157–164. DOI: 10.2174/ 187221208786306289.

Zhang, J., Zhao, C., Hu, P.A., Fu, Y.Q., Wang, Z., Cao, W., Yang, B. and Placido, F. 2013. A UV light enhanced TiO_2/graphene device for oxygen sensing at room temperature. RSC Adv. 3: 22185–22190. DOI: 10.1039/c3ra43480j.

Zhang, J., Zhou, P., Liu, J. and Yu, J. 2014. New understanding of the difference of photocatalytic activity among anatase, rutile and brookite TiO_2. Phys. Chem. Chem. Phys. 16: 20382–20386. DOI: 10.1039/c4cp02201g.

Zhao, G., Li, J., Ren, X., Chen, C. and Wang, X. 2011. Few-layered graphene oxide nanosheets as superior sorbents for heavy metal ion pollution management. Environ. Sci. Technol. 45: 10454–10462. DOI: 10.1021/es203439v.

Zhao, G., Wu, X., Tan, X. and Wang, X. 2011. Sorption of heavy metal ions from aqueous solutions: A review. The Open Colloid. Sci. J. 4: 19–31. DOI: 10.2174/1876530001104010019.

Zhao, X., Jia, Q., Song, N., Zhou, W. and Li, Y. 2010. Adsorption of Pb(II) from an aqueous solution by titanium dioxide/ carbon nanotube nanocomposites: Kinetics, thermodynamics and isotherms. J. Chem. Eng. Data. 55: 4428–4433. DOI: 10.1021/je100586r.

Zhou, W., Liu, X., Cui, J., Liu, D., Li, J., Jiang, H., Wang, J. and Liu, H. 2011. Control synthesis of rutile TiO_2 microspheres, nanoflowers, nanotrees and nanobelts via acidhydrothermal method and their optical properties. 2011. Cryst. Eng. Com. 13: 4557–4563. DOI: 10.1039/C1CE05186E.

CHAPTER 11

Chemical and Biochemical Process Involved in the Degradation of Hazardous Contaminants through Bio and Nanoremediation

Giovanni Arneldi Sumampouw,[1,2] *Antonius Indarto,*[1,3,*] *Veinardi Suendo*[4,5] and *Muhammad Mufti Azis*[6]

1. Introduction

In the earlier chapters, bioremediation, nanoremediation, synthesis and others were described. Here a short review of nanoremediation and bioremediation from many researchers' points of view will be given to form a broad perspective.

1.1 Nano-bioremediation

In recent years, remediation methods gained greater importance, mainly to reduce the concentration of contaminants to below the danger limit as well as to eradicate any negative influences from greenhouse gas emissions (Reddy and Adams, 2015). Remediation faces many problems itself such as effectivity, price, sustainability, etc. Therefore, improvement of the pollution remediation method has attracted researchers to intensively explore this field. In this chapter, our focus is on the nano-bioremediation method.

[1] Department of Chemical Engineering, Institut Teknologi Bandung, Bandung, Indonesia.
[2] Department of Food Engineering, Institut Teknologi Bandung, Jatinangor Campus, Sumedang, Indonesia.
[3] Department of Bioenergy Engineering and Chemurgy, Institut Teknologi Bandung, Jatinangor Campus, Sumedang, Indonesia.
[4] Division of Inorganic and Physical Chemistry, Faculty Mathematics and Natural Sciences, Institut Teknologi Bandung, Bandung, Indonesia.
[5] Research Canter for Nanosciences and Nanotechnology, Institut Teknologi Bandung, Bandung, Indonesia.
[6] Department of Chemical Engineering, Universitas Gadjah Mada, Jogjakarta, Indonesia.
* Corresponding author: antonius.indarto@itb.ac.id

Nano-bioremediation stands for nanoparticles (nanoremediation) and bioremediation. Simply, nano-bioremediation can be described as the combined method of nanotechnology and bioremediation in the remediation field (Pandey, 2018). Some authors prefer adding "nano" first because usually pollutants will first come in contact with nanoparticles, followed by a biological treatment. However, several authors have defined nano-bioremediation as the production of "green nanoparticles" which involve biological activity (Yadav et al., 2017; Tripathi et al., 2018; Sherry et al., 2017). It is more likely to be called bio-nanoparticles which is in line with the earlier definition. Here the first definition will be described in detail.

1.2 Nanoremediation

Nanotechnology applications have been implemented in many areas during the past few decades, not only limited to medicine, textiles, pharmaceutics, optics, electronics and many more (Christian et al., 2013). Nanotechnology itself refers to the utilization of particles with dimensions within a 1–100 nm range to solve a certain problem. Based on earlier research that claimed nanotechnology could be a remediation agent, the remediation field has also been covered by nanotechnology (Singh and Misra 2014; Tratnyek and Johnson, 2006). Nanoparticles mediated remediation is considered to be more effective in the current method of site remediation. Their properties of tiny size and high ratio between surface area and volume may be used for degradation or reduction of hazardous wastes. Moreover, it may be usedfor removal of toxic pollutants both in *in situ* and *ex situ* remediation.

Notably, nano zerovalent iron (nFe^0/nZVI) is the most widely studied nanoparticle for the treatment of environmental contaminants due to its low price (Yadav et al., 2017). In 2012, the price range was about £ 50–150 per kg (Crane and Scott, 2012). Nanoparticles include the two groups either organic or inorganic. Organic nanoparticles may be in the form of carbon nanoparticles (fullerenes) and the inorganic may be magnetic, noble metal and semiconductor nanoparticles.

However, there are some drawbacks , for instance, the issue of reactivity, longevity, transport and fate in the environment of nFe^0. Their tendency in aggregation has caused a major issue for *in-situ* remediation applications. To prevent this issue, iron nanoparticles can be modified in two aspects: (1) preventing the aggregation of nanoparticles and (2) reducing the attachment of nanoparticles to the matrix (Singh and Misra, 2014). In some cases, for example, Gil-Diaz et al. (2016) showed that barley crops which are planted in nZVI treated soil for 4 months showed no overall increase of Fe content (Gil-Díaz et al., 2016). However, one should ensure the use of treated soil or water after remediation to know the most suitable method of remedy.

1.3 Bioremediation

On the other hand, bioremediation is specified as the biological degradation method of organic pollutants to levels below concentration limits established by the regulator. There are several advantages of bioremediation, namely less effort, an efficient process, no dangerous chemicals application, eco-friendly and sustainable. It is widely believed that the bio-based method provides a green, clean and sustainable impact on the environment. However, the limitation is, time-consuming and thetoxicity risk for organisms involved in remediation is an inseparable characteristic of bioremediation (Tripathi et al., 2018).

The process to develop successful bioremediation requires some steps, including finding the right microbes with the capability of degrading the contaminants by isolating them from nature, cultivating those cultures until reaching appropriate populations, followed by performing a study in the laboratory for the catabolic activity and optimizing the process to gain the highest possible degradation efficiency (Anjum et al., 2012).

Many researchers have developed various types of waste removal techniques for the bioremediation process, including the basic technologies such as bio-augmentation, bio-stimulation, bio-attenuation, venting and piles (Abatenh et al., 2017). Azubuike et al. (2016) also mentioned other methods of bioremediation, namely windrow, bioslurping, biosparging, phytoremediation, etc. (Azubuike et al., 2016). Table 11.1 provides an explanation of these bioremediation methods.

There are many other developed techniques for applying these methods, namely biofilters, activated sludge, Integrated Multi-Trophic Aquaculture (IMTA) and constructed wetlands. All of these mentioned methods may be applied using an enzyme, microbe or a combination (Musyoka and Fernandez, 2016). These are not only limited to microorganisms, but animal-mediated soil remediation has also been conducted. Some invertebrates have an ability to adsorb heavy metals from polluted soil and then degrade them. For example, earthworms can effectively accumulate Pb. However, it was noticed that soil pollutants can accumulate and then enter food web which limit the application of this technique (Koul and Taak, 2018).

Table 11.1. Various bioremediation methods (Abatenh et al., 2017; Musyoka and Fernandez 2016; Azubuike et al., 2016).

Method	Description
Biostimulation	Naturally occurring microorganisms (existing bacteria and fungus) are provided with environmental conditions that favor their growth. One can supply the nutrients in the form of fertilizers or growth supplements and trace minerals. Besides, to control the environment (e.g., temperature and oxygen rate) providing the optimum condition for microorganisms to grow using a certain pathway.
Bioattenuation	Eradication of pollutant concentrations from the surroundings and allowing nature to work in its own way. It is carried out within biological processes, physical phenomena and chemical reactions.
Bioaugmentation	The addition of pollutant-degrading microorganisms which is known as biodegradation. A contaminant-degrading microorganism can be added to augment the biodegradative capacity. This could be seen as the next level of biostimulation, because one could isolate the specific microbe from the contaminated site, culture, modified and return it to the site. However, releasing a foreign modified strain to the system may lead to unexpected results, if it is not controlled well.
Venting	Venting oxygen into soil for enhancing indigenous or introduced fungus and bacteria growth.
Biosparging	This may be one of the applications of bioventing. One could inject air into the soil subsurface using a sparger. In this way, air is fed into the saturated zone, thus volatile organic compound could migrate upward to the unsaturated zone where biodegradation takes place.
Piles	Separates some spaces into a smaller area using bio cells (biopiles) to reduce concentrations of petroleum pollutants, thus stimulating microbial respiration which cause increasing their activity for the removal of pollutants.
Rotating Biological Contactor (RDC)	It is a biological filter made of a fixed film disk or film flow bioreactors so it has a higher surface area that can be used by the microbe to attach, grow and consume the organic matter. The fixed-film contactor is submerged (± 40%). It can rotate to provide aeration as well as help with discarding the excess of suspended solids.
Trickling filter	With the same principle to enhance microbial attachment by enlarging the surface area, a trickling filter is constructed with a bed of stones and gravel. Wastewater is trickled or percolated through the stationary bed. Passing wastewater creates biologically active slime.
Denitrifying filter	Designed to encourage the growth of anaerobic bacteria by providing an anaerobic region using a cylindrical anoxic reactor fed with bacterial flocs without dissolved oxygen.
Microbial mats	Microbial mats are a laminated-cohesive microorganism consortium which grows embedded on a matrix, forming a multi-layered sheet.
Bioreactor	A bioreactor is set to provide optimum conditions for microorganisms to grow by mimicking their natural environment. Contaminated water or soil are supplied into this bioreactor with easy control of contaminants removal.
Land farming	It is claimed as the simplest bioremediation technique due to its low cost and smaller equipment requirement. Polluted soils are applied to fixed layer support above the ground surface to allow aerobic degradation of pollutants. However, this technique requires a large operating space and non-optimum of microbial activities due to an uncontrollable environment.
Permeable Reactive Barrier (PRB)	PRB is commonly used by researchers as an *in situ* method for groundwater remediation. This method uses ZVI (Zero-Valent Iron) which is arranged into permanent or semi-permanent reactive barrier. These days many researchers have tried to substitute the barrier to biological agents such as fungus or other microbes.

Higher plants were also studied for removing pollutants by a process called phytoremediation, which is still categorized as bioremediation. Mishra et al. (2009) investigated that rice (*Oryza sativa*), spinach (*Spinaceao oleracea*), eggplant (*Solanum melongena*) and radish (*Raphanus sativus*) are able to accumulate dichlorodiphenyltrichloroethane/DDT and benzene hexachloride from soil (Mishra et al., 2009). For water remediation, some aquatic plants such as *Pistia stratiotes*, *Eichhornia crassipes*, *Typha latifolia*, *Phragmites australis*, *Iris pseudacorus* and *Juncus effuses*, were able to remove pesticides effectively from the water where they were cultivated (Lv et al., 2016; Alencar et al., 2020).

Bioremediation techniques may be conducted *ex situ* and *in situ* depending on the condition. The *ex situ* or *in situ* strategy is chosen considering the contaminants types, degree of contamination, method cost, depth of contamination, geographical location, etc. (Azubuike et al., 2016). Azubuike et al. (2016) classified bioremediation based on site. They categorized biopile, windrow, bioreactor and land farming as *ex situ* and natural attenuation and enhanced processes like bioslurping, bioventing, biosparging and phytoremediation as *in situ* (Azubuike et al., 2016).

Several authors have their own definition on the bioremediation method. In this chapter, techniques are defined in a number of ways to apply the methods. Based on our earlier discussion, it could be classified as *ex situ* or *in situ* based on the methods that were used. For example, bioattenuation methods may be used in a bioreactor (*ex situ*) or directly in the polluted sites (*in situ*).

Sometimes, expected characteristics of agents may not be achieved by biological treatment alone. As an example, mass transfer of the aqueous phase in removal of some hydrophobic compounds, e.g., chlorophenols and Polycyclic Aromatic Hydrocarbon (PAHs), may limit microbial degradation. Keeping this in mind, microbial polymers were introduced. For example, polymers can be used as adsorbents and carriers for phenanthrene in a sand column, where they improve its mass transfer (Wang et al., 2007). Another innovation in bioremediation is dendrimers, which come from Greek words of "dendri" (meaning similar to tree branch) and "meros" (meaning tree part). Analog to the root words, dendrimers are highly branched and monodisperse macromolecules. Dendrimers consist of a central body equipped with some interior branch cells (radial symmetry) and terminal branch cell (peripheral group). These void spaces in dendrimers may be used as a space for microorganisms and nanoparticles to enhance the catalytic activity (Rizwan et al., 2014).

However, natural-based solutions are not only limited to conventional bioremediation. Many researchers have an integrated approach to combine bioremediation with city development. For example, some places such as Amsterdam-Nood in the Netherlands and Green Industrial Heritage Park, are developing methods for remediating brown field using an integrated nature-based technique (Song et al., 2019). Current developments also enable the combination of bioremediation and nanoremediation as they have their own qualities and drawbacks which are summarized in Table 11.2.

Gil-Diaz et al. (2019) suggested the combination of nanotechnology (nZVI) with bioremediation to cope with a high concentration of pollutants in soils. This combination is the best solution compared to some other remediation strategies tested (Gil-Díaz et al., 2019). Vaish and Pathak (2020) claimed that remediation with the Nano-bioremediation (NBR) method is the most advantageous technique for removing pollutants from the environment (Vaish and Pathak, 2020). Table 11.3 lists some contaminants which have been studied for their remediation using nanoparticles, biological agents and NBR.

From an industrial perspective, the recycling of the agent can impact the effectiveness and profitability of a remediation method. Wang et al. (2007) conducted the reusability research of

Table 11.2. Bioremediation vs. nanoremediation (Singh and Misra, 2014).

	Pros	Cons
Bioremediation	• Clean • Green • Sustainable	• Time-consuming • Might be toxic for organisms involved in remediation • Limited to biodegradable compounds
Nanoremediation	Excellent reducing capabilities	Toxicity, fate, and behavior in the environment

Table 11.3. Nanoparticles, biological agents and combined mediated remediation.

Mediator	Contaminants	Results
	Nanoremediation	
Fe/Ni bimetallic (Dong et al., 2018)	Tetracycline (TC)	Removal efficiency of TC by Fe/Ni (5%-wt) decreased a little from 100% after 2 d of aging. However, long aging time decreased the efficiency significantly
Fe_3O_4 capped with CTAB (Elfeky et al., 2017)	Cr(VI)	The system was capable of removing Cr (VI) at a maximum efficiency of 95.77%
TiO_2 (Czech and Rubinowska, 2013)	Diclofenac (DCF)	Degradation efficiency of DCF was 65%
Reduced Graphene Oxide–Silver (rGO-Ag) (Bhunia and Jana, 2014)	Colorless Endocrine Distruptors (Phenol)	Degradation efficiency of phenol was ~ 80%
ZnO (Rajesha et al., 2016)	Benzo-phenone 3	Degradation efficiency of BP-3 was 39.8 and 97% if the electron scavenger existed together with photocatalyst
nZVI (Gao et al., 2016)	Microcystin-lR (drinking water contaminants)	Removal efficiency 28.02%
Fe/Ni (Gao et al., 2016)		Removal efficiency 91.67%
Fe/Pd (Gao et al., 2016)		Removal efficiency 95.14%
	Bioremediation	
Pseudomonas sp.	Oil	Degradation percentage was 70.61%
Bacterial Consortium (Das et al., 2012)	Sulfates	Degradation percentage was up to 80%
	Total dissolved solids	Degradation of total dissolved solids was up to 70%
Cupriavidus sp. (Teng et al., 2017)	Pentachloronitro-benzene in aqueous solution	Removal efficiency reached 73.8% after 5 d
	Pentachloronitro-benzene in soil	Removal efficiency reached 89.3% after 30 d
Nannochloris oculata (Pérez-Legaspi et al., 2016)	Lindane	Seventy three percent of lindane was removed after 4 d of treatment
	NBR	
Pd (0) nanoparticles and *Shewanella onedensis* MR-1 (De Windt et al., 2005)	Polychlorinated biphenyls (PCBs)	The combination was able to dechlorinated around 90% of PCBs and produce less harmful by-products
Polyvinylpyrrolidone (PVP) coated iron oxide nanoparticles with bacterium *Halomonas* sp. (Cao et al., 2020)	Pb and Cd	Metal removal was as high as 100% for both contaminants and the remediation times shortened compared to the individual method
Fe_3O_4 and *Rhodococcus rhodochrous* strain (Hou et al., 2016)	Chlorophenols	*R. rhodocrous* immobilized into κ-crarageenan at a concentration of 9 g/L had higher removal efficiency of chlorophenols and their mixture compared to free cells

carbazole degradation by using Fe_3O_4 which was immobilized with the *Sphingomonas* sp. strain XLDN2-5 cells. The immobilized cells were magnetized so they could be separated from contaminated water using a magnetic field and then reused again. Reusing the immobilized cells up to eight times showed higher removal efficiency compared to non-magnetically immobilized cells and free cells. This better result could appear from the growth of cells inside the magnetic gellan gel beads (Wang et al., 2007). Hou et al. (2016) also reused the magnetically nanoparticle-immobilized cells for degrading chlorophenols. The magnetically immobilized cells of *R. rhodochorus* DSM6263 were able to consume all the chlorophenols until the sixth cycle. Notably, the separation was also easier for this agent since an external magnetic field could be applied for the collection of the biocatalyst from the aqueous phase (Hou et al., 2016).

Plastic pollution is one of the most recent threats in the environment. Not only macro plastics,but also small particles broken to a size of 0.1 μm–5 mm (microplastics) or even smaller at a size of 0.001 μm–0.1 μm (nanoplastics) have gained attention. The source of this material is primarily from households and industries, although it can be formed due to some abiotic factors, e.g.,UV radiation, temperature, microbial degradation, etc. This could have several impacts on the environment and human health, such as a blockage in the digestive tract and abrasion in mucosa if digested by humans. A number of enzymes produced by microbes have been known to degrade polymer, namely polyurethane esterase by *Comamonas acidovorans* to degrade polyester polyurethane and cutinases by *Thermobifidafusca* to degrade polyethylene terephthalate (PET). However, Tiwari et al. (2020) suggested that addition of nanoparticles to some microbial cultures may improve biodegradation of plastics (Tiwari et al., 2020). As an example, SiO_2 may enhance the growth of the plastic degrading bacteria (Pathak and Navneet, 2017).

2. Removal mechanism of nano, Bio and nano-bioremediation

From Table 11.2, it can be seen that nanoremediation is more effective in degrading contaminants than bioremediation which tends to be time-consuming. This characteristic is caused by the removal mechanism. Notably, this not only involves the degradation path but also other parameters. In this chapter, the focus will be on the degradation path. The synthesis, comparison and other fundamental aspects of remediation were explained in the earlier chapter.

2.1 Removal mechanism of nanoremediation

As mentioned earlier, nFe^0 (zero-valent iron nanoparticle) is widely used for the degradation of contaminants, so it will be used as an example. nFe^0 possess a core of Fe^0, equipped with outer shell consisting Fe^{2+} and Fe^{3+} oxides (Khin et al., 2012). The zero-valent iron can react with a contaminant to reduce it using two electrons from the oxidation process of Fe^0 becoming ferrous ions (Fe^{2+}). On the other hand, the iron oxides shell promotes adsorption of contaminants into its shell, thus stimulating electron transfer to reduce the pollutants directly. That is why zero-valent iron nanoparticle is considered as powerful reductants with high surface reactivity and high surface area (Liu et al., 2013; Sheng et al., 2014).

2.1.1 Organic contaminants

Zero-valent iron nanoparticle can reduce the number of halogenated hydrocarbons to less hazardous products such as hydrocarbons, chloride and water. Matheson and Tratynek (1994) proposed three general mechanisms for dehalogenation by Fe^0 as follows:

$$Fe^0 + RX + H^+ \rightarrow Fe^{2+} + RH + X^-$$

$$2Fe^{2+} + RX + H^+ \rightarrow 2\,Fe^{3+} + RH + X^-$$

$$H_2 + RX \rightarrow RH + H^+ + X^-$$

The first reaction is known as dissolving metal reduction. Produced Fe^{2+} becomes the major reductor and provides the electron for the reduction process in the second reaction. The third reaction assumes that hydrogen gas is generated from the corrosion of Fe^0 and is involved in the reduction process. Naja et al. (2008) also stated that while the contaminants are reduced by nFe^0 and Fe^{2+}, hydrogen produced as a result of iron corrosion may reduce the contaminants in the presence of a catalyst. The iron corrosion reaction is described as follows (Gao et al., 2016; Crane and Scott, 2012):

$$Fe^0 + 2H^+ \rightarrow Fe^{2+} + H_2$$

$$Fe^0 + 2H_2O \rightarrow Fe^{2+} + H_2 + 2OH^-$$

$$2Fe^{2+} + 2H_2O \rightarrow 2Fe^{3+} + H_2 + 2OH^-$$

$Fe^{2+} + 2OH^{-} \rightarrow Fe(OH)_2$

$Fe^{3+} + 2H_2O \rightarrow FeOOH + 3H^{+}$

For example, the degradation of chlorinated hydrocarbon using nFe^0 and Fe^{2+} follow these reactions.

$Fe^0 + R - Cl + H^{+} \rightarrow Fe^{2+} + R - H + Cl^{-}$

$Fe^2 + R - Cl + H_2O \rightarrow 2Fe^{3+} + R - H + OH^{-} + Cl^{-}$

As a strong reductor, zero-valent iron will degrade organochloride compounds and Fe^{2+} formed will reduce other organochloride compounds. Naja et al. (2008) conducted RDX (hexahydro-1,3,5-trinitro-1,3,5-triazine) degradation using the zero-valent iron nanoparticle. The proposed degradation path is the same as halogenated hydrocarbon described above, specifically nitro will be reduced to nitrogen and followed by denitrosation of TNX (hexahydro-1,3,5-trinitroso-1,3,5-triazine) formed (Naja et al., 2008). Azo dyes, PAHs and many organic contaminants were known to be effectively degraded by nFe^0.

Not only limited to direct degradation, but several researchers also used light to activate the degradation process, namely photocatalyst (Bhunia and Jana, 2014; Rajesha et al., 2016). The mechanism is quite different from regular degradation. The degradation using the photocatalyst mechanism of ZnO is described in the following reaction (Rajesha et al., 2016):

$ZnO \rightarrow ZnO\ (h^{+} + e^{-})$

$h^{+} + H_2O \rightarrow H^{+} + OH^{*}$

$e^{-} + O_2\ (atm) \rightarrow O_2^{-*}$

$O_2^{-*} + H^{+} \rightarrow HO_2^{*}$

Note: h^{+} is a hole (part in ZnO where e^{-} has been taken away)

It can be seen from the reaction that the last product of this sequence reactions is hydroxyl and peroxide radicals. They may react with organic pollutants near the catalyst in solution or attached on the surface of it. The formation of these two radical increases with the increment of the radiation time (Shwetharani et al., 2015).

Recently, Qian et al. (2020) also proposed a new heterogeneous Fenton-like catalyst containing cobalt to degrade pollutants under neutral and mild conditions. Fenton reagents (H_2O_2 + Fe(II)) have been commonly applied for the decontamination of organic wastewater, which uses the produced OH hydroxyl radicals from H_2O_2 (Qian et al., 2020). The catalyst possesses a porous nanosphere structure with reduced cobalt species state. This involves using the interaction of pollutants to a reduction state of cobalt species to release the electron. These electrons from pollutants are then transported to the surface of the catalyst where H_2O_2 can capture them. H_2O_2 and the electron will react producing hydroxyl radicals (OH). Hence, organic pollutants degradation is achieved through a dual-pathway: (1) there is direct degradation as the pollutant interacts with cobalt and (II) destruction by OH radicals (Zhang et al., 2020).

2.1.2 Inorganic contaminants

If the contaminants are inorganic, several proposed mechanisms were addressed: (1) reduction of contaminants, (2) sorption, (3) complexation, and (4) precipitation/co-precipitation. Singh and Misra (2014) stated that generally the contaminant is removed by a combination of two or more processes (Singh and Misra, 2014). For example, in the remediation of Cr (VI), the hexavalent form will be reduced to its trivalent form and nFe^0 is oxidized to Fe^{3+}. Precipitation of the trivalent form and

Fe^{3+} will then occur on the surface of the nanoparticles. The complete reactions are as follows (Mitra et al., 2017):

$$Cr_2O_7^{2-} (aq) + 2Fe^0 (s) + 14H^+ (aq) \rightarrow 2Cr^{3+} (aq) + 2Fe^{3+} (aq) + 7H_2O$$

$$xCr^{3+} (aq) + (1-x)Fe^{3+} (aq) + 3H_2O \rightarrow Cr_xFe_{(1-x)}(OH)_3 + 3H^+$$

However, Cr (III) cannot further reduce into Cr^0 since it is not thermodynamically viable with Fe^0. This reduction step also occurs for other metals, such as Se(VI) to Se^0 and As(V) to As^0. Unlike other metals removal,U(VI) can be reduced by nZVI as its thermodynamic property enables this reduction reaction into UO_2 (solid) precipitation under favorable environmental conditions as described in this reaction (Tratnyek, 2003):

$$Fe^0 (s) + UO_2(CO_3)_2^{2-} + 2H^+ \rightarrow UO_2 (s) + 2HCO_3^- + Fe^{2+}$$

A study was conducted to observe mechanisms of heavy metals removal using ZnO nanoparticles. ZnO was able to remove some heavy metals using two mechanisms, namely physical adsorption and reduction/oxidation by photo-generated electron-hole pairs. ZnO which were negatively charged mainly due to contribution of OH-groups. These hydroxyl groups were the active adsorptive section because heavy metals tend to react with these groups on the particles surface forming a thin film. The efficiency of adsorption was quite low since it might become saturated after some time. Reduction and oxidation reactions were possible because of electrons and photons. For example, Cu(II) ions were hydrated forming $Cu(H_2O)_6^{2+}$ which then reacted with hydroxyl ions forming a CuO layer on ZnO particles surface by absorption. The adsorption mechanism was dominant for Cu, Cd and Ni removals. On the other hand, reduction was found to be superior on Ag(I), Cr(VI) and Cu(II) ions removal, while oxidation was more effective during the removal of Pb(II) and Mn(II) ions. Removal of Ag(I) and Pb(II) could be improved by UV light exposure because of photocatalytic reaction (Le et al., 2019).

2.1.3 Microbial contaminants

Several metals are toxic for microorganisms. For example, for algae, inorganic mercury exposure may be toxic in these mechanisms because of: (a) blockade of essential biological molecules, (b) essential metal ion from metalloproteins substitution, (c) loss of enzyme specificity, and (d) conformational shift of biomolecules (Gómez-Jacinto et al., 2015). Several nanoparticles for microbial removal are listed in Table 11.4.

Table 11.4. Nanoparticles for microbial degradation (Yadav et al., 2017).

Microorganism	Nanoparticles
Pathogenic bacteria	Silver nanoparticles
E. coli	CeO_2 nanoparticles
E. coli and *Bacillus megaterium*	MgO nanoparticles
Bacillus subtillus	Magnesium nanoparticles
E. coli and *Staphylococcus aureus*	Silver nanoparticles

Silver nanoparticles constantly discharge silver ions which may kill microorganisms through several mechanisms, such as enhancing cytoplasmic membrane permeability that cause bacterial envelope disruption, deactivation of respiratory enzymes, modification of deoxyribonucleic acid (DNA) by interaction with sulfur and phosphorus in it, and inhibition of protein synthesis. Silver nanoparticles themselves are capable of damaging microbes. They can accumulate on the cell wall, leading to cell membrane denaturation. The penetration of silver nanoparticles into the cell wall can change the structure of the cell membrane and burst organelles. Furthermore, the nanoparticles can dephosphorylate protein substrates, thus disturbing bacterial signal transduction that causes cell apoptosis and cell multiplication termination (Yin et al., 2020).

Besides these metal nanoparticles, single-walled carbon nanotubes (SWNTs) were observed to possess anti microbial activity. Direct contact between the cell and SWNTs could cause loss of cellular integrity because of damage to the cell membranes and inactivate the cells subsequently (Kang et al., 2007). Arias and Yang (2009) used Scanning Electron Microscopy (SEM) to observe the individual SWNTs stuck to one side of the cells, while the other side protruded from the cells, thus SWNTs was carried out as needles damage the cell walls. The small size of SWNTs enabled close contact especially when forming aggregates (Arias and Yang, 2009). SWNTs were also reported to induce a stress response to microorganisms. Several genes that are related to oxidative stress (e.g., oxyR and soxRS) were expressed when cells were in contact with carbon nanotubes (Kang et al., 2008). Vecitis et al. (2010) found glutathione production in cells that were exposed to SWNTs increased. This observation proved that the cells were in oxidative stress state since glutathione acts as a redox buffer during oxidative stress (Vecitis et al., 2010).

2.2 Removal mechanism of bioremediation

Unlike nanoremediation, bioremediation uses living organisms, such as fungi, bacteria, algae and some plants to degrade the contaminants. *Escherichia*, *Citrobacter*, *Staphylococcus*, *Bacillus*, *Klebsiella*, *Pseudomonas*, *Rhodococcus* and *Alcaligenes* are commonly used in bioremediation (Kumar and Gopinath, 2016). The resulted products from bioremediation may be milder or less toxic than the degradation product of nanoremediation. For example, degradation of an explosive contaminant, RDX (hexahydro-1,3,5-trinitro1,3,5-triazine) with nZVI produced nitroso by products, which are quite harmful for the environment. However, when RDX was degraded by anaerobic sludge, RDX was completely mineralized into methylenedinitramine, CO_2 and nitrous oxide (N_2O) (Oh et al., 2001). The mechanism of degradation is unique depending on the metabolic path of certain biological agents (Bumpus and Aust, 1987; Wedemeyer, 1967). However, several researchers are trying to propose the typical path of certain pollutants. Here several specific examples will be explained.

2.2.1 Removal of dye

Pandey and Upadhyay (2010) attempted to predict the pathway of textile effluent Direct Orange-102 degradation by *Pseudomonas fluorescens*. Literature has revealed that the end products from dyes biodegradation by *P. fluorescens* were non-toxic in nature (Flaxbart, 1996). They proposed the possible pathway of degradation of Direct Orange-102 by *Pseudomonas fluorescens* in this way: Direct Orange-102 → intermediate compound → 3,7-Diamino-4-hydroxy-napthalene-2-sulfonic acid sodium salt → 7-Amino-3,4-dihydroxy-napthalene-2-sulfonic acid sodium salt or 3-Amino-4,7-dihydroxy-napthalene-2-sulfonic acid sodium salt or 1,3,4,5,6,7,8-Heptahydroxy-napthalene-2-sulfonic acid sodium salt (Pandey and Upadhyay, 2010).

Another study observed *Pseudomonas* sp. and white rod fungi consortium for azo dye Direct Fast Scarlet 4BS removal from dyeing house wastewater. *Pseudomonas* can excrete azo reductase enzyme and white rod fungi releases ligninolytic peroxidases enzyme. Both enzymes were responsible for the 4BS dye decolorization which has two phenyl and naphthyl rings connected by azo bonds. Firstly, the azo bonds were cleaved producing intermediates compounds. The naphthyl ring in the intermediates was then split partly forming a phenyl ring or even split completely to produce aliphatic hydrocarbon intermediates. The final products of the decomposition are simple aliphatic hydrocarbons, alcohols, amines, carbon dioxides, water, nitrogen and ammonia (He et al., 2004).

2.2.2 Removal of heavy metals

Even tough bioremediation has a limitation in the degradation of non-biodegradable contaminants, for example, heavy metals, use microalgae in phycoremediation of heavy metals. Several microalgae have a tolerance to heavy metals. Boron, zinc, cobalt, iron, copper, manganese and molybdenum may be consumed by microalgae as trace elements for cell metabolism and enzymatic process. However, some heavy metals such as mercury, cadmium, chromium, arsenic and lead are still harmful to microalgae (Leong and Chang, 2020; Sun et al., 2015).

Balaji et al. (2016) added that cyanobacterial species, such as *Phormidium*, *Anabaena*, *Spirogyra* and *Oscillatoria* can withstand the Cr(VI) stress and can grow naturally and reproduce in water contaminated by heavy metals. They also added that the ability of adsorption may contribute to their tolerance (Balaji et al., 2016). Plants also have multiple molecular strategies to heavy metal environment, namely (1) metal immobilization, (2) gene regulation, (3) exclusion, (4) metal chelation, and (5) reducing enzyme (Gómez-Jacinto et al., 2015).

Similar to the nanoparticle, the biological-agents mechanism in removing contaminants consist of one or more mechanisms. Observed mechanisms from earlier research were biosorption on the extracellular polymeric substances, bioaccumulation of contaminants inside cells, enzymatic reductase role and reaction between contaminants and electron donor of reducing agents on the biomass surface.

2.2.3 Removal of hydrocarbons

Several proposed mechanisms were applied earlier to decompose various hydrocarbon pollutants (Indarto et al., 2006; 2007; 2008; Indarto, 2012). The hydrocarbon compound structure will determine the ease of biodegradation. The n-alkanes and n-alkyl-aromatics are considered to be the most biodegradable. In contrast, long hydrocarbons (> C22) are considered less biodegradable due to their characteristics (Bossert and Bartha, 1984). Seo et al. (2009) claimed that organic compounds with an aromatic ring are one of the most persistent contaminants in the ecosystem (Seo et al., 2009). The mechanism for each type of hydrocarbons is listed in Table 11.5.

Table 11.5. Biodegradation mechanism of hydrocarbons (summarized from Ivey, 2006).

Type	Mechanisms
N-alkane (Grubbs and Mohaa, 1987)	- Oxidation catalyzed by enzyme begins to form primary fatty alcohol. - The fatty alcohols are oxidized into fatty acid. - β-carbon of the fatty acid is oxidized forming β-keto acid. - Degraded alkane can be achieved by decarboxylation reaction.
Cyclic Alkanes (Bartha, 1986)	- Oxidation of the ring which was supported by the enzyme to produce cycloalkanone. - Cycloalkanone is oxidized by Baeyer-Villiger type reaction to produce lactone. - Generation of n-alkyl fatty acid through hydrolysis. - Fatty acid will follow the n-alkane route.
Aromatics (Leisinger and Brunner, 1986)	- The key factor is to destroy the resonance of the aromatic ring. It could begin by two mechanisms, either catechol or gentisate, that activate the ring and is followed by the ring's cleavage.
Polycyclic Aromatic Hydrocarbons (PAH) (Gibson et al., 1975; Seo et al., 2009)	- As the molecular weight of PAH increases, it will be more difficult for the microbial agent to be soluble in the mixture. - The PAH compounds will be oxidated to produce a dihydrodiol. - The oxidation is followed by ortho cleavage or meta cleavage of the dihydrodiol intermediates and will be converted to tricarboxylic acid.

2.2.4 Removal of agrochemicals

Agrochemicals, including various kinds of pesticides such as insecticides, herbicides and fungicides, are important chemicals for increasing the capacity of food production. However, excessive usage of agrochemicals during the past decades has led to the contamination of soil and bodies of water all around the globe. Hence, agrochemicals removal techniques are needed to remediate the environment, and bioremediation provides a good option. Pesticides can be metabolized by microorganisms to produce less toxic compounds through three phases. The first phase involves oxidation, reduction or hydrolysis of the pesticides to form a more hydrophilic product. Oxidation is the most common step among them, where this reaction is assisted by oxidative enzymes, for example peroxidases, polyphenol oxidases or cytochrome P450s. Phase II of pesticide metabolism is the association of the pesticide or its metabolites to an amino acid, sugar or glutathione. This step increases water solubility further and creates less toxic metabolites. Microorganisms conjugates the pesticides through alkylation, acylation, xylosylation or nitrosation reactions inside or outside their cell bodies. The last phase is conversion of metabolites from phase II into secondary nontoxic conjugates (Hoagland et al., 2000).

An example of microorganisms that degrade pesticides through the hydrolysis process is *Pseudomonas* sp., which proved to have high effectivity of a degrading atrazine, a kind of herbicide. The degradation is mediated by atrazine chlorohydrolase, which is capable of dechlorinating atrazine with hydrolytic reaction to produce hydroxyatrazine. Aminoethyl group of hydroxyatrazine is then removed by hydroxyatrazine ethyl amidohydrolase and becomes N-isopropylammelide. Amidohydrolase produced by the bacteria is responsible to further degrade N-isoproylammelide into cyanuric acid. Finally, cyanuric acid can be converted into CO_2 and NH_3 by a lower pathway (Sadowsky and Wackett, 2000). By this metabolism process, a toxic agrochemical compound can be converted into non-toxic compounds through enzyme activities.

2.3 Degradation mechanism of nano-bioremediation (NBR)

The mechanism is the combination of both mechanisms of nano and bioremediation . Before following several further reactions, contaminants must be adsorbed on the surface of the agent (Vázquez-Núñez et al., 2020). However, the decoupling is not only limited to obtain a combined degradation mechanism. There are other advantages by combining nanoparticles and bioremediation which are listed in Fig. 11.1. The mechanism is simply the combination of each mechanism described earlier):

2.3.1 Providing easier contaminants

NBR provides easier contaminants for bioremediation to degrade. For instance, bulky chlorinated compounds with five or more chlorine atoms in a structure will be difficult for aerobic bacteria to degrade, in contrast, low chlorinated compounds degrade easily in the environment. Therefore nanoremediation will degrade highly chlorinated non-biodegradable compounds to less chlorinated compounds and could be degraded easily by bioremediation (Singh and Misra, 2014).

Alternatively, nanoremediation will completely degrade contaminants in the source. Additionally, it may produce byproducts and these will be degraded by bioremediation simultaneously (Cecchin et al., 2017). Le et al. (2015) studied Aroclor 1248 (PCB) degradation by nZVI and biodegradation using *Burkholderia xenovarans* subsequently. The study results indicated that 89% of the contaminants was degraded by the nZVI remediation. Subsequently, the biphenyls produced from nZVI remediation was removed as high as 90% by bioremediation through the bacterial metabolism. Besides, microorganisms were not damaged by the nZVI (Le et al., 2015).

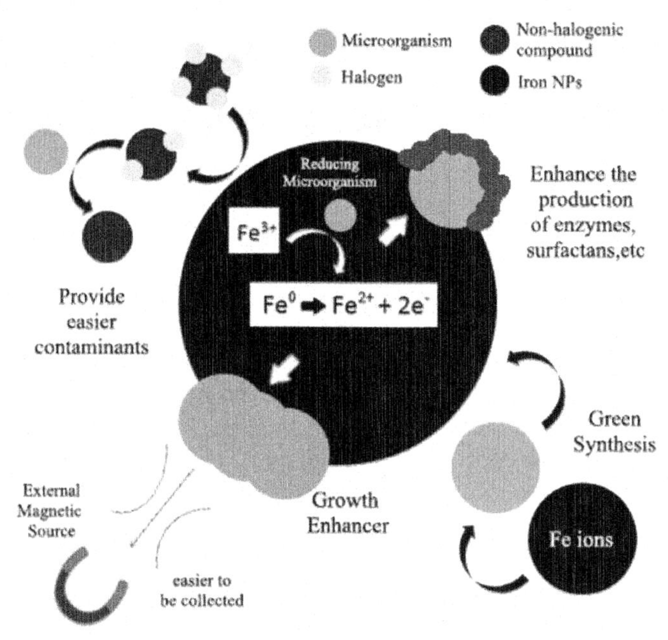

Fig. 11.1. Possible mechanism of decoupling process.

Bokare et al. (2010) conducted the dechlorination of Triclosan using Pd/nFe and followed by laccase derived from *Trametes versicolor*. The reaction mechanisms are as follows:

$Fe^0 + 2H_2O \rightarrow Fe^{2+} + H_2 + 2OH^-$

$H_2 + Pd \rightarrow Pd\text{-}H$

Triclosan + Pd-H → 2-phenoxyphenol + $3Cl^-$

The dechlorination of triclosan on Pd/nFe nanoparticles begins by the adsorption of the chlorinated compound on the surface and reaction with the atomic H on the surface of Pd. Pd was reduced by hydrogen produced from the reduction of water as presented by the first reaction. However, 2-phenoxyphenol produced from this dechlorination has its own problem. In view of structure-reactivity of 2-phenoxyphenol, this compound has an affinity for binding to the E+NAD^+ part in the enzyme, which can be harmful to the microorganisms. From this study, laccase and syringaldehyde were able to oxidize 2-phenoxyphenol to produce polymeric compounds with no harmful effect (Bokare et al., 2010).

2.3.2 Enzymes and metabolic products enhancer

Nanoparticles enhance the production of enzymes or other products from the microorganism. Fang et al. (2012) studied the effect of Fe_3O_4 nanoparticles to indigenous microbes in soil for degrading an herbicide, 2,4-dichlorophenoxyacetic acid (2,4-D). It appeared that this combination of bioremediation and nanoremediation shortened the duration for degrading the herbicide significantly compared to each remediation method alone. This exceptional result led to raising enzyme production by the native microorganisms in the presence of nanoparticles, which was seen by higher enzyme activity. The nanoparticles at a concentration of 20 g/kg soil was able to enhance the activity of amylase by 69.3%, urease by 23.9%, acid phosphatase by 15.4% and catalase by 10.6% (Fang et al., 2012).

Bioremediation is restricted to hydrophobic contaminants, such as hydrocarbon oil. It is widely known that some microorganisms can produce biosurfactants, which are able to decrease surface tension and thus lower the hydrophobicity of organic contaminants. Nanoparticles are able to accelerate the biosurfactants production of the used microorganism agent (Kumari and Singh, 2016). For example, Fe is considered as a main factor for biosurfactants production. Kiran et al. (2014) used iron nanoparticles in *Nocardiopsis* MSA13A for enhancing the biosurfactant production. Consequently, the biosurfactant production increased as high as 80% compared to treatment without iron nanoparticles (Kiran et al., 2014).

2.3.3 Growth enhancer

Nanoparticles can act as a catalyst in growth or metabolism. Many kinds of nanoparticles like Fullerene 60, nanobarium titanate (NBT) and super-magnetic iron oxide nanoparticles (SPION) may reduce the duration of lag phase in the growth of microorganisms and increase the span of exponential and stationary phase. Bhatia et al. (2013) called this a microorganism with a nanoparticle enhancer (Bhatia et al., 2013). For example, a consortium of *Microbacterium*, *Pseudomonas putida* and *Bacterium* Te68R was grown in a medium with NBT and without it. The used consortia were monitored by UV spectrophotometer to measure the bacterial growth cycle. In the presence of NBT, the spectrum shifted in 2 d, while it took 4 d without NBT (Kapri et al., 2009).

Iron oxide nanoparticles (IO-NPs) have also shown to stimulate several microbial strains growth, i.e., *Bacillus subtilis* IC 12488 and *Candida krusei* 963, both in suspension and biofilm.

The stimulation may be the result of the use of iron oxide as a source of iron ions that could be used for metabolism. However, IO-NPs on some scale of concentration could also inhibit growth of other microorganisms tested on suspension and adherent states (*Enterococcus faecalis* ATCC 2921, *Pseudomonas aeruginosa* 1397 and *Escherichia coli* ATCC 25922) (Prodan et al., 2013).

Calcium peroxide (CaO_2) nanoparticles were synthetized to overcome limited Dissolved Oxygen (DO) that inhibited biodegradation process in soil, sediment and groundwater. When this nanoparticle was used to assist biodegradation of diesel, it was able to release oxygen slowly which could maintain the DO level at 4 ppm after 14 d. This enhanced oxygen level could boost diesel degradation to around 60% at d 14, while the control without CaO_2 nanoparticles could only degrade about 10% at the same time. In the first 3 d, the nanoparticles showed an effect of inhibition to the microbes, however they survived and maintainedthe same biomass level after d 3 (Yeh et al., 2018).

Nanoparticles may also produce compounds that can support metabolism and growth of the microorganisms. Xiu et al. (2010) investigated how nZVI affect metabolism of a consortium of methanogenic microbes and dechlorinating culture (*Dehalococcoides* spp.) to degrade trichloroethylene. nZVI could produce hydrogen gas that was consumed by the dechlorinators and methanogens bacteria. The produced hydrogen stimulated methanogens growth significantly, which was a contender for *Dehalococcoides* spp. to consume trichloroethylene. nZVI alone could degrade trichloroethylene, but when combined with dechlorinating culture, it enhanced the degradation enormously although dechlorinating microbes' growth were inhibited by nZVI as they were more sensitive compared to methanogenic bacteria. The inhibition effect was shown by the delay of ethene production . Ethene is the final product of trichloroethylene after it degrades into cis-1,2-dichloroethene and vinyl chloride (Xiu et al., 2010b).

2.3.4 Reusability and recovery

Nanoparticles increase the opportunity to reuse and recover microorganism agents by immobilization. Besides, the immobilized microbial cells possess a higher stability for the removal process (Kumari and Singh, 2016). Immobilization of microorganisms into magnetic nanoparticles offers an easy method for reusing and the recovery by manipulating external magnetic fields. Some desired characteristics of the magnetic nanoparticles are high magnetic susceptibility, retaining superparamagnetism, high chemical stability, low Curie temperature, high coercivity, higher colloidal stability to avoid agglomeration, and narrow size distribution. Iron based nanoparticles, such as Fe_3O_4, Fe_2O_3 and nZVI, have been studied by many researches for this purpose (Giese et al., 2020).

2.3.5 Nanoparticles complementary or support

The addition of microorganisms into nanoparticles often leads to a better performance because of their synergistic mechanisms. A study observed Sulfate Reducing Bacteria (SRB) performance that was coupled with nZVI for groundwater remediation. The results were outstanding, U(VI) was removed at a rate of 98.1%, while ZVI and SRB alone could only remove 17.4 and 67.3%, respectively. They proposed four simultaneous possible mechanisms of this removal (Dong et al., 2019):

(1) nZVI could be used as a reductor for heavy metals (regular reduction path).
(2) The precipitation of heavy metals with the metabolites of SRB. They stated that SRB may reduce sulfate to sulfide (H_2S, HS^-, or S^{2-}) which will react with metal ions.
(3) Corrosion of nZVI produced ferrous ions that could produce iron sulfides after reaction with sulfide. Iron sulfides could adsorb halogenated hydrocarbons and heavy metals into their surfaces.
(4) The adsorption of heavy metals on bacteria surface.

Dong et al. (2019) also conducted an interesting experiment. They coupled the nZVI and Iron Reducing Bacteria (IRB). In the regular mechanism, Fe^0 will be oxidized to produce Fe(II) and oxidized further to produce Fe(III) (Dong et al., 2019). As mentioned earlier, Fe(II) was the major reductant and provided the electron for the reduction process (Matheson and Tratnyek, 1994). The IRB may reduce the produced Fe(III) back to Fe(II). Therefore, the overall degradation efficiency will be enhanced. Amonette et al. (2000) also stated that Fe(II) hydroxides which were adsorbed on the surface of ZVI might increase the electrons density, while the hydroxyl groups could enhance the electron transfer to improve the degradation rates (Amonette et al., 2000). In other words, IRB may reactivate the nZVI.

Further prospects may rely on the development of molecular biology. The performance of a biological agent may be enhanced by the molecular modification by nanoparticles. Chariou et al. (2019) tested the mobility of synthetic nanopesticides made of a virus by incorporating computational modeling into soil column trial. These bio nanoparticles were biodegradable and might be derived from plant viruses. Although the example comes from pesticides, the development of the particle could be a motive for the use of nanobioremediation (Chariou et al., 2019).

It is notable that this synergistic effect is not always applicable to all combinations. The toxic effects or adverse effects may occur if it is not a suitable combination. For example, positive effect in IRB for reducing Fe(III) back to Fe(II) enhanced the degradation. However, IRB may also have negative impacts on the iron environment (Xie et al., 2017), such as affecting the iron corrosion (Gerlach et al., 2000), compound methylation (Si et al., 2015) and the increment of adsorbates mobilization (Cummings et al., 1999).

3. Process Configuration of Integrated Nano-bioremediation (NBR)

Generally, two approaches are used for the application of NBR, namely sequential and concurrent or a combined method. The sequential method is conducted first by implementing nanoremediation, before the contaminants are subjected to bioremediation,while in the concurrent method, the contaminants will be introduced to a system simultaneously where both nanoparticle and biological agents exist. Notably, nanoparticles especially metal and metal oxide may be toxic to microbial agents, so it should be investigated further if one wants to use the concurrent method. For example, *Sphingomonas* sp. PH-07 proved to tolerate nZVI. Therefore, it has the potential to be used in the same system (Kim et al., 2012).

3.1 Sequential method

Bokare et al. (2010) tried to degrade a triclosan contaminant using the sequential method. They used Pd/nFe as the nanoparticle for nanoremediation and followed by laccase derived from *Trametes versicolor* for bioremediation. The remediation by Fe nanoparticles could degrade triclosan by itself. This result is followed by bioremediation with secreted laccase that was able to convert the degraded intermediates into non-toxic compounds (Bokare et al., 2010). This successful attempt led researchers to try the same approach for 2,3,7,8-tetrachlorodibenzo-p-dioxin (2,3,7,8-TeCDD) removal from water. This pollutant hardly degraded if remediation is conducted using biological and chemical treatment alone, due to its extreme recalcitrant property and low bioavailability. Thus, an integration method exploiting nanoremediation with palladized iron nanoparticles (Pd/nFe), followed by oxidative degradation with *Sphingomonas wittichii* RW1 was investigated. 2,3,7,8-TeCDD was fully degraded after 10 hr exposure with Pd/nFe, producing dibenzo-p-dioxin (DD) by dichlorination process. This by-product was then fully metabolized by the microbes to form catechol within 4 hrs. Catechol was used as a substrate for growth and was fully egraded after 18 hr since microbes were applied (Bokare et al., 2012).

Kim et al. (2012) conducted a sequential method to degrade polybrominated diphenyl ethers using nanoscale zero-valent iron and aerobic biodegradation. Nanoscale zero-valent iron was manufactured from $FeCl_3.6H_2O$ and the bacterial strain used in the research was *Sphingomonas* sp. PH-07. The contaminant was introduced to nZVI (100 mg) under the anaerobic condition for 20 d. Sequentially, the reaction mixture was incubated aerobically with PH-07 strain for 4 d. This sequential system achieved a reduction of up to 67% (Kim et al., 2012). A similar contaminant, Aroclor 1248, which is a mixture of 32 polychlorinated biphenyls, was studied in a sequential soil remediation process. The step began with anoxic nanoremediation using bimetallic nanoparticles Pd/nFe, followed by biodegradation utilizing *Burkholderia xenovorans* LB400 in an aerobic condition. The nanoparticles successfully degraded 99% of the trichlorinated biphenyls, 92% of the tetrachlorinated biphenyls, 84% of the pentachlorinated biphenyls and 28% the hexachlorinated biphenyls. The resulted dechlorinated biphenyls were biodegraded by the bacteria producing several end products, such as benzoic acid and cis, cis-muconic acid. This bioremediation succeeded to metabolize 90% of the biphenyls after 24 hr (Le et al., 2015).

Horváthová et al. (2019) compared integration of nanoremediation and bioremediation through two sequences: first bioremediation then followed by nanoremediation (bio-nanoremediation), and the reverse order by performing nanoremediation first before carrying out bioremediation (nano-bioremediation). The experiment was conducted using three species of microbes isolated from soil contaminated with polychlorinated biphenyls/PCBs (*Achromobacter xylosoxidans*, *Stenotrophomonas maltophilia* and *Ochrobactrum anthropic*) and nZVI. They observed that the nano-bioremediation approach gave a better result, with the removal of 75–99% of PCBs after 30 d with 15 d of each treatment. The bio-nanoremediation could only degrade 49–77% of the PCBs after the same period of treatment. The researchers suggested that if bioremediation was done first, the mixture solution already contained PCBs degradation products and polysaccharides, proteins, lipoproteins and other cell structures formed by lysed bacterial cells. These compounds might not be suitable for fresh nZVI particles remediation method, which explained why the bio-nanoremediation approach did not perform better (Horváthová et al., 2019).

3.2 Concurrent method

Ravikumar et al. (2016) conducted concurrent remediation for Cr(VI) contaminants. They used nZVi with immobilized calcium alginate beads (nZVI-C-A) for nanoremediation and *Bacillus subtilis*, *E. coli* and *Acinetobacter junii* for bioremediation. These bacteria consortia were isolated from the chromite mine and grown together in nutrient broth. The microbes flowed through a column filled with nZVI-C-A beads in order to form a biofilm on the surface of the beads. Cr(VI) removal was done in continuous-flow in a glass column. The removal percentage was up to 92% of Cr(VI) and they claimed that this removal was enhanced by the combined technology (Ravikumar et al., 2016).

Yan et al. (2013) also studied the Cr(VI) removal using a concurrent method. Bioreduction of hexavalent chromium into trivalent one was done by Ca-alginate beads impregnated with carbon nanotubes (CNTs). The combination was used for immobilization of *Shewanella oneidensis* MR-1 to support Cr(VI) reduction. The concurrent method performed four times better than each separate technique, primarily because CNTs increase the electron transfer rate (Yan et al., 2013).

Another CNTs and bioremediation combination was shown by Pereira et al. (2014) for dye decolorization. They used CNTs and some other nanomaterials for dye degradation using a redox method and anaerobic metabolism. Reductive transformations of dye by anaerobic sludge (bioremediation) are a slow process due to the limitation of electron transfer. It was assumed that the presence of redox mediators such as Activated Carbon (AC) or CNTs could overcome this condition. After performing some researches, they claimed that the addition of CNTs displayed the best biodegradation rate, outperforming other nanomaterials tested (AC and mesoporous carbons xerogels) (Pereira et al., 2014). However, Zhang et al. (2015) claimed that at higher CNT concentrations, the biodegradation rate is reduced due to the inhibition of bacterial growth and microbial activity. On the other hand, a small amount of CNT was observed to be enough to augment biodegradation efficiency, as it enhances the growth of microorganisms and also degradation genes overexpression (Zhang et al., 2015).

Another approach to mix bioremediation with nanoremediation is by coating the microbes with nanoparticles, rather than immobilizing the bacteria in nanoparticles. Shan et al. (2005) coated *Pseudomonas delafeldii* with magnetic Fe_3O_4 nanoparticles. This coated cell was used to desulfurize dibenzothiopene (DBT). The magnetic nanoparticle was more superior for DBT biodesulfurization compared to cells coated with celite and free cells. The magnetic property of the nanoparticles also enabled reusing of nano bioparticles, and they concluded that it could be used again more than five times, although the remediation time needed to achieve the same performance as the previous cycle became much longer (Shan et al., 2005).

Huang et al. (2017) stated that the coupling of microbe and nZVI for tetracycline degradation is not easy. nZVI is able to degenerate tetracycline partially or fully using reduction and adsorption steps. nZVI can be oxidized into Fe^{2+} and Fe^{3+}which will stimulate Extracellular Polymeric Substances (EPS) production and in turn change microbial consortium structure. The research apparatus consisted of three separate columns and filled with nZVI, mixture of nZVI and microorganism and microorganism (anaerobic activated sludge alone). The microorganisms in the sludge could consume both tetracycline and the intermediate products of tetracycline simultaneously. Moreover, the EPS that was produced by the microbes also contributed in tetracycline removal. Another finding is that the presence of Fe^{2+} and Fe^{3+} may affect the community structure of microbial, some reduced to a great extent, such as *Chitinophagaceae* and *Hyphomicrobiaceae*, while others increased, i.e.,*Comamonadaceae*, *Oxalobacteraceae* and *Chromatiaceae* (Huang et al., 2017).

Cao et al. (2020) also combined the bacterial and nanoscience to remediate a metal contaminant (Cd or Pb). Polyvinylpyrrolidone (PVP) coated iron oxide nanoparticles (NPs) performance was improved by incorporating *Halomonas* sp.,a Gram-negative bacterium. This combination improved the metal remediation rate and also decreased the time needed for metal removal compared to an individual treatment. Interestingly, the bacterial membrane of *Halomonas* sp. was inspected and it contained more Fe rather than Cd or Pb. This may suggest that Fe was used as a nutrient for microbial metabolism (Cao et al., 2020).

Another research of interest was conducted by Xiu et al. (2010). A potential concurrent method was carried out separately with nZVI and *Dehalococcoides* sp. Both nZVI and *Dehalococcoides* sp. can be used to dechlorinate trichloroethylene (TCE), so there is an opportunity to combine this method. However, from research conducted earlier, there was an indication that nZVI can reduce the dechlorination rate because it alters the expression of genes for that purpose. They then used an olefin maleic acid copolymer as a coat for nZVI. This treatment was able to overcome the significant inhibitory effect, even as it increased the expression of genes for dechlorination and thus *Dehaloccoides* sp. participation in the remediation process became more significant (Xiu et al., 2010).

Le et al. (2019) conducted a more natural approach for the combination They used Pd/Fe nanoparticles to treat hexabromocyclododecane (HBCD) in soils with presence of plants. Without additional treatment, planted soil may remove 13% of the HBCD through the phytoremediation process. The removal most probably was done because of (1) the metabolism process of microbes consortium living in rhizosphere, (2) the adsorption mechanism into the plant roots, (3) the accumulation mechanism in the plant tissues (Barac et al., 2004). Plants with nanoparticles achieved the removal rate as high as 41%. Both HBCD removal rate of 13 and 41% were compared with the control where the same soil was not planted; this control was useful for removing any HBCD degradation caused by natural causes and any adsorbed HBCD into soil matrix. After the treatments, around 221–986 mg/g of HBCD were found inside the plant tissues. They also observed the presence of humic acid in HBCD nano-bioremediation from soil and an aqueous system. Compared to the aqueous solution, soil remediation greatly reduced the removal efficiency due to the issues of transfer. For Humic Acid (HAs) cases, apparently it enhanced the HBCD debromination under aqueous conditions but not in the case of soil. It was believed that the HAs can hasten nanoparticles corrosion and aggregation (Le et al., 2019). However, nZVI toxicity towards microorganisms is reduced with the presence of HAs (Chen et al., 2011).

Nano-bioremediation can also be done by immobilized enzyme in nanoparticles. Laccase enzyme derived from *Ganoderma cupreum* was studied for Reactive Violet (RV) 1 dye removal from water after immobilization in amino-functionalized nanosilica particles. The incorporation of laccase into nanosilica increased the stability in storage and after repeated use compared to pure laccase, and also enhanced its thermal stability. High surface area of the nanoparticle could boost the decolorization efficiency of RV 1 dye, with a result of 96.76% removal at 30ºC and pH of 5.0 after 12 hr (Gahlout et al., 2017).

Not only limited to metal nanoparticles, Liu et al. (2018) immobilized bacterium *Arthrobacter globiformis* D47 on the surface of nanocellulose in a fiber form, to create Bacteria-Decorated Nanocellulose (BDN). BDN could rapidly degrade diuron, producing the main metabolite in the form of 3,4-dichloroaniline. After 24 hr, 85% of diuron was degraded and was completely removed after 5 d, with very low concentration of 3,4-dichloroaniline. The cooperation of bioremediation and nanoremediation were achieved by increased nanocellulose adsorption capacity and enzymatic activity (Liu et al., 2018).

4. Bio-Nanoparticles (BNPs)

As mentioned earlier several authors have a different definition of NBR, which is the production of nanoparticles using biological activity for remediation purposes. The produced nanoparticle is defined as bio-nanoparticle and the path is called green synthesis (Pandey, 2018). Bio-nanotechnology in this definition produces nanosized particles through biological approaches combined with chemical and physical principles (Sahayaraj and Rajesh, 2011). The driving force for the development of green synthesis is the high cost of physical and chemical processes. The biological enzymes or released chemicals mainly from plant leaves extracts or juices from plants may reduce the metal ion to nanoparticles. Here only a short review on BNPs will be given, as the earlier chapter has a detailed explanation.

The benefit of synthetizing nanoparticles using plants is their abundance, safe due to the absence of toxicants and variability of metabolites products. In the view of the incubation time for metal ions reduction, plants are better than bacteria and fungi for the synthesis of nanoparticles (Yadav et al., 2017). Yadav et al. (2017) claimed that microbes also show the ability to precipitate metals at a nanometer order. Extracellular secretion of enzymes may produce abundant quantities of nanoparticles with relatively high purity. Similar to microbes, fungi are an excellent example of

Table 11.6. Several plants, microbes, fungi, and yeast for BNPs production.

Type	Name	NPs
Plants	Coriander leaf (Narayanan and Sakthivel, 2008)	Gold
	Pongamia pinnata (Raut et al., 2010)	Silver
	Cinammonum (Premkumar et al., 2018)	Palladium
	Eclipta leaf (Jha et al., 2009)	Silver
	Jatropha curcas (Joglekar et al., 2011)	Lead
	Brassica juncea (Haverkamp et al., 2007)	Gold, silver, and copper alloy
	Aloe vera leaf (Maensiri et al., 2008)	Indium oxide
Microbes	Lactobacillus strains (Garmasheva et al., 2016)	Silver
	Pleurotus sp. (Mazumdar, 2017)	Iron
	Clostridicum thermoaceticum (Cunningham and Lundie, 1993)	Cadmium sulfide
	Diopyros kaki (Song et al., 2010)	Platinum
	Pseudomonas aeruginosa (Husseiny et al., 2007)	Gold
Fungi and yeast	*Rhodospiridium dibocatum* (Seshadri et al., 2011)	Lead sulfide
	Penicillium fellutanum (Kathiresan et al., 2009)	Silver
	Saccharomyces cerevisae (Mourato et al., 2011)	Gold and silver
	Candida glabrata (Krumov et al., 2007)	Cadmium sulfide
	Aspergillus niger (Sagar and Ashok, 2012)	Silver

extracellular enzymes producer. However, due to a larger volume of produced extracellular enzymes, the reduction capacity is also higher.

Plants contain proteins, carbohydrates and coenzyme that have the ability of producing nanoparticles by a metal salt reduction reaction. Some plant extracts have various compounds, such as aldehydes, amides, ascorbic acids, carboxylic acids, flavones, ketones, phenols and terpenoids. These compounds can form metal nanoparticles by reducing metal salts. Some bacteria are also suited for reducing metal ions, with many researches using prokaryotic and actinomycetes for this purpose. Fungi is more efficient for producing metal and metal oxides nanoparticles because of their different intracellular enzymes. In some cases, fungi can produce nanomaterials with higher efficiency (Singh et al., 2018). Several uses of biological agents for nanoparticles production are listed Table 11.6.

5. Nanobioremediation Challenges

Just like other technologies or innovations, the challenges of developing nanobioremediation (NBR) technology are connected. The integration between nanoremediation and bioremediation exclude the drawbacks of each method, especially the disadvantages of applying nanoparticles to the environment. For the sake of future developments, several related challenges or issues should be noted by the stakeholders. Some challenges related to NBR are described here.

5.1 Toxicity

Earlier the toxicity of nanoparticles to the biological agent was mentioned. However, the toxic effect on nanoparticles to the environment (for *in situ* remediations) is also another cause that is attracting the attention of researchers. In recent years, several articles have been published about the toxic effect of nFe^0 on microorganism cells, epithelial cells, fish embryo, etc. Since nZVI could catalyze the formation of hydroxyl radicals (OH^-) from superoxide (O_2^-) and hydrogen peroxide (H_2O_2). These species may induce antioxidant enzymatic activities, peroxidation of membrane lipids, modification of nucleic acids, cell death, etc. (Singh and Misra, 2014).

Nanoparticles also possess a risk of causing health problem for humans. Nanoparticles can enter the human body through several routes, such as ingestion, inhalation and dermal contact. When nanoparticles are consumed through ingestion, most of them will be quickly removed through feces with only a small amount absorbed by the gut lymphatic and distributed to other organs. Although there is currently not enough data to analyze nanoparticles behavior in the food chain (Crisponi et al., 2017), the tiny size and stability of nanoparticles are obviously a sign that they can enter food chain when large scale of nanoremediation and NBR is applied. The higher risk of the health problem is from inhaled nanoparticles, where they can decrease lung efficiency, leading to asthma and other lung chronic diseases. Nanoparticles manufactured from carbon could potentially induce pneumonia, especially for older people (Reijnders, 2006).

Buchman et al. (2019) suggested redesigning the concept of nanomaterials to reduce its toxic effect. Nanoparticles should be designed with a negative surface charge by incorporating ligands (e.g., polyethylene glycol) or changing the morphology, so that nanomaterials would have less interactions and binding with the cell surface. Another alternative is to reduce the toxicity by capping the nanoparticle with a shell material, replacing the toxic compound with less harmful species with similar properties or applying a chelating agent. The production of reactive oxygen species, such as hydroxyl radicals and peroxides, could be suppressed by changing the doping element, adding a shell layer to prevent direct contact with the core or incorporating antioxidant to the surface. However, these suggestions are still a theory assumed from the toxicity mechanism, and should be tested to confirm the reduce toxicity effect (Buchman et al., 2019).

5.2 Disposal

Due to the nanoparticle's role in the NBR catalyst, the disposal and fate in the environment has also become a problem that is similar to the use of nanoparticles because of its toxicity (even though the biological part is safe for the environment). Nanomaterials used for remediation could become pollutants themselves when they are released to the environment. While nanomaterials are released into the body of water (lakes, rivers or oceans), they can get aggregated and sedimented naturally because of dissolved or suspended compounds. However, they can become stable due to ionic power, pH, electrolyte ions that prevent sedimentation. In addition, some modifications to nanoparticles have increased the stability, e.g., surface modification, functionalization and coating (Liu et al., 2014). This event becomes more severe as nanoparticles have a very high mobility property, thus they can spread, reach and impact many places even the ones not visible (Handojo et al., 2020).

One option to this problem is to develop biodegradable nanomaterials (Guerra et al., 2018). For example with nFe^0, it does not only interact with the contaminants but also with the biotic component of that system (Singh and Misra, 2014). Another option is to develop a separation technology to mitigate nanoparticles contamination in the environment. Chemical coagulation with some coagulants, such as polyaluminum chloride or electrocoagulation could be a possible solution. Electro coagulation will produce less sludge compared to chemical coagulation and does not require pH adjustment, but require high conductivity of the treated water suspension and a high cost due to the need of replacing electrodes and high power consumption. Dissolved air flotation is another technology that can remove nanoparticles, often with an aid of a surfactant chemical to boost efficiency (Liu et al., 2014). Therefore, the disposal system must be designed carefully before implementing NBR massively.

5.3 Reusability

In some cases, NBR increases the ease of reusability. The separation of magnetite Fe_3O_4 and *Rhodococcus rhodochrous* strain combination (concurrent method) for chlorophenols degradation which can be separated by using only an external magnetic field . This Fe_3O_4-*R. rhodochrous* combination consumes all the chlorophenols from the first to the sixth cycle and declines after it (Hou et al., 2016). In contrast, Wang et al. (2007) claimed the recycling experiments of Fe_3O_4 nanoparticles immobilized by *Sphingomonas* sp. strain shows that the reusable one has a higher degradation rate due to the effect of the growth of cells in the magnetic gellan gel beads (Wang et al., 2007). It could be concluded that the capacity for degradation is not easy to predict. To emphasize, for industrial implementation, the reusability aspect is not only limited to the ease of separation, but also to the consistency.

5.4 Sustainability

The sustainability aspect of applying NBR cannot be assessed by viewing the results of the remediation which is able to reduce toxicity and hazards from pollutants. It should be monitored as a whole process from its manufacture until application. For example, CNTs production requires a large amount of energy and water, with a bigger impact if they are compared with their toxicity. This high energy demand may also increase greenhouse gases emission for energy generation (Eckelman et al., 2012; Dahlben et al., 2013). In order to conduct a sustainable process, it is important to develop the manufacturing process and material designs to reduce the environmental impacts.

Martins et al. (2017) performed a life cycle assessment for the production of nZVI using a traditional process and green manufacture using oak leaves. This study concluded that the green process produced around 38–50% less environmental impact compared to traditional manufacturing. In the traditional process, electricity generation had the biggest environmental impact, contributing to 45% of the damage. The second largest impact came from sodium borohydride ($NaBH_4$) production, which is used in nZVI manufacturing as a reducing agent for Fe(II) to become zero valency. If electricity was produced in a more sustainable way, the environmental impact of traditional process would decrease quite significantly (Martins et al., 2017).

5.5 Stability

Many of the NBR studies were performed only in the laboratory, where the condition of the process such as temperature, pH, the number of contaminants and the existence of other compound that would interfere with the process which could be maintained easily. However, application in the real field would be more difficult since the condition is harsher and more dangerous. The synthetized nano-bioparticle would be destroyed easily due to this disaggreable condition (Sun, 2019). The degraded nano-bioparticle could even pose a new threat for the environment, because metal leaching from nanoparticles made of metal could become second contaminants (Saputra et al., 2013).

6. Conclusions

Nano-bioremediation is a future nanotechnology-based bioremediation process for contaminant removal where nanoparticles and biological agents are competing with each other. Bioremediation is considerably clean, green and environmentally friendly, but it might be time-consuming and limited to those compounds that are biodegradable. The combination is aimed to conduct a remediation process that possesses a high degradation capacity, less consuming time, sustainable, less toxic and green. The combination may be applied in two ways, namely sequential method and the concurrent method. In the sequential method, the contaminants will be subjected to nanoremediation before bioremediation. While in the concurrent method, the contaminants will be introduced to a system simultaneously where both nanoparticle and biological agents exist. However, further research should be conducted to implement this industrial technology such as the toxicity, fate in the environment, reusability, sustainability and stability aspect.

Existing research is still dealing with hypothetical or single pollutants which is far simpler than real wastewater or soil. This mixture of pollutants can inhibit the growth of some biodegraders of the nano-bioparticle because of a cocktail effect, and thus the study of the efficacy of using nano-bioremediation for this purpose are still required. Studies utilizing genetically modified microorganisms with tremendous effectivity for contaminants removal for the biological part of the nano-bioremediation are still ongoing, where researchers are calculating the risk of large-scale application. Once these research gaps are disclosed, nano-bioremediation will be a promising applicative technology for degrading pollutants.

References

Abatenh, E., Gizaw, B., Tsegaye, Z. and Wassie, M. 2017. The role of microorganisms in bioremediation—A review. Open Journal of Environmental Biology 2(1): 038–046. https://doi.org/10.17352/ojeb.000007.

Alencar, Brenda Thais Barbalho, Victor Hugo Vidal Ribeiro, Cássia Michelle Cabral, Naiane Maria Corrêa dos Santos, Evander Alves Ferreira, Dayana Maria Teodoro Francino, Jose Barbosa dos Santos, Daniel Valadão Silva and Matheus de Freitas Souza. 2020. Use of macrophytes to reduce the contamination of water resources by pesticides. Ecological Indicators 109 (February): 105785. https://doi.org/10.1016/j.ecolind.2019.105785.

Amonette, James E., Darla J. Workman, David W. Kennedy, Jonathan S. Fruchter and Yuri A. Gorby. 2000. Dechlorination of carbon tetrachloride by Fe(II) associated with goethite. Environmental Science & Technology 34(21): 4606–13. https://doi.org/10.1021/es9913582.

Anjum, Reshma, Mashihur Rahman, Farhana Masood and Abdul Malik. 2012. Bioremediation of pesticides from soil and wastewater. pp. 295–328. *In*: Abdul Malik and Elisabeth Grohmann (eds.). Environmental Protection Strategies for Sustainable Development. Dordrecht: Springer Netherlands. https://doi.org/10.1007/978-94-007-1591-2_9.

Arias, L. Renea and Liju Yang. 2009. Inactivation of bacterial pathogens by carbon nanotubes in suspensions. Langmuir 25(5): 3003–12. https://doi.org/10.1021/la802769m.

Azubuike, Christopher Chibueze, Chioma Blaise Chikere and Gideon Chijioke Okpokwasili. 2016. Bioremediation techniques–classification based on site of application: Principles, advantages, limitations and prospects. World Journal of Microbiology and Biotechnology 32(11): 1–18. https://doi.org/10.1007/s11274-016-2137-x.

Balaji, Sundaramoorthy, Thiagarajan Kalaivani, Mohan Shalini, Mohan Gopalakrishnan, Mubarak Ali Rashith Muhammad and Chandrasekaran Rajasekaran. 2016. Sorption sites of microalgae possess metal binding ability towards Cr(VI) from tannery effluents—A kinetic and characterization study. Desalination and Water Treatment 57(31): 14518–29. https://doi.org/10.1080/19443994.2015.1064032.

Barac, Tanja, Safiyh Taghavi, Brigitte Borremans, Ann Provoost, Licy Oeyen, Jan V. Colpaert, Jaco Vangronsveld and Daniel Van Der Lelie. 2004. Engineered endophytic bacteria improve phytoremediation of water-soluble, volatile, organic pollutants. Nature Biotechnology 22(5): 583–88. https://doi.org/10.1038/nbt960.

Bartha, Richard. 1986. Biotechnology of petroleum pollutant biodegradation. Microbial Ecology 12(1): 155–72. https://doi.org/10.1007/BF02153231.

Bhatia, Mayuri, Amandeep Girdhar, Bina Chandrakar and Archana Tiwari. 2013. Implicating nanoparticles as potential biodegradation enhancers: A review. Journal of Nanomedicine and Nanotechnology 4(4): 1–7. https://doi.org/10.4172/2157-7439.1000175.

Bhunia, Susanta Kumar and Nikhil R. Jana. 2014. Reduced graphene oxide-silver nanoparticle composite as visible light photocatalyst for degradation of colorless endocrine disruptors. ACS Applied Materials and Interfaces 6(22): 20085–92. https://doi.org/10.1021/am505677x.

Bokare, Varima, Kumarasamy Murugesan, Jae Hwan Kim, Eun Ju Kim and Yoon Seok Chang. 2012. Integrated hybrid treatment for the remediation of 2,3,7,8-Tetrachlorodibenzo-p-Dioxin. Science of the Total Environment 435-436(October): 563–66. https://doi.org/10.1016/j.scitotenv.2012.07.079.

Bokare, Varima, Kumarasamy Murugesan, Young Mo Kim, Jong Rok Jeon, Eun Ju Kim and Yoon Seok Chang. 2010. Degradation of triclosan by an integrated nano-bio redox process. Bioresource Technology 101(16): 6354–60. https://doi.org/10.1016/j.biortech.2010.03.062.

Bossert, Ingelborg and Richárd Bartha. 1984. Fate of petroleum in soil ecosystems. pp. 435–73. *In*: Atlas, R.M. (ed.). Petroleum Microbiology. New York: Macmillan.

Buchman, Joseph T., Natalie V. Hudson-Smith, Kaitlin M. Landy and Christy L. Haynes. 2019. Understanding nanoparticle toxicity mechanisms to inform redesign strategies to reduce environmental impact. Accounts of Chemical Research 52(6): 1632–42. https://doi.org/10.1021/acs.accounts.9b00053.

Bumpus, J.A. and Aust, S.D. 1987. Biodegradation of DDT [1,1,1-Trichloro-2,2-Bis(4-Chlorophenyl)Ethane] by the white rot fungus phanerochaete chrysosporium. Applied and Environmental Microbiology 53(9): 2001–8. https://doi.org/10.1128/aem.53.9.2001-2008.1987.

Cao, Xiufeng, Amjed Alabresm, Yung Pin Chen, Alan W. Decho and Jamie Lead. 2020. Improved metal remediation using a combined bacterial and nanoscience approach. Science of the Total Environment 704: 135378. https://doi.org/10.1016/j.scitotenv.2019.135378.

Cecchin, Iziquiel, Krishna R. Reddy, Antônio Thomé, Eloisa Fernanda Tessaro and Fernando Schnaid. 2017. Nanobioremediation: Integration of nanoparticles and bioremediation for sustainable remediation of chlorinated organic contaminants in soils. International Biodeterioration and Biodegradation 119: 419–28. https://doi.org/10.1016/j.ibiod.2016.09.027.

Chariou, Paul L., Alan B. Dogan, Alexandra G. Welsh, Gerald M. Saidel, Harihara Baskaran and Nicole F. Steinmetz. 2019. Soil mobility of synthetic and virus-based model nanopesticides. Nature Nanotechnology 14(7): 712–18. https://doi.org/10.1038/s41565-019-0453-7.

Chen, Jiawei, Zongming Xiu, Gregory V. Lowry and Pedro J.J. Alvarez. 2011. Effect of natural organic matter on toxicity and reactivity of nano-scale zero-valent iron. Water Research 45(5): 1995–2001. https://doi.org/10.1016/j.watres.2010.11.036.

Christian, Ferric, Edith, Selly, Dendy Adityawarman and Antonius Indarto. 2013. Application of nanotechnologies in the energy sector: A brief and short review. Frontiers in Energy 7(1): 6–18. https://doi.org/10.1007/s11708-012-0219-5.

Crane, R.A. and Scott, T.B. 2012. Nanoscale zero-valent iron: Future prospects for an emerging water treatment technology. Journal of Hazardous Materials 211-212: 112–25. https://doi.org/10.1016/j.jhazmat.2011.11.073.

Crisponi, Guido, Valeria M. Nurchi, Joanna I. Lachowicz, Massimiliano Peana, Serenella Medici and Maria Antomietta Zoroddu. 2017. Toxicity of nanoparticles: Etiology and mechanisms. 511–46. *In*: Alexandru Mihai Grumezescu (ed.). Antimicrobial Nanoarchitectonics: From Synthesis to Applications. Amsterdam: Elsevier Inc. https://doi.org/10.1016/B978-0-323-52733-0.00018-5.

Cummings, David E., Frank Caccavo, Scott Fendorf and R. Frank Rosenzweig. 1999. Arsenic mobilization by the dissimilatory Fe(III)-reducing bacterium *Shewanella Alga* BrY. Environmental Science and Technology 33(5): 723–29. https://doi.org/10.1021/es980541c.

Cunningham, D.P. and Lundie, L.L. 1993. Precipitation of cadmium by *Clostridium Thermoaceticum*. Applied and Environmental Microbiology 59(1): 7–14. https://doi.org/10.1128/aem.59.1.7-14.1993.

Czech, Bozena and Katarzyna Rubinowska. 2013. TiO2-assisted photocatalytic degradation of diclofenac, metoprolol, estrone and chloramphenicol as endocrine disruptors in water. Adsorption 19(2-4): 619–30. https://doi.org/10.1007/s10450-013-9485-8.

Dahlben, Lindsay J., Matthew J. Eckelman, Ali Hakimian, Sivasubramanian Somu and Jacqueline A. Isaacs. 2013. Environmental life cycle assessment of a carbon nanotube-enabled semiconductor device. Environmental Science and Technology 47(15): 8471–78. https://doi.org/10.1021/es305325y.

Das, Merina Paul, Maharshi Bashwant, Kranthi Kumar and Jayabrata Das. 2012. Control of pharmaceutical effluent parameters through bioremediation. Journal of Chemical and Pharmaceutical Research 4(2): 1061–65.

Dong, Haoran, Long Li, Yue Lu, Yujun Cheng, Yaoyao Wang, Qin Ning, Bin Wang and Lihua Zhang. 2019. Integration of nanoscale zero-valent iron and functional anaerobic bacteria for groundwater remediation: A review. Environment International 124: 265–77. https://doi.org/10.1016/j.envint.2019.01.030.

Dong, Haoran, Zhao Jiang, Junmin Deng, Cong Zhang, Yujun Cheng, Kunjie Hou, Lihua Zhang, Lin Tang and Guangming Zeng. 2018. Physicochemical transformation of Fe/Ni bimetallic nanoparticles during aging in simulated groundwater and the consequent effect on contaminant removal. Water Research 129: 51–57. https://doi.org/10.1016/j.watres.2017.11.002.

Eckelman, Matthew J., Meagan S. Mauter, Jacqueline A. Isaacs and Menachem Elimelech. 2012. New perspectives on nanomaterial aquatic ecotoxicity: Production impacts exceed direct exposure impacts for carbon nanotoubes. Environmental Science and Technology 46(5): 2902–10. https://doi.org/10.1021/es203409a.

Elfeky, Souad A., Shymaa Ebrahim Mahmoud and Ahmed Fahmy Youssef. 2017. Applications of CTAB modified magnetic nanoparticles for removal of chromium (VI) from contaminated water. Journal of Advanced Research 8(4): 435–43. https://doi.org/10.1016/j.jare.2017.06.002.

Fang, Guodong, Youbin Si, Chao Tian, Gangya Zhang and Dongmei Zhou. 2012. Degradation of 2,4-D in soils by Fe3O4 nanoparticles combined with stimulating indigenous microbes. Environmental Science and Pollution Research 19(3): 784–93. https://doi.org/10.1007/s11356-011-0597-y.

Flaxbart, David. 1996. Dictionary of Organic Compounds. Sixth Edition. Volumes 1–9 Edited by J. Buckingham and F. Macdonald (Chapman & Hall). Chapman and Hall: London. 1995. 9144 Pp. $5200.00. ISBN 0-412-54090-8. Journal of the American Chemical Society 118(6): 1580–1580. https://doi.org/10.1021/ja955360e.

Gahlout, Mayur, Darshan M. Rudakiya, Shilpa Gupte and Akshaya Gupte. 2017. Laccase-conjugated amino-functionalized nanosilica for efficient degradation of reactive violet 1 dye. International Nano Letters 7(3): 195–208. https://doi.org/10.1007/s40089-017-0215-1.

Gao, Ying, Feifeng Wang, Yan Wu, Ravendra Naidu and Zuliang Chen. 2016. Comparison of degradation mechanisms of microcystin-LR using nanoscale zero-valent iron (NZVI) and bimetallic Fe/Ni and Fe/Pd nanoparticles. Chemical Engineering Journal 285: 459–66. https://doi.org/10.1016/j.cej.2015.09.078.

Garmasheva, Inna, Nadezhda Kovalenko, Sergey Voychuk, Andriy Ostapchuk, Olena Livins'ka and Ljubov Oleschenko. 2016. Lactobacillus species mediated synthesis of silver nanoparticles and their antibacterial activity against opportunistic pathogens *in vitro*. BioImpacts 6(4): 219–23. https://doi.org/10.15171/bi.2016.29.

Gerlach, Robin, Al B. Cunningham and Frank Caccavo. 2000. Dissimilatory iron-reducing bacteria can influence the reduction of carbon tetrachloride by iron metal. Environmental Science and Technology 34(12): 2461–64. https://doi.org/10.1021/es991200h.

Gibson, David T., Venkatanayarana Mahadevan, Donald M. Jerina, Haruhiko Yagi and Herman J.C. Yeh. 1975. Oxidation of the carcinogens Benzo[a]Pyrene and Benzo[a]Anthracene to dihydrodiols by a bacterium. Science 189(4199): 295–97. https://doi.org/10.1126/science.1145203.

Giese, Ellen C., Debora D.V. Silva, Ana F.M. Costa, Sâmilla G.C. Almeida and Kelly J. Dussán. 2020. Immobilized microbial nanoparticles for biosorption. Critical Reviews in Biotechnology 40(5): 653–66. https://doi.org/10.1080/07388551.2020.1751583.

Gil-Díaz, M., González, A., Alonso, J. and Lobo, M.C. 2016. Evaluation of the Stability of a nanoremediation strategy using barley plants. Journal of Environmental Management 165: 150–58. https://doi.org/10.1016/j.jenvman.2015.09.032.

Gil-Díaz, M., Rodríguez-Valdés, E., Alonso, J., Baragaño, D., Gallego, J.R. and Lobo, M.C. 2019. Nanoremediation and long-term monitoring of brownfield soil highly polluted with As and Hg. Science of the Total Environment 675(July): 165–75. https://doi.org/10.1016/j.scitotenv.2019.04.183.

Gómez-Jacinto, Verónica, Tamara García-Barrera, José Luis Gómez-Ariza, Inés Garbayo-Nores and Carlos Vílchez-Lobato. 2015. Elucidation of the defence mechanism in microalgae chlorella sorokiniana under mercury exposure. Identification of Hg-Phytochelatins. Chemico-Biological Interactions 238(5): 82–90. https://doi.org/10.1016/j.cbi.2015.06.013.

Grubbs, R.B. and Barry A. Mohaa. 1987. *In Situ* Biological Treatment of Troublesome Organics. Fresno.

Guerra, Fernanda, Mohamed Attia, Daniel Whitehead and Frank Alexis. 2018. Nanotechnology for environmental remediation: Materials and applications. Molecules 23(7): 1760. https://doi.org/10.3390/molecules23071760.

Handojo, Lienda, Daniel Pramudita, Dave Mangindaan and Antonius Indarto. 2020. Application of nanoparticles in environmental cleanup: Production, potential risks and solutions. pp. 45–76. *In*: Ram Naresh Bharagava (ed.). Emerging Eco-Friendly Green Technologies for Wastewater Treatment. Singapore: Springer, Singapore. https://doi.org/10.1007/978-981-15-1390-9_3.

Haverkamp, R.G., Marshall, A.T. and Van Agterveld, D. 2007. Pick your carats: Nanoparticles of gold-silver-copper alloy produced *in vivo*. Journal of Nanoparticle Research 9(4): 697–700. https://doi.org/10.1007/s11051-006-9198-y.

He, Fang, Wenrong Hu and Yuezhong Li. 2004. Biodegradation mechanisms and kinetics of azo dye 4BS by a microbial consortium. Chemosphere 57(4): 293–301. https://doi.org/10.1016/j.chemosphere.2004.06.036.

Hoagland, R.E., Zablotowicz, R.M. and Hall, J.C. 2000. Pesticide metabolism in plants and microorganisms: An overview. pp. 2–27. *In*: Christopher Hall, J., Robert E. Hoagland and Robert M. Zablotowicz (eds.). Pesticide Biotransformation in Plants and Microorganisms, 8. Washington D.C.: American Chemical Society. https://doi.org/10.1021/bk-2001-0777.ch001.

Horváthová, Hana, Katarína Lászlová and Katarína Dercová. 2019. Bioremediation vs. nanoremediation: Degradation of polychlorinated biphenyls (PCBS) using integrated remediation approaches. Water, Air, and Soil Pollution 230(8): 1–11. https://doi.org/10.1007/s11270-019-4259-x.

Hou, Jianfeng, Feixia Liu, Nan Wu, Jiansong Ju and Bo Yu. 2016. Efficient biodegradation of chlorophenols in aqueous phase by magnetically immobilized aniline-degrading rhodococcus rhodochrous strain. Journal of Nanobiotechnology 14(1): 1–8. https://doi.org/10.1186/s12951-016-0158-0.

Huang, Lihui, Gaofeng Liu, Guihua Dong, Xueyuan Wu, Chuang Wang and Yangyang Liu. 2017. Reaction mechanism of zero-valent iron coupling with microbe to degrade tetracycline in permeable reactive barrier (PRB). Chemical Engineering Journal 316: 525–33. https://doi.org/10.1016/j.cej.2017.01.096.

Husseiny, M.I., Abd El-Aziz, M., Badr, Y. and Mahmoud, M.A. 2007. Biosynthesis of gold nanoparticles using pseudomonas aeruginosa. Spectrochimica Acta—Part A: Molecular and Biomolecular Spectroscopy 67(3-4): 1003–6. https://doi.org/10.1016/j.saa.2006.09.028.

Indarto, Antonius. 2012. Heterogeneous reactions of HONO formation from NO_2 and HNO_3: A review. Research on Chemical Intermediates 38(3-5): 1029–41. https://doi.org/10.1007/s11164-011-0439-z.

Indarto, Antonius, Jae Wook Choi, Hwaung Lee and Hyung Keun Song. 2006. Treatment of CCl_4 and $CHCl_3$ emission in a gliding-arc plasma. Plasma Devices and Operations 14(1): 1–14. https://doi.org/10.1080/10519990500493833.

Indarto, Antonius, Dae Ryook Yang, Jae Wook Choi, Hwaung Lee and Hyung Keun Song. 2007. CCl_4 decomposition by gliding arc plasma: Role of C2 compounds on products distribution. Chemical Engineering Communications 194(8): 1111–25. https://doi.org/10.1080/00986440701293363.

Indarto, Antonius, Hwaung Lee, Jae Wook Choi and Hyung Keun Song. 2008. Partial oxidation of methane with yttria-stabilized zirconia catalyst in a dielectric barrier discharge. Energy Sources, Part A: Recovery, Utilization and Environmental Effects 30(17): 1628–1636. https://doi.org/ 10.1080/15567030701268385.

Ivey, George A. 2006. Surfactant Enhanced Remediation (SER) Using Ivey-Sol® Surfactant Technology. Vol. 2006.

Jha, Anal K., Kamlesh Prasad, Vikash Kumar and Prasad, K. 2009. Biosynthesis of silver nanoparticles using eclipta leaf. Biotechnology Progress 25(5): 1476–79. https://doi.org/10.1002/btpr.233.

Joglekar, Shriram, Kisan Kodam, Mayur Dhaygude and Manish Hudlikar. 2011. Novel route for rapid biosynthesis of lead nanoparticles using aqueous extract of Jatropha Curcas L. latex. Materials Letters 65(19-20): 3170–72. https://doi.org/10.1016/j.matlet.2011.06.075.

Kang, Seoktae, Mathieu Pinault, Lisa D. Pfefferle and Menachem Elimelech. 2007. Single-walled carbon nanotubes exhibit strong antimicrobial activity. Langmuir 23(17): 8670–73. https://doi.org/10.1021/la701067r.

Kang, Seoktae, Moshe Herzberg, Debora F. Rodrigues and Menachem Elimelech. 2008. Antibacterial effects of carbon nanotubes: Size does matter! Langmuir 24(13): 6409–13. https://doi.org/10.1021/la800951v.

Kapri, Anil, M., Zaidi, G.H. and Reeta Goel. 2009. Nanobarium titanate as supplement to accelerate plastic waste biodegradation by indigenous bacterial consortia. In AIP Conference Proceedings, 1147: 469–74. AIP. https://doi.org/10.1063/1.3183475.

Kathiresan, K., Manivannan, S., Nabeel, M.A. and Dhivya, B. 2009. Studies on silver nanoparticles synthesized by a marine fungus, penicillium fellutanum isolated from coastal mangrove sediment. Colloids and Surfaces B: Biointerfaces 71(1): 133–37. https://doi.org/10.1016/j.colsurfb.2009.01.016.

Khin, Mya Mya, A. Sreekumaran Nair, V. Jagadeesh Babu, Rajendiran Murugan and Seeram Ramakrishna. 2012. A review on nanomaterials for environmental remediation. Energy & Environmental Science 5(8): 8075. https://doi.org/10.1039/c2ee21818f.

Kim, Young Mo, Kumarasamy Murugesan, Yoon Young Chang, Eun Ju Kim and Yoon Seok Chang. 2012. Degradation of polybrominated diphenyl ethers by a sequential treatment with nanoscale zero valent iron and aerobic biodegradation. Journal of Chemical Technology and Biotechnology 87(2): 216–24. https://doi.org/10.1002/jctb.2699.

Kiran, George Seghal, Lipton Anuj Nishanth, Sethu Priyadharshini, Kumar Anitha and Joseph Selvin. 2014. Effect of Fe nanoparticle on growth and glycolipid biosurfactant production under solid state culture by marine nocardiopsis sp. MSA13A. BMC Biotechnology 14(48): 1–10. https://doi.org/10.1186/1472-6750-14-48.

Koul, Bhupendra and Pooja Taak. 2018. *Ex situ* soil remediation strategies. In Biotechnological Strategies for Effective Remediation of Polluted Soils, 39–57. Singapore: Springer. https://doi.org/10.1007/978-981-13-2420-8_2.

Krumov, Nikolay, Stephanie Oder, Iris Perner-Nochta, Angel Angelov and Clemens Posten. 2007. Accumulation of CdS nanoparticles by yeasts in a fed-batch bioprocess. Journal of Biotechnology 132(4): 481–86. https://doi.org/10.1016/j.jbiotec.2007.08.016.

Kumar, S. Raj, and Gopinath, P. 2016. Chapter 2 Nano-bioremediation applications of nanotechnology for bioremediation. pp. 27–48. *In*: Jiaping Paul Chen, Lawrence K. Wang, Mu-Hao Sung Wang, Yung-Tse Hung and Nazih K. Shammas (eds.). Remediation of Heavy Metals in the Environment. Boca Raton: CRC Press. https://doi.org/10.1201/9781315374536-3.

Kumari, Babita and Singh, D.P. 2016. A review on multifaceted application of nanoparticles in the field of bioremediation of petroleum hydrocarbons. Ecological Engineering 97: 98–105. https://doi.org/10.1016/j.ecoleng.2016.08.006.

Le, Anh Thi, Swee Yong Pung, Srimala Sreekantan, Atsunori Matsuda and Dai Phu Huynh. 2019. Mechanisms of removal of heavy metal ions by ZnO particles. Heliyon 5(4): e01440. https://doi.org/10.1016/j.heliyon.2019.e01440.

Le, Thao Thanh, Khanh Hoang Nguyen, Jong Rok Jeon, Arokiasamy J. Francis and Yoon Seok Chang. 2015. Nano/bio treatment of polychlorinated biphenyls with evaluation of comparative toxicity. Journal of Hazardous Materials 287: 335–41. https://doi.org/10.1016/j.jhazmat.2015.02.001.

Le, Thao Thanh, Hakwon Yoon, Min-hui Son, Yu-gyeong Kang and Yoon-seok Chang. 2019. Treatability of hexabromocyclododecane using Pd/Fe nanoparticles in the soil-plant system: Effects of humic acids. Science of the Total Environment 689: 444–50. https://doi.org/10.1016/j.scitotenv.2019.06.290.

Leisinger, Th. and Winfried Brunner. 1986. Poorly degradable substances. pp. 475–513. *In*: Schönberg, W. (ed.). Biotechnology: Microbial Degradations, 8. Weinheim, Germany: VCH Verlagsgesllschaft.

Leong, Yoong Kit and Jo Shu Chang. 2020. Bioremediation of heavy metals using microalgae: Recent advances and mechanisms. Bioresource Technology 303(December 2019): 122886. https://doi.org/10.1016/j.biortech.2020.122886.

Liu, Hong Fang, Tian Wei Qian and Dong Ye Zhao. 2013. Reductive immobilization of perrhenate in soil and groundwater using starch-stabilized ZVI nanoparticles. Chinese Science Bulletin 58(2): 275–81. https://doi.org/10.1007/s11434-012-5425-3.

Liu, Jie, Eden Morales-Narváez, Teresa Vicent, Arben Merkoçi and Guo Hua Zhong. 2018. Microorganism-Decorated nanocellulose for efficient diuron removal. Chemical Engineering Journal 354: 1083–91. https://doi.org/10.1016/j.cej.2018.08.035.

Liu, Y., Tourbin, M., Lachaize, S. and Guiraud, P. 2014. Nanoparticles in wastewaters: Hazards, fate and remediation. Powder Technology 255: 149–56. https://doi.org/10.1016/j.powtec.2013.08.025.

Lv, Tao, Yang Zhang, Mònica E. Casas, Pedro N. Carvalho, Carlos A. Arias, Kai Bester and Hans Brix. 2016. Phytoremediation of imazalil and tebuconazole by four emergent wetland plant species in hydroponic medium. Chemosphere 148(April): 459–66. https://doi.org/10.1016/j.chemosphere.2016.01.064.

Maensiri, S., Laokul, P., Klinkaewnarong, J., Sumalin Phokha, Promarak, V. and Supapan Seraphin. 2008. Indium oxide (In_2O_3) nanoparticles using aloe vera plant extract: Synthesis and optical properties. Optoelectronics and Advanced Materials, Rapid Communications 2(3): 161–65.

Martins, Florinda, Susana Machado, Tomás Albergaria and Cristina Delerue-Matos. 2017. LCA applied to nano scale zero valent iron synthesis. International Journal of Life Cycle Assessment 22(5): 707–14. https://doi.org/10.1007/s11367-016-1258-7.

Matheson, Leah J. and Paul G. Tratnyek. 1994. Reductive dehalogenation of chlorinated methanes by iron metal. Environmental Science and Technology 28(12): 2045–53. https://doi.org/10.1021/es00061a012.

Mazumdar, Harajyoti. 2017. A study on biosynthesis of iron nanoparticles by pleurotus sp. Journal of Microbiology and Biotechnology Research 1(January): 39–49.

Mishra, Virendra K., Alka R. Upadhyay and Tripathi, B.D. 2009. Bioaccumulation of heavy metals and two organochlorine pesticides (DDT and BHC) in crops irrigated with secondary treated waste water. Environmental Monitoring and Assessment 156 (1-4): 99–107. https://doi.org/10.1007/s10661-008-0466-4.

Mitra, Sayak, Avipsha Sarkar and Shampa Sen. 2017. Removal of chromium from industrial effluents using nanotechnology: A review. Nanotechnology for Environmental Engineering 2(1): 1–14. https://doi.org/10.1007/s41204-017-0022-y.

Mourato, Ana, Mário Gadanho, Ana R. Lino and Rogério Tenreiro. 2011. Biosynthesis of crystalline silver and gold nanoparticles by extremophilic yeasts. Bioinorganic Chemistry and Applications 2011. https://doi.org/10.1155/2011/546074.

Musyoka, Sonnia and Fernandez, I. 2016. Types and mechanisms of bioremediation in aquaculture wastes; review. International Journal of Advanced Scientific and Technical Research 1(November): 138–55.

Naja, Ghinwa, Annamaria Halasz, Sonia Thiboutot, Guy Ampleman and Jalal Hawari. 2008. Degradation of hexahydro-1,3,5-Trinitro-1,3,5-Triazine (RDX) using zerovalent iron nanoparticles. Environmental Science and Technology 42(12): 4364–70. https://doi.org/10.1021/es7028153.

Narayanan, K. Badri and Sakthivel, N. 2008. Coriander leaf mediated biosynthesis of gold nanoparticles. Materials Letters 62(30): 4588–90. https://doi.org/10.1016/j.matlet.2008.08.044.

Oh, Byung Taek, Craig L. Just and Pedro J.J. Alvarez. 2001. Hexahydro-1,3,5-Trinitro-1,3,5-Triazine mineralization by zerovalent iron and mixed anaerobic cultures. Environmental Science and Technology 35(21): 4341–46. https://doi.org/10.1021/es010852e.

Pandey, B.V. and Upadhyay, R.S. 2010. Pseudomonas fluorescens can be used for bioremediation of textile effluent direct Orange-102. Tropical Ecology 51(2): 397–403.

Pandey, Garima. 2018. Prospects of nanobioremediation in environmental cleanup. Oriental Journal of Chemistry 34(6): 2838–50. https://doi.org/10.13005/ojc/340622.

Pathak, Vinay Mohan and Navneet. 2017. Review on the Current Status of Polymer Degradation: A Microbial Approach. Bioresources and Bioprocessing. Springer Berlin Heidelberg. https://doi.org/10.1186/s40643-017-0145-9.

Pereira, R.A., Pereira, M.F.R., Alves, M.M. and Pereira, L. 2014. Carbon based materials as novel redox mediators for dye wastewater biodegradation. Applied Catalysis B: Environmental 144: 713–20. https://doi.org/10.1016/j.apcatb.2013.07.009.

Pérez-Legaspi, Ignacio Alejandro, Luis Alfredo Ortega-Clemente, Jesús David Moha-León, Elvira Ríos-Leal, Sergio Curiel Ramírez Gutiérrez and Isidoro Rubio-Franchini. 2016. Effect of the pesticide lindane on the biomass of the microalgae Nannochloris Oculata. Journal of Environmental Science and Health - Part B Pesticides, Food Contaminants, and Agricultural Wastes 51(2): 103–6. https://doi.org/10.1080/03601234.2015.1092824.

Premkumar, J.T., Sudhakar, Abhishek Dhakal, Jeshan Babu Shrestha, Krishnakumar, S. and Balashanmugam, P. 2018. Synthesis of silver nanoparticles (AgNPs) from cinnamon against bacterial pathogens. Biocatalysis and Agricultural Biotechnology 15: 311–16. https://doi.org/10.1016/j.bcab.2018.06.005.

Prodan, Alina Mihaela, Simona Liliana Iconaru, Carmen Mariana Chifiriuc, Coralia Bleotu, Carmen Steluta Ciobanu, Mikael Motelica-Heino, Stanislas Sizaret and Daniela Predoi. 2013. Magnetic properties and biological activity evaluation of iron oxide nanoparticles. Journal of Nanomaterials 2013. https://doi.org/10.1155/2013/893970.

Qian, Yuting, Caidie Qin, Mengmeng Chen and Sijie Lin. 2020. Nanotechnology in soil remediation − applications vs. implications. Ecotoxicology and Environmental Safety 201: 110815. https://doi.org/10.1016/j.ecoenv.2020.110815.

Rajesha, J.B., Alamelu K. Ramasami, Nagaraju, G. and Geetha R. Balakrishna. 2016. Photochemical elimination of endocrine disrupting chemical (EDC) by ZnO nanoparticles, synthesized by gel combustion. Water Environment Research 89(5): 396–405. https://doi.org/10.2175/106143016x14733681696086.

Raut, Rajesh W., Niranjan S. Kolekar, Jaya R. Lakkakula, Vijay D. Mendhulkar and Sahebrao B. Kashid. 2010. Extracellular synthesis of silver nanoparticles using dried leaves of pongamia pinnata (L) pierre. Nano-Micro Letters 2(2): 106–13. https://doi.org/10.5101/nml.v2i2.p106-113.

Ravikumar, K.V.G., Deepak Kumar, Gaurav Kumar, Mrudula, P., Chandrasekaran Natarajan and Amitava Mukherjee. 2016. Enhanced Cr(VI) removal by nanozerovalent iron-immobilized alginate beads in the presence of a biofilm in a continuous-flow reactor. Industrial and Engineering Chemistry Research 55(20): 5973–82. https://doi.org/10.1021/acs.iecr.6b01006.

Reddy, Krishna and Jeffrey Adams. 2015. Sustainable Remediation of Contaminated Sites. New York: Momentum Press.

Reijnders, L. 2006. Cleaner nanotechnology and hazard reduction of manufactured nanoparticles. Journal of Cleaner Production 14(2): 124–33. https://doi.org/10.1016/j.jclepro.2005.03.018.

Rizwan, Md., Man Singh, Chanchal K. Mitra and Roshan K. Morve. 2014. Ecofriendly application of nanomaterials: nanobioremediation. Journal of Nanoparticles 2014(June): 1–7. https://doi.org/10.1155/2014/431787.

Sadowsky, Michael J. and Lawrence P. Wackett. 2000. Genetics of atrazine and s-triazine degradation by Psedomonas sp. strain ADP and other bacteria. pp. 268–82. *In*: Christopher Hall, J., Robert E. Hoagland and Robert M. Zablotowicz (eds.). Pesticide Biotransformation in Plants and Microorganisms. Washington D.C.: American Chemical Society. https://doi.org/10.1021/bk-2001-0777.ch015.

Sagar, Gaikwad and Bhosale Ashok. 2012. Green synthesis of silver nanoparticles using Aspergillus niger and its efficacy against human pathogens. Pelagia Research Library 2(5): 1654–58.

Sahayaraj, Kitherian and Rajesh, S. 2011. Bionanoparticles: Synthesis and antimicrobial applications. pp. 228–44. *In*: Méndez-Vilas, A. (ed.). Science against Microbial Pathogens: Communicating Current Research and Technological Advances. Spain: Formatex Research Center.

Saputra, Edy, Syaifullah Muhammad, Hongqi Sun, Ang, H.M., Tadé, M.O. and Shaobin Wang. 2013. Different crystallographic one-dimensional MnO_2 nanomaterials and their superior performance in catalytic phenol degradation. Environmental Science and Technology 47(11): 5882–87. https://doi.org/10.1021/es400878c.

Seo, Jong Su, Young Soo Keum and Qing X. Li. 2009. Bacterial degradation of aromatic compounds. International Journal of Environmental Research and Public Health 6(1): 278–309. https://doi.org/10.3390/ijerph6010278.

Seshadri, Sachin, K. Saranya and Meenal Kowshik. 2011. Green synthesis of lead sulfide nanoparticles by the lead resistant marine yeast, *Rhodosporidium Diobovatum*. Biotechnology Progress 27(5): 1464–69. https://doi.org/10.1002/btpr.651.

Shan, Guo Bin, Jian Min Xing, Huai Ying Zhang and Hui Zhou Liu. 2005. Biodesulfurization of dibenzothiophene by microbial cells coated with magnetite nanoparticles Applied and Environmental Microbiology 71(8): 4497–4502. https://doi.org/10.1128/AEM.71.8.4497-4502.2005.

Sheng, Gudong, Xiaoyu Shao, Yimin Li, Jianfa Li, Huaping Dong, Wei Cheng, Xing Gao and Yuying Huang. 2014. Enhanced removal of uranium(VI) by nanoscale zerovalent iron supported on Na–bentonite and an investigation of mechanism. The Journal of Physical Chemistry A 118(16): 2952–58. https://doi.org/10.1021/jp412404w.

Sherry Davis, A., Prakash, P. and Thamaraiselvi, K. 2017. Nanobioremediation technologies for sustainable environment. Environmental Science and Engineering (Subseries: Environmental Science), no. 9783319484389: 13–33. https://doi.org/10.1007/978-3-319-48439-6_2.

Shwetharani, R., Fernando, C.A.N. and Geetha R. Balakrishna. 2015. Excellent hydrogen evolution by a multi approach via structure-property tailoring of Titania. RSC Advances 5(49): 39122–30. https://doi.org/10.1039/c5ra04578a.

Si, Youbin, Yan Zou, Xiaohong Liu, Xiongyuan Si and Jingdong Mao. 2015. Mercury methylation coupled to iron reduction by dissimilatory iron-reducing bacteria. Chemosphere 122(March): 206–12. https://doi.org/10.1016/j.chemosphere.2014.11.054.

Singh, Jagpreet, Tanushree Dutta, Ki Hyun Kim, Mohit Rawat, Pallabi Samddar and Pawan Kumar. 2018. Green' synthesis of metals and their oxide nanoparticles: Applications for environmental remediation. Journal of Nanobiotechnology 16(1): 84. https://doi.org/10.1186/s12951-018-0408-4.

Singh, Ritu and Virendra Misra. 2014. Application of zero-valent iron nanoparticles for environmental clean up. In Advanced Materials for Agriculture, Food and Environmental Safety, 9781118773: 385–420. https://doi.org/10.1002/9781118773857.ch14.

Song, Jae Yong, Eun-Yeong Kwon and Beom Soo Kim. 2010. Biological synthesis of platinum nanoparticles using diopyros kaki leaf extract. Bioprocess and Biosystems Engineering 33(1): 159–64. https://doi.org/10.1007/s00449-009-0373-2.

Song, Yinan, Niall Kirkwood, Čedo Maksimović, Xiaodi Zhen, David O'Connor, Yuanliang Jin and Deyi Hou. 2019. Nature based solutions for contaminated land remediation and brownfield redevelopment in cities: A review. Science of the Total Environment 663: 568–79. https://doi.org/10.1016/j.scitotenv.2019.01.347.

Sun, Hongqi. 2019. Grand challenges in environmental nanotechnology. Frontiers in Nanotechnology 1(2): 2. https://doi.org/10.3389/fnano.2019.00002.

Sun, Jing, Jun Cheng, Zongbo Yang, Ke Li, Junhu Zhou and Kefa Cen. 2015. Microstructures and functional groups of nannochloropsis sp. cells with arsenic adsorption and lipid accumulation. Bioresource Technology 194: 305–11. https://doi.org/10.1016/j.biortech.2015.07.041.

Teng, Ying, Xiaomi Wang, Ye Zhu, Wei Chen, Peter Christie, Zhengao Li and Yongming Luo. 2017. Biodegradation of pentachloronitrobenzene by cupriavidus sp. YNS-85 and its potential for remediation of contaminated soils. Environmental Science and Pollution Research 24(10): 9538–47. https://doi.org/10.1007/s11356-017-8640-2.

Tiwari, Neha, Deenan Santhiya and Jai Gopal Sharma. 2020. Microbial remediation of micro-nano plastics: Current knowledge and future trends. Environmental Pollution 265(October): 115044. https://doi.org/10.1016/j.envpol.2020.115044.

Tratnyek, Paul. 2003. Permeable reactive barriers of iron and other zero-valent metals. pp. 371–421. *In*: Matthew A. Tarr (ed.). Chemical Degradation Methods for Wastes and Pollutants: Environmental and Industrial Applications. New York: CRC Press. https://doi.org/10.1201/9780203912553.ch9.

Tratnyek, Paul G. and Richard L. Johnson. 2006. Nanotechnologies for environmental cleanup. Nano Today 1(2): 44–48. https://doi.org/10.1016/S1748-0132(06)70048-2.

Tripathi, Sandeep, Sanjeevi, R., Anuradha, J., Dushyant Singh Chauhan and Ashok K. Rathoure. 2018. Nano-bioremediation: nanotechnology and bioremediation. pp. 202–19. *In*: Ashok K. Rathoure (ed.). Biostimulation Remediation Technologies for Groundwater Contaminants. USA: IGI Global. https://doi.org/10.4018/978-1-5225-4162-2.ch012.

Vaish, Supriya and Bhawana Pathak. 2020. Bionano technological approaches for degradation and decolorization of dye by mangrove plants. pp. 399–412. *In*: Jayanta Kumar Patra, Rashmi Ranjan Mishra and Hrudayanath Thatoi (eds.). Biotechnological Utilization of Mangrove Resources. London: Elsevier. https://doi.org/10.1016/b978-0-12-819532-1.00019-6.

Vázquez-Núñez, Edgar, Carlos Eduardo Molina-Guerrero, Julián Mario Peña-Castro, Fabián Fernández-Luqueño and Ma Guadalupe de la Rosa-Álvarez. 2020. Use of nanotechnology for the bioremediation of contaminants: A review. Processes 8(7): 1–17. https://doi.org/10.3390/pr8070826.

Vecitis, Chad D., Katherine R. Zodrow, Seoktae Kang and Menachem Elimelech. 2010. Electronic-structure-dependent bacterial cytotoxicity of single-walled carbon nanotubes. ACS Nano 4(9): 5471–79. https://doi.org/10.1021/nn101558x.

Wang, Xia, Zhonghui Gai, Bo Yu, Jinhui Feng, Changyong Xu, Yong Yuan, Zhixin Lin and Ping Xu. 2007. Degradation of carbazole by microbial cells immobilized in magnetic gellan gum gel beads. Applied and Environmental Microbiology 73(20): 6421–28. https://doi.org/10.1128/AEM.01051-07.

Wedemeyer, G. 1967. Dechlorination of 1,1,1-Trichloro-2,2-Bis(p-Chlorophenyl)Ethane by aerobacter aerogenes. I. metabolic products. Applied Microbiology 15(3): 569–74. https://doi.org/10.1128/aem.15.3.569-574.1967.

Windt, Wim De, Peter Aelterman and Willy Verstraete. 2005. Bioreductive deposition of palladium (0) nanoparticles on shewanella oneidensis with catalytic activity towards reductive dechlorination of polychlorinated biphenyls. Environmental Microbiology 7(3): 314–25. https://doi.org/10.1111/j.1462-2920.2004.00696.x.

Xie, Yankai, Haoran Dong, Guangming Zeng, Lin Tang, Zhao Jiang, Cong Zhang, Junmin Deng, Lihua Zhang and Yi Zhang. 2017. The interactions between nanoscale zero-valent iron and microbes in the subsurface environment: A review. Journal of Hazardous Materials 321(January): 390–407. https://doi.org/10.1016/j.jhazmat.2016.09.028.

Xiu, Zong Ming, Kelvin B. Gregory, Gregory V. Lowry and Pedro J.J. Alvarez. 2010. Effect of bare and coated nanoscale zerovalent iron on TceA and VcrA gene expression in Dehalococcoides spp. Environmental Science and Technology 44(19): 7647–51. https://doi.org/10.1021/es101786y.

Xiu, Zong Ming, Zhao hui Jin, Tie long Li, Shaily Mahendra, Gregory V. Lowry and Pedro J.J. Alvarez. 2010. Effects of nano-scale zero-valent iron particles on a mixed culture dechlorinating trichloroethylene. Bioresource Technology 101(4): 1141–46. https://doi.org/10.1016/j.biortech.2009.09.057.

Yadav, K.K., Singh, J.K., Gupta, N. and Kumar, V. 2017. A review of nanobioremediation technologies for environmental cleanup: a novel biological approach. Journal of Materials and Environmental Science 8(2): 740–57.

Yan, Fang Fang, Chao Wu, Yuan Yuan Cheng, Yan Rong He, Wen Wei Li and Han Qing Yu. 2013. Carbon nanotubes promote Cr(VI) reduction by alginate-immobilized Shewanella Oneidensis MR-1. Biochemical Engineering Journal 77: 183–89. https://doi.org/10.1016/j.bej.2013.06.009.

Yeh, Chia Shen, Reuben Wang, Wen Chi Chang and Yang hsin Shih. 2018. Synthesis and characterization of stabilized oxygen-releasing CaO_2 nanoparticles for bioremediation. Journal of Environmental Management 212(April): 17–22. https://doi.org/10.1016/j.jenvman.2018.01.068.

Yin, Iris Xiaoxue, Jing Zhang, Irene Shuping Zhao, May Lei Mei, Quanli Li and Chun Hung Chu. 2020. The antibacterial mechanism of silver nanoparticles and its application in dentistry. International Journal of Nanomedicine 15: 2555–62. https://doi.org/10.2147/IJN.S246764.

Zhang, Chengdong, Mingzhu Li, Xu Xu and Na Liu. 2015. Effects of carbon nanotubes on atrazine biodegradation by Arthrobacter sp. Journal of Hazardous Materials 287: 1–6. https://doi.org/10.1016/j.jhazmat.2015.01.039.

Zhang, Xuejian, Junrong Liang, Yong Sun, Fagen Zhang, Chenwei Li, Chun Hu and Lai Lyu. 2020. Mesoporous reduction state cobalt species-doped silica nanospheres: an efficient fenton-like catalyst for dual-pathway degradation of organic pollutants. Journal of Colloid and Interface Science 576: 59–67. https://doi.org/10.1016/j.jcis.2020.05.007.

CHAPTER 12

Environmental Remediation by Novel Nanomaterials and Fungi with High-degradation Capacity of Hazardous Contaminants

Mohd Faheem Khan and *Cormac D. Murphy**

1. Introduction

Continuous advances in the manufacture of chemicals used in myriad applications in multiple sectors result in challenges to the natural environment and the ecosystems it supports. Although the dangers of pollution have been recognized since the start of the industrial revolution, the control and management of pollutants varies across the world. Biological methods have long been acknowledged as the preferred approaches for the removal of pollution over physical (e.g., incineration) and chemical (e.g., UV-treatment) processes. Bioremediation strategies use microorganisms (bacteria, fungi and algae) in *in situ* or *ex situ* modes. *In situ* methods include bioaugmentation, where microorganisms that are capable of degrading a specialized compound are introduced to a contaminated environment, such as soil. *Ex situ* treatment requires that the contaminated material is removed and treated off-site, for example, in a bioreactor.

The application of microorganisms in the remediation of organic pollutants and heavy metals has been explored for years. Microbial enzymes have modified substrate specificities, thus can biotransform xenobiotics that are structurally related to natural substrates. For example, some bacteria belonging to the genera *Pseudomonas* and *Burkholderia,* have evolved a pathway to degrade biphenyl (Bph), which is a naturally occurring compound found after the combustion of lignin (Denef et al., 2004). The enzymes of the pathway can also degrade chlorinated biphenyls (Seeger et al., 1999), which are anthropogenic compounds used earlier in electrical equipment, plasticizers, pigments and fluorinated biphenyls (Hughes et al., 2011). An excellent resource for *in silico* investigations of biodegradation pathways is the EAWAG Biocatalysis/Biodegradation Database (http://eawag-bbd.ethz.ch/index.html), which not only has a repository of biodegradation pathways, but also enables

UCD School of Biomolecular and Biomedical Science, University College Dublin, Belfield, Dublin 4, Ireland.
* Corresponding author: cormac.d.murphy@ucd.ie

a user to predict possible biodegradation pathways of earlier unexplored compounds. However, newer chemicals that are in use and contain highly stable chemical groups are more of a challenge to eliminate alone by biological means. For example, per- and poly-fluorinated alkyl substances (PFAS) are widely used in non-stick coatings, stain repellents and fire-fighting foams, but the presence of many carbon-fluorine bonds makes them biologically recalcitrant. The manufacture and use of PFAS is under close scrutiny and some compounds, such as perfluorooctanoic acid (PFOA) are being phased out. Improved methods of degrading environmentally-damaging pollutants are required, and in this regard the use of nanomaterials has been extensively investigated (Wang et al., 2019), since they have enhanced affinity, higher reactivity, improved selectivity, larger surface area and better disposal capability (Yunus et al., 2012).

In this chapter the role of fungi in the degradation of heavy metal and organic pollutants will be examined together with the role of these microorganisms in the manufacture of important nanomaterials. The combination of nanotechnology and mycoremediation to degrade persistent chemicals will be scrutinized and the potential future research avenues explored.

2. Fungal Degradation of Pollutants

Fungi have been extensively investigated for their abilities to degrade a broad range of xenobiotics and other dangerous pollutants, such as heavy metals, resulting from anthropogenic activity. Fungi can biosorb and/or biotransform pollutants resulting in reduced toxicity, thus have a high potential for remediation applications in contaminated environments. A selection of recent examples of fungal degradation of pollutants is shown in Table 12.1, however the diversity goes beyond this chapter.

Table 12.1. Recent examples of pollutants that are degraded by fungi.

Pollutant Class	Compound	Fungus	Reference
Dye	Turquoise blue	*Coprinus plicatilis*	Akdogan and Topuz (2014)
	Reactive black 5	*Trametes versicolor*	Martinez-Sanchez et al. (2018)
	Malachite green	*Cunninghamella elegans*	Hussain et al. (2017)
Pesticide	Pyrethroid	*Aspergillus* sp.	Kaur and Balomajumder (2020)
	Imidacloprid	*A. terreus*	Mohammed and Badawy (2017)
	Cyhalothrin	*Cunninghamella elegans*	Palmer-Brown et al. (2019)
	Dichlorvos	*Trichoderma atroviride*	Sun et al. (2019)
Pharmaceutical	Piroxicam	*Lentinula edodes*	Muszynska et al. (2019)
	Fluoroquinolones	*Trichoderma* spp.	Manasfi et al. (2020)
	Clofibric acid	*Trametes pubsecens*	Ungureanu et al. (2020)
	Carbamazepine and 17α-ethinylestradiol	*Plerotus* sp. P1	Santos et al. (2012)
Hydrocarbon	Phenanthrene	*Phomopsis liquidambari*	Fu et al. (2018)
	Fluorene	*Pleurotus eryngii* F032	Hadibarata and Kristanti (2014)
	Phenanthrene and pyrene	*Ganoderma lucidum* CCG1	Agrawal et al. (2018)
Polymer	Polyurethane	*Aspergillus tubingensis*	Khan et al. (2017)
	Polyethylene	*Aspergillus* spp.	Saenz et al. (2019)
	Polyethylene	*Penicillium simplicissimum*	Sowmya et al. (2015)
Organohalide	Lindane	*Ganoderma lucidum*	Kaur et al. (2016)
	β-hexachlorohexane	*Penicillium griseofulvus*	Ceci et al. (2015)
	6: 2 Fluorotelomer alcohol	*Phanerochaete chrysosporium*	Merino et al. (2018)
Heavy metal	Cr (VI)	*Cunninghamella elegans*	Hussain et al. (2017)
	Uranium	*Penicillium piscarium*	Coelho et al. (2020)
	Zn (II), Cu (II), Cd (II), Cr (VI) and Ni (II)	*Beauveria bassiana*	Gola et al. (2016)

2.1 Fungal remediation of heavy metals

Fungi have a high tolerance for toxic heavy metals. For example, Srivastava et al. (2011) found five fungal strains in arsenic-contaminated agricultural soils that could tolerate 10 g L^{-1}As (V), including species of *Penicillium*, *Aspergillus* and *Rhizopus*. The mechanisms for metal removal include intracellular compartmentalization in vacuoles (Sharma et al., 2020), metal reduction (Sriharsha et al., 2020), biomineralization (Zhao et al., 2020) and biosorption (Coelho et al., 2020). Immobilized fungi are often applied in the biosorption of metal ions; biomass is usually immobilized on solid supports like TiO_2, SiO_2, carbon nanotubes, Ca-alginate, chitosan microparticles, polymeric nanoparticles, etc., by classical methods such as adsorption, covalent immobilization, entrapment, encapsulation and cross-linking (Velkova et al., 2018). The biomass of *Trichoderma* spp. (*T. viride*, *T. asperellum*, and *T. harzianium*) has been used in both free and immobilized forms for biosorption of several heavy metals from an aqueous solution. For instance, Cr^{6+}, Ni^{2+} and Zn^{2+} were easily removed from an electroplating effluent using Ca-alginate immobilized *T. viride* in a continuous packed-bed column (Kumar et al., 2011). Tan and Ting (2012) used Ca-alginate immobilized heat-inactivated and live biomass of *T. asperellum* for biosorption of Cu^{2+} and found better biosorption with immobilized heat-inactivated cells than live cells. Akhtar et al. (2009) used Ca-alginate immobilized *T. harzianium* for removal of U^{6+} from an aqueous media and recovered up to 98.1–99.3%. Immobilized *Aspergillus niger* biomass on Ca-alginate was applied to the removal of U^{6+} from aqueous solutions in a batch system (Ding et al., 2012). Cu^{2+} can also be removed by immobilized biomass of *Penicillium citrinum*, whereas biosorption of Ni^{2+} and Zn^{2+} was efficiently carried out by thermally deactivated *P. fellutinum* biomass immobilized on sodium bentonite composite (Rashid et al., 2016, Verma et al., 2013). Mahmoud et al. (2015) used the biomass of *Aspergillus ustus*, *F. verticillioides* and *P. funiculosum* immobilized on nanosilica for selective and specific biosorption and removal of Cr^{3+} and Cr^{6+} ions from real wastewater samples.

The composition of fungal biomass impacts on its biosorption characteristics. Tigini et al. (2012) evaluated the ability of *C. elegans* biomass to absorb a range of dye mixtures after culturing the fungus on a media containing different types and quantities of carbon and nitrogen. Biomass from cultures grown in a medium containing starch from potatoes or cereal were shown by Fourier-transform infrared spectroscopy (FTIR) to have high proportions of chitin and chitosan, and were the most effective dye biosorbents. In contrast, biomass from cultures grown on corn steep liquor proportionally contained more protein and was a less effective biosorbent. The polymer chitosan, which is a deacetylated chitin, has proved to be potentiallyuseful in wastewater bioremediation. It is commercially obtained from the shells of crabs and shrimps but can also be obtained from fungi as it is a component of cell walls. Tayel et al. (2016) investigated chitosan isolated from *C. elegans* and determined that it was chemically comparable to that of shrimps. Furthermore, they demonstrated that it effectively removed the metal ions Cu^{2+}, Zn^{2+} and Pb^{2+} from water and had an antimicrobial activity against *E. coli*. Although biosorption of metals is the most predominant application of fungal biomass, there are examples of organic pollutant remediation by biosorption. Luo et al. (2019) used the white rot fungus *Phanerochaete chrysosporium* in the adsorption of the non-steroidal anti-inflammatory drug (NSAID) diclofenac, which is detected in surface waters and sewage treatment plants. The researchers exploited the fungus' tolerance to metal ions and first cultivated it in an Fe-rich medium, to generate iron-rich biomass. This was subsequently used to prepare magnetic biocarbon (Fe/BC) via carbonization and activation with a high specific surface area; the Fe/BC demonstrated high adsorption capacity for diclofenac and could be readily separated from the aqueous solutions using a magnet.

2.2 Fungal enzymes for remediation of organic pollutants

Fungal enzymes, particularly those involved in lignin degradation, are critical factors in a wide variety of xenobiotics that can be transformed by these organisms. Monooxygenases, lignin peroxidases, manganese-dependent peroxidases, versatile peroxidases, laccases and tyrosinases are largely studied for their potential for remediating polluted environments. Figure 12.1 illustrates the fungal pathways and the enzymes involved in the bioremediation of recalcitrant pollutants.

As with other enzyme applications, immobilization enhances important features such as stability, selectivity, ease of enzyme separation, increased reusability, improved storage stability, enhanced thermostability, pH tolerance and organic solvent tolerance (Khan et al., 2019). Immobilization methods include adsorption (enzymes attached to insoluble supports), covalent binding (enzyme and the support matrix that are linked with covalent bonds), entrapment (restricted movement of enzymes in a porous gel), encapsulation (enzyme inside a host semi-permeable membrane or other polymeric materials) and cross-linking (enzyme attachment with each other through ligands) (Sardar and Ahmad 2015). The choice of the immobilization method is crucial to prevent the loss of enzyme activity and maintaining the integrity of the active site. Immobilization of enzymes on nanoparticles (also known as nanozymes) enhances the factors influencing enzyme catalysis, such as increased surface area, improved mass transfer, effective enzyme loading, reduced enzyme unfolding, improved stability and activity (Gupta et al., 2011; Khan et al., 2020).

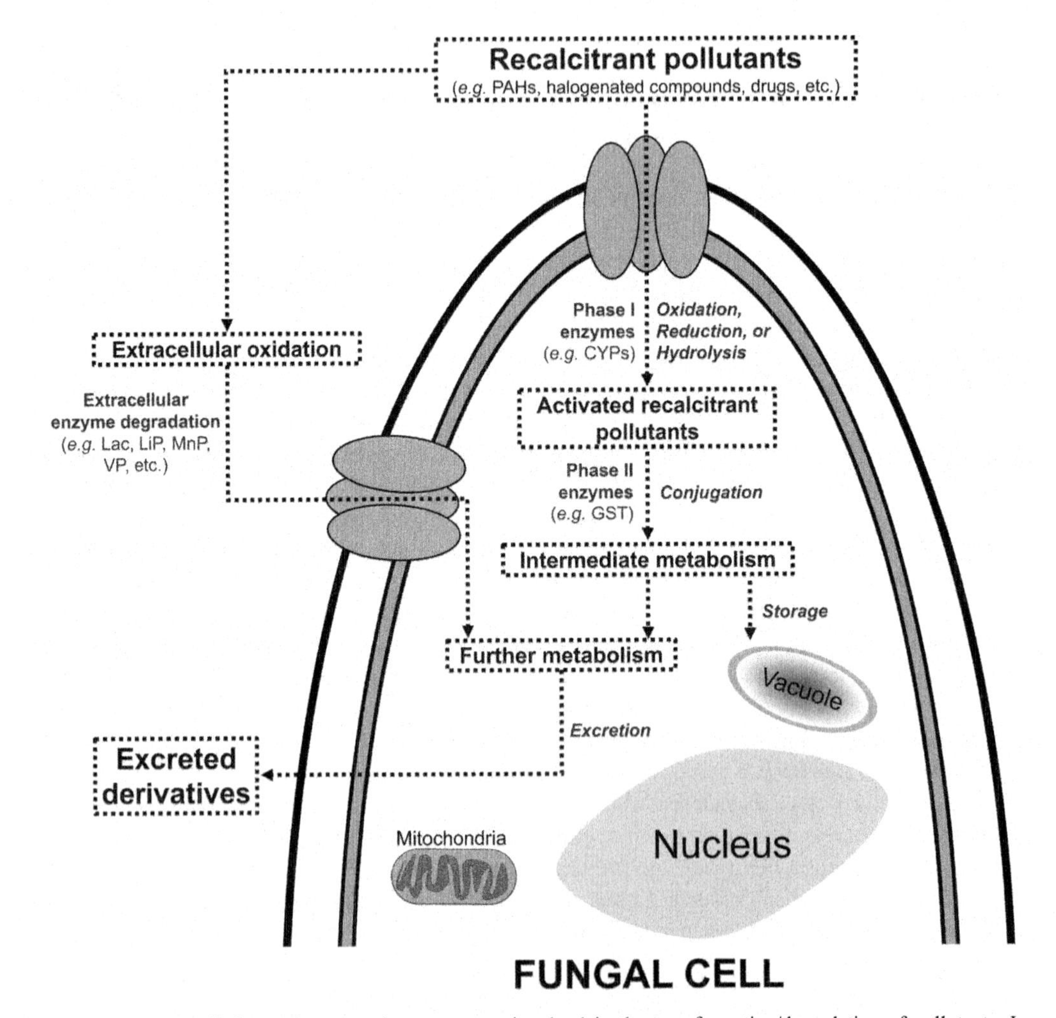

Fig. 12.1. Mycoremediation pathways and enzymes are involved in the transformation/degradation of pollutants. Lac, laccase; LiP, lignin peroxidase; MP, Manganese Peroxidase; VP, Versatile Peroxidase; CYP, cytochrome P450; GST, glutathione S-transferase.

2.2.1 Laccases

Laccases (EC 1.10.3.2) are oxidases that have multi-copper centers in their active sites, which reduce molecular oxygen to water with concurrent oxidation of phenolic compounds such as lignin and humic substances. Over 100 laccases have been identified and isolated from ligninolytic fungal species (*Pleurotus ostreatus*, *Trametes versicolor*, *T. pubescens*, etc.) to date and many are suitable for transformation of toxic xenobiotic compounds (Baldrian, 2006). They have low-substrate specificity thus can oxidize a broad range of substrates including phenols, polyphenols, aromatic amines, phenothiazines, etc. In an earlier study Pozdnyakova et al. (2004) evaluated the degradation efficiency of *Pleurotus ostreatus* laccase for PAH as a substrate and found 95% of anthracene and 14% of fluoranthene were degraded within 2 d at pH 6.0 without the need for redox mediators since laccases have a low redox potential (450–800 mV) compared to ligninolytic peroxidases (> 1 V). Therefore, laccase can transform higher redox potential compounds only when the product is subject to immediate further catalysis or when its redox potential is lowered, for example, by means of chelation (Baldrian, 2006).

In the past decade, immobilized laccases have been investigated with interest as a green biocatalytic tool for environmental remediation due to a broad range in degradation of recalcitrant chemicals that they can transform. For example, Barrios-Estrada et al. (2018) showed 100% degradation of bisphenol A within 24 h using membrane-immobilized laccases from *Pycnoporus sanguineus* and *Trametes versicolor*. Lassouane et al. (2019) reported 99% removal of bisphenol A in only 2 h from an aqueous solution when Ca-alginate-entrapped laccase (from *T. pubescens*) was used. Similarly, Brugnari et al. (2018) used *Pleurotus ostreatus* laccase immobilized on monoaminoethyl-N-aminoethyl (MANAE)-agarose beads that showed complete degradation efficiency of bisphenol A and demonstrated effective reusability for 15 cycles. In another study, Vidal-Limon et al. (2018) synthesized mesoporous nanoparticle-immobilized *Coriollopsis gallica* laccase, which readily detoxified the organochlorine pesticide dichlorophen and concomitantly reduced its cytotoxic effect. Zhang et al. (2017) synthesized dopamine-coated magnetic Fe-NPs for degradation of 4-chlorophenol, which removed 86% of the compound within 2 h. Immobilized laccases are also effective at degrading synthetic dyes and endocrine disrupting compounds. Bayramoglu et al. (2018) revealed 93.0% and 67.5% degradation of Congo Red dye and Bisphenol A, respectively, using laccase immobilized on polymeric microspheres functionalized with cyclic-carbonate in a packed bed reactor. Laccase from *T. harzianum* immobilized in a sol-gel matrix efficiently decolorized malachite green (100%), methylene blue (90%) and Congo Red (60%) after 16, 18 and 20 h, respectively (Bagewadi et al., 2017).

2.2.2 Lignin peroxidase

Lignin peroxidase (EC 1.11.1.14) is a heme-containing extracellular enzyme that oxidatively degrades lignin using H_2O_2. Owing to its high redox potential, the enzyme may be exploited for bioremediation, cosmetic formulation, biorefining, etc. (Falade et al., 2017; Sung et al., 2019). It can oxidize a wide range of environmental pollutants such as phenolic and non-phenolic compounds, xenobiotics and dyes. Notably, the bioremediation of dye effluents from the textile industry is of global importance because of the severe detrimental effects of these toxic dyes on animals. Immobilized lignin peroxidase can effectively degrade/decolorize these harmful dyes in effluents (Bilal et al., 2018). For example, *Ganoderma lucidum* lignin peroxidase immobilized in Ca-alginate beads showed improved decolorization efficiency of Sandal Reactive dyes compared with the free enzyme (Shaheen et al., 2017). Sofia et al. (2016) reported the decolorization of Sandal Fix dyes (95%) by *Schizophyllum commune* lignin peroxidase immobilized on chitosan beads after incubation for 6 h at 30°C. Additionally, the enzyme's kinetic and physicochemical properties were enhanced, and it retained approximately 50% activity after seven cycles, whereas the free enzyme lost all activity after four. Another study showed bisphenol A was degraded up to 90% after incubation with encapsulated lignin peroxidase (with other ligninolytic enzymes) in polyacrylamide hydrogel for 8 h (Gassara et al., 2013).

2.2.3 Manganese peroxidase

Manganese peroxidase (EC 1.11.1.13) is a heme-containing enzyme that can catalyze the oxidation of phenolic compounds and harmful dye pollutants. It is found in lignolytic fungi like *Phanerochaete chrysosporium*, *Anthracophyllum discolor* and *Ganoderma lucidum* where it oxidizes Mn^{2+} to Mn^{3+} in the presence of hydrogen peroxide as part of the lignin-degrading process. Mn peroxidases from *Anthracophyllum discolor* degraded pyrene (> 86%), anthracene (> 65%), fluoranthene (< 15.2%) and phenanthrene (< 8.6%) after immobilization on nanoclay-particles (Acevedo et al., 2010). In another study, Bilal et al. (2016) showed entrapped *Ganoderma lucidum* Mn peroxidases in an agar–agar supporting matrix have improved pH and thermostability and can completely biodegrade and decolorize three synthetic dyes: reactive blue 21, reactive red 195A and reactive yellow 145A.

2.2.4 Tyrosinase

Tyrosinases (EC 1.14.18.1) are distributed across many microorganisms, plants and animals. They are copper-containing enzymes that use oxygen to catalyze monophenolase and diphenolase reactions, thus a phenol such as tyrosine is oxidized to catechol (dopa) in the monophenolase step and to a quinone in the diphenolase step. The edible mushroom *Agaricus bisporus* is an inexpensive and convenient source of tyrosinase that has been examined for its potential in bioremediation (Zaidi et al., 2014). Ba et al. (2018) used co-immobilized mushroom tyrosinase and laccase aggregates on a cross-linking matrix in a hybrid bioreactor for degradation of a large number of waste pharmaceuticals such as acetaminophen, atenolol, bezafibrate, caffeine, carbamazepine, ciprofloxacin, fenofibrate, ibuprofen, indomethacin, ketoprofen, ofloxacin, naproxen, mefenamic acid and trimethoprim. They found complete removal of these waste pharmaceuticals within 5 d in a continuous operation. Mushroom tyrosinase can also remove phenol from wastewater very efficiently when immobilized on modified iron oxide magnetic nanoparticles (up to 78% in 60 min incubation) and on glutaraldehyde activated polyacrylonitrile beads (up to 90% in 12 hr incubation) (Abdollahi et al., 2018; Wu et al., 2017). The same mushroom tyrosinase immobilized on glutaraldehyde activated diatom-biosilica particles degrades phenols (84%) and other phenolic compounds like *p*-cresol (74%) and phenylacetate (90%) (Bayramoglu et al., 2013). Kampmann et al. (2014) developed a batch catalyst system of immobilized *A. bisporus* tyrosinase on silica alginate beads, which showed almost 100% degradation of bisphenol A in 11 repeated cycles.

2.2.5 Cytochrome P450 monooxygenase (CYP)

This class of heme-containing enzyme (EC 1.14.14.1) is found in plants, animals and microorganisms. Fungal CYPs catalyze the biotransformation of a range of xenobiotic compounds, requiring reducing power from NADPH, electrons which are transferred via a separate reductase. CYPs catalyze a range of reactions including hydroxylation, demethylation, epoxidation, N-oxidation and deamination. Fungal CYPs are characteristically membrane-bound (Cresnar and Petric, 2011) and are challenging to investigate *in vitro*, so their involvement in xenobiotic biotransformation is often indirectly inferred from inhibition of the whole cell biocatalysis using CYP-specific inhibitors, such as 1-aminobenzotriazole and piperonyl butoxide. They have been implicated in the biodegradation of a range of pollutants including the insecticide lindane (Xiao and Kondo, 2020), the neonicotinoid acetamiprid (Wang et al., 2019) and PAHs such as benzo[a]pyrene (Ostrem Loss et al., 2019). *Cunninghamella* spp. are known to have extensive CYP activity and can metabolize a range of environmental pollutants such as pesticides including the organophosphorous pesticides diazinon (Zhao et al., 2020) and fenitrothion

(Zhu et al., 2017) and the fungicide cyazofamid (Lee et al., 2016). These fungi are also known for their ability to biotransform drug molecules such as carvediol, bomhexine and etonogestrel (Baydoun et al., 2016; Dube and Kumar, 2017; Zawadzka et al., 2017) yielding metabolites that are analogous to those formed in mammals, thus have the potential for determining the environmental impact of drugs and their metabolites in the environment. Owing to the instability of CYP enzymes and the need for an additional reductase, they are not routinely immobilized in the same manner as the other enzymes described above. However, *Cunninghamella elegans*, which can biotransform a range of xenobiotics (Murphy, 2015), has numerous CYPs (Palmer-Brown et al., 2019) and can grow as a biofilm, which is a natural form of immobilization (Amadio et al., 2013). The biofilm can degrade xenobiotics more efficiently than planktonic cells, for example, Mitra et al. (2013) cultivated *C. elegans* biofilm on a hydrophobic surface and demonstrated that fluoranthene biotransformation was 22-fold greater in the biofilm compared with planktonic cells. The fungal biofilm is also readily reusable, unlike planktonic cells, Hussain et al. (2017) cultivated *C. elegans* in an Erlenmeyer flask containing a steel spring and found that it could simultaneously remove malachite green and Cr (VI) from water; the biofilm could be recycled 19 times without loss of activity. Some fungi form biofilms in combinations with other microbes, and these have been shown to be effective degraders of pollutants. For example, Perera et al. (2019) demonstrated that a mixed community isolated after enrichment culture of soil samples from a municipal landfill in medium containing hexadecane, comprised a fungus (*Aspergillus*) and a bacterium (*Bacillus*). Electron microscopy showed that the microbes existed as a biofilm and by measuring hexadecane concentrations by GC-MS a more effective biodegradation of the substrate was observed in the co-culture (> 99%) compared with the individual isolates (approximately 53% and 10% by fungus and bacteria, respectively). A 'mycoalga' biofilm comprising the fungus *Mucor indicus* and the alga *Chlorella vulgaris* was developed to remove nutrient pollution in aquaculture wastewater. The biofilm removed both nitrogen and phosphorus and was easily harvested to be used as a feed supplement (Barnharst et al., 2018).

3. Nanomaterials

The study of nano-sized substances, dealing with their size, structure-dependent properties and their manipulations for industrial, biomedical and environmental applications, encompass the discipline of 'Nanotechnology'. These nano-scale substances can be nanoparticles (NPs) or nanomaterial (NMs); they have dimensions within the range of 1 nm to 1 μm and a specific surface area/volume ratio greater than 60 m^2 cm^{-3} (Wohlleben et al., 2017). They can be further categorized depending on the number of external dimensions that are in the nanoscale: nanoparticles have three, nanorod/fibers have two and nanoplates have one. Naturally-occurring NPs are ubiquitous in the environment and are formed by biogeochemical processes; these heterogeneous NPs are distinctive from synthetic or engineered NPs (ENPs) that are manufactured for specific purposes. NPs and NMs, often functionalized, are commonly used for remediation applications as they show highly favorable properties towards *in situ* as well as *ex situ* applications. For instance, various metal-based nanomaterials are useful *in situ* and *ex situ* applications for removal of heavy metals and degradation of toxic organic pollutants from ground water, wastewater, soil, sediment or other contaminated environmental sources (Guerra et al., 2018). They can remove contaminants via absorption, adsorption, photocatalysis, filtration or redox reactions (Naghdi et al., 2016). NM, such as nanoscale metal oxides, nanofibers and carbon nanotubes are used as nanoadsorbants (Qu et al., 2013) as they have a high specific surface area, accessible absorption/absorption sites, tunable surface chemistry and are reusable. TiO_2, ZnO_2 and fullerene-based nanomaterials are photocatalytically active and are used in solar disinfection systems (Raizada et al., 2019). Wastewater treatment employs membrane technology using NPs of Ag, Au or TiO_2 that interact with polymeric membranes (Ladner et al., 2012) and halogenated organic compounds, oils and metals are the target pollutants that can be degraded by redox reaction based-catalysis, predominantly involving nanoscale zerovalent iron (nZVI or nanoiron) (Gholami-Shabani et al., 2018).

3.1 Naturally occurring environmental nanoparticles

Naturally occurring nanomaterials include humic acids, fulvic acids, aerosols and volcanic ashes and are ubiquitous in the environment. They are principally formed by decaying processes caused by (micro)biological or chemical changes. Anthropogenic activities, such as combustion, industrialization, land use, deforestation and solid waste generation, contribute to environmental nanoparticles (Lespes et al., 2020).

Natural NPs have applications in the field of biomedicines, cosmetics, agriculture and waste removal. Non-biological naturally occurring inorganic micro-/nanoparticles include calcium carbonate, silicate, silica, alumina, zeolite, basanite, etc. These inorganic NPs have been used for water purification (desalination and defluorination), bioimaging, therapeutics, bone regeneration, etc. (Park et al., 2011; Setiawan et al., 2017). Inorganic NPs are also produced naturally by microorganisms in the environment. For example, Bansal et al. (2005) demonstrated that the fungus *Fusarium oxysporum* produces extracellular protein that hydrolyzes the anionic complexes SiF_6^{2-} and TiF_6^{2-} to produce SiO_2 and TiO_2 NPs, respectively. When microbes such as *Pseudomonas*, *Serratia*, *Thiobacillus* and *Stenotrophomonas* are exposed to Ag^+, Au^{3+}, S^{2-} and SeO_3^{2-} inorganic salts, detoxifying oxidative or reductive pathways are triggered, which results in the formation of NPs (Griffin et al., 2018). These natural nanoparticles formed by microbes are the emerging nanofactories, which may be exploited in environmental cleanup applications (Yadav et al., 2017).

3.2 Synthetic (or engineered) nanoparticles

Synthetic or engineered nanoparticles (ENPs) are synthesized artificially with specific characteristics, such as optical, electronic, magnetic, mechanical, chemical and biological properties, engineered for tailor-made applications. Unlike the naturally occurring nanoparticles that are formed heterogeneously and disseminated in the environment, ENPs are formed as pure powders or suspensions of nanoparticles that are homogeneous in size, shape and structure. The ENPs are classified into four main groups (i) metal-based nanoparticles (such as pure metals like Fe, Ag, Au, Pt and Ti metal-derived compounds like oxides, hydroxides, phosphates, sulfides and halides and metal chalcogenides like quantum dots); (ii) carbonaceous nanomaterials (such as carbon nanotubes, nanofibers, fullerenes (C60), graphene and graphene oxide); (iii) nanopolymers (or polymer with branched units like dendrimers); and (iv) composites (such as nanoclay, nanosponge and other bio-organic/inorganic complex).

4. Green Synthesis of Nanoparticles by Fungi

The broad applications of NPs require effective synthetic methods to meet the demand and there are variations in the sustainability of different approaches. For example, chemical synthesis of NPs requires reducing agents, such as sodium borohydride ($NaBH_4$), which is toxic and generates hydrogen gas, which is hazardous. Instinctively, a more sustainable approach involves biological reducing agents; a life cycle analysis of nZVI synthesis by Martins et al. (2017) concluded that the environmental impacts could be reduced by 50% using a green synthesis approach, and that the chemical synthesis is more expensive. Bioreduction of metals, using plant extracts or microorganisms, to synthesize metal nanoparticles (MNPs) has been intensively investigated (Silva et al., 2016). Fungi from the genera *Aspergillus*, *Fusarium*, *Penicillium*, *Rhizopus*, *Trichoderma* and *Verticillium* have been examined for their ability to synthesize MNP, in particular those of Ag and Au. For instance, a wide range of shape and size morphology of Ag and Au NPs have been obtained by *Penicillium duclauxii* and *Verticillium luteoalbum*, respectively (Almaary et al., 2020; Khandel and Shahi, 2018). The work-flow involved in NP synthesis is illustrated in Fig. 12.2, and involves the addition of the metal salt to either the fungal mycelium directly or a cell free extract. Fungal extracts are preferred for ease of recovery, since intracellularly-located MNPs require a more involved extraction process. The MNPs

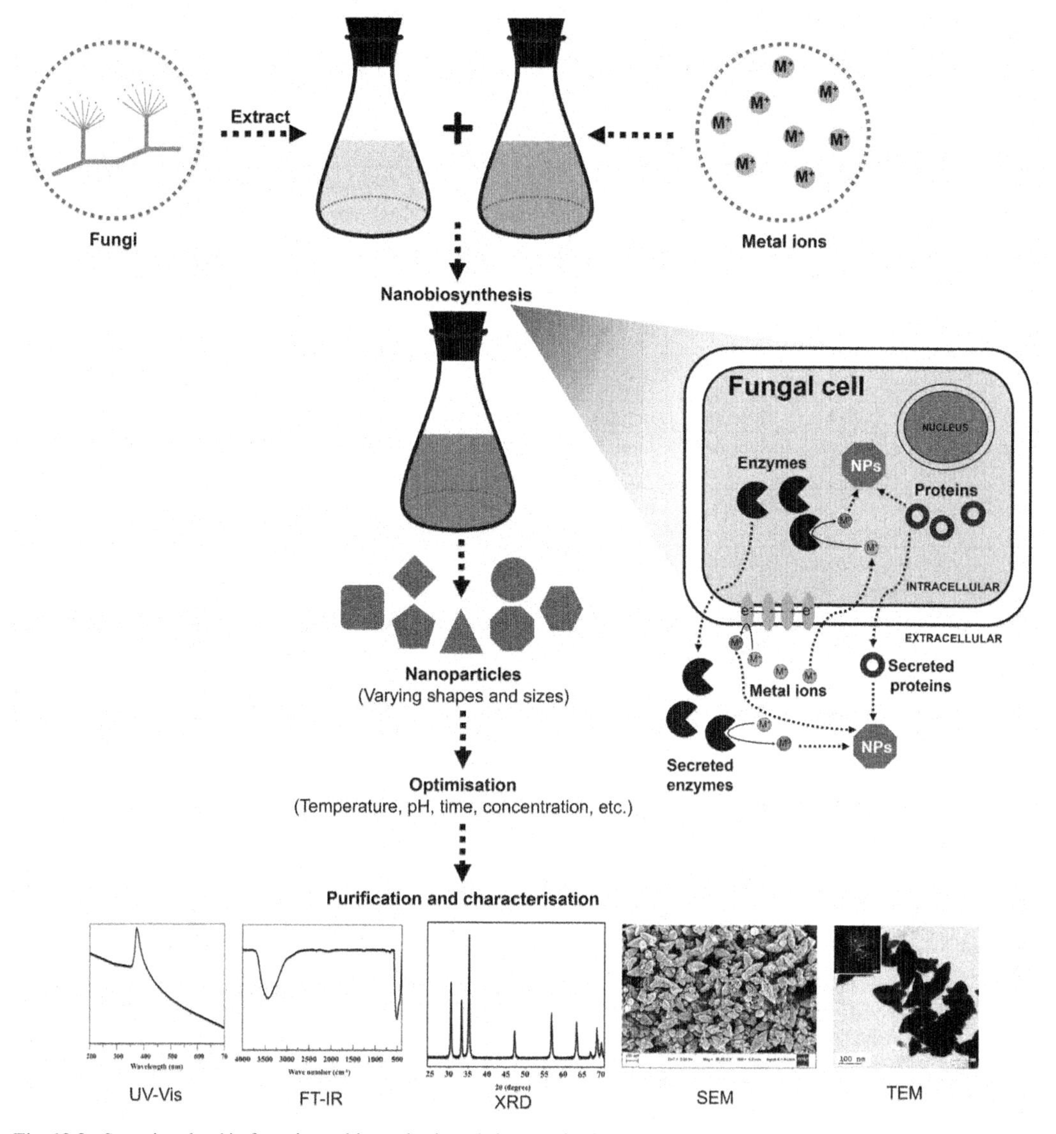

Fig. 12.2. Steps involved in fungal nanobiosynthesis and characterisation. Parts of this figure are reprinted from International Journal of Macromolecules, 136, Khan et al., A strategic approach of enzyme engineering by attribute ranking and enzyme immobilization on zinc oxide nanoparticles to attain thermostability in mesophilic *Bacillus subtilis* lipase for detergent formulation, 66–82, Copyright (2019), with permission from Elsevier.

formed are characterized using a range of techniques, including UV-visible spectrophotometry, electron microscopy, X-Ray Diffraction (XRD), Fourier-Transform Infrared spectroscopy (FTIR) and Dynamic Light Scattering (DLS). Some examples of MNP synthesis are given in Table 12.2, which also highlights their potential applications. There is a large diversity in the morphology of the MNPs that are produced by fungi, and reproducibility is a challenge, since the morphology is affected not only by the species of fungus, but by media composition, metal ion concentration, pH, temperature and incubation time (Silva et al., 2016). For example, Balakumaran et al. (2016) screened extracts from 65 fungal isolates from soil and found that one, later identified as *A. terreus*, produced stable NPs of Au and Ag after 24 h. The conditions of production were optimized with respect to the concentration of biomass, pH and temperature, and it was found that extracts from 10 g biomass/100 ml added to 1 mM metal ion, pH 7 and 30°C gave the most stable MNPs. One major

Table 12.2. Examples of different nanoparticles synthesized using fungi.

Metal	Fungus	Biomass or Extract	Size	Shape	Application	Reference
Ag	*Trichoderma atroviride*	Extract	10–15 nm	Spherical	Antimicrobial	Abdel-Azeem et al. (2020)
	Pleurotus giganteus	Extracts	2–20 nm	Spherical	Antibacterial	Debnath et al. (2019)
	Trichoderma viride	Extract	100–250 nm	Spherical	Antifungal	Manikandaselvi et al. (2020)
	Fomes fomentarius	Extract	Varied	Varied	Antibacterial/ cytotoxic	Rehman et al. (2020)
Au	*Fusarium solani*	Extract	40–45 nm	Needles; flowers	Cytotoxic	Clarance et al. (2020)
	Aspergillus WL-Au	Mycelium	Varied	Rod-shaped	None	Qu et al. (2020)
	Phonopsis XP-8	Mycelium	15 nm	Spherical	Dye degradation	Xu et al. (2019)
CdS	*Aspergillus niger*	Mycelium	5 nm	Spherical	Antimicrobial, cytotoxic	Alsaggaf et al. (2020)
Co_3O_4	*Aspergillus brasiliensis*	Extract	20–27 nm	Quasi-spherical	Antimicrobial	Omran et al. (2020)
Pt	*Neurospora crassa*	Biomass and extract	20–110 nm	Spherical	None	Castro-Longoria et al. (2012)
ZrO_2	*Penicillium* spp.	Extract	< 100 nm	Spherical	Antibacterial	Golnaraghi Ghomi et al. (2019)
Fe_3O_4	*Aspergillus niger* YESM1	Extract	2–16 nm	Spherical	Adsorption of Cr (VI)	Mahanty et al. (2019)
$PbSO_4$	*Aspergillus* spp.	Biomass (periplasmic space)	5–20 nm	Spherical	None	Pavani et al. (2012)
Cu	*Stereum hirsutum*	Extract	5–20 nm	Spherical	None	Cuevas et al. (2015)
NiO	*Hypocrea lixii*	Biomass and extract	1.25–3.8 nm	Spherical	None	Salvadori et al. (2015)
Ti	*Fomes fomentarius*	Extract	Varied	Varied	Antibacterial/ cytotoxic	Rehman et al. (2020)

advantage of using fungi to synthesize MNPs is the ease of recovery, which often requires a simple water wash (Vahabi and Dorcheh, 2014).

4.1 Biosynthesis mechanism

Although there are numerous reports of fungi applied to the synthesis of MNPs, relatively few of the studies include an assessment of the mechanism involved in their biosynthesis. It is generally accepted that a range of biomolecules that fungi produce is involved in the synthesis of MNPs, but the specific molecules are often unidentified. However some research does attempt to shed light on the mechanism and reveals that enzymatic and non-enzymatic routes are involved in metal reduction. One example of the latter is the production of Au-NPs in *Mucor plumbeus* biomass reported by Maliszewska et al. (2017), who showed that by electron microscopy the Ag^{3+} initially absorbed to the mycelial surface, where it is reduced and transported into the cytoplasm. Furthermore, following X-ray Photoelectron Spectroscopy (XPS) analysis of the surface of the mycelium and FTIR analysis of the Au-NPs, it was concluded that the polysaccharide chitosan is probably the main reducing agent. Molnar et al. (2018) investigated the synthesis of Au-NPs by different fractions from cultures of thermophilic fungi: culture supernatant, aqueous mycelial extract and an intracellular fraction. They observed AuNPs synthesized in all fractions. Furthermore, when the culture supernatant of one of the fungi, *Thermoascus thermophilus*, was further fractionated according to molecular size, Au^{3+} reduction was only observed in the fraction containing molecules with a mass < 3 kDa (small molecules), but that capping required larger (> 3 kDa) molecules.

4.1.1 Fungal proteins involved in NP synthesis

Proteins and peptides act as both reducing agents to produce MNPs and stabilizing/capping agents by coating the particles, and fungi are attractive for the synthesis of MNPs as they produce numerous extracellular proteins. The mechanism can be linked to the enzymatic activity or can occur independently of enzyme catalysis. Anil Kumar et al. (2007) used purified nitrate reductase from *Fusarium oxysporum* to synthesize AgNP and demonstrated that the reduction of Ag^{2+} required the enzymatic oxidation of NADPH. The reaction mixture also contained 4-hydroxyquinoline, which acted as an electron shuttle to transfer electrons from the coenzyme to the Ag^{2+} and phytochelatin, which capped the growing nanoparticles. Xylanases from the fungi *A. niger* and *T. longibrachiatum* were used to synthesize spherical and flower-shaped AuNPs that were highly bioactive (antibacterial, anticoagulant and thrombolytic (Elegbede et al., 2020). The same enzymes were used to produce silver-gold alloy nanoparticles (Ag-AuNPs), which in addition to having antibacterial activity, could degrade the dyes malachite green and methylene blue (Elegbede et al., 2019). These syntheses did not require the addition of the enzyme's normal substrate, thus were not reliant on the enzyme's catalytic cycle. Vetchinkina et al. (2017) investigated the synthesis of Au, Ag, Se and Si NPs by the mushrooms *Lentinus edodes*, *P. ostreatus*, *Ganoderma lucidum* and *Grifola frondosa*, using mycelia and purified phenol oxidase enzymes from *L. edodes*. Electron microscopy revealed that MNPs were present both intracellularly and extracellularly. The biosynthesis of AuNPs was studied in more detail by incubation purified intracellular Mn-peroxidase, laccase and tyrosinase with $HAuCl_4$, yielding AuNPs that were spherical (Mn-peroxidase) or irregular spheroids (laccase and tyrosinase). The authors proposed that Au^{3+} reduction occurs at the protophorphyrin IX prosthetic group in Mn-peroxidase and by-passes the catalytic cycle. In the case of laccase and tyrosinase, it was proposed that the reduction of Au^{3+} was indirectly associated with enzyme activity and was catalyzed by H_2O_2 that forms from O_2 in the absence of a phenolic substrate. Faramarzi and Forootanfar (2011) showed that the laccase from *Paraconiothyrium* variabile produced AuNPs more rapidly when inactivated by heat and stated that this might be the result of reducing functional groups being exposed on denaturation.

5. Nanoparticles Applied to Remediation

ENPs have the potential for environmental (bio)remediation (Tripathi et al., 2018), and Table 12.3 summarizes their applications. The factors involved in the selection of the best ENPs to mitigate a particular pollutant in a specific environmental include (a) complete knowledge of applicability of ENPs to be used for remediation, (b) understanding the contaminant to be removed, (c) approachability to the remediation site, (d) amount of ENPs required for efficient remediation, and (e) recycling or recovery of the used ENPs after remediation. The importance of different ENPs in bioremediation applications will be described next.

5.1 Use of metal and metal oxide nanoparticles in remediation

Metallic NPs have several applications such as in biomedical, pharmaceutical, biosensing, bioremediation, etc., and can be synthesized either from pure metals (Fe, Mg, Ca, Mn, Ag, Zn, Au, Pt and Ti,) or from their derived compounds like oxides, hydroxides, phosphates, sulfides, halides, etc., (Elizabeth et al., 2019). Nanoiron or nanoscale Zero Valent Iron (nZVI) is the most predominant metallic nanoparticle used to remove inorganic and organic pollutants from the environment, owing to its excellent magnetic and catalytic properties. The benefits of using nZVI for environmental remediation include: cost-effectiveness, safety in comparison to other chemical oxidants, longevity (3–15 years), sustainability, reusability, reduced number of toxic products, it can complement

Table 12.3. Use of different nanoparticles in remediation (Guerra et al., 2018; Tripathi et al., 2018).

Type	Remediation Process Involved	Examples	Size	Application in Bioremediation
Metal and metal oxides	Adsorption and redox reaction	Zero valent iron NPs (nZVI)	50–200 nm	Detoxification of pesticides, herbicides, organochlorine, polychlorinated biphenyls, etc.
		Silver and gold NPs	10–20 nm	Water disinfection and kill pathogens
		Titanium oxide (TiO_2)	~ 5 nm	Removal of dye and organic waste from textile effluent, water disinfection and kill pathogenic microbes
		Zinc oxide (ZnO)	~ 20 nm	Removal of dyes from waste water
		Bimetallic NPs	20–40 nm	Remove and degrade chlorinated and brominated contaminants from water and soil
Carbon-based nanomaterials	Absorption, adsorption, membranes filtration, antimicrobial agents, and environmental sensors	Carbon nanotubes	1–30 μm	Use in water purification and filtration systems, sorption of organophosphates and heavy metals, degrade PFAs and herbicides, removes electronics, computers, plastics waste, etc.
		Fullerenes	~ 0.72 nm	Absorbs and adsorbs organic and organometallic compounds
Nanopolymers	Sorption, filtration encapsulation, chemical and biological catalysis	Chitosan NPs	10–1000 nm	Biodegradation of organic compounds (Bisphenol A; 2,6-Dimethoxyphenol; organic dyes, etc.) and industrial waste water treatment
		CLEAs	1–100 μm	Organic dye degradation, degradation of acetaminophen and other organic drugs
		Poly (ethylene) glycol modified urethane acrylate (PMUA)	~ 80 nm	Sorption of hydrophobic organic contaminants and increases solubility of PAHs
		Dendrimer	1.5–10 nm	Degrades organic compounds and arsenic pollutants
Other nanoparticles	Sorption, chemical catalysis	Nanoclay	Up to 1000 nm	Sorption of PAHs (pyrene, anthracene, fluoranthene, phenanthrene)
		Nanosponge (highly cross-linked cyclodextrin-based polymers)	250–1000 nm	Removal of Triclopyr (3,5,6-trichloro-2-pyridinyloxyacetic acid)
		Anion-SAMMS (Self-Assembled Monolayer on Mesoporous Supports), Thiol-SAMMS and HOPO-SAMMS	2–20 nm	Remove inorganic ions, heavy metal ions, actinides and lanthanides

bioremediation and is not impacted by Soil Oxidant Demand (SOD) (Grieger et al., 2019). Examples of the application of nZVI include the detoxification of toxic metals such as As (III) and As (V) (Kanel et al., 2006). In an earlier study Ponder et al. (2000) used modified nZVI in "ferragels" that can readily separate and immobilize toxic Cr (VI) and Pb (II) from water; the metals are reduced to Cr (III) and Pb (0) while iron is oxidized goethite (α-FeOOH). A large number of organic pollutants can be remediated with nanoiron, in particular halogenated compounds that are rapidly remediated by adsorption or reductive dehalogenation (Gholami-Shabani et al., 2018). For example, Cao et al. (2017) investigated the removal of the halogenated (chlorinated and fluorinated) antibiotic florfenicol by sulfidized-nZVI and observed dechlorination and defluorination. Interestingly, no reaction occurred with unsulfidized nZVI. In contrast, Lawal and Choi (2018) used the bimetallic Pd/nZVI to remove perfluoroactanoic acid, but did not observe fluoride ions, implying that removal was by adsorption. Metal oxides NPs, such as iron oxide (Fe_3O_4) NPs, have the potential for wastewater remediation because of their superparamagnetic properties, strong adsorption capacity and chemical inertness. He et al. (2017) loaded Fe_3O_4/biochar nanocomposites with photosynthetic bacteria for wastewater treatments and showed that the magnetic property enables easy recovery for repeated use with efficient biodegradation of hazardous compounds in contaminated water. Likewise, magnetic Fe_3O_4/CeO_2 nanocomposite was considered as a good heterogeneous Fenton-like catalyst for degradation of 4-chlorophenol (Xu and Wang, 2012). In an interesting turn to the application of nZVI, Li et al. (2020) found that the fungus *Neurospora crassa* precipitated iron carbonate when cultured in medium containing Fe^{2+} and urea. On heating to 900°C a composite of nZVI/fungal biomass was formed that was coralline in appearance. The 'iron coral' proved to be more effective at removing CCl_4 from an aqueous solution than either carbonized fungal biomass or nZVI alone.

Other metal/metal oxide nanoparticles that have been studied for toxic metal removal include MnO, ZnO and MgO. Manganese oxide (MnO) nanoparticles have a high surface area, polymorphic structure, contain nanoporous or nanotunnel manganese oxides and hydrous manganese oxide and adsorb heavy metal ions on their external surface followed by intraparticle diffusion (Zhang et al., 2018). They are used to eliminate toxic metals such as As (V), As (III), Pb (II), Cd (II) and Zn (II) from contaminated water samples (Gholami-Shabani et al., 2018). Zinc oxide (ZnO) nanoparticles also have the advantage of a large surface area, chemo- and thermostability and antimicrobial properties. One potential application is to remove heavy metals from wastewater (Co^{2+}, Ni^{2+}, Cu^{2+}, Cd^{2+}, As^{3+}, Pb^{2+} and Hg^{2+}) owing to their microporous characteristics, which display high adsorption affinity (Tyagi et al., 2017). Similarly, magnesium oxide (MgO) nanoparticles can eliminate heavy metals from polluted water such as Pb (II) and Cd (II) (Kumar et al., 2013). Yang et al. (2017) reported that UVC-activated TiO_2 NPs can enhance biodegradation of heavy hydrocarbons in soils impacted by crude oil spills. Their results suggest that adding ultraviolet radiation C (UVC)-activated TiO_2 NPs to soil slurries can transform heavy hydrocarbons into biodegradable byproducts. Ag NP-based materials and bimetallic Au/Ag NPs have been used to treat dyes such as crystal violet; rhodamine B, methylene blue, methyl orange and other nitro-derivative dyes (Fiorati et al., 2020).

5.2 Use of carbon-based nanomaterials in remediation

Carbon containing nanomaterials can be synthesized in diverse morphologies such as hollow tubes (nanotubes), ellipsoids or spheres (e.g., fullerenes (C60)) or sheets (graphene/graphene oxide). These carbon-based nanomaterials are being explored for the elimination of metals and organics from soils and groundwater (Wang et al., 2013). Bina et al. (2012) used multi-walled, single-walled and hybrid carbon nanotubes (CNTs) for absorption of ethylbenzene from an aqueous solution and found single-walled CNTs (99.5%) performed better than multi-walled (91.7%) and hybrid CNTs (97.6%). However, multi-walled CNTs are more efficient in removing Ni^{2+} ions and cationic dyes from an aqueous solution (Gong et al., 2009; Kandah and Meunier, 2007). The nanocomposite of CNTs and calcium alginate removes up to 69.9% Cu^{2+} even at acidic pH (Li et al., 2010). Other heavy metal ions like Cd^{2+}, Pb^{2+}, As^{3+} Cr^{6+} and Hg^{2+} are also removed by CNTs from an aqueous media (Gupta et al., 2015). Graphene oxide rapidly adsorbs both cationic dyes (like crystal violet, methylene blue, rhodamine B, malachite green, etc.) and anionic dyes (acid orange 8, direct red 23, alizarin red S, etc.), from water (Baby et al., 2019). Zhang et al. (2018) used modified graphene oxide functionalized with 4-sulfophenylazo for heavy metal remediation from an aqueous media and found higher adsorption capacity for Pb^{2+} (689 mg g^{-1}), Cd^{2+} (267 mg g^{-1}), and Cr^{3+} (191 mg/g) than Cu^{2+} (59 mg g^{-1}) and Ni^{2+} (66 mg g^{-1}). Due to their high biocompatible and physicochemical properties, carbon-based materials have great potential for their use in environment remediation and water purification.

5.3 Use of nanopolymers in remediation

Synthetic nanopolymers like dendrimers (PAMAM) or poly (amidoamine) contain functional groups like amines, carboxylates, hydroxamates, etc. They are chelating agents that are used in ultrafiltration to bind metal ions and enable encapsulation of a wide-range of cations like Ag^+, Au^+, Cu^{2+}, Ni^{2+}, Fe^{2+}, Fe^{3+}, Zn^{2+} and U^{6+} in water purification systems. Thus, dendrimer-based nanomaterials have been used to develop ultrafiltration and microfiltration processes for the recovery of dissolved ions from water (Guerra et al., 2018). Similarly, the bioactive polymer chitosan is also a promising material to absorb heavy metal (Pb^{2+} and Cu^{2+} up to a concentration of 100–300 ppm) from an aqueous media and soil matrices. Fungal mycelia of *Cunninghamella elegans* produces chitosan, which can be treated with sodium tripolyphosphate for the synthesis of chitosan nanoparticles that have the potential in eliminating heavy metal contaminants from soil (Alsharari et al., 2018). Cross-Linked Enzyme Aggregates (CLEA) offer advantages of catalytic properties, stability, reusability of enzymes, non-toxic, cheap and easy synthesis. CLEAs are successfully used for decolorization and detoxification of dyes (malachite green, Remazol brilliant blue R and Reactive Black, using laccase-CLEAs nanoclusters), elimination of endocrine-disrupting chemicals (bisphenol A, nonylphenol, triclosan, etc.,) and bioconversion of agro-industrial waste (lignocellulosic biomass) (Yamaguchi et al., 2018). A variety of enzymes are satisfactorily tested with the CLEA immobilization method and thus, are establishing their potential as an upcoming tool to optimize biocatalytic processes in environmental pollution control (Velasco-Lozano et al., 2016).

5.4 Other nanoparticles in bioremediation

Natural nanoclays are a good source of nanomaterial for sorption or dispersion of toxic substances, but require engineering to modify their physicochemical properties to enable them to be used for remediation. Modified nanoclays, such as heat-activated nanoclay, acid-activated nanoclay or alkali-activated nanoclay, nanoclay saturated with metal ions like Ca^{2+}, Na^+ or Mg^{2+} and Fe-based redox-modified clay, were found to be nontoxic or less harmful when applied directly to the environment. These nanoclays have successfully removed organic and inorganic pollutants, such as heavy metal (e.g., cadmium, arsenic, lead), PAHs, oil and pesticides from water, soil and sediments (Biswas et al., 2019).

5.5 Nanobioremediation

One potential limitation of using metal nanoparticles in the removal of pollutants is that alone they will not mineralize organic compounds. Microorganisms can completely degrade pollutants, but the degradation is typically slow. The concept of combining rapid degradation of organic pollutants by nanoparticles with microorganisms that will mineralize more easily with metabolizable intermediates has been investigated. For example, Aroclor 1248 is a mixture of polychlorinated biphenyl (PCB) congeners that are recalcitrant in the environment owing to a high degree of chlorination. Le et al. (2015) used Pd/nFe bimetallic nanoparticles to dechlorinate Aroclor 1248 and the bacterium *Burkholderia xenovorans* LB400 to biodegrade the resulting biphenyl. Reductive dechlorinationof the PCB congeners is achieved by H_2 production from the nZVI and atomic H produced at the Pd surface, with 89% of the chlorine removed over 6 d; the bacterium can degrade biphenyl with the production of benzoic acid. The authors confirmed the reduction in Aroclor toxicity by exposing *E. coli* to the compound before and after nanobiotreatment and measuring the amount of reactive oxygen species, bacterial numbers and glutathione peroxidase activity. A later investigation by Horvathova et al. (2019) examined both a 'bionano' and a 'nanobio' approach to PCB degradation, i.e., microbial treatment followed by nZVI or *vice versa.* In this study Delor 103 (a mixture of di- to hexa-chlorobiphenyls) was incubated with nZVI and three different bacterial strains (*Alcaligenes xylosoxidans*, *Stenotrophomonas maltophilia* and *Ochrobactrum anthropi*) which were isolated from a sediment close to a former PCB manufacturer. The nanobio-treatment proved to be more effective at removing PCBs in both artificially contaminated minimal medium and in typically contaminated sediment. The addition of triton X-100 to the experiments in sediment also improved the degree of biodegradation, possibly by solubilizing PCBs that were sequestered in the sediment particles.

Fungal laccase has been used in combination with nanoparticles to enable the degradation of environmental pollutants. For instance, Bokare et al. (2010) used Pd/nZVI in combination with laccase from *T. versicolor* to degrade triclosan. The nanoparticles dechlorinated the substrate yielding 2-phenoxyphenol, which was subsequently polymerized to a non-toxic product by the enzyme. A similar approach was used by Dai et al. (2015), who investigated the degradation of the halogenated phenolic compounds triclosan, tetrabromobisphenol A and 2-bromo-4-fluorophenol using a combination of trimetallic reduction with [Fe|Ni|Cu] nanoparticles and *T. versicolor* laccase. The nanoparticles allowed the dehalogenation of the substrates and the enzyme catalyzed the oxidation of the resulting phenols yielding quinones, hydroquinones and corresponding radicals, which spontaneously polymerized. Dai et al., also designed a plug-flow reactor using the same sequence of trimetallic reduction and laccase, both immobilized on sponge, to treat simulated wastewater containing the three halophenolics. Over 95% transformation efficiency was observed, and the nanoparticles and laccase retained their activities for 256 hr, suggesting the strong potential for practical wastewater treatment.

5.6 Nanobiosensors for detection of pollutants

One challenge for monitoring pollution in different environments is the detection of myriad compounds. Although high performance instrumentation, such as LC- and GC-MS, is routinely used in many countries, there are limitations owing to the cost and expertise. Other cost-efficient and portable detection methods are required and in this case biosensors are tools that can also be exploited for detection of the pollutants (such as pesticides, herbicides, PAHs, heavy metals, pharmaceuticals, toxic intermediates, etc.,) that are released into the environment by means of industrial effluents, agricultural wastes or domestic wastes. A biosensor is usually comprised of three components: (i) detector, (ii) transducer; (iii) signal processing system. The detector identifies the stimulus, then the transducer converts it into a useful output and finally a signal processing system involves amplification of received output followed by its display (Perumal and Hashim, 2014). Nanobiosensors consist of a nanomaterial-based platform, which is directly associated with a biological entity for the detection of a compound. The biological entity can be cells or enzymes attached to an electrode (Čvančarová et al., 2020). Nanomaterials with a biological entity can detect environmental pollutants with high sensitivity and specificity. Thus, nanofabrication is an important consideration for the desired and specific manufacturing of a sensing device for detecting a metabolite. Like other biosensors, nanobiosensors rely on the principle of electrochemistry, which uses redox reactions and are categorized as potentiometric biosensor (measures oxidation or reduction potential), conductometric nanobiosensor (measures changes in the ionic strength) and amperometric nanobiosensor (measures current as a function of time) (Perumal and Hashim, 2014).

Fungal and yeast cells have been used as whole-cell biosensors for the detection of many target metabolites such as endocrine disruptors (Chamas et al., 2017). On the other hand, fungal enzymes like laccase and tyrosinase have been investigated for assimilating various phenolic compounds. Table 12.4 lists examples of pollutants that can be detected using fungal enzyme-based nanobiosensors. Biocatalyst nanocomposites used in nanobiosensors showed high enzyme loading, improved catalytic activity and stability with high specificity and selectivity of sensing metabolite compared to the corresponding whole cell. A range of applications utilizing nanobiocatalysts in a single sensing device are these days in demand which can observe the different pollutants equally well in a single operation resulting in energy, cost and saving time.

Laccase, particularly from *T. versicolor*, is used as it is relatively stable and catalyzes electron transfer reactions in the absence of a cofactor. Work by Liu et al. (2006) demonstrated that laccase

Table 12.4. Examples of fungal enzyme-based nanobiosensors for pollutant detection.

Pollutant for Detection	Immobilized Enzyme	Nanomaterial used for Sensing	Reference
Bisphenol A	Mushroom tyrosinase	Multi-walled carbon nanotubes	Yin et al. (2010)
		Nanographene	Wu et al. (2012)
		Sol–gel TiO_2 modified with multi-walled carbon nanotubes, polycationic polymer and Nafion	Kochana et al. (2015)
		Graphene-Au nanocomposite	Pan et al. (2015)
		Diazonium-functionalized boron doped diamond electrode modified with multi-walled carbon nanotubes	Zehani et al. (2015)
Catechol	Mushroom tyrosinase	Chitosan-carbon coated nickel nanoparticles	Yang et al. (2012)
	Laccase from *Trametes versicolor*	Electrospun Cu/carbon composite nanofibers	Fu et al. (2014)
		Au nanoparticles crosslinked zein ultrafine fibers	Chen et al. (2015)
		Graphene oxide/palladium–copper alloyed nanocages	Mei et al. (2015)
Catechol and hydroquinone	Laccase from *Trametes versicolor*	Multi-walled carbon nanotubes	Qu et al. (2013)
	Mushroom tyrosinase	Mesoporous carbon nitride	Zhou et al. (2014)
Pyrocatechol	Laccase from *Trametes versicolor*	Osmium tetroxide and multi walled carbon nanotubes	Das et al. (2014)
Phenolic compounds	Mushroom tyrosinase	Liposome bioreactor and chitosan nanocomposite	Guan et al. (2013)
		Graphitized ordered mesoporous carbon and cobaltosic oxide nanorod	Wang et al. (2014)
	Laccase from *Trametes versicolor*	Graphene oxide-glycol chitosan nanohybrid	Boujakhrout et al. (2016)

entrapped in a carbon nanotube-chitosan composite had improved access to substrates such as ABTS (2,2′-azino-bis (3-ethylbenzothiazoline-6-sulfonic acid)) and catechol. The resulting sensor had a limit of detection for catechol of 0.66 μM. More recently Othman and Wollenberger (2020)immobilized a chemically modified laccase from *T. hirsute* on screen printed carbon electrodes with carboxy-functionalized multi-walled carbon nanotubes that was more sensitive for catechols, retained 77% of its response after 10 successive cycles and did not lose its activity after 20 d storage. Laccase immobilized on a nanocomposite of graphene oxide/ferrous-ferric oxide nanoparticles and chitosan were used for the electrochemical detection of bisphenol A (limit of detection = 18 nM) and can be stored for 1 mon (Fernandes et al., 2020). 4-Chlorophenol was detected using laccase immobilized on a nanocomposite of ZnO and chitosan with a limit of detection of 0.7 μM (Mendes et al., 2017).

Mushroom tyrosinase can oxidize the endocrine-disrupting chemical bisphenol A in the presence of oxygen, yielding o-diquinone, thus has been applied to the development of a biosensor for this important environmental pollutant. Wu et al. (2020) designed a biosensor of tyrosinase nanocapsules (nTyr) that contained single enzyme molecules surrounded by a polyacrylamide shell. The nTyr was reached in two steps: protein acrylolation anchored the polymerizable groups to the surface of the enzyme and initiation of polymerization in an aqueous solution containing the monomers and cross-linker (Fig. 12.3). The resulting biosensor had high electrocatalytic activity with enhanced stability (i.e., reduced dissociation) and had a limit of detection of 12 nM.

5.7 Nanoparticle toxicity

The use of NPs or NMs in various industrial sectors results in their distribution in the environment, which can have an adverse impact on microbes, plants, animals and humans. Three aspects are required when assessing the environmental impacts of nanomaterials: (i) nanomaterial mobility in the environment (for example, from soil to drinking water or plants or food); (ii) nanomaterial toxicity

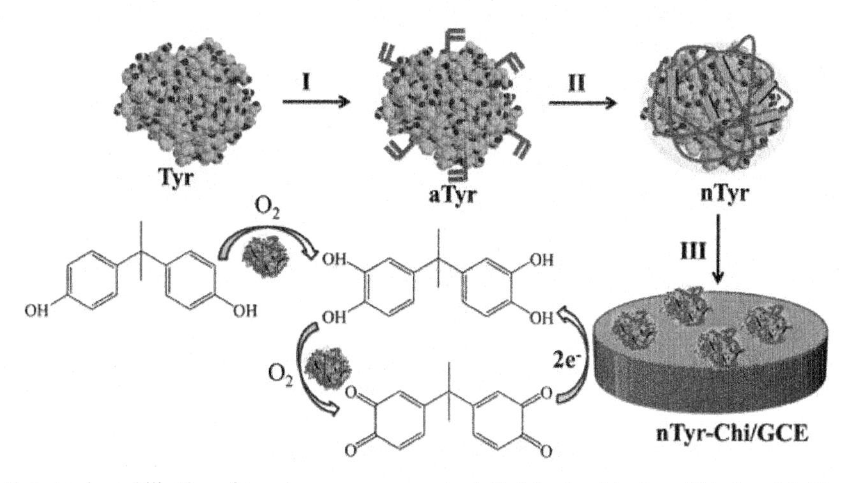

Fig. 12.3. Two step immobilization of tyrosinase as nanocapsules (nTyr) for the detection of bisphenol A. The text can be seen for a description of the procedure. Reprinted from Biosensors and Bioelectronics, 165, Wu et al. (2020) Tyrosinase nanocapsule based nano-biosensor for ultrasensitive and rapid detection of bisphenol A with excellent stability in different application scenarios, Pages 112407, Copyright (2020), with permission from Elsevier.

to the organisms living in soils, water and sediments; and (iii) extent of nanomaterial exposure and their mechanisms involved (Ray et al., 2009).

Nanoparticles can cause oxidative stress resulting in the production of reactive oxygen and nitrogen species (ROS/RNS) which lead to damaging membrane integrity, cytoskeletal damage, mitochondrial damage, ER and Golgi stress, proteolysis, lipid peroxidation, nucleic acid defragmentation, interruption of energy transduction, release of harmful and toxic components, etc. (Klaine et al., 2008). Figure 12.4A illustrates different cytotoxicities caused by nanoparticles. To gain access to the different subcellular compartments (like mitochondria, nucleus, ER, Golgi bodies, peroxisomes, etc.), nanoparticles enter the cells through endocytic pathways and connect to the endolysosomal network using motor proteins and cytoskeleton and cause cytotoxicity (Shang et al., 2014).

Compared with animal cells, plants, algae and fungi cells have cell walls as the primary site or barrier for NPs interaction; however, their mechanisms of nanotoxicity are poorly understood (Navarro et al., 2008). Cell walls have pores which provide permeability (sieving properties) for NPs to enter into the cells followed by their internalization (Ovecka et al., 2005). The toxicity of NPs towards bacteria has been studied extensively with conflicting results: species such as *E. coli*, *S. aureus* and *Dehalococcoides* are inhibited by NPs, whereas growth of methanogens and sulfate-reducing bacteria is stimulated (Cecchin et al., 2017). Fungi appear to show a much higher tolerance to nZVI toxicity than bacteria (Lefevre et al., 2016), for example, Diao and Yao (2009) demonstrated that 10 mg/mL nZVI completely inactivated *Bacillus subtilis* and *Pseudomonas aeruginosa*, whereas *Aspergillus versicolor* was unaffected. Nevertheless, NPs can have negative effects on fungi: Otero-González et al. (2013) showed that oxygen uptake by *S. cerevisiae* was not impacted by TiO_2 NPs at concentrations of 1000 mg/L, but Mn_2O_3 NPs at the same concentration caused the oxygen uptake to fall nearly 70%. Shah et al. (2010) studied the impact of nZVI and Cu NPs on lignocellulolytic enzymes production by fungus *Trametes versicolor.* The production of Mn-peroxidase, cellobiohydrolase, β-xylosidase and β-glucosidase was decreased by both NPs while laccase was not influenced by Fe NPs, but was affected by Cu NPs when incubated for 24 hr.

The physicochemical properties of nanoparticles that impact their toxicity are summarized in Fig. 12.4B. The nanomaterial size can interfere with biological functions depending on its ability to enter the host (cellular uptake), the effectiveness of endocytic processing and the physiological responses in the cell (Li et al., 2015). However, there is no clear relationship between NP size and toxicity. Park et al. (2011) compared the toxic effects of variable sized Ag NPs on L929 fibroblasts in terms of genotoxicity, cytotoxicity and systemic toxicity and reported that 20 nm NPs were more

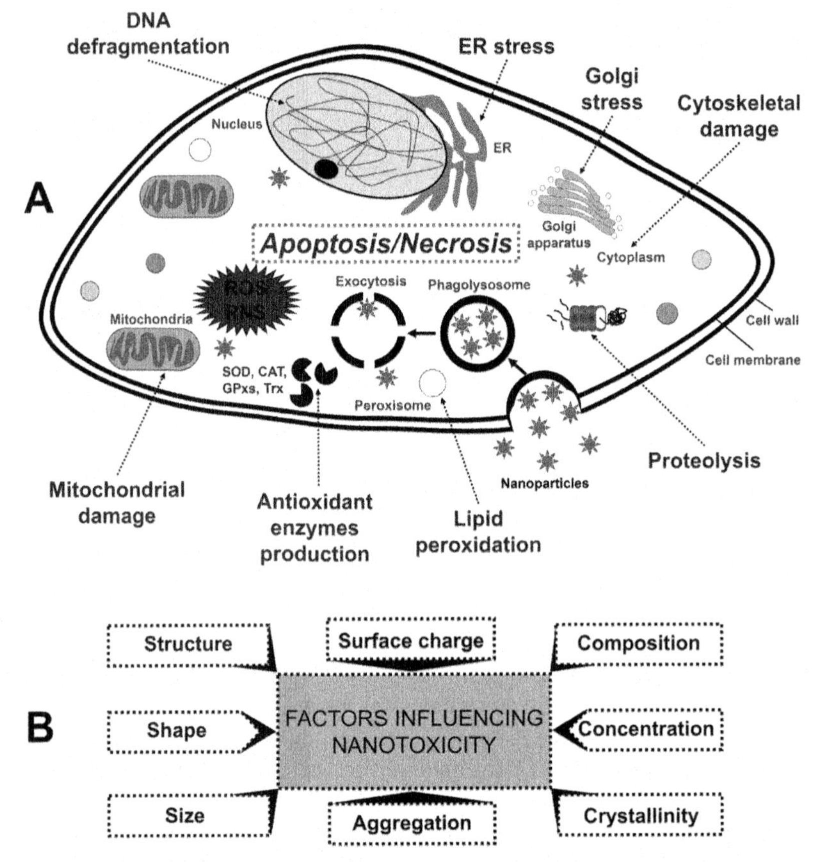

Fig. 12.4. (A) Cytotoxicity caused by nanoparticles. (B) Different factors influencing nanotoxicity.

toxic than the larger NPs. In contrast, Yin et al. (2005) examined the *in vitro* effects of nickel ferrite NPs on a Neuro-2A cell line and the data inferred cytotoxicity was independent of nanoparticle size.

Nanomaterials are found in a number of shapes and structures,such as nanospheres, nanorods, nanofibers, nanosheets, etc., and the differences in structure influences toxicity. Klingenfuss (2014) reported that TiO_2 NPs have different types, sizes, structures and crystallinities and the variation in these physicochemical properties can potentially affect their bioavailability to soil microorganisms, such as rhizobial bacteria and mycorrhizal fungi. Their major impact is on wrapping of NPs during phagocytosis. It has been observed that nanorods or nanofibers are relatively less toxic than spherical nanoparticles because of slow endocytic processing (Gatoo et al., 2014). Another property that influences toxicity is surface charge as this impacts on the adsorption, binding to surface proteins and transmembrane permeability. Reports show that positively charged NPs are more toxic than negatively charged and neutral NPs, thus Si NPs that are positively charged owing to surface amino groups ($NP–NH_2$) are more cytotoxic than neutral ($NP-N_3$) and negatively charged (NP-COOH) types (Bhattacharjee et al., 2010). Additionally, nanoparticle functionalization (or surface modification) can alter NPs bioavailability and toxicity against mycorrhiza/rhizobia. For example, Burke et al. (2015) observed carboxylic acid-functionalized negatively charged Fe_3O_4 NPs showed an inhibitory effect on nodulation in soybean, whereas amine-functionalized positively charged Fe_3O_4 NPs improved nodulation. Similarly, polyvinyl pyrrolidone functionalized Ag NPs (PVP-Ag NPs) reduced arbuscular mycorrhizal fungal colonization than Ag_2S NPs applied in the soil at same rate 100 mg kg^{-1} (Judy et al., 2015). Nanoparticle solubility is also an important factor, which is controlled by chemical composition and crystallinity. Griffitt et al. (2008) reported that soluble forms of Ag NPs and Cu NPs caused toxicity in aquatic organisms, whereas TiO_2 NPs of same size and shape did not show any cytotoxicity issues, which also emphasizes the role of chemical composition on NPs' toxicity. With regard to crystallinity of NPs, Gurr et al. (2005) observed that in the absence of light rutile TiO_2 NPs have a crystalline structure that induced various oxidative damages to bronchial epithelial cells, whereas amorphous anatase TiO_2 NPs did not. Likewise, nanoparticle aggregates of fullerenes (C_{60}) induce oxidative responses through the generation of superoxide anions and lipid peroxidation that leads to cytotoxicity (Nel et al., 2006). The most crucial factor that determines the toxicities caused by nanoparticles is concentration. Zhang et al. (2010) compared the toxicity of graphene and carbon nanotubes on neural phaeochromocytoma-derived PC12 cells and observed both the nanomaterials induce shape- and concentration-dependent cytotoxicity. Nanoparticles like Ag, TiO_2, and ZnO NPs are known for antibacterial and fungal activity in the soil microflora and high concentrations can deform and damage fungal hyphae and bacterial cells (Kumar et al., 2011; Shinde, 2015). However, some of the reported responses by microorganisms to NPs are contrary, for example, Feng et al. (2013) investigated the effect of Ag and FeO NPs on mycorrhizal clover and observed a concentration-dependent inhibition with FeONPs, while with AuNPs there was inhibition at 0.01 mg/kg but not at concentrations over 0.1 mg/kg. Sarabia-Castillo and Fernández-Luqueño (2016) studied the effect on nodulation by exposure of TiO_2, ZnO, and Fe_2O_3 NPs on *Rhizobium leguminosarum-Pisum sativum* L. symbiosis at same concentration (6 g L^{-1}) and exposure time (35 d). Their results revealed Fe_2O_3 NPs do not affect nodulation whereas ZnO NPs and TiO_2 NPs negatively affect the nodule development. Overall, the toxic effects of NPs are complex and context dependent.

6. Conclusions, Future Directions and Final Remarks

In this chapter the ability of fungi and fungal enzymes to biotransform/biodegrade xenobiotics was explored, and the application of fungi in the biosynthesis of a range of nanomaterials described. Given the global importance of both areas of research, it is likely that in the near future studies will continue in them. The combination of fungi and nanotechnology has found success in relation to biosensors, but there is less literature regarding nano-fungal applications in environmental clean-up. This would appear to be an area that could be further exploited, since fungi are effective pollutant degraders, can make NPs under mild conditions, and are less impacted than bacteria from the toxic effects of NMs. The emergence of pollutants that are recalcitrant is of immediate concern and the application of fungi and nanoparticles to their remediation is highly relevant. The challenges faced with difficult-to-treat chemicals are likely to be successfully overcome with a combined approach, using several technologies and mixed, rather than pure, cultures.

Acknowledgement

We thank the Irish Research Council for the provision of a Government of Ireland Postdoctoral Fellowship to MFK (IRC/GOI-PD/1064).

References

Abdel-Azeem, A., Nada, A.A., O'Donovan, A., Thakur, V.K. and El Kelish, A. 2020. Mycogenic silver nanoparticles from endophytic *Trichoderma atroviride* with antimicrobial activity. J. Renew. Mater. 8: 171–185. DOI: 10.32604/jrm.2020.08960.

Abdollahi, K., Yazdani, F., Panahi, R. and Mokhtarani, B. 2018. Biotransformation of phenol in synthetic wastewater using the functionalized magnetic nano-biocatalyst particles carrying tyrosinase. 3 Biotech. 8: 419. DOI: 10.1007/s13205-018-1445-2.

Acevedo, F., Pizzul, L., Castillo, M.D., Gonzalez, M.E., Cea, M., Gianfreda, L. and Diez, M.C. 2010. Degradation of polycyclic aromatic hydrocarbons by free and nanoclay-immobilized manganese peroxidase from *Anthracophyllum discolor*. Chemosphere 80: 271–278. DOI: 10.1016/j.chemosphere.2010.04.022.

Agrawal, N., Verma, P. and Shahi, S.K. 2018. Degradation of polycyclic aromatic hydrocarbons (phenanthrene and pyrene) by the ligninolytic fungi *Ganoderma lucidum* isolated from the hardwood stump. Bioresour Bioprocess. 5. DOI: 10.1186/s40643-018-0197-5.

Akdogan, H.A. and Topuz, M.C. 2014. Decolorization of turquoise blue HFG by *Coprinus plicatilis* for water bioremediation. Bioremediat. J. 18: 287–294. DOI: 10.1080/10889868.2014.933171.

Akhtar, K., Khalid, A.M., Akhtar, M.W. and Ghauri, M.A. 2009. Removal and recovery of uranium from aqueous solutions by Ca-alginate immobilized *Trichoderma harzianum*. Bioresour. Technol. 100: 4551–4558. DOI: 10.1016/j.biortech.2009.03.073.

Almaary, K.S., Sayed, S.R.M., Abd-Elkader, O.H., Dawoud, T.M., El Orabi, N.F. and Elgorban, A.M. 2020. Complete green synthesis of silver-nanoparticles applying seed-borne *Penicillium duclauxii*. Saudi. J. Biol. Sci. 27: 1333–1339. DOI: 10.1016/j.sjbs.2019.12.022.

Alsaggaf, M.S., Elbaz, A.F., El Badawy, S. and Moussa, S.H. 2020. Anticancer and antibacterial activity of cadmium sulfide nanoparticles by *Aspergillus niger*. Adv. Polym. Technol. 2020. DOI: 10.1155/2020/4909054.

Alsharari, S.F., Tayel, A.A. and Moussa, S.H. 2018. Soil emendation with nano-fungal chitosan for heavy metals biosorption. Int. J. Biol. Macromol. 118: 2265–2268. DOI: 10.1016/j.ijbiomac.2018.07.103.

Amadio, J., Casey, E. and Murphy, C.D. 2013. Filamentous fungal biofilm for production of human drug metabolites. Appl. Microbiol. Biotechnol. 97: 5955–5963. DOI: 10.1007/s00253-013-4833-x.

Anil Kumar, S., Abyaneh, M.K., Gosavi, S.W., Kulkarni, S.K., Pasricha, R., Ahmad, A. and Khan, M.I. 2007. Nitrate reductase-mediated synthesis of silver nanoparticles from $AgNO_3$. Biotechnol. Lett. 29: 439–445. DOI: 10.1007/s10529-006-9256-7.

Ba, S., Haroune, L., Soumano, L., Bellenger, J.P., Jones, J.P. and Cabana, H. 2018. A hybrid bioreactor based on insolubilized tyrosinase and laccase catalysis and microfiltration membrane remove pharmaceuticals from wastewater. Chemosphere 201: 749–755. DOI: 10.1016/j.chemosphere.2018.03.022.

Baby, R., Saifullah, B. and Hussein, M.Z. 2019. Carbon nanomaterials for the treatment of heavy metal-contaminated water and environmental remediation. Nanoscale Res. Lett. 14: 341. DOI: 10.1186/s11671-019-3167-8.

Bagewadi, Z.K., Mulla, S.I. and Ninnekar, H.Z. 2017. Purification and immobilization of laccase from *Trichoderma harzianum* strain HZN10 and its application in dye decolorization. J. Genet. Eng. Biotechnol. 15: 139–150. DOI: 10.1016/j.jgeb.2017.01.007.

Balakumaran, M.D., Ramachandran, R., Balashanmugam, P., Mukeshkumar, D.J. and Kalaichelvan, P.T. 2016. Mycosynthesis of silver and gold nanoparticles: Optimization, characterization and antimicrobial activity against human pathogens. Microbiol. Res. 182: 8–20. DOI: 10.1016/j.micres.2015.09.009.

Baldrian, P. 2006. Fungal laccases—occurrence and properties. FEMS Microbiol Rev. 30: 215–242. DOI: 10.1111/j.1574-4976.2005.00010.x.

Bansal, V., Rautaray, D., Bharde, A., Ahire, K., Sanyal, A., Ahmad, A. and Sastry, M. 2005. Fungus-mediated biosynthesis of silica and titania particles. J. Mater. Chem. 15: 2583–2589. DOI: 10.1039/b503008k.

Barnharst, T., Rajendran, A. and Hu, B. 2018. Bioremediation of synthetic intensive aquaculture wastewater by a novel feed-grade composite biofilm. Int. Biodeterior. Biodegradation 126: 131–142. DOI: 10.1016/j.ibiod.2017.10.007.

Barrios-Estrada, C., Rostro-Alanis, M.J., Parra, A.L., Belleville, M.P., Sanchez-Marcano, J., Iqbal, H.M.N. and Parra-Saldivar, R. 2018. Potentialities of active membranes with immobilized laccase for Bisphenol A degradation. Int. J. Biol. Macromol. 108: 837–844. DOI: 10.1016/j.ijbiomac.2017.10.177.

Baydoun, E,, Wahab, A.T., Shoaib, N., Ahmad, M.S., Abdel-Massih, R., Smith, C., Naveed, N. and Choudhary, M.I. 2016. Microbial transformation of contraceptive drug etonogestrel into new metabolites with *Cunninghamella blakesleeana* and *Cunninghamella echinulata*. Steroids 115: 56–61. DOI: 10.1016/j.steroids.2016.08.003.

Bayramoglu, G., Akbulut, A. and Arica, M.Y. 2013. Immobilization of tyrosinase on modified diatom biosilica: Enzymatic removal of phenolic compounds from aqueous solution. J. Hazard. Mater. 244–245: 528–536. DOI: 10.1016/j.jhazmat.2012.10.041.

Bayramoglu, G., Karagoz, B. and Arica MY. 2018. Cyclic-carbonate functionalized polymer brushes on polymeric microspheres: Immobilized laccase for degradation of endocrine disturbing compounds. J. Ind. Eng. Chem. 60: 407–417. DOI: 10.1016/j.jiec.2017.11.028.

Bhattacharjee, S., de Haan, L.H., Evers, N.M., Jiang, X., Marcelis, A.T., Zuilhof, H., Rietjens, I.M. and Alink, G.M. 2010. Role of surface charge and oxidative stress in cytotoxicity of organic monolayer-coated silicon nanoparticles towards macrophage NR8383 cells. Part. Fibre Toxicol. 7: 25. DOI: 10.1186/1743-8977-7-25.

Bilal, M., Asgher, M., Shahid, M. and Bhatti, H.N. 2016. Characteristic features and dye degrading capability of agar-agar gel immobilized manganese peroxidase. Int. J. Biol. Macromol. 86: 728–740. DOI: 10.1016/j.ijbiomac.2016.02.014.

Bilal, M., Rasheed, T., Iqbal, H.M.N. and Yan, Y. 2018. Peroxidases-assisted removal of environmentally-related hazardous pollutants with reference to the reaction mechanisms of industrial dyes. Sci. Total Environ. 644: 1–13. DOI: 10.1016/j.scitotenv.2018.06.274.

Bina, B., Pourzamani, H., Rashidi, A. and Amin, M.M. 2012. Ethylbenzene removal by carbon nanotubes from aqueous solution. J. Environ. Public. Health 2012: 817187. DOI: 10.1155/2012/817187.

Biswas, B., Warr, L.N., Hilder, E.F., Goswami, N., Rahman, M.M., Churchman, J.G., Vasilev, K., Pan, G. and Naidu, R. 2019. Biocompatible functionalisation of nanoclays for improved environmental remediation. Chem. Soc. Rev. 48: 3740–3770. DOI: 10.1039/c8cs01019f.

Bokare, V., Murugesan, K., Kim, Y.M., Jeon, J.R., Kim, E.J. and Chang, Y.S. 2010. Degradation of triclosan by an integrated nano-bio redox process. Bioresour. Technol. 101: 6354–6360. DOI: 10.1016/j.biortech.2010.03.062.

Boujakhrout, A., Jimenez-Falcao, S., Martinez-Ruiz, P., Sanchez, A., Diez, P., Pingarron, J.M. and Villalonga, R. 2016. Novel reduced graphene oxide-glycol chitosan nanohybrid for the assembly of an amperometric enzyme biosensor for phenols. Analyst 141: 4162–4169. DOI: 10.1039/c5an02640g.

Brugnari, T., Pereira, M.G., Bubna, G.A., de Freitas, E.N., Contato, A.G., Correa, R.C.G., Castoldi, R., de Souza, C.G.M., Polizeli, M., Bracht, A. et al. 2018. A highly reusable MANAE-agarose-immobilized *Pleurotus ostreatus* laccase for degradation of bisphenol A. Sci. Total Environ. 634: 1346–1351. DOI: 10.1016/j.scitotenv.2018.04.051.

Burke, D.J., Pietrasiak, N., Situ, S.F., Abenojar, E.C., Porche, M., Kraj, P., Lakliang, Y. and Samia, A.C. 2015. Iron oxide and titanium dioxide nanoparticle effects on plant performance and root associated microbes. Int. J. Mol. Sci. 16: 23630–23650. DOI: 10.3390/ijms161023630.

Cao, Z., Liu, X., Xu, J., Zhang, J., Yang, Y., Zhou, J., Xu, X. and Lowry, G.V. 2017. Removal of antibiotic florfenicol by sulfide-modified nanoscale zero-valent iron. Environ. Sci. Technol. 51: 11269–11277. DOI: 10.1021/acs.est.7b02480.

Castro-Longoria, E., Moreno-Velazquez, S.D., Vilchis-Nestor, A.R., Arenas-Berumen, E. and Avalos-Borja, M. 2012. Production of platinum nanoparticles and nanoaggregates using *Neurospora crassa*. J. Microbiol. Biotechnol. 22: 1000–1004. DOI: 10.4014/jmb.1110.10085.

Cecchin, I., Reddy, K.R., Thome, A., Tessaro, E.F. and Schnaid, F. 2017. Nanobioremediation: Integration of nanoparticles and bioremediation for sustainable remediation of chlorinated organic contaminants in soils. Int. Biodeterior. Biodegradation 119: 419–428. DOI: 10.1016/j.ibiod.2016.09.027.

Ceci, A., Pierro, L., Riccardi, C., Pinzari, F., Maggi, O., Persiani, A.M., Gadd, G.M. and Petrangeli, Papini, M. 2015. Biotransformation of beta-hexachlorocyclohexane by the saprotrophic soil fungus *Penicillium griseofulvum*. Chemosphere 137: 101–107. DOI: 10.1016/j.chemosphere.2015.05.074.

Chamas, A., Pham, H.T.M., Baronian, K. and Kunze, G. 2017. Biosensors based on yeast/fungal cells. pp. 351–371. *In*: Sibirny, A. (ed.). Biotechnology of Yeasts and Filamentous Fungi. Springer. DOI: 10.1007/978-3-319-58829-2_12.

Chen, X.D., Li, D.W., Li, G.H., Luo, L., Ullah, N., Wei, Q.F. and Huang, F.L. 2015. Facile fabrication of gold nanoparticle on zein ultrafine fibers and their application for catechol biosensor. Appl. Surf. Sci. 328: 444–452. DOI: 10.1016/j.apsusc.2014.12.070.

Clarance, P., Luvankar, B., Sales, J., Khusro, A., Agastian, P., Tack, J.C., Al Khulaifi, M.M., Al-Shwaiman, H.A., Elgorban, A.M., Syed, A. et al. 2020. Green synthesis and characterization of gold nanoparticles using endophytic fungi *Fusarium solani* and its *in vitro* anticancer and biomedical applications. Saudi. J. Biol. Sci. 27: 706–712. DOI: 10.1016/j.sjbs.2019.12.026.

Coelho, E., Reis, T.A., Cotrim, M., Rizzutto, M. and Correa, B. 2020. Bioremediation of water contaminated with uranium using *Penicillium piscarium*. Biotechnol. Prog. 36: e30322. DOI: 10.1002/btpr.3032.

Cresnar, B. and Petric, S. 2011. Cytochrome P450 enzymes in the fungal kingdom. Biochim Biophys Acta. 1814: 29–35. DOI: 10.1016/j.bbapap.2010.06.020.

Cuevas, R., Duran, N., Diez, M.C., Tortella, G.R. and Rubilar, O. 2015. Extracellular biosynthesis of copper and copper oxide nanoparticles by *Stereum hirsutum*, a native white-rot fungus from Chilean forests. J. Nanomater. 2015. DOI: 10.1155/2015/789089.

Čvančarová, M., Shahgaldian, P. and Corvini, P.F.-X. 2020. Enzyme-based nanomaterials in bioremediation. pp. 345–372. *In*: Filip, J., Cajthaml, T., Najmanová, P., Černík, M. and Zbořil, R. (eds.). Advanced Nano-Bio Technologies for Water and Soil Treatment. Springer. DOI: 10.1007/978-3-030–29840-1_16.

Da, Y., Song, Y., Wang, S. and Yuan, Y. 2015. Treatment of halogenated phenolic compounds by sequential tri-metal reduction and laccase-catalytic oxidation. Water Res. 71: 64–73. DOI: 10.1016/j.watres.2014.12.047.

Das, P., Barbora, L., Das, M. and Goswami, P. 2014. Highly sensitive and stable laccase based amperometric biosensor developed on nano-composite matrix for detecting pyrocatechol in environmental samples. Sens. Actuators B. Chem. 192: 737–744. DOI: 10.1016/j.snb.2013.11.021.

Debnath, G., Das, P. and Saha, A.K. 2019. Green synthesis of silver nanoparticles using mushroom extract of *Pleurotus giganteus*: Characterization, antimicrobial, and alpha-amylase inhibitory activity. Bionanoscience 9: 611–619. DOI: 10.1007/s12668-019-00650-y.

Denef, V.J., Park, J., Tsoi, T.V., Rouillard, J.M., Zhang, H., Wibbenmeyer, J.A., Verstraete, W., Gulari, E., Hashsham, S.A. and Tiedje, J.M. 2004. Biphenyl and benzoate metabolism in a genomic context: Outlining genome-wide metabolic networks in *Burkholderia xenovorans* LB400. Appl. Environ. Microbiol. 70: 4961–4970. DOI: 10.1128/AEM.70.8.4961-4970.2004.

Diao, M.H. and Yao, M.S. 2009. Use of zero-valent iron nanoparticles in inactivating microbes. Water Res. 43: 5243–5251. DOI: 10.1016/j.watres.2009.08.051.

Ding, D.X., Tan, X., Hu, N., Li, G.Y., Wang, Y.D. and Tan, Y. 2012. Removal and recovery of uranium (VI) from aqueous solutions by immobilized *Aspergillus niger* powder beads. Bioprocess. Biosyst. Eng. 35: 1567–1576. DOI: 10.1007/s00449-012-0747-8.

Dube, A.K. and Kumar, M.S. 2017. Biotransformation of bromhexine by *Cunninghamella elegans*, *C. echinulata* and *C. blakesleeana*. Braz. J. Microbiol. 48: 259–267. DOI: 10.1016/j.bjm.2016.11.003.

Elegbede, J.A., Lateef, A., Azeez, M.A., Asafa, T.B., Yekeen, T.A., Oladipo, I.C., Aina, D.A., Beukes, L.S. and Gueguim-Kana E.B. 2020. Biofabrication of gold nanoparticles using xylanases through valorization of corncob by *Aspergillus niger* and *Trichoderma longibrachiatum*: Antimicrobial, antioxidant, anticoagulant and thrombolytic activities. Waste and Biomass Valorization 11: 781–791. DOI: 10.1007/s12649-018-0540-2.

Elegbede, J.A., Lateef, A., Azeez, M.A., Asafa, T.B., Yekeen, T.A., Oladipo, I.C., Hakeem, A.S., Beukes, L.S. and Gueguim-Kana, E.B. 2019. Silver-gold alloy nanoparticles biofabricated by fungal xylanases exhibited potent biomedical and catalytic activities. Biotechnol. Prog. 35: e2829. DOI: 10.1002/btpr.2829.

Elizabeth, P.-S., Néstor, M.-M. and David, Q.-G. 2019. Nanoparticles as dental drug-delivery systems. pp. 567–593. *In*: Subramani, K. and Ahmed, W. (eds.). Nanobiomaterials in Clinical Dentistry. Elsevier. DOI: 10.1016/B978-0-12-815886-9.00023-1.

Falade, A.O., Nwodo, U.U., Iweriebor, B.C., Green, E., Mabinya, L.V. and Okoh, A.I. 2017. Lignin peroxidase functionalities and prospective applications. Microbiologyopen 6: e00394. DOI: 10.1002/mbo3.394.

Faramarzi, M.A. and Forootanfar, H. 2011. Biosynthesis and characterization of gold nanoparticles produced by laccase from *Paraconiothyrium variabile*. Colloids Surf. B. Biointerfaces 87: 23–27. DOI: 10.1016/j.colsurfb.2011.04.022.

Feng, Y., Cui, X., He, S., Dong, G., Chen, M., Wang, J. and Lin, X. 2013. The role of metal nanoparticles in influencing arbuscular mycorrhizal fungi effects on plant growth. Environ. Sci. Technol. 47: 9496–9504. DOI: 10.1021/es402109n.

Fernandes, P.M.V., Campina, J.M. and Silva, A.F. 2020. A layered nanocomposite of laccase, chitosan, and Fe3O4 nanoparticles-reduced graphene oxide for the nanomolar electrochemical detection of bisphenol A. Mikrochim. Acta 187: 262. DOI: 10.1007/s00604-020-4223-x.

Fiorati, A., Bellingeri, A., Punta, C., Corsi, I. and Venditti, I. 2020. Silver nanoparticles for water pollution monitoring and treatments: Ecosafety challenge and cellulose-based hybrids solution. Polymers 12: 1635. DOI: 10.3390/polym12081635.

Fu, J., Qiao, H., Li, D., Luo, L., Chen, K. and Wei, Q. 2014. Laccase biosensor based on electrospun copper/carbon composite nanofibers for catechol detection. Sensors 14: 3543–3556. DOI: 10.3390/s140203543.

Fu, W., Xu, M., Sun, K., Hu, L., Cao, W., Dai, C. and Jia, Y. 2018. Biodegradation of phenanthrene by endophytic fungus *Phomopsis liquidambari in vitro* and *in vivo*. Chemosphere 203: 160–169. DOI: 10.1016/j.chemosphere.2018.03.164.

Gassara, F., Brar, S.K., Verma, M. and Tyagi, R.D. 2013. Bisphenol A degradation in water by ligninolytic enzymes. Chemosphere 92: 1356–1360. DOI: 10.1016/j.chemosphere.2013.02.071.

Gatoo, M.A., Naseem, S., Arfat, M.Y., Dar, A.M., Qasim, K. and Zubair, S. 2014. Physicochemical properties of nanomaterials: Implication in associated toxic manifestations. Biomed. Res. Int. 2014: 498420. DOI: 10.1155/2014/498420.

Gholami-Shabani, M., Gholami-Shabani, Z., Shams-Ghahfarokhi, M. and Razzaghi-Abyaneh, M. 2018. Application of nanotechnology in mycoremediation: Current status and future prospects. pp. 89–116. *In*: Prasad, R., Kumar, V., Kumar, M. and Wang, S. (eds.). Fungal Nanobionics: Principles and Applications. Springer. DOI: 10.1007/978-981-10-8666-3_4.

Gola, D., Dey, P., Bhattacharya, A., Mishra, A., Malik, A., Namburath, M. and Ahammad, S.Z. 2016. Multiple heavy metal removal using an entomopathogenic fungi *Beauveria bassiana*. Bioresour. Technol. 218: 388–396. DOI: 10.1016/j.biortech.2016.06.096.

Golnaraghi Ghomi, A.R., Mohammadi-Khanaposhti, M., Vahidi, H., Kobarfard, F., Ameri Shah Reza, M. and Barabadi, H. 2019. Fungus-mediated extracellular biosynthesis and characterization of zirconium nanoparticles using standard *Penicillium* species and their preliminary bactericidal potential: A novel biological approach to nanoparticle synthesis. Iran. J. Pharm. Res. 18: 2101–2110. DOI: 10.22037/ijpr.2019.112382.13722.

Gong, J.L., Wang, B., Zeng, G.M., Yang, C.P., Niu, C.G., Niu, Q.Y., Zhou, W.J. and Liang, Y. 2009. Removal of cationic dyes from aqueous solution using magnetic multi-wall carbon nanotube nanocomposite as adsorbent. J. Hazard. Mater. 164: 1517–1522. DOI: 10.1016/j.jhazmat.2008.09.072.

Grieger, K., Hjorth, R., Carpenter, A.W., Klaessig, F., Lefevre, E., Gunsch, C., Soratana, K., Landis, A.E., Wickson, F. and Hristozov, D. 2019. Sustainable environmental remediation using NZVI by managing benefit-risk trade-offs. pp. 511–562. *In*: Phenrat, T. and Lowry, G. (eds.). Nanoscale Zerovalent Iron Particles for Environmental Restoration. Springer. DOI: 10.1007/978-3-319-95340-3_15.

Griffin, S., Masood, M.I., Nasim, M.J., Sarfraz, M., Ebokaiwe, A.P., Schafer, K.H., Keck, C.M. and Jacob, C. 2018. Natural nanoparticles: A particular matter inspired by nature. Antioxidants 7: 3. DOI: 10.3390/antiox7010003.

Griffitt, R.J., Luo, J., Gao, J., Bonzongo, J.C. and Barber, D.S. 2008. Effects of particle composition and species on toxicity of metallic nanomaterials in aquatic organisms. Environ. Toxicol. Chem. 27: 1972–1978. DOI: 10.1897/08-002.1.

Guan, H.N., Liu, X.F. and Wang, W. 2013. Encapsulation of tyrosinase within liposome bioreactors for developing an amperometric phenolic compounds biosensor. J. Solid. State. Electr. 17: 2887–2893. DOI: 10.1007/s10008-013-2181–5.

Guerra, F.D., Attia, M.F., Whitehead, D.C. and Alexis, F. 2018. Nanotechnology for environmental remediation: Materials and applications. Molecules 23: 1760. DOI: 10.3390/molecules23071760.

Gupta, M.N., Kaloti, M., Kapoor, M. and Solanki, K. 2011. Nanomaterials as matrices for enzyme immobilization. Artif. Cell. Blood. Sub. 39: 98–109. DOI: 10.3109/10731199.2010.516259.

Gupta, S., Bhatiya, D. and Murthy, C.N. 2015. Metal removal studies by composite membrane of polysulfone and functionalized single-walled carbon nanotubes. Sep. Sci. Technol. 50: 421–429. DOI: 10.1080/01496395.2014.973516.

Gurr, J.R., Wang, A.S.S., Chen, C.H. and Jan, K.Y. 2005. Ultrafine titanium dioxide particles in the absence of photoactivation can induce oxidative damage to human bronchial epithelial cells. Toxicology 213: 66–73. DOI: 10.1016/j.tox.2005.05.007.

Hadibarata, T. and Kristanti, R.A. 2014. Potential of a white-rot fungus *Pleurotus eryngii* F032 for degradation and transformation of fluorene. Fungal. Biol. 118: 222–227. DOI: 10.1016/j.funbio.2013.11.013.

He, S., Zhong, L., Duan, J., Feng, Y., Yang, B. and Yang, L. 2017. Bioremediation of wastewater by iron oxide-biochar nanocomposites loaded with photosynthetic bacteria. Front. Microbiol. 8: 823. DOI: 10.3389/fmicb.2017.00823.

Horvathova, H., Laszlova, K. and Dercova, K. 2019. Bioremediation vs. nanoremediation: Degradation of polychlorinated biphenyls (PCBs) using integrated remediation approaches. Water Air Soil Pollut. 230. DOI: 10.1007/s11270-019-4259-x.

Hughes, D., Clark, B.R. and Murphy, C.D. 2011. Biodegradation of polyfluorinated biphenyl in bacteria. Biodegradation 22: 741–749. DOI: 10.1007/s10532-010-9411-7.

Hussain, S., Quinn, L., Li, J.J., Casey, E. and Murphy, C.D. 2017. Simultaneous removal of malachite green and hexavalent chromium by *Cunninghamella elegans* biofilm in a semi-continuous system. Int. Biodeterior. Biodegradation 125: 142–149. DOI: 10.1016/j.ibiod.2017.09.003.

Judy, J.D., Kirby, J.K., Creamer, C., McLaughlin, M.J., Fiebiger, C., Wright, C., Cavagnaro, T.R. and Bertsch, P.M. 2015. Effects of silver sulfide nanomaterials on mycorrhizal colonization of tomato plants and soil microbial communities in biosolid-amended soil. Environ. Pollut. 206: 256–263. DOI: 10.1016/j.envpol.2015.07.002.

Kampmann, M., Boll, S., Kossuch, J., Bielecki, J., Uhl, S., Kleiner, B. and Wichmann, R. 2014. Efficient immobilization of mushroom tyrosinase utilizing whole cells from *Agaricus bisporus* and its application for degradation of bisphenol A. Water. Res. 57: 295–303. DOI: 10.1016/j.watres.2014.03.054.

Kandah, M.I. and Meunier, J.L. 2007. Removal of nickel ions from water by multi-walled carbon nanotubes. J. Hazard. Mater. 146: 283–288. DOI: 10.1016/j.jhazmat.2006.12.019.

Kanel, S.R., Greneche, J.-M. and Choi, H. 2006. Arsenic (V) removal from groundwater using nano scale zero-valent iron as a colloidal reactive barrier material. Environ. Sci. Technol. 40: 2045–2050.

Kaur, H., Kapoor, S. and Kaur, G. 2016. Application of ligninolytic potentials of a white-rot fungus *Ganoderma lucidum* for degradation of lindane. Environ. Monit. Assess. 188: 588. DOI: 10.1007/s10661-016-5606-7.

Kaur, P. and Balomajumder, C. 2020. Effective mycoremediation coupled with bioaugmentation studies: An advanced study on newly isolated *Aspergillus* sp. in Type-II pyrethroid-contaminated soil. Environ. Pollut. 261: 114073. DOI: 10.1016/j.envpol.2020.114073.

Khan, M.F., Kundu, D., Gogoi, M., Shrestha, A.K., Karanth, N.G. and Patra, S. 2020. Enzyme-responsive and enzyme immobilized nanoplatforms for therapeutic delivery: An overview of research innovations and biomedical applications. pp. 165–200. *In*: Yata, V., Ranjan, S., Dasgupta, N. and Lichtfouse, E. (eds.). Nanopharmaceuticals: Principles and Applications. Springer Vol 3. DOI: 10.1007/978-3-030-47120-0_6.

Khan, M.F., Kundu, D., Hazra, C. and Patra, S. 2019. A strategic approach of enzyme engineering by attribute ranking and enzyme immobilization on zinc oxide nanoparticles to attain thermostability in mesophilic *Bacillus subtilis* lipase for detergent formulation. Int. J. Biol. Macromol. 136: 66–82. DOI: 10.1016/j.ijbiomac.2019.06.042.

Khan, S., Nadir, S., Shah, Z.U., Shah, A.A., Karunarathna, S.C., Xu, J., Khan, A., Munir, S. and Hasan, F. 2017. Biodegradation of polyester polyurethane by *Aspergillus tubingensis*. Environ. Pollut. 225: 469–480. DOI: 10.1016/j.envpol.2017.03.012.

Khandel, P. and Shahi, S.K. 2018. Mycogenic nanoparticles and their bio-prospective applications: Current status and future challenges. J. Nanostructure. Chem. 8: 369–391. DOI: 10.1007/s40097-018-0285-2.

Klaine, S.J., Alvarez, P.J., Batley, G.E., Fernandes, T.F., Handy, R.D., Lyon, D.Y., Mahendra, S., McLaughlin, M.J. and Lead J.R. 2008. Nanomaterials in the environment: Behavior, fate, bioavailability, and effects. Environ. Toxicol. Chem. 27: 1825–1851. DOI: 10.1897/08-090.1.

Klingenfuss, F. 2014. Testing of TiO_2 Nanoparticles on Wheat and Microorganisms in a Soil Microcosm [master thesis]. University of Gothenburg, Gothenburg.

Kochana, J., Wapiennik, K., Kozak, J., Knihnicki, P., Pollap, A., Wozniakiewicz, M., Nowak, J. and Koscielniak, P. 2015. Tyrosinase-based biosensor for determination of bisphenol A in a flow-batch system. Talanta 144: 163–170. DOI: 10.1016/j.talanta.2015.05.078.

Kumar, K.Y., Muralidhara, H.B., Nayaka, Y.A., Balasubramanyam, J. and Hanumanthappa, H. 2013. Hierarchically assembled mesoporous ZnO nanorods for the removal of lead and cadmium by using differential pulse anodic stripping voltammetric method. Powder Technol. 239: 208–216. DOI: 10.1016/j.powtec.2013.02.009.

Kumar, N., Shah, V. and Walker, V.K. 2011. Perturbation of an arctic soil microbial community by metal nanoparticles. J Hazard. Mater. 190: 816–822. DOI: 10.1016/j.jhazmat.2011.04.005.

Kumar, R., Bhatia, D., Singh, R., Rani, S. and Bishnoi, N.R. 2011. Sorption of heavy metals from electroplating effluent using immobilized biomass *Trichoderma viride* in a continuous packed-bed column. Int. Biodeterior. Biodegradation 65: 1133–1139. DOI: 10.1016/j.ibiod.2011.09.003.

Ladner, D.A., Steele, M., Weir, A., Hristovski, K. and Westerhoff, P. 2012. Functionalized nanoparticle interactions with polymeric membranes. J. Hazard. Mater. 211–212: 288–295. DOI: 10.1016/j.jhazmat.2011.11.051.

Lassouane, F., Ait-Amar, H., Amrani, S. and Rodriguez-Couto, S. 2019. A promising laccase immobilization approach for Bisphenol A removal from aqueous solutions. Bioresour. Technol. 271: 360–367. DOI: 10.1016/j.biortech.2018.09.129.

Lawal, W.A. and Choi, H. 2018. Feasibility study on the removal of perfluorooctanoic acid by using palladium-doped nanoscale zerovalent iron. J. Environ. Eng. 144: 04018115. DOI: 10.1061/ (Asce)Ee.1943-7870.0001468.

Le, T.T., Nguyen, K.H., Jeon, J.R., Francis, A.J. and Chang, Y.S. 2015. Nano/bio treatment of polychlorinated biphenyls with evaluation of comparative toxicity. J. Hazard. Mater. 287: 335–341. DOI: 10.1016/j.jhazmat.2015.02.001.

Lee, H., Kim, E., Shin, Y., Lee, J.H., Hur, H.G. and Kim, J.H. 2016. Identification and formation pattern of metabolites of cyazofamid by soil fungus *Cunninghamella elegans*. Appl. Biol. Chem. 59: 9–14. DOI: 10.1007/s13765-015-0127-6.

Lefevre, E., Bossa, N., Wiesner, M.R. and Gunsch, C.K. 2016. A review of the environmental implications of *in situ* remediation by nanoscale zero valent iron (nZVI): Behavior, transport and impacts on microbial communities. Sci. Total Environ. 565: 889–901. DOI: 10.1016/j.scitotenv.2016.02.003.

Lespes, G., Faucher, S. and Slaveykova, V.I. 2020. Natural nanoparticles, anthropogenic nanoparticles, where is the frontier? Front. Environ. Sci. 8: 71.

Li, Q.W., Liu, D.Q., Wang, T.Z., Chen, C.M. and Gadd, G.M. 2020. Iron coral: Novel fungal biomineralization of nanoscale zerovalent iron composites for treatment of chlorinated pollutants. Chem. Eng. J. 402. DOI: 10.1016/j.cej.2020.126263.

Li, X.M., Liu, W., Sun, L.W., Aifantis, K.E., Yu, B., Fan, Y.B., Feng, Q.L., Cui, F.Z. and Watari, F. 2015. Effects of physicochemical properties of nanomaterials on their toxicity. J. Biomed. Mater. Res. A. 103: 2499–2507. DOI: 10.1002/jbm.a.35384.

Li, Y., Liu, F., Xia, B., Du, Q., Zhang, P., Wang, D., Wang, Z. and Xia, Y. 2010. Removal of copper from aqueous solution by carbon nanotube/calcium alginate composites. J. Hazard. Mater. 177: 876–880. DOI: 10.1016/j.jhazmat.2009.12.114.

Liu, Y., Qu, X., Guo, H., Chen, H., Liu, B. and Dong, S. 2006. Facile preparation of amperometric laccase biosensor with multifunction based on the matrix of carbon nanotubes-chitosan composite. Biosens. Bioelectron 21: 2195–2201. DOI: 10.1016/j.bios.2005.11.014.

Luo, H., Zhang, Y., Xie, Y., Li, Y., Qi, M., Ma, R., Yang, S. and Wang, Y. 2019. Iron-rich microorganism-enabled synthesis of magnetic biocarbon for efficient adsorption of diclofenac from aqueous solution. Bioresour. Technol. 282: 310–317. DOI: 10.1016/j.biortech.2019.03.028.

Mahanty, S., Bakshi, M., Ghosh, S., Gaine, T., Chatterjee, S., Bhattacharyya, S., Das, S., Das, P. and Chaudhuri, P. 2019. Mycosynthesis of iron oxide nanoparticles using manglicolous fungi isolated from Indian sundarbans and its application for the treatment of chromium containing solution: Synthesis, adsorption isotherm, kinetics and thermodynamics study. Environ. Nanotechnol. Monit. Manag. 12: 100276. DOI: 10.1016/j.enmm.2019.100276.

Mahmoud, M.E., Yakout, A.A., Abdel-Aal, H. and Osman, M.M. 2015. Speciation and selective biosorption of Cr (III) and Cr (VI) using nanosilica immobilized-fungi biosorbents. J. Environ. Eng. 141: 04014079. DOI: 10.1061/(Asce) Ee.1943-7870.0000899.

Maliszewska, I., Tylus, W., Checmanowski, J., Szczygiel, B., Pawlaczyk-Graja, I., Pusz, W. and Baturo-Ciesniewska, A. 2017. Biomineralization of gold by *Mucor plumbeus*: The progress in understanding the mechanism of nanoparticles' formation. Biotechnol. Prog. 33: 1381–1392. DOI: 10.1002/btpr.2531.

Manasfi, R., Chiron, S., Montemurro, N., Perez, S. and Brienza, M. 2020. Biodegradation of fluoroquinolone antibiotics and the climbazole fungicide by *Trichoderma* species. Environ. Sci. Pollut. Res. Int. 27: 23331–23341. DOI: 10.1007/s11356-020-08442-8.

Manikandaselvi, S., Sathya, V., Vadivel, V., Sampath, N. and Brindha, P. 2020. Evaluation of bio control potential of AgNPs synthesized from *Trichoderma viride*. Adv. Nat. Sci.-Nanosci. 11. DOI: 10.1088/2043-6254/ab9d16.

Martinez-Sanchez, J., Membrillo-Venegas, J., Martinez-Trujillo, A. and Garcia-Rivero, A.M. 2018. Decolorization of reactive black 5 by immobilized *Trametes versicolor*. Rev. Mex. Ing. Quim. 17: 107–121. DOI: 10.24275/uam/izt/dcbi/revmexingquim/2018v17n1/Martinez.

Martins, F., Machado, S., Albergaria, T. and Delerue-Matos, C. 2017. LCA applied to nano scale zero valent iron synthesis. Int. J. Life Cycle Assess. 22: 707–714. DOI: 10.1007/s11367-016-1258-7.

Mei, L.P., Feng, J.J., Wu, L., Zhou, J.Y., Chen, J.R. and Wang, A.J. 2015. Novel phenol biosensor based on laccase immobilized on reduced graphene oxide supported palladium-copper alloyed nanocages. Biosens. Bioelectron. 74: 347–352. DOI: 10.1016/j.bios.2015.06.060.

Mendes, R.K., Arruda, B.S., de Souza, E.F., Nogueira, A.B., Teschke, O., Bonugli, L.O. and Etchegaray, A. 2017. Determination of chlorophenol in environmental samples using a voltammetric biosensor based on hybrid nanocomposite. J. Braz. Chem. Soc. 28: 1212–1219. DOI: 10.21577/0103-5053.20160282.

Merino, N., Wang, M., Ambrocio, R., Mak, K., O'Connor, E., Gao, A., Hawley, E.L., Deeb, R.A., Tseng, L.Y. and Mahendra, S. 2018. Fungal biotransformation of 6: 2 fluorotelomer alcohol. Remediation 28: 59–70. DOI: 10.1002/rem.21550.

Mitra, S., Pramanik, A., Banerjee, S., Haldar, S., Gachhui, R. and Mukherjee, J. 2013. Enhanced biotransformation of fluoranthene by intertidally derived *Cunninghamella elegans* under biofilm-based and niche-mimicking conditions. Appl. Environ. Microbiol. 79: 7922–7930. DOI: 10.1128/AEM.02129-13.

Mohammed, Y.M.M. and Badawy, M.E.I. 2017. Biodegradation of imidacloprid in liquid media by an isolated wastewater fungus *Aspergillus terreus* YESM3. J. Environ. Sci. Health. B 52: 752–761. DOI: 10.1080/03601234.2017.1356666.

Molnar, Z., Bodai, V., Szakacs, G., Erdelyi, B., Fogarassy, Z., Safran, G., Varga, T., Konya, Z., Toth-Szeles, E., Szucs, R. et al., 2018. Green synthesis of gold nanoparticles by thermophilic filamentous fungi. Sci. Rep. 8: 3943. DOI: 10.1038/s41598-018-22112-3.

Murphy, C.D. 2015. Drug metabolism in microorganisms. Biotechnol. Lett. 37: 19–28. DOI: 10.1007/s10529-014-1653-8.

Muszynska, B., Dabrowska, M., Starek, M., Zmudzki, P., Lazur, J., Pytko-Polonczyk, J. and Opoka, W. 2019. *Lentinula edodes* mycelium as effective agent for piroxicam mycoremediation. Front. Microbiol. 10: 313. DOI: 10.3389/fmicb.2019.00313.

Naghdi, M., Taheran, M., Sarma, S.J., Brar, S.K., Ramirez, A.A. and Verma, M. 2016. Nanotechnology to remove contaminants. pp. 101–128. *In*: Ranjan, S., Dasgupta, N. and Lichtfouse, E. (eds.). Nanoscience in Food and Agriculture 1. Sustainable Agriculture Reviews Vol 20. Springer. DOI: 10.1007/978-3-319-39303-2_4.

Navarro, E., Baun, A., Behra, R., Hartmann, N.B., Filser, J., Miao, A.J., Quigg, A., Santschi, P.H. and Sigg, L. 2008. Environmental behavior and ecotoxicity of engineered nanoparticles to algae, plants, and fungi. Ecotoxicology 17: 372–386. DOI: 10.1007/s10646-008-0214-0.

Nel, A., Xia, T., Madler, L. and Li, N. 2006. Toxic potential of materials at the nanolevel. Science 311: 622–627. DOI: 10.1126/science.1114397.

Omran, B.A., Nassar, H.N., Younis, S.A., El-Salamony, R.A., Fatthallah, N.A., Hamdy, A., El-Shatoury, E.H. and El-Gendy, N.S. 2020. Novel mycosynthesis of cobalt oxide nanoparticles using *Aspergillus brasiliensis* ATCC 16404-optimization, characterization and antimicrobial activity. J. Appl. Microbiol. 128: 438–457. DOI: 10.1111/jam.14498.

Ostrem Loss, E.M., Lee, M.K., Wu, M.Y., Martien, J., Chen, W., Amador-Noguez, D., Jefcoate, C., Remucal, C., Jung, S., Kim, S.C. et al. 2019. Cytochrome P450 monooxygenase-mediated metabolic utilization of benzo[a]pyrene by *Aspergillus* species. mBio. 10. DOI: 10.1128/mBio.00558-19.

Otero-González, L., García-Saucedo, C., Field, J.A. and Sierra-Álvarez, R. 2013. Toxicity of TiO_2, ZrO_2, Fe0, Fe_2O_3, and Mn_2O_3 nanoparticles to the yeast, *Saccharomyces cerevisiae*. Chemosphere 93: 1201–1206.

Othman, A.M. and Wollenberger, U. 2020. Amperometric biosensor based on coupling aminated laccase to functionalized carbon nanotubes for phenolics detection. Int. J. Biol. Macromol. 153: 855–864. DOI: 10.1016/j.ijbiomac.2020.03.049.

Ovecka, M., Lang, I., Baluska, F., Ismail, A., Illes, P. and Lichtscheidl, I.K. 2005. Endocytosis and vesicle trafficking during tip growth of root hairs. Protoplasma 226: 39–54. DOI: 10.1007/s00709-005-0103-9.

Palmer-Brown, W., de Melo Souza, P.L. and Murphy, C.D. 2019. Cyhalothrin biodegradation in *Cunninghamella elegans*. Environ. Sci. Pollut. Res. Int. 26: 1414–1421. DOI: 10.1007/s11356-018-3689-0.

Palmer-Brown, W., Miranda-CasoLuengo, R., Wolfe, K.H., Byrne, K.P. and Murphy, C.D. 2019. The CYPome of the model xenobiotic-biotransforming fungus *Cunninghamella elegans*. Sci. Rep. 9: 9240. DOI: 10.1038/s41598-019-45706-x.

Pan, D., Gu, Y., Lan, H., Sun, Y. and Gao, H. 2015. Functional graphene-gold nano-composite fabricated electrochemical biosensor for direct and rapid detection of bisphenol A. Anal. Chim. Acta 853: 297–302. DOI: 10.1016/j.aca.2014.11.004.

Park, M.V., Neigh, A.M., Vermeulen, J.P., de la Fonteyne, L.J., Verharen, H.W., Briede, J.J., van Loveren, H. and de Jong, W.H. 2011. The effect of particle size on the cytotoxicity, inflammation, developmental toxicity and genotoxicity of silver nanoparticles. Biomaterials 32: 9810–9817. DOI: 10.1016/j.biomaterials.2011.08.085

Park, Y.B., Mohan, K., Al-Sanousi, A., Almaghrabi, B., Genco, R.J., Swihart, M.T. and Dziak, R. 2011. Synthesis and characterization of nanocrystalline calcium sulfate for use in osseous regeneration. Biomed. Mater. 6: 055007. DOI: 10.1088/1748-6041/6/5/055007.

Pavani, K.V., Kumar, N.S. and Sangameswaran, B.B. 2012. Synthesis of lead nanoparticles by *Aspergillus* species. Pol. J. Microbiol. 61: 61–63. DOI: 10.33073/pjm-2012-008.

Perera, M., Wijayarathna, D., Wijesundera, S., Chinthaka, M., Seneviratne, G. and Jayasena, S. 2019. Biofilm mediated synergistic degradation of hexadecane by a naturally formed community comprising *Aspergillus flavus* complex and *Bacillus cereus* group. BMC Microbiol. 19: 84. DOI: 10.1186/s12866-019-1460-4.

Perumal, V. and Hashim, U. 2014. Advances in biosensors: Principle, architecture and applications. J. Appl. Biomed. 12: 1–15. DOI: 10.1016/j.jab.2013.02.001.

Ponder, S.M., Darab, J.G. and Mallouk, T.E. 2000. Remediation of Cr (VI) and Pb (II) aqueous solutions using supported, nanoscale zero-valent iron. Environ. Sci. Technol. 34: 2564–2569. DOI: DOI 10.1021/es9911420.

Pozdnyakova, N.N., Rodakiewicz-Nowak, J. and Turkovskaya, O.V. 2004. Catalytic properties of yellow laccase from *Pleurotus ostreatus* D1. J. Mol. Catal. B. Enzym. 30: 19–24. DOI: 10.1016/j.molcatb.2004.03.005.

Qu, J.Y., Lou, T.F., Kang, S.P. and Du, X.P. 2013. Simultaneous determination of catechol and hydroquinone using a self-assembled laccase biosensor based on nanofilm. Sensor Lett. 11: 1567–1572. DOI: 10.1166/sl.2013.3017.

Qu, X., Alvarez, P.J. and Li, Q. 2013. Applications of nanotechnology in water and wastewater treatment. Water Res. 47: 3931–3946. DOI: 10.1016/j.watres.2012.09.058.

Qu, Y., Lian, S., Shen, W., Li. Z., Yang, J. and Zhang, H. 2020. Rod-shaped gold nanoparticles biosynthesized using Pb (2+)-induced fungus *Aspergillus* sp. WL-Au. Bioprocess Biosyst. Eng. 43: 123–131. DOI: 10.1007/s00449-019-02210-w.

Raizada, P., Sudhaik, A. and Singh, P. 2019. Photocatalytic water decontamination using graphene and ZnO coupled photocatalysts: A review. Materials Science for Energy Technologies 2: 509–525.

Rashid, A., Bhatti, H.N., Iqbal, M. and Noreen, S. 2016. Fungal biomass composite with bentonite efficiency for nickel and zinc adsorption: A mechanistic study. Ecol. Eng. 91: 459–471. DOI: 10.1016/j.ecoleng.2016.03.014.

Ray, P.C., Yu, H. and Fu, P.P. 2009. Toxicity and environmental risks of nanomaterials: Challenges and future needs. J. Environ. Sci. Health. C. Environ. Carcinog. Ecotoxicol. Rev. 27: 1–35. DOI: 10.1080/10590500802708267.

Rehman, S., Farooq, R., Jermy, R., Asiri, S.M., Ravinayagam, V., Jindan, R.A., Alsalem, Z., Shah, M.A., Reshi. Z., Sabit, H. et al. 2020. A wild fomes fomentarius for biomediation of one pot synthesis of titanium oxide and silver nanoparticles for antibacterial and anticancer application. Biomolecules 10. DOI: 10.3390/biom10040622.

Rizwan, M., Singh, M., Mitra, C.K. and Morve, R. K. 2014. Ecofriendly application of nanomaterials: Nanobioremediation. Journal of Nanoparticles 2014. DOI: 10.1155/2014/431787.

Saenz, M., Borodulina, T., Diaz, L. and Banchon, C. 2019. Minimal conditions to degrade low density polyethylene by *Aspergillus terreus* and *niger*. J. Ecol. Eng. 20: 44–51. DOI: 10.12911/22998993/108699.

Salvadori, M.R., Ando, R.A., Nascimento, C.A. and Correa, B. 2015. Extra and intracellular synthesis of nickel oxide nanoparticles mediated by dead fungal biomass. PLoS One 10: e0129799. DOI: 10.1371/journal.pone.0129799.

Santos, I.J.S., Grossman, M.J., Sartoratto, A., Ponezi, A.N. and Durrant, L.R. 2012. Degradation of the recalcitrant pharmaceuticals carbamazepine and 17 alpha-ethinylestradiol by ligninolytic fungi. pp. 169–174. *In*: Ibic2012: International Conference on I ndustrial Biotechnology. 10.3303/cet1227029.

Sarabia-Castillo, C. and Fernández-Luqueño, F. 2016. TiO_2, ZnO, and Fe_2O_3 nanoparticles effect on *Rhizobium leguminosarum-Pisum sativum* L. symbiosis. 3rd Biotechnology Summit 2016, Ciudad Obregón, Sonora, Mexico, 24–28 October 2016, 144–149.

Sardar, M. and Ahmad, R. 2015. Enzyme immobilization: an overview on nanoparticles as immobilization matrix. Biochem. Anal. Biochem. 4: 1–8. DOI: 10.4172/2161-1009.1000178.

Seeger, M., Zielinski, M., Timmis, K.N. and Hofer, B. 1999. Regiospecificity of dioxygenation of di- to pentachlorobiphenyls and their degradation to chlorobenzoates by the bph-encoded catabolic pathway of *Burkholderia* sp. strain LB400. Appl. Environ. Microbiol. 65: 3614–3621. DOI: 10.1128/AEM.65.8.3614-3621.1999.

Setiawan, H., Khairani, R., Rahman, M.A., Septawendar, R., Mukti, R.R., Dipojono, H.K. and Purwasasmita, B.S. 2017. Synthesis of zeolite and γ-alumina nanoparticles as ceramic membranes for desalination applications. J. Aust. Ceram. Soc. 53: 531–538. DOI: 10.1007/s41779-017-0064-4.

Shah, V., Dobiasova, P., Baldrian, P., Nerud, F., Kumar, A. and Seal, S. 2010. Influence of iron and copper nanoparticle powder on the production of lignocellulose degrading enzymes in the fungus *Trametes versicolor*. J. Hazard. Mater. 178: 1141–1145. DOI: 10.1016/j.jhazmat.2010.01.141.

Shaheen, R., Asgher, M., Hussain, F. and Bhatti, H.N. 2017. Immobilized lignin peroxidase from *Ganoderma lucidum* IBL-05 with improved dye decolorization and cytotoxicity reduction properties. Int. J. Biol. Macromol. 103: 57–64. DOI: 10.1016/j.ijbiomac.2017.04.040.

Shang, L., Nienhaus, K. and Nienhaus, G.U. 2014. Engineered nanoparticles interacting with cells: size matters. J. Nanobiotechnology 12: 5. DOI: 10.1186/1477-3155-12-5.

Sharma, K.R., Giri, R. and Sharma, R.K. 2020. Lead, cadmium and nickel removal efficiency of white-rot fungus *Phlebia brevispora*. Lett. Appl. Microbiol. DOI: 10.1111/lam.13372.

Shinde, S.S. 2015. Antimicrobial activity of ZnO nanoparticles against pathogenic bacteria and fungi. Sci. Med. Central. 3: 1033.

Silva, L.P., Bonatto, C.C. and Polez, V.L.P. 2016. Green synthesis of metal nanoparticles by fungi: Current trends and challenges. pp. 71–89. *In*: Prasad, R. (ed.). Advances and Applications Through Fungal Nanobiotechnology. Springer. DOI: 10.1007/978-3-319-42990–8_4.

Sofia, P., Asgher, M., Shahid, M. and Randhawa, M.A. 2016. Chitosan beads immobilized *Schizophyllum commune* ibl-06 lignin peroxidase with novel thermo stability, catalytic and dye removal properties. J. Anim. Plant. Sci. 26: 1451–1463.

Sowmya, H.V., Ramalingappa, Krishnappa, M. and Thippeswamy, B. 2015. Degradation of polyethylene by *Penicillium simplicissimum* isolated from local dumpsite of Shivamogga district. Environ. Dev. Sustain. 17: 731–745. DOI: 10.1007/s10668-014-9571-4.

Sriharsha, D.V., Lokesh Kumar, R. and Janakiraman, S. 2020. Absorption and reduction of chromium by fungi. Bull. Environ. Contam. Toxicol. 105: 645–649. DOI: 10.1007/s00128-020-02979-7.

Srivastava, P.K., Vaish, A., Dwivedi, S., Chakrabarty, D., Singh, N. and Tripathi, R.D. 2011. Biological removal of arsenic pollution by soil fungi. Sci. Total Environ. 409: 2430–2442. DOI: 10.1016/j.scitotenv.2011.03.002.

Sun, J., Yuan, X., Li, Y., Wang, X. and Chen, J. 2019. The pathway of 2,2-dichlorovinyl dimethyl phosphate (DDVP) degradation by *Trichoderma atroviride* strain T23 and characterization of a paraoxonase-like enzyme. Appl. Microbiol. Biotechnol. 103: 8947–8962. DOI: 10.1007/s00253-019-10136-2.

Sung, H.J., Khan, M.F. and Kim, Y.H. 2019. Recombinant lignin peroxidase-catalyzed decolorization of melanin using in-situ generated H_2O_2 for application in whitening cosmetics. Int. J. Biol. Macromol. 136: 20–26. DOI: 10.1016/j.ijbiomac.2019.06.026.

Tan, W.S. and Ting, A.S. 2012. Efficacy and reusability of alginate-immobilized live and heat-inactivated *Trichoderma asperellum* cells for Cu (II) removal from aqueous solution. Bioresour. Technol. 123: 290–295. DOI: 10.1016/j.biortech.2012.07.082.

Tayel, A.A., Gharieb, M.M., Zaki, H.R. and Elguindy, N.M. 2016. Bio-clarification of water from heavy metals and microbial effluence using fungal chitosan. Int. J. Biol. Macromol. 83: 277–281. DOI: 10.1016/j.ijbiomac.2015.11.072.

Tigini, V., Prigione, V., Donelli, I., Freddi, G. and Varese, G.C. 2012. Influence of culture medium on fungal biomass composition and biosorption effectiveness. Curr. Microbiol. 64: 50–59. DOI: 10.1007/s00284-011-0017-z.

Tripathi, S., Sanjeevi, R., Anuradha, J., Chauhan, D.S. and Rathoure, A.K. 2018. Nano-bioremediation: nanotechnology and bioremediation. pp. 202–219. *In*: Rathoure, A.K. (ed.). Biostimulation Remediation Technologies for Groundwater Contaminants. IGI Global. DOI: 10.4018/978-1-5225-4162–2.ch012.

Tyagi, I., Gupta, V., Sadegh, H., Ghoshekandi, R.S. and Makhlouf, A.S.H. 2017. Nanoparticles as adsorbent; a positive approach for removal of noxious metal ions: A review. Sci. Technol. Dev. 34: 195–214. DOI: 10.3923/std.2015.195.214.

Ungureanu, C.V., Favier, L. and Bahrim, G.E. 2020. Improving biodegradation of clofibric acid by *Trametes pubescens* through the design of experimental tools. Microorganisms. 8. DOI: 10.3390/microorganisms8081243.

Vahabi, K. and Dorcheh, S.K. 2014. Biosynthesis of silver nano-particles by *Trichoderma* and its medical applications. pp. 393–404. *In*: Gupta, V.K., Schmoll, M., Herrera-Estrella, A., Upadhyay, R.S., Druzhinina, I. and Tuohy, M.G. (eds.). Biotechnology and biology of *Trichoderma*. Elsevier. DOI: 10.1016/B978-0-444-59576-8.00029-1.

Velasco-Lozano, S., López-Gallego, F., Mateos-Díaz, J.C. and Favela-Torres, E. 2016. Cross-linked enzyme aggregates (CLEA) in enzyme improvement—A review. Biocatalysis 1: 166–177. DOI: 10.1515/boca-2015-0012.

Velkova, Z., Kirova, G., Stoytcheva, M., Kostadinova, S., Todorova, K. and Gochev, V. 2018. Immobilized microbial biosorbents for heavy metals removal. Eng. Life Sci. 18: 871–881. DOI: 10.1002/elsc.201800017.

Verma, A., Shalu, Singh, A., Bishnoi, N.R. and Gupta, A. 2013. Biosorption of Cu (II) using free and immobilized biomass of *Penicillium citrinum*. Ecol. Eng. 61: 486–490. DOI: 10.1016/j.ecoleng.2013.10.008.

Vetchinkina, E.P., Loshchinina, E.A., Vodolazov, I.R., Kursky, V.F., Dykman, L.A. and Nikitina, V.E. 2017. Biosynthesis of nanoparticles of metals and metalloids by basidiomycetes. Preparation of gold nanoparticles by using purified fungal phenol oxidases. Appl. Microbiol. Biotechnol. 101: 1047–1062. DOI: 10.1007/s00253-016-7893-x.

Vidal-Limon, A., Garcia Suarez, P.C., Arellano-Garcia, E., Contreras, O.E. and Aguila, S.A. 2018. Enhanced degradation of pesticide dichlorophen by laccase immobilized on nanoporous materials: A cytotoxic and molecular simulation investigation. Bioconjug. Chem. 29: 1073–1080. DOI: 10.1021/acs.bioconjchem.7b00739.

Wang, J., Ohno, H., Ide, Y., Ichinose, H., Mori, T., Kawagishi, H. and Hirai, H. 2019. Identification of the cytochrome P450 involved in the degradation of neonicotinoid insecticide acetamiprid in *Phanerochaete chrysosporium*. J. Hazard. Mater. 371: 494–498. DOI: 10.1016/j.jhazmat.2019.03.042.

Wang, S.B., Sun, H.Q., Ang, H.M. and Tade, M.O. 2013. Adsorptive remediation of environmental pollutants using novel graphene-based nanomaterials. Chem. Eng. J. 226: 336–347. DOI: 10.1016/j.cej.2013.04.070.

Wang, X., Lu, X.B., Wu, L.D. and Chen, J.P. 2014. Direct electrochemical tyrosinase biosensor based on mesoporous carbon and Co_3O_4 nanorods for the rapid detection of phenolic pollutants. Chemelectrochem. 1: 808–816. DOI: 10.1002/celc.201300208.

Wang, Y.N., O'Connor, D., Shen, Z.T., Lo, I.M.C., Tsang, D.C.W., Pehkonen, S., Pu, S.Y. and Hou, D.Y. 2019. Green synthesis of nanoparticles for the remediation of contaminated waters and soils: Constituents, synthesizing methods, and influencing factors. J. Clean. Prod. 226: 540–549. DOI: 10.1016/j.jclepro.2019.04.128.

Wohlleben, W., Mielke, J., Bianchin, A., Ghanem, A., Freiberger, H., Rauscher, H., Gemeinert, M. and Hodoroaba, V.D. 2017. Reliable nanomaterial classification of powders using the volume-specific surface area method. J. Nanopart. Res. 19: 61. DOI: 10.1007/s11051-017-3741-x.

Wu, L., Deng, D., Jin, J., Lu, X. and Chen, J. 2012. Nanographene-based tyrosinase biosensor for rapid detection of bisphenol A. Biosens. Bioelectron. 35: 193–199. DOI: 10.1016/j.bios.2012.02.045.

Wu, L., Lu, X., Niu, K., Dhanjai and Chen, J. 2020. Tyrosinase nanocapsule based nano-biosensor for ultrasensitive and rapid detection of bisphenol A with excellent stability in different application scenarios. Biosens Bioelectron. 165: 112407. DOI: 10.1016/j.bios.2020.112407.

Wu, Q., Xu, Z.Q., Duan, Y.J., Zhu, Y.C., Ou, M.R. and Xu, X.P. 2017. Immobilization of tyrosinase on polyacrylonitrile beads: Biodegradation of phenol from aqueous solution and the relevant cytotoxicity assessment. RSC Adv. 7: 28114–28123. DOI: 10.1039/c7ra03174b.

Xiao, P. and Kondo, R. 2020. Potency of *Phlebia* species of white rot fungi for the aerobic degradation, transformation and mineralization of lindane. J. Microbiol. 58: 395–404. DOI: 10.1007/s12275-020-9492-x.

Xu, L.J. and Wang, J.L. 2012. Magnetic nanoscaled Fe_3O_4/CeO_2composite as an efficient fenton-like heterogeneous catalyst for degradation of 4-chlorophenol. Environ. Sci. Technol. 46: 10145–10153. DOI: 10.1021/es300303f.

Xu, X., Yang, Y., Zhao, X., Zhao, H., Lu, Y., Jiang. C., Shao, D. and Shi, J. 2019. Recovery of gold from electronic wastewater by *Phomopsis* sp. XP-8 and its potential application in the degradation of toxic dyes. Bioresour. Technol. 288: 121610. DOI: 10.1016/j.biortech.2019.121610.

Yadav, K., Singh, J., Gupta, N. and Kumar, V. 2017. A review of nanobioremediation technologies for environmental cleanup: A novel biological approach. J. Mater. Environ. Sci. 8: 740–757.

Yamaguchi, H., Kiyota, Y. and Miyazaki, M. 2018. Techniques for preparation of cross-linked enzyme aggregates and their applications in bioconversions. Catalysts 8: 174. DOI: 10.3390/catal8050174.

Yang, L., Xiong, H., Zhang, X. and Wang, S. 2012. A novel tyrosinase biosensor based on chitosan-carbon-coated nickel nanocomposite film. Bioelectrochemistry 84: 44–48. DOI: 10.1016/j.bioelechem.2011.11.001.

Yang, Y., Javed, H., Zhang, D., Li, D., Kamath, R., McVey, K., Sra, K. and Alvarez, P.J. 2017. Merits and limitations of TiO 2-based photocatalytic pretreatment of soils impacted by crude oil for expediting bioremediation. Front. Chem. Sci. Eng. 11: 387–394. DOI: 10.1007/s11705-017-1657-8.

Yin, H., Too, H.P. and Chow, G.M. 2005. The effects of particle size and surface coating on the cytotoxicity of nickel ferrite. Biomaterials 26: 5818–5826. DOI: 10.1016/j.biomaterials.2005.02.036.

Yin, H., Zhou, Y., Xu, J., Ai. S., Cui, L. and Zhu, L. 2010. Amperometric biosensor based on tyrosinase immobilized onto multiwalled carbon nanotubes-cobalt phthalocyanine-silk fibroin film and its application to determine bisphenol A. Anal. Chim. Acta 659: 144–150. DOI: 10.1016/j.aca.2009.11.051.

Yunus, I.S., Harwin, Kurniawan, A., Adityawarman, D. and Indarto, A. 2012. Nanotechnologies in water and air pollution treatment. Environ. Technol. Rev. 1: 136–148. DOI: 10.1080/21622515.2012.733966.

Zaidi, K.U., Ali, A.S., Ali, S.A. and Naaz, I. 2014. Microbial tyrosinases: Promising enzymes for pharmaceutical, food bioprocessing, and environmental industry. Biochem. Res. Int. 2014: 854687. DOI: 10.1155/2014/854687.

Zawadzka, K., Bernat, P., Felczak, A. and Lisowska, K. 2017. Microbial detoxification of carvedilol, a beta-adrenergic antagonist, by the filamentous fungus *Cunninghamella echinulata*. Chemosphere 183: 18–26. DOI: 10.1016/j.chemosphere.2017.05.088.

Zehani, N., Fortgang, P., Saddek Lachgar, M., Baraket, A., Arab, M., Dzyadevych, S.V., Kherrat, R. and Jaffrezic-Renault, N. 2015. Highly sensitive electrochemical biosensor for bisphenol A detection based on a diazonium-functionalized boron-doped diamond electrode modified with a multi-walled carbon nanotube-tyrosinase hybrid film. Biosens. Bioelectron. 74: 830–835. DOI: 10.1016/j.bios.2015.07.051.

Zhang, C.Z., Chen, B., Bai, Y. and Xie, J. 2018. A new functionalized reduced graphene oxide adsorbent for removing heavy metal ions in water via coordination and ion exchange. Sep. Sci. Technol. 53: 2896–2905. DOI: 10.1080/01496395.2018.1497655.

Zhang, D., Deng, M.F., Cao, H.B., Zhang, S.P. and Zhao, H. 2017. Laccase immobilized on magnetic nanoparticles by dopamine polymerization for 4-chlorophenol removal. Green Energy Environ. 2: 393–400. DOI: 10.1016/j.gee.2017.04.001.

Zhang, Y., Ali, S.F., Dervishi, E., Xu, Y., Li, Z., Casciano, D. and Biris, A.S. 2010. Cytotoxicity effects of graphene and single-wall carbon nanotubes in neural phaeochromocytoma-derived PC12 cells. ACS Nano. 4: 3181–3186. DOI: 10.1021/nn1007176.

Zhao, M.A., Gu, H., Zhang, C.J., Jeong, H., Kim, E.O.H. and Zhu, Y.Z. 2020. Metabolism of insecticide diazinon by *Cunninghamella elegans* ATCC36112. RSC Adv. 10: 19659–19668. DOI: 10.1039/d0ra02253e.

Zhao, W.W., Zhu, G,Q,, Daugulis, A.J., Chen, Q., Ma, H.Y., Zheng, P., Liang, J. and Ma, X.K. 2020. Removal and biomineralization of Pb2+ in water by fungus *Phanerochaete chrysoporium*. J. Clean. Prod. 260. DOI: 10.1016/j.jclepro.2020.120980.

Zhou, Y., Tang, L., Zeng, G., Chen, J., Cai, Y., Zhang, Y., Yang, G., Liu, Y., Zhang, C. and Tang, W. 2014. Mesoporous carbon nitride based biosensor for highly sensitive and selective analysis of phenol and catechol in compost bioremediation. Biosens. Bioelectron. 61: 519–525. DOI: 10.1016/j.bios.2014.05.063.

Zhu, Y.Z., Fu, M., Jeong, I.H., Kim, J.H. and Zhang CJ. 2017. Metabolism of an insecticide fenitrothion by *Cunninghamella elegans* ATCC36112. J. Agric. Food Chem. 65: 10711–10718. DOI: 10.1021/acs.jafc.7b04273.

CHAPTER 13

Nanomaterials and Edible or Nonedible Crops for Dissipate Pollutants

Using Nanotechnology to Bioremediate the Environment

García-Mayagoitia Selvia,[1] *Fernández-Luqueño Fabián,*[1,*]
Medina-Pérez Gabriela,[2] *López-Valdez Fernando,*[3]
Campos-Montiel Rafael[2] and *Mariana Miranda-Arámbula*[3]

1. Introduction

A healthy environment is essential for the survival of all living beings. However, the large growth in the human population has posed numerous environmental concerns, such as polluting the air, land and water bodies. Our ecosystem is constantly under threat as a result of rising industrialization and urbanization. Among the list of the world's most serious environmental concerns, water scarcity, soil degradation and air pollution are at the top of the list (Keshav et al., 2021). The environment is well known to face high polluted levels at the highest concentrations as never seen before. Besides, different technologies have been developed during the last years to face this reality; the uncontrolled distribution of contaminants to the environment. Therefore, nanosized materials have been seen as a new technology with the capacity to degrade, remove or immobilize hazardous materials, which could jeopardize human and environmental health (Medina-Perez et al., 2019).

The tiny particle size, large specific surface area and high surface activity of nanomaterials make them ideal for combining with heavy metal ions by ion exchange or adsorption. They can also reduce the capacity of heavy metal ions to migrate through soil. As a result, nanoparticles are becoming

[1] Sustainability of Natural Resources and Energy Programs, Cinvestav-Saltillo, Coahuila. CP 25900, Mexico.
[2] ICAP – Institute of Agricultural Sciences, Autonomous University of the State of Hidalgo.Tulancingo de Bravo, Hidalgo. C.P. 43000, Mexico.
[3] Agricultural Biotechnology and Agronanobiotechnology Group. Research Center for Applied Biotechnology (CIBA) - Instituto Politécnico Nacional, Tlaxcala, C.P. 90700, Mexico.
* Corresponding author: cinves.cp.cha.luqueno@gmail.com

increasingly important and are being widely used in the treatment of soil contamination. However, due to the fact that soils serve as the final sink for nanomaterials and are key contamination sources, soil-microbiota and plants are among the most important eco-receptors for nanoparticles exposure, and some of these nanoparticles are hazardous to soil organisms and plants (Pérez-Hernández et al., 2020).

2. Environmental Remediation through Nanotechnologies

Many different remediation technologies are available when pollution is found on the soil, water or air. On the other hand, most soil remediation procedures convert soil pollutants into less harmful forms or simply stabilize contaminants in the soil matrix and thus are incapable of completely removing dangerous chemicals from the soil. A phytoremediation approach for soil remediation, on the other hand, is a good plant-based alternative technique for soil remediation that has the potential to remove, degrade, sequester and stabilize a broad spectrum of contaminants. Phytoremediation has been acclaimed as a promising, cost-effective and environmentally benign technique since it is powered by solar energy and is unlikely to create secondary contamination. Besides, the capacity of plants to absorb contaminants and the bioavailability of pollutants are both crucial factors in the effectiveness of phytoremediation.

However, the biological availability of several hazardous compounds that are now in use and the accumulation capacity of some plants are both restricted. The use of integrated remediation strategies, including phytoremediation and NMs, is an emerging technology that is of help in tackling these challenges. In part because of the diverse reactions of plants to NMs and the diverse influence of NMs on soil contaminants, the synergistic and antagonistic effects of integrated remediation approaches in the decontamination of polluted soils was discovered. The scientific community described at length the impact of NMs on the phytoremediation of organic pollutants and heavy metals polluted soils (Sarabia-Castillo et al., 2020; Vazquez-Nuñez et al., 2020).

3. Synthesis of Novel Materials

Nanotechnology is increasingly valuable for detecting contamination and assisting in the remediation of pollution through novel and advanced techniques for the prevention, treatment and cleaning of waste sites (soil, sediment, solid waste, air and water) (Wiesner and Bottero, 2017; Mehndiratta et al., 2013); describe how nanomaterials (NM) can function in various roles in pollution control through three different ways: source reduction, pollutant degradation and pollutant detection. Among the documented findings is that NMs remediate toxic metals and organic pollutants due to their properties such as the interaction of metal ions with oxohydroxyl, surface absorption, spontaneous precipitation, reduction, photodegradation, magnetic separation and separation by permeable reactive barriers (Das et al., 2018). Recently, nanoscale materials are being explored for bioremediation, including nanoscale zeolites, metal oxides, bimetallic Pd/FeO, NP, zinc oxide, carbon nanotubes, chitosan and graphene nanofilms. Various organic nanomaterials, such as dendrimers, modified dendrimers, carbon nanotubes, calcium alginate and multi-walled carbon nanotubes are used for metal removal, radionuclide remediation, toluene hydrogenation, crude oil degradation and phenol degradation. In addition, a large number of inorganic NPs, such as the NPs of TiO_2, polysulfone-zero valence coated iron, polyacrylic acid) oxide, chitosan-gelatin/graphene oxide and Ag-iron oxide/fly ash have been used for arsenic, uranium, inorganic nitrate and remediation of polychlorinated biphenyls. There are socio-environmental considerations that should not be ignored. However, nanotechnology also offers alternatives for the recovery of water, air and soil. Nonetheless, there are still many doubts about the massive use of nanomaterials, given that eventually, they will also become waste. Additionally, some questions are still unanswered, specifically in terms of toxicity; since they could affect the health of people and the environment (Handford et al., 2014).

3.1 Chemical conventional methods of nanoparticle synthesis

There are two groups of methods to obtain nanomaterials. The first one is "Top-down," which consists of dividing mass solids into smaller portions. These procedures include (a) Chemical vapor deposition,which consists of disintegration of volatile compounds in a vacuum chamber; this process can also be carried out near a solid surface to obtain material in the form of nanoparticles. (b) Thermic evaporation; the material is heated in a vacuum chamber to evaporate and then condense on a cold sheet under controlled conditions. (c) Laser ablation; a high-power laser produces metallic atomic vapors that carry inert gas. Later, they are deposited in substrates like crystalline oxides under ultra vacuum conditions. (d) Sputtering magnetron; extraction of atoms from the surface of an electrode due to the momentous exchange with ions that bombard the surface atoms (Tulinski and Jurczyk, 2017). The second is "Bottom-up", this method creates tools by manipulating atoms or molecules to produce small objects. This approach is more common in nanoparticle synthesis. These groups of procedures include condensation inert gas, sol-gel method, hydrothermal and solvothermal synthesis, coprecipitation, inert gas condensation, ion sputtering scattering, microemulsion, microwave, pulse laser ablation, sonochemical, spark discharge, template synthesis and biological synthesis (Zanella, 2012; Hayashi and Hakuta, 2010; Tulinski and Jurczyk, 2017).

3.2 Green synthesis

The United Nations Sustainable Development Goals 2030 recently stated that any new material or product should be functional, cost-effective, safe and sustainable. Nanotechnology is one of the technologies that could enable such green growth (Gottardo et al., 2021). The need to sustain remediation processes and current advances in environmental nanotechnology is trying to keep aside traditional methods of synthesizing nanoparticles from chemical and physical methods. The synthesis of nanoparticles was recently approached through green chemistry or biological synthesis using vascular plants and microorganisms such as bacteria, yeasts, algae, fungi and actinomycetes (Salem and Fouda, 2021). The importance of biological synthesis has grown because traditional methods have proved costly, toxic and unsustainable. Obtaining nanoparticles from potential biological systems of plants or microbes has recently been developed. In this respect, nanoparticles have been synthesized through microorganisms and plant extracts, where the primary process is reduction/oxidation reaction (Yadav et al., 2017).

3.3 Smart nanomaterials

By definition, smart materials are those that can react when they are externally chemical and physically stimulated (pH, temperature, moisture, force, electric fields, magnetic fields or the interaction with a specific molecule). In nature, there are several examples of this stimulated environmental change in material configuration, one of them is the camouflage of some insect species via neurally controlled chromatophore apparatus (Yoshida and Lahann, 2008). There are already smart materials in modern society: piezoelectric, thermoresponsive, shape memory alloys, polychromic, chromogenic or halochromic, polymers and ferrofluids. These materials can be categorized as active or passive. Active materials possess the capacity to transduce energy (piezoelectric materials, memory alloys, etc.), while passive materials do not possess the capacity to transducer energy. Both of them act as sensors. Smart nanometric materials have found application in different fields of research (energy generation, 3D printing, environmental remediation, textiles and healthcare products). Smart nanomaterials are classified into several groups: Selvan et al. (2010) mentioned five categories due to their function or structure: (1) Remote-actuated; (2) Environmentally-responsive (pH, temperature, light, oxidation-reduction, certain chemicals and others); (3) Miniaturized (nanoscaled devices and technologies); (4) Hybrids (biotic–abiotic, organic-inorganic); and (5) Transforming (structures that can change irreversibly).

The European Chemicals Agency (ECHA) in 2019 published a registry of 48 products defined as smart materials considered as second and eight of the third generation of nano-applications in technology, most of the second generation materials have been developed for medical applications, a quarter are concerned with electronic devices, only three products are currently applied in environmental issues such as pollution remediation and gas detection, experimentally used to treat surface water, seawater, soil or air.

3.4 Nanohybrids

Most often, standard materials (metals, ceramics or plastics) cannot fulfill all the required functions; so materials science developed mixtures between them. This way, these new materials can show superior properties to their pure counterparts (Table 13.1). Many of the resulting materials show improvement in mechanical properties. A well-known example is reinforced polymers with inorganic fiber. The structural building blocks in these materials incorporated into the matrix are predominantly natural inorganic and show size ranges from micrometer to millimeter. Therefore, its heterogeneous composition is quite visible to the human eye. It soon became apparent that the decrease in the size

Table 13.1. Examples of nanomaterials currently used for remediation.

Type of Material	Examples	Functionality	Reference
Polymeric smart materials	Polyurethane foams possessed high absorption, Cellulose nanoparticles (cellulose nanofibrils and nanocrystals), dendrimers	High absorption capacity Effective in detecting and removing hazardous contaminants, capable of detecting and removing hazardous contaminants	Dongre et al., 2019; Okesola and Smith, 2016; Mastronardi et al., 2014; Liu et al., 2010; Roy and Gupta, 2003; Davison, 2002; Shrivastava and Tomar, 2020
Nanostructures of semiconductors	ZnO nanostructures; semiconductor–metal nanocomposites, SnO_2–based semiconductor systems, TiO_2 nanocomposites	Detect and destroy harmful chemical contaminants	Wang et al., 2013; Ansari et al., 2013; Kamat and Meisel, 2003
Sorbent nanocomposites	Zeolite coated with layered double hydroxide, biochar-bentonite composite for the removal of antibiotic, Hybrid silicate-hydrochar composite	Efficient capacity to adsorb harmful ions and to minimize environmental pollution during wastewater management	Ashiq et al., 2019; Yamada et al., 2006; Deng et al., 2020
Microbial calcite	*Sporosarcinapasteurii* can induce $CaCO_3$ precipitation in the surroundings in response to environmental cues such as high pH and available nutrients and minerals	Self-healing concrete, surface crack remediation, and durability enhancement	Bang et al., 2010
Carbonous nanomaterials	Graphene oxide, nanotubes, nanorods	Found as a single layer of carbon atoms arranged as honeycomb lattice with oxygen, advanced removal agent and had potential to remove pollution	Wang et al., 2013; Pendolino and Armata, 2017
Metal ions and clusters	Nanomaterials like titania, carbon nanotubes, zero-valent iron, dendrimers and silver nanotubes, Alpha Fe_2O_3	was used in environmental remediation and water purification by absorbance of heavy metals and other pollutants	Ghasemzadeh et al., 2014; Ryder and Tan, 2014; Soler and Sánchez, 2014
Nanospheres	Removing pollutants from wastewater. Nanospheres had to adopt capacity and found to possess efficient antibacterial properties	Hierarchical SnO_2 was prepared by efficient methods and was utilized as a photocatalyst for photocatalytic degradation of certain dyes like mytheme blue and helps in environmental remediation	Nima et al., 2020; Yang et al., 2019

of the inorganic units at the same level as organic building blocks could lead to more homogeneous materials, allowing a fine adjustment of the properties of the material at the molecular and nanoscale levels, generating new materials that retain the characteristics of both initial phases or even new properties. Makishima (2004) classified by far the most accepted for hybrid materials, according to the scale in which the union occurs and the type of the component substances, (1) *Composites*: Mix of different materials dispersed in a matrix at level micrometric, (2) *Nanocomposites*: A mixture of materials of the same type at a sub-micrometric level. (3) *Hybrids*: Mix different types of materials at a sub-micrometric level; and (4) *Nanohybrides*: A mixture at the atomic or molecular level of different materials through chemical bonds. There are several examples of inorganic-organic hybrids, such as layered silicates combined with polymers; clays (aluminosilicates) combined with a polymer such as nylon; silica particles dispersed in a polymeric matrix, etc. Aich et al. (2014) extensively reviewed information on the application potential of nanohybrids; (a) electronic: solar and fuel cells; (2) environmental: contaminant sorption, membrane technology, catalytic–photocatalytic–electrocatalytic applications and antimicrobial–antibacterial processes and devices and (3) medical cancer treatment, health protection, etc.

Some examples or nanohybrids are in Fig. 13.1: (1) Carbon-carbon, such as nanobud (fullerenes covalently bound to a single-wall carbon nanotube) or peapod (fullerenes encapsulated inside a single-wall carbon nanotube), and (iii) nano-onion (multi-shelled fullerenes); (b) Carbon–metallic, Metal–metal, (d) Organic molecule-coated (Aich et al., 2014). Plazas-Tuttle et al., 2015 presented a classification for adaptive nanohybrids based on the environmental stimuli that they respond to (1) pH-responsive; (2) thermo-responsive; (3) photo-responsive; and (4) multi-stimuli-responsive.

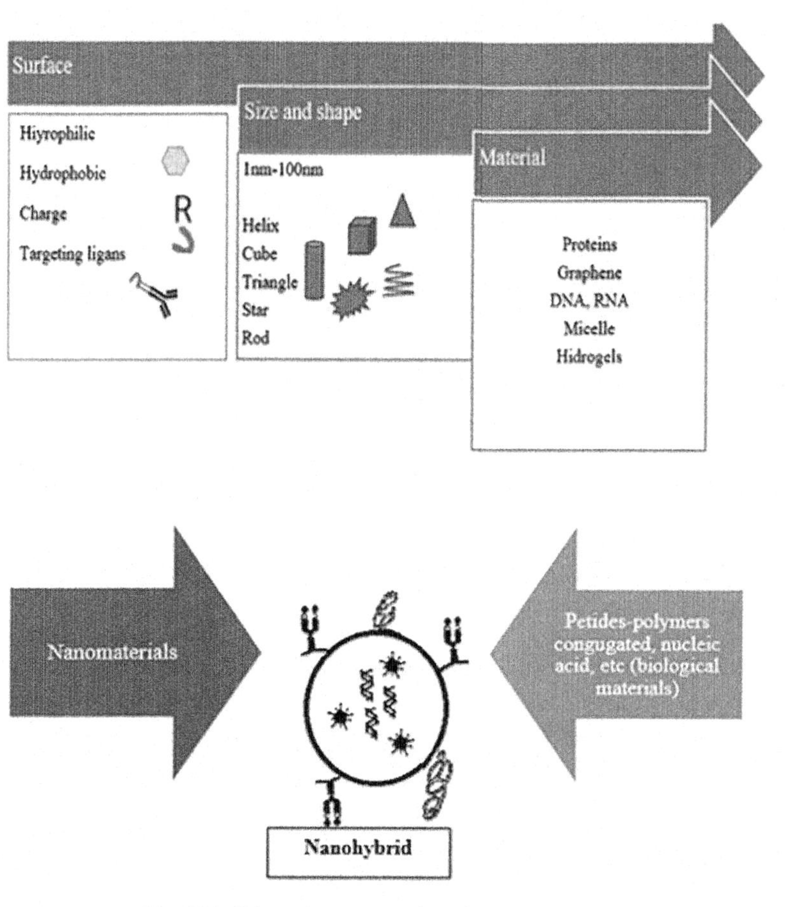

Fig. 13.1. Schematic representation of a nanohybrid.

3.5 Bionanomaterials

Bionanomaterials are defined as materials of molecular origin that are made up of total or partial biological compounds (proteins, enzymes, oligosaccharides, DNA, RNA, viruses) with dimensions of nano order. As a result, these molecules offer novel applications, such as sensors, adhesives, fibers, generate energy, etc. For example, a biosensor consists of a sensitive element and an interface, thus being designed to detect the presence of specific analytes. Therefore, many biodegradable polymers and naturally derived nanomaterials have been extensively used in various biomedical, pharmaceuticals, industrial, packaging and agriculture fields (Honek, 2013; Singh, 2011). Complementary base pairing in DNA integration has been exploited to develop target molecule detectors or modified nanopores. Multiplexed biosensors detect and quantify multi-analytes in a miniaturized device, useful in detecting pathogens. In addition, it has also been possible to bind gold nanoparticles to biomolecules. Like other nanomaterials, bionanomaterials have physical, chemical, biological, mechanical, optical, electrical, catalytic, etc., properties, the dimensions also determine their functionality to a great extent.

Particular properties such as shape, size, chemistry, surface properties and agglomeration allow the design of functional structures, such as binding with proteins and receptors, administering a drug, disinfectant, antibiotic, etc. (Wang et al., 2017). The surface area per volume ratio and the particle size distribution are the two main relational factors. Parameters are required to study the morphology of synthesized bionanomaterials (Boverhof et al., 2015; Bleeker et al., 2013). The size of bionanomaterials broadly defines their application. In the same way, they influence the volume, size and characteristic quantum effects that determine their mechanical properties. When these bionanoparticles are mixed in a common material, they refine the grain and form an intergranular or intergranular structure. Intragranular, improving the grain boundary and increasing the mechanical properties of common materials. For example, when 3% nano palm oil was added to epoxy compounds, their impact and tensile strength increased (Khalil et al., 2013).

3.6 Nanomachines

Micro and nanomachines have recently been developed, and are considered one of the most significant advances that can be propelled by different mechanisms (autoelectrophoresis, bubble propulsion, diffusion phoresis) (Gao and Wang, 2014; Burdick et al., 2008; Mei et al., 2008) and responding to environmental stimuli at the same time (light, magnetic fields, ultrasounds) (Gao and Wang, 2014; Zhang et al., 2019). These investigations were initiated in the biomedical area. However, these nanomachines will have a potential application in environmental bioremediation. Micro/nanomotors are applied for detection and environmental monitoring, such as monitoring the quality of water and air by taking measurements of chemical, biological or physical variables; the main advantage of these devices is high sensitivity and the timing of fast responses (Zhang et al., 2019). By mimicking biological behavior, nanomachines could use search strategies to trace chemical traces back to their source optimizing the demand and consumption of biological fuels (Zhao et al., 2018; Ma et al., 2016; Zhang et al., 2014; Gao et al., 2012). Another important application is the detection of toxic compounds, sometimes within inaccessible areas or hostile environments, for example, those nanomotors whose acceleration is modified concerning the presence of silver ions. Silver-based acceleration has formed the basis for motion-based detection of DNA hybridization relative to silver nanoparticle tags (Zhang et al., 2019; Gao et al., 2012) reported that Janus Ir/Silica micromotors can run on extremely low levels of hydrazine fuel (down to 1 ppb).

4. Biotechnologies to Phytoremediate Hazardous Contaminants

The definition of "phytoremediation", used by some authors, is cleaning or controlling wastes, especially hazardous wastes, mainly by using green plants. Different phytoremediation types, including phreatophytes, are used to manage plumes in groundwater pollutants and contaminated vadose zones.

Photoautotrophs organisms are vascular plants, green algae, cyanobacteria and fungi, which are also involved in biomass synthesis and maintenance, direct metabolism, storage, detoxification or control of pollutants (McCutcheon and Jørgensen, 2008). This benign technique improves the biological quality of soil, air and water, but showed practical limitations as the slow remediation process is subjected to biological cycles (Cameselle and Gouveia, 2019).

For some phytoremediation processes such as detoxification and contaminants accumulation or storage in plants, it is essential to know the metabolism of plants. In saprophytic fungi, this metabolism process is called glycosylation. The contaminant molecules are sequestrated, forming a glycoprotein group by adding a glycosyl group; by this process, the molecules can be transported and stored or transformed by plant cells. Heterotrophs organisms are responsible for the metabolism or mineralization of organic contaminants, which can also accumulate metals and other elements using local accumulations of non-living organic matter and oxidized inorganic compounds (McCutcheon and Jørgensen, 2008).

In different phytoremediation types, the metabolism processes of some microorganisms such as bacteria and cyanobacteria are included in the bioremediation process. The actual biotechnology recognizes their importance in this kind of waste management. There are different types of phytoremediation, including phytodegradation, phytostimulation, biocontainment, rhizofiltration, phytovolatilization, phytoextraction phytoslurry, phytophotolysis and phytostabilization (McCutcheon and Jørgensen, 2008). In Fig. 13.2 most of the standard phytoremediation techniques are shown and how the heavy metals move from roots to leaves to be released in different forms by the plants.

There is research on biotechnologies for the remediation of contaminated air, water and soil, looking for methods that could enhance the plant's purification rates (Wei et al., 2021). These new techniques, including the nanophytoremediation, are involved in using biotechnological resources and nanomaterials to improve the phytoremediation, obtaining higher levels of uptake of contaminants in a short time of the remedial process (Jesitha and Harikumar, 2018).

The biotechnological phytoremediation process also involves the importance of microorganisms in remediation techniques. The interaction between the fungal endophytes or endophytic fungi with their hosts influences the host's overall productivity, growth enhancement and the efficiency of carbon metabolism processes. In various cases, the strategy of host-endophyte symbiotic co-evolution is developed against hazardous contaminants like Heavy Metals (HM). Novel biotechnology is the use

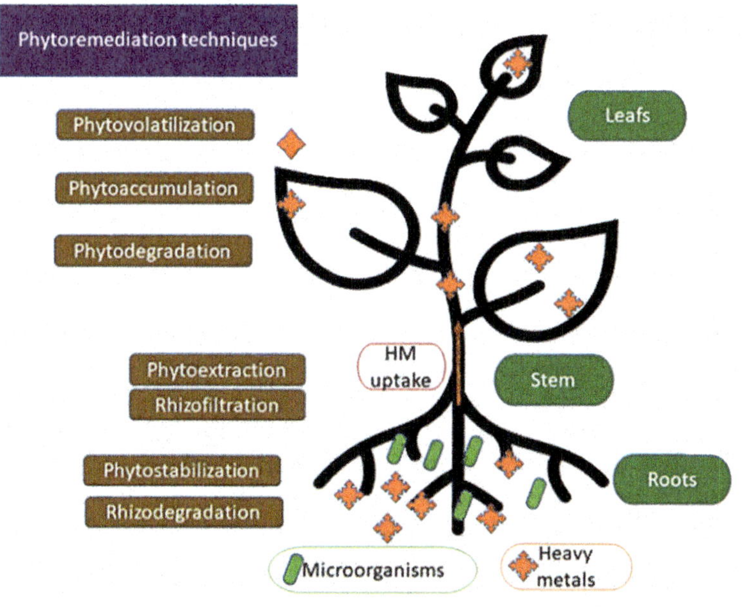

Fig. 13.2. Standard phytoremediation techniques for heavy metals remediation.

of these endophytic fungi as a potent agent of assisted-phytoremediation. These fungi are reported to enhance the plant's ability to sequester, detoxify, accumulate or extract environmentally hazardous pollutants as bio-hazardous contaminants; some examples of these contaminants are the metals and metalloids pollutants, carcinogenic agents, industrial organic waste materials, inorganic pesticides, herbicides, hydrocarbon-based elements and chlorinated products, this biotechnology is called endophyte-assisted phytoremediation technique (Nandy et al., 2020).

Many hazardous contaminants responsible for air pollution come from different sources such as industrial production, vehicles, fossil fuel burning and volatile gases escaping from activities inside houses. These toxic gases can cause ecosystem deterioration and human health hazards, including climate change (Wei et al., 2021). The Particulate Matter (PM), carbon monoxide (CO), Nitrogen Oxides (NOx), Volatile Organic Compounds (VOC), Polycyclic Aromatic Hydrocarbons (PAH), heavy metals (HM) and sulfur dioxide (SO_2) are typical pollutants from anthropogenic origins, also the result of their secondary atmospheric reactions such as peroxyacetyl nitrates (PAN) and ozone (O_3). Another kind of air contaminant is the Persistent Organic Pollutants (POPs), which are not released in large amounts but persevere in the environment for long periods due to their environmental stability (Gawronski et al., 2017).

Phytoremediation is an under exploited solution to enhance air quality. Plants represent a sustainable solution for air contamination. An important characteristic of plants is that they can improve air quality, taking CO_2, releasing O_2 simultaneously through light-dependent photosynthesis and increase air humidity by water vapor transpired from leaves via the stomata, which are microscopic leaf pores (Brilli et al., 2018). The amount of gas exchange depends on many factors such as temperature, humidity, carbon dioxide concentration, light intensity, specific photosynthetic system and air pollution. The adaptation that most plants have to go through includes developing different Photosynthetic Systems (PS), such as C3, C4 photosynthesis, Crassulacean Acid Metabolism (CAM) and facultative CAM. Plants with specific PS occupy different ecosystems, depending mainly on the climate (Gawronski et al., 2017).

Facultative CAM plant species are beneficial for air phytoremediation, as their stomata open at night and take up CO_2 along with pollutants. During the day, they attain other steps of photosynthesis. They are helpful for indoor phytoremediation and recommended for development in vast green roofs, with hot conditions during the day and plants carrying out the gas exchange at night (Gawronski et al., 2017). It is also reported that bacteria colonize the waxy surface of leaves, an average of 10^6–10^7 microorganisms cm^2; these bacteria are dynamically tangled in the reduction of available organic contaminants as toluene, contributing to air remediation, however it is challenging to determine the role of bacteria in remediation (Gawronski et al., 2017).

There is some research in the efficacy of air biofiltration mechanisms, which explains the importance of using living green walls. It was demonstrated that indoor vegetation could remove volatile organic compounds (VOC). In this experiment, a forced-air system was used to draw contaminated air through a green wall based on a soilless growing medium containing activated carbon; the combination of substrate media and botanical component with the biofiltration system led to significant VOC reduction, averaging 57%, indicating a high level of VOC removal efficiency for the active green wall biofilter (Torpy et al., 2018).

Another research in which the use of home plants growing in an adapted growing medium and soil. Both media were abundant in microorganisms, which additionally contribute to the remediation of VOC, using these pollutants as a resource of carbon and energy. The gas conversion with soil is promoted when it is poured with water, which helps to drive air from the soil, followed by a phase of dryness, allowing a new batch of air (contaminated air) to go through the soil pores. Aerial parts and the root zone from the plants removed formaldehyde, using two species, *Ficus japonica* and *Ficus benjamina*, measured day and night. Remarkably, both plants remove formaldehyde equally during the day, while at night, the removal in the soil was done by soil microorganisms as rhizobacteria (Gawronski et al., 2017).

In everyday life, humans are exposed to many artificial organic compounds, such as polychlorinated biphenyls (PCB), that are common complexes used as heat transfer liquids. The PCB enters plants throughout the epidermis. A part of these compounds remain trapped within the cuticle. It is probably metabolized by cytochrome P-450 when it penetrates the cells, as reported elsewhere, in cells of pine needles and eucalyptus leaves. Winter melon (*Benincasa hispida*) can take the di(1-ethylhexyl) phthalate from the air through its leaves, shoots and even flowers, making these species comparable to *Brassica chinensis* and *Brassica campestris* species, effective for phytoremediation, but is a valuable vegetable in warmer regions. It cannot be used for consumption when used for soil remediation (Gawronski et al., 2017).

Most heavy metals (HM) come from the combustion of fossil fuels and are associated with larger raw particles from active mining and mine tailings. Due to this, these particles are present in the air as Particulate Matter (PM). Some HM dominating or present in significant proportions are lead, zinc, cadmium, arsenic, mercury, chromium VI and copper. Certain metals are very toxic and not necessary for plants. The presence of HM in the epidermis from plants should be treated with care as they are taken up from the soil very quickly and translocated to the above-ground parts of plants. Some examples are Cd and metalloids as As. It is essential to note that most plants accumulate PM on their surface. In other words, the plants act as biofilters involved in air phytoremediation (Gawronski et al., 2017).

When the CO rises in the environment, it concerns humans because 750 ppm is toxic for them. On the other hand, plants can tolerate higher concentrations of CO. The source of CO usually comes from the combustion of organic matter in conditions with limited oxygen. The plants metabolize CO by CO_2 oxidation or reduction and further incorporate it into amino acids inside the cells. In studies of 35 woody plant species, more than 15 were seen to have a high ability to remediate CO. Some species of trees are: *A. saccharum*, *A. saccharinum*, *G. triacanthos*, *P. resinosa*, *P. nigra*, *F. pennsylvanica* and two shrubs' species *S. vulgaris* and *Hydrangea* sp. (Gawronski et al., 2017).

Nitrogen (N) is essential in protein block building and is the next most important element after carbon for plants. Nonetheless, plants can only absorb N when it is already fixed to them for uptake. In developed areas, the most dominant is N-containing pollutants such as NOx (most NO_2 gas) due to vehicular combustion of fossil fuels. It was reported that some plants could use N in the form of NO_2. Nearly 200 herbaceous and woody species were investigated due to the ability of NO_2 uptake, reporting more than 600-fold differences in NO_2 between the uptake and assimilation. Some efficient species in remediation include *M. kobus*, *E. viminalis*, *E. grandis*, *E. globulus*, *P. nigra*, *Populus* sp., *R. pseudoacacia*, *S. japonicum*, *P. cerasoides* (Gawronski et al., 2017).

Phytovolatilization implies the uptake of HM and other contaminants by roots from plants and the translocation to the atmosphere via the evapotranspiration process. This metabolization of pollutants transforms them into a gaseous state. Later they are released to the environment. It is notable that volatilizing organic compounds (VOC) such as trichloroethylene (TCE) might emerge at high rates by plants. Some HM can be volatilized in this process (Khan et al., 2020).

In most cases, water pollution comes from industrial wastewater release or from chemical fertilizers and pesticides applied to farmland and produced in sewage, leading to plant eutrophication, decreased freshwater resources, pollution of aquatic ecosystems and causing human diseases (Wei et al., 2021). Many challenges for wastewater phytoremediation exist, such as the impact of metal specification, impacts of ions and organics coexisting, choice of plants species and its farming conditions, phytoremediation time lapses, biomass wastes removal, the plants capacity for accumulation, pest and disease attack on plants, metallic ions bioavailability, invasive plants, agronomic practices and amendments for soil. Plant selection is based on root type, roots depth, plant growth ratio, transpiration ratio, sources of seeds and plants, allelopathy and type of plants, biomass production and harvesting process (Jeevanantham et al., 2019).

Sharma (2021) studied the importance of bacteria associated with plants, enhancing the efficiency of HM phytoremediation from polluted sites and described the advances in wastewater treatment applications. Bacteria-assisted phytoremediation is a strategy that is cost-effective and essential for metal sequestration mechanisms, having high metal biosorption capacities. For the phytoremediation process of metals reduction in wastewater, it is helpful to apply some microorganisms consortia such as Rhizobacteria that are potential growth-promoting, mixed with degrading bacteria and endophytic bacteria. Bacteria genera as *Acidovorax*, *Alcaligenes*, *Bacillus*, *Mycobacterium*, *Paenibacillus*, *Pseudomonas* and *Rhodococcus*, were extensively used in phytoremediation techniques. Plant growth and the biomass directly influence phytoremediation, and is enhanced by microbes, mainly *Bacillus* and *Pseudomonas*, encouraging plant growth (Sharma, 2021).

An environmental problem is wastewaters containing contaminants that are androgenic, mutagenic and carcinogenic compounds, which are toxic even after remediation treatments, including a variety of contaminating metals like Ni, Fe, Zn, Cr, Cd, As, Cu, Mn and Mg, and others from industry such as paper, tannery and distillery. Some other contaminants are androgenic, mutagenic and carcinogenic compounds. The importance of the interaction between Metal-Binding Proteins (MBP) and microbial in supporting the phytoremediation these proteins is reported to improve the accumulation and tolerance to metals like Fe, Cr, Zn, As, Cd, Ni, and Pb, by plants, enhancing the microorganism's role in the binding protein synthesis. Many protection mechanisms exist that enable microorganisms to provide plants protection from HM stress, like compartmentalization, exclusion, complexity rendering and the synthesis of binding proteins. HM-binding proteins, including phytochelatins (PC) and metallothioneins (MT), are rich in cysteine residues and are considered enhancers to the tolerance or resistance in hyperaccumulating plants in the presence of elevated chelating molecules of plant cells like MT and PC compounds. Studying the role of MBP and understanding their mechanisms in cells are helpful in gene engineering of bacteria and obtaining the overexpression of these proteins to enhance their characteristics against HM, making them an option for industrial wastewater phytoremediation and restoration (Sharma et al., 2021).

A technique that helps to degrade organic xenobiotics such as chlorinated hydrocarbons and herbicides is photodegradation. These contaminants affect shallow ground waters, soils and sediments. The process involves plants taking the contaminant and transforming or converting them into less toxic metabolites. Malathion and crufomate are organophosphorus pesticides, photo transformed by aquatic plants like parrot feather, duckweed, elodea. This remediation could benefit other possible applications in large areas like landfills, petrochemical, fuel storage sites and affected agrochemicals. The importance of this technique is that the quantities of the pollutant will be less toxic for plants and be accessible to plant roots uptake (Khan et al., 2020).

Soil pollution is mainly referred to the toxic substances exceeding in the soil and changing its characteristics, affecting crops and animal production, as well as drinking and groundwater, which affect human health. The HM, sewage and urban industrial waste, including large amounts of fertilizers and pesticides applied to farmland, are the primary source of soil pollution (Wei et al., 2021). Even during the efforts of the last 30 yr, there is still no reliable technology to focus on the problems of soil contamination. However, it is still costly and difficult to remove contaminants from soil. Modern techniques are expensive and involve huge quantities of chemicals and energy. Others require special conditions that could change soil characteristics, and sometimes these changes are permanent (Cameselle and Gouveia, 2019).

One plant specie, *Brassica rapa*, can develop and grow in contaminated soil mixed with PAH and metals like Cd, Cr and Pb. This phytoremediation technique is combined with the electro-phytoremediation test. In this research, around the *B. rapa*, 1 ACV/cm potential gradient, resulting on PAH as anthracene and phenanthrene was effectively eliminated, but was not the same for the removal of metal, which was small. From these results, it is suggested that the decisive effect was from the electricity, showing an effect on plant growth and in the production of biomass, using the Alternating Current (AC) for large scale applications, making it the most suitable electric field (Cameselle and Gouveia, 2019).

In the search for phytoremediation of HM, some studies used bamboo. Bamboo species present a high capacity to absorb HM and develop a high ability to adapt to metalliferous environments. Tissues from bamboo can accumulate large amounts of HM in the rhizome and culm, mainly in the cell wall, vacuole and cytoplasm. *Phyllostachys praecox* seen in metal contaminated soils, have high durability, allowing a significant uptake and accumulation of HM. Some management strategies can be helpful to improve phytoremediation with bamboo; these plants are used as HM tolerant species, and when intercropped with hyperaccumulators, are helpful for the application of fertilizers (Bian et al., 2020).

Another essential factor to consider in the research and development of phytotechnology is the effects of climate change. These consequences sometimes indicate that soil remediation failed due the extreme abiotic stress. The harmful effects of climate change are on the production of plants, harvest, phenology and overall efficiency (Nguyen-Sy et al., 2019). It is reported that some important crops, such as rice, wheat, sorghum, maize, chili, pepper, soybean, tobacco and okra, are affected by high-temperature stress. Some of these crops are economically important plants that are highly effective in remediation metal-infiltrated soil. Some transgenic techniques, particularly the CRISPR-Cas9 genome editing system, are involved in making these plants resilient to high temperatures, which will enhance the phytoremediation of HM in the soil, making them a promising biotechnology technique (Sarma et al., 2021).

Many environmental factors are present in soil, such as the initial concentration of metals, pH, texture, temperature, microorganisms and the contaminants' chemical nature, could affect metal uptake in plants. In soil, the phytoextraction technique is a slow process by itself, in remediation of HM, it may require some chemical agents such as ethylenediamine tetra-acetic-acid (EDTA), N-(2-hydroxyethyl)-ethylenediaminetetraacetic acid (HEDTA), diethylenetriamine penta-acetic (acid DTPA) or other synthetic compounds containing properties for acidifying the soil like ammonium sulfate (NH_4SO_4) or ammonium nitrate (NH_4NO_3), which accelerate the phytoextraction process and promote the attachment of HM through plant roots and their forward translocation to aerial parts. After this treatment, it is essential to eliminate the hazards of using other techniques such as *ex situ* phytoextraction and the periodic application of these chemicals (Khan et al., 2020).

As mentioned earlier, some negative effects are the poor quality of soil, air pollution, erosion and flood, due to abandoned mining sites. An example is bauxite mines, which are necessary for the high demand of alumina, which leads to an environmental problem. An option reported is the use of Jatropha curcas to remediate these mining sites. *J. curcas*, used as a phytotranslocation technique, showed that it could grow on this mined soil quickly and is easier to propagate (Rahim et al., 2019; Narayanan et al., 2021).

A process that combines nanotechnology with plants to degrade hazardous materials present in the environment is called phytonanotechnology. Many nanoparticles (NP) are developed to clean up organic pollutants (PAH, PPCP, PCB and organic solvents). A few studies described that the combination of NP with some plant varieties revealed positive results in decontamination of polluted soil. In phytonanotechnology, the NP used should be less toxic and permit the augmentation of plant growth hormones. Sometimes NP increased the plant height and germination biomass. Some challenges using nanomaterials is that they should not be in a higher concentration as they could be toxic for plants. Nutrient absorption can also be inhibited while the protein content, chlorophyll, carotenoids and biomass are lowered (Verma et al., 2021).

Researchers reported that the use of nanomaterials (NM) could improve the HM accumulation by three possible methods: (i) Cell wall permeability improvement, (ii) NM and HM co-transportation, and (iii) Transporter gene expression regulation. Using graphene oxide (GO) in phytoremediation, has shown that it is helpful for removing HM from soil and increasing the As uptake in wheat; GO is also helpful for improving the reduction biomass from plants, the roots quantity exposed to toxins, changing the urea cycle and fab metabolism, transforming from arsenate to arsenite, when oxidative stress is reduced in wheat seedlings compared to HM exposure without remediation (Hu et al., 2014; Kumar et al., 2019).

A study which evaluated the efficiency of nanoparticle nanoscale Zero-Valent Iron (nZVIs) in combination with *Tradescantia spathacea* and *Alternanthera dentata*, claimed that this technique of nanophytoremediation effectively remediates soil contaminated with HM. The obtained results showed that HM uptake was higher in roots in comparison to shoots. *T. spathacea* was used in the phytoremediation technique, accumulated 47% of Pb and 45.3% of Cd, and with nanophytoremediation, the uptake was enhanced, obtained 84.4% of Pb and 64.8% of Cd. In the results using *A. dentata*, it was seen as 60.5% for Pb and 52.5% for Cd; with nanophytoremediation, the uptake accumulated was 73.7% for Pb and 71.3% for Cd. These results proved the efficiency and reinforced the possibility of using nanophytoremediation to remediation soil contaminated with HM and pesticide residues (Jesitha and Harikumar, 2018).

Titanium dioxide (TiO_2), NPs application, enhanced the Cd accumulation from polluted soils and reduced Cd toxicity by improving soybeans, photosynthesis rate and plant growth (Singh et al., 2015). The iron oxide NMs uptake using wheat plants mitigate the toxicity of Cd and Cr in crops. The efficiency of the remediation process depends on the kind of metal contamination, such as in the contamination rates, as the NM used and the plant combination for phytoremediation (Kumar et al., 2019).

There have been many biotechnological techniques in the last years to improve the phytoremediation process, including the importance of microorganisms and their biochemical interactions with plants. Some bacteria and fungi had been bioengineered to assist plant remediation, enhancing the process in remediation of soil, water and air. However there is not enough information about nanomaterials combined with phytoremediation, as most nanoparticles include metals that can capture pollutants but are sometimes unavailable for the plant's uptake. This is still being researched for new nanomaterials. Many research reports have expressed the importance of the uptake of contaminants and their transformation, evidence of obtaining healthy soils, clean water and air, using greener techniques and solving pollution problems in shorter periods.

5. Novel Materials and Plants Working Together to Dissipate Pollutants

Plants have been used for remediation for soils and waters contaminated in several strategies such as wetlands. Pollution of soils, air and waters is the common denominator in an unnatural equation. So, it is our responsibility to clean or restore these matrices. Phytoremediation is an accepted and essential method for treating contaminated soils and waters. As stated by Song et al. (2019), many studies have been performed to enhance the phytoremediation efficiency by several strategies.

Guan et al. (2019) used a Plant Microbial Fuel Cell (PMFC) as a novel technology that involves plants, microbes and electrochemical elements to create renewable energy. They remediated soils polluted by Cr (VI). Comparing different plants and several electrode materials for electricity generation and Cr (VI) removal under different soil Cr (VI) concentrations, reaching a removal efficiency of Cr (VI) of 99%. The total Cr was also reduced, using *Chinese pennisetum*, where the soil was from actual metal-contaminated sites (Guan et al., 2019). They concluded that using PMFCs to remediate contaminated soils is promising, and effects can be attributed to bioelectrochemical processes and plant uptake (Guan et al., 2019).

Another critical subject is germination, this stage is essential for the life cycle of plants and is susceptible to the presence of soil contaminants (Kaur et al., 2017), it is important as it has been used for testing at this stage of the plant for remediation purposes, selecting the possible candidates for phytoremediation. The germination studies are very relevant as they are used as ecotoxicological tools. As a general observation from results where comparing cultivated crop versus native species, it was found that native species were better suited to site conditions. Kaur et al. (2017) recommended selecting candidates for phytoremediation focused on native plant species and their characteristics as species phylogeny, plant morphological and functional characters and tolerance to environmental stresses.

For heavy metal removal, microbial-phytoremediation can be an excellent alternative strategy. Using plants and symbiotic microorganisms (frequently endophytic microorganisms) to remove, transfer or stabilize heavy metals to reduce their concentration or toxicity in the soil (Tang and Ni, 2021). Among the heavy metals or chemical elements considered the most toxic, mercury, lead, arsenic, chromium and cadmium, where the last is highly resistant to corrosion, damages the soil causing irreversible pollution (Tang and Ni, 2021). On the other hand, Arbuscular Mycorrhizal Fungi (AMF) can develop a connection between the soil and plant roots and offers an additional effect to improve the plant's absorption of heavy metals. Hu et al. (2013) compared the efficiencies of phytoextraction by cadmium hyperaccumulation in weeds as *Sedum alfredii* (Hance.) and *Lolium perenne* (L.) that were evaluated with AMF (*Glomus caledonium* 90036 (Gc) and *Glomus mosseae* M47V (Gm)). As a result, both AMF increased the roots colonization rates significantly. Concerning phytoextraction efficiency, it was highlighted that the weed *Sedum alfredii* with 78% led to total Cd extraction.

Perfluoroalkyl and polyfluoroalkyl (PFAs) are substances from an extensive family of artificial compounds (derived from hydrocarbons in which fluorine ions have replaced the protons) that have been used as surface coating and protectant formulations due to their specific surfactant properties. They are artificial persistent organic compounds with a high risk to health and the environment due to their resistance to degradation, the tendency for global transport and ability to bioaccumulate (Jiao et al., 2020). These types of compounds family can be taken up by the foliar and roots. According to Jiaoet al. (2020), studies of kinetics of perfluoro octane sulfonic acid (PFOS) uptake through wheat roots (*Triticum aestivum* L.) seedlings under hydroponics arrangement and uptake of PFOS and perfluorooctanoic acid (PFOA) by maize (*Zea mays* L. cv. TY2), where PFOS use the passive process. On the other hand, PFOA was reported to use a different pathway (Jiao et al., 2020). In the case of foliar uptake, volatized organic compounds are carried out by air and later by aerial deposition. According to Jiao et al. (2020), long-chain PFCAs (C9–C12) were detected in the tree leaves: oriental plane (*Platanus orientalis* Linn), Chinese pine (*Pinus tabulaeformis* Carr.) and white poplar (*Populus alba* sp.), where they found that their total concentrations were above those of short-chain perfluoroalkyl carboxylic acids (PFCAs, C5–C7) by 1.2–50 folds, and ionized and neutral PFAs were also found. It was stated that this was due to predominant foliar uptake of the long-chain compounds or more rapid conversion from their precursors. Concerning plant metabolism of PFAs, it should be kept in mind that there are many PFAs, about 4,730 commercially produced compounds (Jiao et al., 2020), and in a way similar to PAHs, there are many compounds that are not susceptible to transformation or degradation by metabolic pathways of plants or microorganisms, probably due to their structure and arrangement of polyfluoroalkyl chains. However, plant roots can promote biodegradation via a synergistic activity of plant root exudates and microorganisms in the rhizosphere zone. Plant performance can be improved by additional agents such as microbiota or organic elements like sewage sludge. For more information on pathways of plant metabolism, the review of Jiao et al. (2020) can be consulted. More attention and claims of research is a global position of PFAs in agricultural soils and methods of quantification in foods such as vegetables, and these types of chemical compounds could be a little more critical than PAHs due to a lack of knowledge, as they are not visible in the matrix soil as PAHs.

Recently, nanomaterials have been attracting the attention of researchers to find new applications for several technological areas, for example, to improve performance devices. The nanomaterials present interesting characteristics due to their properties that are shown by their nano-size (1–100 nm). Some phytoremediation research has been detailed by Song et al. (2019). An interesting compilation was made by them, for example, nanoparticles (NPs) that were studied to remove some organics compounds as (a) trinitrotoluene, (b) endosulfan, and (c) trichloroethylene using (a) panicum (*Panicum* maximum Jacq.) with zero-valent iron (nZVI), for (b1) Chittaratha (*Alpinia calcarata* Roscoe), (b2) Tulsi (*Ocimum sanctum* L.) with nZVI, and (b3) lemongrass (*Cymbopogon citratus* (DC.) Stapf.), and (c) Eastern cottonwood (*Populus deltoides* Bartr.) with fullerene nanoparticles, reaching removal efficiency for (a) 85.7 to 100% after 120 d, for (b1) 81.2 to 100%, (b2) 20.76 to 76.28%, (b3) 65.08 to 86.16%, and 26 to 82% (with 2 and 15 mg $\cdot$ L^{-1}) fullerene NPs, respectively. Concerning metallic contaminants, the most studied have been Pb, Cd, and As, mainly, plants as ryegrass (*Lolium perenne* L.), sunflower (*Helianthus annuus* L.), rye (*Secale montanum* Guss.), white popinac (*Leucaena leucocephala* (Lam.) de Wit) and maize (*Zea mays* L.) were used for Pb; soybean (*Glycine max* (L.) Merr.), sunflower (*Helianthus annuus* L.), ryegrass (*Lolium perenne* L.), ramie (*Boehmeria nivea* (L.) Gaudich), rye (*Secale montanum* Guss.), white popinac (*Leucaena leucocephala* (Lam.) de Wit) and maize (*Zea mays* L.) were used for Cd; *Isatis cappadocica* (Desv.), ryegrass (*Lolium perenne* L.), sunflower (*Helianthus annuus* L.) and rye (*Secale montanum* Guss.) for as (Song et al., 2019). Several NPs were used in these studies: Nano-hydroxyapatite and nanocarbon black, silicon NP, Salicylic acid NPs, TiO_2NP, nZVI, nano-silica, ZnONPs, AgNPs, for example, and the results were variables (Song et al., 2019) depending on conditions, plants, NPs and contaminants, and according to them, nanotechnology provides a compelling alternative method or strategy with some additional effects as increasing pollutant phytoavailability and promoting plant growth.

Nanomaterials (NMs) as single-walled carbon nanotubes (SWCNTs) and multi-walled carbon nanotubes (MWCNTs) are characterized by several layers of carbon with extraordinary mechanical and electrical properties. This way, nanomaterials can have a substantial potential for phytoremediation due to high capacities of adsorption andabsorption to help plants remove or phytoextract processes of contaminants into the soil matrix. Nanomaterials such as SWCNTs can affect the physiological process of plants as phenotype, genotype (seed germination or DNA damage) and increase the ROS with high variability response for different vegetal species (various stages of plant oncogenesis and varieties). On the other hand, MWCNTs improve the germination process in tomato and rice crops, speed germination of barley, soybean, maize and mustard seeds, stimulate the biomass accumulation (dosage 100 mg L^{-1}), among other, these materials have shown toxic effects at 10–600 mg L^{-1} in suspensions cells of *A. thaliana*, diminishing the chlorophyll content; in rice, increasing the forming of ROS and decreasing the cellular viability. The MWCNTs can form smaller and large agglomerates as an essential factor determining its toxicity (Zaytseva and Neumann, 2016). The study of Gong et al. (2019) demonstrated that MWCNTs at adequate levels can improve the phytoremediation efficiency of some heavy metals such as Cd (promotes oxidative damage in plants by accumulation) in contaminated river sediments. They experimented with ramie (*Boehmeria nivea* (L.) Gaudich) seedlings that were cultivated in Cd-contaminated river sediments at 100, 500, 1,000 and 5,000 mg MWCNTs · kg^{-1}, and found that at 500 mg MWCNTs · kg^{-1} promoted the accumulation and translocation of Cd in these plantlets and alleviated Cd-induced toxicity by stimulating plant growth, reducing oxidative stress, activating antioxidant enzyme activities and increasing specific antioxidant content (Gong et al., 2019). Plant growth inhibition, Cd accumulation reduction and oxidative damage aggravation were found at 5,000 mg MWCNTs · kg^{-1} (Gong et al., 2019). These results showed that one needs to find adequate doses for the application of any nanomaterial as MWCNTs to improve or reach the best conditions for phytoremediation efficiency and at the same time, consider the inevitable release of MWCNTs at high levels would exacerbate metal-induced toxicity to plants (Gong et al., 2019).

A recent strategy combines earlier ones,such as bioremediation, phytoremediation and the use of NMs for some contaminant removal in soil. An example of this is the work of Daryabeigi Zand et al. (2020) who evaluated *Trifolium repens* seedling at several concentrations of NPs (0, 150, 300, 500 and 1,000 mg nZVI · kg^{-1}) and PGPRs. Their results were nZVI and PGPRs positively affect plant establishment and growth in contaminated soil with Sb. Additionally, they found high accumulation of Sb in the shoot than in roots, and nZVI significantly increased the accumulation capacity of Sb (3.8 mg per pot) in this crop. It was also reported that at 1,000 mg nZVI · kg^{-1}, it negatively impacted phytoremediation performance and plant growth. At the same time, using both nZVI and PGPR as an integrated strategy can reach beneficial effects on photosynthesis. However, only the most negligible concentration of nZVI was required with this strategy.

Most research is at the basic level of exploration or experimentation, but has not been applied as yet, combined or integrated strategies as plants, microorganisms and nanomaterials, or strategy innovations are still at the beginning, which requires a significant opportunity for conducting investigations.

6. Conclusion

Pollutants of many classes and types have been discovered in soils, inflicting severe damage. Certain pollutants can alter the biological balance of soil ecosystems and constitute a risk to human health as they are persistent and stay in soils for years or decades. Organic pollutants like chlorinated solvents, polycyclic aromatic hydrocarbon (PAH) insecticides, organophosphorus pesticides and inorganic pollutants like heavy metals and microbiological contaminants such as bacteria and viruses are all found in soil.

Phytoremediation and microremediation are more environmentally friendly and cost-effective than most standard chemical and physical cleanup approaches. Biotechnology's efficacy in regenerating soils degraded with organic and inorganic contaminants has been demonstrated over a long period through extensive research.

Nanotechnology, which is contributing significantly to solve these problems, provides innovative and practical solutions to a wide range of environmental problems. As a result, more attention should be paid to the biological toxicity of nanomaterials used in environmental cleanup technology. Finally, one of the most significant factors inhibiting the widespread use of NPs in the environmental sector is their cost. As a result, the capacity to correctly synthesis NPs using environmentally benign technologies at a reasonable cost will be critical in the future growth of nanotechnology. Although NPs can break down, eliminate and dissipate pollutants, additional research is needed to thoroughly comprehend how they do so. It is also crucial to look into the impact of NPs on the ecosystem on the ground.

Furthermore, integrating computers and other disciplines will allow one to more accurately apply NPs and their environmental effects, enabling to better promote NP technology for pollution stress mitigation in the environment.

Acknowledgments

This research was founded by the projects 'CienciaBásica SEP-CONACyT-151881', 'FONCYT-COAHUILA COAH-2019-C13-C006', and 'FONCYT-COAHUILA COAH-2021-C15-C095', by the Sustainability of Natural Resources and Energy Program (Cinvestav-Saltillo), and by Cinvestav Zacatenco. To Instituto Politécnico Nacional for financial support (Project SIP-20161934, 20172065, and 20181453). G. M.-P., F. F.-L., and F. L.-V. received grant-aided support from 'Sistema Nacional de Investigadores (SNI)', Mexico. To Institute of Agricultural Sciences of the Autonomous University of the State of Hidalgo (ICAP-UAEH).

References

Aich, Nirupam, Jaime Plazas-Tuttle, Jamie R. Lead and Navid B. Saleh. 2014. A critical review of nanohybrids: Synthesis, applications and environmental implications. Environmental Chemistry 11(6): 609–623.

Ansari, Sajid Ali, Mohammad Mansoob Khan, Shafeer Kalathil, Ambreen Nisar, Jintae Lee and Moo Hwan Cho. 2013. Oxygen vacancy induced band gap narrowing of ZnO nanostructures by an electrochemically active biofilm. Nanoscale 5(19): 9238–9246.

Ashiq, Ahmed, Nadeesh M. Adassooriya, Binoy Sarkar, Anushka Upamali Rajapaksha, Yong Sik Ok and Meththika Vithanage. 2019. Municipal solid waste biochar-bentonite composite for the removal of antibiotic ciprofloxacin from aqueous media. Journal of Environmental Management 236: 428–435.

Bang, S.S., Lippert, J.J., Yerra, U., Mulukutla, S. and Ramakrishnan, V. 2010. Microbial calcite, a bio-based smart nanomaterial in concrete remediation. International Journal of Smart and Nano Materials 1(1): 28–39.

Bian, F., Zhong, Z., Zhang, X., Yang, C. and Gai, X. 2020. Bamboo – An untapped plant resource for the phytoremediation of heavy metal contaminated soils. Chemosphere 246: 125750. http: //dx.doi.org/10.1016/j.chemosphere.2019.125750.

Bleeker, Eric A.J., Wim, H. de Jong, Robert E. Geertsma, Monique Groenewold, Evelyn H.W. Heugens, Marjorie Koers-Jacquemijns, Dik van de Meent, Jan R. Popma, Anton G. Rietveld and Susan W.P. Wijnhoven. 2013. Considerations on the EU definition of a nanomaterial: Science to support policy making. Regulatory Toxicology and Pharmacology 65(1): 119–125.

Boverhof, Darrell R., Christina M. Bramante, John H. Butala, Shaun F., Clancy, Mark Lafranconi, Jay West and Steve C. Gordon. 2015. Comparative assessment of nanomaterial definitions and safety evaluation considerations. Regulatory Toxicology and Pharmacology 73(1): 137–150.

Brilli, F., Fares, S., Ghirardo, A., de Visser, P., Calatayud, V., Muñoz, A., Annesi-Maesano, I., Sebastiani, F., Alivernini A., Varriale, V. and Menghini, F. 2018. Plants for sustainable improvement of indoor air quality. Trends Plant Sci. 23(6): 507–512.

Burdick, Jared, Rawiwan Laocharoensuk, Philip M. Wheat, Jonathan D. Posner and Joseph Wang. 2008. Synthetic nanomotors in microchannel networks: Directional microchip motion and controlled manipulation of cargo. Journal of the American Chemical Society 130(26): 8164–8165.

Cameselle, C. and Gouveia, S. 2019. Phytoremediation of mixed contaminated soil enhanced with electric current. J. Hazard Mater. 361: 95–102.

Daryabeigi Zand, A., Tabrizi, A.M. and Heir, A.V. 2020. The influence of association of plant growth-promoting rhizobacteria and zero-valent iron nanoparticles on removal of antimony from soil by Trifolium repens. Environmental Science and Pollution Research 27(34): 42815–42829. https: //doi.org/10.1007/s11356-020-10252-x.

Das, Surajit, Jaya Chakraborty, Shreosi Chatterjee and Himanshu Kumar. 2018. Prospects of biosynthesized nanomaterials for the remediation of organic and inorganic environmental contaminants. Environmental Science: Nano 5(12): 2784–2808.

Davison, Brian H. 2002. Green Biopolymers for Improved Decontamination of Metals from Surfaces: Sorptive Characterization and Coating Properties. Oak Ridge National Lab. (ORNL), Oak Ridge, TN (United States).

Deng, Jiaqin, Xiaodong Li, Xue Wei, Yunguo Liu, Jie Liang, Biao Song, Yanan Shao and Wei Huang. 2020. Hybrid silicate-hydrochar composite for highly efficient removal of heavy metal and antibiotics: Coadsorption and mechanism. Chemical Engineering Journal 387: 124097.

Dongre, Rajendra S., Kishor Kumar Sadasivuni, Kalim Deshmukh, Akansha Mehta, Soumen Basu, Jostna S. Meshram, Mariam Al Ali Al-Maadeed and Alamgir Karim. 2019. Natural polymer based composite membranes for water purification: A review. Polymer-Plastics Technology and Materials 58(12): 1295–1310.

Gao, Wei and Joseph Wang. 2014. The environmental impact of micro/nanomachines: A review. Acs Nano 8(4): 3170–3180.

Gao, Wei, Aysegul Uygun and Joseph Wang. 2012. Hydrogen-bubble-propelled zinc-based microrockets in strongly acidic media. Journal of the American Chemical Society 134(2): 897–900.

Gawronski, S.W., Gawronska, H., Lomnicki, S., Sæbo, A. and Vangronsveld, J. 2017. Plants in air phytoremediation. Phytoremediation 319–346. http: //dx.doi.org/10.1016/bs.abr.2016.12.008.

Ghasemzadeh, Gholamreza, Mahdiye Momenpour, Fakhriye Omidi, Mohammad R. Hosseini, Monireh Ahani and Abolfazl Barzegari. 2014. Applications of nanomaterials in water treatment and environmental remediation. Frontiers of Environmental Science & Engineering 8(4): 471–482.

Gong, X., Huang, D., Liu, Y., Zeng, G., Wang, R., Xu, P., Zhang, C., Cheng, M., Xue, W. and Chen, S. 2019. Roles of multiwall carbon nanotubes in phytoremediation: Cadmium uptake and oxidative burst in Boehmeria nivea (L.) Gaudich. Environmental Science: Nano. 6(3): 851–862. https: //doi.org/10.1039/C8EN00723C.

Gottardo, Stefania, Agnieszka Mech, Jana Drbohlavova, Aleksandra Malyska, Søren Bøwadt, Juan Riego Sintes and Hubert Rauscher. 2021. Towards safe and sustainable innovation in nanotechnology: State-of-play for smart nanomaterials. NanoImpact 100297.

Guan, C.-Y., Tseng, Y.-H., Tsang, D.C.W., Hu, A. and Yu, C.-P. 2019. Wetland plant microbial fuel cells for remediation of hexavalent chromium contaminated soils and electricity production. Journal of Hazardous Materials 365: 137–145. https: //doi.org/10.1016/j.jhazmat.2018.10.086.

Handford, Caroline E., Moira Dean, Maeve Henchion, Michelle Spence, Christopher T. Elliott and Katrina Campbell. 2014. Implications of nanotechnology for the agri-food industry: opportunities, benefits and risks. Trends in Food Science & Technology 40(2): 226–241.

Hayashi, Hiromichi and Yukiya Hakuta. 2010. Hydrothermal synthesis of metal oxide nanoparticles in supercritical water. Materials 3(7): 3794–3817.

Honek, John F. 2013. Bionanotechnology and Bionanomaterials: John Honek Explains the Good Things that can come in very Small Packages. BioMed Central.

Hu, J., Wu, S., Wu, F., Leung, H.M., Lin, X. and Wong, M.H. 2013. Arbuscular mycorrhizal fungi enhance both absorption and stabilization of Cd by Alfred stonecrop (Sedum alfredii Hance) and perennial ryegrass (Lolium perenne L.) in a Cd-contaminated acidic soil. Chemosphere 93(7): 1359–1365. https: //doi.org/10.1016/j.chemosphere.2013.07.089.

Hu, X., Kang, J., Lu, K., Zhou, R., Mu, L. and Zhou, Q. 2014. Graphene oxide amplifies the phytotoxicity of arsenic in wheat. Sci. Rep. 4: 6122.

Jeevanantham, S., Saravanan, A., Hemavathy, R.V., Senthil Kumar, P., Yaashikaa, P.R. and Yuvaraj, D. 2019. Removal of toxic pollutants from water environment by phytoremediation: A survey on application and future prospects. Environ. Technol. Inno. 13: 264–276. http: //dx.doi.org/10.1016/j.eti.2018.12.007.

Jesitha, K. and Harikumar, P.S. 2018. Application of nano-phytoremediation technology for soil polluted with pesticide residues and heavy metals. Phytoremediation 415–439. http: //dx.doi.org/10.1007/978-3-319-99651-6_18.

Jiao, X., Shi, Q. and Gan, J. 2020. Uptake, accumulation and metabolism of PFASs in plants and health perspectives: A critical review. Critical Reviews in Environmental Science and Technology 0(0): 1–32. https: //doi.org/10.1080/1064 3389.2020.1809219.

Kamat, Prashant, V. and Dan Meisel. 2003. Nanoscience opportunities in environmental remediation. Comptes Rendus Chimie 6 (8–10): 999–1007.

Kaur, N., Erickson, T.E., Ball, A.S. and Ryan, M.H. 2017. A review of germination and early growth as a proxy for plant fitness under petrogenic contamination—Knowledge gaps and recommendations. Science of the Total Environment 603: 728–744. https: //doi.org/10.1016/j.scitotenv.2017.02.179.

Keshav, K.S., Akash, S. and Sarita, R. 2021. A study on nanomaterials for water purification. Materials Today: Proceedings. Mater Today Proc. In Press. https: //doi.org/10.1016/j.matpr.2021.07.116.

Khalil, H.P.S. Abdul, Fizree, H.M., Bhat, A.H., Jawaid, M. and Abdullah, C.K. 2013. Development and characterization of epoxy nanocomposites based on nano-structured oil palm ash. Composites Part B: Engineering 53: 324–333.

Khan, M., Shaheen, S., Ali, S., Yi, Z., Cheng, L., Samrana, Khan M.D., Azam, M., Rizwan, M., Afzal, M. et al., 2020. In situ phytoremediation of metals. concepts and strategies in plant sciences. Phytoremediation. Springer, 103–121. http: //dx.doi.org/10.1007/978-3-030-00099-8_4.

Kumar, S., Prasad, S., Yadav, K.K., Shrivastava, M., Gupta, N., Nagar, S., Bach, Q.-V., Kamyab, H., Khan, S.A., Yadav, S. and Malav, L.C. 2019. Hazardous heavy metals contamination of vegetables and food chain: Role of sustainable remediation approaches—A review. Environ. Res. 179(Pt A): 108792.

Liu, Yan-Ju, Xin Lan, Hai-Bao Lu and Jin-Song Leng. 2010. Recent progresses in polymeric smart materials. International Journal of Modern Physics B 24(15-16): 2351–2356.

Ma, Xing, Xu Wang, Kersten Hahn and Samuel Sánchez. 2016. Motion control of urea-powered biocompatible hollow microcapsules. Acs Nano 10(3): 3597–3605.

Makishima, A. 2004. Possibility of hybrid materials. Ceram. Jap. 39(2): 90–91.

Mastronardi, Emily, Amanda Foster, Xueru Zhang and Maria C. DeRosa. 2014. Smart materials based on DNA aptamers: Taking aptasensing to the next level. Sensors 14(2): 3156–3171.

McCutcheon, S.C. and Jørgensen, S.E. 2008. Phytoremediation. Encyclopedia of Ecology, 568–582. http: //dx.doi.org/10.1016/b978-0-444-63768-0.00069-x.

Medina-Pérez, G., Fernández-Luqueño, F., Vazquez-Nuñez, E., López-Valdez, F., Prieto-Mendez, J., Madariaga-Navarrete, A. and Miranda-Arámbula, M. 2019. Remediation of polluted soils using nanotechnologies: Environmental benefits and risks. Pol. J. Environ. Stud. 28(3): 1013–1030.

Mehndiratta, Poorva, Arushi Jain, Sudha Srivastava and Nidhi Gupta. 2013. Environmental pollution and nanotechnology. Environment and Pollution 2(2): 49.

Mei, Yongfeng, Gaoshan Huang, Alexander A. Solovev, Esteban Bermúdez Ureña, Ingolf Mönch, Fei Ding, Thomas Reindl, Ricky K.Y. Fu, Paul K. Chu and Oliver G. Schmidt. 2008. Versatile approach for integrative and functionalized tubes by strain engineering of nanomembranes on polymers. Advanced Materials 20(21): 4085–4090.

Nandy, S., Das, T., Tudu, C.K., Pandey, D.K., Dey, A. and Ray, P. 2020. Fungal endophytes: Futuristic tool in recent research area of phytoremediation. S. Afr. J. Bot. 134: 285–295. http: //dx.doi.org/10.1016/j.sajb.2020.02.015.

Narayanan, M., Natarajan, D., Kandasamy, G., Kandasamy, S., Shanmuganathan, R. and Pugazhendhi, A. 2021. Phytoremediation competence of short-term crops on magnesite mine tailing. Chemosphere 270: 128641.

Nguyen-Sy, T., Cheng, W., Tawaraya, K., Sugawara, K. and Kobayashi, K. 2019. Impacts of climatic and varietal changes on phenology and yield components in rice production in Shonai region of Yamagata Prefecture, Northeast Japan for 36 years. Plant Prod Sci. 22(3): 382–394. http: //dx.doi.org/10.1080/1343943x.2019.1571421.

Nima, Ambika Madhusoodanan, Philomina Amritha, Vidhya Lalan and Ganesanpotti Subodh. 2020. Green synthesis of blue fluorescent carbon nanopheres from the pith of tapioca (Manihot esculenta) stem for Fe (III) detection. Journal of Materials Science: Materials in Electronics 32(23): 21767–21778.

Okesola, Babatunde O. and David K. Smith. 2016. Applying low-molecular weight supramolecular gelators in an environmental setting–self-assembled gels as smart materials for pollutant removal. Chemical Society Reviews 45(15): 4226–4251.

Pendolino, Flavio and Nerina Armata. 2017. Graphene Oxide in Environmental Remediation Process: Springer.

Pérez-Hernández, H., Fernández-Luqueño, F., Huerta-Lwanga, E., Mendoza-Vega, J. and Álvarez-Solís, J.D. 2020. Effect of engineered nanoparticles on soil biota: Do they improve the soil quality and crop production or jeopardize them? Land Degrad Dev. 31(16): 2213–2230.

Plazas-Tuttle, Jaime, Lewis S. Rowles, Hao Chen, Joseph H. Bisesi, Tara Sabo-Attwood and Navid B. Saleh. 2015. Dynamism of stimuli-responsive nanohybrids: Environmental implications. Nanomaterials 5(2): 1102–1123.

Rahim, F.A.A. and Zainuddin, T.H.T.Z. 2019. Jatropha curcas as a potential plant for bauxite phytoremediation. IOP Conference Series: Earth and Environmental Science 308: 012006. http: //dx.doi.org/10.1088/1755-1315/308/1/012006.

Roy, Ipsita and Munishwar Nath Gupta. 2003. Smart polymeric materials: Emerging biochemical applications. Chemistry & Biology 10(12): 1161–1171.

Ryder, Matthew R. and Jin-Chong Tan. 2014. Nanoporous metal organic framework materials for smart applications. Materials Science and Technology 30(13): 1598–1612.

Salem, Salem S. and Amr Fouda. 2021. Green synthesis of metallic nanoparticles and their prospective biotechnological applications: an overview. Biological Trace Element Research 199(1): 344–370.

Sarabia-Castillo, C.R., Pérez-Moreno, A., Galindo-Ortiz, Fraga-Pecina, N., Pérez-Hernández, H., Medina-Pérez, G. and Fernández-Luqueño F. 2020. Phytonanotechnology and environmental remediation. pp. 159–185. In: Thajuddin, N. and Mathew Silvy (eds.). Phytonanotechnology: Challenges and prospects. Elsevier. ISBN: 9780128223482. 336 pp.

Sarma, H., Islam, N.F., Prasad, R., Prasad, M.N.V., Ma, L.Q. and Rinklebe, J. 2021. Enhancing phytoremediation of hazardous metal(loid)s using genome engineering CRISPR–Cas9 technology. J. Hazard Mater. 414: 125493. http: //dx.doi.org/10.1016/j.jhazmat.2021.125493.

Schmidt. 2008. Versatile approach for integrative and functionalized tubes by strain engineering of nanomembranes on polymers. Advanced Materials 20(21): 4085–4090.

Selvan, Subramanian Tamil, Timothy Thatt Yang Tan, Dong Kee Yi and Nikhil R. Jana. 2010. Functional and multifunctional nanoparticles for bioimaging and biosensing. Langmuir 26(14): 11631–11641.

Sharma, P. 2021. Efficiency of bacteria and bacterial assisted phytoremediation of heavy metals: An update. Bioresour. Technol. 328: 124835.
Sharma, P., Pandey, A.K., Udayan, A. and Kumar, S. 2021. Role of microbial community and metal-binding proteins in phytoremediation of heavy metals from industrial wastewater. Bioresource Technology 326: 124750. http: //dx.doi.org/10.1016/j.biortech.2021.124750.
Shrivastava, Sushmita and Rajesh Singh Tomar. 2020. Smart materials: An effective tool for bioremediation. Materials Today: Proceedings 29: 508–511.
Singh, Ravindra P. 2011. Prospects of nanobiomaterials for biosensing. International Journal of Electrochemistry 2011.
Singh, S., Parihar, P., Singh, R., Singh, V.P. and Prasad, S.M. 2015. Heavy metal tolerance in plants: Role of transcriptomics, Proteomics, Metabolomics and Ionomics. Front. Plant. Sci. 6: 1143.
Soler, Lluís and Samuel Sánchez. 2014. Catalytic nanomotors for environmental monitoring and water remediation. Nanoscale 6(13): 7175–7182.
Song, B., Xu, P., Chen, M., Tang, W., Zeng, G., Gong, J., Zhang, P. and Ye, S. 2019. Using nanomaterials to facilitate the phytoremediation of contaminated soil. Critical Reviews in Environmental Science and Technology 49(9): 791–824. https: //doi.org/10.1080/10643389.2018.1558891.
Tang, X. and Ni, Y. 2021. Review of remediation technologies for cadmium in soil. E3S Web of Conferences 233: 01037. https: //doi.org/10.1051/e3sconf/202123301037.
Torpy, F., Clements, N., Pollinger, M., Dengel, A., Mulvihill, I., He, C. and Irga, P. 2018. Testing the single-pass VOC removal efficiency of an active green wall using methyl ethyl ketone (MEK). Air Qual. Atmos. Health 11(2): 163–170.
Tulinski, Maciej and Mieczyslaw Jurczyk. 2017. Nanomaterials synthesis methods. Metrology and Standardization of Nanotechnology: Protocols and Industrial Innovations, 75–98.
Vázquez-Núñez, E., Molina-Guerrero, C.E., Peña-Castro, J.M., Fernández-Luqueño, F. and De La Rosa-Álvarez, M.G. 2020. Use of nanotechnology for the bioremediation of contaminants: A Review. Processes 8(7): 1–17.
Verma, A., Roy, A. and Bharadvaja, N. 2021. Remediation of heavy metals using nanophytoremediation. Advanced Oxidation Processes for Effluent Treatment Plants, 273–296. Elsevier Inc. http: //dx.doi.org/10.1016/b978-0-12-821011-6.00013-x.
Wang, Shaobin, Hongqi Sun, Ha-Ming Ang and Tadé, M.O. 2013. Adsorptive remediation of environmental pollutants using novel graphene-based nanomaterials. Chemical Engineering Journal 226: 336–347.
Wang, Xiumei, Murugan Ramalingam, Xiangdong Kong and Lingyun Zhao. 2017. Nanobiomaterials: Classification, Fabrication and Biomedical Applications: John Wiley & Sons.
Wei, Z., Van Le, Q., Peng, W., Yang, Y., Yang, H., Gu, H., Lam, S.S. and Sonne, C. 2021. A review on phytoremediation of contaminants in air, water and soil. J Hazard Mater. 403: 123658.
Wiesner, Mark R. and Jean-Yves Bottero. 2017. Environmental Nanotechnology: Applications and Impacts of Nanomaterials: McGraw-Hill Education.
Yadav, K.K., Singh, J.K., Gupta, N. and Kumar, V.J.J.M.E.S. 2017. A review of nanobioremediation technologies for environmental cleanup: A novel biological approach. J. Mater. Environ. Sci. 8(2): 740–757.
Yamada, Hirohisa, Yujiro Watanabe and Kenji Tamura. 2006. Development of environmental purification materials with smart functions. Nukleonika 51(suppl. 1): 61–67.
Yang, Xueli, Hao Li, Tai Li, Zezheng Li, Weifeng Wu, Chaoge Zhou, Peng Sun, Fangmeng Liu, Xu Yan and Yuan Gao. 2019. Highly efficient ethanol gas sensor based on hierarchical SnO2/Zn2SnO4 porous spheres. Sensors and Actuators B: Chemical 282: 339–346.
Yoshida, Mutsumi and Joerg Lahann. 2008. Smart nanomaterials. Acs Nano 2(6): 1101–1107.
Zanella, Rodolfo. 2012. Metodologías para la síntesis de nanopartículas: controlando forma y tamaño. Mundo Nano. Revista Interdisciplinaria en Nanociencias y Nanotecnología 5(1).
Zaytseva, O. and Neumann, G. 2016. Carbon nanomaterials: Production, impact on plant development, agricultural and environmental applications. Chemical and Biological Technologies in Agriculture 3(1): 17. https: //doi.org/10.1186/s40538-016-0070-8.
Zhang, Hua, Wentao Duan, Mengqian Lu, Xi Zhao, Sergey Shklyaev, Lei Liu, Tony Jun Huang and Ayusman Sen. 2014. Self-powered glucose-responsive micropumps. Acs Nano 8(8): 8537–8542.
Zhang, Yabin, Ke Yuan and Li Zhang. 2019. Micro/nanomachines: From functionalization to sensing and removal. Advanced Materials Technologies 4(4): 1800636.
Zhao, Xi, Kayla Gentile, Farzad Mohajerani and Ayusman Sen. 2018. Powering motion with enzymes. Accounts of Chemical Research 51(10): 2373–2381.

CHAPTER 14

Conducting Polymers Based Nanocomposites for the Environmental Pollutants Detection

Thatchanamoorthy Thenrajan and *Jeyaraj Wilson**

1. Introduction

Conducting Polymers (CPs) are believed to be valuable materials in the field of electrochemical science as they involve different classes of applications such as medicine, biology and material science, etc. The progress in their nanostructured preparation due to their tunable properties in the electrochemical performance has increased the attention of researchers widening its tertiary in many fields compared to the other existing materials (De Alvarenga et al., 2020). The π-conjugated polymers due to their attractive magnetic and optical nature also gained interest in organic electronics. However an absence in dispersibility and mechanical strength hinders their use in some biomedical applications. Composite formation of CPs with other materials alters their texture and configuration thus producing a platform in device fabrication perspectives Hence, combining CPs with classical materials like graphene, metal/metal oxide, carbon-based composites will yield striking properties combined with advantages in major biomedical and energy applications, i.e., modified electrochemical behavior is very suitable for sensor applications; due to good biocompatibility CPs use in bone tissue engineering and bio coatings; by their impressive optical properties CPs used as fluorescent sensors; owing to high mobility and easy binding affinity CPs were selected for targeted drug delivery and bioactuators. Generally metal oxide, graphene-based materials were widely used as sensor elements; and CPs namely Polyaniline (PANI), Polypyrrole (PPy), Poly (Ethylene dioxythiophene) (PEDOT), Polythiophene (PT) attracted researchers in the detection of neurotransmitters, vitamins, amino acids, food, beverage, industry and environmental application, etc. In this chapter, the CP composites for environmental pollutant detection will be described. In monitoring the environmental hazardous elements, as a result of their electrostatic nature, charged wastes in liquid wastes can be removed by using the assessed CPs deposited on conductive substrates. This can be performed by the application of an external electrical field opposing the pollutant charge which takes place by the electrochemical ion exchange. This will lead to electrostatic attraction of charged wastes and makes

Polymer El, electronics Laboratory, Department of Bioelectronics and Biosensors, Alagappa University, Karaikudi - 630 003, Tamil Nadu.

* Corresponding author: wilson.j2008@yahoo.com

their removal easier. In the case of uncharged CPs, this can be done by the reaction between the primary groups of CPs like amine complexes or nitrogen with the metal ion or other functional groups of pollutants will effectually eliminate the polluted matter through chemical bond formation. Thus, CPs provides a strong platform for the removal, adsorption and detection of the pollutants compared to other materials.

The increase in population and rapid urbanization has led to tremendous growth along with huge wastes. Pollutants are divided as biodegradable and non-biodegradable wastes. Biodegradable pollutants often consist of organic wastes which are easily broken down in the microbial action in normal surroundings whereas, the non-biodegradable pollutants are not readily decomposed and remain in the environment causing harmful effects to land, air and aquatic systems. A large amount of these organic pollutants is emitted by industries which can damage the marine ecosystem and the use of chemical manure in agriculture will lead to stacking of polluted waste, in turn creating side effects when such cultivated crops are consumed by humans. Addressing this issue is very serious these days with the growth in population and with the help of science for these side effects. This chapter on the CPs-based composites for sensing of environmental pollutants is presented to understand this critical situation. Some details of the major pollutants in recent times along with the future scope of the CPs in the scientific society (Fig. 14.1) are also included.

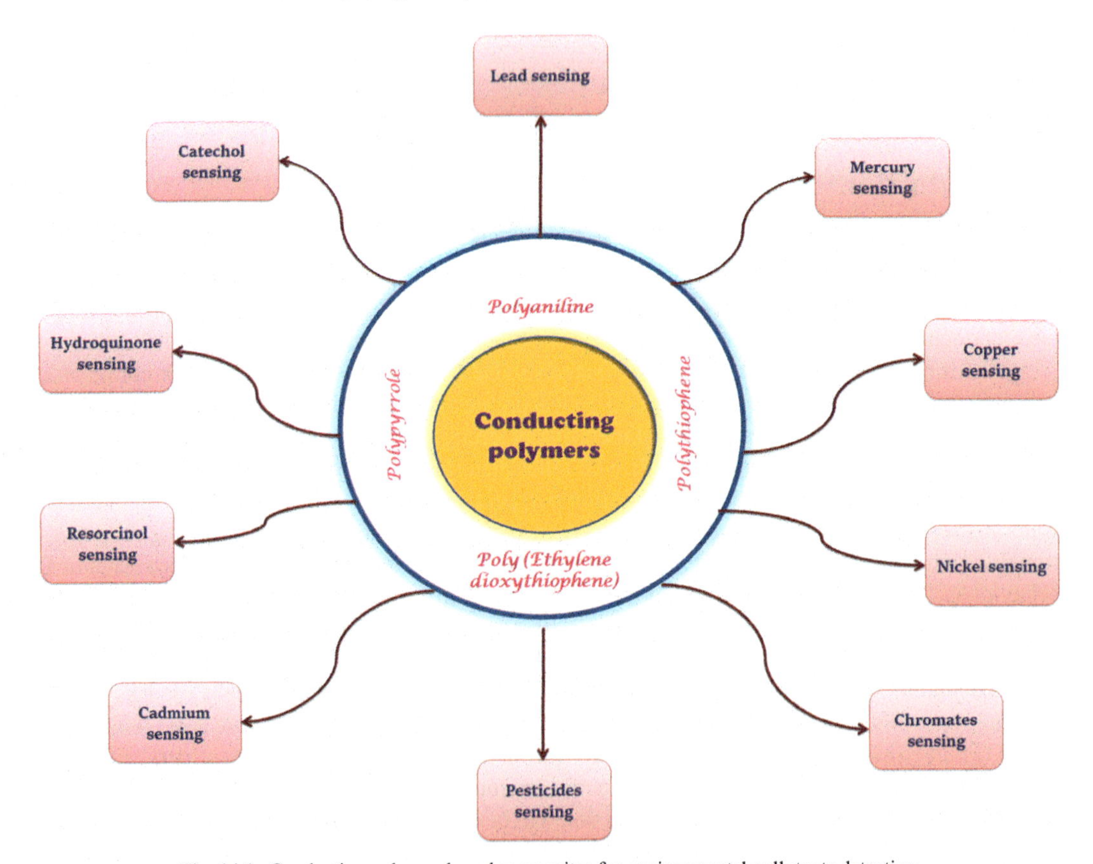

Fig. 14.1. Conducting polymer-based composites for environmental pollutants detection.

2. History

Since many suitable materials were used for the effective sensing of numerous environmental pollutants, the role of CPs finds an important place, i.e., metal/metal oxide composites, graphene-based materials, carbon-based based materials and others produced excellent results, however some drawbacks do appear and hence promoting the use of CPs for better performance are required. Some of the limitations are addressed below.

Metal/metal oxide composites has gained significant attention in the field of biosensors towards various analytes detection since the electrochemical sensors are in less in cost, have a fast response and have an easy operating procedure with real sample analysis. They are easily attracted by the target as a result of its existing metal ion and oxide acting as their primary functional group. Particularly, zinc, iron, copper, aluminum-based metal oxides have been involved for environmental pollutants detection with better results. Due to their doped and mixed composites they are used in many sensor applications such as optical, humidity, conductometric, etc. Their working mode on ionic conductivity creates internal heat which needs to be rectified and they often function in an AC mode which can prevent or normalize their temperature. Some drawbacks could appear with the loss of anti-intereference, low limit detection and accuracy restricts its use. Hence some problems limit their behavior in many applications (Kishore Kumar et al., 2020). In order to enhance their electrochemical response, these materials are made of composites with other metal or metal oxides in a small amount in order to achieve an efficient sensing platform. While zinc oxide, iron oxide, tungsten, etc., based composites were extensively used for biomedical applications, other metal oxides such as gold, silver, copper, aluminum, platinum composites were replaced because of their high cost. Challenges like reduction in material preparation temperature, interferent nature and long-term stability in all environments for designing devices have led researchers to create a need for alternative materials.

Graphene-based materials are gaining interest owing to their fascinating properties and conductivity in the field of biosensors. Graphene is a carbon allotrope with sp^2-hybridized carbon atoms and its family is mainly filled with graphene derivaties originating from pure graphene which are Graphene Oxide (GO), reduced Graphene Oxide (rGO), and graphene-based quantum dots. Their main attraction lies in the modification of both surface and conducting property, easy functionalization for better interaction, high surface to volume ratio and huge electron conductivity. However, pure graphene is costly, easily reactive with heat and temperature, has high electron density, dispersibility are some concerns restricting its use. Graphene-based derivates and composites were used in energy-storage and sensor applications due to their unique actions. Particularly in the field of biosensors, literature has reported graphene-based materials for effective sensing of biomolecules, metal ions and pollutants. Despite such benefits and influence in many areas, it lacks in a few unavoidable ways. The goal remains to produce functionalized graphene composites and derivatives which should be able to create high sensitivity and preciseness at cheaper rates. Furthermore, the prepared materials should withstand in all systems and surroundings providing good reproducibility, long lasting stability and flexibility. In electrochemical sensors, this particular material is still lagging due to its fouling, unnecessary noise and non-specific binding during electrode fabrication/coating and agglomeration when formed composites occur with organic polymers. While investigating its biocompatibility, graphene-based nanomaterials are often found to be toxic for human health which needs to be resolved. This not only affects the human body but the deposition of graphene products may stimulate large eradication of small fishes in the aquatic system which finally leads to their entire destruction. Therefore understanding the behavior in the ecosystem should be clearly addressed in the near future to expand its use in all applications (Kumunda et al., 2021). Eliminating all the defects and the emergence of new materials which replace graphene is urgently required.

Some important aspects can be comparatively rectified and performed well with CPs. First their preparation strategies require minimum time and that the reagents used are cost effective. The fabrication of CPs is much simpler when compared with the other materials and their efficiency in a wide range of detection with the lowest limit have led to their popularity. The conduction in CPs is mainly due to its unbounded or delocalized π bonds which undergoes binding with the neighboring atoms generating charges which are essential for the conversion of insulator to metal. This event is achieved by production of polarons, bipolarons, solitons, etc. The CPs are now synthesized in nanostructured forms like nanotubes, nanowires, nanocubes, etc., which provide a high surface area and create necessary binding sites for the diffusion or electrostatic interaction of the target analyte.

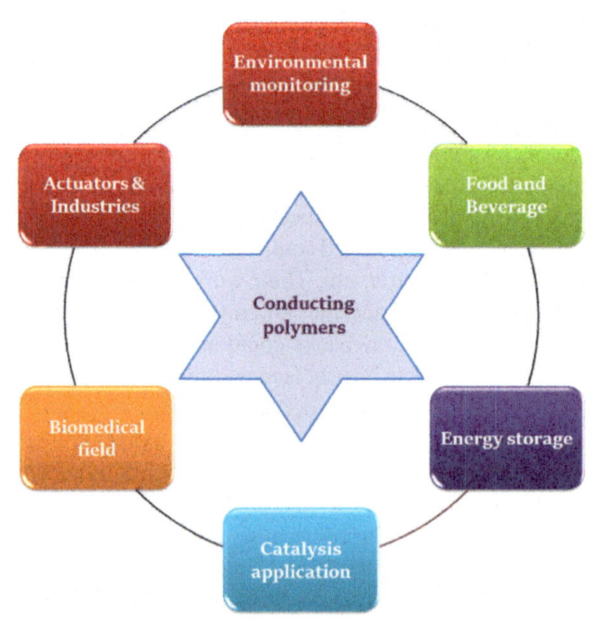

Fig. 14.2. Conducting polymers and its applications.

The composite formation of CPs will reduce the inter domain spacing of the prepared material which in turn proliferates high binding affinity towards target biomolecules. This will enhance the selectivity and sensitivity of the hybrid. Similarly, CPs show better performance than the other existing materials in temperature-related problems in chemresistors (Namsheer and Rout, 2021). The NH_3 attached on the surface of the CPs act as the primary functional group providing necessary interaction with organic and inorganic compounds which is suitable for environmental-pollutant sensing. The role of CPs is not any less than the other materials described earlier. The individual CPs environmental pollutant sensing data are described below in detail (Fig. 14.2).

3. Different Types of Environmental Pollutants

3.1 Hydroquinone

Hydroquinone (HQ) a major industrial pollutant, while it has an important role in pharmaceuticals, cosmetics, oil refinery plants, but some major defects are a result of its discharge in the environment. Its concentration in high levels is toxic to both flora and fauna ecosystem ultimately leading to death. It also leads to spells of fainting, lethargy, skin-disease, head aches, even at a low level. Due to its dangerous effects and low degradability in nature, the US Environmental Protection Agency (EPA) and the European Union (EU) have listed HQ as one of the hazardous pollutants.

3.2 Catechol

Catechol (CC) is a derivative of benzene and is used widely in a number of fields such as oil and photographic industries, including as an inhibitor during a polymerization reaction. It is generally toxic in nature with a strong odor and acts as one of the major pollutants in the marine environment. Due to its hazardous effects on humans, it is classified as Group 2 B carcinogenic agent by the International Agency for Research on Cancer (IARC). Increase in its level leads to problems in the nervous system along with infections related to the gastrointestinal tract, such as a decrease in the function of the liver and deformation of the renal tube causing damage to internal body organs. To protect human from such diseases, the detection of CC is important as its discharge is in a large amount from industries but remains as non-degradable waste in land and water producing health issues to the surroundings consuming/living in it as claimed by EPA.

3.3 Resorcinol

Resorcinol (RC) (1,2-benzenediol) has attracted attention in the food, pharmaceutical and cosmetic industries and is also used as antioxidants. Being a phenolic compound it has some disadvantages such as, irritation in the eyes and skin if exposed directly and also causes gastrointestinal and oral function defects when inhaled or consumed (Ameen et al., 2017). It is generally detected simultaneously with HQ and CC as they are industrial pollutants appearing from some common industries. These phenolic compounds are generally found in contaminated water and also in other items which are prepared using them.

3.4 Lead

Lead (Pb) is one of the endangered metal ion species causing immense problems to human health including the major internal systems. Its prolonged toxicity is generally referred as Pb poisoning. This serious issue that needs to be addressed in human health as exposure leads to the reduction in the fertility rate, fetuses death, neurological disorders, heart and intestine related problems, ultimately leading to mortality in the worst cases (Wasi et al., 2013). Its sources of exposure usually occur when people work with Pb-related products and Pb used product preparation industries. For example, occupations like manufacturing of ships, defense materials, grids for batteries, books, pigments generally use Pb contained/related products which will affect one directly/indirectly. Direct exposure is a continuous process which occurs when a certain part of agricultural land is contaminated with Pb. Food and crops cultivated on them will contain Pb and people consuming them will be affected. On the other hand, water flowing from these areas through pipes will also get affected, poisoning the entire ecosystem during their flow. Other exposures includes inhaling or direct contact through the mouth, skin and eyes and can cause severe health issues. Therefore monitoring this metal pollutant in the environment is urgently required these days.

3.5 Cadmium

Cadmium (Cd) poisoning is also a dangerous threat in global health leading to a number of problems associated with the human body. If exposed to Cd, it could be related to cardio, nervous, reproductive, respiratory, ortho and nephrology related functions to be drastically affected even causing cancer. Considering its serious threat, the IARC has described Cd as a carcinogenic agent in Group 1 for humans. In addition, the infection caused by Cd in an aquatic system leads to mortality in mammalian cells, and at excessive levels in cytoplasmic and nuclear space, it causes death (Wasi et al., 2013). When the contamination is below 1 ppm, it can even cause bacterial and algae death which shows its hazardous nature making its detection very important (Sall et al., 2020).

3.6 Mercury

Mercury (Hg) is considered one harmful heavy metal when environmental outcomes are taken into account. The largest pollution of Hg is from coal mines and its units while other sources are identified as natural and manmade emissions. The emission of gaseous Hg from coal mines generally affects the air and thereby contaminating the surrounding areas. In general, plants absorb Hg from air which settles on the outer portion of leaves and trunks. Hg containing low activity will contaminate the soil and in turn affect plants growing on them when they consume water and nutrients flowing/deposited inside. Even a very small amount of Hg is very dangerous since it is toxic and accumulated inside the body. Since it contains thiol binding agents, it will cause difficulty in leaving the human body as it binds well with the thiol residues in protein.

3.7 Copper

Copper (Cu) is usually found in organic complexes and inorganic salts such as Cu (I) or Cu (II). It is one of the most widely used metals in daily life with a number of applications in major industries. It is also used for bacterial enzymes, proteins involved in transport of electrons during the redox process, However its consumption and exposure creates serious defects in human health. Most of its oral form of intake is from duodenal mucosa of about 30–50% through Cu-binding proteins. Its high level of exposure leads to blocking of bacterial growth of microbes and serves as an important component of bactericides in agriculture. Its accumulation in land and water affects living organisms and crops which could lead to problems for human health.

3.8 Nickel

Nickel (Ni) composites are mainly involved in modem industries for refining, electroplating and production of long-lasting Ni–Cd batteries. The injection of Ni containing aerosols in the environment is largely due to the solid waste incineration which could affect the surroundings (Pichon and Chapuis-Hugon, 2008). Its main form of infection in the human body is through direct exposure from industries by inhaling and studies have pointed out that respiratory problems are caused by Ni. Severe exposure also leads to tumors tested in animal models after examining different modes of Ni infection.

3.9 Chromates

Chromium (Cr) is a metal found in deposits of nature and its compounds have multifarious industrial uses. Large scale industries and combustion of crude oil form different sources, and are the two main reasons of Cr emission in the atmosphere. The advantages of Cr develops from the preparation of metal alloys and coatings to paints, cements and paper industries when they are over used causing defects if mishandled. Their pollution is mainly found in land, air and water ecosystems since they are emitted widely from steel industries, leather processing units and wood preservation plants along with sediments in the soil. Generally, Cr occurs in oxidation states from +2 to +6 with +3 and +6 which is the main component for the presence of a carcinogenic effect of some Cr derivatives. In particular, the hexavalent state is very poor since it is mutagenic and toxic 100 times more than the trivalent form.

3.10 Pesticides based pollutants

Nowadays the main source for pest eradication in modern agricultural methods is due to the invention of pesticides as they are often used by many people in farming and are also classified on their prepared content. If their use is limited no harm can occur, but an increase in their level for controlling pests will ultimately lead to serious issues. This is a matter of grave concern related to problems throughout the world, where many sectors are involved in pollution against public health and the protection of animals (Vikrant et al., 2018). Moreover, water systems connected with these lands have also been completely contaminated by these hazardous factors including canals and ponds. When consumed, it creates a toxic effect creating nerve disorders, cardio diseases and other related problems including cancer induced by various forms of pesticide agents such as dichlorodiphenyltrichloroethane DDT, dichlorodiphenyldichloroethylene (DDE), endosulfan, parathion, malathion, chlordane and atrazine (Wasi et al., 2013). Not only pesticides, insecticides also play a major role of pesticides used about 38% in the world in different forms for pest control over the last five decades (Pundir et al., 2019) fruits, vegetables and processed food as health and environmental hazardous compounds. Thus, detection of these harmful OP pesticides at an ease with high sensitivity and selectivity is the need of hour. Bio-sensing technology meet these requirements and has been employed at a large scale for detection. The present review is aimed mainly to provide the overview of the past and recent advances occurred in the field of biosensor technology employed for the detection of these OP compounds. The review describes the principle and strategy of various OP biosensors including electrochemical (amperometric, potentiometric. Fat soluble pesticides are used more as they can accumulate and stay in the body causing harmful effects leading to increase of food chain levels. Farmers use these pesticides regularly in their lands for the controlling pests usually developing skin and eye related issues. Long time exposure leads to nervous problems, hormones mimicking ultimately leads to death and due to pesticide poisoning nearly one million workers have been affected by this (Rawtani et al., 2018).

In order to discover these hazardous compounds at an early stage for global safety, various materials were applied with satisfactory results. In this chapter, CPs are described for their efficiency and outstanding advantages compared with other materials.

4. Conducting Polymer-based Composites

4.1 Polyaniline based composites

Polyaniline (PANI) has been widely used as efficient materials in synthesizing composites for energy and sensor applications, possessing numerous properties like cheaper rate, convenient tunable nature and large scale production (Zhang et al., 2020). Due to the π–π interactions, PANI based polymers having a benzene ring structure readily forms composites with aromatic compounds with remarkable properties and thus is used in extraction of solid phase wastes from a liquid media. In composite formation with many materials, it has also been widely used for the detection of other pollutants. Detection of 4-aminophenol by electrochemical method involving graphene and carbon as composite materials with PANI was reported earlier. In this work, 4-aminophenol reacts with the adsorbed oxygen on the PANI/G/CNT oxidizing to quinoneimine plus H^+ ion, released free electrons (Rahman et al., 2018). The same GO was made to form composite with Molecularly Imprinted Polymer (MIP) along with PANI for the effective sensing of *p*-nitrophenol where the possible interaction is mainly due to the dimerization reaction (Saadati et al., 2018). Similarly, for the same analyte MIP composite of PANI and polyvinyl sulfonic acid (PVSA) was also investigated. The effects due to the 'Zipper effect' led to good sensing results (Roy et al., 2013). With cotton fibers, flexible sensors made with PANI were designed for superior selectivity towards ammonia (NH_3). This occurred due to the reactions between the PANI and NH_3 that contributed equally for effective detection results (Zhang et al., 2020). Carbon nanotubes (CNTs) were also used with PANI as an efficient composite for the environmental pollutant malathion detection on graphite as an electrode. The CNTs are responsible for improving the sensitivity and the results are due to the oxidation process of malathion enhancing the reduction of PANI which in turn increases the peak current (Ebrahim et al., 2014). Nanostructured films from polystyrene (PS) formed core shell particles with PANI were examined for the sensor behavior of dry gas flow, ethanol vapor, hydrogen chloride and ammonia where the performance was greatly influenced by PANI shells packed in the films. Protonation due to HCL doping also produced radical cations leading to superior outcomes of the composite (Yang and Liau, 2009). Similarly, PANI coating for protection of steel reinforcements in concrete was studied along with the corrosion inhibition. The results indicated the appreciable corrosion resistant of PANI in ambient systems (Saravanan et al., 2007). A superparamagnetic attapulgite/Fe_3O_4/PANI (ATP/Fe_3O_4/PANI) nanocomposite was successfully synthesized for the detection of benzoylurea in environmental water samples. The report demonstrated excellent sensing owing to the π-π interactions and electrostatic attraction in the determination of benzoylurea by the composite (Yang et al., 2016). PANi-α-Fe_2O_3–MoS_2-DNA hybrid composite for the detection of HQ, CC, RC and nitric oxide in water samples where DNA played a key role in enhancing the conductive behavior of the composite. Combining the interaction of amine and electrostatic interaction with the OH of the pollutants gives excellent sensitivity in simultaneous sensing (Thivya and Wilson, 2019). Generally, with metal/metal oxide and composites,

PANI forms both an electrostatic interaction and a covalent bond with the target analyte as a result of the existing metal ion and NH_3 groups; while with other CPs and natural polymer composites it forms inter hydrogen bonding and a covalent bond due to the presence of OH and NH_3 groups. This makes it more efficient in sensing a platform than the other materials (Table 14.1).

Table 14.1. Representation of various composites with PANI for different analytes.

	Composites	Analytes	Possible Interaction
PANI	graphene/carbon nanotubes	4-aminophenol	reaction of aminophenol with the adsorbed oxygen on the PANI/G/CNT
	molecularly imprinted polymer (MIP)/GO	*p*-nitrophenol	production of a radical species due to phenol group oxidation leading to the dimerization reaction
	molecularly imprinted polymer (MIP)/polyvinyl sulfonic acid (PVSA)	*p*-nitrophenol	Zipper effect
	cotton	ammonia	reaction between NH_3 and PANI
	single walled carbon nanotubes	malathion	alternate redox process between the material and analyte
	polystyrene	dry gas flow, ethanol vapor, hydrogen chloride, and ammonia	formation of radical cations produced by HCl doping
	superparamagnetic attapulgite/Fe_3O_4	benzoylurea	π–π interactions and electrostatic attraction with benzoylurea
	α-Fe_2O_3–MoS_2-DNA	HQ, CC, RC and nitric oxide	amine and electrostatic interaction with OH group of environment pollutants

4.2 Polypyrrole based composites

Polypyrrole (PPy), is known for its features such as being hydrophilic, easily oxidized, water soluble and its mass change depends on the relative humidity of air (Syritski et al., 1999). It has good environmental stability and its surface modifications can be achieved by doping with appropriate selection of materials in preparation time. Its properties of conductivity, redox activity and simple synthesis strategies have led to its importance in the research of sensor fabrications. Due to these parameters, applications such as batteries, supercapacitors, fuel cells etc., were explored with PPy composites emerging in a commercial field. Hollow PPy composite synthesized for detection of toxic herbicide was achieved with good results and also effectively in real sample analysis (Abraham and Vasantha, 2020). PPy made composite with mercaptoacetic acid (PPy/MAA) was used for removal of noble metal ions (Ag^+). This metal ion loaded composite (PPy/MAA/Ag^0) was found to have antimicrobial activity and was involved in the reduction of 4-nitrophenol and NO_2 in gas-sensing applications (Das et al., 2017). PPy films deposited on QCM disks were investigated for the fabrication of environmental sensors (Syritski et al., 1999). Graphene quantum dots prepared with molecularly imprinted PPy as composite was used for detection of bisphenol A with satisfactory results (Tan et al., 2016). Using carbon fibers as substrates, doped PPy films were used for sensitive and selective nitrate sensing (Bendikov et al., 2005). PPy was also used in humidity sensors where it showed satisfactory results. Another report using alumina as substrates, metal oxide TiO_2 nanoparticles with PPy composite thin films were fabricated as resistive-type humidity sensors (Su and Huang, 2007). Functionalized PPy core–shell nanofibers mats were proposed for the extraction of trace polar analytes disulfonated (acid yellow 9) and monosulfonated azo dyes from environmental water samples (Qi et al., 2015). Silver molybdate (Ag_2MoO_4) with PPy as a nanocomposite platform was investigated as both photocatalyst electrocatalyst for simultaneous pollutant sensing of heavy metals such as Methylene Blue (MB), (Cr(VI)), ciprofloxacin (CIP) and azomycin (Abinaya et al., 2019). Many successful applications have shown the efficacy of PPy gas sensors for electronic nose fabrications for environmental monitoring and also for the detection of various analytes like beverage samples, cattle wastes and sewage (Ameer and Adeloju, 2005). More involvement and improvement in this area of research by will provide a stable platform for immunosensors which will predominantly give rise to point of care and hands on portable sensor devices (Ramanavičius et al., 2006). Due to its good catalytic activity, stability in the presence of electrolytes, PPy provides the necessary abilities for effective composite formation and sensor performance (Table 14.2).

Table 14.2. Representation of various composites with PPy for different analytes.

	Composites	Analytes	Possible Interaction
PPy	mercaptoacetic acid	4-nitrophenol , metal ion (Ag^+) and NO_2	ionic interaction between AgNPs and 4-nitrophenolate by PPy-MAA matrix
	graphene quantum dots	bisphenol A	reversibe binding of the composite imprinted sites to Bisphenol A
	alkyl aryl sulfonate (AAS) film	fenuron	electrostatic interaction between the phosphate anions of composite with nitrogen atoms in fenuron
	TiO_2 nanoparticles	humidity	electrostatic Interaction of ions (Ag+) and TiO_2
	functionalized core–shell electrospun nanofibers	disulfonated and monosulfonated azo dyes	π–π interaction and electrostatic attraction
	silver molybdate	methylene blue (MB), ciprofloxacin (CIP) and azomycin	formation of hydroxyl radicals

4.3 Poly (Ethylene dioxythiophene) based composites

Among various CPs, Poly(3,4-ethylenedioxythiophene) (PEDOT) also plays a vital role due of its high conductivity, excellent environmental stability, easy processing, efficient synthesis, low cost and commercial availability. PEDOT has attracted large research efforts worldwide and found widespread applications in different areas of electrochemical sensors, supercapacitors, organic light emitting diodes, solar cells, display devices and corrosion protection materials. Another striking property of PEDOT is its unique ability to catalyze some electrode reactions, such as iodine-iodide redox reaction, water splitting, hydrogen peroxide redox reaction and oxygen reduction reaction. As reported recently, the formation of polaronic states in PEDOT, leads to the decrease of the HOMO-LUMO gap to enhance the reactivity of the system. Many studies use composite materials containing PEDOT as one component, such as various PEDOT/metal nanoparticles, PEDOT/carbon-based materials, etc., to constitute modified electrode materials to study different electro-active molecules (Tian et al., 2019). The PEDOT film prepared with better electrocatalytic behavior and sensing performance exhibited significant selectivity and sensitivity towards the detection of BPA. The sensor ability of PEDOT: poly(styrene sulfonate) PSS was studied for its efficiency in quinonic compounds reduction. Herea PEDOT:PSS immobilized tyrosinase sensing platform was fabricated, which showed good results in achieving a wide linear range and low detection limit for the estimation of phenolic elements (Moczko et al., 2012). Multiwall Carbon Nanotubes-doped Poly (3, 4-ethylenedioxythiophene) was reported for the environmental pollutant Bisphenol A. PEDOT has a high affinity for organic molecules, while MWCNT can provide an excellent electron-transport channel thus these combinations produced good sensing results (Tian et al., 2019). At the same time PPy, molecularly imprinted PEDOT was also synthesized for the detection of 2,4-dichlorophenol on paper electrode using carbon fiber. The effective interaction of MIP with the target analyte produced a good sensor performance. The measure of strength of interaction between the target molecule was given by β, the imprinting factor, which was a larger value of about 2.9384 indicating greater selectivity of the sensor (Maria et al., 2020). Recently, a PEDOT:PSS functionalized ZnO inkjet-printed paper sensor was evaluated by hydrazine sensing involving nafion as a binding element was performed. When the pH of the solution is close to the pKa value of hydrazine, it diffuses through nafion and interaction with ZnO giving a good performance (Beduk et al., 2020). These paper-based sensors will be very useful as point of care devices in future device fabrications. Moreover, the high conductivity, low redox potential and other excellent properties make the PEDOT a suitable material for the sensor field application (Table 14.3).

Table 14.3. Representation of some composites with PEDOT for different analytes.

	Composites	Analytes	Possible Interaction
PEDOT	Tyrosinase immobilized on PSS	phenolic compounds	covalent and non-covalent interactions
	Carbon fiber paper electrode	2,4-dichlorophenol	covalent or non-covalent bond
	Poly(styrene sulfonate) functionalized with zinc oxide (ZnO)	hydrazine	hydrazine diffusion through nafion and interaction with ZnO

4.4 Polythiophene based composites

Polythiophene (PT) and its derivatives have not only attracted significant attention in electrochemical science but also in optical sensor applications as a result of its diverse properties like material formation and high mechanical stability. In addition, PT shows easy operation and tunable properties, durability and strong absorption in visible regions (Chandra et al., 2017). It has been used in many areas including chemical, optical sensors, light-emitting diodes, solar cells, rechargeable batteries, supercapacitors and biosensors because of its high environmental and thermal stabilities. Especially in sensor applications, PT is made composite with many materials owing to its excellent properties such as outstanding electrocatalytic activity, maximum conductivity and strong surface adsorptive ability. The ability of PTs gas sensing comes from its preparation in both chemical and electrochemical means which is widely used by researchers for determining various gases and Volatile Organic Compounds (VOC). It includes a number of compounds responsible for pollutants in the air when interacting with PT causing changes in their physical appearance and electrical behavior, but easily evaporates at ambient conditions. The detection of VOCs (methanol, dichloromethane, hexane, toluene, chloroform, tetrahydrofuran, water, etc.), emission levels is currently of wide interest (Gonçalves et al., 2010). Polymers band gap may be altered as a result of the change in polaron densities owing to the interaction of the organic compound and gas molecules. Some disadvantages such as swelling, reduction of conductivity may occur when CPs are exposed to the vapors of solvent chemicals. As PT based derivatives have a fluorescent nature they have been used in a fluorescence sensor. The detection of Cu^{2+}, Pb^{2+} and Cd^{2+} metal ions using PT along with nucleotides and amino acids was investigated (Maiti et al., 2009). PT incorporated with pendant terpyridine groups led to simultaneous detection of Co and Cu ions showed good sensing results. Generally, to avoid the interferents activity, electrode surface deposited with metal ions along with polymer films for the metal ion detection will attain selective results (Savan et al., 2014). Gas sensing properties of PT supported tin-doped titanium nanocomposites were evaluated in different hazardous gases. Adsorption capability of the composite can be increased with PT coated metal oxide surface exposed to LPG as an electron transfer between the materials and the analyte leading to effective sensing (Chandra et al., 2017). Methanol sensing using the PT/α-Fe_2O_3 nanocomposite platform was performed by Farid A. Harraz et al. (2020) the synergetic effect produced more active sites which leads to enhance the adsorption/diffusion of the gas molecule with the composite. Due to the fluorescent nature, there is more scope in optical sensors for VOCs detection and adsorption of pollutants in liquid matter (Table 14.4).

Table 14.4. Representation of various composites with PPy for different analytes.

	Composites	Analytes	Possible Interaction
PT	nucleotides and amino acids	Cu^{2+}, Pb^{2+} and Cd^{2+}	electrostatic interaction
	α-Fe_2O_3	methanol	adsorption-diffusion of methanol molecules on the modified electrode surface
	PT derivatives	LPG	electrostatic interaction
	pendant terpyridine groups	Co and Cu ions	electrostatic interaction
	tin doped titanium nanocomposite	LPG	exchange of electrons between the LPG and oxide surface

5. Conclusion with Future Scope

With the intervention of nanotechnology significant progress has been made in the preparation route, surface modifications and other significant properties of CPs. From supercapacitors to biomedical applications has led to good results and given sufficient support to global research on CPs. Nanostructured Cps has improved the molecular structure, crystallinity, conjugate length and electrical conductivity. MIPs are cheap and very useful for natural detection. The porous 3D nanostructured CPs provides high surface area and affinity to analytes of interest, thus playing an excellent immobilization platform. However, some of the defects in CPs need to be rectified in the future. Reduction in size to form 1D structures, control of deposition rate and increased solubility should be achieved for improving CPs. Properties like irradiation, sonication during synthesis are required to have good reproducibility than carbon-based materials. On continuous measurement, the analytical signal fades with time due to the swelling of the polymer layer. The stability, resistant to high temperature, dispersibility also needs to be addressed to bring the CPs towards device fabrication and point of care devices in future. It is expected to design a flexible high-performance substrate which could be designed with CPs nanostructured materials. Thus CPs-based composites could offer a large scope for future generation sensor instruments. It is hoped that this chapter provides the necessary details on the CPs for environmental pollutants detection along with the rectification of its defects.

References

Abinaya, M., Rajakumaran, R., Chen, S.M., Karthik, R. and Muthura, V. 2019. *In situ* synthesis, characterization, and catalytic performance of polypyrrole polymer-incorporated Ag2MoO4 nanocomposite for detection and degradation of environmental pollutants and pharmaceutical drugs. ACS Appl. Mater. Interfaces 11: 38321–38335. https://doi.org/10.1021/acsami.9b13682.

Abraham, D.A. and Vasantha, V.S. 2020. Hollow polypyrrole composite synthesis for detection of trace-level toxic herbicide. ACS Omega 5: 21458–21467. https://doi.org/10.1021/acsomega.0c01870.

Ameen, S., Kim, E.B., Akhtar, M.S. and Shin, H.S. 2017. Electrochemical detection of resorcinol chemical using unique cabbage like ZnO nanostructures. Mater. Lett. 209: 571–575. https://doi.org/10.1016/j.matlet.2017.08.100.

Ameer, Q. and Adeloju, S.B. 2005. Polypyrrole-based electronic noses for environmental and industrial analysis. Sensors Actuators, B. Chem. 106: 541–552. https://doi.org/10.1016/j.snb.2004.07.033.

Beduk, T., Bihar, E., Surya, S.G., Robles, A.N.C., Inal, S. and Salama, K.N. 2020. A paper-based inkjet-printed PEDOT:PSS/ZnO sol-gel hydrazine sensor.

Bendikov, T.A., Kim, J. and Harmon, T.C. 2005. Development and environmental application of a nitrate selective microsensor based on doped polypyrrole films. Sensors Actuators, B. Chem. 106: 512–517. https://doi.org/10.1016/j.snb.2004.07.018.

Chandraa, M.R., Siva Prasada Reddy, P., Raoa, T.S., Pammic, S.V.N., Siva Kumar, K., Vijay Babub, K., Kiran Kumare, Ch. and Hemalatha, K.P.J. 2017. Enhanced visible-light photocatalysis and gas sensor properties of polythiophene supported tin doped titanium nanocomposite. J. Phys. Chem. Solids 105: 99–105. https://doi.org/10.1016/j.jpcs.2017.02.014.

Das, R., Giri, S., King Abia, A.L., Dhonge, B. and Maity, A. 2017. Removal of noble metal ions (Ag+) by mercapto group-containing polypyrrole matrix and reusability of its waste material in environmental applications. ACS Sustain Chem. Eng. 5: 2711–2724. https://doi.org/10.1021/acssuschemeng.6b03008.

De Alvarenga, G., Hryniewicz, B.M., Jasper, I., Silva, R.J., Klobukoski, V., Costa, F.S., Cervantes, T.N.M., Amaral, C.D.B., Schneider, J.T., Toledo, L.B., Zamora, P.P., Valerio, T.L., Soares, F., Silva, B.J.G. and Vidotti, M. 2020. Recent trends of micro and nanostructured conducting polymers in health and environmental applications. J. Electroanal. Chem. 879: 114754. https://doi.org/10.1016/j.jelechem.2020.114754.

Ebrahima, S., El-Raeyb, R., Hefnawya, A., Ibrahimb, H., Solimana, M. and Abdel-Fattah, T.M. 2014. Electrochemical sensor based on polyaniline nanofibers/single wall carbon nanotubes composite for detection of malathion. Synth. Met. 190: 13–19. https://doi.org/10.1016/j.synthmet.2014.01.021.

Gonçalves, V.C., Nunes, B.M., Balogh, D.T. and Olivati, C.A. 2010. Detection of volatile organic compounds using a polythiophene derivative. Phys. Status Solidi. Appl. Mater. Sci. 207: 1756–1759. https://doi.org/10.1002/pssa.200983723.

Harraza, F.A., Faisal, M., Jalalaha, M., Almadiyc, A.A., Al-Sayaria, S.A. and Al-Assiri, M.S. 2020. Conducting polythiophene/α-Fe2O3 nanocomposite for efficient methanol electrochemical sensor. Appl. Surf. Sci. 508: 145226. https://doi.org/10.1016/j.apsusc.2019.145226.

Kishore Kumar, D., Raghava Reddy, K., Sadhu,V., Shetti, N.P., Venkata Reddy, Ch., Chouhan, R.S. and Naveen, S. 2020. Metal oxide-based nanosensors for healthcare and environmental applications. INC.

Kumunda, C., Adekunle, A.S., Mamba, B.B., Hlongwa, N.W. and Nkambule, T.T.I. 2021. Electrochemical detection of environmental pollutants based on graphene derivatives: A review. Front. Mater. 7: https://doi.org/10.3389/fmats.2020.616787.

Maiti, J., Pokhrel, B., Boruah, R. and Dolui, S.K. 2009. Polythiophene based fluorescence sensors for acids and metal ions. Sensors Actuators, B Chem. 141: 447–451. https://doi.org/10.1016/j.snb.2009.07.008.
Maria, C.G.A., Akshaya, K.B., Risona, S., Varghesea, A. and Georgea, L. 2020. Molecularly imprinted PEDOT on carbon fiber paper electrode for the electrochemical determination of 2,4-dichlorophenol. Synth. Met. 261: 116309. https://doi.org/10.1016/j.synthmet.2020.116309.
Moczko, E., Istamboulie, G., Calas Blanchard, C., Rouillon, R. and Noguer, T. 2012. Biosensor employing screen-printed PEDOT:PSS for sensitive detection of phenolic compounds in water. J. Polym. Sci. Part A Polym. Chem. 50: 2286–2292. https://doi.org/10.1002/pola.26009.
Namsheer, K. and Rout, C.S. 2021. Conducting polymers: A comprehensive review on recent advances in synthesis, properties and applications. RSC Adv. 11: 5659–5697. https://doi.org/10.1039/d0ra07800j.
Pichon, V. and Chapuis-Hugon, F. 2008. Role of molecularly imprinted polymers for selective determination of environmental pollutants—A review. Anal. Chim. Acta 622: 48–61. https://doi.org/10.1016/j.aca.2008.05.057.
Pundir, C.S., Malik, A. and Preety. 2019. Bio-sensing of organophosphorus pesticides: A review. Biosens. Bioelectron 140. https://doi.org/10.1016/j.bios.2019.111348.
Qi, F., Li, X., Yang, B., Rong, F. and Xu, Q. 2015. Disks solid phase extraction based polypyrrole functionalized core-shell nanofibers mat. Talanta 144: 129–135. https://doi.org/10.1016/j.talanta.2015.05.040.
Rahman, M.M., Hussein, M.A., Alamry, K.A., Al-Shehry, F.M. and Asiri, A.M. 2018. Polyaniline/graphene/carbon nanotubes nanocomposites for sensing environmentally hazardous 4-aminophenol. Nano-Structures and Nano-Objects 15: 63–74. https://doi.org/10.1016/j.nanoso.2017.08.006.
Ramanavičius, A., Ramanavičiene, A. and Malinauskas, A. 2006. Electrochemical sensors based on conducting polymer-polypyrrole. Electrochim. Acta 51: 6025–6037. https://doi.org/10.1016/j.electacta.2005.11.052.
Rawtani, D., Khatri, N., Tyagi, S. and Pandey, G. 2018. Nanotechnology-based recent approaches for sensing and remediation of pesticides. J. Environ. Manage 206:749–762. https://doi.org/10.1016/j.jenvman.2017.11.037.
Roy, A.C., Nishaa, V.S., Dhanda, C., Azahar Alia, Md. and Malhotra, B.D. 2013. Molecularly imprinted polyaniline-polyvinyl sulphonic acid composite based sensor for para-nitrophenol detection. Anal. Chim. Acta 777: 63–71. https://doi.org/10.1016/j.aca.2013.03.014.
Saadati, F., Ghahramani, F., Shayani-jam, H., Piri, F. and Yaftian, M.R. 2018. Synthesis and characterization of nanostructure molecularly imprinted polyaniline/graphene oxide composite as highly selective electrochemical sensor for detection of p-nitrophenol. J. Taiwan. Inst. Chem. Eng. 86: 213–221. https://doi.org/10.1016/j.jtice.2018.02.019.
Sall, M.L., Fall, B., Diédhiou, I., Hadji Dièye, El., Lo, M., Diagne Diaw, A.K., Sall, D.G., Raouafi, N. and Fall, M. 2020. Toxicity and electrochemical detection of lead, cadmium and nitrite ions by organic conducting polymers: A review. Chem. Africa 3: 499–512. https://doi.org/10.1007/s42250-020-00157-0.
Saravanan, K., Sathiyanarayanan, S., Muralidharan, S., Syed Azim, S. and Venkatachari, G. 2007. Performance evaluation of polyaniline pigmented epoxy coating for corrosion protection of steel in concrete environment. Prog. Org. Coatings 59: 160–167. https://doi.org/10.1016/j.porgcoat.2007.03.002.
Savan, E.K., Koytepe, S., Pasahan, A., Erdogdu, G. and Seckin, T. 2014. Amperometric simultaneous measurement of copper and cobalt ions with polythiophene incorporating pendant terpyridine groups. Polym. - Plast. Technol. Eng. 53: 1817–1824. https://doi.org/10.1080/03602559.2014.935405.
Su, P.G. and Huang, L.N. 2007. Humidity sensors based on TiO2 nanoparticles/polypyrrole composite thin films. Sensors Actuators, B. Chem. 123: 501–507. https://doi.org/10.1016/j.snb.2006.09.052.
Syritski, V., Reut, J., Opik, A. and Idla, K. 1999. Environmental QCM sensors coated with polypyrrole. Synth. Met. 102: 1326–1327. https://doi.org/10.1016/S0379-6779(98)01047-9.
Tan, F., Cong, L., Li, X., Zhao, Q., Zhao, H., Quan, X. and Chen, J. 2016. An electrochemical sensor based on molecularly imprinted polypyrrole/graphene quantum dots composite for detection of bisphenol A in water samples. Sensors Actuators, B Chem. 233: 599–606. https://doi.org/10.1016/j.snb.2016.04.146.
Thivya, P. and Wilson, R.R.J. 2019. Environmental pollutants simultaneous determination: DNA catalyst mediated polyaniline biocomposite nanostructures. Biocatal. Agric. Biotechnol. 21: 101352. https://doi.org/10.1016/j.bcab.2019.101352.
Tian, Q., Xua, J., Zuoa, Y., Lia, Y., Zhanga, J., Zhoua, Y., Duana, X., Lub, L., Jiaa, H., Xua, Q. and Yu, Y. 2019. Three-dimensional PEDOT composite based electrochemical sensor for sensitive detection of chlorophenol. J. Electroanal. Chem. 837: 1–9. https://doi.org/10.1016/j.jelechem.2019.01.055.
Vikrant, K., Tsang, D.C.W., Raza, N., Giri, B.S., Kukkar, D. and Kim, K. 2018. Potential utility of metal-organic framework-based platform for sensing pesticides. ACS Appl. Mater. Interfaces 10: 8797–8817. https://doi.org/10.1021/acsami.8b00664.
Wasi, S., Tabrez, S. and Ahmad, M. 2013. Toxicological effects of major environmental pollutants: An overview. Environ. Monit. Assess 185: 2585–2593. https://doi.org/10.1007/s10661-012-2732-8.
Yang, L.Y. and Liau, W. Bin. 2009. Environmental responses of nanostructured polyaniline films based on polystyrene-polyaniline core-shell particles. Mater. Chem. Phys. 115: 28–32. https://doi.org/10.1016/j.matchemphys.2008.10.074.
Yang, X., Qiao, K., Ye, Y., Yang, M., Li, J., Gao, H., Zhang, S., Zhou, W. and Lu, R. 2016. Facile synthesis of multifunctional attapulgite/Fe3O4/polyaniline nanocomposites for magnetic dispersive solid phase extraction of benzoylurea insecticides in environmental water samples. Anal. Chim. Acta 934: 114–121. https://doi.org/10.1016/j.aca.2016.06.027.
Zhang, W., Zhang, X., Wua, Z., Abdurahman, K., Cao, Y., Duan, H. and Jia, D. 2020. Mechanical, electromagnetic shielding and gas sensing properties of flexible cotton fiber/polyaniline composites. Compos. Sci. Technol. 188: 107966. https://doi.org/10.1016/j.compscitech.2019.107966.

Index

About the Editors

Dr. Fabián Fernández-Luqueño is a Full Professor at the Center for Research and Advanced Studies of the National Polytechnic Institute (Cinvestav Saltillo), Mexico. He received a Bachelor's degree in engineering in soil science, his M.Sc. in edaphology, and a Ph.D. in Biotechnology. His research area is plant-environment interactions, bio- and nano-remediation, agricultural nanotechnology, and the effect of nanomaterials on crops, soils, and microorganisms. Fabián has authored more than 65 scientific papers, 12 science articles, 17 books, and 35 book chapters, and has been a reviewer of more than 20 high-impact peer-reviewed journals (Ranked in JCR-Q1) and associate editor from Heliyon (IF = 3.772) & Terra Latinoamericana (Scopus). He has been a dissertation advisor for undergraduate, bachelor, master, and Ph.D. degrees. Fabián is a member of Researchers-National Systems in level II, Mexican Academy of Sciences, and is currently vice-president of the Mexican Society of Soil Science.

Dr. Fernando López-Valdez is a full Professor at the Research Centre for Applied Biotechnology (Centro de Investigación en Biotecnología Aplicada), Instituto Politécnico Nacional, Mexico. He is a Bachelor grade in Chem. Engineering (UV, México). M.Sc. Degree in Biotechnology (Bioprocesses) and Ph.D. in Biotechnology (Agricultural Biotech.) from CINVESTAV, Mexico. Dr. Fernando López-Valdez has been a Postgraduate professor for more than 20 years. He is also a Author or co-author of over 20 book chapters, two books as Editor, over 20 articles papers (JCR or peer-reviewed). He has lectured for science divulgation on Biotechnology, Agricultural, and Environmental Biotechnology fields. Currently, he is the Associate Editor for Terra Latinoamericana Journal, Mexican Society of Soil Science (SMCS).

Dr. Gabriela Medina-Pérez is a Doctor of Science from the Scientific and Technological Development Program for Society with an emphasis on Agricultural Nanotechnology from the Center for Research and Advanced Studies of the National Polytechnic Institute (CINVESTAV-IPN). She is a Master in Food Sciences from the Autonomous University of the State of Hidalgo (UAEH) and an Agroindustrial Engineer from the Institute of Agricultural Sciences of UAEH (ICAP-UAEH). She is a member of the National System of Researchers (SNI) with a recent entry at the candidate level. An active member of the International Network of Government Scientific Advice (INGSA) since 2018. She is currently a full-time research professor in the academic area of agro-industrial engineering and food engineering at the ICAP-UAEH. She is also the general coordinator of the institute's outreach. She has participated as lead author and co-author in several indexed articles, book chapters, and conference proceedings on bioactive compounds, nanotechnology, agricultural biotechnology, and sustainable development.